T0224305

Remote Sensing of Plant Biodiversity

Jeannine Cavender-Bares • John A. Gamon
Philip A. Townsend
Editors

Remote Sensing of Plant Biodiversity

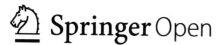

Springer Open

Editors
Jeannine Cavender-Bares
Ecology, Evolution & Behavior
University of Minnesota
Saint Paul, MN, USA

John A. Gamon
University of Nebraska–Lincoln,
School of Natural Resources
Lincoln, NE, USA

University of Alberta,
Departments of Earth & Atmospheric
Sciences and Biological Sciences
Edmonton, AB, Canada

Philip A. Townsend
Department of Forest and Wildlife Ecology
University of Wisconsin–Madison
Madison, WI, USA

ISBN 978-3-030-33159-7   ISBN 978-3-030-33157-3   (eBook)
https://doi.org/10.1007/978-3-030-33157-3

© The Editor(s) (if applicable) and The Author(s) 2020, Corrected Publication 2020
This book is an open access publication.
Open Access This book is licensed under the terms of the Creative Commons Attribution 4.0
International License (http://creativecommons.org/licenses/by/4.0/), which permits use, sharing,
adaptation, distribution and reproduction in any medium or format, as long as you give appropriate credit
to the original author(s) and the source, provide a link to the Creative Commons license and indicate if
changes were made.
The images or other third party material in this book are included in the book's Creative Commons
license, unless indicated otherwise in a credit line to the material. If material is not included in the book's
Creative Commons license and your intended use is not permitted by statutory regulation or exceeds the
permitted use, you will need to obtain permission directly from the copyright holder.
The use of general descriptive names, registered names, trademarks, service marks, etc. in this publication
does not imply, even in the absence of a specific statement, that such names are exempt from the relevant
protective laws and regulations and therefore free for general use.
The publisher, the authors, and the editors are safe to assume that the advice and information in this book
are believed to be true and accurate at the date of publication. Neither the publisher nor the authors or the
editors give a warranty, express or implied, with respect to the material contained herein or for any errors
or omissions that may have been made. The publisher remains neutral with regard to jurisdictional claims
in published maps and institutional affiliations.

Cover artwork was designed by Daniel Tschanz.

This Springer imprint is published by the registered company Springer Nature Switzerland AG
The registered company address is: Gewerbestrasse 11, 6330 Cham, Switzerland

Foreword

At last, here it is. For some time now, the world has needed a text providing both a new theoretical foundation and practical guidance on how to approach the challenge of biodiversity decline in the Anthropocene. This is a global challenge demanding global approaches to understand its scope and implications. Until recently, we have simply lacked the tools to do so. We are now entering an era in which we can realistically begin to understand and monitor the multidimensional phenomenon of biodiversity at a planetary scale. This era builds upon three centuries of scientific research on biodiversity at site to landscape levels, augmented over the past two decades by airborne research platforms carrying spectrometers, lidars, and radars for larger-scale observations. Emerging international networks of fine-grain in-situ biodiversity observations complemented by space-based sensors offering coarser-grain imagery—but global coverage—of ecosystem composition, function, and structure together provide the information necessary to monitor and track change in biodiversity globally.

This book is a road map on how to observe and interpret terrestrial biodiversity across scales through plants—primary producers and the foundation of the trophic pyramid. It honors the fact that biodiversity exists across different dimensions, including both phylogenetic and functional. Then, it relates these aspects of biodiversity to another dimension, the spectral diversity captured by remote sensing instruments operating at scales from leaf to canopy to biome. The biodiversity community has needed a Rosetta Stone to translate between the language of satellite remote sensing and its resulting spectral diversity and the languages of those exploring the phylogenetic diversity and functional trait diversity of life on Earth. By assembling the vital translation, this volume has globalized our ability to track biodiversity state and change. Thus, a global problem meets a key component of the global solution.

The editors have cleverly built the book in three parts. Part 1 addresses the theory behind the remote sensing of terrestrial plant biodiversity: why spectral diversity relates to plant functional traits and phylogenetic diversity. Starting with first principles, it connects plant biochemistry, physiology, and macroecology to remotely sensed spectra and explores the processes behind the patterns we observe. Examples from the field demonstrate the rising synthesis of multiple disciplines to create a new cross-spatial and spectral science of biodiversity.

Part 2 discusses how to implement this evolving science. It focuses on the plethora of novel in-situ, airborne, and spaceborne Earth observation tools currently and soon to be available while also incorporating the ways of actually making biodiversity measurements with these tools. It includes instructions for organizing and conducting a field campaign. Throughout, there is a focus on the burgeoning field of imaging spectroscopy, which is revolutionizing our ability to characterize life remotely.

Part 3 takes on an overarching issue for any effort to globalize biodiversity observations, the issue of scale. It addresses scale from two perspectives. The first is that of combining observations across varying spatial, temporal, and spectral resolutions for better understanding—that is, what scales and how. This is an area of ongoing research driven by a confluence of innovations in observation systems and rising computational capacity. The second is the organizational side of the scaling challenge. It explores existing frameworks for integrating multi-scale observations within global networks. The focus here is on what practical steps can be taken to organize multi-scale data and what is already happening in this regard. These frameworks include essential biodiversity variables and the Group on Earth Observations Biodiversity Observation Network (GEO BON).

This book constitutes an end-to-end guide uniting the latest in research and techniques to cover the theory and practice of the remote sensing of plant biodiversity. In putting it together, the editors and their coauthors, all preeminent in their fields, have done a great service for those seeking to understand and conserve life on Earth—just when we need it most. For if the world is ever to construct a coordinated response to the planetwide crisis of biodiversity loss, it must first assemble adequate—and global—measures of what we are losing.

Woody Turner
Earth Science Division
NASA Headquarters,
Washington, DC, USA

Contents

About the Authors

Christiana Ade is a PhD student in the Civil and Environmental Engineering Department at the University of California Merced. She is passionate about using remote sensing to study changes in wetland vegetation diversity and water quality, and evaluate how these changes may impact key Earth system cycles.

Julián Aguirre-Santoro is an Assistant Professor at the Natural Sciences Institute of the National University of Colombia. His research focuses on taxonomy, phylogenetics, biogeography, and macroevolution of Neotropical plant lineages.

Doreen Boyd is a Professor in the School of Geography, University of Nottingham, UK. Her research focuses include active and passive remote sensing, ecology and conservation, and human rights intersections with environmental change.

Betsabe de la Barreda-Bautista is a research associate at the School of Geography, University of Nottingham (UK), working on remote sensing for environmental-human dynamics, principally monitoring how vegetation responds to climatological and anthropogenic events.

Nicholas Basinger is an Assistant Professor of Weed Science at the University of Georgia. His research focuses on weed ecology, biology, and diversification of the agroecosystem and non-cropland ecosystems to reduce the impacts of weedy species.

Erik Bolch is a Master's student in Environmental Systems at the University of California Merced. His research focuses utilizing UAS imaging spectroscopy to study biodiversity and invasive plants.

Jan Bumberger heads the working group "Environmental Sensor and Information Systems" and is the scientific coordinator for research data management at the Helmholtz Centre for Environmental Research – UFZ, Germany. His research focuses on scalable sensor network technologies as well as the calibration and validation of remote sensing data with broadband electromagnetic and optical spectral methods.

Ana Carnaval was born and raised in Brazil. She received her Master's degree in Zoology from the Museu Nacional do Rio de Janeiro and has a PhD in Evolutionary Biology from The University of Chicago. She is an Associate Professor at the City College of CUNY, and her research focuses on the historical biogeography of the Atlantic forest fauna and flora.

Jeannine Cavender-Bares is a Distinguished McKnight University Professor in the Department of Ecology, Evolution and Behavior at the University of Minnesota. Her research focuses on the ecology and evolution of plant function, applying phylogenetics and spectral data to community ecology, and remote sensing of biodiversity.

Jennifer Costanza is a Research Assistant Professor in the Department of Forestry and Environmental Resources at North Carolina State University. Her research focuses on forest dynamics, the ecological effects of global change, land change modeling, and disturbance ecology.

Kyla Dahlin is an Assistant Professor in the Department of Geography, Environment, and Spatial Sciences and the Ecology, Evolutionary Biology, and Behavior Program at Michigan State University. Her research aims to better understand and quantify ecosystem processes and disturbance responses through the application of emerging technologies, including air- and spaceborne remote sensing, spatial statistics, and process-based modeling.

Néstor Fernández is a conservation biology researcher interested in the advancement of biodiversity conservation solutions through the integration of multiple sources of earth observations in order to quantify and model changes in terrestrial ecosystem dynamics and in species distributions.

Andrew O. Finley is a Professor at the Departments of Forestry and Geography, Environment, and Spatial Sciences at Michigan State University. His research interests lie in developing methodologies for monitoring and modeling environmental processes, Bayesian statistics, spatial statistics, and statistical computing.

Keith Gaddis is the Senior Support Scientist for the National Aeronautics and Space Administration's Biological Diversity and Ecological Forecasting Programs. He is an ecologist and biogeographer by training with expertise using remote sensing and genetics to address questions in ecology, evolution, and conservation biology.

John Gamon is a Professor in Earth and Atmospheric Sciences and Biological Sciences at the University of Alberta, and also conducts airborne remote sensing research in vegetation biodiversity and productivity at the School of Natural Resources at the University of Nebraska, Lincoln. His interests include ecophysiology, ecosystem metabolism, biodiversity, ecoinformatics, and sustainability.

Gary Geller has a PhD in Biology from UCLA where he studied the interaction of plants with their environment, focusing on modeling plant architecture. At NASA/JPL he combines that experience with system engineering and has worked with the NASA Ecological Forecasting program for the last 15 years, including two years seconded to the Group on Earth Observations in Geneva.

Hamed Gholizadeh is an Assistant Professor in the Geography Department at Oklahoma State University. His research focuses on using proximal, airborne, and spaceborne remote sensing to study biodiversity.

Renato Goldenberg is a Full Professor in the Department of Botany at Universidade Federal do Paraná, Brazil. His research focuses on systematics of flowering plants.

John Grady was a postdoctoral fellow at Michigan State University and Bryn Mawr College at the time of writing this chapter and is now a postdoctoral fellow at the National Great Rivers Research and Educational Center. He is broadly interested in how traits affect interactions and emergent features of ecology.

Ryan P. Hanavan is a Research Entomologist and the Remote Sensing Program Manager with the Forest Health Assessment and Applied Sciences Team with the US Forest Service. His research focuses on early detection remote sensing techniques for insect and disease threats.

Erin Hestir is an Associate Professor in Civil and Environmental Engineering at the University of California Merced. Her research focuses on aquatic ecosystems under threat from competing pressures to meet societal needs for water and food security while sustaining biodiversity and other ecosystem services.

Marco Heurich is Head of the Department of Visitor Management and National Park Monitoring of Bavarian Forest National Park and Associate Professor of Wildlife Ecology at the University of Freiburg. His research focuses on exploring movement ecology and biodiversity by using a variety of remote sensing techniques from camera traps to satellites.

Martina Hobi is a Group Leader in the Forest Resources and Management Research Unit at the Swiss Federal Institute for Forest, Snow, and Landscape Research WSL. She specializes in the analysis of forest dynamical processes using a combination of terrestrial, dendroecological, and remote sensing data.

Sarah Hobbie is a Distinguished McKnight University Professor in the Department of Ecology, Evolution and Behavior at the University of Minnesota. She studies global change impacts on ecosystems, urban ecosystem ecology, and human impacts on urban water resources.

Stéphane Jacquemoud is a Professor of Remote Sensing in the Institut de Physique du Globe de Paris at the University of Paris. His research focuses include remote sensing of natural surfaces, leaf and soil radiative transfer models, and imaging spectroscopy.

Miriam Kaehler has worked as a visiting scholar at the Institute of Systematic Botany at the New York Botanical Garden, and as a post-doc research fellow at Universidade de São Paulo and Universidade Federal do Paraná, both in Brazil. Her research focuses on molecular systematics and taxonomy of Bignoniaceae, as well as plant inventory.

Shruti Khanna is a Senior Environmental Scientist at the California Department of Fish and Wildlife. Her research focuses on understanding invasion ecology of nonnative plants through remote sensing.

Doug King is a Distinguished Research Professor in the Department of Geography and Environmental Studies, Carleton University, Ottawa, Canada. His research focuses on high-resolution UAV to satellite scale imaging for vegetation composition, structure, productivity and health modeling, mapping, and monitoring in forest, wetland, agriculture, and tundra environments.

Mayara Krasinski Caddah is a Professor in the Department of Botany at the Federal University of Santa Catarina, Brazil. Her research focuses on plant taxonomy and evolution, applying phylogenetics and population genetics to genera and species delimitation.

Daniel Kükenbrink is a PhD candidate at the Remote Sensing Laboratories of the University of Zurich. Three-dimensional forest reconstruction and the radiative transfer in forest ecosystems are among his main research interests.

Andrew Latimer is a Professor in the Department of Plant Sciences at the University of California Davis. He studies how plant populations and communities respond to change, including sudden, major disturbance such as fire and drought, as well as more gradual changes in climate.

Angela Lausch is head of the working group functional landscape ecology and Remote Sensing at the Helmholtz Centre for Environmental Research – UFZ, and Privat Dozent (PD) at the Humboldt University to Berlin. She uses hyperspectral remote sensing on airborne, drone, and laboratory platforms at the UFZ. Her research focuses on remote sensing, trait and functional ecology, bio- and geodiversity, data science, linked open data, and the semantic web.

Reik Leiterer is a Senior Scientist at the University of Zurich, Switzerland. His aim is to enable more robust and operational environmental monitoring using data-driven methods.

Lúcia Lohmann is a Professor in the Department of Botany at the University of São Paulo, Brazil. Her research is highly integrative, combining components of classic taxonomy, phylogenetics, molecular biology, ecology, evolution, and biogeography to address questions associated with the assembly and evolution of the Neotropical biota.

Mike Madritch is a Professor in the Department of Biology at Appalachian State University. His research focuses on linking aboveground plant communities with belowground microbial communities and ecosystem functioning.

Paul Magdon is a Senior Scientist at the Chair of Forest Inventory and Remote Sensing at the University of Göttingen, Germany. In his research he addresses methodological questions, relevant for the development of remote sensing–assisted forest and biodiversity monitoring systems.

Sparkle Malone is an Assistant Professor in the Department of Biological Sciences at Florida International University. Her primary research focus is to improve our understanding of how climate and disturbance regimes influence spatial and temporal variability in ecosystem structure and function using remote sensing, eddy covariance, and spatial and temporal models.

Roberta Martin is an Associate Professor in the School for Geographical Sciences and Urban Planning and the Center for Global Discovery and Conservation Science at the Arizona State University. Her research incorporates field and laboratory studies of trees and corals with aircraft- and satellite-based maps to understand the biodiversity and health of forests and reefs.

Kyle C. McDonald is Terry Elkes Professor in the Earth and Atmospheric Sciences Department at the City College of New York, City University of New York. Professor McDonald studies terrestrial ecosystems with satellite-borne remote sensing instruments.

Jose Eduardo Meireles is an Assistant Professor in the School of Biology and Ecology at the University of Maine. His research focuses on plant phylogenetics, population genetics, and evolutionary ecology.

Fabián A. Michelangeli is a Curator in the Institute of Systematic Botany at The New York Botanical Garden. His research focuses on tropical plant taxonomy, systematics, and their evolution.

Felix Morsdorf is a Group Leader in the Geography Department at the University of Zurich. He uses laser scanners, drones, and physical modeling to derive diversity-related metrics of vegetation structure.

Scott Ollinger is a Professor in the Department of Natural Resources and the Environment at the University of New Hampshire. His research interests span a variety of topics within the fields of ecology and biogeochemistry including carbon and nitrogen cycling, forest productivity and succession, plant-soil interactions, remote sensing, ecosystem modeling, and the effects of multiple environmental stressors on forests.

Brian O'Meara studies macroevolution by developing and applying phylogenetic models. He is a Professor in the Department of Ecology and Evolutionary Biology at the University of Tennessee, Knoxville.

Stephanie Pau is an Associate Professor in the Department of Geography at Florida State University. Her research focuses on the diversity, structure, and function of forest and savanna systems, often combining field surveys with remote sensing imagery and statistical modeling.

Ryan Pavlick is a Research Technologist at the Jet Propulsion Laboratory, California Institute of Technology; working on observing and understanding Earth's biodiversity, terrestrial ecosystems and carbon cycle from space.

Andrea Paz has an MSc in Biology and is a PhD candidate at the City University of New York, USA. She is interested in understanding the influence of environments in determining species distributions and diversity patterns in the Neotropics.

Henrique Miguel Pereira is a Professor of Biodiversity Conservation at the German Centre for Integrative Biodiversity Research (iDiv) at the Martin-Luther University Halle-Wittenberg and the co-chair of the Group on Earth Observations Biodiversity Observation Network. His research uses a combination of field work, long-term monitoring, and modeling to understand the drivers of global biodiversity change.

Jesús N. Pinto-Ledezma is a Grand Challenge in Biology Postdoctoral Fellow in the Department of Ecology, Evolution and Behavior at the University of Minnesota. He works with both macroecological and evolutionary theory and methods with the aim of understanding the processes that underlie biodiversity patterns at different temporal and spatial scales.

Erika Podest is a scientist with the Carbon Cycle and Ecosystems Group at NASA's Jet Propulsion Laboratory in Pasadena, California. Her research focuses on the use of microwave remote sensing for characterizing wetland inundation dynamics and vegetation growing season length in the northern high latitudes.

Jennifer Pontius is a Research Ecologist with the US Forest Service and Director of the Environmental Science Program at the University of Vermont. Her research focuses on developing novel remote sensing methods to assess forest health, structure, and productivity to better inform the management of forested ecosystems.

Quentin Read was a postdoctoral fellow at Michigan State University and Bryn Mawr College at the time of writing this chapter and is now a postdoctoral fellow at the National Socio-Ecological Synthesis Center. A community ecologist by training, he is now working on quantifying the environmental and ecological impacts of food waste in the United States.

Martin Reader is a PhD student in the Department of Geography at the University of Zurich. He is examining the impacts of human modification on river delta biodiversity and ecosystem services.

Sydne Record is an Associate Professor in the Department of Biology at Bryn Mawr College. Her research focuses on modeling the past, present, and future states of the natural world to better understand the services that ecosystems afford to society.

Marcelo Reginato is a Professor in the Department of Botany at the Universidade Federal do Rio Grande do Sul, Brazil. His research focuses include plant systematics, biogeography, and macroevolution.

Duccio Rocchini is Professor in Biology and Ecology at the University of Trento, Italy. His research focuses on biodiversity analysis at multiple spatial scales, ecological informatics, plant community ecology, remote sensing, spatial ecology, spatio-ecological modeling by open-source software, and species distribution modeling.

Maria J. Santos is a Professor of Earth System Science in the Department of Geography, University of Zurich. Her research focuses on understanding coupled social-ecological systems and their impacts on biodiversity and ecosystem services, using field, archival, and remote sensing data, GIS, and statistical modeling.

Michael E. Schaepman is a Professor of Remote Sensing and Vice President of Research at the University of Zurich, Switzerland. Following a focus on optical engineering and designing imaging spectrometers, he moved on to signal processing and radiative transfer modeling and is currently working on remotely sensing ecological genomics.

Gabriela Schaepman-Strub is a Professor of Earth System Science in the Department of Evolutionary Biology and Environmental Studies at the University of Zurich, Switzerland. Her research focuses on radiation-vegetation interaction from leaf to canopy scale, Arctic biodiversity and ecosystem functions, and land surface energy fluxes under climate change.

David Schimel is a Senior Research Scientist at the Jet Propulsion Laboratory, California Institute of Technology; leading research focused on interactions between climate and ecosystems, combining models and observations. His team is working on satellite missions that involve spectroscopic detection of carbon cycle dynamics and biodiversity on Earth.

Fabian Schneider is a postdoc at the NASA Jet Propulsion Laboratory, California Institute of Technology. His research interests lie in remote sensing of plant functional traits, biodiversity, and ecosystem functioning to improve our understanding of the interactions between plant communities and global change.

Franziska Schrodt is an Assistant Professor in the School of Geography, University of Nottingham (UK). She combines ecological data on (functional, phylogenetic, and taxonomic) biodiversity and geodiversity with novel data sources (e.g., remotely sensed) and analysis techniques (e.g., machine learning) to further develop our understanding of ecosystem dynamics and potential impacts on human well-being.

Karsten Schulz is Professor of Hydrology and Integrated Water Management at the University of Natural Resources and Life Sciences Vienna, Austria. By integrating information from various remote sensing platforms, his research aims to improve the prediction and understanding of water, energy, and solute fluxes in the soil-plant-atmosphere continuum at different spatial scales.

Anna K. Schweiger is a Postdoctoral Associate at the Institut de recherche en biologie végétale, Université de Montréal. Anna is currently particularly interested in developing theory and methods for using spectra of plants and spectral diversity of landscapes to study plant community composition and the ecosystem benefits of biodiversity.

Shawn P. Serbin is a Scientist in the Environmental and Climate Sciences Department at Brookhaven National Laboratory, Upton, New York. Shawn's research aims to improve the understanding of the processes driving the fluxes and pools of carbon, water, and energy in terrestrial ecosystems and utilizes a range of tools including remote sensing to develop novel scaling methods and model-data synthesis approaches.

Philip Townsend is a Distinguished Professor in the Department of Forest and Wildlife Ecology at the University of Wisconsin, Madison. His research focuses include physiological remote sensing, imaging spectroscopy, forest ecosystem ecology, and watershed hydrology.

Woody Turner has spent a career promoting the use of satellite imagery to answer questions about the distribution and abundance of life on Earth and providing natural resource managers with remote sensing solutions to their day-to-day challenges.

Susan Ustin is a Distinguished Professor in the Department of Land, Air and Water Resources at the University of California Davis. Her research focuses include remote sensing of plant functional properties, ecosystem processes, imaging spectroscopy, and mapping of plant species and biodiversity.

Ran Wang is a Research Assistant Professor at the University of Nebraska – Lincoln, and visiting scientist at University of Alberta, where he received his PhD. He uses proximal, airborne, and satellite remote sensing to study vegetation biodiversity and productivity.

Chris Williams is the Geospatial Analysis Lead at the British Geological Survey. His background is in glaciology, terrain and environmental data analysis, and geo-environmental product development.

Adam Wilson is an Assistant Professor in the Geography Department and the Department of Environment and Sustainability at the University at Buffalo. His research investigates the spatial patterns and processes of biodiversity and ecosystem function using remote sensing and field observations together with mechanistic and statistical modeling to understand how ecosystems change through space and time.

Phoebe Zarnetske is an Associate Professor in the Department of Integrative Biology at Michigan State University. Her research uses a combination of observational data, experiments, and modeling to connect observed patterns of biodiversity and community composition with underlying mechanisms.

Brian Zutta is a researcher at Universidad Alas Peruanas and an assistant project scientist at the University of California Los Angeles. He recently coordinated the development of the national measurement, reporting and verification and forest monitoring system for the National Forest Conservation Program (PNCB) of the Ministry of the Environment (MINAM) in Lima, Peru.

About the Editors

Jeannine Cavender-Bares is a Distinguished McKnight University Professor in the Department of Ecology, Evolution and Behavior at the University of Minnesota. She earned a Master's at the Yale School of Forestry and Environmental Studies and a PhD in Biology at Harvard University. Her research focuses on the ecology and evolution of plant function, applying phylogenetics and spectral data to community ecology, and remote sensing of biodiversity. She is committed to advancing international efforts for global monitoring and assessment of biodiversity and ecosystem services to aid management efforts towards sustainability.

John A. Gamon is a Professor in the Departments of Earth and Atmospheric Sciences and Biological Sciences at the University of Alberta. He also conducts research in Quantitative Remote Sensing at the Center for Advanced Land Management Information Technologies (CALMIT) in the School of Natural Resources at the University of Nebraska, Lincoln. He earned his Master's and his PhD in Botany at the University of California, Davis, and did his postdoctoral research on Remote Sensing and Ecophysiology at the Carnegie Institution in Stanford, CA. His research focuses include photosynthesis, ecosystem function, productivity, biodiversity, ecoinformatics, and sustainability.

Philip A. Townsend is a Distinguished Professor in the Department of Forest and Wildlife Ecology at the University of Wisconsin, Madison. He earned his PhD in Geography at the University of North Carolina, Chapel Hill. His research focuses include physiological remote sensing, imaging spectroscopy, ecosystem ecology, and watershed hydrology.

Chapter 1
The Use of Remote Sensing to Enhance Biodiversity Monitoring and Detection: A Critical Challenge for the Twenty-First Century

Jeannine Cavender-Bares, John A. Gamon, and Philip A. Townsend

1.1 Introduction

Improved detection and monitoring of biodiversity is critical at a time when Earth's biodiversity loss due to human activities is accelerating at an unprecedented rate. We face the largest loss of biodiversity in human history, a loss which has been called the "sixth mass extinction" (Leakey 1996; Kolbert 2014), given that its magnitude is in proportion to past extinction episodes in Earth history detectable from the fossil record. International efforts to conserve biodiversity (United Nations 2011) and to develop an assessment process to document changes in the status and trends of biodiversity globally through the Intergovernmental Science-Policy Platform on Biodiversity and Ecosystem Services (Díaz et al. 2015) have raised awareness about the critical need for continuous monitoring of biodiversity at multiple spatial scales across the globe. Biodiversity itself—the variation in life found among ecosystems and organisms at any level of biological organization—cannot practically be observed everywhere. However, if habitats, functional traits, trait diversity, and the spatial turnover of plant functions can be remotely sensed, the potential exists to globally inventory the diversity of habitats and traits associated with terrestrial biodiversity. To face this challenge, there have been recent calls for

J. Cavender-Bares (✉)
Department of Ecology, Evolution & Behavior, University of Minnesota,
Saint Paul, MN, USA
e-mail: cavender@umn.edu

J. A. Gamon
Department of Earth & Atmospheric Sciences, University of Alberta, Edmonton, AB, Canada

Department of Biological Sciences, University of Alberta, Edmonton, AB, Canada

CALMIT, School of Natural Resources, University of Nebraska – Lincoln, Lincoln, NE, USA

P. A. Townsend
Department of Forest and Wildlife Ecology, University of Wisconsin, Madison, WI, USA

© The Author(s) 2020
J. Cavender-Bares et al. (eds.), *Remote Sensing of Plant Biodiversity*,
https://doi.org/10.1007/978-3-030-33157-3_1

1

a global biodiversity monitoring system (Jetz et al. 2016; Proença et al. 2017; The National Academy of Sciences 2017). A central theme of this volume is that remote sensing (RS) will play a key role in such a system.

1.2   Why a Focus on Plant Diversity?

Plants and other photosynthetic organisms form the basis of almost all primary productivity on Earth, and their diversity and function underpin virtually all other life on this planet. Plants—collectively called vegetation—regulate the flow of critical biogeochemical cycles, including those for water, carbon, and nitrogen. They affect soil chemistry and other properties, which in turn affect the productivity and structure of ecosystems. Given the importance of plant diversity for providing the ecosystem services on which humans depend—including food production and the regulating services that maintain clean air and freshwater supply (Millennium Ecosystem Assessment 2005; IPBES 2018a)—it is critical that we monitor and understand plant biodiversity from local to global scales, encompassing genetic variation within and among species to the entire plant tree of life (Cavender-Bares et al. 2017; Jetz et al. 2016; Turner 2014).

Of the 340,000 known seed plants on Earth and the 60,000 known tree species (Beech et al. 2017), 1 out of every 5 seed plants and 1 out of every 6 tree species are threatened (Kew Royal Botanic Gardens 2016). Vulnerability to threats ranging from climate change to disease varies among species and lineages because of evolved differences in physiology and spatial proximity to threats. Across all continents, the largest threat to terrestrial biodiversity is land use change due to farming and forestry, while climate change, fragmentation, and disease loom as ever-increasing threats (IPBES 2018a; b; c; d). Many plant species are at risk for extinction due to a combination of global change factors, including drought stress, exotic species, pathogens, land use change, altered disturbance regimes, application of chemicals, and overexploitation.

1.3   The Promise of Remote Sensing to Detect Plant Diversity

Different plants have evolved to synthesize different mixes of chemical compounds arranged in contrasting anatomical forms to support survival and growth. In addition, the structures of plant canopies correspond to different growth strategies in response to climate, environment, or disturbance. Differences among individual plants, populations, and lineages result from contrasting evolutionary histories, genetic backgrounds, and environmental conditions. Because these differences are readily expressed in aboveground physiology, biochemistry, and structure, many of these properties can be detected using spectral reflectance from leaves and plant canopies (Fig. 1.1). Plant pigments absorb strongly electromagnetic radiation in the

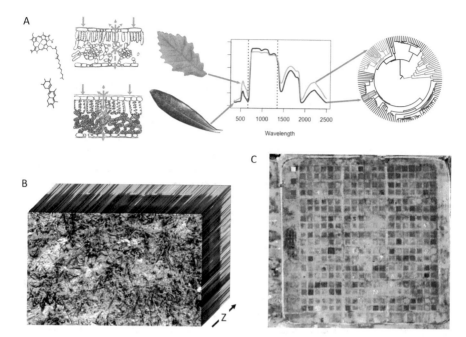

Fig. 1.1 (**a**) The chemical, structural, and anatomical attributes of plants influence the way they interact with electromagnetic energy to generate spectral reflectance profiles that reveal information about plant function and are tightly coupled to their evolutionary origins in the tree of life. (Adapted from Cavender-Bares et al. 2017.) Imaging spectroscopy offers the potential to remotely detect patterns in diversity and chemical composition and vegetation structure that inform our understanding of ecological processes and ecosystem functions. Examples are shown from the Cedar Creek Ecosystem Science Reserve long-term biodiversity experiment. (**b**) The image cube (0.5 m × 1 m) at 1 mm spatial resolution (400–1000 nm) detects sparse vegetation early in the season in which individual plants can be identified. The "Z-dimension" (spectral dimension) illustrates different spectral reflectance properties for different scene elements, including different species. At this spatial resolution, plant diversity is predicted from remotely sensed spectral diversity (Wang et al. 2018). (**c**) AVIRIS NextGen false color image of the full experiment at 1 m spatial resolution (400–2500 nm). Each square is a 9 × 9 m plot with a different plant composition and species richness. Wang et al. (2019) mapped chemical composition and a suite of other functional traits and their uncertainties in all of the experimental plots. By combining spectral data at different scales, proximal and remote imagery can be used to examine the scale dependence of the spectral diversity–biodiversity relationship in detail (e.g., Wang et al. 2018; Gamon et al. Chap. 16)

visible wavelengths (400–700 nm), while other chemical compounds and structural attributes of plants that tend to be conserved through evolutionary history affect longer wavelengths. The patterns of light absorbed, transmitted, and reflected at different wavelengths from vegetation reveal leaf and canopy surface properties, tissue chemistry, and anatomical structures and morphological attributes of leaves, whole plants, and canopies. Thus, technological advances for assessing optical properties of plants provide profound opportunities for detecting functional traits of organisms at different levels of biological organization. These advances are occurring at multiple spatial scales, with technologies ranging from field spectrometers and airborne systems to emerging satellite systems. As a consequence, there is high potential to detect and monitor plant diversity—and other forms of diversity—across a range of spatial scales, and to do so iteratively and continuously, particularly if multiple methods can be properly coordinated.

Calls for a global biodiversity observatory (Fig. 1.2) that can detect and monitor functional plant diversity from space (Jetz et al. 2016; Proença et al. 2017; The National Academy of Sciences 2017; Geller et al., Chap. 20) have been met with widespread support. Forthcoming satellite missions, including the Surface Biology and Geology (SBG) mission in planning stages at the US National Aeronautics and Space Administration (NASA) and related missions in Europe and Japan (Schimel et al., Chap. 19), will make unprecedented spectroscopic data available to scientists, management communities, and decision-makers, but at relatively coarse spatial scales. At the same time, rapid progress is being made with field spectroscopy using unmanned aerial vehicles (UAVs) and other airborne platforms that are offering novel ways to use RS to advance our understanding of the linkages between optical (e.g., spectral or structural) diversity and multiple dimensions of biodiversity (e.g., species, functional, and phylogenetic diversity) at finer spatial scales (Fig. 1.1). These advances present a timely and tremendously important opportunity to detect changes in the Earth's biodiversity over large regions of the planet. One can fairly ask whether user communities are ready to make use of the data. Effective interpretation and application of remotely sensed data to determine the status and trends of plant biodiversity and plant functions across the tree of life—with linkages to all other living organisms—requires integration across vastly different knowledge arenas. Critically, it requires integration with in-situ direct and indirect measures of species distributions, their evolutionary relationships, and their functions. Approaches for integration are the primary focus of this book.

A central requirement to advance monitoring of biodiversity at the global scale is to decipher the sources of variation that contribute to spectral variation, both from a biological perspective and from a physical perspective. Distinct fields of biology have developed a range of methodologies for understanding plant ecological and evolutionary processes that underlie these sources of variation. Similarly, radiative transfer models have been developed largely based on the physics of light interacting with vegetation canopy elements and the atmosphere. These models have yet to capture the full range of plant traits, often preferring to represent "average" vegetation conditions for a region instead of the variation present, so are not yet ready for the task. All of these methods have unique approaches to analyzing complex, multidimensional data sets, and neither the analytical approaches nor the data structures have been brought together in a systematic or comprehensive manner. A common language among disciplines (including biology, physiological ecology, landscape ecology, genetics, phylogenetics, geography, spectroscopy, and radiative transfer) related to the RS of biodiversity is currently lacking. This book provides a framework for how biodiversity, focusing particularly on plants, can be detected using proximally and remotely sensed hyperspectral data (with many contiguous spectral bands) and other tools, such as lidar (with its ability to detect structure). The chapters in this book present a range of perspectives and approaches on how RS can be

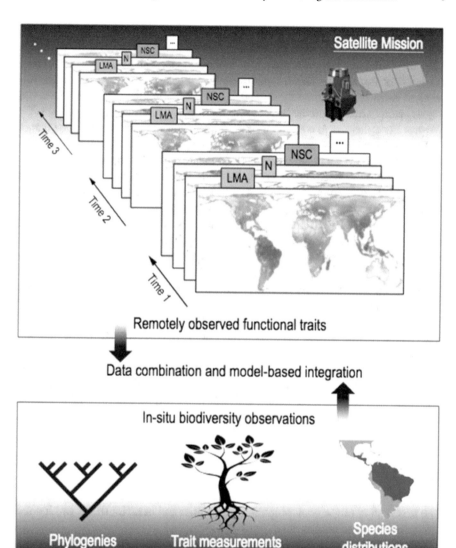

Fig. 1.2 An envisioned global biodiversity observatory in which remotely sensed high spectral resolution spectroscopic data from satellites is integrated with biodiversity observations through natural history studies, phylogenetic systematics, functional trait measurements, and species distribution data. The figure is adapted from Jetz et al. 2016 based on the National Center for Ecological Analysis and Synthesis Working Group "Prospects and priorities for satellite monitoring of global terrestrial biodiversity"

integrated to detect and monitor the status and trends of plant diversity, as well the biodiversity of other organisms and life processes that depend on plants above- and belowground. Biological and computational experts from three disciplines—RS and leaf optics, plant functional biology, and systematics—present insights to advance our understanding of how to link spectral and other kinds of RS data with functional traits, species distributions, and the tree of life for biodiversity detection. The authors detail the approaches and conditions under which efforts to detect plant biodiversity are likely to succeed, being explicit about the advantages and disadvantages of each. A theme running through many chapters is the challenge of moving across spatial scales from the leaf level to the canopy, ecosystem, and global scale. We provide a glossary that allows a common language across disciplines to emerge.

Here we explore the prospects for integrating components from each of these fields to remotely detect biodiversity and articulate the major challenges in our ability to directly link spectral data of vegetation to species diversity, functional traits, phylogenetic information, and functional biodiversity at the global scale. RS offers the potential to fill in data gaps in biodiversity knowledge locally and globally, particularly in remote and difficult-to-access locations, and can help define the larger spatial and temporal background needed for more focused and effective local or regional studies. It also may increase the likelihood of capturing temporal variation, and it allows monitoring of biodiversity at different spatial scales with different platforms and approaches. In essence, it provides the context within which changing global biodiversity patterns can be understood. The concept of "optical surrogacy" (Magurran 2013)—in which the linkage of spectral measurements to associated patterns and processes is used—may be useful in predicting ecosystem processes and characteristics that themselves are not directly observable (Gamon 2008; Madritch et al. 2014; Fig. 1.3). In a broad sense, such relationships between various expressions of biodiversity and optical (spectral) diversity provide a fundamental principle for "why RS works" as a metric of biodiversity and why so many different methods at different scales can provide useful information.

1.4   The Contents of the Book

The first section of the book presents the potential and basis for direct and indirect remote detection of biodiversity.

Cavender-Bares et al. (Chap. 2) present an overview of biodiversity itself, including the in-situ methods and metrics for measuring biodiversity, particularly plant diversity. The chapter provides a layperson's overview of the elements, methods, and metrics for detecting and analyzing biodiversity and points to the potential of spectral data, collected at multiple spatial and biological scales, to enhance the study of biodiversity. In doing so, it bridges an ecological and evolutionary understanding of the diversity of life, considering both its origins and consequences.

Serbin and Townsend (Chap. 3) describe various approaches for measuring plant and ecosystem function using spectroscopy, providing both the historical development of past advances and the potential of these approaches looking forward. The chapter explains why we are able to retrieve functional traits from spectra, which traits can be retrieved, and where spectra show features important for different aspects of plant function. It also raises the challenge of scaling plant function from leaves to canopies and landscapes.

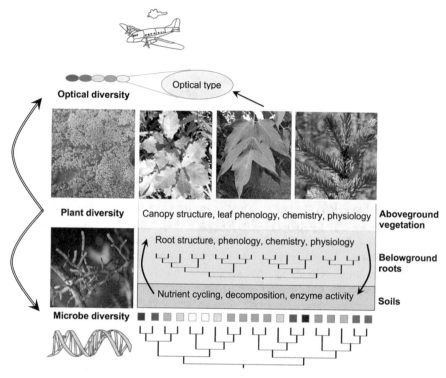

Fig. 1.3 Optical methods for detecting the functional, structural, and chemical components of vegetation, which are tightly coupled to the genetic and phylogenetic backgrounds of plants, are linked to belowground processes and the structure and function of microbial communities

Morsdorf et al. (Chap. 4) then present the Laegeren forest site in Switzerland as a virtual laboratory. They demonstrate how spectroscopy can be operationalized for RS of functional diversity to explain plant biodiversity patterns and ecosystem functions. The Laegeren site is one of the best-studied sites in the world for this purpose and is used as a case study to explain ground truthing and what can be learned from landscape-level detection of functional diversity.

Martin (Chap. 5) summarize the experiences with "spectronomics"—a framework aimed at integrating chemical, phylogenetic, and spectral RS data—using airborne imagery to detect forest composition and function in wet tropical forests. In these vast, largely inaccessible landscapes that harbor enormous taxonomic variation, approaches that rely solely on field-based observations are infeasible, illustrating an essential role for RS. As pioneers in using spectroscopy to detect plant chemistry, function, and biodiversity in tropical forests around the world, these researchers highlight some of the major lessons they have learned.

Pontius et al. (Chap. 6) consider how biodiversity can be protected given current threats to forest and vegetation conditions and present approaches for detailed and accurate detection of forest disturbance and decline. They review current techniques used to assess and monitor forest ecosystem condition and disturbance and outline a general approach for earlier, more detailed and accurate decline assessment. They also discuss the importance of engaging land managers, practitioners, and decision-makers in these efforts to ensure that the products developed can be utilized by stakeholders to maximize their impact.

Meireles et al. (Chap. 7) provide a framework to explain how spectral reflectance data from plants is tightly coupled to the tree of life and demonstrate how spectra can reveal evolutionary processes in plants. They clarify that many spectral features in plants are inherited and are thus very similar among close relatives—in other words, they are highly phylogenetically conserved. Simulations reveal that spectral information of plants appears to follow widely used evolutionary models, making it possible to link plant spectra to the tree of life in a predictive manner. As a consequence, methods developed in evolutionary biology to understand the tree of life can now benefit the RS community. The chapter provides evidence that evolutionary lineages may be easier than individual species to detect through RS methods, particularly if they are combined with other approaches for estimating which species and lineages have the potential to be present in a given location. A caveat is that spatial resolution of satellite spectral data will limit such inferences, but leaf- and canopy-level spectra (obtainable from proximal and airborne sensing) can contribute enormously to our understanding of these fundamental links between spectral patterns and gene sequences.

Madritch et al. (Chap. 8) link aboveground plant biodiversity and productivity to belowground processes. They explain the functional mechanisms—which can be revealed by remotely sensed spectral data—that influence interactions of plant hosts with insects and soil organisms, in turn influencing ecosystem functions, such as decomposition and nutrient cycling. The chapter provides an example of using the concept of surrogacy, in which the biochemical linkage of spectral measurements to associated patterns and processes aboveground is used to provide estimates of soil and microbial processes belowground that are not directly observable via RS.

The next three chapters focus on linking satellite-based remotely sensed data to biodiversity prediction. Pinto-Ledezma and Cavender-Bares (Chap. 9) present an example of how currently available satellite-based RS products can be used to generate next-generation species distribution models to predict where species and lineages are likely occur and the habitats they may have access to and persist in under altered climates in the future. They compare RS-based methods for generating predictive models with widely used approaches that use meteorologically derived climate variables. They demonstrate the advantages of RS-based models in regions where meteorological data is only sparsely available. Such predictive modeling that harnesses species occurrence data and temporal information about the biotic environment may make spectral methods of species and evolutionary lineage detection more tractable.

Building on the availability of satellite RS data with near-global coverage to predict biodiversity, Record et al. (Chap. 10) explore how RS illuminates the relationship between biodiversity and geodiversity—the variety of abiotic features and processes that provide the template for the development of biodiversity. They introduce a variety of globally available geodiversity measures and examine how they can be combined with biodiversity data to understand how biodiversity responds to geodiversity. The authors use the analogy of the "stage" that defines the patterns of life to some degree, often measured as habitat heterogeneity, a key driver of species diversity. They illustrate the approach by examining the relationship between biodiversity and geodiversity with tree biodiversity data from the US Forest Inventory and Analysis Program and geodiversity data from remotely sensed elevation from the Shuttle Radar Topography Mission (SRTM). In doing so, they outline the challenges and opportunities for using RS to link biodiversity to geodiversity.

Paz et al. (Chap. 11) present an approach for using RS data to predict patterns of plant diversity and endemism in the tropics within the Brazilian Atlantic rainforest. They examine how RS environmental data from tropical regions can be used to support biodiversity prediction at multiple spatial, temporal, and taxonomic scales.

Bolch et al. (Chap. 12) summarize the range of approaches that can be used to optimize detection of invasive alien species (IAS), which pose severe threats to biodiversity. These approaches emphasize the ability to detect individual plant species that have distinct functional properties. The chapter presents current RS capabilities to detect and track invasive plant species across terrestrial, riparian, aquatic, and human-modified ecosystems. Each of these systems has a unique set of issues and species assemblages with its own detection requirements. The authors examine how RS data collection in the spectral, spatial, and temporal domains can be optimized for a particular invasive species based on the ecosystem type and image analysis approach. RS approaches are enhancing studies of the invasion processes and enabling managers to monitor invasions and predict the spread of IAS.

The next three chapters of the book explore how components of diversity can be detected spectrally and remotely with a focus on optical detection methods and technical challenges. Lausch et al. (Chap. 13) delve into the complexity of monitoring vegetation diversity and explain how no single monitoring approach is sufficient on its own. The chapter introduces the range of Earth observation (EO) techniques available for assessing vegetation diversity, covering close-range EO platforms, spectral approaches, plant phenomics facilities, ecotrons, wireless sensor networks (WSNs), towers, air- and spaceborne EO platforms, UAVs, and approaches that integrate air- and spaceborne EO data. The chapter presents the challenges with these approaches and concludes with recommendations and future directions for monitoring vegetation diversity using RS.

Ustin and Jacquemoud (Chap. 14) provide the physical basis for detecting the optical properties of leaves based on how they modify the absorption and scattering of energy to reveal variation in function. The chapter provides considerable detail on how the combination of absorption and scattering properties of leaves together creates the shape of their reflectance spectrum. It also reviews and summarizes the most common interactions between leaf properties and light and the physical processes that regulate the outcomes of these interactions.

Schweiger (Chap. 15) describes a set of best practices for planning field campaigns and collecting and processing data, focusing on spectral data of terrestrial plants collected across various levels of measurements, from leaf to canopy to airborne. These approaches also generally apply to RS of aquatic systems, soil, and the atmosphere and to active RS systems, such as lidar, thermal, and satellite data collection. Schweiger discusses how goals for data collection can be broadly classified into model calibration, model validation, and model interpretation.

The final chapters of the book move to the issues of temporal and spectral scale and integration across scales. Gamon et al. (Chap. 16) present a thorough examination of the challenges in spectral methods for detecting biodiversity posed by issues of spatial, temporal, and spectral dimensions of scale. They explain why the size of the organism relative to the pixel size of detection has consequences for spectral detection of different components of biodiversity and draw on a rich history of literature on scaling effects, including geostatistical approaches for sampling across spatial scales. The chapter emphasizes the importance of developing biodiversity monitoring systems that are "scale-aware" as well as the value of an integrated, multi-scale sampling approach.

Schrodt et al. (Chap. 17) outline how environmental and socioeconomic data can be integrated with biodiversity and RS data to expand knowledge of ecosystem functioning and inform biodiversity conservation decisions. They present the concepts, data, and methods necessary to assess plant species and ecosystem properties across spatial and temporal scales and provide a critical discussion of the major challenges.

Fernández et al. (Chap. 18) provide a framework for understanding Essential Biodiversity Variables (EBVs) to integrate in-situ biodiversity observations and RS through modeling. They argue that open and reproducible workflows for data integration are critical to ensure traceability and reproducibility to allow each EBV to be updated as new data and observation systems become available. The chapter makes the case that the development of a globally coordinated system for biodiversity monitoring will require the mobilization of and integration of in-situ biodiversity data not yet publicly available with emerging RS technologies, novel biodiversity models, and informatics infrastructures.

Schimel et al. (Chap. 19) discuss the prospects and pitfalls for RS of biodiversity at the global scale, focusing on imaging spectroscopy and NASA's Surface Biology and Geology mission concept.

Finally, Geller et al. (Chap. 20) provide an epilogue to the book and present a vision for a global biodiversity monitoring system that is flexible and accessible to a range of user communities. Such a system will require a coordinated effort among space agencies, the RS community, and biologists to bring information about the status and trends in biodiversity, ecosystem functions, and ecosystem services together so that different data streams inform each other and can be integrated. The chapter explains that the Group on Earth Observations Biodiversity Observation Network (GEO BON), the International Long Term Ecological Research Site (ILTER) network, the US National Ecological Observatory Network (NEON), and a variety of sponsors and other organizations are working to enhance coordination and to develop guidelines and standards that will serve this vision.

Indeed, a rapidly advancing global movement has emerged with a shared vision to develop the capacity to monitor the status and trends in the Earth's biodiversity. The authors of this book have sought to contribute to that shared vision through their varied perspectives and experiences. Collectively, the chapters present a range of approaches and knowledge that can transform the ability of humanity to detect and interpret the changing functional biodiversity of planet Earth.

1.5   The Origins of the Book

Before closing, we offer a note of acknowledgment on how this book came into existence. The editors, who themselves have contrasting backgrounds spanning several disciplines, were collaboratively funded, starting in 2013, by the US National Science Foundation (NSF) and the NASA Dimensions of Biodiversity program on the project *Linking remotely sensed optical diversity to genetic, phylogenetic and functional diversity to predict ecosystem processes* (DEB-1342872,1342778). We worked together in several field sites for 5 years to advance our own understanding approaches for remote detection of plant biodiversity. The importance and high potential for rapid advances, as well as the need for the involvement of numerous experts, were obvious from the start. With support from the National Institute for Mathematical and Biological Synthesis (NIMBioS) for the working group on Remote Sensing of Biodiversity, we brought together many of the experts represented within the book, including a symposium in the fall of 2018 where authors shared their work and provided feedback to each other. Further interactions were fostered by the National Center for Ecological Analysis and Synthesis (NCEAS), annual meetings of the NASA Biological Diversity and Ecological Forecasting Program, the Keck Institute for Space Studies, the NSF Research Coordination Network on Biodiversity across scales, and bioDISCOVERY, an international research program fostering collaborative interdisciplinary activities on biodiversity and ecosystem science. bioDISCOVERY, which is part of Future Earth and is hosted and supported by the University Zürich's University Research Priority Program on Global Change and Biodiversity, provided generous financial support for the editing process. Additional financial support was provided by the NSF RCN (DEB: 1745562) and the Keck Institute for Space Studies. Mary Hoff served as technical editor for all of the chapters in the book. We express gratitude to all of these institutions and our many colleagues who contributed to the conception of this work along the way.

References

Beech E, Rivers M, Oldfield S, Smith PP (2017) GlobalTreeSearch: the first complete global database of tree species and country distributions. J Sustain For 36:454–489

Cavender-Bares J, Gamon JA, Hobbie SE, Madritch MD, Meireles JE, Schweiger AK, Townsend PA (2017) Harnessing plant spectra to integrate the biodiversity sciences across biological and spatial scales. Am J Bot 104:1–4. https://doi.org/10.3732/ajb.1700061

Díaz S, Demissew S, Carabias J, Joly C, Lonsdale M, Ash N, Larigauderie A, Adhikari JR, Arico S, Báldi A, Bartuska A, Baste IA, Bilgin A, Brondizio E, Chan KMA, Figueroa VE, Duraiappah A, Fischer M, Hill R, Koetz T, Leadley P, Lyver P, Mace GM, Martin-Lopez B, Okumura M, Pacheco D, Pascual U, Pérez ES, Reyers B (2015) The IPBES conceptual framework—connecting nature and people. Curr Opin Environ Sustain 14:1–16

Gamon JA (2008) Tropical remote sensing – opportunities and challenges. In: Kalacska M, Sanchez-Azofeifa GA (eds) Hyperspectral remote sensing of tropical and subtropical forests. CRC Press. Taylor and Francis Group, Boca Raton, pp 297–304

IPBES (2018a) Summary for policymakers of the regional and subregional assessment of biodiversity and ecosystem services for Europe and Central Asia of the Intergovernmental Science-Policy Platform on Biodiversity and Ecosystem Services. IPBES Secretariat, Bonn

IPBES (2018b) Summary for policymakers of the regional assessment report on biodiversity and ecosystem services for Africa of the Intergovernmental Science-Policy Platform on Biodiversity and Ecosystem Services. IPBES Secretariat, Bonn

IPBES (2018c) Summary for policymakers of the regional assessment report on biodiversity and ecosystem services for Asia and the Pacific of the Intergovernmental Science-Policy Platform on Biodiversity and Ecosystem Services. IPBES Secretariat, Bonn

IPBES (2018d) Summary for policymakers of the regional assessment report on biodiversity and ecosystem services for the Americas of the Intergovernmental Science-Policy Platform on Biodiversity and Ecosystem Services. IPBES Secretariat, Bonn

Jetz W, Cavender-Bares J, Pavlick R, Schimel D, Davis FW, Asner GP, Guralnick R, Kattge J, Latimer AM, Moorcroft P, Schaepman ME, Schildhauer MP, Schneider FD, Schrodt F, Stahl U, Ustin SL (2016) Monitoring plant functional diversity from space. Nature Plants 2:16024

Kew Royal Botanic Gardens (2016) State of the World's plants. Board of Trustees of the Royal Botanic Gardens, Kew, London, UK

Kolbert E (2014) The sixth extinction: an unnatural history. Henry Holt & Company, New York

Leakey R, Roger L (1996) The sixth extinction: patterns of life and the future of humankind. Weidenfeld and Nicolson, London

Madritch MD, Kingdon CC, Singh A, Mock KE, Lindroth RL, Townsend PA (2014) Imaging spectroscopy links aspen genotype with below-ground processes at landscape scales. Philos Trans R Soc Lond B Biol Sci 369:20130194

Magurran AE (2013) Measuring biological diversity. Wiley-Blackwell, Malden

Millennium Ecosystem Assessment (2005) Ecosystems and human well-being: current state & trends assessment. Island Press, Washington, D.C.

Proença V, Martin LJ, Pereira HM, Fernandez M, McRae L, Belnap J, Böhm M, Brummitt N, García-Moreno J, Gregory RD, Honrado JP, Jürgens N, Opige M, Schmeller DS, Tiago P, van Swaay CAM (2017) Global biodiversity monitoring: from data sources to essential biodiversity variables. Biol Conserv 213:256–263

Turner W (2014) Sensing biodiversity. Science 346:301

The National Academy of Sciences (2017) Thriving on our changing planet: a decadal strategy for earth observation from space. The National Academies Press, Washington, D.C.

United Nations (2011) Resolution 65/161. Convention on biological diversity

Wang R, Gamon JA, Cavender-Bares J, Townsend PA, Zygielbaum AI (2018) The spatial sensitivity of the spectral diversity–biodiversity relationship: an experimental test in a prairie grassland. Ecol Appl 28:541–556

Wang Z, Townsend PA, Schweiger AK, Couture JJ, Singh A, Hobbie SE, Cavender-Bares J (2019) Mapping foliar functional traits and their uncertainties across three years in a grassland experiment. Remote Sensing of Environment 221:405–416

Open Access This chapter is licensed under the terms of the Creative Commons Attribution 4.0 International License (http://creativecommons.org/licenses/by/4.0/), which permits use, sharing, adaptation, distribution and reproduction in any medium or format, as long as you give appropriate credit to the original author(s) and the source, provide a link to the Creative Commons license and indicate if changes were made.

The images or other third party material in this chapter are included in the chapter's Creative Commons license, unless indicated otherwise in a credit line to the material. If material is not included in the chapter's Creative Commons license and your intended use is not permitted by statutory regulation or exceeds the permitted use, you will need to obtain permission directly from the copyright holder.

Chapter 2
Applying Remote Sensing to Biodiversity Science

Jeannine Cavender-Bares, Anna K. Schweiger, Jesús N. Pinto-Ledezma, and Jose Eduardo Meireles

A treatment of the topic of biodiversity requires consideration of what biodiversity is, how it arises, what drives its current patterns at multiple scales, how it can be measured, and its consequences for ecosystems. Biodiversity science, by virtue of its nature and its importance for humanity, intersects evolution, ecology, conservation biology, economics, and sustainability science. These realms then provide a basis for discussion of how remote detection of biodiversity can advance our understanding of the many ways in which biodiversity is studied and impacts humanity. We start with a discussion of how biodiversity has been defined and the ways it has been quantified. We briefly discuss the nature and patterns of biodiversity and some of the metrics for describing biodiversity, including remotely sensed spectral diversity. We discuss how the historical environmental context at the time lineages evolved has left "evolutionary legacy effects" that link Earth history to the current functions of plants. We end by considering how remote sensing (RS) can inform our understanding of the relationships among ecosystem services and the trade-offs that are often found between biodiversity and provisioning ecosystem services.

J. Cavender-Bares (✉) · J. N. Pinto-Ledezma
Department of Ecology, Evolution and Behavior, University of Minnesota,
Saint Paul, MN, USA
e-mail: cavender@umn.edu

A. K. Schweiger
Department of Ecology, Evolution and Behavior, University of Minnesota,
Saint Paul, MN, USA

Institut de recherche en biologie végétale, Université de Montréal, Montréal, QC, Canada

J. E. Meireles
Department of Ecology, Evolution and Behavior, University of Minnesota,
Saint Paul, MN, USA

School of Biology and Ecology, University of Maine, Orono, ME, USA

© The Author(s) 2020
J. Cavender-Bares et al. (eds.), *Remote Sensing of Plant Biodiversity*,
https://doi.org/10.1007/978-3-030-33157-3_2

2.1   What Is Biodiversity?

Biodiversity encompasses the totality of variation in life on Earth, including its ecosystems, the species generated through evolutionary history across the tree of life, the genetic variation within them, and the vast variety of functions that each organism, species, and ecosystem possess to access and create resources for life to persist. Changes in the Earth's condition, including the actions of humanity, have consequences for the expression of biodiversity and how it is changing through time.

2.2   The Hierarchical Nature of Biodiversity

Since Darwin (1859), we have understood that biodiversity is generated by a process of descent with modification from common ancestors. As a result, biological diversity is organized in a nested hierarchy that recounts the branching history of species (Fig. 2.1a). Individual organisms are nested within populations, which are nested within species and within increasingly deeper clades. This hierarchy ultimately represents the degree to which species are related to each other and often conveys when in time lineages split (Fig. 2.1a).

Evolution results in the accumulation of changes in traits that causes lineages to differ. The degree of trait divergence between taxa is expected to be proportional to the amount of time they have diverged from a common ancestor. As a consequence, distantly related taxa are expected to be phenotypically more dissimilar (Fig. 2.1b).

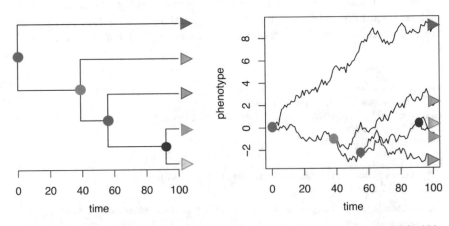

Fig. 2.1 (**a**) The hierarchical organization of biodiversity. Species (triangles) are nested within phylogenetic lineages (clades) due to shared ancestry. All species within a lineage have common ancestor (filled circles). (**b**) Differences in the phenotypes (or trait values) of species (triangles) tend to increase with time since divergence from a common ancestor, shown by the orange circle. The divergence points at which species split are shown in the simulation, and the filled circles indicate the common ancestor of each lineage

Since spectral signals are integrated measures of phenotype, spectra should be more dissimilar among distantly related groups than among close relatives (Cavender-Bares et al. 2016b; McManus et al. 2016; Schweiger et al. 2018). This expectation can be seen from models of evolution in which traits change over time following a random walk (Brownian motion process; Fig. 2.1b) (O'Meara et al. 2006; Meireles et al., Chap. 7). In cases of convergent evolution—where natural selection causes distant relatives to evolve similar functions in similar environments—however, phenotypes can be more similar than expected under Brownian motion.

This hierarchy of life is relevant to RS of plant diversity because certain depths of the tree of life may be more accurately detected than others at different spatial resolutions and geographic regions. For example, it could be easier to detect deeper levels in the hierarchy (such as genera or families) in hyper-diverse communities than in shallow levels (such as species) because deep splits tend to have greater trait divergence. Meireles et al. (Chap. 7) further explain why and how phylogenetic information can be leveraged to detect plant diversity.

2.3 The Making of a Phenotype: Phylogeny, Genes, and the Environment

The phenotype of an organism is the totality of its attributes, and it is quantified in terms of its myriad functions and traits. The phenotype of an organism is a product of the interaction between the information encoded in its genes—the genotype—and the environment over the course of development. Understanding the relative influence of gene combinations, environmental conditions, and ontogenetic stage is an active area of investigation across different disciplines (Diggle 1994; Sultan 2000; Des Marais et al. 2013; Palacio-López et al. 2015).

Although genotypes often play a critical role in determining phenotypic outcomes, many processes can result in mismatches between genotype and phenotype. One of the most well documented of these processes is known as phenotypic plasticity—when organisms with the same genotype display different phenotypes, usually in response to different environmental conditions (Bradshaw 1965; Scheiner 1993; Des Marais et al. 2013). Plasticity can also result in distinct genotypes developing similar phenotypes when growing under the same environmental conditions.

A similar story can be told about the relationship of phenotypic similarity and phylogenetic relatedness. As we have seen earlier, closely related taxa are expected to be more similar to each other than distantly related taxa. However, convergent evolution can lead to plants from different branches of the tree of life to evolve very similar traits—such as succulents, which are found within both euphorbia and very distantly related cacti taxa.

The fact that phenotypes can be, but not necessarily are, directly related to specific genotypes and phylogenetic history should be considered when remotely sensing biodiversity. Only phenotypes can be remotely sensed directly. Genetic and phylogenetic information can only be inferred from spectra to the degree that absorption features of plant chemical or structural characteristics at specific wavelengths relate to phenotypic information. However, the effects plant traits have on spectra are only partially understood. Identifying the regions of the spectrum that are influenced by specific traits is complicated by overlapping absorption features and subtle differences in plant chemical, structural, morphological, and anatomical characteristics that simultaneously influence the shape of the spectral response (Ustin and Jacquemoud, Chap. 14).

2.4   Patterns in Plant Diversity

One of the most intensively studied patterns in biodiversity is the latitudinal gradient, in which low-latitude tropical regions harbor more species, genera, and families of organisms than high-latitude regions. In particular, wet tropical areas tend to reveal higher diversity of organisms than colder and drier climates (Fig. 2.2). Humboldt (1817) documented these patterns quite clearly for plant diversity. Naturalists since then have sought to explain these patterns.

Tropical biomes have existed longer than more recent biomes, such as deserts, Mediterranean climates, and tundra, which expanded as the climate began to cool some 35 million years ago. Tropical biomes also cover more land surface area than other biomes. Tropical species thus have had more time and area (integrated over the time since their first appearance) for species to evolve and maintain viable populations (Fine and Ree 2006). Lineages that originally evolved in the tropics may also have been less able to disperse out of the tropics and to evolve new attributes adapted to cold or dry climates—due to phylogenetic conservatism—restricting their ability to diversify (Wiens and Donoghue 2004). However, not all lineages follow this latitudinal gradient. Ectomycorrhizal fungi, for example, show higher diversity at temperate latitudes, where they likely have higher tree host density (Tedersoo and Nara 2010). Moreover, other measures of diversity do not necessarily follow these patterns. Variation in functional attributes of species, for example, follow different patterns depending on the trait (Cavender-Bares et al. 2018; Echeverría-Londoño et al. 2018; Pinto-Ledezma et al. 2018b). Specific leaf area, one of the functional traits that is highly aligned with the leaf economic spectrum (discussed below), shows higher variation at high latitudes than low latitudes across the Americas. In contrast, seed size shows higher variation at low latitudes (Fig. 2.2b).

At regional scales, variation in the environment, as discussed by Record et al. (Chap. 10), sets the stage for variation in biodiversity because species have evolved to inhabit and can adapt to different environments, which allows them to partition resources and occupy different niches created by environmental variation. Thus, habitat diversity begets biodiversity, and remotely sensed measures of environmental variation have long been known to predict biodiversity patterns (Kerr et al. 2001).

Land area is another long-observed predictor of species diversity, first described for species within certain guilds on island archipelagoes (Diamond and Mayr 1976).

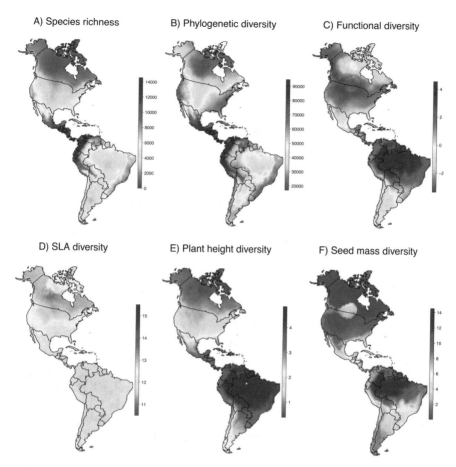

Fig. 2.2 Map of the Americas showing plant species richness, phylogenetic diversity, and functional diversity. Species richness and phylogenetic and functional diversities were estimated based on available information from the Botanical Information and Ecology Network (BIEN) database (Enquist et al. 2016, https://peerj.com/preprints/2615/). Distribution of functional diversity (trait mean) for three functional traits (**d–f**) was log-transformed for plotting purposes. (**a**) Species richness; (**b**) phylogenetic diversity; (**c**) first principal component of functional trait means; (**d**) specific leaf area (mm²/mg); (**e**) plant height (m); and (**f**) seed mass (mg). Diversity metrics were calculated from an estimated presence-absence matrix (PAM) for all vascular plant species at 1 degree spatial resolution (PAM dimension = 5353 pixels × 98,291 species) using range maps and predicted distributions. Functional diversity is based on the first principal component of a principal component analysis (PCA) of species means for the three functional traits

These observations led to the generalization that richness (number of species, S) increases with available land area (A), giving rise to the well-known species-area relationships, in which the log of species number increases linearly with the log of the area available:

$$\log(S) = z \times \log(A) + c, \tag{2.1}$$

or simply,

$$S = cA^{z}, \tag{2.2}$$

where c is the y-intercept of the log-log relationship and z is the slope.

2.5   Functional Traits, Community Assembly, and Evolutionary Legacy Effects on Ecosystems

2.5.1   Functional Traits and the Leaf Economic Spectrum

There is a long history of using functional traits to understand ecological processes, including the nature of species interactions, the assembly of species into ecological communities, and the resulting functions of ecosystems. Species with different functions are likely to have different performance in different environments and to use resources differently, allowing them to partition ecological niches. They are thus less likely to compete for the same resources, promoting their long-term coexistence. An increased focus on trait-based methodological approaches to understanding the relationship between species functional traits and the habitats or ecological niches was spurred by the formalization of the leaf economic spectrum (LES) (Wright et al. 2004). The LES shows that relationships exist among several key traits across a broad range of species and different climates (Reich et al. 1997; Wright et al. 2004) and that simple predictors, such as specific leaf area (SLA, or its reciprocal leaf mass per area, LMA) and leaf nitrogen content, represent a major axis of life history variation. This axis ranges from slow-growing ("conservative") species that tolerate low-resource environments to fast-growing ("acquisitive") species that perform well in high-resource environments (Reich 2014). Variations in relatively easy-to-measure plant traits are tightly coupled to hard-to-measure functions, such as leaf lifespan and growth rate, which reveal more about how a plant invests and allocates resources over time to survive in different kinds of environments. High correlations of functional traits provide strong evidence for trait coordination across the tree of life. The variation in plant function across all of its diversity is relatively constrained and can be explained by a few major axes of trait information (Díaz et al. 2015). Conveniently, traits such as SLA and N are readily detectable via spectroscopy. Other traits—such as leaf lifespan or photosynthetic rates—that are harder to measure but are correlated with these readily detectable traits can thus be inferred, permitting greater insight into ecological processes.

2.5.2 Plant Traits, Community Assembly, and Ecosystem Function

Considerable evidence supports the perspective that plant traits influence how species sort along environmental gradients and are linked to abiotic environmental filters that prevent species without the appropriate traits from persisting in a given location. Traits thus influence the assembly of species in communities—and consequently, the composition, structure, and function of ecosystems. Variation in traits among individual plants and species within communities indicates differences in resource use strategies of plants, which have consequences for ecosystem functions, such as productivity and resistance to disturbance, disease, and extreme environmental conditions. Moreover, the distribution of plant traits within communities influences resource availability for other trophic levels, above- and belowground, which affects community structure and population dynamics in other trophic levels. A major goal of functional ecology is to develop predictive rules for the assembly of communities based on an understanding of which traits or trait combinations (e.g., the leaf-height-seed (LHS) plant ecology strategy, sensu Westoby 1998) are important in a given environment, how traits are distributed within and among species, and how those traits relate to mechanisms driving community dynamics and ecosystem function (Shipley et al. 2017). This predictive framework requires selecting relevant traits; describing trait variation and incorporating this variation into models; and scaling trait data to community- and ecosystem-level processes (Funk et al. 2017). Selecting functional traits for ecological studies is not trivial. Depending on the question, individual traits or trait combinations can be selected that contribute to a mechanistic understanding of the critical processes examined. One can distinguish *response traits*, which influence a species response to its environment, and *effect traits*, which influence ecosystem function (Lavorel and Garnier 2002). These may or may not be different traits. Disturbance or global change factors that influence whether a species can persist within a habitat or community based on its response traits may impact ecosystem functions in complex ways (Díaz et al. 2013). Plant traits are at the heart of understanding how the evolutionary past influences ongoing community assembly processes and ecosystem function (Fig. 2.3). Traits also influence species interactions, which contribute to continuing evolution. Remotely sensed plant traits, if detected and mapped (Serbin and Townsend, Chap. 3; Morsdorf et al., Chap. 4) at the appropriate pixel size and spatial extent (Gamon et al., Chap. 16), can provide a great deal of insight into these different processes (Fig. 2.4).

2.5.3 Phylogenetic, Functional, and Spectral Dispersion in Communities

The rise of phylogenetics in community ecology was based on the idea that functional similarity due to shared ancestry should be predictive of environmental sorting and limiting similarity. These processes depend on physiological tolerances

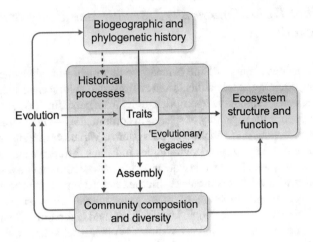

Fig. 2.3 Plant traits that have evolved over time influence how plants assemble into communities, which shapes ecosystem structure and function. Traits reflect biogeographic and environmental legacies and evolve in response to changing environments. They play a central role in ecological processes influencing the distribution of organisms and community assembly. A range of traits influence the way plants reflect light, such that many traits can be mapped continuously across large spatial extents with imaging spectroscopy. The remote detection of plant traits provides incredible potential to observe and understand patterns that reveal information about community assembly, changes in ecosystem function, and how legacies from the past shape community structure and ecosystem processes today. (Reprinted from Cavender-Bares et al. 2019, with permission)

in relation to environmental gradients and intensity of competition as a consequence of shared resource requirements (Webb 2000a, 2002). The underlying conceptual framework was formalized in terms of functional traits in individual case studies (Cavender-Bares et al. 2004; Verdu and Pausas 2007). The tendency to oversimplify the interpretation of phylogenetic patterns in communities, whereby phylogenetic overdispersion was equated with the outcome of competitive exclusion and phylogenetic clustering was interpreted as evidence for environmental sorting, led to a series of studies investigating the importance of scale (Cavender-Bares et al. 2006; Swenson et al. 2006) and the role of Janzen-Connell-type mechanisms, i.e., density-dependent mortality due to pathogens and predators (Gilbert and Webb 2007; Parker et al. 2015). Further developments revealed that the relationship between patterns and ecological processes is context-dependent—in particular, with respect to spatial scale (Emerson and Gillespie 2008; Cavender-Bares et al. 2009; Gerhold et al. 2015). Later studies revisited assumptions about the nature of competition and expected evolutionary and ecological outcomes (Mayfield and Levine 2010). Likewise, interpreting spectral dispersion will depend on the spatial resolution and pixel (grain) size of remotely sensed imagery relative to plant size (Marconi et al. 2019) as well as on the consideration of specific spectral regions and their functional importance. When traits and spectral regions are highly phylogenetically conserved (see Meireles et al., Chap. 7), trait, phylogenetic, and spectral data provide equivalent information. However, when some traits and spectral regions are

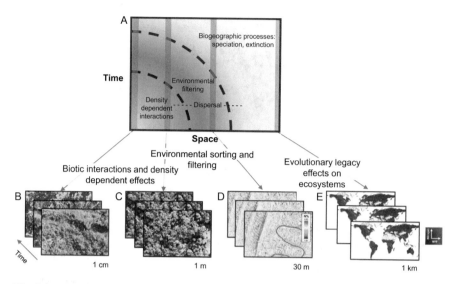

Fig. 2.4 (**a**) Biological processes change with spatial and temporal scale as do the patterns they give rise to. (Adapted from Cavender-Bares et al. 2009.) Detection and interpretation of those patterns will shift with spatial resolution (pixel size) and extent (**b–e**). (**a**) At high spatial resolutions (1 cm pixel size)—that allow detection of individual herbaceous plants and their interactions—and relatively restricted spatial extents in which the abiotic environment is fairly homogeneous, spectral dissimilarity among pixels may indicate complementarity of contrasting functional types. (**b**) The grain size sufficient to detect species interactions is likely to shift with plant size. For example, the interactions of trees in the Minnesota oak savanna and their vulnerability to density-dependent diseases, such as oak wilt (*Bretziella fagacearum*), can be studied at a 1 m pixel size. (**c**) At somewhat larger spatial resolution (30 m pixel sizes) and extent, environmental sorting—which includes interactions of species with both the biotic and abiotic environments—may be detected by comparing spectral similarity of neighbors and comparing mapped functional traits to environmental variation. Images adapted from Singh et al. (2015). The ability to detect change through time may be especially important in understanding species interactions and ecological sorting processes in relation to the biotic and abiotic environment. (**d**) At the global scale, it may be possible to detect the evolutionary legacy effects. For instance, regions with similar climate and geology can differ in vegetation composition and ecosystem function as a consequence of differences in which lineages evolved in a given biogeographic region and their historical migration patterns. Shown are mapped values of %N and NPP based on Moderate Resolution Imaging Spectroradiometer (MODIS) data. (Adapted from Cavender-Bares et al. 2016a)

conserved, but others related to species interactions or with the abiotic environment vary considerably among close relatives, there is the potential to tease apart spectral signals that may relate to species interactions.

Spatial patterns of spectral similarity and dissimilarity also have the potential to provide meaningful information about ecological processes and the forces that dominate community assembly at a particular scale. For example, to the extent that spectral similarity of neighboring plants can be determined, high spectral similarity might indicate that functionally and/or phylogenetically similar individuals are sorting into the same environment, while spectral dissimilarity might indicate that quite distinct individuals are able to coexist if they exhibit complementarity by

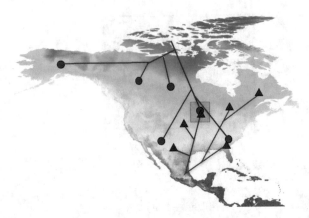

Fig. 2.5 Evolutionary legacy effects as a consequence of biogeographic origin. Two lineages are shown that have contrasting origins, one from the tropics and one from high latitudes. Both diversified and expanded to colonize intermediate latitudes such that their descendants sometimes co-occur. Lineages with ancestors from contrasting climatic environments likely differ in functional traits that reflect their origins and thus may assemble in contrasting microenvironments within the local communities where they co-occur. (Adapted from Cavender-Bares et al. 2016a)

partitioning resources. Approaches that use spectral detection of patterns that might be interpreted within this framework will need to pay close attention to the pixel-to-plant size ratio—or the grain size at which biological diversity varies (Gamon et al., Chap. 16; Serbin and Townsend, Chap. 3; Schimel et al., Chap. 19)— as well as to the spatial extent at which density-dependent processes and environmental sorting pressures are strongest. Often these processes are expected to dominate at different spatial scales, such that competition and Janzen-Connell-type mechanisms operate at very local scales, while environmental sorting may be more important at landscape scales. Other factors, such as the geographic locations and environmental conditions under which lineages diversified, may impact spectral patterns of phylogenetic, functional trait, and spectral similarity at continental scales (Fig. 2.5). At the same time, spectral similarity will be driven by similar ecological forces, since both genetic and phylogenetic compositions, as well as environmental factors, drive phenotypic variation that can be spectrally detected.

2.6   Evolutionary Legacy Effects on Ecosystems

Ecological communities are formed by resident species (incumbents) and colonizer species. Incumbents may have originated in the study region (or at least have had considerable time to adapt to their biotic and abiotic environment), whereas colonizers evolved elsewhere and subsequently dispersed into the region. However, the processes that determine species distributions and the assembly of ecological communities are complex. Species within communities experience unique combinations of evolutionary constraints and innovations due to legacies of their biogeographic origins and the environmental conditions in which they evolved (Cavender-Bares et al. 2016a; Pinto-Ledezma et al. 2018a). Historical contingencies play a role in which lineages can take advantage of opportunities to diversify following climate change or other disturbances and environmental transitions. The rate of species range expansion and contraction and the evolution of species functional traits that allow species to establish and persist in some regions or under particular environmental conditions but not elsewhere are shaped by biogeographic history (Moore et al. 2018). For example, when species from two distinct lineages—one that evolved in tropical climates and the other that evolved in temperate climates—colonize a new environment, they are predicted to persist in contrasting microhabitats as a consequence of niche conservatism (Ackerly 2003; Harrison 2010; Cavender-Bares et al. 2016a). These evolutionary legacies—collectively referred to as "historical factors" (Ricklefs and Schluter 1993)—operate at different spatial and temporal scales that leave their imprints on species current functional attributes and distributions and consequently on ecosystem function itself (Fig. 2.4c, Cavender-Bares et al. 2016a). RS approaches can help reveal how the deep past has influenced current biodiversity patterns and ecosystem function by decoupling climate and geological setting from ecosystem function. Current and forthcoming RS instruments (Lausch et al., Chap. 13; Schimel et al., Chap. 19) enable the monitoring of plant productivity, dynamics of vegetation growth, seasonal changes in chemical composition, and other ecosystem properties independently of climate and geology. These technologies thus provide opportunities to detect how biodiversity is sorted across the globe and to determine how variable ecosystem functions can be in the same geological and environmental setting. Both are important for developing robust predictive models of how lineages respond to current and future environmental conditions with important consequences for managing ecosystems in the Anthropocene.

2.7 Quantifying Multiple Dimensions of Biodiversity

Several major dimensions of biodiversity have emerged in the literature that capture different aspects of the variation of life. *Taxonomic diversity* focuses on differences between species or between higher-order clades, such as genera or families. Estimating the numbers and/or abundances of different taxa across units of area captures this variation. *Phylogenetic diversity* captures the evolutionary distances between species or individuals, represented in terms of millions of years since divergence from a common ancestor or molecular distances based on accumulated mutations since divergence. *Functional diversity* focuses on the variation among species as a consequence of measured differences in their functional traits, frequently calculated as a multivariate metric but also calculated for individual trait variation. *Spectral diversity* captures the variability in spectral reflectance from vegetation (or from other surfaces), either measured and calculated among individual plants or, more commonly, calculated among pixels or among other meaningful spatial units.

Biodiversity metrics can have different components, including (1) taxonomic units; (2) abundance, frequency, or biomass of those units and their degree of evenness; and (3) the dispersion or distances between those units in trait, evolutionary, or spectral space. Myriad metrics quantify the major dimensions and components of diversity. Here we briefly describe several frequently used metrics; the equation for each metric and the source citation that provides the full details are given in Table 2.1.

Table 2.1 A brief summary of metrics that are commonly used to estimate the diversity of different facets/dimensions of plant diversity

Metric	Equation	Definition	Reference
Whittaker alpha	α	Number of species found in a sample or particular area, generally expressed as species richness	Whittaker (1960)
Whittaker beta	γ/α	Variation of species composition between two samples. Can be interpreted as the effective number of distinct compositional units in the region	Whittaker (1960)
Whittaker gamma	γ	Overall diversity (number of species) within a region	Whittaker (1960)
Shannon's H	$-\sum_{i=1}^{s} p_i \ln p_i$	Metric that characterizes species diversity in a sample. Assumes that all species are represented in the sample and that individuals within species were sampled randomly	Shannon (1948)
Simpson's D	$\dfrac{1}{\sum_{i=1}^{S} P_i^2}$	Metric that characterizes species diversity in a sample. Contrary to Shannon's H, Simpson's D captures the variance of the species abundance distribution	Simpson (1949)
Faith's PD (phylogenetic diversity)	$\sum_{e \in z(T)} \lambda e$	Sum of the lengths of all phylogenetic branches (from the root to the tip) spanned by a set of species	Faith (1992)
PSV (phylogenetic species variability)	$\dfrac{ntrC - \sum C}{n(n-1)} = 1 - \bar{c},$	Measures the variability in an unmeasured neutral trait or the relative amount of unshared edge length	Helmus (2007)
PSR (phylogenetic species richness)	nPSV	The deviation from species richness, penalized by close relatives	Helmus (2007)

(continued)

Table 2.1 (continued)

Metric	Equation	Definition	Reference
PSE (phylogenetic species evenness)	$\dfrac{m\,diag(C)'\,M - M'CM}{m^2 - \bar{m}_i m}$	PSV metric modified to account for relative species abundance or simply abundance-weighted PSV	Helmus (2007)
$^qPD(T)$ (phylogenetic branch diversity)	$\left[\sum_{i=1}^{B} L_i x\left(\dfrac{a_i}{\bar{T}}\right)^q\right]^{\frac{1}{1-q}}$	Hill number (the effective total branch length) of the average time of a tree's generalized entropy over evolutionary time intervals	Chao et al. (2010)
$^qD(T)$ (phylogenetic Hill numbers)	$\left(\dfrac{^qPD(\bar{T})}{\bar{T}}\right)$	Effective number of species or lineages	Chao et al. (2010)
FRic (functional richness)	Quickhull algorithm	Quantity of functional space filled by the sample. The number of species within the sample must be higher than the number of functional traits	Barber et al. (1996), Villeger et al. (2008)
FEve (functional evenness)	$\dfrac{\displaystyle\sum_{i=1}^{S-1} \min\left(PEW_i, \dfrac{1}{S-1}\right) - \dfrac{1}{S-1}}{1 - \dfrac{1}{S-1}}$	Quantifies the abundance distribution in functional trait space	Villeger et al. (2008)
FDiv (functional divergence)	$\dfrac{\Delta d + \overline{dG}}{\Delta\lvert d\rvert + \overline{dG}}$	Metric that measures the spread of species abundance across trait space	Villeger et al. (2008)
$^qD(TM)$ (functional trait dispersion)	$1 + (S-1) \times {}^qE(T) \times M'$	Metric that quantify the effective number of functionally distinct species for a given level of species dispersion	Scheiner et al. (2017)
βsor (Sørensen pairwise dissimilarity)	$\dfrac{b+c}{2a+b+c}$	Compares the shared species relative to the mean number of species in a sample. Bray-Curtis dissimilarity is a special case of Sørensen dissimilarity that accounts for species abundance	Sørensen (1948), Baselga (2010)
βsim (Simpson pairwise dissimilarity)	$\dfrac{\min(b,c)}{a + \min(b,c)}$	Similar to Sørensen dissimilarity but independent of species richness	Simpson (1943), Lennon et al. (2001), Baselga (2010)
βjac (Jaccard index)	$\dfrac{a}{a+b+c}$	Metric that compares the shared species to the total number of species in all samples	Jaccard (1900)

2.7.1 The Spatial Scale of Diversity: Alpha, Beta, and Gamma Diversity

Diversity metrics are designed to capture biological variation at different spatial extents. Alpha diversity (α) represents the diversity within local communities, which are usually spatial subunits within a region or landscape. Whittaker first defined beta diversity (β) as the variation in biodiversity among local communities and gamma diversity (γ) as the total biodiversity in a region or a region's species pool (Whittaker 1960).

$$\beta = \frac{\gamma}{\alpha},$$ (2.3)

where β is beta diversity, γ gamma diversity, and α alpha diversity.

Other authors have defined beta diversity differently (see Tuomisto 2010), including using variance partitioning methods (Legendre and De Cáceres 2013).

2.7.2 Taxonomic Diversity

Species richness is the number of species for a given area. It does not include abundance of individuals within species. However, the relative abundances, frequency, and biomass of species within a community matter in terms of capture rarity and evenness. Abundance-weighted metric, such as Simpson's diversity index (D), incorporates both richness and evenness. A set of indices based on Hill numbers—a unified standardization method for quantifying and comparing species diversity across samples, originally presented by Mark Hill (1973)—were refined by Chao et al. (2005, 2010). These are generalizable to all of the dimensions of diversity and consider the number of species and their relative abundances within a local community. Hill numbers require the specification of the diversity order (q), which determines the sensitivity of the metric to species relative abundance. Different orders of q result in different diversity measures; for example, $q = 0$ is simply species richness, $q = 1$ gives the exponential of Shannon's entropy index, and $q = 2$ gives the inverse of Simpson's concentration index.

2.7.3 Phylogenetic Diversity

Phylogenetic diversity (PD) considers the extent of shared ancestry among species (Felsenstein 1985). For example, a plant community composed of two species that diverged from a common ancestor more recently is less phylogenetically diverse than a community of two species that diverged less recently. Faith's (1992) metric of PD sums the branch lengths among species within a community (from the root of the phylogeny to the tip). One feature of this metric is that it scales with species richness because as new species are added into the community, new branch lengths are also added. Other metrics were subsequently developed that calculate the mean evolutionary distances among species independently of the number of species [e.g., mean phylogenetic distance (MPD, Webb 2000b; Webb et al. 2002) or phylogenetic species variability (PSV), Helmus 2007]. Helmus (2007) developed two more phylogenetic diversity metrics that scale either with richness or by incorporating species abundances. Phylogenetic species richness (PSR) increases with the number of species, but reduces the effect of species richness proportionally to their degree of shared ancestry. Phylogenetic species evenness (PSE) is similar to PSV but includes abundances by adding individuals as additional tips descending from a single species node, with branch lengths of 0. Chao et al. (2010) defined the phylogenetic Hill number, $^{q}D(T)$, as the effective number of equally abundant and equally distinct lineages and phylogenetic branch diversity, $^{q}PD(T)$, as the effective total lineage length from the root node (i.e., the total evolutionary history of an assemblage) (Chao et al. 2014).

Phylogenetic endemism is another aspect of biodiversity that can be estimated from phylogenetic information and range maps of species (Faith et al. 2004). Phylogenetic endemism can be simply defined as the quantity of PD restricted to a given geographic area. This metric thus focuses on geographic areas, rather than on species, to discern areas of high endemism based on evolutionary history for conservation purposes.

2.7.4 Functional Diversity

Widely used metrics of functional diversity consider the area or volume of trait space occupied by a community of species, the distances of each species to the center of gravity of those traits, and the trait distances between species (Mouillot et al. 2013). Functional attribute diversity (FAD) is a simple multivariate metric calculated as the sum of species pairwise distances of all measured continuous functional traits (Walker et al. 1999). Villeger et al. (2008) developed a series of functional diversity metrics that incorporate trait dispersion and distance among species as well as species abundances, including functional richness (FRic), functional divergence (FDiv), and functional evenness (FEve). Building on the framework of Villeger et al. (2008), Laliberté and Legendre (2010) developed functional dispersion (FDis), a functional diversity metric that is independent of species richness and can include species relative abundances (Table 2.1).

Scheiner's functional trait dispersion [$^{q}D(TM)$, Scheiner et al. 2017] calculates the effective number of species (or units) that are as distinct as the most distinct species (or unit) in that community. $^{q}D(TM)$ decomposes diversity estimates into three components: the number of units (S), functional evenness [$^{q}E(T)$, the extent to which units are equally dispersed], and mean dispersion [M', the average distance or the distinctiveness of these units]. Functional diversity measured as $^{q}D(TM)$ is maximized when there are more units in a community that are more equitably distributed (or less clumped) and more dispersed (or positioned further apart) in space. Like Chao's approach, $^qD(TM)$ includes Hill numbers (^{q}), which allow weighting of abundances: small and large q values emphasize rare and common species, respectively. Like many other biodiversity metrics, $^qD(TM)$ can be calculated from pairwise distances among species or individuals; thus, the metric can be applied to estimate different dimensions of biodiversity, including functional, phylogenetic (Scheiner 2012; Presley et al. 2014) and spectral components (Schweiger et al. 2018).

Briefly, functional trait dispersion [$^qD(TM)$] is calculated as:

$$^q D(TM) = 1 + (S-1) \times {}^q E(T) \times M^{'} \tag{2.4}$$

where:

S = species richness
$E(T)$ = trait evenness
M' = trait dispersion
q = Hill number

2.7.5   Spectral Diversity

Like taxonomic, functional, and phylogenetic diversity, spectral diversity can be calculated in many different ways. Spectral alpha diversity metrics include the coefficient of variation of spectral indices (Oindo and Skidmore 2002) or spectral bands among pixels (Hall et al. 2010; Gholizadeh et al. 2018, 2019; Wang et al. 2018, the convex hull volume (Dahlin 2016) and the convex hull area (Gholizadeh et al. 2018) of pixels in spectral feature space, the mean distance of pixels from the spectral centroid (Rocchini et al. 2010), the number of spectrally distinct clusters or "spectral species" in ordination space (Féret and Asner 2014), and spectral variance (Laliberté et al. 2019). Schweiger et al. (2018) applied $^qD(TM)$ to species mean spectra and to individual pixels extracted at random from high-resolution proximal RS data. The second approach is independent of species identity and uses the same number of pixels per community for analysis. In this manner, the problem of diversity scaling with the number of species in a community is eliminated, and greater differences in reflectance spectra among pixels result in increased spectral diversity. Conceptually, spectral diversity metrics are versatile and can be tailored to match taxonomic or phylogenetic units, e.g., by using mean spectra for focal taxa, or to resemble functional diversity by selecting spectral bands that align with known absorption features for specific chemical traits or spectral indices that capture plant characteristics of known ecological importance. If measured at the appropriate scale (see Gamon et al. Chap. 16), spectral diversity can integrate the variation captured by other metrics of diversity and similarly predicts ecosystem function (Fig. 2.6).

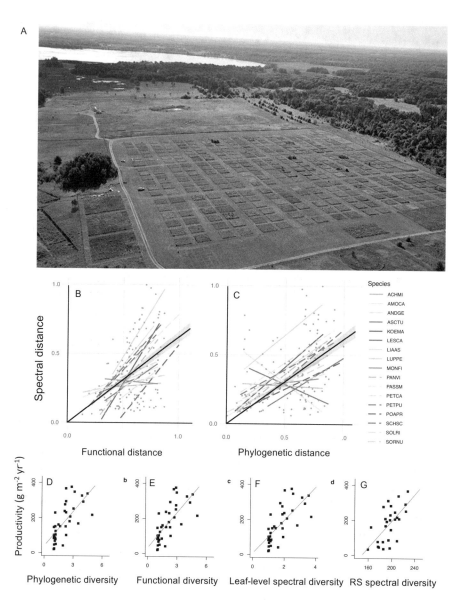

Fig. 2.6 (**a**) Aerial photo of the Cedar Creek long term biodiversity experiment (BioDIV) (Courtesy of Cedar Creek Ecosystem Science Reserve). (**b**) Pairwise phylogenetic and (**c**) functional distances for the 17 most abundant prairie-grassland species in BioDIV are well-predicted by their spectral distances based on leaf-level spectral profiles (400–2500 nm). (**d**) Phylogenetic, (**e**) functional, and (**f**) leaf-level spectral diversities based on Scheiner's $^qD(TM)$ metric all predict ecosystem productivity in BioDIV. (**g**) Independent of information about species identities or their abundances, remotely sensed spectral diversity detected at high spatial resolution (1 mm) also predicts productivity. All graphs are redrawn from Schweiger et al. 2018. Species abbreviations in **b** and **c** are as follows: ACHMI = *Achillea millefolium* L., AMOCA = *Amorpha canescens* Pursh, ANDGE = *Andropogon gerardii* Vitman, ASCTU = *Asclepias tuberosa* L., KOEMA = *Koeleria macrantha* (Ledeb.) Schult., LESCA = *Lespedeza capitata* Michx., LIAAS = *Liatris aspera* Michx., LUPPE = *Lupinus perennis* L., MONFI = *Monarda fistulosa* L., PANVI = *Panicum virgatum* L., PASSMI = *Pascopyrum smithii* (Rydb.) Á. Löve, PETCA = *Petalostemum candidum* (Willd.), PETPU = *Petalostemum purpureum* (Vent.) Rydb., POAPR = *Poa pratensis* L., SCHSC = *Schizachyrium scoparium* Michx., SOLRI = *Solidago rigida* L., SORNU = *Sorghastrum nutans* (L.) Nash

2.7.6 Beta Diversity Metrics

Whittaker's 1960 definition of beta diversity (Eq. 2.3) quantified the degree of differentiation among communities in relation to environmental gradients. Under this definition, beta diversity is defined as the ratio between regional (gamma) and local (alpha) diversities (Eq. 2.3) and measures the number of different communities in a region and the degree of differentiation between them (Whittaker 1960; Jost 2007). Indices such as Bray-Curtis dissimilarity and Jaccard and Sørensen indices evaluate similarity of communities based on the presence or abundance of species within them. Metrics of similarity used for species have been adapted for phylogenetic and functional trait distances (Bryant et al. 2008; Graham and Fine 2008; Kembel et al. 2010; Cardoso et al. 2014) and can equally be applied to spectral information (Gamon et al., Chap. 16).

While the ratio between regional and local communities provides a simple means to estimate beta diversity, there are many different ways to calculate taxonomic, functional, and phylogenetic beta diversity that can be grouped into pairwise and multiple-site metrics (reviewed in Baselga 2010). Notably, beta diversity can be partitioned into components that capture species replacement—the "turnover component"—caused by the exchange of species among communities and differences in the number of species, the "nestedness component," caused by differences in the number of species among communities. The turnover component can be interpreted as the difference between two community assemblages that contain contrasting subsets of species from a regional source pool, while the nestedness component represents the difference in species composition between two communities due to attrition of species in one assemblage relative to the other (Baselga 2010; Cardoso et al. 2014). Examining these different components of beta diversity for multiple dimensions of plant diversity provides a means to discern the role of historical and ongoing environmental sorting processes in the distribution of plant diversity at continental extents (Pinto-Ledezma et al. 2018b). In contrast to traditional diversity metrics, spectral diversity (alpha and beta) is only beginning to receive attention in biodiversity studies (Rocchini et al. 2018). Although different approaches have been proposed (Schmidtlein et al. 2007; Féret and Asner 2014; Rocchini et al. 2018; Laliberté et al. 2019), the estimation and mapping of dissimilarities in spectral composition (i.e., the variation among pixels) is similar to traditional estimations of beta diversity. For example, Laliberté et al. (2019) adapted the total community composition variance approach (Legendre and De Cáceres 2013) to estimate spectral diversity as spectral variance, partitioning the spectral diversity of a region (gamma diversity) into additive alpha and beta diversity components.

2.8 Links Between Plant Diversity, Other Trophic Levels, and Ecosystem Functions

Plant diversity has consequences for other trophic levels, sometimes reducing herbivory on focal species (Castagneyrol et al. 2014), but also increasing the diversity of insects and their predators in an ecosystem (Dinnage et al. 2012; Lind et al. 2015).

The distribution of plant traits within communities influences resource availability for other trophic levels above- and belowground, which affects community assembly and population dynamics across trophic levels. Diversity of neighbors surrounding focal trees can both increase and decrease pathogen and herbivore pressure on them (Grossman et al. 2019). Thus, while we know that plant diversity impacts other trophic levels, consistent rules across the globe that explain how and why these impacts occur remain elusive. An increasing number of studies reveal that plant diversity influences belowground microbial diversity and composition (Madritch et al. 2014; Cline et al. 2018). While these relationships are significant, they may explain limited variation given the number of other factors that influence microbial diversity and potentially due to a mismatch in sampling scales. Ultimately, it appears that chemical composition and productivity of aboveground components of ecosystems that can be remotely sensed are critical drivers of belowground processes, including microbial diversity (Madritch et al., Chap. 8).

Biodiversity loss is known to substantially decrease ecosystem functioning and ecosystem stability (Cardinale et al. 2011; O'Connor et al. 2017). Yet, the nature and scale of biodiversity-ecosystem function relationships remains a central question in biodiversity science. The issue is one that is ready to be tackled across scales using RS technology. The long-term biodiversity experiment at Cedar Creek Ecosystem Science Reserve (Tilman 1997) (Fig. 2.6), for example, has revealed the increasing effects of biodiversity on productivity over time (Reich et al. 2012) and that phylogenetic and functional diversity are highly predictive of productivity (Cadotte et al. 2008; Cadotte et al. 2009). Remotely sensed spectral diversity also predicts productivity (Sect. 2.9). Increased stability has also been linked to both higher plant richness (Tilman et al. 2006) and phylogenetic diversity (Cadotte et al. 2012) in this experiment. Tree diversity experiments show similar effects of increasing productivity with diversity (Tobner et al. 2016; Grossman et al. 2017) (Fig. 2.7), and these same trends emerge as the dominant pattern in forest plots globally

Fig. 2.7 The Forest and Biodiversity (FAB) experiment at the Cedar Creek Ecosystem Science Reserve shows overyielding (**a**)—greater productivity than expected in species-rich communities compared to monocultures—also called the net biodiversity effect (NBE). Curves show 90% predictions from multiple linear regression models (yellow 2013–2014; blue 2014–2015). (Redrawn from Grossman et al. 2018.) Photos (**b, c**) show juvenile trees grown in mixtures with varying neighborhood composition. The first phase of the experiment, shown here, includes three 600 m² blocks, each consisting of 49 plots (9.25 m²) planted in a grid with 0.5 m spacing

(Liang et al. 2016). Hundreds of rigorous biodiversity experiments have been designed and conducted to tease apart effects of changing numbers of species (richness) from effects of changing identities of species (composition) (O'Connor et al. 2017; Grossman et al. 2018; Isbell et al. 2018). Complementarity among diverse plant species that vary in their functional attributes and capture and respond to resources differently is the primary explanation for increasing productivity with diversity (Williams et al. 2017). Nevertheless, both the nature of biodiversity-ecosystem function (BEF) relationships and their causal mechanisms remain variable and scale dependent in natural systems. In the Nutrient Network global grassland experiments, in which communities have assembled naturally, the relationship between diversity and productivity is variable (Adler et al. 2011). In tropical forest plots around the globe, at spatial extents of 0.04 ha or less, the biodiversity-productivity relationship is strong. However, as scales increase to 0.25 or 1.0 ha, the relationship is no longer consistently positive and can frequently be negative (Chisholm et al. 2013). These varied relationships at contrasting spatial scales may result from nonlinear, hump-shaped relationships between biodiversity and ecosystem function across resource availability gradients as the nature of species interactions and their level of complementarity shift (Jaillard et al. 2014). RS methods—including imaging spectroscopy and LiDAR—that can detect both the diversity and the structure and function of ecosystems (Martin, Chap. 5; Atkins et al. 2018) can discern these relationships across spatial extents and biomes in natural systems. They thus have high potential to enhance our understanding of the scale and context dependence of linkages between biodiversity and ecosystem function (Grossman et al. 2018).

2.9   Incorporating Spectra into Relationships Between Biodiversity and Ecosystem Function

Detection of spectral diversity, in particular, offers the potential to contribute to the quantification of BEF relationships at large scales (Schweiger et al. 2018) and is thus worth discussing in more detail. The variability captured by spectral diversity in a given ecosystem depends on the way the spectral diversity is calculated, as well as its spatial and spectral resolution (Sect. 2.7.5; Gamon et al., Chap. 16). From a functional perspective, spectral profiles measured at the leaf level depend on the chemical, structural, morphological, and anatomical characteristics of leaves (Ustin and Jacquemoud, Chap. 14). Variation in spectra and spectral diversity can be used to test hypotheses about how specific traits influence ecosystem function, community composition, and other characteristics of ecosystems, when using spectral bands or spectral indices with known associations with specific plant traits (Serbin and Townsend, Chap. 3). Moreover, spectral bands and indices can be weighted based on prior information about the relative contribution of individual traits to specific ecosystem characteristics. However, while the absorption features of some chemical traits are known, the effects of other, particularly nonchemical, plant traits on spectra are less well understood, in part due to overlapping spectral features and challenges associated with accurately describing nonchemical traits (Ustin and Jacquemoud, Chap. 14). Using the full spectral profile of plants in spectral diversity calculations provides a means to integrate chemical, structural, morphological, and anatomical variation and to acknowledge the many ways plants differ from one another.

It is certainly more complicated to decipher the biological meaning of spectral diversity calculated from spectral profiles than from measures of biodiversity that are based on a specific set of plant traits or spectral bands or indices with known links to specific traits. However, the variance that is explained by models based on spectral profiles can be partitioned into known and unknown sources of variation. This provides a means to assess the relative contribution of traits with known spectral characteristics and traits that are less well understood spectrally or that are of yet-unrecognized importance. At the canopy level, when spectra are measured from a distance, the question of what spectra and spectral diversity represent is further complicated by the influences that plant architecture, soil, and other materials have on the spectral characteristics of image pixels (Wang et al. 2018; Gholizadeh et al. 2018). Again, the degree to which these characteristics matter for a particular ecosystem needs to be evaluated in the particular context of the study. Some ecosystem components such as shade, soil, rock, or debris, which influence remotely sensed spectra, are biologically meaningful because they influence light availability and microclimate and provide resources for other trophic levels.

The association between plant spectra and traits can be illustrated by plotting spectral distances against functional distances or dissimilarity, as illustrated using species from the Cedar Creek biodiversity experiment (Fig. 2.6d). Given that functional differences among species are expected to increase with evolutionary divergence time (Fig. 2.1b), positive relationships are also expected among spectral and phylogenetic distances. The observed associations among spectral, functional, and phylogenetic dissimilarity (Fig. 2.6a, b) allow biodiversity metrics based on any of these dimensions of biodiversity to explain a similar proportion of the total variability in aboveground productivity (Fig. 2.6c–e), which is known to increase with the functional diversity of the plant community in this system (Cadotte et al. 2009). The species in the biodiversity experiment at Cedar Creek are relatively functionally dissimilar and distantly related, such that spectral, functional, and phylogenetic diversity also predict species richness (not shown). One advantage of spectral diversity is that the metric can be calculated from remotely sensed image pixels without depending on information about the distribution and abundance of species in an area, their functional traits, or phylogenetic relationships (Schweiger et al. 2018). By extracting a random number of high-resolution image pixels in each plant community, Schweiger et al. (2018) found that remotely sensed spectral diversity explained the biodiversity effect on aboveground productivity about as well as spectral diversity calculated using leaf-level spectra (Fig. 2.6).

2.10   Links Between Biodiversity and Ecosystem Services

Humans benefit from ecosystem functions and biodiversity. The benefits we derive from nature, often called ecosystem services, are a product of the biodiversity—assembled over millions of years—and ecosystem properties of a given region, or the whole Earth (Daily 1997). Daily (1997) defines ecosystem services as "the conditions and processes through which natural ecosystems, and the species that make them up, sustain and fulfill human life." Ecosystem services, referred to by the Intergovernmental Science-Policy Platform on Biodiversity and Ecosystem Services (IPBES) as "nature's contributions to people" (Díaz et al. 2018), are a socioecological concept that emerged from the Millennium Ecosystem Assessment (2005) and include provisioning, regulating, supporting, and cultural services. Some ecosystem service categories include direct benefits of biodiversity—through the use and spiritual values that humans establish with elements of biodiversity and ecosystems—and indirect benefits through the contributions of biodiversity to critical regulating ecosystem functions. The diversities of functional traits of plants make up the primary productivity of life on Earth and are essential to the ecosystem services on which all life depends. Assessment of ecosystem services depends on understanding both the ecosystem functions on which ecosystem services are derived and how services are valued by humans (Schrodt et al., Chap. 17). Modeling efforts that incorporate remotely sensed data can be used to describe ecosystem functions and quantify the services they generate (Sharp et al. 2018). (For modeling tools that enable mapping and valuing ecosystem services, see https://naturalcapitalproject.stanford.edu/invest/.)

2.11   Trade-Offs Between Biodiversity and Ecosystem Services

Biodiversity—as well as many regulating services to which biodiversity contributes and upon which it depends—frequently shows a negative trade-off with provisioning ecosystem services, such as agricultural production (Haines-Young and Potschin 2009). The nature of these trade-offs depends on the biophysical context, including the climate, soils, hydrology, and geology, and will differ among regions. A trade-off curve represents the limits set by these biophysical constraints and can be thought of in economic terms as an "efficiency frontier" that sets the boundaries on possible combinations of biodiversity (or regulating services) and provisioning services (Polasky et al. 2008). Combinations above the curve are not possible; outcomes beneath the curve provide fewer total benefits than what is actually possible from the environment. Quantifying the biodiversity and ecosystem service potential from land and how they trade off are critical to efficient management of ecosystems. Current RS tools and forthcoming technologies are well-poised to decrease uncertainty in estimates of biodiversity—ecosystem service trade-offs—and can

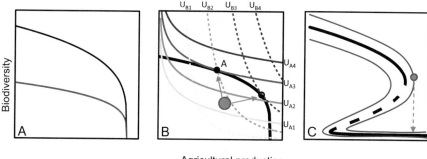

Agricultural production

Fig. 2.8 RS technologies that enable detection of biodiversity and ecosystem functions aid in modeling their trade-offs. (**a**) The "efficiency frontier," or biophysical constraints that limit biodiversity and crop production, depends on the specific climatic, historical, and resource context of the land area and on the growth or replenishment rate of the natural system. These constraints can vary among ecosystems (red vs black curves). (**b**) Where along the efficiency frontier we want to manage for depends on human values. The superimposed curves show isolines of equal utility (U_{A1-4} or U_{B1-4}) for two different stakeholders (A and B) who have sharply different willingness to give up natural habitat for crop production and vice versa. Utility—or benefits to each stakeholder—increases moving from yellow to dark red. The two points at which the highest utility curve for each stakeholder intersects with the efficiency frontier represent the greatest feasible benefit to the stakeholder (points A and B). Often ecosystems are managed well below the efficiency frontier (green circle). RS may enable detection of components of biodiversity and ecosystem services at relevant spatial scales that can inform stakeholders about improved outcomes and aid negotiation among stakeholders. (**c**) Some trade-offs can have thresholds and tipping points that, once traversed, may result in a degraded alternative state. RS approaches that can aid in predicting uncertainties and temporal variability in trade-offs, indicated by thin blue lines, can help maximize ecosystem service benefits without overshooting thresholds that risk pushing the system into a degraded state. (Adapted from Cavender-Bares et al. 2015b)

contribute meaningfully to decision-making and resource management (Chaplin-Kramer et al. 2015; de Araujo Barbosa et al. 2015; Schrodt et al., Chap. 17).

Where along the efficiency frontier we wish to target our management efforts depends on human preferences. These can differ strongly among different stakeholders that have contrasting priorities (Cavender-Bares et al. 2015a, b). Distinguishing the biophysical limits of ecosystems from contrasting stakeholder preferences for what they want from ecosystems is a critical contribution to participatory processes that enable dialogue and progress toward sustainability (Cavender-Bares et al. 2015b; King et al. 2015). RS technologies that can enhance detection of biodiversity as well as both regulating and provisioning ecosystem services—and changes in these at multiple scales—can thus increase clarity in decision-making processes in the face of rapid global change (Fig. 2.8).

Acknowledgments We thank Shan Kothari for providing valuable feedback on an earlier version of the manuscript. The authors acknowledge funding from a NSF and NASA Dimensions grant (DEB 1342778), as well as the Cedar Creek NSF Long-Term Ecological Research grant (DEB

1234162) for support of some of the data presented here. We also thank the National Institute for Mathematical and Biological Synthesis (NIMBioS) working group on "Remotely Sensing Biodiversity," the Keck Institute for Space Studies working group "Unlocking a New Era in Biodiversity Science," and the NSF Research Coordination Network "Biodiversity Across Scales" (DEB 1745562).

References

Ackerly DD (2003) Community assembly, niche conservatism and adaptive evolution in changing environments. Int J Plant Sci 164:S165–S184

Adler PB, Seabloom EW, Borer ET, Hillebrand H, Hautier Y, Hector A, Harpole WS, O'Halloran LR, Grace JB, Anderson TM, Bakker JD, Biederman LA, Brown CS, Buckley YM, Calabrese LB, Chu C-J, Cleland EE, Collins SL, Cottingham KL, Crawley MJ, Damschen EI, Davies KF, DeCrappeo NM, Fay PA, Firn J, Frater P, Gasarch EI, Gruner DS, Hagenah N, Hille Ris Lambers J, Humphries H, Jin VL, Kay AD, Kirkman KP, Klein JA, Knops JMH, La Pierre KJ, Lambrinos JG, Li W, MacDougall AS, McCulley RL, Melbourne BA, Mitchell CE, Moore JL, Morgan JW, Mortensen B, Orrock JL, Prober SM, Pyke DA, Risch AC, Schuetz M, Smith MD, Stevens CJ, Sullivan LL, Wang G, Wragg PD, Wright JP, Yang LH (2011) Productivity is a poor predictor of plant species richness. Science 333:1750–1753

Atkins JW, Fahey RT, Hardiman BS, Gough CM (2018) Forest canopy structural complexity and light absorption relationships at the subcontinental scale. J Geophys Res Biogeosci 123:1387–1405

Barber CB, Dobkin DP, Huhdanpaa H (1996) The quickhull algorithm for convex hulls. ACM Trans Math Softw 22(4):469–483

Baselga A (2010) Partitioning the turnover and nestedness components of beta diversity: partitioning beta diversity. Glob Ecol Biogeogr 19:134–143

Bradshaw A (1965) Evolutionary significance of phenotypic plasticity in plants. Adv Genet 13:115–155

Bryant JA, Lamanna C, Morlon H, Kerkhoff AJ, Enquist BJ, Green JL (2008) Microbes on mountainsides: contrasting elevational patterns of bacterial and plant diversity. Proc Natl Acad Sci 105:11505–11511

Cadotte MW, Cardinale BJ, Oakley TH (2008) Evolutionary history and the effect of biodiversity on plant productivity. Proc Natl Acad Sci 105:17012–17017

Cadotte MW, Cavender-Bares J, Tilman D, Oakley TH (2009) Using phylogenetic, functional and trait diversity to understand patterns of plant community productivity. PLoS One 4:e5695

Cadotte MW, Dinnage R, Tilman D (2012) Phylogenetic diversity promotes ecosystem stability. Ecology 93:S223–S233

Cardinale BJ (2011) Biodiversity improves water quality through niche partitioning. Nature 472:86–89

Cardoso P, Rigal F, Carvalho JC, Fortelius M, Borges PAV, Podani J, Schmera D (2014) Partitioning taxon, phylogenetic and functional beta diversity into replacement and richness difference components. J Biogeogr 41:749–761

Castagneyrol B, Jactel H, Vacher C, Brockerhoff EG, Koricheva J (2014) Effects of plant phylogenetic diversity on herbivory depend on herbivore specialization. J Appl Ecol 51:134–141

Cavender-Bares J, Ackerly DD, Baum DA, Bazzaz FA (2004) Phylogenetic overdispersion in Floridian oak communities. Am Nat 163:823–843

Cavender-Bares J, Keen A, Miles B (2006) Phylogenetic structure of floridian plant communities depends on taxonomic and spatial scale. Ecology 87:S109–S122

Cavender-Bares J, Kozak KH, Fine PVA, Kembel SW (2009) The merging of community ecology and phylogenetic biology. Ecol Lett 12:693–715

Cavender-Bares J, Balvanera P, King E, Polasky S (2015a) Ecosystem service trade-offs across global contexts and scales. Ecol Soc 20:22

Cavender-Bares J, Polasky S, King E, Balvanera P (2015b) A sustainability framework for assessing trade-offs in ecosystem services. Ecol Soc 20:17

Cavender-Bares J, Ackerly DD, Hobbie SE, Townsend PA (2016a) Evolutionary legacy effects on ecosystems: biogeographic origins, plant traits, and implications for management in the era of global change. Annu Rev Ecol Evol Syst 47:433–462

Cavender-Bares J, Meireles JE, Couture JJ, Kaproth MA, Kingdon CC, Singh A, Serbin SP, Center A, Zuniga E, Pilz G, Townsend PA (2016b) Associations of leaf spectra with genetic and phylogenetic variation in oaks: prospects for remote detection of biodiversity. Remote Sens 8:221. https://doi.org/10.3390/rs8030221

Cavender-Bares J, Arroyo MTK, Abell R, Ackerly D, Ackerman D, Arim M, Belnap J, Moya FC, Dee L, Estrada-Carmona N, Gobin J, Isbell F, Köhler G, Koops M, Kraft N, Macfarlane N, Mora A, Piñeiro G, Martínez-Garza C, Metzger J-P, Oatham M, Paglia A, Peri PL, Randall R, Weis J (2018) Status and trends of biodiversity and ecosystem functions underpinning nature's benefit to people. In: Regional and subregional assessments of biodiversity and ecosystem services: regional and subregional assessment for the Americas. IPBES Secretariat, Bonn

Cavender-Bares J (2019) Diversification, adaptation, and community assembly of the American oaks (Quercus), a model clade for integrating ecology and evolution. New Phytol 221:669–692

Chao A, Chazdon RL, Colwell RK, Shen T-J (2005) A new statistical approach for assessing similarity of species composition with incidence and abundance data. Ecol Lett 8:148–159

Chao A, Chiu C-H, Jost L (2010) Phylogenetic diversity measures based on Hill numbers. Philos Trans R Soc Lond Ser B Biol Sci 365:3599–3609

Chao A, Chiu C-H, Jost L (2014) Unifying species diversity, phylogenetic diversity, functional diversity, and related similarity and differentiation measures through Hill numbers. Annu Rev Ecol Evol Syst 45:297–324

Chaplin-Kramer R, Sharp RP, Mandle L, Sim S, Johnson J, Butnar I, Milà i Canals L, Eichelberger BA, Ramler I, Mueller C, McLachlan N, Yousefi A, King H, Kareiva PM (2015) Spatial patterns of agricultural expansion determine impacts on biodiversity and carbon storage. Proc Natl Acad Sci 112:7402

Chisholm RA, Muller-Landau HC, Abdul Rahman K, Bebber DP, Bin Y, Bohlman SA, Bourg NA, Brinks J, Bunyavejchewin S, Butt N, Cao H, Cao M, Cárdenas D, Chang L-W, Chiang J-M, Chuyong G, Condit R, Dattaraja HS, Davies S, Duque A, Fletcher C, Gunatilleke N, Gunatilleke S, Hao Z, Harrison RD, Howe R, Hsieh C-F, Hubbell SP, Itoh A, Kenfack D, Kiratiprayoon S, Larson AJ, Lian J, Lin D, Liu H, Lutz JA, Ma K, Malhi Y, McMahon S, McShea W, Meegaskumbura M, Razman SM, Morecroft MD, Nytch CJ, Oliveira A, Parker GG, Pulla S, Punchi-Manage R, Romero-Saltos H, Sang W, Schurman J, Su S-H, Sukumar R, Sun IF, Suresh HS, Tan S, Thomas D, Thomas S, Thompson J, Valencia R, Wolf A, Yap S, Ye W, Yuan Z, Zimmerman JK (2013) Scale-dependent relationships between tree species richness and ecosystem function in forests. J Ecol 101:1214–1224

Cline LC, Hobbie SE, Madritch MD, Buyarski CR, Tilman D, Cavender-Bares JM (2018) Resource availability underlies the plant-fungal diversity relationship in a grassland ecosystem. Ecology 99:204–216

Dahlin KM (2016) Spectral diversity area relationships for assessing biodiversity in a wildland-agriculture matrix. Ecol Appl 26(8):2758–2768

Daily GC (1997) Nature's services. Island Press, Washington, D.C.

Darwin C (1859) On the origin of species. Murray, London

de Araujo Barbosa CC, Atkinson PM, Dearing JA (2015) Remote sensing of ecosystem services: a systematic review. Ecol Indic 52:430–443

Des Marais D, Hernandez K, Juenger T (2013) Genotype-by-environment interaction and plasticity: exploring genomic responses of plants to the abiotic environment. Annu Rev Ecol Evol Syst 44:5–29

Diamond JM, Mayr E (1976) Species-area relation for birds of the Solomon Archipelago. Proc Natl Acad Sci U S A 73:262–266

Díaz S, Purvis A, Cornelissen JHC, Mace GM, Donoghue MJ, Ewers RM, Jordano P, Pearse WD (2013) Functional traits, the phylogeny of function, and ecosystem service vulnerability. Ecol Evol 3:2958–2975

Díaz S, Kattge J, Cornelissen JHC, Wright IJ, Lavorel S, Dray S, Reu B, Kleyer M, Wirth C, Colin Prentice I, Garnier E, Bönisch G, Westoby M, Poorter H, Reich PB, Moles AT, Dickie J, Gillison AN, Zanne AE, Chave J, Joseph Wright S, Sheremet'ev SN, Jactel H, Baraloto C, Cerabolini B, Pierce S, Shipley B, Kirkup D, Casanoves F, Joswig JS, Günther A, Falczuk V, Rüger N, Mahecha MD, Gorné LD (2015) The global spectrum of plant form and function. Nature 529:167

Díaz S, Pascual U, Stenseke M, Martín-López B, Watson RT, Molnár Z, Hill R, Chan KMA, Baste IA, Brauman KA, Polasky S, Church A, Lonsdale M, Larigauderie A, Leadley PW, van Oudenhoven APE, van der Plaat F, Schröter M, Lavorel S, Aumeeruddy-Thomas Y, Bukvareva E, Davies K, Demissew S, Erpul G, Failler P, Guerra CA, Hewitt CL, Keune H, Lindley S, Shirayama Y (2018) Assessing nature's contributions to people. Science 359:270

Diggle P (1994) The expression of andromonoecy in *Solanum hirtum* (Solanaceae)—phenotypic plasticity and ontogenetic contingency. Am J Bot 81:1354–1365

Dinnage R, Cadotte MW, Haddad NM, Crutsinger GM, Tilman D (2012) Diversity of plant evolutionary lineages promotes arthropod diversity. Ecol Lett 15:1308–1317

Echeverría-Londoño S, Enquist BJ, Neves DM, Violle C, Boyle B, Kraft NJB, Maitner BS, McGill B, Peet RK, Sandel B, Smith SA, Svenning J-C, Wiser SK, Kerkhoff AJ (2018) Plant functional diversity and the biogeography of biomes in North and South America. Front Ecol Evol 6:219

Emerson BC, Gillespie RG (2008) Phylogenetic analysis of community assembly and structure over space and time. Trends Ecol Evol 23:619–630

Enquist BJ, Condit R, Peet RK, Schildhauer M, Thiers BM (2016) Cyberinfrastructure for an integrated botanical information network to investigate the ecological impacts of global climate change on plant biodiversity. https://doi.org/10.7287/peerj.preprints.2615v2

Faith DP (1992) Conservation evaluation and phylogenetic diversity. Biol Conserv 61:1–10

Faith DP, Reid CAM, Hunter J (2004) Integrating phylogenetic diversity, complementarity, and endemism for conservation assessment. Conserv Biol 18:255–261

Felsenstein J (1985) Phylogenies and the comparative method. Am Nat 125:1–15

Féret J-B, Asner GP (2014) Mapping tropical forest canopy diversity using high-fidelity imaging spectroscopy. Ecol Appl 24:1289–1296

Fine P, Ree R (2006) Evidence for a time integrated species area effect on the latitudinal gradient in tree diversity. Am Nat 168:796–804

Funk JL, Larson JE, Ames GM, Butterfield BJ, Cavender-Bares J, Firn J, Laughlin DC, Sutton-Grier AE, Williams L, Wright J (2017) Revisiting the Holy Grail: using plant functional traits to understand ecological processes: plant functional traits. Biol Rev 92:1156–1173

Gerhold P, Cahill JF, Winter M, Bartish IV, Prinzing A (2015) Phylogenetic patterns are not proxies of community assembly mechanisms (they are far better). Funct Ecol 29:600–614

Gholizadeh H, Gamon JA, Zygielbaum AI, Wang R, Schweiger AK, Cavender-Bares J (2018) Remote sensing of biodiversity: soil correction and data dimension reduction methods improve assessment of α-diversity (species richness) in prairie ecosystems. Remote Sens Environ 206:240–253

Gholizadeh H, Gamon JA, Townsend PA, Zygielbaum AI, Helzer CJ, Hmimina GY, Yu R, Moore RM, Schweiger AK, Cavender-Bares J (2019) Detecting prairie biodiversity with airborne remote sensing. Remote Sens Environ 221:38–49

Gilbert GS, Webb CO (2007) Phylogenetic signal in plant pathogen-host range. Proc Natl Acad Sci U S A 104:4979–4983

Graham C, Fine P (2008) Phylogenetic beta diversity: linking ecological and evolutionary processes across space and time. Ecol Lett 11:1265–1277

Grossman JJ, Cavender-Bares J, Hobbie SE, Reich PB, Montgomery RA (2017) Species richness and traits predict overyielding in stem growth in an early-successional tree diversity experiment. Ecology 98:2601–2614

Grossman JJ, Vanhellemont M, Barsoum N, Bauhus J, Bruelheide H, Castagneyrol B, Cavender-Bares J, Eisenhauer N, Ferlian O, Gravel D, Hector A, Jactel H, Kreft H, Mereu S, Messier C, Muys B, Nock C, Paquette A, Parker J, Perring MP, Ponette Q, Reich PB, Schuldt A, Staab M, Weih M, Zemp DC, Scherer-Lorenzen M, Verheyen K (2018) Synthesis and future research directions linking tree diversity to growth, survival, and damage in a global network of tree diversity experiments. Environ Exp Bot 152:68

Grossman JJ, Cavender-Bares J, Reich PB, Montgomery RA, Hobbie SE (2019) Neighborhood diversity simultaneously increased and decreased susceptibility to contrasting herbivores in an early stage forest diversity experiment. J Ecol 107:1492–1505

Haines-Young R, Potschin M (2009) The links between biodiversity, ecosystem services and human well-being. In: Raffaelli D (ed) Ecosystem ecology: a new synthesis. Cambridge University Press, Cambridge

Hall K, Johansson LJ, Sykes MT, Reitalu T, Larsson K, Prentice HC (2010) Inventorying management status and plant species richness in semi-natural grasslands using high spatial resolution imagery. Appl Veg Sci 13(2):221–233

Harrison S, Damschen EI, Grace JB (2010) Ecological contingency in the effects of climatic warming on forest herb communities. Proc Natl Acad Sci USA 107:19362–19367

Helmus MR (2007) Phylogenetic measures of biodiversity. Am Nat 169:E68–E83

Hill MO (1973) Diversity and evenness: a unifying notation and its consequences. Ecology 54:427–432

Humboldt, A (1817) *De distributione geographica plantarum secundum coeli temperiem et altitudinem montium: prolegomena* (On the Distribution of Plants) CE, Lutetiae Parisiorum, In: Libraria Graeco-Latino-Germanica

Isbell F, Cowles J, Dee LE, Loreau M, Reich PB, Gonzalez A, Hector A, Schmid B (2018) Quantifying effects of biodiversity on ecosystem functioning across times and places. Ecol Lett 21:763–778

Jaccard P (1900) Contribution au problème de l'immigration post-glaciaire de la flore alpine. Bull Soc Vaud Sci Nat 36:87–130

Jaillard B, Rapaport A, Harmand J, Brauman A, Nunan N (2014) Community assembly effects shape the biodiversity-ecosystem functioning relationships. Funct Ecol 28:1523–1533

Jost L (2007) Partitioning diversity into independent alpha and beta components. Ecology 88:2427–2439

Kembel SW, Ackerly DD, Blomberg SP, Cornwell WK, Cowan PD, Helmus MR, Morlon H, Webb CO (2010) Picante: R tools for integrating phylogenies and ecology. Bioinformatics 26:1463–1464

Kerr JT, Southwood TRE, Cihlar J (2001) Remotely sensed habitat diversity predicts butterfly species richness and community similarity in Canada. Proc Natl Acad Sci 98:11365–11370

King E, Cavender-Bares J, Balvanera P, Mwampamba TH, Polasky S (2015) Trade-offs in ecosystem services and varying stakeholder preferences: evaluating conflicts, obstacles, and opportunities. Ecol Soc 20:25

Laliberté E, Legendre P (2010) A distance-based framework for measuring functional diversity from multiple traits. Ecology 91(1):299–305

Legendre P, De Cáceres M (2013) Beta diversity as the variance of community data: dissimilarity coefficients and partitioning. Ecol Lett 16(8):951–963

Laliberté E, Schweiger AK, Legendre P (2019) Partitioning plant spectral diversity into alpha and beta components. Ecol Lett https://doi.org/10.1111/ele.13429

Lavorel S, Garnier E (2002) Predicting changes in community composition and ecosystem functioning from plant traits: revisiting the Holy Grail. Funct Ecol 16:545–556

Lennon JJ, Koleff P, Greenwood JJD, Gaston KJ (2001) The geographical structure of British bird distributions: diversity, spatial turnover and scale. J Anim Ecol 70:966–979

Liang J, Crowther TW, Picard N, Wiser S, Zhou M, Alberti G, Schulze E-D, McGuire AD, Bozzato F, Pretzsch H, de-Miguel S, Paquette A, Hérault B, Scherer-Lorenzen M, Barrett CB, Glick HB, Hengeveld GM, Nabuurs G-J, Pfautsch S, Viana H, Vibrans AC, Ammer C, Schall P,

Verbyla D, Tchebakova N, Fischer M, Watson JV, Chen HYH, Lei X, Schelhaas M-J, Lu H, Gianelle D, Parfenova EI, Salas C, Lee E, Lee B, Kim HS, Bruelheide H, Coomes DA, Piotto D, Sunderland T, Schmid B, Gourlet-Fleury S, Sonké B, Tavani R, Zhu J, Brandl S, Vayreda J, Kitahara F, Searle EB, Neldner VJ, Ngugi MR, Baraloto C, Frizzera L, Bałazy R, Oleksyn J, Zawiła-Niedźwiecki T, Bouriaud O, Bussotti F, Finér L, Jaroszewicz B, Jucker T, Valladares F, Jagodzinski AM, Peri PL, Gonmadje C, Marthy W, O'Brien T, Martin EH, Marshall AR, Rovero F, Bitariho R, Niklaus PA, Alvarez-Loayza P, Chamuya N, Valencia R, Mortier F, Wortel V, Engone-Obiang NL, Ferreira LV, Odeke DE, Vasquez RM, Lewis SL, Reich PB (2016) Positive biodiversity-productivity relationship predominant in global forests. Science 354:aaf8957
Lind EM, Vincent JB, Weiblen GD, Cavender-Bares JM, Borer ET (2015) Trophic phylogenetics: evolutionary influences on body size, feeding, and species associations in grassland arthropods. Ecology 96:998–1009
Madritch MD, Kingdon CC, Singh A, Mock KE, Lindroth RL, Townsend PA (2014) Imaging spectroscopy links aspen genotype with below-ground processes at landscape scales. Philos Trans R Soc Lond B Biol Sci 369:20130194
Marconi S, Graves S, Weinstein B, Bohlman S, White E (2019) Rethinking the fundamental unit of ecological remote sensing: Estimating individual level plant traits at scale. bioRxiv:556472
Mayfield MM, Levine JM (2010) Opposing effects of competitive exclusion on the phylogenetic structure of communities. Ecol Lett 13:1085–1093
McManus MK, Asner PG, Martin ER, Dexter GK, Kress JW, Field BC (2016) Phylogenetic structure of foliar spectral traits in tropical forest canopies. Remote Sens 8:196. https://doi.org/10.3390/rs8030196
Millennium Ecosystem Assessment (2005) Ecosystems and human well-being: biodiversity synthesis. World Resources Institute, Washington, D.C.
Moore TE, Schlichting CD, Aiello-Lammens ME, Mocko K, Jones CS (2018) Divergent trait and environment relationships among parallel radiations in Pelargonium (Geraniaceae): a role for evolutionary legacy? New Phytol 219:794–807
Mouillot D, Graham NAJ, Villéger S, Mason NWH, Bellwood DR (2013) A functional approach reveals community responses to disturbances. Trends Ecol Evol 28:167–177
Oindo BO, Skidmore AK (2010) Interannual variability of NDVI and species richness in Kenya. Int J Remote Sens 23(2):285–298
O'Connor MI, Gonzalez A, Byrnes JEK, Cardinale BJ, Duffy JE, Gamfeldt L, Griffin JN, Hooper D, Hungate BA, Paquette A, Thompson PL, Dee LE, Dolan KL (2017) A general biodiversity–function relationship is mediated by trophic level. Oikos 126:18–31
O'Meara BC, Ané C, Sanderson MJ, Wainwright PC (2006) Testing for different rates of continuous trait evolution using likelihood. Evolution 60:922–933
Palacio-López K, Beckage B, Scheiner S, Molofsky J (2015) The ubiquity of phenotypic plasticity in plants: a synthesis. Ecol Evol 5:3389–4000
Parker IM, Saunders M, Bontrager M, Weitz AP, Hendricks R, Magarey R, Suiter K, Gilbert GS (2015) Phylogenetic structure and host abundance drive disease pressure in communities. Nature 520:542–544
Pinto-Ledezma JN, Jahn AE, Cueto VR, Diniz-Filho JAF, Villalobos F (2018a) Drivers of phylogenetic assemblage structure of the Furnariides, a widespread clade of lowland neotropical birds. Am Nat 193:E41–E56
Pinto-Ledezma JN, Larkin DJ, Cavender-Bares J (2018b) Patterns of beta diversity of vascular plants and their correspondence with biome boundaries across North America. Front Ecol Evol 6:194
Polasky S, Nelson E, Camm J, Csuti B, Fackler P, Lonsdorf E, Montgomery C, White D, Arthur J, Garber-Yonts B, Haight R, Kagan J, Starfield A, Tobalske C (2008) Where to put things? Spatial land management to sustain biodiversity and economic returns. Biol Conserv 141:1505–1524
Presley SJ, Scheiner SM, Willig MR (2014) Evaluation of an integrated framework for biodiversity with a new metric for functional dispersion. PLoS One 9:e105818

Reich PB (2014) The world-wide 'fast-slow' plant economics spectrum: a traits manifesto. J Ecol 102:275–301

Reich PB, Walters MB, Ellsworth DS (1997) From tropics to tundra: global convergence in plant functioning. Proc Natl Acad Sci 94:13730–13734

Reich PB, Tilman D, Isbell F, Mueller K, Hobbie SE, Flynn DFB, Eisenhauer N (2012) Impacts of biodiversity loss escalate through time as redundancy fades. Science 336:589–592

Ricklefs RE, Schluter D (1993) In: Ricklefs RE, Schluter D (eds) Species diversity: regional and historical influences. University of Chicago Press, Chicago, pp 350–363

Rocchini D, Balkenhol N, Carter GA, Foody GM, Gillespie TW, He KS, Kark S, Levin N, Lucas K, Luoto M, Nagendra H, Oldeland J, Ricotta C, Southworth J, Neteler M (2010) Remotely sensed spectral heterogeneity as a proxy of species diversity: Recent advances and open challenges. Ecol Inform 5(5):318–329

Rocchini D, Luque S, Pettorelli N, Bastin L, Doktor D, Faedi N, Feilhauer H, Féret JB, Foody GM, Gavish Y, Godinho S, Kunin WE, Lausch A, Leitão PJ, Marcantonio M, Neteler M Ricotta C, Schmidtlein S, Vihervaara P, Wegmann M, Nagendra H (2018) Measuring β-diversity by remote sensing: A challenge for biodiversity monitoring. Methods Ecol Evol 9(8):1787–1798

Scheiner SM (1993) Genetics and evolution of phenotypic plasticity. Annu Rev Ecol Syst 24:35–68

Scheiner SM (2012) A metric of biodiversity that integrates abundance, phylogeny, and function. Oikos 121:1191–1202

Scheiner SM, Kosman E, Presley SJ, Willig MR (2017) Decomposing functional diversity. Methods Ecol Evol 8:809–820

Schmidtlein S, Zimmermann P, Schüpferling R, Weiß C (2007) Mapping the floristic continuum: Ordination space position estimated from imaging spectroscopy. J Veg Sci 18(1):131–140

Schweiger AK, Cavender-Bares J, Townsend PA, Hobbie SE, Madritch MD, Wang R, Tilman D, Gamon JA (2018) Plant spectral diversity integrates functional and phylogenetic components of biodiversity and predicts ecosystem function. Nat Ecol Evol 2:976–982

Simpson GG (1943) Mammals and the nature of continents. Am J Sci 241:1–31

Simpson EH (1949) Measurement of diversity. Nature 163(4148):688–688

Singh A, Serbin SP, McNeil BE, Kingdon CC, Townsend PA (2015) Imaging spectroscopy algorithms for mapping canopy foliar chemical and morphological traits and their uncertainties. Ecol Appl 25:2180–2197

Shannon CE (1948) A mathematical theory of communication. Bell Syst Tech J 27(3):379–423

Sharp R, Tallis HT, Ricketts T, Guerry AD, Wood SA, Chaplin-Kramer R, Nelson E, Ennaanay D, Wolny S, Olwero N, Vigerstol K, Pennington D, Mendoza G, Aukema J, Foster J, Forrest J, Cameron D, Arkema K, Lonsdorf E, Kennedy C, Verutes G, Kim CK, Guannel G, Papenfus M, Toft J, Marsik M, Bernhardt J, Griffin R, Glowinski K, Chaumont N, Perelman A, Lacayo M, Mandle L, Hamel P, Vogl AL, Rogers L, Bierbower W, Denu D, Douglass J (2018) In: T. N. C. Project (ed) InVEST 3.5.0. User's Guide. Stanford University, University of Minnesota, The Nature Conservancy, and World Wildlife Fund

Shipley B, Belluau M, Kühn I, Soudzilovskaia NA, Bahn M, Penuelas J, Kattge J, Sack L, Cavender-Bares J, Ozinga WA, Blonder B, van Bodegom PM, Manning P, Hickler T, Sosinski E, Pillar VDP, Onipchenko V, Poschlod P (2017) Predicting habitat affinities of plant species using commonly measured functional traits. J Veg Sci 28:1082–1095

Sørensen TA (1948) A method of establishing groups of equal amplitude in plant sociology based on similarity of species content, and its application to analyses of the vegetation on Danish commons. K Dan Vidensk Selsk Biol Skr 5:1–34

Sultan S (2000) Phenotypic plasticity for plant development, function and life history. Trends Plant Sci 5:537–542

Swenson NG, Enquist BJ, Pither J, Thompson J, Zimmerman JK (2006) The problem and promise of scale dependency in community phylogenetics. Ecology 87:2418–2424

Tedersoo L, Nara K (2010) General latitudinal gradient of biodiversity is reversed in ectomycorrhizal fungi. New Phytol 185:351–354

Tilman D (1997) Community invasibility, recruitment limitation, and grassland biodiversity. Ecology 78:81–92

Tilman D, Reich PB, Knops JMH (2006) Biodiversity and ecosystem stability in a decade-long grassland experiment. Nature 441:629–632

Tobner CM, Paquette A, Gravel D, Reich PB, Williams LJ, Messier C (2016) Functional identity is the main driver of diversity effects in young tree communities. Ecol Lett 19:638–647

Tuomisto H (2010) A diversity of beta diversities: straightening up a concept gone awry. Part 1. Defining beta diversity as a function of alpha and gamma diversity. Ecography 33:2–22

Verdu M, Pausas JG (2007) Fire drives phylogenetic clustering in Mediterranean Basin woody plant communities. J Ecol 95:1316–1323

Villeger S, Mason NWH, Mouillot D (2008) New multidimensional functional diversity indices for a multifaceted framework in functional ecology. Ecology 89:2290–2301

Walker B, Kinzig A, Langridge J (1999) Plant attribute diversity, resilience, and ecosystem function: the nature and significance of dominant and minor species. Ecosystems 2:95–113

Wang R, Gamon JA, Cavender-Bares J, Townsend PA, Zygielbaum AI (2018) The spatial sensitivity of the spectral diversity–biodiversity relationship: an experimental test in a prairie grassland. Ecol Appl 28:541–556

Webb CO (2000a) Exploring the phylogenetic structure of ecological communities: an example for rain forest trees. Am Nat 156:145–155

Webb CO (2000b) Exploring the phylogenetic structure of ecological communities: an example for rain forest trees. Am Nat 156:145–155

Webb CO, Ackerly DD, McPeek MA, Donoghue MJ (2002) Phylogenies and community ecology. Annu Rev Ecol Syst 33:475–505

Westoby M (1998) A leaf-height-seed (LHS) plant ecology strategy scheme. Plant Soil 199:213–227

Whittaker RH (1960) Vegetation of the Siskiyou Mountains, Oregon and California. Ecol Monogr 30:279–338

Wiens JJ, Donoghue MJ (2004) Historical biogeography, ecology and species richness. Trends Ecol Evol 19:639–644

Williams LJ, Paquette A, Cavender-Bares J, Messier C, Reich PB (2017) Spatial complementarity in tree crowns explains overyielding in species mixtures. Nat Ecol Evol 1:0063

Wright IJ, Reich PB, Westoby M, Ackerly DD, Baruch Z, Bongers F, Cavender-Bares J, Chapin T, Cornelissen JH, Diemer M (2004) The worldwide leaf economics spectrum. Nature 428:821–827

Open Access This chapter is licensed under the terms of the Creative Commons Attribution 4.0 International License (http://creativecommons.org/licenses/by/4.0/), which permits use, sharing, adaptation, distribution and reproduction in any medium or format, as long as you give appropriate credit to the original author(s) and the source, provide a link to the Creative Commons license and indicate if changes were made.

The images or other third party material in this chapter are included in the chapter's Creative Commons license, unless indicated otherwise in a credit line to the material. If material is not included in the chapter's Creative Commons license and your intended use is not permitted by statutory regulation or exceeds the permitted use, you will need to obtain permission directly from the copyright holder.

Chapter 3
Scaling Functional Traits from Leaves to Canopies

Shawn P. Serbin and Philip A. Townsend

3.1 Introduction

Fossil energy use and land use change are the dominant drivers of the accelerating increase in atmospheric CO_2 concentration and the principal causes of global climate change (IPCC 2018; IPBES 2018). Many of the observed and projected impacts of rising CO_2 concentration and increased anthropogenic pressures on natural resources portend increasing risks to global terrestrial biomes, including direct impacts on biodiversity, yet the uncertainty surrounding the forecasting of biodiversity change, future climate, and the fate of terrestrial ecosystems by biodiversity and Earth system models (ESMs) is unacceptably high, hindering informed policy decisions at national and international levels (Jetz et al. 2007; Friedlingstein et al. 2014; Rice et al. 2018). As such, the impact of our changing climate and altered disturbance regimes on terrestrial ecosystems is a major focus of a number of disciplines, including the biodiversity, remote sensing (RS), and global change research communities.

Here we provide an overview of approaches to scale and map plant functional traits and diversity across landscapes with a focus on current approaches, leveraging on best practices provided by Schweiger (Chap. 15), benefits and issues with general techniques for linking and scaling traits and spectra, and other key considerations that need to be addressed when utilizing RS observations to infer plant functional traits across diverse landscapes.

S. P. Serbin (✉)
Brookhaven National Laboratory, Environmental and Climate Sciences Department, Upton, NY, USA
e-mail: sserbin@bnl.gov

P. A. Townsend
Department of Forest and Wildlife Ecology, University of Wisconsin, Madison, WI, USA

© The Author(s) 2020
J. Cavender-Bares et al. (eds.), *Remote Sensing of Plant Biodiversity*,
https://doi.org/10.1007/978-3-030-33157-3_3

3.1.1   Plant Traits and Functional Diversity

The importance of characterizing leaf and plant functional traits across scales is tied to the crucial role these traits play in mediating ecosystem structure, functioning, and resilience or response to perturbations (Lavorel and Garnier 2002; Reich et al. 2003; Wright et al. 2004; Reich 2014; Funk et al. 2017). The structural, biochemical, physiological, and phenological properties of plants regulate the growth and performance or fitness of plants and their ability to propagate or survive in diverse environments. As such, these traits are used to characterize the axes of variation that define broad plant functional types (PFTs), which in turn describe global vegetation patterns and properties (Ustin and Gamon 2010; Díaz et al. 2015), particularly in ESMs (Bonan et al. 2002; Wullschleger et al. 2014). Our focus here will be on leaf traits related to nutrition and defense that broadly fit within the concept of the leaf economics spectrum (LES, Wright et al. 2004), because these are most amendable to measurements using spectral methods. Other traits relating to reproductive strategies, hydraulics, physiology (though see Serbin et al. 2015), wood characteristics, etc. may be inferred from the traits described here, especially when combined with climate, soils, topography, or other data that generally are not directly detectable using RS.

Leaf nutritional properties and morphology are strong predictors of the photosynthetic capacity, plant growth, and biogeochemical cycling of terrestrial ecosystems (Aber and Melillo 1982; Green et al. 2003; Wright et al. 2004; Díaz et al. 2015). With respect to litter turnover and nutrient cycling, leaf traits that correspond to the distribution and magnitude of structural carbon and chemical compounds such as lignin and cellulose are used to infer the recalcitrant characteristics of canopy foliage (Madritch et al., Chap. 8). Capturing the spatial variation in these traits can therefore provide critical information on the nutrient cycling potential of ecosystems (Ollinger et al. 2002). On the other hand, leaf mass per area (LMA)—the ratio of a leaf's dry mass to its surface area—and its reciprocal, specific leaf area (SLA), correspond to a fundamental trade-off of leaf construction costs versus light-harvesting potential (Niinemets 2007; Poorter et al. 2009). The amount of foliar nitrogen within a leaf, on a mass (N_{mass}, %) or area (N_{area}, g/m^2) basis, strongly regulates the photosynthetic capacity of leaves given its fundamental role in the light-harvesting pigments of leaves (chlorophyll a and b) and photosynthetic machinery, namely, the enzyme RuBisCo (Field and Mooney 1986; Evans and Clarke 2018). Other traits, such as the concentration or content of water and accessory pigments, are important indicators of plant health and stress (Ustin et al. 2009). Moreover, the covariation of traits is also a primary focus of ecological and biodiversity research given strong trade-offs defining different leaf form and function (Díaz et al. 2015). For example, across the spectrum of plant functional diversity (Wright et al. 2004), foliar nitrogen and LMA form a key axis of variation that describes end-members between "cheap" thinner, low-LMA leaves with high leaf nitrogen, higher photosynthetic rates and faster turnover versus thick, expensive leaves with high LMA, low nitrogen, slower turnover, and longer leaf life spans. Other traits with strong evidence for detection in the literature relate to plant allocation strategies (e.g., starch and sugar content) or defense compounds, such as phenolics (e.g., Asner et al. 2015; Kokaly and Skidmore 2015; Couture et al. 2016; Ely et al. 2019).

Despite the importance of characterizing leaf and plant functional traits across global biomes, the plasticity and high functional diversity of these traits makes this apparently simple goal extremely challenging (Reich et al. 1997; Wu et al. 2017; Osnas et al. 2018), and as such global coverage has been historically limited to specific biomes (Schimel et al. 2015). Leaf traits can vary strongly within and across species (Serbin et al. 2014; Osnas et al. 2018) and are strongly mediated by an array of biotic and abiotic factors (Díaz et al. 2015; Neyret et al. 2016; Butler et al. 2017). Within a canopy, for example, functional traits typically show high variation with average light condition and quality (Niinemets 2007; Neyret et al. 2016) where lower canopy leaves tend to be thinner and have lower photosynthetic rates and altered pigment pools to account for the lower light quality. Plant traits can also change across local resource gradients, including with variations in water, nutrient availability, and disturbance legacy (Singh et al. 2015; Butler et al. 2017; Enquist et al. 2017). Importantly, this pattern can be confounded by species composition, which is generally the strongest driver of trait variation.

RS has provided new avenues to explore trait variation at larger scales and continuously across landscapes (Fig. 3.1). For example, Dahlin et al. (2013) observed that leaf functional traits were more strongly mediated by plant community composition than environment across a water-limited Mediterranean ecosystem, explaining 46–61% of the variation on the landscape. Likewise, McNeil et al. (2008) found that 93% of variation in nutrient cycling in northern hardwood forests of the US Adirondacks could be explained by species identity. Yet the presence or absence of specific plant species is, in part, a consequence of habitat sorting processes and the adaptive mechanisms of plants that influence the environments in which they can persist, including their modification of traits in response to local conditions (Reich et al. 2003). Mapping species or communities to infer traits is impractical at anything other than the local scale due to the presence of more than 200,000 plant species on Earth. Dispersal and other stochastic processes also play a role. Across broad environmental gradients, traits display much larger variation, where climate, topography, and edaphic conditions drive changes in plant community composition and structure, which, in turn, drive the patterns of potential and realized plant traits in any one location (Díaz et al. 2015; Butler et al. 2017). Finally, factors such as convergent evolution may make some species spectrally similar, while phenology and phenotypic variation may make the same species look different across locations.

Temporal regulation of traits is a key factor driving changes in functional properties and the resulting functioning of the ecosystem. Seasonal changes in traits can be significant (e.g., Yang et al. 2016) and can strongly regulate vegetation functioning (e.g., Wong and Gamon 2015). Moreover, during the lifetime of a leaf, traits can change significantly (e.g., Wilson et al. 2001; Niinemets 2016), and in evergreen species, leaf age has been shown to be a strong covariate with functional trait values (e.g., Chavana-Bryant et al. 2017; Wu et al. 2017). Age-dependent and phenological changes in leaf traits can, in turn, have significant impacts on ecosystem functioning

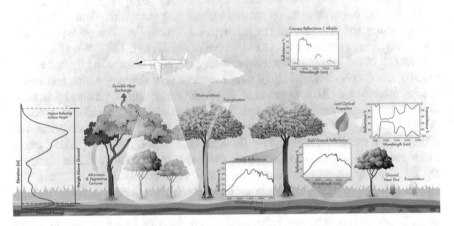

Fig. 3.1 There is a strong coupling between vegetation composition, structure and function, and the signatures observed by remote sensing instrumentation. Passive optical, thermal, and active sensing systems can be used to identify and map a range of phenomena, including minor to major variation in vegetation properties, health, and status across a landscape. Specifically, high spectral resolution imaging spectroscopy data can be used to infer functional traits of the vegetation through the measurement of canopy-scale optical properties which are driven by variation in leaf biochemistry and morphology, as well as overall canopy structure

(Wu et al. 2016). Given the role plant traits play in community assembly, characterizing the distribution, spatial patterns, and seasonality of traits is crucial for improved prediction of biodiversity change and ecosystem responses to global change.

Numerous plant trait databases have been developed to store information on the variation in functional traits across space and time (e.g., Wright et al. 2004; Kattge et al. 2011; LeBauer et al. 2018) needed to inform biodiversity and ecological modeling research. However, repeated direct measurement of plant traits is logistically challenging, which limits the geographic and temporal coverage of trait variation in these databases. Moreover, capturing plant trait variation through time is critical, but currently lacking from most observations (but with notable exceptions, e.g., Stylinski et al. 2002; Yang et al. 2016) given a host of additional technical and monetary challenges. In particular, efforts to collect direct, repeat samples of functional traits in remote areas, such as high-latitude ecosystems and the remote tropics, can be severely hindered by access and other logistical considerations.

On the other hand, RS can provide the critical unifying observations to link in-situ measurements of plant traits to the larger spatial and temporal scales needed to improve our understanding of global functional and plant biodiversity (Fig. 3.1, Table 3.1). As such, a strong interest in the use of RS to characterize foliar functional traits and their diversity has emerged from three key areas: research in RS of leaf optical properties (Jacquemoud et al. 2009), the concept of the leaf

Table 3.1 List of key foliar functional traits that can be estimated from imaging spectroscopy

Functional characterization[a]	Trait	Example of functional role	Example Citations
Primary	Foliar N (% dry mass or area based)	Critical to primary metabolism (e.g., Rubisco).	Johnson et al. (1994), Gastellu-Etchegorry et al. (1995), Mirik et al. (2005), Martin et al. (2008), Gil-Pérez et al. (2010), Gökkaya et al. (2015), Kalacska et al. (2015), Singh et al. (2015)
	Foliar P (% dry mass)	DNA, ATP synthesis	Mirik et al. (2015), Mutanga and Kumar (2007), Gil-Pérez et al. (2010), Asner et al. (2015)
	Sugar (% dry mass)	Carbon soiree	Asner and Martin (2015)
	Starch (% dry mass)	Storage compound, carbon reserve	Matson et al. (1994)
	Chlorophyll-total (ng g^{-1})	Light-harvesting capability	Johnson et al. (1994), Zarco-Tejada et al., (2000a); Moorthy et al., (2008); Gil-Pérez et al. (2010), Zhang et al. (2008)
	Carotenoids (ng g^{-1})	Light harvesting, antioxidants	Datt (1998), Zarco-Tejada et al. (2000a)
	Other pigments (e.g., anthocyanins; ng g^{-1})	Photoprotection. NPQ	van den Berg and Perkins (2005)
	Water content (% fresh mass)	Plant water status	Gao and Goetz (1995), Gao (1996), Serrano et al., (2000), Asner et al. (2015)
Physical	Leaf mass per area (g m^{-2})	Measure of plant resource allocation strategies	Asner et al. (2015), Singh et al. (2015)
	Fiber (% dry mass)	Structure, decomposition	Mirik et al. (2005), Singh et al. (2015)
	Cellulose (% dry mass)	Structure, decomposition	Gastellu-Etchegorry et al. (1995), Thulin et al. (2014), Singh et al. (2015)
	Lignin (% dry mass)	Structure, decomposition	Singh et al. (2015)
Metabolism	Vcmax (μmol m^{-2} s^{-1})	Rubisco-limited photosynthetic capacity	Serbin et al. (2015)
	Photochemical Reflectance Index (PRI)	Indicator of non-photochemical quenching (NPQ) and photosynthetic efficiency, xanthophyll cycle	Gamon et al. (1992), Asner et al. (2004)
	Fv/Fm	Photosynthetic capacity	Zarco-Tejada et al. (2000c)

(continued)

Table 3.1 (continued)

Functional characterization[a]	Trait	Example of functional role	Example Citations
Secondary	Bulk phenolics (% dry mass)	Stress responses	Asner et al. (2015)
	Tannins (% dry mass)	Defenses, nutrient cycling, stress responses	Asner et al. (2015)

[a]Categories of functional characterization are for organizational purposes only: *Primary* refers to compounds that are critical to photosynthetic metabolism; *Physical* refers to non-metabolic attributes that are also important indicators of photosynthetic activity and plant resource allocation; *Metabolism* refers to measurements used to describe rate limits on photosynthesis; and *Secondary* refers compounds that are not directly related to plant growth, but indirectly related to plant function through associations with nutrient cycling, decomposition, community dynamics, and stress responses

economics spectrum (Wright et al. 2004), and the development of global-scale foliar trait databases (Kattge et al. 2011). Within the signals observed by passive optical, thermal, and active sensing systems, such as light detection and ranging (lidar) platforms, is a whole host of underlying leaf chemical, physiological, and plant structure information that drives the spatial and temporal variation in RS observations (Ollinger 2011; Figs. 3.1, 3.2, and 3.3). As a result, RS provides the only truly practical approach to observing spatial and temporal variation in plant traits, canopy structure, ecosystem functioning, and biodiversity in absence of being able to map all species or communities everywhere (Schimel et al. 2015; Jetz et al. 2016). RS observations can provide the synoptic view of terrestrial ecosystems and capture changes on the landscape from disturbances and necessary temporal coverage via multiple repeat passes or targeted collection at specific phenological stages, yielding information needed to fill critical gaps in trait observations across global biomes (Cavender-Bares et al. 2017; Schimel et al., Chap. 19).

3.1.2   Historical Advances in Remote Sensing of Vegetation

Over the last four-plus decades, passive optical RS has been used as a key tool for characterizing and monitoring the composition, structure, and functioning of terrestrial ecosystems across space and time. For example, spectral vegetation indices (SVIs), such as the normalized difference vegetation index (NDVI), have been used to capture broad-scale plant seasonality or phenology and changes in composition, monitor plant pigmentation and stress, and track changes in productivity through time and in response to environmental change (e.g., Goward and Huemmrich 1992; Kasischke et al. 1993; Myneni and Williams 1994; Gamon et al. 1995; Ahl et al. 2006; Mand et al. 2010). Platforms, such as the Advanced Very High Resolution Radiometer (AVHRR), originally designed for atmospheric research, have been

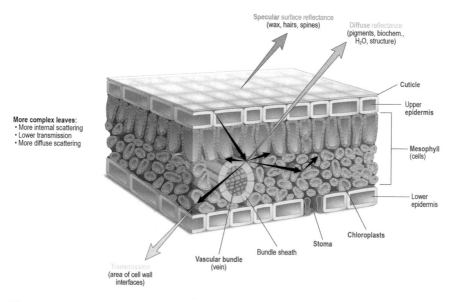

Fig. 3.2 The internal structure and biochemistry of leaves within a canopy control the optical signatures observed by remote sensing instrumentation. The amount of incident radiation that is reflected by, transmitted through, or absorbed by leaves within a canopy is regulated by these structural and biochemical properties of leaves. For example, leaf properties such as a thick cuticle layer, high wax, and/or a large amount of leaf hairs can significantly influence the amount of first-surface reflectance (that is the reflected light directly off the outer leaf layer that does not interact with the leaf interior), causing less solar radiation to penetrate into the leaf. The thickness of the mesophyll layer associated with other properties, such as thicker leaves, can cause higher degree of internal leaf scattering, less transmittance through the leaf, and higher absorption in some wavelengths. Importantly, the diffuse reflectance out of the leaf is that modified by internal leaf properties and contains useful for mapping functional traits

leveraged to capture changes in plant "greenness" based on the ratio of red absorption in leaves (signal of pigmentation levels and change) to near-infrared reflectance (tied to internal cellular structure and water content) to monitor changes in plant vigor and change (e.g., Tucker et al. 2001; Zhou et al. 2001; Goetz et al. 2005; Goetz et al. 2006). With the advent of focused Earth-observing (EO) sensors, such as the Landsat constellation, the science and use of optical RS observations for monitoring plant properties and functioning increased substantially (e.g., Chen and Cihlar 1996; Turner et al. 1999; Townsend 2002; Jones et al. 2007; Sonnentag et al. 2007; Drolet et al. 2008; Foster et al. 2008; Peckham et al. 2008; Yilmaz et al. 2008). Since the earliest uses, optical RS observations from the leaf to suborbital to satellite EO platforms have been heavily leveraged in the plant sciences, RS, and biodiversity communities (e.g., Jacquemoud et al. 1995; Roberts et al. 2004; Ustin et al. 2004; Gitelson et al. 2006; Hilker et al. 2008; Pettorelli et al. 2016; Cavender-Bares et al. 2017).

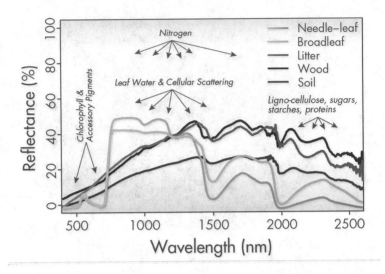

Fig. 3.3 High spectral resolution measurements of leaves and plant canopies enable the indirect, non-contact measurement of key structural and chemical absorption features that are associated with the physiological and biochemical properties of plants

3.1.3 Remote Sensing as a Tool for Scaling and Mapping Plant Traits

The use of leaf-level spectroscopy to understand plant functioning via biochemistry dates to the early twentieth century with papers describing light absorption and reflectance (Shull 1929; McNicholas 1931; Rabideau et al. 1946; Clark 1946; Krinov 1953). Billings and Morris (1951) made a direct linkage to differing ecological strategies of plants, in particular demonstrating that visible and near-infrared reflectance of species growing in different environments is directly linked to strategies associated with thermoregulation. Similarly, Gates et al. (1965) connected the interaction of light with leaves to internal leaf pigments and leaf structure (Fig. 3.2.) and how this relates to larger ecological processes.

By the 1970s, work with spectrophotometers at the US Department of Agriculture (USDA) led to the use of spectral methods for constituent characterization—near-infrared spectroscopy (NIRS) to predict moisture, protein, fat, and carbohydrate content of feed (Norris and Hart 1965; Norris et al. 1976; Shenk et al. 1981; Davies 1998; Workman and Weyer 2012), generally using linear regression on dry samples. In the 1980s and 1990s, field and laboratory studies used these earlier spectrometer systems to develop relationships and approaches to link leaf optical properties and underlying biochemical and structural properties, including variations in leaf moisture condition (Hunt and Rock 1989). For example, Elvidge (1990) utilized spectroscopy to describe optical properties of dried plant materials in the 0.4–2.5 micron range that enable detection of plant biochemistry from spectroscopy. Similarly, Curran (1989) summarized spectral features across this same spectral range that could be used in RS of plants, identifying not just the specific absorption features associated with pigments but also features related to harmonics and overtones related to molecular bonds of hydrogen (H) with carbon (C), nitrogen (N), and oxygen (O) in organic compounds (e.g., Fig. 3.3). In addition, by the late 1980s, researchers began to utilize novel, experimental airborne imaging spectrometer systems to map vegetation canopy chemistry in diverse landscapes. Using an early-generation NASA imaging spectrometer, the airborne imaging spectrometer (AIS, Vane and Goetz 1988), these studies illustrated the capacity to map landscape variation in foliar biochemical properties, including nitrogen and lignin (Peterson et al. 1988; Wessman et al. 1988; Wessman et al. 1989). AIS was the precursor to the Airborne Visible/Infrared Imaging Spectrometer (AVIRIS, Vane 1987). Following on this work, several others explored the impacts of leaf functional traits on reflectance properties of plant canopies and the ability to retrieve canopy chemistry, leveraging several important airborne campaigns including the Oregon Transect Ecosystem Research (OTTER) project and the Accelerated Canopy Chemistry Program (ACCP) (e.g., Card et al. 1988; Peterson et al. 1988; Matson et al. 1994; Bolster et al. 1996; Martin and Aber 1997).

These early studies became the basis for studies using imaging spectrometry to infer nutrient use and cycling in natural ecosystems (e.g., Martin and Aber 1997; Ollinger et al. 2002; Ollinger and Smith 2005). By the 1990s, the promise of spectroscopy for ecological characterization led to the increased use of handheld portable spectrometers in the field (e.g., instruments from Analytical Spectral Devices, GER, Spectra Vista Corporation, Spectral Evolution, Ocean Optics, LiCor, and PP Systems), as well as research that led to the use of narrowband SVIs for characterizing rapid changes in leaf function in response to the environment and leaf physiology (e.g., photochemical reflectance index, PRI, Gamon et al. 1992; Penuelas et al. 1995; Gamon et al. 1997). The review by Cotrozzi et al. (2018) provides a more detailed summary of the history of spectroscopy for plant studies, while Table 3.1 provides a summary of the key functional traits observable with spectroscopic RS approaches. As a consequence of studies at the leaf level and using early imaging spectrometers, a host of airborne sensor systems emerged, such as AVIRIS (Green et al. 1998), HyMap (Cocks et al. 1998), Airborne Prism Experiment (APEX, Schaepman et al. 2015), the Carnegie Airborne Observatory (CAO, Asner et al. 2012), AVIRIS-Next Generation (Miller et al. 2018; Thompson et al. 2018), and the US National Ecological Observatory Network (NEON) imaging spectrometer (Kampe et al. 2010) in the twenty-first century. The NASA prototype satellite EO-1 (Middleton et al. 2013) included the Hyperion sensor as an early test of the capacity to make hyperspectral measurements from space, leading to the development of a number of spaceborne missions planned for the early 2020s (Schimel et al., Chap. 19).

3.1.4 Key Considerations for the Use of Imaging Spectroscopy Data for Scaling and Mapping Plant Functional Traits

One of the chief challenges to effectively using imaging spectroscopy has been the acquisition of data of sufficient resolution, quality, and consistency for broad application in vegetation studies (Table 3.2). This necessitates measurements in the shortwave infrared (SWIR, 1100–2500 nm) in addition to the visible and near infrared (VNIR, 400–1100 nm). While VNIR wavelengths are most sensitive to pigments and overall canopy health, longer wavelengths are required to retrieve many biochemicals and LMA (Serbin et al. 2014; Kokaly and Skidmore 2015; Serbin et al. 2015; Singh et al. 2015). Spectral resolution is critical as well, with 10 nm band spacing and 10 nm full-width half maximum (FWHM) generally considered essential to identify traits detected based on narrow absorption features. Even finer resolution is required to detect spectral features that rely on narrow (<0.5 nm) atmospheric windows, such as solar-induced fluorescence (SIF, Yang et al. 2018). Other key considerations include sufficient signal-to-noise ratio (SNR) to identify important spectral features, accounting for both coherent and random noise related to detector sensitivity, dark current, and stray light. Additional sensor characteristics important to using imaging spectroscopy include spectral distortion. Most sensors are push-broom sensors, in which an image is constructed via the forward movement of the platform. Spatial samples are measured in the X-dimension (pixels) of the detector array and spectral wavelengths in the Y-dimension. Nonuniformity may arise due to differences in detectors in both dimensions, meaning that different detectors in the X-dimension see different central wavelengths (smile) and offsets in the Y-dimension lead to band-to-band misregistration (keystone). All of these effects can influence the ability to detect traits reliably within one scene or across multiple scenes using common algorithms. Full understanding of detector (and thus image) uniformity as well as the measurement point-spread function in 3-D (spatial X [detector X], spatial Y [platform movement], and spectral [detector Y]) is critical to accurate retrievals.

All RS data require some level of post-processing. Imaging spectroscopy is no different; prior to implementing algorithms for trait retrieval (Sect. 3.2.2), additional efforts must be undertaken to ensure consistent measurements in consistent units such that retrievals from imagery from multiple sources, dates, locations, etc. can be compared. Minimally, pixel measurements should be converted to radiances (w m^{-2} sr^{-1} nm^{-1}) based on laboratory calibrations and regular vicarious measurements of stable targets. With proper instrument characterization, keystone, smile, and other radiometric artifacts can be reduced. Subsequently, atmospheric corrections to convert radiance to reflectance (percent) are essential for cross-site studies. The approaches to atmospheric correction are numerous and tailored to particular environments, e.g., terrestrial vs. aquatic systems. Even within terrestrial applications, approaches differ among airborne data products (e.g., NASA's AVIRIS-Classic and AVIRIS-NG sensors vs. NEON AOP) and do not necessarily yield consistent reflectance imagery. Finally, new approaches that take advantage of

Table 3.2 Traceability matrix for a global imaging spectroscopy misson for terrestrail ecosystem functioning and biogeochemical processes

Science target	Science objectives	Functional characterization	Trait	Spectal range and sampling	Other measurement characteristics	Example citations
Theme III: Marine and Terrestrial Ecosystems and Natural Resource Management New essential measurements of the biochemical, physiological and functional attributes of the Earth's terrestrial vegetation	O1. To deliver new quantification of biogeochemical cycles, ecosystem functioning and factors that influence vegetation health and ecosystem services O2. To advance Earth system modes with improved process representation and quantification.	Primary biochemical content	Foliar N (% dry mass or area based)	450–2450 nm @ ≤15 nm	Seasonal cloud free measurement for ≤ 80% terrestrial vegetation areas. Radiometric range and sampling to capture range of vegetation signals from tropical to high latitude summers. Signals-to-Noise Ratio consistent with tropical to high latitude vegetation (e.g., red region, >500:1). At least three years of measurement to capture inter-annual variability and seasonally as robust baseline for ≥80 of the terrestrial ecosystems.	Johnson et al. (1994), Gastellu-Eichegorry et al. (1995), Mirik et al. (2005), Martin et al. (2008), Gil-Pérez et al. (2010), Gökkaya et al. (2015), Kalacska et al. (2015), Singh et al. (2015)
			Foliar P (% dry mass)	450–2450 nm @ ≤15 nm		Mirik et al. (2005), Mutangao and Kumar (2007), Gil-Pérez et al. (2010), Asner et al. (2015)

(continued)

Table 3.2 (continued)

Science target	Science objectives	Functional characterization	Trait	Spectal range and sampling	Other measurement characteristics	Example citations
			Sugar (% dry mass)	1500–2400 nm @ ≤15 nm		Asner and Martin (2015)
			Starch (% dry mass)	1500–2400 nm @ ≤15 nm		Matson et al. (1994)
			Chlorophyll-total (mg g^{-1})	450–740 nm @ ≤ 10 nm		Johnson et al. (1994), Zarco-Tejada et al. (2000a), Gil-Perez et al. (2010), Zhang et al. (2008), Kalacska et al. (2015)
			Carotenoids (mg g^{-1})	450–740 nm @ ≤10 nm		Datt (1998), Zarco-Tejada et al. (2000a)
			Other pigments (e.g., anthocyanins; mg g^{-1})	980 nm ± 40, 1140 ± 50 @ ≤20 nm		van den Berg and Perkins (2005)
			Water content (% fresh mass)	1100–2400 nm @ ≤20 nm		Gao and Goetz (1995), Gao (1996), Thompson et al. (2015), Asner et al. (2016)
			Leaf mass per area (g m^{-2})	1500–2400 nm @ ≤20 nm		Asner et al. (2015), Singh et al. (2015)
			Fiber (% dry mass)	1500–2400 nm @ ≤20 nm		Mirik et al. (2005), Singh et al. (2015)

			References
Physical/ structural content	Cellulose (% dry mass)	1500–2400 nm @ ≤20 nm	Gastellu-Etchegorry et al. (1995), Thulin et al. (2014), Singh et al. (2015)
	Lignin (% dry mass)	1500–2400 nm @ ≤15 nm	Johnson et al. (1994), Gastellu-Etchegorry et al. (2014), Thulin et al. (2014), Singh et al. (2015)
Metabolism	Vcmax (μmol m^{-2} s^{-1})	450–2450 nm @ ≤15 nm	Serbin et al. (2015)
	Photochemical Reflectance Index (PRI).	450 to 650 nm @ ≤10 nm	Gamon et al. (1992), Asner et al. (2004)
	Fraction of absorbed photosynthetically active radiation by chlororphyll, fAP ARchl.	450 to 800 nm @ ≤20 nm	
Secondary biochemcial content	Bulk phoeolics (% dry mass)	1100–2400 nm @ ≤10 nm	Asner et al. (2015)
	Tannins (% dry mass)	1100–2400 nm @ 10 nm	Asner et al. (2015)
Required for atmosopheric correction	Water vapor	980 nm ± 50, 1140 ± 50 @ ≤20 nm	Thompson et al. (2015), Gao et al. (1993)
	Cirrus clouds	940 nm ± 30, 1140 ± 40 @ ≤20 nm	
	Aerosos	450–1200 nm @ ≤20 nm	

advances in computing capacities and newer optimal estimation (OE) approaches for radiative transfer retrieval of atmospheric parameters are poised to transform atmospheric correction in the 2020s (Thompson et al. 2018).

Following atmospheric correction, scene-dependent corrections are often required, including corrections for different illumination and reflectance due to sun-target-sensor geometry, i.e., the bidirectional reflectance distribution function (BRDF). Current methods to correct for across-track (and along-track) illumination variation account for differences in vegetation structure and density, either through continuous functions (Schläpfer et al. 2015; Weyermann et al. 2015) or using land-cover stratification (Jensen et al. 2018). However, BRDF corrections are also rapidly changing and likely will be improved by new OE methods. As well, methods requiring land cover stratification are generally limited to local studies, whereas broad-scale implementation across biomes and through time will be most stable as long as scene-specific stratification is not required.

In addition to BRDF, corrections for topographic illumination are required (Singh et al. 2015). However, such corrections can result in poor performance for highly shaded slopes; they enhance noise on shaded slopes while suppressing signal on illuminated slopes. In addition, differential illumination may still remain in images due to multiple sensor artifacts as well as effects of vegetation structure (Knyazikhin et al. 2013). These effects can be effectively addressed using vector normalization (Feilhauer et al. 2010; Serbin et al. 2015) or continuum removal (e.g., Dahlin et al. 2013). Such approaches largely address structure-induced reflectance effects of broadleaf and graminoid canopies, with minor variances remaining in conifers. The residual effect of canopy structure on trait mapping largely relates to an inability to fully account for within-canopy scattering of diffuse radiation, especially in conifer forests.

Finally, when integrating data from multiple sources to map canopy traits, users must address wavelength calibrations. Different sensors may have different band centers, and these may change (on airborne devices) as they are recalibrated from time to time. This requires image resampling, which is data and processing intensive and—to be done precisely—requires good knowledge of spectral response functions or model recalibration to new wavelengths.

3.2   Linking Plant Functional Traits to Remote Sensing Signatures

All materials interact with light energy in different and characteristic ways. With respect to terrestrial ecosystems, spectroscopic RS leverages spectroradiometers, which measure the intensity of light energy reflected from or transmitted through leaves, plant canopies, or other materials (e.g., wood, soil, Fig. 3.3). The absorbing and scattering properties of the individual elements (e.g., leaves, twigs, stems) within the canopy or surface (soil) are defined by their physical and 3-D structure as well as chemical constituents or bonds (Figs. 3.2 and 3.3), which drives the vari-

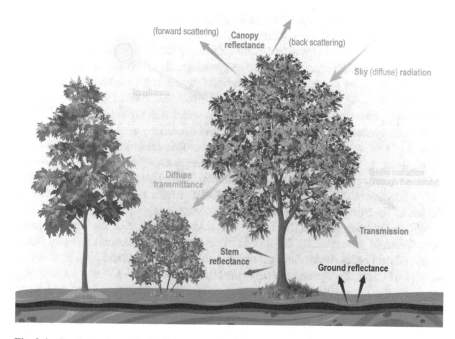

Fig. 3.4. Similar to those of a leaf, the properties of vegetation canopies strongly control the optical signatures observed by passive remote sensing instrumentation (Ollinger 2011). Specifically, the height and three-dimensional shape of the individual plants comprising the canopy as well as their leaf area index (LAI), leaf optical properties and stem and soil optical properties regulate the amount of incident radiation that reflects back from and transmits through a canopy. In addition, canopy properties and sun-sensor geometry can modify the shape and strength of the reflectance signature of vegetation canopies, which requires careful consideration when developing methods to map leaf functional traits

ability observed in reflectance spectra (Figs. 3.1 and 3.4). Thus, the underlying variation in plant canopy structure, function, and leaf traits in turn drives the optical properties and spectral signatures detected by RS platforms (Ollinger 2011). As such, the capacity to infer plant health, status, stress, and leaf and plant functional traits with optical RS observations is tied to the physical principle that plant physiological properties, structure, and distribution of foliage within plant canopies are reflected in the RS signatures of leaves within a canopy (Curran 1989; Kokaly et al. 2009; Ollinger 2011).

3.2.1   Spectroscopy and Plant Functional Traits

With the advent of laboratory and field spectrometer instrumentation, the leaf to landscape-scale RS of vegetation traits and functional properties began in earnest in the early 1980s (Sect. 3.1.3). As stated in Sect. 3.1.4, there are a host of important considerations with the use of leaf and imaging spectroscopy for scaling plant functional traits. In addition, the underlying drivers of vegetation optical properties are complex and numerous (Ustin et al. 2004; Ollinger 2011). For example, in the visible range (~0.4–0.75 microns) of the electromagnetic (EM) spectrum, the strong absorption of solar energy by photosynthetic pigments in healthy, green foliage dominates the optical properties of leaves (Ustin et al. 2009; Figs. 3.2. and 3.3). Importantly, knowledge of leaf pigment pools and fluxes provides key insight into plant photosynthesis, environmental stress, and overall vigor. As such a significant amount of research has focused on the retrieval of foliar primary and accessory pigments using spectroscopic and other RS measurements (e.g., Jacquemoud et al. 1996; Richardson et al. 2002; Sims and Gamon 2002; Ustin et al. 2009; Féret et al. 2017). Blackburn (2007) and Ustin et al. (2009) provide more detailed reviews on the use of spectroscopy to remotely sense pigments in higher plants.

Within the near-infrared (NIR, ~0.8–1.2 microns) portion of the EM spectrum, optical signals are generally dominated by scattering from internal leaf structures, structural properties, water, and leaf epidermal layer (Figs. 3.2 and 3.3). In addition, strong leaf water absorption features in the NIR, centered on ~0.97 and 1.1 microns, are often used to remotely sense vegetation water content (e.g., Hunt and Rock 1989; Gao and Goetz 1995; Sims and Gamon 2003; Stimson et al. 2005; Colombo et al. 2008). Much of the early research into the use of spectroscopic RS focused on leaf and canopy water content retrieval given its importance in plant function and as an important indicator of moisture (Fig. 3.5.) and other stress. In attached, fresh leaf material, water also dominates the spectral absorption features of the SWIR (1.3–2.5 micron) portion of the EM (Hunt and Rock 1989; Sims and Gamon 2003); as a result, spectral optical properties are strongly regulated by leaf and canopy water content in this region (Fig. 3.5). Along with water absorption, a number of other biochemical and structural trait absorption features exist in the SWIR wavelength region (Fig. 3.3), including cellulose, lignin, structural carbon, and nutrients and proteins (Curran 1989; Elvidge 1990; Kokaly et al. 2009; Ollinger 2011; Ely et al., 2019). Removal of water from leaf materials can sometimes enhance the detection of these absorption features (e.g., see Serbin et al. 2014 and references within; Fig. 3.5). However, at the canopy scale, a number of studies have demonstrated the capacity to retrieve these foliar biochemical properties in the SWIR region (e.g., Wessman et al. 1988; Martin and Aber 1997; Townsend et al. 2003; Kokaly et al. 2009; Asner et al. 2015; Singh et al. 2015), perhaps because of the increased signal due to multiple scattering within canopies (Baret et al. 1994).

In addition to the underlying leaf biochemical and structural characteristics, leaf orientation, display, and distribution in a canopy are also strong drivers of plant optical properties (Ollinger, 2011; Fig. 3.4). Decreasing the leaf area of a canopy generally results in a higher reflectance signal from elements deeper within the canopy, including twigs, branches, stems, and soil/litter layer (Asner 1998; Asner et al. 2000; Ollinger 2011). Canopies with flat, horizontal leaves tend to have higher NIR reflectance than those with more erect, vertical leaves, depending on the sun-sensor geometry. Leaf anatomy and average leaf angle vary widely across species

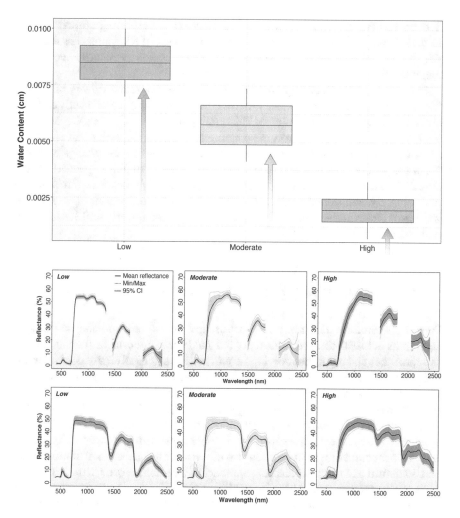

Fig. 3.5. Together, leaf optical properties and canopy architecture regulate the remote sensing signatures observed in remote sensing data. In addition, changes in leaf internal biochemistry or structure (i.e., functional traits) as a result of biotic or abiotic factors can change these signatures over space and time. For example, a prolonged drought can cause changes in leaf internal water content and potentially a redistribution of internal pigmentation. We can simulate the potential changes in optical signatures associated with a drought using a leaf and canopy-scale radiative transfer models (RTM), in this case PROSPECT-5b (Féret JB et al. 2008) and SAIL (Verhoef and Bach 2007), to illustrate the changes in leaf an canopy spectra over the course of a low, moderate, and high drought event. Here we modified pigment and water content from low to high for a range of canopies, as represented by different LAIs, and for canopy-scale reflectance, we incorporated the sensor characteristics of AVIRIS-classic (Green et al. 1998) to illustrate what the canopy reflectance might look like from that sensor. (For illustration purposes only)

(Falster and Westoby 2003), with consequences for interpreting optical RS signatures (Ollinger 2011). Thus, when considering the use of RS approaches for mapping leaf traits, careful consideration of vegetation structure, collection characteristics, and sensor design is important.

Phenology, leaf seasonality, and leaf age are also important drivers of optical properties for a number of reasons. First, leaf traits can change significantly over the lifetime of a leaf (e.g., Wilson et al. 2001; Niinemets 2016; Chavana-Bryant et al. 2017; Wu et al. 2017), and the corresponding leaf optical properties will change in concert (Yang et al. 2016). Average leaf angle distribution can also change with leaf age or seasonally from younger, recently expanded leaves to fully expanded (Raabe et al. 2015), which can have significant impacts on canopy reflectance (Huemmrich 2013). Finally, atmospheric, insect, or other stressors typically change the chemical makeup of leaves and so their optical properties (e.g., Couture et al. 2013; Ainsworth et al. 2014; Cotrozzi et al. 2018).

3.2.2 Approaches for Linking Traits and Spectral Signatures

Despite the promise and utility of spectroscopy for the retrieval and mapping of plant traits across space and time, there has not been consensus or standardization of approaches and algorithm development in the RS and biodiversity communities. This is not entirely unexpected given the complexity of connecting traits and RS observations across the various scales of interest, from leaves to individual trees, communities, and landscapes (Schweiger, Chap. 15). In addition, early approaches (e.g., Peterson et al. 1988) were often later deemed inappropriate and often replaced by other techniques (e.g., Grossman et al. 1996). Access to more powerful, improved, and cheaper computing resources has also allowed for the exploration of more complex statistical and machine-learning approaches (see Schweiger, Chap. 15).

Two primary approaches have been utilized to link RS observations to functional traits—empirical, statistically based techniques and radiative transfer modeling (RTM; see also Meireles et al., Chap. 7; Ustin, Chap. 14).

3.2.2.1 Empirical Scaling Approaches

With respect to empirical techniques, the use of SVIs was one of the earliest methods to explore the capacity to link a range of plant functional traits to vegetation spectra. Typically, with this approach a single SVI is linked with a trait of interest, such as leaf pigments or water content, to develop a simple statistical relationship between the trait of interest and corresponding variation in optical properties (e.g., Sims and Gamon 2003; Gitelson 2004; Colombo et al. 2008; Feret et al. 2011). The derived model is then used to estimate trait values for new leaves using only spectral measurements. This approach typically assumes the researcher has an *a priori* understanding of the links between the trait and resulting variation in the electromagnetic spectrum and thus selects specific wavelengths, and therefore SVI, for their analysis. An alternative approach is to explore the spectra and trait space to identify new or previously unknown SVIs that maximize the correspondence between optical properties and traits of interest (e.g., Inoue et al. 2008), akin to a data mining exercise. A challenge of this approach can be interpretation of the selected SVIs, where the resulting vegetation indices may not contain wavelengths with known absorption features relating to the trait of interest. The same general approach can also leverage multiple SVIs, provided the research avoids highly correlated portions of the spectrum (Grossman et al., 1996), to attempt to capture how variation in the trait of interest is reflected in various portions of the EM spectrum to other sites and plant species. However, a limitation to the use of SVIs has been the ability to generalize across broad canopy architectures, species, and environments due to the often site-specific modeling results or potential signal saturation issues with some SVIs (Shabanov et al. 2005; Glenn et al. 2008).

Continuous spectral wavelet transforms have been used to reduce the dimensionality of spectral data prior to developing simple statistical models (e.g., Blackburn and Ferwerda 2008). Wavelets are functions that are used to decompose a full, complex signal into simpler component sub-signals. When used with spectral data, the full reflectance signature can be decomposed in a way that allows the resulting wavelet coefficients assigned to each sub-signal to be related to concentrations of chemical constituents or other traits of interest, through standard statistical modeling approaches (e.g., linear regression). Previous studies have explored the use of wavelet methods to retrieve a host of functional traits, including pigments, water, and nitrogen content (e.g., Blackburn and Ferwerda 2008; Cheng et al. 2011; Li et al. 2018; Wang et al. 2018). Continuum removal together with band-depth analysis (Kokaly and Clark 1999) has also been utilized as a means to retrieve the chemical composition of leaves. In this approach, continuum removal lines are fit through the absorption features of interest based on those regions not in the areas of interest, then the original spectra are divided by corresponding values of the continuum removal line. The band centers can then be found by finding the minimum of the continuum-removed spectra. Normalization of the band centers is often used to standardize the values across samples. These data are then used to develop models to predict functional traits at the leaf and canopy scales, including foliar nitrogen and recalcitrant properties, such as the amount of lignin and cellulose (Kokaly et al. 2009).

In addition to the empirical SVI approach, as discussed in Schweiger (Chap. 15), partial least-squares regression (PLSR) modeling has been used extensively in the development of spectra-trait models for measuring, scaling, and mapping plant functional traits (e.g., Ollinger et al. 2002; Townsend et al. 2003; Asner and Martin 2008; Martin et al. 2008; Dahlin et al. 2013; Singh et al. 2015; Ely et al. 2019). A key attribute of PLSR is the capacity to utilize the entire measured portion of the EM spectrum as predictors (i.e., X matrix) without requiring a priori selection of wavelengths or SVIs (Wold et al. 1984; Geladi and Kowalski 1986; Wold et al. 2001). PLSR avoids collinearity (i.e., spectral autocorrelation across wavelengths) in the predictor variables (i.e., reflectance wavelengths), even if predictors exceed the number of observations (Geladi and Kowalski 1986; Wold et al. 2001; Carrascal et al. 2009). This is done through singular value decomposition (SVD), which reduces the X matrix down to relatively few non-correlated latent components. While PLSR was originally used in chemometrics, the features and benefits of PLSR also fit well within the goals of connecting spectral signatures to leaf functional traits. PLSR leverages the fact that different portions of the EM spectrum change in concert with various nutritional, structural, and morphological properties of leaves and canopies—in other words, leveraging the known covariance between variations in leaf optical properties and leaf traits (Ollinger 2011). Importantly, PLSR also allows for univariate or multivariate modeling where multiple predictands (i.e., Y matrix) can be modeled simultaneously with the same spectral matrix to account for the covariance between X and Y but also among the various Y (response) variables (Wold et al. 1984; Geladi and Kowalski 1986; Wold et al. 2001). Wolter et al. (2008) review of the use of PLSR in RS research, and Carrascal et al. (2009) summarize its use in ecology, as well as key features of PLSR.

Several approaches and implementations of PLSR have been used within the overarching "plant trait mapping" paradigm, including various spectral transformations and the use of prescreening of wavelengths or down-selection of suitable of pixels (e.g., Townsend et al. 2003; Feilhauer et al. 2010; Schweiger, Chap. 15; Asner et al. 2015). In a typical PLSR implementation (e.g., Fig. 3.6), foliar samples are first collected from vegetation canopies and processed to obtain the functional traits of interest. For leaf-scale algorithms, the optical properties of the leaves are typically measured in situ or within a small window (2–4 hours) prior to further

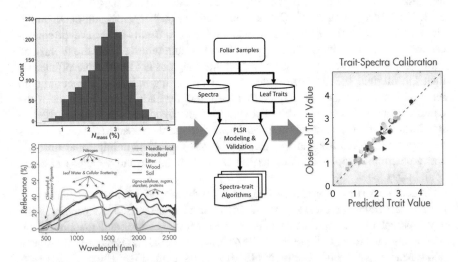

Fig. 3.6. A simple example illustrating how leaf functional traits and optical properties (e.g. reflectance) are combined in an empirical partial least-squares regression (PLSR) modeling approach to develop spectra-trait algorithms. The input traits and reflectance spectra are combined and used to train and test a PLSR model, using either cross-validation and/or independent validation (e.g., Serbin et al. 2014), and the resulting model can then be applied to other spectral measurements to estimate the traits of interest

processing. Leaf and/or image spectra for the pixel containing the plots or sample locations are then linked with these functional trait measurements to develop the PLSR algorithm. Typically, for models utilizing imaging spectroscopy data, plot-scale estimates of traits are derived using measurements of basal area, leaf area by species, or other means to produce a weighted average of each trait by dominant species within given ground area (e.g., McNeil et al. 2008; Singh et al. 2015). The algorithm is evaluated using internal validation during model development (e.g., cross-validation) and/or using a set of training and validation data to build and test the model predictive capacity across a range of similar samples and optical properties. Some approaches utilize additional steps to characterize the uncertainties associated with the sample collection, measurements, and other issues (e.g., instrument noise) in the PLSR modeling step. For example, Serbin et al. (2014) and Singh et al. (2015) introduced a novel PLSR approach that can account for uncertainty in the prediction of trait values, which has later been used by other groups (Asner et al. 2015). Image-scale algorithms are often used to derive functional trait maps (e.g., Fig. 3.7) to explore the spatial and/or temporal patterns of traits across the landscapes of interest (e.g., Ollinger et al. 2002; McNeil et al. 2008).

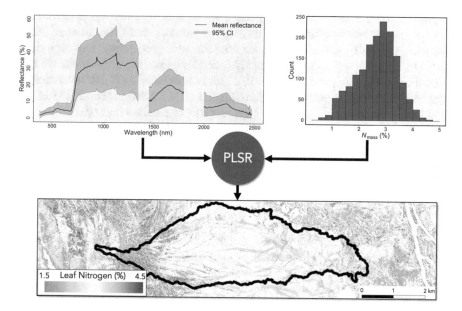

Fig. 3.7. Much like developing a leaf-scale PLSR model for estimating leaf functional traits, such as leaf nitrogen concentration (Fig. 3.6), we can also utilize high spectral resolution imaging spectroscopy data, such as that from NASA AVIRIS to build models applicable at the canopy to landscape scales (e.g., Dahlin et al. 2013; Singh et al. 2015). Here we show a simple illustration of the linkage between functional traits scaled to the canopy, for example based on a weighted average of the dominant species in the plot, connected with the reflectance signature of these canopies. Once linked, we can develop PLSR algorithms conceptually similar to that of leaves resulting in canopy-scale spectra-trait models capable of mapping functional traits across the broader landscape

While the PLSR approach produces algorithms that "weight" wavelengths by their importance in the prediction (Wold et al. 2001) of the functional traits of interest (e.g., Serbin et al. 2014), some researchers have also explored modifications to the standard PLSR approach that provide additional reductions in data dimensionality. For example, Li et al. (2008) coupled PLSR with a genetic algorithm (GA) approach to select a smaller subset of wavelengths to use in the final PLSR model for predicting leaf water content, measured as equivalent water thickness (EWT). DuBois et al. (2018) combined the SVI and PLSR approach by using all two-band AVIRIS wavelength combinations to model the relationship between spectral reflectance and ecosystem carbon fluxes across a water-limited environment. To date, the spectra-trait PLSR modeling approach has shown the capacity to characterize the widest array of leaf functional traits using the optical properties of plants across a broad range of species and ecosystems (e.g., Dahlin et al. 2013; Asner et al. 2014; Asner et al. 2015; Serbin et al. 2015; Singh et al. 2015; Couture et al. 2016).

Similar to the PLSR approach, researchers have leveraged various machine-learning approaches to connect RS observations to functional traits (e.g., Féret et al. 2018). Schweiger (Chap. 15) describes two commonly used machine-learning approaches in RS; several other approaches have also been used to model trait variation as a function of spectral measurements. More recently, Gaussian processes regression (GPR) has been recommended as superior to other machine-learning approaches for trait mapping from imaging spectroscopy data (Verrelst et al. 2012; Verrelst et al. 2016). GPR is a nonlinear nonparametric probabilistic approach similar to kernel ridge regression that directly generates uncertainty (or confidence) levels for the prediction (Wang et al. 2019). This is in contrast to PLSR uncertainties, generally assessed through permutation (Singh et al. 2015; Serbin et al. 2015). PLSR and GPR yield very similar results, both in terms of absolute trait predictions and relative scaling of uncertainties (Wang et al. 2019). PLSR is much more computationally efficient, and results are readily interpretable in terms of wavelength quantitative contribution to prediction (see Fig. 3.1 in Schimel et al., Chap. 19), whereas GPR only identifies relatively important wavelengths.

The challenge with most machine-learning approaches is that some level of data reduction is required for optimal performance. Standard approaches, such as principle component analysis (PCA) or minimum noise fraction (MNF) transformations, may reduce data dimensionality. However, features important to trait estimation may be buried in lower principle components, as high contrast variation (albedo, greenness, water content) dominate scene properties. In contrast, PLSR rotates the data into latent vectors optimized to the empirical dependent variables, which generally yields strong models for calibration data but can lead to poor model performance when confronted with new data that differ considerably from the model-building data sets.

3.2.2.2 Radiative Transfer Models and Scaling Functional Traits

An alternative to statistical, field-based, and empirical approaches for connecting leaf and canopy optical properties with plant functional traits, RTMs can be used either at the leaf and canopy scales to directly retrieve leaf traits (e.g., Colombo et al. 2008; Darvishzadeh et al. 2008; Feret et al. 2011; Banskota et al. 2015; Shiklomanov et al. 2016) or in hybrid approaches where statistical algorithms are developed based on RTM simulations (e.g., Asner et al. 2011). RTMs encapsulate our best mechanistic understanding of the coordination among leaf properties, canopy structure, and resulting spectral signatures at the leaf and canopy scales, but abstracted to operate with different degrees of complexity and assumptions (Bacour et al. 2002; Nilson et al. 2003; Kobayashi and Iwabuchi 2008; See also Morsdorf et al., Chap. 4; Ustin and Jacquemoud, Chap. 14).

At the leaf scale, RTMs were generally spawned from earlier work that identified the relationships between fresh and dried leaf reflectance and a range of foliar traits, including pigments, water content, nitrogen, dry matter, cellulose, and lignin. The realization that leaf optical properties were fundamentally tied to the concentration and distribution of leaf traits led to the development of models that could closely mimic the spectral patterns across the shortwave spectral region (0.4–2.5 microns) based on select leaf properties, such as chlorophyll and water content, as well as structural variables. By far the most widely and commonly used leaf-level RTM is the PROSPECT model (Jacquemoud and Baret 1990; Feret et al. 2008), which simulates leaf directional-hemispherical reflectance (R) and transmittance (T), allowing for the calculation of leaf absorption (1-R+T) based on leaf biochemical and morphological properties, primary and accessory pigments, water content, LMA, or dry matter content, brown material, and an approximation of the thickness of the internal leaf mesophyll layer (Féret et al. 2008; Féret et al. 2017). PROSPECT then simulates leaf optical properties based on a generalized plate model describing leaves as a stack of N homogenous absorbing layers that are calculated based on the values of input leaf traits and their corresponding spectral absorption coefficient. Other prominent leaf models include the Leaf Incorporating Biochemistry Exhibiting Reflectance and Transmittance Yields (LIBERTY) model (Dawson et al. 1998) and LEAFMOD (Ganapol et al. 1998). In particular, LIBERTY is notable given its original application focusing on improving the modeling of needle-leaf evergreen conifer species and their leaf optical properties based on several leaf traits, similar to PROSPECT, but also including foliar lignin and nitrogen content.

Moving to the canopy scale, RTMs are far more numerous with a wide variety of complexities, assumptions, and requirements (Verhoef and Bach 2007; Widlowski et al. 2015; Kuusk 2018). Most canopy RTMs leverage leaf-scale models, such as PROSPECT, to provide the leaf optical properties (i.e., leaf single-scattering albedo) needed to simulate canopy directional-hemispherical reflectance across select wavelengths, simulated spectral bands, or specific SVIs. Generally, the soil boundary layer is either prescribed or simulated using a simple model of soil BRDF (e.g., Hapke model, Verhoef and Bach 2007), and stem or woody material reflectance and transmittance (when used) is prescribed. Canopy RTMs can be separated into two main classes, homogenous and heterogenous models. Homogenous models assume the canopy to be horizontally unlimited and treated as a turbid medium of sufficiently large number of phytoelements (leaves, stems, other materials). For example, the Ross–Nilson model of plate medium (Ross 1981) assumes these elements to be composed of small bi-Lambertian "plates" described by their reflectance and transmittance properties with a specific leaf angle distribution (LAD). Leaves are small compared to the full canopy medium, with no self-shading, and transmittance is a function of optical properties and leaf area index (LAI). Additional canopy parameters were added, including the hot-spot and canopy clumping to describe sun-sensor illumination effects and the inhomogeneity of the canopy elements (Kuusk 2018). Early SAIL models also fall into this classification (e.g., Verhoef 1984). On the other hand, heterogenous canopy RTM models, including 3-D models, address the fact that vegetation canopies are heterogenous (e.g., gaps between crowns, spatial structure, differing canopy architectures) but range widely in their complexity and implementations. These models provide enhanced detail in the modeling of vegetation canopies but are necessarily more complex. Often these models require additional information to model vegetation "scenes," which can include information on tree crown shape, stem location, and other properties (e.g., hot spot, clumping) in addition to leaf optical properties, sun-sensor geometry, and LAI. These models range from 3-D Monte Carlo ray-tracing models, such as FLIGHT (North 1996) and FLiES (Kobayashi and Iwabuchi 2008), to analytical and hybrid approaches using a variety of canopy structure schemes including geometric optical (GO) representation of individual plants where tree placement follows a statistical distribution and leaf and stem scattering elements are homogenously distributed (e.g., Kuusk and Nilson 2000; Nilson et al. 2003). For example, multiple stream, including four-stream, two-layer models often utilize simplifying assumptions, to model canopies as homogenous and continuous (i.e., "slab canopies"), but which are composed of a large number of small scattering elements (leaves, sometimes leaves and stems) with arbitrary inclination angles (e.g., 4SAIL2, Verhoef and Bach 2007). The scattering elements and the soil can be prescribed with specific optical properties using observed data or based on a leaf RTM, such as PROSPECT (Jacquemoud et al. 2009). In addition, some models can divide complex scenes into smaller cells to perform the radiative transfer calculations (e.g., DART, Gastellu-Etchegorry et al. 2015) where the level of simulation detail is based on the size of the cells and the degree of detail built into the model scene components. See the review by Kuusk (2018) for more details regarding canopy RTMs and their design, diversity, assumptions, and approaches.

The use of RTMs allows for the estimation of leaf and canopy traits using simulated canopy reflectance, without some of the limitations or challenges of empirical approaches (3.3.1), such as the requirement of field sampling, scaling leaf traits to the canopy, and other issues such as the timing of field and imagery collections. Furthermore, RTMs can provide a more mechanistic connection between traits and reflectance allowing for potentially broader application than empirical approaches in areas were ground sampling may be sparse (e.g., remote regions such as the Arctic or the tropics). In addition, RTMs provide the opportunity to prototype inversion approaches across a range of remote sensing platforms and evaluate the trade-offs between different sensor designs, spectral resolutions, and temporal coverage (Shiklomanov et al. 2016), enabling the development of cross-platform retrieval algorithms.

Depending on the application, and RTM complexity, inversion can be conducted at the pixel or larger patch scales (i.e., collections of relatively homogenous areas of vegetation) to characterize spatial and temporal patterns in plant functional (e.g., pigments) and structural (e.g., LAI) properties. In RTM inversion, the leaf-scale model is often the focus, where the goal is to invert the canopy and leaf models jointly to extract estimated foliar traits based on observed canopy reflectance (e.g., Colombo et al. 2008). Many other studies have focused on retrieving canopy-scale parameters, such as LAI (e.g., Darvishzadeh et al. 2008; Banskota et al. 2015). Early approaches leveraged RTM inversions that focused on numerical optimization techniques to minimize the difference between modeled and observed reflectance across similar wavelengths (e.g., Jacquemoud et al. 1995). Other methods have utilized look-up table (LUT) inversion (e.g., Weiss et al. 2000) where a range of simulated canopy reflectance patterns are generated in advanced by varying leaf and canopy inputs across predetermined values. These simulated spectra are then compared to observations where either a single or select number of closely matching modeled spectra, and their associated inputs, are selected as the solution to the inversion. Bayesian RTM inversion methods have also been utilized (e.g., Shiklomanov et al. 2016) as a means to retrieve leaf and canopy properties as joint posterior probability distributions through iterative sampling of the input parameter space. The use of RTMs ranges from retrieval of vegetation functional and structural traits to the characterization of landscape functional diversity (Kattenborn et al. 2017; Kattenborn et al. 2019).

3.3 Important Considerations, Caveats, and Future Opportunities

3.3.1 Field Sampling and Scaling Considerations

There are several important considerations and best practices when developing algorithms for the remote estimation of plant traits (see Schweiger, Chap. 15). We will only briefly touch on these here. A key first step is to consider the scope of the research and area of interest, focusing specifically on considerations such as local climate conditions, terrain, vegetation, and canopy access. Specifically, the spatial locations, site, and canopy access (e.g., is it possible to reach canopy foliage?); vegetation composition and canopy architecture; timing of collection; and methods for sample retrieval are key to identify prior to field campaigns in order to maximize the utility of the field samples for conversion of RS signatures to accurate trait maps. Furthermore, it may be important to consider what approach may be best to characterize the vegetation canopy architecture and/or composition to facilitate scaling of each trait to the pixel or plot scale (e.g., using basal area, LAI). This may strongly depend on the dominant vegetation types, where more open canopies may require a different approach to a closed canopy, or on the spatial resolution of the imagery. Observational data range is a primary consideration (see Schweiger, Chap. 15), and sample locations should be chosen to cover the range of canopy types and vegetation communities that will fall within the RS observations. The timing of the field sampling should be as close to the RS collection date as possible, as an optimal approach, but at least be selected to match the phenological stage of the vegetation during the imagery collection, if leveraging sample campaigns in following year(s).

A number of different methods have been used to collect plant functional traits to link with RS imagery (e.g., Wang et al. 2019). Common approaches for the collection of canopy leaf samples include the use of slingshot, pruning pole, and shotgun (Lausch et al., Chap. 13), but also include line-launcher and air cannon (e.g., Serbin et al. 2014); simpler tools and hand shears are often used for accessible, shorter canopies. Regardless of the sample collection approach, harvested leaves should be reasonably intact and minimally damaged in order to avoid any issues with changes in leaf chemistry from physical damage or stress. In addition, leaves should be immediately measured for leaf optical properties and fresh mass, if these are of interest, then stored in humidified and sealed bags and placed in a cool, dark place prior to transport for further processing. Processing should then be completed within 2–4 hours of sampling—though a much shorter time between sample and measurement or different sample storage and handling (e.g., flash freezing in liquid nitrogen) may be needed for specific biochemical traits. Typically top-of-canopy, sunlit samples have been the main focus; however, more recent work has also begun to focus on collection of canopy and subcanopy samples (e.g., Serbin et al. 2014; Singh et al. 2015). This provides the ability to evaluate the depth in the canopy needed to link traits with image, which may vary by vegetation type or LAI.

3.3.2   Evaluating Functional Trait Maps and the Need to Quantify Uncertainties

Maps of plant functional traits are useful for a wide variety of applications. From an ecological perspective, maps of plant traits across broad biotic and abiotic gradients can be used to explore the drivers of plant trait variation in relation to climate, soils, and vegetation types (e.g., McNeil et al. 2008). Modeling activities can leverage these trait maps as either inputs for model parameterization across space and time (Ollinger and Smith 2005) or to evaluate prognostic plant trait predictions. However, to maximize the utility of functional trait maps a detailed understanding of the their uncertainties across space and time is required.

In the earliest functional trait mapping work, predictive model uncertainties were limited to the "goodness of fit" and overall model root mean square error (RMSE) statistics provided by the modeling approach (e.g., Wessman et al. 1988; Martin and Aber 1997; Townsend et al. 2003). While this information is helpful to understand the accuracy of the model fit, that level of accuracy assessment is insufficient for characterizing the uncertainty of the trait maps themselves. Mapping efforts should instead provide an accounting of the trait measurement, scaling, and algorithm uncertainties and provide this information in the resulting trait map data products. However, detailed error propagation is not trivial, particularly with respect to empirical modeling approaches, and is an ongoing and active area of research in the RS sciences and not discussed in detail here. On the other hand, efforts to provide product uncertainties do exist. Serbin et al. (2014) and Singh et al. (2015) illustrate how to incorporate data and modeling uncertainties at the leaf and canopy scales in the mapping of plant functional traits. This approach captures the uncertainties stemming from the leaf-level estimation of traits (Serbin et al. 2014) and the modeling of plot-level spectra and trait values (Singh et al. 2015) using a similar PLSR and uncertainty analysis approach. The result is an ensemble of PLSR models to apply to new RS data providing mean and error metrics for every pixel in the image. However, even approaches such as these fail to incorporate and propagate the uncertainties stemming from the atmospheric correction workflow given the challenge of extract the information needed to enable this on a pixel-by-pixel or even a scene-by-scene basis. Future work will be required to focus on capturing this information and providing it to the end-user who conducts the trait mapping efforts.

Uncertainty in RTM approaches have generally been derived based on inversion approaches applied to imagery. For example, as described in Sect. 3.2.2.2, a commonly used approach to the inversion of RTM simulations for the RS of functional traits is the use of LUTs. Some LUT approaches provide results based on the "best fit" of the model inversion results to the RS observations. However, this only provides an assessment of error where field measurements can be used to evaluate the retrieved values. Given the challenge of equifinality in RTM approaches, later efforts have used an ensemble of best fit results to provide a mean and distribution of values that provide a good fit of modeled reflectance to observed (e.g., Weiss et al. 2000; Banskota et al. 2015). Using this approach allows for the description of pixel-level uncertainty based on the best fit ensembles; however, these need to be combined with an accuracy assessment to get a true uncertainty of the functional trait retrievals. More recent approaches have leveraged Bayesian inversion approaches that provide output that is not a point estimate for each parameter but rather the joint probability distribution that includes estimates of parameter uncertainties and covariance structure (Shiklomanov et al. 2016). Regardless of the approach, the key is that the derived products provide a reasonable assessment of trait uncertainty across the spatial and temporal domain (where appropriate).

3.3.3   Current and Future Opportunities in the Use of Remote Sensing to Characterize Functional Traits and Biodiversity

The ability to map foliar functional traits from imaging spectroscopy greatly expands the potential for understanding patterns of vegetation function and functional diversity both locally and broadly across biomes, especially in comparison to the challenges of fully characterizing spatial and temporal (across seasons and between years) variation using field data (e.g., the TRY database). With forthcoming spaceborne sensors (see Schimel et al., Chap. 19) and continental-scale experiments

like the US National Ecological Observatory Network (NEON), we are able to test
relationships among traits and characterize functional diversity at unprecedented
scales. For example, NEON is collecting imaging spectroscopy data at 1 m resolution and waveform lidar data almost annually for 30 years at 81 10 km × 10 km sites
covering 20 biomes defined for the USA. With the addition of lidar, which enables
measuring traits such as plant area index, canopy height, canopy volume, and
aboveground biomass (of forests), a broad suite of traits can be leveraged to test
relationships that have been published in the literature (e.g., the leaf economics
spectrum) and are generally tested now at global scales using extensive—but still
not comprehensive—databases such as TRY. With spaceborne imaging, phenological variation in traits (e.g., Yang et al. 2016) can be further explored. For example,
preliminary mapping of key functional traits across all NEON biomes in the USA
shows the leaf economics spectrum relationship between LMA and nitrogen for forest and grassland ecosystems east of the US Rocky Mountains (Fig. 3.8.) in comparison to the data set used for the original LES studies, GLOPNET (Global Plant
Trait Network, Wright et al. 2004; Reich et al. 2007). Importantly, the use of data
from RS platforms, such as NEON, AVIRIS, and upcoming spaceborne sensors (see
Schimel et al., Chap. 19), enables the filling of critical research gaps and global
coverage in remote regions, as suggested by Jetz et al. (2016) and Schimel et al.
(2015). The relationship does not differ significantly from published relationships
but does suggest a breadth of the relationship as well as outliers for a number of
observations many orders of magnitude higher than is possible from field databases.
Field databases are still required for basic science studies, as well as inventory, calibration, and validation, but RS offers new possibilities for baseline characterization
of Earth's functional diversity and thus testing new hypotheses about the drivers of
such variation, using the range of traits detectable from RS (Tables 3.1 and 3.2).

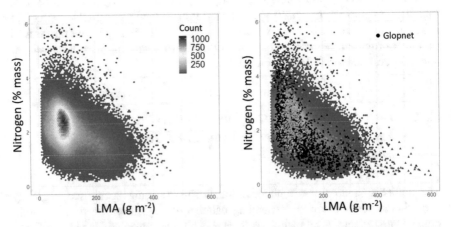

Fig. 3.8. LMA versus nitrogen for NEON for GLOPNET observations (black dots, truncated to
observations with LMA <600) vs. pixel predictions derived for NEON sites east of the US Rocky
Mountains (color gradient). Color gradient is density of pixel observations based on 333,500 pixel
values randomly extracted from 447 flight NEON flight lines in 18 sites across 6 biomes

Furthermore, coupling of spectral and functional trait databases (e.g., ecosis.org) will facilitate more rapid development and testing of new functional algorithms or the expansion of the scope of inference of existing models. In addition, the inclusion of high spectral resolution sensors on unmanned aerial systems (UASs, Shiklomanov et al. 2019) provides the opportunity to leverage similar scaling approaches as presented in this chapter with UAS observations to provide unprecedented temporal coverage and targeted spatial sampling that can be used to understand ecosystem in new detail or aid in the scaling from the plant to grid cell. In all, functional trait maps from imaging spectroscopy will supplement data and approaches presented by Butler et al. (2017) or Moreno-Martínez et al. (2018) for broad-scale trait characterization.

Acknowledgments The authors would like to thank Anna Schweiger, Erin Hestir, and Jeannine Cavender-Bares for their careful reviews, input, and suggestions on earlier versions of this chapter as part of the and the National Institute of Mathematical Biology and Synthesis Working Group on Remotely Sensing Biodiversity. Special thanks to Tiffany Bowman and Yelena Belyavina for assistance with graphics. S.P.S was supported by the Next-Generation Ecosystem Experiments (NGEEs) in the Arctic and tropics that are supported by the Office of Biological and Environmental Research in the Department of Energy, Office of Science, and through the United States Department of Energy contract No. DE-SC0012704 to Brookhaven National Laboratory. P.T. acknowledges support from NSF Emerging Frontiers Macrosystems Biology and NEON-Enabled Science (MSB-NES) grant 1638720, USDA McIntire-Stennis WIS01809 and Hatch WIS01874, NASA Biodiversity Program grant 80NSSC17K0677, and AIST program grant 80NSSC17K0244.

References

Aber JD, Melillo JM (1982) Nitrogen immobilization in decaying hardwood leaf litter as a function of initial nitrogen and lignin content. Can J Bot 60:2261–2269

Ahl DE, Gower ST, Burrows SN, Shabanov NV, Myneni RB, Knyazikhin Y (2006) Monitoring spring canopy phenology of a deciduous broadleaf forest using modis. Remote Sens Environ 104:88–95

Ainsworth EA, Serbin SP, Skoneczka JA, Townsend PA (2014) Using leaf optical properties to detect ozone effects on foliar biochemistry. Photosynth Res 119:65–76

Asner GP (1998) Biophysical and biochemical sources of variability in canopy reflectance. Remote Sens Environ 64:234–253

Asner GP, Nepstad D, Cardinot G, Ray D (2004) Drought stress and carbon uptake in an Amazon forest measured with spaceborne imaging spectroscopy. Proc Natl Acad Sci U S A 101:6039–6044

Asner GP, Knapp DE, Boardman J, Green RO, Kennedy-Bowdoin T, Eastwood M, Martin RE, Anderson C, Field CB (2012) Carnegie Airborne Observatory-2: increasing science data dimensionality via high-fidelity multi-sensor fusion. Remote Sens Environ 124:454–465

Asner GP, Martin RE (2008) Spectral and chemical analysis of tropical forests: scaling from leaf to canopy levels. Remote Sens Environ 112:3958–3970

Asner PG, Martin ER (2015) Spectroscopic remote sensing of non-structural carbohydrates in forest canopies. Remote Sens 7(4)

Asner GP, Martin RE, Anderson CB, Knapp DE (2015) Quantifying forest canopy traits: imaging spectroscopy versus field survey. Remote Sens Environ 158:15–27

Asner GP, Brodrick PG, Anderson CB, Vaughn N, Knapp DE, Martin RE (2016) Progressive forest canopy water loss during the 2012-2015 California drought. Proc Natl Acad Sci U S A 113:E249–E255

Asner GP, Martin RE, Carranza-Jiménez L, Sinca F, Tupayachi R, Anderson CB, Martinez P (2014) Functional and biological diversity of foliar spectra in tree canopies throughout the Andes to Amazon region. New Phytol 204:127–139

Asner GP, Martin RE, Knapp DE, Tupayachi R, Anderson C, Carranza L, Martinez P, Houcheime M, Sinca F, Weiss P (2011) Spectroscopy of canopy chemicals in humid tropical forests. Remote Sens Environ 115:3587–3598

Asner GP, Wessman CA, Bateson CA, Privette JL (2000) Impact of tissue, canopy, and landscape factors on the hyperspectral reflectance variability of arid ecosystems. Remote Sens Environ 74:69–84

Bacour C, Jacquemoud S, Tourbier Y, Dechambre M, Frangi JP (2002) Design and analysis of numerical experiments to compare four canopy reflectance models. Remote Sens Environ 79:72–83

Banskota A, Serbin SP, Wynne RH, Thomas VA, Falkowski MJ, Kayastha N, Gastellu-Etchegorry JP, Townsend PA (2015) An LUT-based inversion of DART model to estimate forest LAI from hyperspectral data. IEEE J Sel Top Appl Earth Obs Remote Sens 8:3147–3160

Baret F, Vanderbilt VC, Steven MD, Jacquemoud S (1994) Use of spectral analogy to evaluate canopy reflectance sensitivity to leaf optical properties. Remote Sens Environ 48:253–260

Billings WD, Morris RJ (1951) Reflection of visible and infrared radiation from leaves of different ecological groups. Am J Bot 38(5):327–331

Blackburn GA (2007) Hyperspectral remote sensing of plant pigments. J Exp Bot 58:855–867

Blackburn GA, Ferwerda JG (2008) Retrieval of chlorophyll concentration from leaf reflectance spectra using wavelet analysis. Remote Sens Environ 112:1614–1632

Bolster KL, Martin ME, Aber JD (1996) Determination of carbon fraction and nitrogen concentration in tree foliage by near infrared reflectance: a comparison of statistical methods. Can J For Res 26:590–600

Bonan GB, Levis S, Kergoat L, Oleson KW (2002) Landscapes as patches of plant functional types: an integrating concept for climate and ecosystem models. Global Biogeochem Cy 16:5-1–5-23

Butler EE, Datta A, Flores-Moreno H, Chen M, Wythers KR, Fazayeli F, Banerjee A, Atkin OK, Kattge J, Amiaud B, Blonder B, Boenisch G, Bond-Lamberty B, Brown KA, Byun C, Campetella G, Cerabolini BEL, Cornelissen JHC, Craine JM, Craven D, de Vries FT, Díaz S, Domingues TF, Forey E, González-Melo A, Gross N, Han W, Hattingh WN, Hickler T, Jansen S, Kramer K, Kraft NJB, Kurokawa H, Laughlin DC, Meir P, Minden V, Niinemets Ü, Onoda Y, Peñuelas J, Read Q, Sack L, Schamp B, Soudzilovskaia NA, Spasojevic MJ, Sosinski E, Thornton PE, Valladares F, van Bodegom PM, Williams M, Wirth C, Reich PB (2017) Mapping local and global variability in plant trait distributions. Proc Natl Acad Sci 114:E10937–E10946

Card DH, Peterson DL, Matson PA, Aber JD (1988) Prediction of leaf chemistry by the use of visible and near infrared reflectance spectroscopy. Remote Sens Environ 26:123–147

Carrascal LM, Galván I, Gordo O (2009) Partial least squares regression as an alternative to current regression methods used in ecology. Oikos 118:681–690

Cavender-Bares J, Gamon JA, Hobbie SE, Madritch MD, Meireles JE, Schweiger AK, Townsend PA (2017) Harnessing plant spectra to integrate the biodiversity sciences across biological and spatial scales. Am J Bot 104:966–969

Chavana-Bryant C, Malhi Y, Wu J, Asner GP, Anastasiou A, Enquist BJ, Cosio Caravasi EG, Doughty CE, Saleska SR, Martin RE, Gerard FF (2017) Leaf aging of Amazonian canopy trees as revealed by spectral and physiochemical measurements. New Phytol 214:1049–1063

Chen JM, Cihlar J (1996) Retrieving leaf area index of boreal conifer forests using landsat TM images. Remote Sens Environ 55:153–162

Cheng T, Rivard B, Sanchez-Azofeifa A (2011) Spectroscopic determination of leaf water content using continuous wavelet analysis. Remote Sens Environ 115:659–670

Clark W (1946) Photography by infrared: its principles and applications: J. Wiley & sons, Incorporated

Cocks T, Jensen R, Stewart A, Wilson I, and Shields T (1998) The HyMap airborne hyperspectral sensor: the system, calibration and performance. In: Proceedings of 1st EARSeL Workshop on Imaging Spectroscopy, Zurich, Switzerland, pp 37–42

Colombo R, Merom M, Marchesi A, Busetto L, Rossini M, Giardino C, Panigada C (2008) Estimation of leaf and canopy water content in poplar plantations by means of hyperspectral indices and inverse modeling. Remote Sens Environ 112:1820–1834

Cotrozzi L, Townsend PA, Pellegrini E, Nali C, Couture JJ (2018) Reflectance spectroscopy: a novel approach to better understand and monitor the impact of air pollution on Mediterranean plants. Environ Sci Pollut Res 25:8249–8267

Couture J, Singh A, Rubert-Nason KF, Serbin SP, Lindroth RL, Townsend PA (2016) Spectroscopic determination of ecologically relevant plant secondary metabolites. Methods Ecol Evol (in press).

Couture JJ, Serbin SP, Townsend PA (2013) Spectroscopic sensitivity of real-time, rapidly induced phytochemical change in response to damage. New Phytol 198:311–319

Curran PJ (1989) Remote-sensing of foliar chemistry. Remote Sens Environ 30:271–278

Dahlin KM, Asner GP, Field CB (2013) Environmental and community controls on plant canopy chemistry in a Mediterranean-type ecosystem. Proc Natl Acad Sci 110:6895–6900

Darvishzadeh R, Skidmore A, Schlerf M, Atzberger C (2008) Inversion of a radiative transfer model for estimating vegetation LAI and chlorophyll in a heterogeneous grassland. Remote Sens Environ 112:2592–2604

Datt B (1998) Remote sensing of chlorophyll a, chlorophyll b, chlorophyll a+b, and total carotenoid content in eucalyptus leaves. Remote Sens Environ 66(2):111–121

Davies T (1998) The history of near infrared spectroscopic analysis: Past, present and future "From sleeping technique to the morning star of spectroscopy". Analusis 26(4):17–19

Dawson TP, Curran PJ, Plummer SE (1998) LIBERTY – Modeling the Effects of Leaf Biochemical Concentration on Reflectance Spectra. Remote Sens Environ 65:50–60

Díaz S, Kattge J, Cornelissen JHC, Wright IJ, Lavorel S, Dray S, Reu B, Kleyer M, Wirth C, Colin Prentice I, Garnier E, Bönisch G, Westoby M, Poorter H, Reich PB, Moles AT, Dickie J, Gillison AN, Zanne AE, Chave J, Joseph Wright S, Sheremet'ev SN, Jactel H, Baraloto C, Cerabolini B, Pierce S, Shipley B, Kirkup D, Casanoves F, Joswig JS, Günther A, Falczuk V, Rüger N, Mahecha MD, Gorné LD (2015) The global spectrum of plant form and function. Nature 529:167

Drolet GG, Middleton EM, Huemmrich KF, Hall FG, Amiro BD, Barr AG, Black TA, McCaughey JH, Margolis HA (2008) Regional mapping of gross light-use efficiency using MODIS spectral indices. Remote Sens Environ 112:3064–3078

DuBois S, Desai AR, Singh A, Serbin SP, Goulden ML, Baldocchi DD, Ma S, Oechel WC, Wharton S, Kruger EL, Townsend PA (2018) Using imaging spectroscopy to detect variation in terrestrial ecosystem productivity across a water-stressed landscape. Ecol Appl 28:1313–1324

Elvidge CD (1990) Visible and near-infrared reflectance characteristics of dry plant materials. Int J Remote Sens 11:1775–1795

Ely KS, Burnett AC, Lieberman-Cribbin W, Serbin S, and Rogers A (2019) Spectroscopy can predict key leaf traits associated with source–sink balance and carbon–nitrogen status. J Exp Bot. 70:1789–1799

Enquist BJ, Bentley LP, Shenkin A, Maitner B, Savage V, Michaletz S, Blonder B, Buzzard V, Espinoza TEB, Farfan-Rios W, Doughty CE, Goldsmith GR, Martin RE, Salinas N, Silman M, Díaz S, Asner GP, Malhi Y (2017) Assessing trait-based scaling theory in tropical forests spanning a broad temperature gradient. Global Ecol Biogeogr 26:1357–1373

Evans JR, Clarke VC (2018) The nitrogen cost of photosynthesis. J Exp Bot 70:7–15

Falster DS, Westoby M (2003) Leaf size and angle vary widely across species: what consequences for light interception? New Phytol 158:509–525

Feilhauer H, Asner GP, Martin RE, Schmidtlein S (2010) Brightness-normalized partial least squares regression for hyperspectral data. J Quant Spectrosc Radiat Transf 111:1947–1957

Feret J-B, Francois C, Gitelson A, Asner GP, Barry KM, Panigada C, Richardson AD, Jacquemoud S (2011) Optimizing spectral indices and chemometric analysis of leaf chemical properties using radiative transfer modeling. Remote Sens Environ 115:2742–2750

Féret JB, Francois C, Asner GP, Gitelson AA, Martin RE, Bidel LPR, Ustin SL, le Maire G, Jacquemoud S (2008) PROSPECT-4 and 5: Advances in the leaf optical properties model separating photosynthetic pigments. Remote Sens Environ 112(6):3030–3043

Féret JB, Gitelson AA, Noble SD, Jacquemoud S (2017) PROSPECT-D: towards modeling leaf optical properties through a complete lifecycle. Remote Sens Environ 193:204–215

Féret JB, le Maire G, Jay S, Berveiller D, Bendoula R, Hmimina G, Cheraiet A, Oliveira JC, Ponzoni FJ, Solanki T, de Boissieu F, Chave J, Nouvellon Y, Porcar-Castell A, Proisy C, Soudani K, Gastellu-Etchegorry JP, Lefèvre-Fonollosa MJ (2018) Estimating leaf mass per area and equivalent water thickness based on leaf optical properties: Potential and limitations of physical modeling and machine learning. In: Remote Sens Environ

Field C, Mooney HA (1986) The photosynthesis-nitrogen relationship in wild plants. In: Givnish T (ed) On the economy of plant form and function. Cambridge University Press, Cambridge, pp 22–55

Foster JR, Townsend PA, Zganjar CE (2008) Spatial and temporal patterns of gap dominance by low-canopy lianas detected using EO-1 Hyperion and Landsat Thematic Mapper. Remote Sens Environ 112:2104–2117

Friedlingstein P, Meinshausen M, Arora VK, Jones CD, Anav A, Liddicoat SK, Knutti R (2014) Uncertainties in CMIP5 climate projections due to carbon cycle feedbacks. J Clim 27:511–526

Funk JL, Larson JE, Ames GM, Butterfield BJ, Cavender-Bares J, Firn J, Laughlin DC, Sutton-Grier AE, Williams L, Wright J (2017) Revisiting the Holy Grail: using plant functional traits to understand ecological processes. Biol Rev 92:1156–1173

Gamon JA, Field CB, Goulden ML, Griffin KL, Hartley AE, Joel G, Penuelas J, Valentini R (1995) Relationships between NDVI, canopy structure, and photosynthesis in 3 Californian vegetation types. Ecol Appl 5:28–41

Gamon JA, Penuelas J, Field CB (1992) A narrow-waveband spectral index that tracks diurnal changes in photosynthetic efficiency. Remote Sens Environ 41:35–44

Gamon JA, Serrano L, Surfus JS (1997) The photochemical reflectance index: an optical indicator of photosynthetic radiation use efficiency across species, functional types, and nutrient levels. Oecologia 112:492–501

Ganapol BD, Johnson LF, Hammer PD, Hlavka CA, Peterson DL (1998) LEAFMOD: a new within-leaf radiative transfer model. Remote Sens Environ 63:182–193

Gao BC (1996) NDWI—A normalized difference water index for remote sensing of vegetation liquid water from space. Remote Sens Environ 58(3):257–266

Gao BC, Goetz AFH (1995) Retrieval of equivalent water thickness and information related to biochemical components of vegetation canopies from AVIRIS data. Remote Sens Environ 52(3):155–162

Gao BC, Heidebrecht KB, Goetz AFH (1993) Derivation of scaled surface reflectances from AVIRIS data. Remote Sens Environ, 44:165–178

Gastellu-Etchegorry JP, Zagolski F, Mougtn E, Marty G, Giordano G (1995). An assessment of canopy chemistry with AVIRIS—a case study in the Landes Forest, South-west France. Int J Remote Sens 16(3):487–501

Gastellu-Etchegorry J-P, Yin T, Lauret N, Cajgfinger T, Gregoire T, Grau E, Feret J-B, Lopes M, Guilleux J, Dedieu G, Malenovský Z, Cook BD, Morton D, Rubio J, Durrieu S, Cazanave G, Martin E, Ristorcelli T (2015) Discrete anisotropic radiative transfer (DART 5) for modeling airborne and satellite spectroradiometer and LIDAR acquisitions of natural and urban landscapes. Remote Sens 7:1667–1701

Gates DM, Keegan HJ, Schleter JC, Weidner VR (1965) Spectral properties of plants. Appl Opt 4(1):11–20

Geladi P, Kowalski BR (1986) Partial least-squares regression - A tutorial. Anal Chim Acta 185:1–17

Gil-Pérez B, Zarco-Tejada PJ, Correa-Guimaraes A, Relea-Gangas E, Navas-Gracia LM, Hernández-Navarro S, Sanz-Requena JF, Berjón A, Martín-Gil J (2010). Vitis-Journal of Grapevine Research 49(4):167–173

Gitelson AA (2004) Wide dynamic range vegetation index for remote quantification of biophysical characteristics of vegetation. J Plant Physiol 161:165–173

Gitelson AA, Vina A, Verma SB, Rundquist DC, Arkebauer TJ, Keydan G, Leavitt B, Ciganda V, Burba GG, Suyker AE (2006) Relationship between gross primary production and chlorophyll content in crops: Implications for the synoptic monitoring of vegetation productivity. J Geophys Res-Atmos 111:13

Glenn EP, Huete AR, Nagler PL, Nelson SG (2008) Relationship Between Remotely-sensed Vegetation Indices, Canopy Attributes and Plant Physiological Processes: What Vegetation Indices Can and Cannot Tell Us About the Landscape. Sensors (Basel) 8:2136–2160

Goetz SJ, Bunn AG, Fiske GJ, Houghton RA (2005) Satellite-observed photosynthetic trends across boreal north america associated with climate and fire disturbance. Proc Natl Acad Sci U S A 102:13521–13525

Goetz SJ, Fiske GJ, Bunn AG (2006) Using satellite time-series data sets to analyze fire disturbance and forest recovery across Canada. Remote Sens Environ 101:352–365

Gökkaya K, Thomas V, Noland TL, McCaughey H, Morrison I, Treitz P (2015) Prediction of macronutrients at the canopy level using spaceborne imaging spectroscopy and LiDAR data in a mixedwood boreal forest. Remote Sens 7:9045–9069

Goward SN, Huemmrich KF (1992) Vegetation canopy PAR absorptance and the normalized difference vegetation index – An assessment using the SAIL model. Remote Sens Environ 39:119–140

Green DS, Erickson JE, Kruger EL (2003) Foliar morphology and canopy nitrogen as predictors of light-use efficiency in terrestrial vegetation. Agric For Meteorol 115:165–173

Green RO, Eastwood ML, Sarture CM, Chrien TG, Aronsson M, Chippendale BJ, Faust JA, Pavri BE, Chovit CJ, Solis MS, Olah MR, Williams O (1998) Imaging spectroscopy and the Airborne Visible Infrared Imaging Spectrometer (AVIRIS). Remote Sens Environ 65:227–248

Grossman YL, Ustin SL, Jacquemoud S, Sanderson EW, Schmuck G, Verdebout J (1996) Critique of stepwise multiple linear regression for the extraction of leaf biochemistry information from leaf reflectance data. Remote Sens Environ 56:182–193

Hilker T, Coops NC, Wulder MA, Black TA, Guy RD (2008) The use of remote sensing in light use efficiency based models of gross primary production: A review of current status and future requirements. Sci Total Environ 404:411–423

Huemmrich KF (2013) Simulations of seasonal and latitudinal variations in leaf inclination angle distribution: implications for remote sensing. J Adv Remote Sens 02:9

Hunt ER, Rock BN (1989) Detection of changes in leaf water content using near-infrared and middle-infrared reflectances. Remote Sens Environ 30:43–54

Inoue Y, Penuelas J, Miyata A, Mano M (2008) Normalized difference spectral indices for estimating photosynthetic efficiency and capacity at a canopy scale derived from hyperspectral and CO2 flux measurements in rice. Remote Sens Environ 112:156–172

IPCC (2018) Summary for Policymakers. In: Global Warming of 1.5°C. An IPCC Special Report on the impacts of global warming of 1.5°C above pre-industrial levels and related global greenhouse gas emission pathways, in the context of strengthening the global response to the threat of climate change, sustainable development, and efforts to eradicate poverty [Masson-Delmotte, V., P. Zhai, H.-O. Pörtner, D. Roberts, J. Skea, P.R. Shukla, A. Pirani, W. Moufouma-Okia, C. Péan, R. Pidcock, S. Connors, J.B.R. Matthews, Y. Chen, X. Zhou, M.I. Gomis, E. Lonnoy, T. Maycock, M. Tignor, and T. Waterfield (eds.)]. World Meteorological Organization, Geneva, Switzerland, 32 pp.

IPBES (2018) Summary for policymakers of the regional assessment report on biodiversity and ecosystem services for the Americas of the Intergovernmental Science-Policy Platform on Biodiversity and Ecosystem Services. IPBES secretariat, Bonn, Germany

Jacquemoud S, Baret F (1990) PROSPECT – a model of leaf optical-properties of spectra. Remote Sens Environ 34:75–91

Jacquemoud S, Baret F, Andrieu B, Danson FM, Jaggard K (1995) Extraction of vegetation biophysical parameters by inversion of the PROSPECT + SAIL models on sugar beet canopy reflectance data. Application to TM and AVIRIS sensors. Remote Sens Environ 52:163–172

Jacquemoud S, Ustin SL, Verdebout J, Schmuck G, Andreoli G, Hosgood B (1996) Estimating leaf biochemistry using the PROSPECT leaf optical properties model. Remote Sens Environ 56:194–202

Jacquemoud S, Verhoef W, Baret F, Bacour C, Zarco-Tejada PJ, Asner GP, Francois C, Ustin SL (2009) PROSPECT + SAIL models: a review of use for vegetation characterization. Remote Sens Environ 113:S56–S66

Jensen DJ, Simard M, Cavanaugh KC, Thompson DR (2018) Imaging spectroscopy BRDF correction for mapping Louisiana's coastal ecosystems. IEEE Trans Geosci Remote Sens 56:1739–1748

Jetz W, Cavender-Bares J, Pavlick R, Schimel D, Davis FW, Asner GP, Guralnick R, Kattge J, Latimer AM, Moorcroft P, Schaepman ME, Schildhauer MP, Schneider FD, Schrodt F, Stahl U, Ustin SL (2016) Monitoring plant functional diversity from space. Nature Plants 2:16024

Jetz W, Wilcove DS, Dobson AP (2007) Projected impacts of climate and land-use change on the global diversity of birds. PLoS Biol 5:e157

Johnson LF, Hlavka CA, Peterson DL (1994) Multivariate analysis of AVIRIS data for canopy biochemical estimation along the oregon transect. Remote Sens Environ 47(2):216–230

Jones LA, Kimball JS, McDonald KC, Chan STK, Njoku EG, Oechel WC (2007) Satellite microwave remote sensing of boreal and arctic soil temperatures from AMSR-E. IEEE Trans Geosci Remote Sens 45:2004–2018

Kalacska M, Lalonde M, Moore TR (2015) Estimation of foliar chlorophyll and nitrogen content in an ombrotrophic bog from hyperspectral data: Scaling from leaf to image. Remote Sens Environ 169:270–279

Kampe TU, Johnson BR, Kuester M, Keller M (2010) NEON: the first continental-scale ecological observatory with airborne remote sensing of vegetation canopy biochemistry and structure. J Appl Remote Sens 4:043510

Kasischke ES, French NHF, Harrell P, Christensen NL, Ustin SL, Barry D (1993) Monitoring of wildfires in boreal forests using large-area AVHRR NDVI composite image data. Remote Sens Environ 45:61–71

Kattenborn T, Fassnacht FE, Pierce S, Lopatin J, Grime JP, Schmidtlein S (2017) Linking plant strategies and plant traits derived by radiative transfer modelling. J Veg Sci 28:717–727

Kattenborn T, Fassnacht FE, Schmidtlein S (2019) Differentiating plant functional types using reflectance: which traits make the difference? Remote Sens Ecol Conserv 5(1):5–19

Kattge J, Díaz S, Lavorel S, Prentice IC, Leadley P, Bönisch G, Garnier E, Westoby M, Reich PB, Wright IJ, Cornelissen JHC, Violle C, Harrison SP, Van Bodegom PM, Reichstein M, Enquist BJ, Soudzilovskaia NA, Ackerly DD, Anand M, Atkin O, Bahn M, Baker TR, Baldocchi D, Bekker R, Blanco CC, Blonder B, Bond WJ, Bradstock R, Bunker DE, Casanoves F, Cavender-Bares J, Chambers JQ, Chapin Iii FS, Chave J, Coomes D, Cornwell WK, Craine JM, Dobrin BH, Duarte L, Durka W, Elser J, Esser G, Estiarte M, Fagan WF, Fang J, Fernández-Méndez F, Fidelis A, Finegan B, Flores O, Ford H, Frank D, Freschet GT, Fyllas NM, Gallagher RV, Green WA, Gutierrez AG, Hickler T, Higgins SI, Hodgson JG, Jalili A, Jansen S, Joly CA, Kerkhoff AJ, Kirkup D, Kitajima K, Kleyer M, Klotz S, Knops JMH, Kramer K, Kühn I, Kurokawa H, Laughlin D, Lee TD, Leishman M, Lens F, Lenz T, Lewis SL, Lloyd J, Llusià J, Louault F, Ma S, Mahecha MD, Manning P, Massad T, Medlyn BE, Messier J, Moles AT, Müller SC, Nadrowski K, Naeem S, Niinemets Ü, Nöliert S, Nüske A, Ogaya R, Oleksyn J, Onipchenko VG, Onoda Y, Ordoñez J, Overbeck G, Ozinga WA, Patiño S, Paula S, Pausas JG, Peñuelas J, Phillips OL, Pillar V, Poorter H, Poorter L, Poschlod P, Prinzing A, Proulx R, Rammig A, Reinsch S, Reu B, Sack L, Salgado-Negret B, Sardans J, Shiodera S, Shipley B, Siefert A, Sosinski E, Soussana JF, Swaine E, Swenson N, Thompson K, Thornton P, Waldram M, Weiher E, White M, White S, Wright SJ, Yguel B, Zaehle S, Zanne AE, Wirth C (2011) TRY – a global database of plant traits. Glob Chang Biol 17:2905–2935

Knyazikhin Y, Schull MA, Stenberg P, Mõttus M, Rautiainen M, Yang Y, Marshak A, Latorre Carmona P, Kaufmann RK, Lewis P, Disney MI, Vanderbilt V, Davis AB, Baret F, Jacquemoud S, Lyapustin A, Myneni RB (2013) Hyperspectral remote sensing of foliar nitrogen content. Proc Natl Acad Sci 110:E185–E192

Kobayashi H, Iwabuchi H (2008) A coupled 1-D atmosphere and 3-D canopy radiative transfer model for canopy reflectance, light environment, and photosynthesis simulation in a heterogeneous landscape. Remote Sens Environ 112:173–185

Kokaly RF, Asner GP, Ollinger SV, Martin ME, Wessman CA (2009) Characterizing canopy biochemistry from imaging spectroscopy and its application to ecosystem studies. Remote Sens Environ 113:S78–S91

Kokaly RF, Clark RN (1999) Spectroscopic determination of leaf biochemistry using band-depth analysis of absorption features and stepwise multiple linear regression. Remote Sens Environ 67:267–287

Kokaly RF, Skidmore AK (2015) Plant phenolics and absorption features in vegetation reflectance spectra near 1.66μm. Int J Appl Earth Obs Geoinf 43:55–83

Krinov EL (1953) Spectral reflectance properties of natural formations. National Research Council of Canada (Ottawa) Technical Translations TT-439

Kuusk A (2018) 3.03 - Canopy radiative transfer modeling. In: Liang S (ed). Comprehensive Remote Sensing. Oxford: Elsevier, 9–22, https://doi.org/10.1016/B978-0-12-409548-9.10534-2

Kuusk A, Nilson T (2000) A directional multispectral forest reflectance model. Remote Sens Environ 72:244–252

Lavorel S, Garnier E (2002) Predicting changes in community composition and ecosystem functioning from plant traits: revisiting the Holy Grail. Funct Ecol 16:545–556

LeBauer D, Kooper R, Mulrooney P, Rohde S, Wang D, Long SP, Dietze MC (2018) BETYdb: a yield, trait, and ecosystem service database applied to second-generation bioenergy feedstock production. GCB Bioenergy 10(1):61–71

Li D, Wang X, Zheng H, Zhou K, Yao X, Tian Y, Zhu Y, Cao W, Cheng T (2018) Estimation of area- and mass-based leaf nitrogen contents of wheat and rice crops from water-removed spectra using continuous wavelet analysis. Plant Methods 14:76

Li L, Cheng YB, Ustin S, Hu XT, Riaño D (2008) Retrieval of vegetation equivalent water thickness from reflectance using genetic algorithm (GA)-partial least squares (PLS) regression. Adv Space Res 41:1755–1763

Mand P, Hallik L, Penuelas J, Nilson T, Duce P, Emmett BA, Beier C, Estiarte M, Garadnai J, Kalapos T, Schmidt IK, Kovacs-Lang E, Prieto P, Tietema A, Westerveld JW, Kull O (2010) Responses of the reflectance indices PRI and NDVI to experimental warming and drought in European shrublands along a north-south climatic gradient. Remote Sens Environ 114:626–636

Martin ME, Aber JD (1997) High spectral resolution remote sensing of forest canopy lignin, nitrogen, and ecosystem processes. Ecol Appl 7:431–443

Martin ME, Plourde LC, Ollinger SV, Smith ML, McNeil BE (2008) A generalizable method for remote sensing of canopy nitrogen across a wide range of forest ecosystems. Remote Sens Environ 112:3511–3519

Matson P, Johnson L, Billow C, Miller J, Pu RL (1994) Seasonal patterns and remote spectral estimation of canopy chemistry across the Oregon transect. Ecol Appl 4:280–298

McNeil BE, Read JM, Sullivan TJ, McDonnell TC, Fernandez IJ, Driscoll CT (2008) The spatial pattern of nitrogen cycling in the Adirondack Park, New York. Ecol Appl 18:438–452

McNicholas HJ (1931) The visible and ultraviolet absorption spectra of carotin and xanthophyll and the changes accompanying oxidation. Bureau of Standards Journal of Research 7(1):171. Research Paper 337 (RP337)

Middleton EM, Ungar SG, Mandl DJ, Ong L, Frye SW, Campbell PE, Landis DR, Young JP, Pollack NH (2013) The Earth observing one (EO-1) satellite mission: over a decade in space. IEEE J Sel Top Appl Earth Obs Remote Sens 6:243–256

Miller CE, Green RO, Thompson DR, Thorpe AK, Eastwood M, Mccubbin IB, Olson-Duvall W, Bernas M, Sarture CM, Nolte S, Rios LM, Hernandez MA, Bue BD, Lundeen SR (2019) ABoVE: Hyperspectral Imagery from AVIRIS-NG, Alaskan and Canadian Arctic, 2017-2018. ORNL DAAC, Oak Ridge, Tennessee, USA. https://doi.org/10.3334/ORNLDAAC/1569

Mirik M, Norland JE, Crabtree RL, Biondini ME (2005) Hyperspectral one-meter-resolution remote sensing in yellowstone National Park, Wyoming: I. Forage nutritional values. Rangel Ecol Manag 58:452–458

Moorthy I, Miller JR, Noland TL (2008) Estimating chlorophyll concentration in conifer needles with hyperspectral data: An assessment at the needle and canopy level. Remote Sens Environ 112(6):2824–2838

Moreno-Martínez Á, Camps-Valls G, Kattge J, Robinson N, Reichstein M, van Bodegom P, Kramer K, Cornelissen JHC, Reich P, Bahn M, Niinemets Ü, Peñuelas J, Craine JM, Cerabolini BEL, Minden V, Laughlin DC, Sack L, Allred B, Baraloto C, Byun C, Soudzilovskaia NA, Running SW (2018) A methodology to derive global maps of leaf traits using remote sensing and climate data. Remote Sens Environ 218:69–88

Mutanga O, Kumar L (2007) Estimating and mapping grass phosphorus concentration in an African savanna using hyperspectral image data. Int J Remote Sens 28(21):4897–4911

Myneni RB, Williams DL (1994) On the relationship between FAPAR and NDVI. Remote Sens Environ 49:200–211

Neyret M, Bentley LP, Oliveras I, Marimon BS, Marimon-Junior BH, Almeida de Oliveira E, Barbosa Passos F, Castro Ccoscco R, dos Santos J, Matias Reis S, Morandi PS, Rayme Paucar G, Robles Cáceres A, Valdez Tejeira Y, Yllanes Choque Y, Salinas N, Shenkin A, Asner GP, Díaz S, Enquist BJ, Malhi Y (2016) Examining variation in the leaf mass per area of dominant species across two contrasting tropical gradients in light of community assembly. Ecology and Evolution 6:5674–5689

Niinemets U (2007) Photosynthesis and resource distribution through plant canopies. Plant Cell Environ 30:1052–1071

Niinemets Ü (2016) Leaf age dependent changes in within-canopy variation in leaf functional traits: a meta-analysis. J Plant Res 129:313–338

Nilson T, Kuusk A, Lang M, Lükk T (2003) Forest reflectance modeling: theoretical aspects and applications. Ambio 32:535–541

North PRJ (1996) Three-dimensional forest light interaction model using a Monte Carlo method. IEEE Trans Geosci Remote Sens 34:946–956

Norris KH, Hart JR (1965) Direct spectrophotometric determination of moisture content of grain and seeds. Proceedings of the 1963 International Symposium on Humidity and Moisture, Reinhold, New York, vol. 4, pp 19–25

Norris KH, Barnes RF, Moore JE, Shenk JS (1976) Predicting forage quality by infrared replectance spectroscopy. J Anim Sci 43(4):889–897

Ollinger SV (2011) Sources of variability in canopy reflectance and the convergent properties of plants. New Phytol 189:375–394

Ollinger SV, Smith ML (2005) Net primary production and canopy nitrogen in a temperate forest landscape: An analysis using imaging spectroscopy, modeling and field data. Ecosystems 8:760–778

Ollinger SV, Smith ML, Martin ME, Hallett RA, Goodale CL, Aber JD (2002) Regional variation in foliar chemistry and N cycling among forests of diverse history and composition. Ecology 83:339–355

Osnas JLD, Katabuchi M, Kitajima K, Wright SJ, Reich PB, Van Bael SA, Kraft NJB, Samaniego MJ, Pacala SW, Lichstein JW (2018) Divergent drivers of leaf trait variation within species, among species, and among functional groups. Proc Natl Acad Sci 115:5480–5485

Peckham SD, Ahl DE, Serbin SP, Gower ST (2008) Fire-induced changes in green-up and leaf maturity of the Canadian boreal forest. Remote Sens Environ 112:3594–3603

Penuelas J, Filella I, Gamon JA (1995) Assessment of photosynthetic radiation-use efficiency with spectral reflectance. New Phytol 131:291–296

Peterson DL, Aber JD, Matson PA, Card DH, Swanberg N, Wessman C, Spanner M (1988) Remote-sensing of forest canopy and leaf biochemical contents. Remote Sens Environ 24:85–108

Pettorelli N, Wegmann M, Skidmore A, Mücher S, Dawson TP, Fernandez M, Lucas R, Schaepman ME, Wang T, O'Connor B, Jongman RHG, Kempeneers P, Sonnenschein R, Leidner AK, Böhm M, He KS, Nagendra H, Dubois G, Fatoyinbo T, Hansen MC, Paganini M, de Klerk HM, Asner GP, Kerr JT, Estes AB, Schmeller DS, Heiden U, Rocchini D, Pereira HM, Turak E, Fernandez N, Lausch A, Cho MA, Alcaraz-Segura D, McGeoch MA, Turner W, Mueller A, St-Louis V,

Penner J, Vihervaara P, Belward A, Reyers B, Geller GN (2016) Framing the concept of satellite remote sensing essential biodiversity variables: challenges and future directions. Remote Sens Ecol Conser 2:122–131

Poorter H, Niinemets U, Poorter L, Wright IJ, Villar R (2009) Causes and consequences of variation in leaf mass per area (LMA): a meta-analysis. New Phytol 182:565–588

Raabe K, Pisek J, Sonnentag O, Annuk K (2015) Variations of leaf inclination angle distribution with height over the growing season and light exposure for eight broadleaf tree species. Agric For Meteorol 214-215:2–11

Rabideau GS, French CS, Holt AS (1946) The absorption and reflection spectra of leaves, chloroplast suspensions, and chloroplast fragments as measured in an Ulbricht sphere. Am J Bot 33(10):769–777

Reich PB (2014) The world-wide 'fast–slow' plant economics spectrum: a traits manifesto. J Ecol 102:275–301

Reich PB, Walters MB, Ellsworth DS (1997) From tropics to tundra: Global convergence in plant functioning. Proc Natl Acad Sci U S A 94:13730–13734

Reich PB, Wright IJ, Cavender-Bares J, Craine JM, Oleksyn J, Westoby M, Walters MB (2003) The evolution of plant functional variation: traits, spectra, and strategies. Int J Plant Sci 164:S143–S164

Reich PB, Wright IJ, Lusk CH (2007) Predicting leaf physiology from simple plant and climate attributes: a global GLOPNET analysis. Ecol Appl 17:1982–1988

Rice J, Seixas CS, Zaccagnini ME, BedoyaGaitán M, Valderrama N, Anderson CB, Arroyo MTK, Bustamante M, Cavender-Bares J, Diaz-de-Leon A, Fennessy S, Márquez JRG, Garcia K, Helmer EH, Herrera B, Klatt B, Ometo JP, Osuna VR, Scarano FR, Schill S, and Farinaci JS (2018) IPBES, The regional assessment report on biodiversity and ecosystem services for the Americas. Bonn, Germany

Richardson AD, Duigan SP, Berlyn GP (2002) An evaluation of noninvasive methods to estimate foliar chlorophyll content. New Phytol 153:185–194

Roberts DA, Ustin SL, Ogunjemiyo S, Greenberg J, Dobrowski SZ, Chen JQ, Hinckley TM (2004) Spectral and structural measures of northwest forest vegetation at leaf to landscape scales. Ecosystems 7:545–562

Ross J (1981) Optical properties of phytoelements. In: Ross J (ed) The radiation regime and architecture of plant stands. Springer Netherlands, Dordrecht, pp 175–187

Schaepman ME, Jehle M, Hueni A, D'Odorico P, Damm A, Weyermann J, Schneider FD, Laurent V, Popp C, Seidel FC, Lenhard K, Gege P, Küchler C, Brazile J, Kohler P, De Vos L, Meuleman K, Meynart R, Schläpfer D, Kneubühler M, Itten KI (2015) Advanced radiometry measurements and Earth science applications with the Airborne Prism Experiment (APEX). Remote Sens Environ 158:207–219

Schimel D, Pavlick R, Fisher JB, Asner GP, Saatchi S, Townsend P, Miller C, Frankenberg C, Hibbard K, Cox P (2015) Observing terrestrial ecosystems and the carbon cycle from space. Glob Chang Biol 21:1762–1776

Schläpfer D, Richter R, Feingersh T (2015) Operational BRDF effects correction for wide-field-of-view optical scanners (BREFCOR). IEEE Trans Geosci Remote Sens 53:1855–1864

Serbin SP, Singh A, Desai AR, Dubois SG, Jablonski AD, Kingdon CC, Kruger EL, Townsend PA (2015) Remotely estimating photosynthetic capacity, and its response to temperature, in vegetation canopies using imaging spectroscopy. Remote Sens Environ 167:78–87

Serbin SP, Singh A, McNeil BE, Kingdon CC, Townsend PA (2014) Spectroscopic determination of leaf morphological and biochemical traits for northern temperate and boreal tree species. Ecol Appl 24:1651–1669

Serrano L, Ustin SL, Roberts DA, Gamon JA, Penuelas J (2000) Deriving water content of chaparral vegetation from AVIRIS data. Remote Sens Environ 74(3):570–581

Shabanov NV, Huang D, Yang WZ, Tan B, Knyazikhin Y, Myneni RB, Ahl DE, Gower ST, Huete AR, Aragao L, Shimabukuro YE (2005) Analysis and optimization of the MODIS leaf area index algorithm retrievals over broadleaf forests. IEEE Trans Geosci Remote Sens 43:1855–1865

Shenk JS, Landa I, Hoover MR, Westerhaus MO (1981) Description and evaluation of a near infrared reflectance spectro-computer for forage and grain analysis1. Crop Sci 21:355–358

Shiklomanov AN, Dietze MC, Viskari T, Townsend PA, Serbin SP (2016) Quantifying the influences of spectral resolution on uncertainty in leaf trait estimates through a Bayesian approach to RTM inversion. Remote Sens Environ 183:226–238

Shiklomanov A, Bradley BA, Dahlin K, Fox A, Gough C, Hoffman FM, Middleton E, Serbin S, Smallman L, Smith WK (2019) Enhancing global change experiments through integration of remote-sensing techniques. Front Ecol Environ 17(4):215–224

Shull CA (1929) A Spectrophotometric Study of Reflection of Light from Leaf Surfaces. Bot Gaz 87(5):583–607

Sims DA, Gamon JA (2002) Relationships between leaf pigment content and spectral reflectance across a wide range of species, leaf structures and developmental stages. Remote Sens Environ 81:337–354

Sims DA, Gamon JA (2003) Estimation of vegetation water content and photosynthetic tissue area from spectral reflectance: a comparison of indices based on liquid water and chlorophyll absorption features. Remote Sens Environ 84:526–537

Singh A, Serbin SP, McNeil BE, Kingdon CC, Townsend PA (2015) Imaging spectroscopy algorithms for mapping canopy foliar chemical and morphological traits and their uncertainties. Ecol Appl 25:2180–2197

Sonnentag O, Chen JM, Roberts DA, Talbot J, Halligan KQ, Govind A (2007) Mapping tree and shrub leaf area indices in an ombrotrophic peatland through multiple endmember spectral unmixing. Remote Sens Environ 109:342–360

Stimson HC, Breshears DD, Ustin SL, Kefauver SC (2005) Spectral sensing of foliar water conditions in two co-occurring conifer species: Pinus edulis and Juniperus monosperma. Remote Sens Environ 96:108–118

Stylinski CD, Gamon JA, Oechel WC (2002) Seasonal patterns of reflectance indices, carotenoid pigments and photosynthesis of evergreen chaparral species. Oecologia 131:366–374

Thompson DR, Gao B-C, Green RO, Roberts DA, Dennison PE, Lundeen SR (2015) Atmospheric correction for global mapping spectroscopy: ATREM advances for the HyspIRI preparatory campaign. Remote Sens Environ 167:64–77

Thompson DR, Natraj V, Green RO, Helmlinger MC, Gao B-C, Eastwood ML (2018) Optimal estimation for imaging spectrometer atmospheric correction. Remote Sens Environ 216:355–373

Thulin S, Hill MJ, Held A, Jones S, Woodgate P (2014) Predicting levels of crude protein, digestibility, lignin and cellulose in temperate pastures using hyperspectral image data. Am J Plant Sci 5:997–1019

Townsend PA (2002) Estimating forest structure in wetlands using multitemporal SAR. Remote Sens Environ 79:288–304

Townsend PA, Foster JR, Chastain RA, Currie WS (2003) Application of imaging spectroscopy to mapping canopy nitrogen in the forests of the central Appalachian Mountains using Hyperion and AVIRIS. IEEE Trans Geosci Remote Sens 41:1347–1354

Tucker CJ, Slayback DA, Pinzon JE, Los SO, Myneni RB, Taylor MG (2001) Higher northern latitude normalized difference vegetation index and growing season trends from 1982 to 1999. Int J Biometeorol 45:184–190

Turner DP, Cohen WB, Kennedy RE, Fassnacht KS, Briggs JM (1999) Relationships between leaf area index and landsat Tm spectral vegetation indices across three temperate zone sites. Remote Sens Environ 70:52–68

Ustin SL, Gamon JA (2010) Remote sensing of plant functional types. New Phytol 186:795–816

Ustin SL, Gitelson AA, Jacquemoud S, Schaepman M, Asner GP, Gamon JA, Zarco-Tejada P (2009) Retrieval of foliar information about plant pigment systems from high resolution spectroscopy. Remote Sens Environ 113:S67–S77

Ustin SL, Roberts DA, Gamon JA, Asner GP, Green RO (2004) Using imaging spectroscopy to study ecosystem processes and properties. Bioscience 54:523–534

Vane G (1987) First results from the airborne visible/infrared imaging spectrometer (AVIRIS): SPIE.

Vane G, Goetz AFH (1988) Terrestrial imaging spectroscopy. Remote Sens Environ 24(1):1–29

van den Berg AK, Perkins TD (2005) Nondestructive estimation of anthocyanin content in autumn sugar maple leaves. HortScience HortSci 40(3): 685-686.

Verhoef W (1984) Light scattering by leaf layers with application to canopy reflectance modeling: the SAIL model. Remote Sens Environ 16:125–141

Verhoef W, Bach H (2007) Coupled soil-leaf-canopy and atmosphere radiative transfer modeling to simulate hyperspectral multi-angular surface reflectance and TOA radiance data. Remote Sens Environ 109:166–182

Verrelst J, Alonso L, Camps-Valls G, Delegido J, Moreno J (2012) Retrieval of vegetation biophysical parameters using gaussian process techniques. IEEE Trans Geosci Remote Sens 50:1832–1843

Verrelst J, Rivera JP, Gitelson A, Delegido J, Moreno J, Camps-Valls G (2016) Spectral band selection for vegetation properties retrieval using Gaussian processes regression. Int J Appl Earth Obs Geoinf 52:554–567

Wang J, Chen Y, Chen F, Shi T, Wu G (2018) Wavelet-based coupling of leaf and canopy reflectance spectra to improve the estimation accuracy of foliar nitrogen concentration. Agric For Meteorol 248:306–315

Wang Z, Townsend PA, Schweiger AK, Couture JJ, Singh A, Hobbie SE, Cavender-Bares J (2019) Mapping foliar functional traits and their uncertainties across three years in a grassland experiment. Remote Sens Environ 221:405–416

Weiss M, Baret F, Myneni RB, Pragnere A, Knyazikhin Y (2000) Investigation of a model inversion technique to estimate canopy biophysical variables from spectral and directional reflectance data. Agronomie 20:3–22

Wessman CA, Aber JD, Peterson DL (1989) An evaluation of imaging spectrometry for estimating forest canopy chemistry. Int J Remote Sens 10:1293–1316

Wessman CA, Aber JD, Peterson DL, Melillo JM (1988) Remote-sensing of canopy chemistry and nitrogen cycling in temperate forest ecosystems. Nature 335:154–156

Weyermann J, Kneubühler M, Schläpfer D, Schaepman ME (2015) Minimizing reflectance anisotropy effects in airborne spectroscopy data using Ross–Li model inversion with continuous field land cover stratification. IEEE Trans Geosci Remote Sens 53:5814–5823

Widlowski J-L, Mio C, Disney M, Adams J, Andredakis I, Atzberger C, Brennan J, Busetto L, Chelle M, Ceccherini G, Colombo R, Côté J-F, Eenmäe A, Essery R, Gastellu-Etchegorry J-P, Gobron N, Grau E, Haverd V, Homolová L, Huang H, Hunt L, Kobayashi H, Koetz B, Kuusk A, Kuusk J, Lang M, Lewis PE, Lovell JL, Malenovský Z, Meroni M, Morsdorf F, Mõttus M, Ni-Meister W, Pinty B, Rautiainen M, Schlerf M, Somers B, Stuckens J, Verstraete MM, Yang W, Zhao F, Zenone T (2015) The fourth phase of the radiative transfer model intercomparison (RAMI) exercise: actual canopy scenarios and conformity testing. Remote Sens Environ 169:418–437

Wilson KB, Baldocchi DD, Hanson PJ (2001) Leaf age affects the seasonal pattern of photosynthetic capacity and net ecosystem exchange of carbon in a deciduous forest. Plant Cell Environ 24:571–583

Wold S, Ruhe A, Wold H, Dunn WJ (1984) The collinearity problem in linear-regression – The partial least-squares (PLS) regression approach to generalized inverses. SIAM J Sci Stat Comput 5:735–743

Wold S, Sjostrom M, Eriksson L (2001) PLS-regression: a basic tool of chemometrics. Chemom Intell Lab Syst 58:109–130

Wolter PT, Townsend PA, Sturtevant BR, Kingdon CC (2008) Remote sensing of the distribution and abundance of host species for spruce budworm in Northern Minnesota and Ontario. Remote Sens Environ 112:3971–3982

Workman J, Weyer L (2012) Practical guide and spectral atlas for interpretive near-infrared spectroscopy. CRC Press, Boca Raton, 326. https://doi.org/10.1201/b11894

Wong CYS, Gamon JA (2015) The photochemical reflectance index provides an optical indicator of spring photosynthetic activation in evergreen conifers. New Phytol 206:196–208

Wright IJ, Reich PB, Westoby M, Ackerly DD, Baruch Z, Bongers F, Cavender-Bares J, Chapin T, Cornelissen JHC, Diemer M, et al. (2004) The worldwide leaf economics spectrum. Nature 428(6985):821–827

Wu J, Albert LP, Lopes AP, Restrepo-Coupe N, Hayek M, Wiedemann KT, Guan K, Stark SC, Christoffersen B, Prohaska N, Tavares JV, Marostica S, Kobayashi H, Ferreira ML, Campos KS, da Silva R, Brando PM, Dye DG, Huxman TE, Huete AR, Nelson BW, Saleska SR (2016) Leaf development and demography explain photosynthetic seasonality in Amazon evergreen forests. Science 351:972–976

Wu J, Chavana-Bryant C, Prohaska N, Serbin SP, Guan K, Albert LP, Yang X, Leeuwen WJD, Garnello AJ, Martins G, Malhi Y, Gerard F, Oliviera RC, Saleska SR (2017) Convergence in relationships between leaf traits, spectra and age across diverse canopy environments and two contrasting tropical forests. New Phytol 214:1033–1048

Wullschleger SD, Epstein HE, Box EO, Euskirchen ES, Goswami S, Iversen CM, Kattge J, Norby RJ, van Bodegom PM, Xu X (2014) Plant functional types in Earth system models: past experiences and future directions for application of dynamic vegetation models in high-latitude ecosystems. Ann Bot 114:1–16

Yang X, Shi H, Stovall A, Guan K, Miao G, Zhang Y, Zhang Y, Xiao X, Ryu Y, Lee J-E (2018) FluoSpec 2-an automated field spectroscopy system to monitor canopy solar-induced fluorescence. Sensors (Basel) 18:2063

Yang X, Tang J, Mustard JF, Wu J, Zhao K, Serbin S, Lee J-E (2016) Seasonal variability of multiple leaf traits captured by leaf spectroscopy at two temperate deciduous forests. Remote Sens Environ 179:1–12

Yilmaz MT, Hunt ER, Jackson TJ (2008) Remote sensing of vegetation water content from equivalent water thickness using satellite imagery. Remote Sens Environ 112:2514–2522

Zarco-Tejada, Pablo J (2000a) Hyperspectral remote sensing of closed forest canopies: estimation of chlorophyll fluorescence and pigment content. Ph.D. Thesis, York University, Toronto Ontario, Canada

Zarco-Tejada PJ, Miller JR, Mohammed GH, Noland TL, Sampson PH (2000b) Chlorophyll fluorescence effects on vegetation apparent reflectance: II. laboratory and airborne canopy-level measurements with hyperspectral Data. Remote Sens Environ 74(3):596–608

Zarco-Tejada PJ, Miller JR, Mohammed GH, Noland TL, Sampson PH (2000c) Optical indices as bioindicators of forest condition from hyperspectral CASI data. Remote Sensing in the 21st Century: Economic and Environmental Applications, 517–522

Zhang Y, Chen JM, Miller JR, Noland TL (2008) Leaf chlorophyll content retrieval from airborne hyperspectral remote sensing imagery. Remote Sens Environ 112(7):3234–3247

Zhou LM, Tucker CJ, Kaufmann RK, Slayback D, Shabanov NV, Myneni RB (2001) Variations in northern vegetation activity inferred from satellite data of vegetation index during 1981 to 1999. J Geophys Res-Atmos 106:20069–20083

Open Access   This chapter is licensed under the terms of the Creative Commons Attribution 4.0 International License (http://creativecommons.org/licenses/by/4.0/), which permits use, sharing, adaptation, distribution and reproduction in any medium or format, as long as you give appropriate credit to the original author(s) and the source, provide a link to the Creative Commons license and indicate if changes were made.

The images or other third party material in this chapter are included in the chapter's Creative Commons license, unless indicated otherwise in a credit line to the material. If material is not included in the chapter's Creative Commons license and your intended use is not permitted by statutory regulation or exceeds the permitted use, you will need to obtain permission directly from the copyright holder.

Chapter 4
The Laegeren Site: An Augmented Forest Laboratory

Combining 3-D Reconstruction and Radiative Transfer Models for Trait-Based Assessment of Functional Diversity

Felix Morsdorf, Fabian D. Schneider, Carla Gullien, Daniel Kükenbrink, Reik Leiterer, and Michael E. Schaepman

4.1 Introduction

Global change is altering biodiversity in an unprecedented manner (Parmesan and Yohe 2003), and its impact on humankind may be large (Chapin III et al. 2000; Isbell et al. 2017). Forests are of special relevance because they hold most of the terrestrial biomass (Bar-On et al. 2018), are a hot spot of biodiversity (Wilson et al. 2012), and are subject to climate- and human-induced changes (Gardner 2010; Hansen et al. 2013). To monitor and potentially mitigate changes in biodiversity, Pereira et al. (2013) defined a set of essential biodiversity variables (EBVs), which should be comprehensive, concise, and standardized. Originally, most of these EBVs were to be measured in situ within ecosystems, but because forest plots are particularly scarce in the regions where change is happening the fastest (Chave et al. 2014), remote sensing (RS) has been acknowledged as a vital component to contribute to the aims of EBVs in the form of RS-enabled EBVs (RS-EBVs; Pettorelli et al. 2016; O'Connor et al. 2015). More specifically, RS technologies such as imaging spectroscopy and laser scanning have been attributed with the potential to play an important role in providing the necessary information for RS-EBVs, be it at regional, national, or global scale (Skidmore et al. 2015; Jetz et al. 2016).

Still in its early stage is the design and use of the EBV framework to include and combine RS-EBVs with in-situ measurements. In-situ measurements are often based on point measurements of individual species, whereas RS-EBVs are area-

F. Morsdorf (✉) · F. D. Schneider · C. Gullien · D. Kükenbrink · R. Leiterer
· M. E. Schaepman
Remote Sensing Laboratories, Department of Geography, University of Zurich,
Zurich, Switzerland
e-mail: felix.morsdorf@geo.uzh.ch

© The Author(s) 2020
J. Cavender-Bares et al. (eds.), *Remote Sensing of Plant Biodiversity*,
https://doi.org/10.1007/978-3-030-33157-3_4

based, with spatial characteristics depending on sensor resolution and coverage, similar to the concept of grain and extent in ecology (Turner 1989).

For large-scale assessments (i.e., regional, continental, global), cost and effort of fieldwork is a limiting factor with respect to in-situ observations. Data from the newest generation of optical satellites (e.g., Landsat 8 and Sentinel-2) have high potential for a global biodiversity assessment due to their high spatial resolution (10–30 m), multispectral information, and temporal coverage, with repeat passes within 5–6 days, depending on the area of interest. Nevertheless, due to their recent launch, these sensors do not provide a long time series, and the complementarity of lower resolution satellite data or airborne or terrestrial RS data in combination with in-situ observation is beneficial to map changes at decadal or longer timescales.

All optical RS approaches use reflected light of the vegetation canopy to infer information about its state (Schaepman et al. 2009; Homolová et al. 2013). Leaf-level biochemistry (e.g., traits such as chlorophyll and water content) has strong links with leaf reflectance and transmittance (Jacquemoud and Baret 1990). However, when light interacts with the canopy, a multitude of scattering and absorption processes have to be considered (North 1996), taking place at different levels (e.g., leaf, tree, canopy; Niinemets et al. 1998) of the canopy. Thus, passive optical observational approaches of forested ecosystems are susceptible to the effects of forest structure because directional effects associated with illumination and observation geometry may interact with signals related to leaf-level biochemistry (Hilker et al. 2008; Knyazikhin et al. 2013). Consequently, the reflectance signal at the canopy level is influenced by both vegetation structure and leaf-level physiology, and disentangling those based on passive optical data alone remains a difficult problem (Kotz et al. 2004). The effect of vegetation structure on RS indices and products (e.g., RS-EBVs) is difficult to assess, and its impact on current observations and predictions may be large. The validation of advanced wall-to-wall RS products becomes increasingly difficult because of spatiotemporal mismatches of in-situ observations with RS data. Hence, we need a framework to be able to upscale and validate leaf-level physiological traits to the level of RS data to test potential observables for RS-informed EBVs.

Radiative transfer (RT) modeling has been used for several decades to simulate and understand the signals in passive optical data (Myneni et al. 1995, 1997; Meroni et al. 2004; Lewis and Disney 2007; Gastellu-Etchegorry et al. 1996). In addition, RT models (RTMs) have been used with existing medium- to low-resolution spaceborne missions for the retrieval of products such as leaf area index (LAI) or fraction of absorbed photosynthetic radiation (fAPAR) through inversion (Myneni et al. 1997; Running et al. 2004). One particular issue with RTMs of vegetation is their parameterization. While modeling approaches simulating low-resolution data [such as Moderate Resolution Imaging Spectroradiometer (MODIS) or MEdium Resolution Imaging Spectrometer (MERIS)] mainly used one-dimensional parameterizations of the vegetation (Jacquemoud 1993; Huemmrich 2001; Verhoef and Bach 2007), higher-resolution sensors will need 3-D parameterization to account for effects like shadowing and multiple scattering (Asner and Warner 2003; Disney et al. 2006; Widlowski et al. 2015). The first RTMs incorporating 3-D forest structure were called geometric-optical radiative transfer (GORT)-type models (Ni et al. 1999).

While they were better at modeling directional effects than 1-D models, they still lacked multiple scattering and did not have full energy balance closure of incoming and outgoing radiation across all spectral domains. More advanced models use Monte Carlo ray tracing (MCRT) to add multiple scattering and provide a sound physical representation of the photon's interaction with vegetation canopies (Disney et al. 2006). While the inclusion of more physical processes (e.g., multiple scattering) certainly improves MCRT-type models over simpler approaches, their parameterization and benchmarking remains an issue. A large effort in testing RT models was undertaken in the course of the radiation transfer modeling intercomparison (RAMI) exercise, where different models were tested using a set of artificial scenes of different complexity, including 3-D scenes, to see if the models produced comparable results (Widlowski et al. 2008, 2015). However, this benchmarking remained relative (i.e., representing actual forest patches that could be validated with real-world Earth observation (EO) data acquired over the same area was not an aim of the RAMI exercise). One reason, among others, for this was the lack of suitable technologies and methods to capture and represent the 3-D vegetation structure at small scales (e.g., branches, leaves, and/or shoots).

Today, laser scanning is an established tool for retrieving quantitative measures of canopy structure (Nelson 1997; Lefsky et al. 1999; Næsset 2002; Morsdorf et al. 2004; Popescu et al. 2002; Morsdorf et al. 2006, 2010; Nelson 2013; Wulder et al. 2012). Airborne (ALS)-, terrestrial (TLS)-, and unmanned aerial vehicle (UAV)-based laser scanning (Morsdorf et al. 2017) provide a direct means to assess vegetation structure by combining the known position and orientation of the sensor with the time of flight of a laser pulse to produce a point cloud of exact 3-D coordinates. Measurements can be made across scales (e.g., stand, tree, branch, and leaf level) with finer scales often captured by close-range laser scanning (Morsdorf et al. 2018). The amount of structural detail contained in the point cloud can be overwhelming, and the extraction of meaningful information remains a challenge (Wulder et al. 2013; Morsdorf et al. 2018). Due to large data sets, automated methods for the extraction of either semantic information, such as single-tree detection based on ALS (Hyyppa et al. 2001; Morsdorf et al. 2004; Kaartinen et al. 2012; Wang et al. 2016) or tree geometry reconstruction from TLS (Cote et al. 2009; Raumonen et al. 2013) or the derivation of biophysical variables such as LAI (Morsdorf et al. 2006), are preferable over manual and/or empirical approaches.

Figure 4.1 shows an example of a single-tree-based 3-D reconstruction using ALS- and TLS-derived information. Using the 3-D information derived by ALS and TLS, one can reconstruct a virtual representation of the forest that will be used by the RT model to simulate the radiative regime of the canopy. Such an approach can be utilized to upscale measurements of leaf biochemistry to the canopy scale and to validate imaging spectroscopy-derived RS-EBVs across larger regions. In addition, this approach uses a set of three physiological and tree morphological functional traits, derived from imaging spectroscopy and laser scanning, respectively, to showcase the potential of these technologies to map the functional diversity of forests and to provide relevant information for RS-enabled EBVs.

Here we describe how we (i) designed and implemented an observational scheme to gather in-situ and structural data across several scales to simulate the 3-D radiative

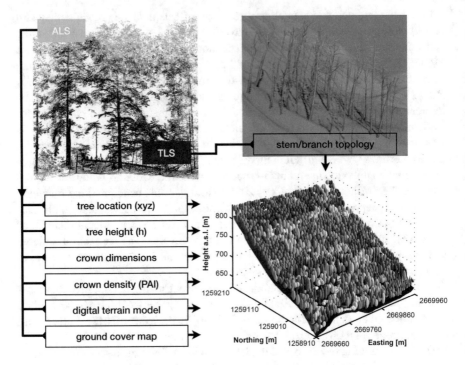

Fig. 4.1 Workflow of the 3-D reconstruction using ALS and TLS measurements

regime of the forest, (ii) tested the simulation by comparing simulated and actual RS data in their spectral and spatial information dimension, and (iii) use the approach to demonstrate how remotely sensed functional traits can be used to compute regional-scale functional richness, showcasing the information content of RS-EBVs.

4.2   The Laegeren Site: Description and History

The Laegeren site is located at N 47° 28′, 49″ and E 8° 21′, 05″ at 680 m a.s.l. on the southern slope of the Laegeren mountain, approximately 15 km northwest of Zürich, Switzerland (Fig. 4.2). The southern slope of the Laegeren marks the boundary of the Swiss Plateau, which is bordered by the Jura and the Alps. Since 1986, a 45-m-tall flux tower has provided micrometeorological data at high temporal resolution. Since April 2004, CO_2 and H_2O flux measurements are a routinely contribution to the FLUXNET/CarboEurope-IP network (Eugster et al. 2007). The mean annual temperature is 8°C. The mean annual precipitation is 1200 mm, and the growing season lasts 170–190 days. The natural vegetation cover around the tower is a mixed beech forest. The western part is dominated by broad-leaved trees, mainly beech (*Fagus sylvatica L.*) and ash (*Fraxinus excelsior L.*). In the eastern part, beech and

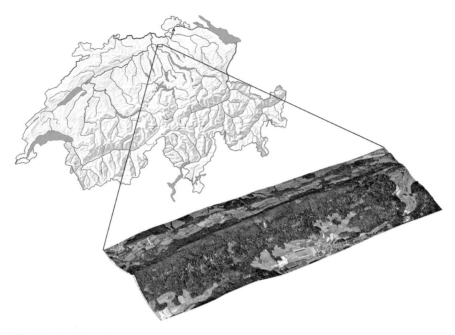

Fig. 4.2 Location of the Laegeren site within Switzerland

Norway spruce (*Picea abies (L.) Karst.*) are dominant. The forest stand has a relatively high diversity of species, ages, and diameters (Eugster et al. 2007). The ground cover mainly consists of bare soil, boulders, and litter, while the sparse understory vegetation is dominated by herbs and shrubs. Average canopy height (CH) is 24.9 m, with a maximum of 49 m, and the stem density is 270 stems per ha.

4.3   Data

4.3.1   In-Situ Data

Ground data with varying spatial, spectral, and temporal resolution allow for the 3-D reconstruction of the Laegeren, its attribution with leaf optical properties (LOPs), and generation of a reference database for parameterization and validation purposes. We used multitemporal TLS on a 60 m × 60 m plot (Sect. 4.3.2.2) and an extensive forest inventory for an area of 300 m × 300 m, which is extended to 300 m × 900 m for the simulation of EO data. In the inventory data, the type and accurate position of the trees, as well as their crown dimension and offset due to leaning stems, social position, and vertical stratification of the crown, were recorded (Sect. 4.3.1.2). In addition, the occurrence and characterization of the understory was mapped in the field and interpolated to a 2 m × 2 m grid using an ALS-based classification (Leiterer et al. 2013).

4.3.1.1 Measurements of Leaf Optical Properties

To obtain the optical properties of tree foliage, we used an integrating sphere coupled with an Analytical Spectral Devices (ASD) FieldSpec-3. Measurements of hemispherical and directional reflectance and transmittance, both from the abaxial and adaxial side of the leaves, were taken. To take into account the vertical variability of LOPs, we sampled in three different crown parts (top, middle, bottom), representing different lighting conditions in the canopy (e.g., sunlit, transitional, shaded). Deciduous leaves were collected from ten individual trees of five species (*Acer pseudoplatanus* spp., *Fagus excelsior, F. sylvatica, Ulmus glabra*, and *Tilia platyphyllos*). Measurements of aerosol optical depth (AOD) and precipitable amount of water (PAW) were provided by the aerosol robotic network (AERONET) as level 2.0 quality-assured data. For details of the sampling and measurement scheme, see Schneider et al. (2014).

4.3.1.2 Forest Inventory

An exhaustive forest inventory was carried out, individually addressing all single trees with a diameter at breast height (DBH) above 20 cm on the 300 m × 300 m site. Variables recorded for each tree included DBH, species, social status, and crown shift (i.e., an estimation of the magnitude and horizontal direction of the crown center in respect to the foot of the stem). The latter is of particular relevance on the Laegeren site because many trees have leaning stems due to topography and shallow soils. A geodetic tachymeter was used for the surveying, enabling fast and accurate electronic tree location measurements. Using all measured points, a polygonal traverse was calculated resulting in the *x, y, z* coordinates for each measurement position with an error range of millimeters for the TLS measurements and a maximum of 10 cm in *x, y*, and *z* for all other field measurements. The relative locations were transformed to absolute Swiss national coordinates using three differentially corrected global positioning system (GPS) base points, which were placed in canopy gaps. See Fig. 4.3 for a visualization of the tree inventory.

4.3.2 RS Data

4.3.2.1 Airborne Laser Scanning

To provide 3-D structure information across the whole study area, we relied on two airborne laser scanning campaigns, using a RIEGL LMS-Q680i scanner under leaf-on conditions and a RIEGL LMS-Q560 scanner under leaf-off conditions. Flight strips have an overlap of approximately 50%. Full-waveform features, namely, echo width and intensity, were extracted from the data using the software RiANALYZE and were assigned to the individual returns in the multiple-echo point cloud.

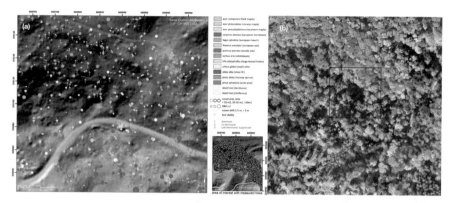

Fig. 4.3 Subset of single-tree ground inventory (**a**) and UAV-based RGB imagery acquired in fall (**b**). The black box in (**b**) denotes the subset presented in (**a**). The gray structure southwest of the bounding box is the flux tower, and the small inset shows the total extent of the single-tree ground inventory

The point cloud was filtered to classify ground and vegetation points, and the ground points were subsequently interpolated to a raster of 1 m resolution. For a detailed description of the digital terrain model (DTM) generation, see Leiterer et al. (2013). DTM accuracy was assessed using more than 500 TLS-measured road surface and bare soil points (see Sect. 4.3.2.2), which were related to the national land survey and resulted in a mean height uncertainty of about ±0.25 m. For each point of the full point cloud, the height above ground was calculated by subtracting the interpolated DTM value from the corresponding echo height above sea level, providing the vertical distance of the vegetation echoes to the terrain underneath.

4.3.2.2   Terrestrial Laser Scanning

On a subset of about 60 m × 60 m, a ground-based TLS survey was carried out using a Riegl VZ1000 instrument. A total of 40 scans on 20 scan locations were taken because each location had to be covered by two scans due to the VZ1000's camera scanning pattern (Morsdorf et al. 2018). About 50 reflective targets were placed within the scene and later used for co-registration of the scans. For co-registering RiSCAN Pro was used, and we used the ALS data to subsequently globally adjust (rotate and translate) the unified TLS point cloud. Due to the high and dense canopy, TLS needs to be complemented by laser data from above the canopy, providing more information in the upper part, either by ALS or UAV-based laser scanners. For biomass retrievals, the occlusion of upper canopy material in TLS data might be less of a problem because stems generally taper off toward the top. However, if simulation of the radiative regime and subsequent comparison with EO data gathered with a top-of-canopy perspective is the aim, TLS in denser forests needs to be complemented with laser scanning data from above the canopy (Morsdorf et al. 2017, 2018) (Fig. 4.4).

Fig. 4.4 Terrestrial laser scan of a beech-dominated part of the study area. Transect measures about 30 m (width) × 4 m (depth). One can observe a general thinning of the point cloud toward the top due to occlusion

4.3.3 Multispectral and Imaging Spectroscopy Data

Imaging spectroscopy data were acquired under clear sky conditions using the APEX imaging spectrometer (Schaepman et al. 2015). The average flight altitude was 4500 m a.s.l. resulting in an average ground pixel size of 2 m. APEX measured at-sensor radiances in 316 spectral bands ranging from 372 nm to 2540 nm. APEX data were processed to hemispherical-conical reflectance factors in the APEX processing and archiving facility (Hueni et al. 2009). Level 1 (L1) calibrated radiances were obtained by inverting the instrument model, applying coefficients established during calibration, and characterization at the APEX Calibration Home Base (CHB) in Oberpfaffenhofen, Germany. The position and orientation of each pixel in 3-D space was based on automatic geocoding in PARGE v3.269, using the swis-sALTI3D DTM. L1 data were then converted to hemispherical conical reflectance factors (HCRFs, Schaepman-Strub et al. 2006) by employing ATCOR4 v7.0 in the smile aware mode. The APEX data were complemented with other passive optical data of varying spatial and spectral resolution to build up an EO data set (Fig. 4.5). This EO data set enables cross-comparisons between the 3-D RT modeled and the actual, measured top-of-atmosphere (TOA) reflectance values at different spectral and spatial resolutions and thus an absolute evaluation of the 3-D reconstructed forest scenes and the RTM parameterization. The EO data acquired during the 2010–2014 growing seasons covers a variety of spectral and spatial resolutions:

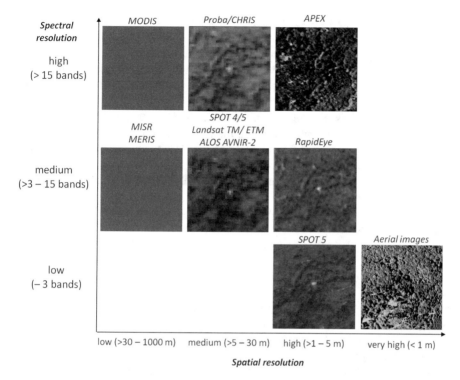

Fig. 4.5 The spatial and spectral scales covered by Earth observation (EO) data gathered for validation and up- and downscaling purposes

imaging spectrometer data from APEX (2 m × 2 m, see above for details) as well as multispectral data from RapidEye (5 m × 5 m, 4 scenes), SPOT HRG (10 m × 10 m, 5 scenes), PROBA CHRIS (17 m × 17 m, 1 scene), Landsat TM/ ETM+/OLI (30 m × 30 m, 37 scenes), ENVISAT MERIS (300 m × 300 m, 8 scenes), and Aqua/Terra MODIS (250 m × 250 m, monthly). We use APEX data for the spectral validation and a RapidEye scene for the spatial validation of our 3-D RTM approach.

4.4 Methods

4.4.1 In-Situ Data Processing

4.4.1.1 Optical Properties

LOPs were calculated separately for deciduous and coniferous trees. A linear spectral forward mixing was applied to calculate the reflectance and transmittance spectra of sunlit, transitional, and shaded leaves and needles. Because the spectra were found to match well with those in literature, the data were used directly instead of a forward simulation of a LOP model (Feret et al. 2008). This was done to reduce the number of parameters and associated uncertainties. The broadleaf species composition used for spectral mixing was derived from the forest inventory information and is dominated by beech (about 50%), with lesser contributions from maple, elm, linden, and ash.

One particular issue of the Laegeren site is its large variation in the spectral background. Because we had multitemporal full-waveform lidar data available for the Laegeren site, we used this information to classify the ground into distinct classes (gravel, litter, soil) and assigned matching spectra from our field measurements to these classes (Leiterer et al. 2013). As Schneider et al. (2014) showed, using several understory classes instead of a homogenous (black) background makes simulated top-of-canopy (TOC) and top-of-atmosphere (TOA) reflectance values more realistic.

4.4.1.2   3-D Reconstruction

Two different approaches for 3-D reconstruction of the vegetation structure were implemented and tested. The first approach relied on a single-tree identification and the second one on a direct computation of plant area index (PAI) values inside a voxel cell. Voxels are basically 3-D pixels, dividing the 3-D space into equal-sized cubes. The single-tree detection (individual tree crown, ITC) method used was based on Morsdorf et al. (2004), which derives tree location, height, and crown diameter to reconstruct the forest in 3-D based on simple geometric primitives like rotational paraboloids. However, as with most local maxima detection-based ITC methods, its performance within the mixed forest stands of the Laegeren site was suboptimal, with tree detection rates of only 50–70%. This is much lower than what can be expected for conifer forests, where rates of up to 90% can be achieved (Kaartinen et al. 2012; Wang et al. 2016). Conifers generally have conical crowns with one distinct peak (treetop), greatly facilitating their detection as local maxima in a digital surface model (DSM). The main difference between the voxel-grid and ITC approaches is the added level of semantics (Morsdorf et al. 2018) in the single-tree case, which might be relevant for some species- and individual-focused experiments (i.e., when trying to link EO-based traits with genetic information of the individual tree). If the aim of the 3-D reconstruction is an accurate simulation of the radiative regime, single-tree identification adds a layer of unnecessary complexity, so the voxel-grid approach led to better results (Schneider et al. 2014) and was subsequently used for upscaling of the trait information (Schneider et al. 2017).

4.4.1.3   Linking Field and RS Data

The perspective of forest inventory is from within or beneath the canopy and the main sampling unit is the tree, quantified as diameter at breast height (DBH). RS, on the other hand, has a top-down perspective on the canopy, and the sampling unit

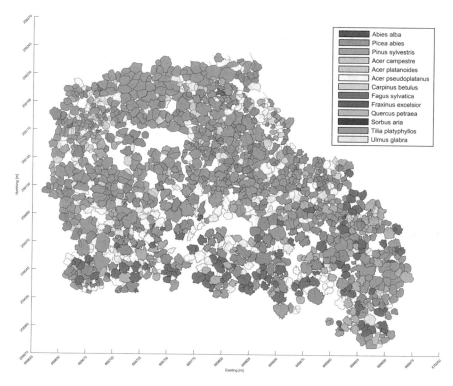

Fig. 4.6 Map of crown polygons determined from a combination of ALS and maximum leaf senescence (Fall) UAV data, linked with species information derived from the stem-referenced field inventory. Only the combination of these crown outlines and the stem map (see Fig. 4.3) allowed for individual specific computation of physiological and morphological traits

is normally a pixel. Linking these different perspectives can be difficult under any circumstances; the Laegeren site is situated on a steep slope, and trees have irregular crown shapes and different growing directions, which further complicates information matching. The first necessary step is to convert the pixel-based EO data to tree-based data by using 2-D polygons of crown boundaries. Considering the low success rates of ALS-based ITC, we manually delineated tree crowns based on UAV imagery acquired in the fall (Fig. 4.3) and matched each crown with the forest inventory data. The field inventory provided valuable additional information, such as magnitude and direction of the crown shift for trees with leaning stems, which hinders a direct stem and crown location matching based on location (Fig. 4.6).

4.4.2 Radiative Transfer Modeling

The RTM used to upscale and validate leaf-level traits such as chlorophyll and leaf water content is Discrete Anisotropic Radiative Transfer (DART; Gastellu-Etchegorry et al. 2015). Generally, a DART scene is built out of voxels with a predefined size.

To simulate vegetation such as grass or tree crowns, voxels can be filled by turbid media parameterized with PAI and leaf optical properties (LOPs). Further details of the DART model and examples of DART simulations can be found in Gastellu-Etchegorry et al. (2015). We use flux tracking in reflectance mode with the sun and the atmosphere as the only radiation sources and used DART version 5.6.0 (v739). Optical properties described in Sect. 4.4.1.1 and the forest reconstruction described in Sect. 4.4.1.2 are used to parameterize the forest canopy, background, and terrain in DART. For details of model parameterization, see Schneider et al. (2014); for details on the model-based upscaling of leaf-level traits, see Schneider et al. (2017). For the modeling results shown in Sect. 4.5.1.1, we used sun and observation angles as in the actual APEX and RapidEye acquisitions, respectively. We evaluate the performance of the combined 3-D reconstruction and RT simulation approach in two ways: spectrally, by comparing averaged simulated spectra on the core (i.e., covered by TLS measurements) site with those obtained by the APEX instrument, and, spatially, by comparing simulated bands of RapidEye over an area of 900 m × 300 m with DART-simulated reflectance at those particular wavelength regions.

4.4.3   Validation of Trait Predictions Using the RTM Approach

Three functional traits were derived from ALS data, canopy height (CH), PAI, and foliage height diversity (FHD), forming a set of morphological traits. These three were chosen because they are ecologically relevant and can be easily derived from airborne laser scanning data. Three additional functional traits—chlorophylls (CHL), carotenoids (CAR), and equivalent water thickness (EWT)—were chosen and computed using specific band ratios from the IS data (Schneider et al. 2017), forming a set of physiological traits. Both CH and CHL have been identified as primary observables for RS-EBVs, so their validation and scaling is particularly relevant. The traits were computed at a spatial aggregation unit of 6 m; for the ALS data, all echo values within a 6 m × 6 m grid cell were used for the computation, while for the IS data only sunlit pixels within the grid cell were retained for subsequent index computation. The shadow mask used for extracting sunlit pixels was derived from a DSM based on the ALS data and the solar illumination angle at the time of the IS overflight. For more details on the selected traits and their computation, please refer to Schneider et al. (2017).

The physiological traits used in this study are by definition leaf-level parameters, which need to be upscaled or averaged to be representative for the tree or canopy level. On the other hand, the morphological traits can be directly estimated from ALS data for any spatial unit. However, the chosen spatial scale and context might change how the data are interpreted (e.g., tree height needs to be estimated using single-tree information, whereas vegetation height can be derived at all different scales at the stand or plot level. See Fig. 4.7 for a map of the computed physiological and morphological traits.

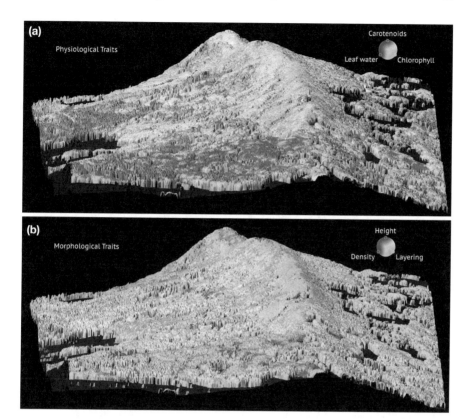

Fig. 4.7 Physiological (**a**) and morphological (**b**) traits derived from IS and ALS. For the ALS-based morphological traits, density is plant area index (PAI) and layering foliage height diversity (FHD)

4.4.4 Computation of Functional Richness

To showcase how the RTM-validated EO traits can be used for spatially explicit diversity assessments, we compute the functional richness within the 3-D trait space using a spatial subset of pixels (e.g., in a 60 m × 60 m box containing 100 pixels). In the case of the morphological traits, the 3-D trait space is spanned by the axes CH, PAI, and FHD, whereas for the physiological traits the trait space is spanned by the axes CHL, CAR, and EWT. The richness within the trait space is based on volume of a 3-D convex hull of all pixels' trait values (i.e., the larger the variation of the respective traits, the larger the volume of the convex hull). As an example, if all trait values were the same, the richness would be zero because no volume would be spanned in the 3-D trait space. Computing the richness using a pixel-based approach has the advantage of resolving both inter- and intraspecific variation of the traits, with the latter being potentially as large as the former (e.g., as observed in our leaf spectra). For details on the definition and computation of richness and other diversity-related metrics in the scope of this work, please refer to Schneider et al. (2017).

4.5 Results and Discussion

4.5.1 Forward Simulation of Passive Optical Imagery and Comparison With EO Data

4.5.1.1 Spectral Validation

Figure 4.8 compares the spectral response of a 20 m × 20 m subplot within the Laegeren site simulated by the DART RTM with the average APEX spectrum of the same area. In contrast to Schneider et al. (2014), the improved version of the DART model used in this study shows good agreement (within the standard deviation for the 10 × 10 pixel areas) for all wavelengths, including the visible domain. The version of DART used in this study (5.6.0, v739) has a more sophisticated parameterization of the atmosphere than the older version, improving the spectral response in the visible domain (Grau and Gastellu-Etchegorry 2013; Yin et al. 2013; Gastellu-Etchegorry et al. 2015). The very good agreement of simulated and measured spectra across all bands shows that our approach of combining a 3-D reconstruction of the forest and LOPs of leaves and needles was successful in capturing the dominant scattering components of this natural system. In the near-infrared domain of the spectra, this is likely due to ALS and TLS providing accurate physical representations of 3-D canopy structure, whereas in the visible domain the quality of the LOPs

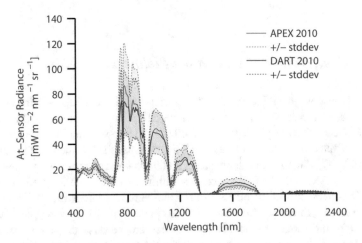

Fig. 4.8 Simulated spectral response by DART for subplot S1 in comparison with APEX data acquired over the same area. The standard deviation is computed from the single pixels in the 20 m × 20 m plot

and the representation of the atmosphere are contributing more to this excellent result. Thus, we have shown that the measured physiological trait variation at leaf level can be upscaled to canopy level (as observed by IS instruments). We used this to forward validate IS-derived physiological traits that are the basis of the functional richness computation in Sect. 4.5.2. This RTM-based link is a key component of our validation framework because typically field-measured spectra and the traits based on spectral indices cannot be assumed to be representative for the respective signals measured at the imaging sensor above the canopy.

4.5.1.2   Spatial Validation

Figure 4.9 shows a comparison of the spatial patterns in both simulated and actual RapidEye imagery obtained over the Laegeren site in leaf-on conditions. When sub-sampled to the 5 m resolution of RapidEye, our approach produces very similar spatial patterns, properly resolving shadows and highlights due to forest structure and underlying topography effectively contained in the ALS data and transferred to

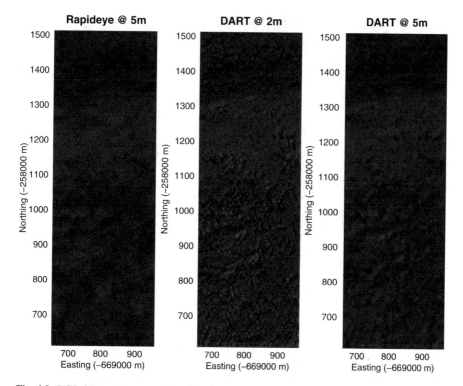

Fig. 4.9 RGB false-color composite using the RapidEye bands 5 (R), 3 (G), and (2), both for the simulated image using DART (center panel) and actual RapidEye data acquired over the site (left panel). The right panel shows a smoothed version of the DART simulation to accommodate for the lower resolution (5 m) of the RapidEye imaging sensor

the RT model by the PAI voxel grid. If LOPs can be assumed to be uniform across a site or region, the presented approach can be used for larger areas, only relying on ALS data to parameterize the RT model and using the LOPs measured at a subset of the site or taken from spectral libraries. ALS data are generally available at regional and national scales, effectively bridging the gap between point-based field inventories and global scale satellite imagery. Using such larger-scale simulated EO data, retrieval methods for RS-EBV primary observables such as CHL and vegetation height can be tested and validated. The 3-D simulation environment we established explicitly or implicitly contains all these variables in an easily retrievable format.

4.5.2 Functional Diversity of Laegeren Site

Figure 4.10 shows the spatial distribution of the richness as computed from the three morphological and physiological traits. The most prominent pattern is the strong topographic effect of the Laegeren mountain ridge, which is equally present in both

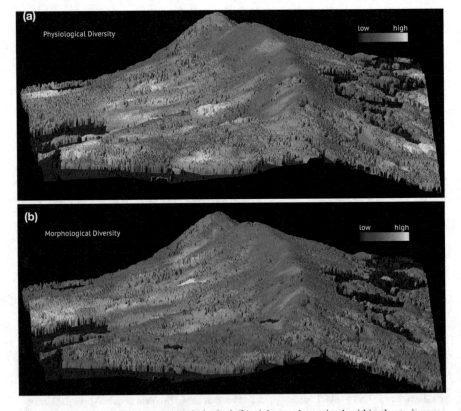

Fig. 4.10 Physiological (**a**) and morphological (**b**) richness determined within the trait space spanned by three traits

richness maps. The ridge and the associated steep slopes affect many environmental variables, which could act as filter for niche space and thus diversity. Higher altitude is linked with decreased temperature, whereas the slope is contributing to lesser soil depths and water availability and increased incoming radiation, at least for the part south of the ridge. Thus, the environmental conditions are harsher close to the ridge, which might explain the decrease of functional richness we observe in this context. In the lower regions, morphological and physiological richness exhibit differing spatial patterns. We assume that changes of the morphological richness in lower parts parallel to ridge are caused by the different stand management regimes and associated stand ages and structures. For the physiological richness, differences in species composition seem to be a dominant effect, with the conifer-dominated stands having a lower functional richness and the old-growth mixed stands being functionally richer.

4.6   Conclusion and Outlook

Modern RS technologies increasingly face a validation paradox—i.e., it is very difficult to provide ground-based validation data that match the spatial (resolution and extent), temporal, and thematic characteristics of modern EO data sets. As an example, ALS-derived tree height is assumed to be more accurate than field measurements, but it cannot be proved using field data alone. By using laser scanning-derived 3-D structure together with LOPs in an RT model approach, we have shown a way to overcome such mismatches and provide a framework that could be established across a range of sites around the globe to prototype and validate EO-based data and products in the future. Such a forward validation will as well pave the way for products that are not measurable in the field, but still might be relevant in the context of ecosystem function and diversity. The RTM approach provides a physical and mechanistic way to learn about the information content of EO data, and a combination of this approach with recent developments in the machine learning domain could provide interesting perspectives.

With the trait-based functional richness assessment, we demonstrated how a spatially extended monitoring using the complementary technologies imaging spectroscopy and lidar would work and what kind of insights into ecosystem functioning it could generate. In addition, the trait maps and the derived functional richness could be used for spatiotemporal gap filling of in-situ observational networks such as the global forest biodiversity initiative, complementing the diversity information that these provide.

In the future, these data streams in conjunction with the EBV concept (Fernández et al., Chap. 18) will give policy-makers around the world useful tools to assess and report on the biodiversity. To speed up this process, the European Space Agency funded the GlobDiversity project starting in 2017 in the tradition of similar projects for some of the essential climate variables. The project's goal is to demonstrate the capability and utility of producing a set of selected RS-EBV data sets in different regions and biomes around the globe and with high spatial resolution (10–30 m) using the newest-generation satellite data, such as Sentinel-2 and Landsat 8. In addition, the project shall suggest in a reference document how to describe RS-EBVs and how they could be engineered and validated. We believe that the 3-D reconstruction and RT modeling approach highlighted in this chapter could be applied across a global range of sites to fulfill this task.

Acknowledgments The contributions of F.M., F.D.S., and M.E.S. are supported by the University of Zurich Research Priority Program on Global Change and Biodiversity, and the contribution of D.K. was with support from the European Union's 7th Framework Program (FP7/ 2014–2018) under EUFAR2 contract no. 312609.

References

Asner GP, Warner AS (2003) Canopy shadow in ikonos satellite observations of tropical forests and savannas of. Environment 87(4):521–533

Bar-On YM, Phillips R, Milo R (2018) The biomass distribution on earth. Proc Natl Acad Sci 115(25):6506

Chapin FS III, Zavaleta ES, Eviner VT, Naylor RL, Vitousek PM, Reynolds HL, Hooper DU, Lavorel S, Sala OE, Hobbie SE, Mack MC, Dıaz S (2000) Consequences of changing biodiversity. Nature 405(6783):234–242

Chave J, Réjou-Méchain M, Burquez A, Chidumayo E, Colgan MS, Delitti WB, Duque A, Eid T, Fearnside PM, Goodman RC, Henry M, Martínez-Yrízar A, Mugasha WA, Muller-Landau HC, Mencuccini M, Nelson BW, Ngomanda A, Nogueira EM, Ortiz-Malavassi E, Pelissier R, Ploton P, Ryan CM, Saldarriaga JG, Vieilledent G (2014) Improved allometric models to estimate the aboveground biomass of tropical trees. Glob Chang Biol 20(10):3177–3190

Cote JF, Widlowski JL, Fournier RA, Verstraete MM (2009) The structural and radiative consistency of three-dimensional tree reconstructions from terrestrial lidar. Remote Sens Environ 113(5):1067–1081

Disney M, Lewis P, Saich P (2006) 3d modeling of forest canopy structure for remote sensing simulations in the optical and microwave domains. Remote Sens Environ 100(1):114–132

Eugster W, Zeyer K, Zeeman M, Michna P, Zingg A, Buchmann N, Emmeneg- ger L (2007) Nitrous oxide net exchange in a beech dominated mixed forest in Switzerland measured with a quantum cascade laser spectrometer. Biogeosci Discuss 4:1167–1200

Feret JB, Francois C, Asner GP, Gitelson AA, Martin RE, Bidel LP, Ustin SL, le Maire G, Jacquemoud S (2008) Prospect-4 and 5: advances in the leaf optical properties model separating photosynthetic pigments. Remote Sens Environ 112(6):3030–3043

Gardner T (2010) Monitoring forest biodiversity: improving conservation through ecologically responsible management, vol 9781849775106. Earthscan, London

Gastellu-Etchegorry J, Zagolski F, Romier J (1996) A simple anisotropic reflectance model for homogeneous multilayer canopies. Remote Sens Environ 57(1):22–38

Gastellu-Etchegorry JP, Yin T, Lauret N, Cajgfinger T, Gregoire T, Grau E, Feret JB, Lopes M, Guilleux J, Dedieu G, Malenovský Z, Cook BD, Morton D, Rubio J, Durrieu S, Cazanave G, Martin E, Ristorcelli T (2015) Discrete anisotropic radiative transfer (DART 5) for modeling airborne and satellite spectroradiometer and lidar acquisitions of natural and urban landscapes. Remote Sens 7(2):1667–1701

Grau E, Gastellu-Etchegorry JP (2013) Radiative transfer modeling in the earth-atmosphere system with {DART} model. Remote Sens Environ 139:149–170

Hansen MC, Potapov PV, Moore R, Hancher M, Turubanova SA, Tyukavina A, Thau D, Stehman SV, Goetz SJ, Loveland TR, Kommareddy A, Egorov A, Chini L, Justice CO, Townshend JRG (2013) High-resolution global maps of 21st-century forest cover change. Science 342(6160):850–853

Hilker T, Coops NC, Hall FG, Black TA, Wulder MA, Nesic Z, Krishnan P (2008) Separating physiologically and directionally induced changes in pri using brdf models. Remote Sens Environ 112(6):2777–2788

Homolová L, Malenovský Z, Clevers JGPW, Garćia-Santos G, Schaepman ME (2013) Review of optical-based remote sensing for plant trait mapping. Ecol Complex 15:1–16

Huemmrich KF (2001) The geosail model: a simple addition to the sail model to describe discontinuous canopy reflectance. Remote Sens Environ 75(3):423–431

Hueni A, Biesemans J, Meuleman K, Dell'Endice F, Schlapfer D, Odermatt D, Kneubuehler M, Adriaensen S, Kempenaers S, Nieke J, Itten K (2009) Structure, components, and interfaces of the airborne prism experiment (apex) processing and archiving facility. IEEE Trans Geosci Remote Sens 47(1):29–43

Hyyppa J, Kelle O, Lehikoinen M, Inkinen M (2001) A segmentation-based method to retrieve stem volume estimates from 3-d tree height models produced by laser scanners. IEEE Trans Geosci Remote Sens 39:969–975

Isbell F, Gonzalez A, Loreau M, Cowles J, Dıaz S, Hector A, Mace GM, Wardle DA, O'Connor MI, Duffy JE, Turnbull LA, Thompson PL, Larigauderie A (2017) Linking the influence and dependence of people on biodiversity across scales. Nature 546:65–72

Jacquemoud S (1993) Inversion of the prospect + sail canopy reflectance model from aviris equivalent spectra: theoretical study. Remote Sens Environ 44(2–3):281–292

Jacquemoud S, Baret F (1990) Prospect: a model of leaf optical properties spectra. Remote Sens Environ 34(2):75–91

Jetz W, Cavender-Bares J, Pavlick R, Schimel D, Davis F, Asner G, Guralnick R, Kattge J, Latimer A, Moorcroft P, Schaepman M, Schildhauer M, Schneider F, Schrodt F, Stahl U, Ustin S (2016) Monitoring plant functional diversity from space. Nat Plants 2(3):16024

Kaartinen H, Hyyppa J, Yu X, Vastaranta M, Hyyppa H, Kukko A, Holopainen M, Heipke C, Hirschmugl M, Morsdorf F, Næsset E, Pitkanen J, Popescu S, Solberg S, Wolf BM, Wu JC (2012) An international comparison of individual tree detection and extraction using airborne laser scanning. Remote Sens 4:950–974

Knyazikhin Y, Lewis P, Disney MI, Mottus M, Rautiainen M, Stenberg P, Kaufmann RK, Marshak A, Schull MA, Latorre Carmona P, Vanderbilt V, Davis AB, Baret F, Jacquemoud S, Lyapustin A, Yang Y, Myneni RB (2013) Reply to Ollinger et al.: remote sensing of leaf nitrogen and emergent ecosystem properties. Proc Natl Acad Sci 110(27):E2438

Kotz B, Schaepman M, Morsdorf F, Bowyer P, Itten K, Allgöwer B (2004) Radiative transfer modeling within a heterogeneous canopy for estimation of forest fire fuel properties. Remote Sens Environ 92(3):332–344

Lefsky MA, Cohen WB, Acker SA, Parker GG, Spies TA, Harding D (1999) Lidar remote sensing of the canopy structure and biophysical properties of Douglas-fir western hemlock forests. Remote Sens Environ 70:339–361

Leiterer R, Mücke W, Morsdorf F, Hollaus M, Pfeifer N, Schaepman M (2013) Operational forest structure monitoring using airborne laser scanning. Photogrammetrie Fernerkundung Geoinformation 3:173–184

Lewis P, Disney M (2007) Spectral invariants and scattering across multiple scales from within-leaf to canopy. Remote Sens Environ 109(2):196–206

Meroni M, Colombo R, Panigada C (2004) Inversion of a radiative transfer model with hyperspectral observations for LAI mapping in poplar plantations. Remote Sens Environ 92(2):195–206

Morsdorf F, Meier E, Kotz B, Itten KI, Dobbertin M, Allgower B (2004) Lidar-based geometric reconstruction of boreal type forest stands at single tree level for forest and wildland fire management. Remote Sens Environ 92(3):353–362, forest Fire Prevention and Assessment

Morsdorf F, Kotz B, Meier E, Itten K, Allgower B (2006) Estimation of LAI and fractional cover from small footprint airborne laser scanning data based on gap fraction. Remote Sens Environ 104(1):50–61

Morsdorf F, Marell A, Koetz B, Cassagne N, Pimont F, Rigolot E, Allgower B (2010) Discrimination of vegetation strata in a multi-layered mediterranean forest ecosystem using height and intensity information derived from airborne laser scanning. Remote Sens Environ 114(7):1403–1415

Morsdorf F, Eck C, Zgraggen C, Imbach B, Schneider F, Kükenbrink D (2017) UAV-based LiDAR acquisition for the derivation of high-resolution forest and ground information. Lead Edge 36(7):566–570

Morsdorf F, Kükenbrink D, Schneider FD, Abegg M, Schaepman ME (2018) Close-range laser scanning in forests: towards physically based semantics across scales. Interface Focus 8(2):20170046

Myneni R, Nemani R, Running S (1997) Estimation of global leaf area index and absorbed PAR using radiative transfer models. IEEE Trans Geosci Remote Sens 35:1380–1393

Myneni RB, Maggion S, Iaquinta J, Privette JL, Gobron N, Pinty B, Kimes DS, Verstraete MM, Williams DL (1995) Optical remote sensing of vegetation: Modeling, caveats, and algorithms. Remote Sens Environ 51(1):169–188

Næsset E (2002) Predicting forest stand characteristics with airborne scanning laser using a practical two-stage procedure and field data. Remote Sens Environ 80(1):88–99

Nelson R (1997) Modeling forest canopy heights: the effects of canopy shape. Remote Sens Environ 60:327–334

Nelson R (2013) How did we get here? An early history of forestry lidar. Can J Remote Sens 39(s1):S6–S17

Ni W, Li X, Woodcock C, Caetano M, Strahler A (1999) An analytical hybrid gort model for bidirectional reflectance over discontinuous plant canopies. IEEE Trans Geosci Remote Sens 37:987–999

Niinemets Ü, Kull O, Tenhunen J (1998) An analysis of light effects on foliar morphology, physiology, and light interception in temperate deciduous woody species of contrasting shade tolerance. Tree Physiol 18(10):681–696

North P (1996) Three-dimensional forest light interaction model using a monte carlo method. IEEE Trans Geosci Remote Sens 34(4):946–956

O'Connor B, Secades C, Penner J, Sonnenschein R, Skidmore A, Burgess ND, Hutton JM (2015) Earth observation as a tool for tracking progress towards the aichi biodiversity targets. Remote Sens Ecol Conserv 1(1):19–28

Parmesan C, Yohe G (2003) A globally coherent fingerprint of climate change impacts across natural systems. Nature 421(6918):37–42

Pereira H, Ferrier S, Walters M, Geller G, Jongman R, Scholes R, Bruford M, Brummitt N, Butchart S, Cardoso A, Coops N, Dulloo E, Faith D, Freyhof J, Gregory R, Heip C, Hoft R, Hurtt G, Jetz W, Karp D, McGeoch M, Obura D, Onoda Y, Pettorelli N, Reyers B, Sayre R, Scharlemann J, Stuart S, Turak E, Walpole M, Wegmann M (2013) Essential biodiversity variables. Science 339(6117):277–278

Pettorelli N, Wegmann M, Skidmore A, Mücher S, Dawson T, Fernandez M, Lucas R, Schaepman M, Wang T, O'Connor B, Jongman R, Kempeneers P, Sonnenschein R, Leidner A, Bohm M, He K, Nagendra H, Dubois G, Fatoyinbo T, Hansen M, Paganini M, de Klerk H, Asner G, Kerr J, Estes A, Schmeller D, Heiden U, Rocchini D, Pereira H, Turak E, Fernandez N, Lausch A, Cho M, Alcaraz-Segura D, McGeoch M, Turner W, Mueller A, St-Louis V, Penner J, Vihervaara P, Belward A, Reyers B, Geller G (2016) Framing the concept of satellite remote sensing essential biodiversity variables: challenges and future directions. Remote Sensing in Ecology and Conservation 2(3):122–131

Popescu SC, Wynne RH, Nelson RF (2002) Estimating plot-level tree heights with lidar: local filtering with a canopy-height based variable window size. Comput Electron Agric 37(1–3):71–95

Raumonen P, Kaasalainen M, Akerblom M, Kaasalainen S, Kaartinen H, Vastaranta M, Holopainen M, Disney M, Lewis P (2013) Fast automatic precision tree models from terrestrial laser scanner data. Remote Sens 5(2):491

Running S, Nemani R, Heinsch F, Zhao M, Reeves M, Hashimoto H (2004) A continuous satellite-derived measure of global terrestrial primary production. Bioscience 54(6):547–560

Schaepman ME, Ustin SL, Plaza AJ, Painter TH, Verrelst J, Liang S (2009) Earth system science related imaging spectroscopy - an assessment. Remote Sens Environ 113(Suppl. 1):S123–S137

Schaepman ME, Jehle M, Hueni A, D'Odorico P, Damm A, Weyermann J, Schnei- d FD, Laurent V, Popp C, Seidel FC, Lenhard K, Gege P, Küchler C, Brazile J, Kohler P, Vos LD, Meuleman K, Meynart R, Schlapfer D, Kneubühler M, Itten KI (2015) Advanced radiometry measurements and earth science applications with the airborne prism experiment (apex). Remote Sens Environ 158:207–219

Schaepman-Strub G, Schaepman M, Painter T, Dangel S, Martonchik J (2006) Reflectance quantities in optical remote sensing–definitions and case studies. Remote Sens Environ 103(1):27–42

Schneider FD, Leiterer R, Morsdorf F, Gastellu-Etchegorry JP, Lauret N, Pfeifer N, Schaepman ME (2014) Simulating imaging spectrometer data: 3d forest modeling based on lidar and in situ data. Remote Sens Environ 152:235–250

Schneider FD, Morsdorf F, Schmid B, Petchey OL, Hueni A, Schimel DS, Schaepman ME (2017) Mapping functional diversity from remotely sensed morphological and physiological forest traits. Nat Commun 8(1):1441

Skidmore A, Pettorelli N, Coops N, Geller G, Hansen M, Lucas R, Mücher C, O'Connor B, Paganini M, Pereira H, Schaepman M, Turner W, Wang T, Weg- mann M (2015) Environmental science: agree on biodiversity metrics to track from space. Nature 523(7561):403–405

Turner M (1989) Landscape ecology: the effect of pattern on process. Annu Rev Ecol Syst 20:171–197

Verhoef W, Bach H (2007) Coupled soil-leaf-canopy and atmosphere radiative transfer modeling to simulate hyperspectral multi-angular surface reflectance and {TOA} radiance data. Remote Sens Environ 109(2):166–182

Wang Y, Hyyppa J, Liang X, Kaartinen H, Yu X, Lindberg E, Holmgren J, Qin Y, Mallet C, Ferraz A, Torabzadeh H, Morsdorf F, Zhu L, Liu J, Alho P (2016) International benchmarking of the individual tree detection methods for modeling 3-d canopy structure for silviculture and forest ecology using airborne laser scanning. IEEE Trans Geosci Remote Sens 54(9):5011–5027

Widlowski JL, Robustelli M, Disney M, Gastellu-Etchegorry JP, Lavergne T, Lewis P, North P, Pinty B, Thompson R, Verstraete M (2008) The rami online model checker (romc): a web-based benchmarking facility for canopy reflectance models. Remote Sens Environ 112(3):1144–1150

Widlowski JL, Mio C, Disney M, Adams J, Andredakis I, Atzberger C, Brennan J, Busetto L, Chelle M, Ceccherini G, Colombo R, Cote JF, Eenmae A, Essery R, Gastellu-Etchegorry JP, Gobron N, Grau E, Haverd V, Homolova L, Huang H, Hunt L, Kobayashi H, Koetz B, Kuusk A, Kuusk J, Lang M, Lewis PE, Lovell JL, Malenovsky Z, Meroni M, Morsdorf F, Mottus M, Ni-Meister W, Pinty B, Rautiainen M, Schlerf M, Somers B, Stuckens J, Verstraete MM, Yang W, Zhao F, Zenone T (2015) The fourth phase of the radiative transfer model intercomparison (rami) exercise: actual canopy scenarios and conformity testing. Remote Sens Environ 169:418–437

Wilson J, Peet R, Dengler J, Partel M (2012) Plant species richness: the world records. J Veg Sci 23(4):796–802

Wulder M, Coops N, Hudak A, Morsdorf F, Nelson R, Newnham G, Vastaranta M (2013) Status and prospects for lidar remote sensing of forested ecosystems. Can J Remote Sens 39(s1):S1–S5

Wulder MA, White JC, Nelson RF, Næsset E, Orka HO, Coops NC, Hilker T, Bater CW, Gobakken T (2012) Lidar sampling for large-area forest characterization: a review. Remote Sens Environ 121:196–209

Yin T, Gastellu-Etchegorry JP, Lauret N, Grau E, Rubio J (2013) A new approach of direction discretization and oversampling for 3d anisotropic radiative transfer modeling. Remote Sens Environ 135:213–223

Open Access This chapter is licensed under the terms of the Creative Commons Attribution 4.0 International License (http://creativecommons.org/licenses/by/4.0/), which permits use, sharing, adaptation, distribution and reproduction in any medium or format, as long as you give appropriate credit to the original author(s) and the source, provide a link to the Creative Commons license and indicate if changes were made.

The images or other third party material in this chapter are included in the chapter's Creative Commons license, unless indicated otherwise in a credit line to the material. If material is not included in the chapter's Creative Commons license and your intended use is not permitted by statutory regulation or exceeds the permitted use, you will need to obtain permission directly from the copyright holder.

Chapter 5
Lessons Learned from Spectranomics: Wet Tropical Forests

Roberta E. Martin

5.1 Introduction

One of the major challenges for biodiversity science is how to measure biodiversity at spatial scales relevant for conservation and management (Turner 2014). Supported by technological, computational, and modeling advances, along with increased data availability, remote sensing (RS) has become an essential tool for ecologists and land managers because it provides data on the optical properties of the Earth's surface at landscape to global scales (Jetz et al. 2016). At the same time, increasing awareness of how little we know about the species inhabiting our planet has led to a surge in ground-based activities to catalog what's out there and establish baselines such as Conservation International's Rapid Assessment Program and/or community aggregated information needed for biodiversity assessment (Myers et al. 2000). In addition, advances in genetic analysis, physiological experiments, and trait-based studies have advanced our understanding of functional biodiversity (Cavender-Bares et al. 2006; Kress et al. 2009; Baraloto et al. 2012). Despite these knowledge gains, linking the information from these disparate sources in a useful manner presented a new hurdle. In 2007 the Spectranomics approach was launched to address this challenge using canopy functional traits and their resultant spectral properties.

R. E. Martin (✉)
School of Geographical Sciences and Urban Planning, Arizona State University, Tempe, AZ, USA

Center for Global Discovery and Conservation Science, Arizona State University, Tempe, AZ, USA
e-mail: Roberta.Martin@asu.edu

© The Author(s) 2020
J. Cavender-Bares et al. (eds.), *Remote Sensing of Plant Biodiversity*,
https://doi.org/10.1007/978-3-030-33157-3_5

Plants play a foundational role in establishing and maintaining ecosystem function, biogeochemical cycling, hydrological cycling, and biodiversity (Mooney et al. 1996; Schimel et al. 2013). More specifically, canopy plants (those that occupy the sun-facing portion of a landscape) serve as dominant primary producers through the capture and utilization of light. Their structures also provide habitat for vast numbers of species living in the shadows. To maintain this premier position in a forest ecosystem, plants have evolved a vast array of strategies for growth, defense, and longevity, largely manifested as chemical and/or structural adjustments in their leaves (Reich et al. 2003; Wright et al. 2004; Diaz et al. 2016). The molecular arrangement of these foliar properties generates an optical reflectance spectrum that can be measured at a variety of scales with spectroscopy (Curran 1989; Jacquemoud and Ustin 2001; Ustin et al. 2009; Ustin and Jacquemoud, Chap. 14). The ultimate result is a massive number of tree species coalescing into forest communities of varying complexity, with unique taxonomic compositions and functional roles that can potentially be mapped across a forested landscape (Reichstein et al. 2014 and others).

Despite understanding the important role different canopy species and communities of species play in creating and maintaining biodiversity, the measurement, mapping, and monitoring of forest canopy composition and functional diversity has remained a challenge. Current Earth-observing satellite technology is limited to detecting changes in vegetation cover as well as major differences in vegetation type and photosynthesis (Running et al. 1994; Tucker and Townshend 2000) and does not easily reveal compositional differences or changes over time (Turner et al. 2003). Tropical forest canopy diversity is especially underexplored because spatial and temporal variation often exceeds our ability to adequately utilize field-based approaches (Marvin et al. 2014). Airborne imaging spectroscopy can provide an intermediate solution; however, a fundamental prerequisite for determining whether species diversity or a particular species might be successfully mapped is an assessment of chemical uniqueness and diversity among plant taxa. This is important because the spectroscopy of canopies is driven primarily by the chemical composition of the foliage (Curran 1989; Asner et al. 2015).

5.2 Spectranomics Approach

The Spectranomics approach was developed to link plant canopy functional traits to their spectral properties with the objective of providing time-varying, scalable methods for remote sensing (RS) of forest biodiversity (Asner and Martin 2009). In the pool of potentially important plant functional traits, foliar chemicals stand out as core physiologically based predictors of plant adaptation to environmental conditions (Díaz et al. 1998; Wright et al. 2010). We selected a suite of 23 canopy chemical traits based on their strong ecological and evolutionary relevance, spatial variation in species and communities, and measurable spectral properties. These traits consist of those that (i) mediate or are indicative of photosynthesis and carbon uptake (chlorophyll a and b, carotenoids, nitrogen, $\delta^{13}C$, and $\delta^{15}N$; non-soluble carbohydrates); (ii) are related to structure (leaf mass per area and water content, lignin, cellulose, and hemicellulose) and chemical defense (phenols and tannins); and (iii) are defining general metabolic processes (macro- and micronutrients; here calcium, magnesium, phosphorus, potassium and boron, iron, manganese, zinc) (Table 5.1). The distribution and variation of these traits in plant canopy leaves evolve as a function of stoichiometric relationships among constituents in response to biotic and abiotic pressures and are often formulated differently at the species level (Díaz et al. 1998). This evolved chemical makeup of plant canopies and its similarity and uniqueness among species, which we call *chemical phylogeny*, is an essential component of Spectranomics (Fig. 5.1a).

Table 5.1 Summary statistics for 22 foliar chemical traits and leaf mass per area (LMA) from top-of-canopy leaves collected from 12,012 individual trees at sites across the wet tropics as part of the Spectranomics Program

	Mean	Standard deviation	Minimum	Maximum	Median	Skew	Kurtosis
Light capture and growth							
Chlorophyll *a* (mg g^{-1})	4.67	1.96	0.01	26.71	4.40	2.47	1.76
Chlorophyll *b* (mg g^{-1})	1.73	0.78	0.01	11.32	1.62	0.86	0.58
Carotenoids (mg g^{-1})	1.38	0.53	0.01	7.87	1.31	0.80	0.59
Nitrogen (%)	2.01	0.70	0.35	6.15	1.91	1.21	0.91
NSC (%)	46.20	11.56	12.73	86.33	45.72	31.40	25.61
$\delta^{13}C$ (‰)	−30.59	1.89	−36.40	−19.90	−30.70	−32.90	−34.00
$\delta^{15}N$ (‰)	1.38	2.70	−10.30	10.50	1.40	−2.00	−4.00
Structure and defense							
LMA (g m^{-2})	113.63	44.44	15.65	622.36	105.33	66.92	51.68
Water (%)	58.40	8.28	9.17	90.79	57.57	48.92	44.87
Carbon (%)	49.30	3.31	31.60	65.00	49.70	45.00	41.80
Lignin (%)	24.15	9.94	0.25	81.08	23.47	11.67	7.65
Cellulose (%)	17.82	5.75	1.08	56.60	17.30	10.89	8.41
Hemicellulose (%)	11.63	5.03	0.00	49.31	11.25	5.72	2.84
Phenols (mg g^{-1})	101.11	53.63	0.00	358.19	101.71	27.02	10.54
Tannins (mg g^{-1})	46.45	26.82	−0.64	238.79	43.14	15.27	5.66
Macronutrients							
Calcium (%)	0.96	0.81	0.00	8.36	0.74	0.18	0.08
Magnesium (%)	0.26	0.15	0.02	2.71	0.23	0.11	0.08
Phosphorus (%)	0.12	0.07	0.02	0.86	0.10	0.05	0.04
Potassium (%)	0.76	0.45	0.13	5.64	0.65	0.35	0.25
Micronutrients							
Boron (µg g^{-1})	27.19	23.48	1.16	321.89	20.03	8.35	5.31
Iron (µg g^{-1})	80.58	206.63	7.13	9470.68	47.78	26.55	19.47
Manganese (µg g^{-1})	304.32	512.14	3.03	7331.67	103.80	19.75	11.55
Zinc (µg g^{-1})	17.08	44.47	1.65	2535.98	11.77	6.49	4.62

NSC nonstructural carbohydrates

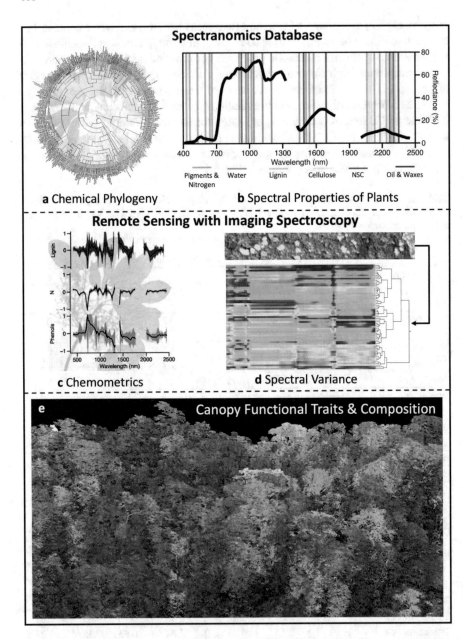

Fig. 5.1 The essential interactive elements of the Spectranomics Database include phylogenetic, chemical, and spectral information on canopy species. (**a**) Assays of 23 foliar chemical traits combined are collected, organized, and analyzed phylogenetically, producing a new tree of life based on the relatedness of functional trait signatures. This generic phylogeny shows the chemical relatedness of thousands of species in the Spectranomics Database. (**b**) An example of a remotely sensed canopy reflectance spectrum of one species is shown along with indicators of key chemical contributions to the spectrum (Curran 1989; Ustin et al. 2009; Kokaly et al. 2009). (**c**) Chemometric

Another component of Spectranomics is the spectral properties of plant canopies (Fig. 5.1b). Canopy spectra are derived from the way plant foliage interacts with solar radiation, and this interaction is strongly determined by foliar chemicals. Across the full solar spectrum, from the ultraviolet to the visible to the near-infrared and the shortwave-infrared regions of the electromagnetic spectrum (350–3500 nm), plants have many common and yet also unique patterns of interaction with solar energy. Chemometric studies determine how these chemicals relate to reflectance spectra, and the methods today range from traditional spectroscopic assays and newer machine learning approaches (Wold et al. 2001; Serbin et al. 2014; Feilhauer et al. 2015). Spectral properties also provide a tantalizing pathway forward to scale up from leaves to landscapes (Ustin et al. 2004) to the planetary level (Jetz et al. 2016), but only if we can accurately and repeatedly measure and interpret the spectra of plants over increasingly larger portions of Earth (Fig. 5.1c–e).

The realization of Spectranomics rests in a number of choices made early on to attempt to reduce unwanted sources of variation combined with extensive sampling. We focused on humid tropical forests for their high diversity and relative freedom from extreme phenological changes brought about by seasonal cycles such as those experienced in temperate regions but may not completely eliminate smaller phenological variation that might arise in reaction to drought or solar variations. We targeted only mature, fully sunlit, top-of-canopy leaves (trees and lianas) to limit variation attributable to intra-canopy shade and ontogeny and to best relate leaf properties to airborne and satellite-based spectral measurements. Prior to Spectranomics, our work and that of many others did not follow a strategically consistent, integrated method for global spectral-functional trait database building needed to reveal canopy plant functional spectral-chemical patterns at the biospheric scale.

We have collected, cataloged, and stored more than 13,000 canopy tree and liana specimens, in over 3 million tissue samples, representing about 10,000 species biased to humid tropical ecosystems (Fig. 5.2a). For perspective, this number approaches the total number of tree species in the Amazon basin (roughly 11,000; Hubbell et al. 2008), a value that would put the global tropical tree inventory at 30,000 species if we liberally extrapolate to the entire Neotropics plus the African and Asian-Oceanic tropics. The Spectranomics database focuses only on species found in the canopy, meaning they are in full sunlight and are observable from above. Since roughly 30–60% of tree species in a tropical forest plot makes it to the canopy (e.g., Bohlman 2015), the current Spectranomics database contains at least

---

Fig. 5.1 (continued) equations are derived to quantitatively relate canopy functional traits (chemicals) to spectral data. Example relationships are shown for foliar lignin, nitrogen (N), and polyphenols. The x-axis indicates spectral wavelengths of 400–2500 nm; the y-axes indicate relative importance of the spectrum to each example chemical constituent shown. (**d**) An example of spectra from individual crowns clustered based on their spectral variation. (**e**) A 3-D view of a portion of lowland Amazonian forest canopy. Different colors indicate different species detected based on 15 chemical traits using airborne imaging spectroscopy

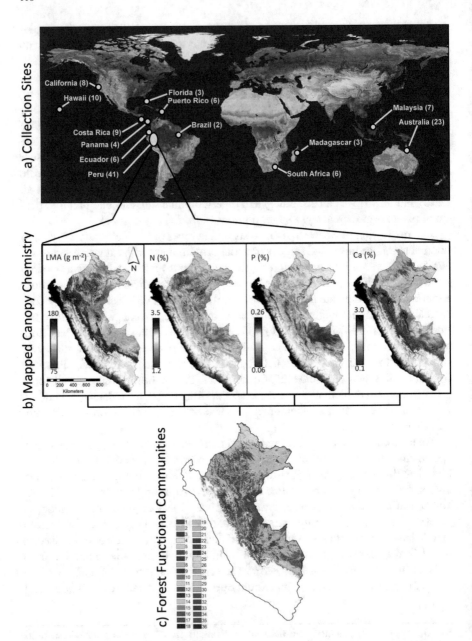

Fig. 5.2 An illustration of functional biodiversity mapping from foliar traits. (**a**) The 2018 global distribution of 128 forest landscapes contributing to the Spectranomics database. (**b**) Example maps of four foliar traits generated for the Andes-to-Amazon region of Peru using airborne imaging spectroscopy and modeling (Asner et al. 2017). (**c**) Map of 36 forest functional communities derived from a classification based on seven forest canopy traits derived from airborne imaging spectroscopy

half of the known tropical forest canopy species worldwide with measured foliar traits (Table 5.1). From investigations of these data and the fundamental patterns they uncover, Spectranomics has evolved into a new pathway to biological and ecological discovery, as well as a new tool for conservation-relevant mapping, particularly in high-diversity tropical forests.

5.3 Lessons Learned from Spectranomics

As the Spectranomics database has grown through the years, new relationships among plant phylogeny, canopy chemical traits, and spectral properties have emerged that reveal patterns at nested biogeographic scales. The extent of sampling across continents, along regional environmental gradients, and within local tree communities, coupled with consistent methods and analysis, has provided for quantitative testing of these relationships at multiple scales such that they can now be used to forecast the functional traits and biodiversity components that can be remotely mapped and monitored with spectral RS instrumentation.

5.3.1 Nested Geography of Canopy Chemical Traits in Humid Tropical Forest

Humid tropical forests cover over 20 million km of land area, span an enormous range of environmental conditions from hot lowland forests to cool montane rainforests along equatorial tree line at almost 3500 m on a variety of geological substrates, and support thousands of tree species. The high degree of complexity of this region provided an ideal setting to develop and use Spectranomics to test how environment and phylogeny interact to sort the spectral-chemical diversity of forest canopies. Based on results from multiple field studies throughout this region (Martin et al. 2007; Asner and Martin 2011; Asner et al. 2014b; McManus Chauvin et al. 2018) as well as their collective analysis (Asner and Martin 2016), we discovered that canopy chemical trait diversity of humid tropical forests occurs in a nested pattern driven by long-term adjustment of tree communities to large-scale environmental factors, particularly geologic substrate and climate. More specifically, geographic variation at the soil order level, expressing broad changes in fertility, underpins major shifts in foliar phosphorus (P) and calcium (Ca) (Fig. 5.2). Additionally, elevation-dependent shifts in average community leaf dry mass per area (LMA), chlorophyll, and carbon allocation (including nonstructural carbohydrates) are most strongly correlated with changes in foliar Ca. We also found that chemical diversity within communities is driven by differences between species rather than by plasticity within species. Finally, elevation- and soil-dependent changes in nitrogen (N), LMA, and leaf carbon allocation are mediated by canopy compositional turnover, whereas foliar P and Ca are driven more by changes in site conditions than by phylogeny. In short, Spectranomics led us to understand that canopy functional traits can be nested regionally by environmental setting but expressed locally within any given environment by their evolutionary origin.

5.3.2 Spectral Properties of Humid Tropical Forest Canopies

In concert with chemical trait collections, we measured the spectral properties of canopy foliage from thousands of humid tropical tree canopies and determined that all 23 chemical traits can be remotely sensed to varying degrees (Asner et al. 2011; Chadwick and Asner 2016; Martin et al. 2018). Utilizing leaf-level spectral-chemical relationships, we discovered that the spectral properties of canopy foliage closely tracked canopy functional trait responses to macro-environmental changes such as broad differences in soil fertility (Asner and Martin 2011; Asner et al. 2012b). Similar to the functional trait findings, we discovered that the spectral properties of foliage within communities along elevation gradients were largely determined by phylogenetic identity (Asner et al. 2014a). Consequently, canopy functional traits and spectral properties tracked one another at nested ecological scales, a result that suggests what we might find if we collected map-based spectral data over a much larger geographic area using RS instrumentation.

When coupled with DNA analyses, Spectranomics data indicate that forest canopies show strong phylogenetic organization of their foliar spectral properties, particularly in the shortwave-infrared (1500–2500 nm) wavelength region (McManus et al. 2016). This finding suggests that mapping of forest canopies with airborne imaging spectroscopy may provide spatial insight to the genetic distribution and genealogy of forest canopy taxa. Growth-form-specific studies using the Spectranomics approach revealed that lianas (woody vines) maintain functional traits and spectral properties unique from their host tree canopies (Asner and Martin 2012). Lianas are important drivers and limiters of biodiversity and carbon cycling in tropical forests (Schnitzer and Bongers 2011), and these measured differences predicted and underpinned the subsequent mapping of lianas in tropical forests using airborne imaging spectroscopy (Marvin et al. 2016).

Spectranomics data have been collected and archived under stringent field and analytical standards, which has facilitated the development new quantitative linkages between canopy foliar spectroscopy and canopy functional traits (Feilhauer et al. 2010, 2015; Féret et al. 2011, 2017). Spectral modeling studies showed that full-spectrum (350–3500 nm) data provided retrieval capability for three times the number of chemicals as 350–1300 nm data from less expensive, more common visible to near-infrared spectrometers. These studies also pointed to the need for sampling fully sunlit foliage in higher-density portions of tree crowns to minimize the effect of canopy structure on chemical trait retrievals. These findings were key guiding components in the development of laser-guided imaging spectroscopy that links Spectranomics field surveys to remotely sensed spectra to generate consistent canopy chemical trait retrieval at multiple geographic scales.

5.3.3 Spectranomics for Biodiversity Mapping

The Spectranomics fieldwork pointed toward two particular forecasts. First, Spectranomics suggested that spectral mapping from current aircraft and future satellites will reveal where whole forest communities are functionally similar and where they are unique. Second, Spectranomics suggested that spectral RS will reveal the presence and patterning of specific canopy species, within communities and across environmental gradients, based on their functional trait "signatures."

Both forecasts were subsequently proven correct during mapping studies. Numerous landscape-scale studies now show that location of particular forest canopy species and their evolved canopy functional traits mirror soil nutrient resources mediated by topography, parent material, and climate (Higgins et al. 2014; Chadwick and Asner 2016; Balzotti et al. 2016). These findings demonstrate that Spectranomics directly connects plants to ecosystem processes such as biogeochemical cycles, which form an essential link to the rest of the Earth system. At a larger scale, a 2016 report on Andean and Amazonian forests mapped with airborne imaging spectroscopy confirmed the forecasted ecological shifts in forest canopy functional composition, sorted geographically by large-scale environmental factors including elevation, geology, soils, and climate (Fig. 5.2b, c; Asner et al. 2017). While the Spectranomics database provided a field-based preview of how communities of species would differ from one another, the mapping step provided a first synoptic view of the geographic distribution. Importantly, the mapping phase also revealed numerous new combinations of functional traits that had not been detected in the field program. The new canopy functional trait maps are a key stepping-stone to biogeographic assembly, not only of the functional diversity of the Andes-to-Amazon but also of the biological diversity of the region. The approach from Peru is currently being applied in Ecuador as well as Malaysian Borneo.

The second forecast from Spectranomics—which coexisting species within communities can maintain relatively unique canopy functional traits and spectral properties—has been explored and confirmed in a series of studies using airborne and space-based imaging spectroscopy. From Hawaii to Panama, and from Africa to the Amazon, hundreds of target species have been mapped based on their spectral signatures, underpinned by a knowledge of their functional traits (Fig. 5.3; Carlson et al. 2007; Papeş et al. 2010; Colgan et al. 2012; Baldeck and Asner 2014; Baldeck et al. 2015; Graves et al. 2016). Further, the new concept of "spectral species" was developed to map species richness (alpha diversity) and compositional turnover (beta diversity) in forest landscapes without the need to detect individual species (Féret and Asner 2014).The separability of the spectral species is determined by their canopy functional traits.

More broadly, Spectranomics has enabled a different kind of interaction between field or laboratory studies of plants and RS of functional and biological diversity of ecosystems. The forecasting capability made possible with the Spectranomics database has been central to planning whether and how to undertake spectral mapping activities in different regions and under what environmental conditions the RS technology will yield new insight. In turn, this has transformed the interaction

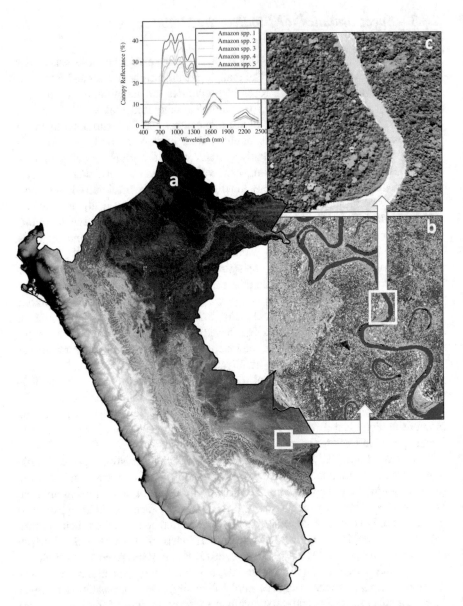

Fig. 5.3 Three scale-dependent views of the Peruvian Andes-Amazon region derived from airborne imaging spectroscopy using data and information from Spectranomics. (**a**) Peru-wide map shows the distribution of functionally distinct forests. Different colors indicate varying combinations of remotely sensed canopy foliar nitrogen (N), phosphorus (P), and leaf mass per area (LMA) (Asner et al. 2016). (**b**) Zoom image from the Peru-wide map indicates major changes in canopy N, P, and LMA with a lowland Amazonian forest (Asner et al. 2015a). Red indicates higher N + P and lower LMA relative to yellow and blue. (**c**) Individual species detections within the zoom box of panel b, derived using species-specific canopy spectra (Féret and Asner 2013; Baldeck et al. 2015)

between field and RS work from the traditional approach of mapping and ground truthing to one based on botanical, ecological, and biophysical knowledge in the interpretation of remotely sensed data.

This interaction between Spectranomics and RS also provided the scientific guidance, and initial funding, for a new class of mapping instruments, starting with a next-generation, high-fidelity visible-to-shortwave infrared (VSWIR) imaging spectrometer, built by the California Institute of Technology's Jet Propulsion Laboratory (JPL) for the Global Airborne Observatory, formerly the Carnegie Airborne Observatory (Asner et al. 2012a). JPL then built an identical instrument for NASA's Airborne Visible/Infrared Imaging Spectrometer (AVIRIS; http://aviris.jpl.nasa.gov) program, as well as several copies for the US National Ecological Observatory Network (NEON, https://www.neonscience.org; Kampe et al. 2011).

5.3.4 Scientific and Conservation Opportunities

An important outgrowth of Spectranomics is an emerging opportunity to partner discovery-based science with applied environmental conservation at large geographic scales. Conservation and management actions are usually limited in scope and effectiveness by numerous interacting financial, logistical, cultural, and political factors. An increasing ability to map canopy diversity may provide an avenue to identify the location and essential components of high-value conservation targets. Moreover, near-real-time scientific discovery from spectral RS can lead to more tactical conservation decision-making. Our specific experience is that, as land use pressures expand, intensify, and change over time, a mapping capability built upon the details of forest canopy function and composition, rather than just forest cover, supports improved conservation discussions and planning. This type of approach is needed to identify current and potential threats to, as well as current protections and opportunities for new protection of, species, communities, and ecosystems. The evolving biodiversity mapping capabilities made possible through Spectranomics are providing a tool set to support the current portfolio of Global Airborne Observatory activities (e.g., http://www.theborneopost.com/2016/04/06/3d-mapping-to-decide-on-land-use/).

The Spectranomics approach is starting to catch on in the scientific community, as highlighted in chapters throughout this book as well as new programs such as NEON and Canada's recently announced Spectranomics program for boreal forests (the Canadian Airborne Biodiversity Observatory; http://www.caboscience.org/), but there is much more to do to bring our approach to the global level. First, more scientists could get involved through building plant canopy trait laboratories and databases, paired with a specific style of leaf-level spectral measurements in the field. Currently, many functional trait and spectral measurement protocols are incompatible with the Spectranomics approach. For example, many foliar trait studies have involved the collection of samples in understory or shaded settings, in part because this foliage is easier to reach, yet spectral RS is most sensitive to canopy-level foliar chemical and structural traits (Jacquemoud et al. 2009). Additionally, most field-based trait studies do not include the use of a high-fidelity field spectrometer, which must be applied on fresh foliage to ensure connectivity to biotic and environmental conditions. Moreover, high-fidelity imaging spectrometers needed for mapping, such as the Global Airborne Observatory or AVIRIS, demand stringent and consistent field and laboratory trait measurement practices. Most of these issues can be remedied by incorporating one or more of the protocols provided on the Spectranomics website (https://gdcs.asu.edu/labs/martinlab/spectranomics). More could be done to boost capacity throughout the science community to generate data suitable for Spectranomics-type applications. Community-wide efforts to develop a global biodiversity monitoring system (Geller, Chap. 20) will greatly enhance humanity's ability to monitor and manage biodiversity for sustainability in the Anthropocene.

Acknowledgments I thank the Spectranomics scientific co-founder and programmatic funder, Dr. Greg Asner, for the inspiration to write this piece. I also thank the numerous scientists, engineers, technicians, students, and supporters of the Spectranomics Project, which has been made possible by the John D. and Catherine T. MacArthur Foundation.

References

Asner GP, Martin RE (2009) Airborne spectranomics: mapping canopy chemical and taxonomic diversity in tropical forests. Front Ecol Environ 7:269–276. https://doi.org/10.1890/070152

Asner GP, Martin RE (2011) Canopy phylogenetic, chemical and spectral assembly in a lowland Amazonian forest. New Phytol 189:999–1012. https://doi.org/10.1111/j.1469-8137.2010.03549.x

Asner GP, Martin RE (2012) Contrasting leaf chemical traits in tropical lianas and trees: implications for future forest composition. Ecol Lett 15:1001–1007

Asner GP, Martin RE (2016) Convergent elevation trends in canopy chemical traits of tropical forests. Glob Chang Biol 22:2216–2227. https://doi.org/10.1111/gcb.13164

Asner GP, Martin RE, Knapp DE, Tupayachi R, Anderson C, Carranza L, Martinez P, Houcheime M, Sinca F, Weiss P (2011) Spectroscopy of canopy chemicals in humid tropical forests. Remote Sens Environ 115:3587–3598. https://doi.org/10.1016/j.rse.2011.08.020

Asner GP, Knapp DE, Boardman J, Green RO, Kennedy-Bowdoin T, Eastwood M, Martin RE, Anderson C, Field CB (2012a) Carnegie Airborne Observatory-2: increasing science data dimensionality via high-fidelity multi-sensor fusion. Remote Sens Environ 124:454–465. https://doi.org/10.1016/j.rse.2012.06.012

Asner GP, Martin RE, Bin SA (2012b) Sources of canopy chemical and spectral diversity in lowland bornean forest. Ecosystems 15:504–517. https://doi.org/10.1007/s10021-012-9526-2

Asner GP, Martin RE, Carranza-Jiménez L, Sinca F, Tupayachi R, Anderson CB, Martinez P (2014a) Functional and biological diversity of foliar spectra in tree canopies throughout the Andes to Amazon region. New Phytol 204:127–139. https://doi.org/10.1111/nph.12895

Asner GP, Martin RE, Tupayachi R, Anderson CB, Sinca F, Carranza-Jimenez L, Martinez P (2014b) Amazonian functional diversity from forest canopy chemical assembly. Proc Natl Acad Sci 111:5604–5609. https://doi.org/10.1073/pnas.1401181111

Asner GP, Anderson CB, Martin RE, Tupayachi R, Knapp DE, Sinca F (2015a) Landscape biogeo-chemistry reflected in shifting distributions of chemical traits in the Amazon forest canopy. Nat Geosci 8:567–575. https://doi.org/10.1038/ngeo2443

Asner GP, Martin RE, Anderson CB, Knapp DE (2015b) Quantifying forest canopy traits: imaging spectroscopy versus field survey. Remote Sens Environ 158:15–27. https://doi.org/10.1016/j.rse.2014.11.011

Asner GP, Ustin SL, Townsend PA, Martin RE and Chadwick KD (2015) Forest biophysical and biochemical properties from hyperspectral and LiDAR remote sensing. In: Remote sensing handbook. Land resources monitoring, modeling, and mapping with remote sensing, pp 429–448. https://doi.org/10.1201/b19322-22

Asner GP, Knapp DE, Anderson CB, Martin RE, Vaughn N (2016) Large-scale climatic and geo-physical controls on the leaf economics spectrum. Proc Natl Acad Sci 113:E4043–E4051. https://doi.org/10.1073/pnas.1604863113

Asner GP, Martin RE, Knapp DE, Tupayachi R, Anderson CB, Sinca F, Vaughn NR, Llactayo W (2017) Airborne laser-guided imaging spectroscopy to map forest trait diversity and guide conservation. Science 355:385–389. https://doi.org/10.1126/science.aaj1987

Baldeck C, Asner G (2014) Single-species detection with airborne imaging spectroscopy data: a comparison of support vector techniques. IEEE J Sel Top Appl Earth Obs Remote Sens 8:2501–2512

Baldeck CA, Asner GP, Martin RE, Anderson CB, Knapp DE, Kellner JR, Wright SJ (2015) Operational tree species mapping in a diverse tropical forest with airborne imaging spectros-copy. PLoS One 10:e0118403. https://doi.org/10.1371/journal.pone.0118403

Balzotti CS, Asner GP, Taylor PG, Cleveland CC, Cole R, Martin RE, Nasto M, Osborne BB, Porder S, Townsend AR (2016) Environmental controls on canopy foliar nitrogen distributions in a Neotropical lowland forest. Ecol Appl 26:2449–2462. https://doi.org/10.1002/eap.1408

Baraloto C, Hardy OJ, Paine CET, Dexter KG, Cruaud C, Dunning LT, Gonzalez M-A, Molino J-F, Sabatier D, Savolainen V, Chave J (2012) Using functional traits and phylogenetic trees to examine the assembly of tropical tree communities. J Ecol 100:690–701. https://doi.org/10.1111/j.1365-2745.2012.01966.x

Bohlman SA (2015) Species diversity of canopy versus understory trees in a Neotropical Forest: implications for Forest structure, function and monitoring. Ecosystems 18:658–670. https://doi.org/10.1007/s10021-015-9854-0

Carlson KM, Asner GP, Hughes RF, Ostertag R, Martin RE, Hughes FR, Ostertag R, Martin RE (2007) Hyperspectral remote sensing of canopy biodiversity in Hawaiian lowland rainforests. Ecosystems 10:536–549. https://doi.org/10.1007/s10021-007-9041-z

Cavender-Bares J, Keen A, Miles B (2006) Phylogenetic structure of Floridian plant communities depends on taxonomic and spatial scale. Ecology 87:S109–S122

Chadwick KD, Asner GP (2016) Organismic-scale remote sensing of canopy foliar traits in low-land tropical forests. Remote Sens 8:87. https://doi.org/10.3390/rs8020087

Colgan MS, Baldeck CA, baptiste FJ, Asner GP (2012) Mapping savanna tree species at eco-system scales using support vector machine classification and BRDF correction on air-borne hyperspectral and LiDAR data. Remote Sens 4:3462–3480. https://doi.org/10.3390/rs4113462

Curran PJ (1989) Remote sensing of foliar chemistry. Remote Sens Environ 30:271–278. https://doi.org/10.1016/0034-4257(89)90069-2

Díaz S, Cabido M, Casanoves F, Diaz S, Cabido M, Casanoves F (1998) Plant functional traits and environmental filters at a regional scale. J Veg Sci 9:113–122. https://doi.org/10.2307/3237229

Diaz S, Kattge J, Cornelissen JH, Wright IJ, Lavorel S, Dray S, Reu B, Kleyer M, Wirth C, Prentice IC, Garnier E, Bonisch G, Westoby M, Poorter H, Reich PB, Moles AT, Dickie J, Gillison AN, Zanne AE, Chave J, Wright SJ, Sheremet'ev SN, Jactel H, Baraloto C, Cerabolini B, Pierce S, Shipley B, Kirkup D, Casanoves F, Joswig JS, Gunther A, Falczuk V, Ruger N, Mahecha MD, Gorne LD (2016) The global spectrum of plant form and function. Nature 529:167–171. https://doi.org/10.1038/nature16489

Feilhauer H, Asner GP, Martin RE, Schmidtlein S (2010) Brightness-normalized partial least squares regression for hyperspectral data. J Quant Spectrosc Radiat Transf 111:1947–1957. https://doi.org/10.1016/j.jqsrt.2010.03.007

Feilhauer H, Asner GP, Martin RE (2015) Multi-method ensemble selection of spectral bands related to leaf biochemistry. Remote Sens Environ 164:57–65. https://doi.org/10.1016/j.rse.2015.03.033

Féret JB, Asner GP (2013) Tree species discrimination in tropical forests using airborne imaging spectroscopy. IEEE Trans Geosci Remote Sens 51:73–84

Feret J-B, Asner GP (2014) Microtopographic controls on lowland Amazonian canopy diversity from imaging spectroscopy. Ecol Appl 24:1297–1310

Féret J-B, Asner GP (2014) Mapping tropical forest canopy diversity using high-fidelity imaging spectroscopy. Ecol Appl 24:1289–1296. https://doi.org/10.1890/13-1824.1

Féret J-B, François C, Gitelson A, Asner GP, Barry KM, Panigada C, Richardson AD, Jacquemoud S (2011) Optimizing spectral indices and chemometric analysis of leaf chemical properties using radiative transfer modeling. Remote Sens Environ 115:2742–2750. https://doi.org/10.1016/j.rse.2011.06.016

Féret J-B, Gitelson AA, Noble SD, Jacquemoud S (2017) PROSPECT-D: towards modeling leaf optical properties through a complete lifecycle. Remote Sens Environ 193:204–215. https://doi.org/10.1016/J.RSE.2017.03.004

Graves S, Asner G, Martin R, Anderson C, Colgan M, Kalantari L, Bohlman S (2016) Tree species abundance predictions in a tropical agricultural landscape with a supervised classification model and imbalanced data. Remote Sens 8:161. https://doi.org/10.3390/rs8020161

Higgins MA, Asner GP, Martin RE, Knapp DE, Anderson C, Kennedy-Bowdoin T, Saenz R, Aguilar A, Joseph Wright S (2014) Linking imaging spectroscopy and LiDAR with floristic composition and forest structure in Panama. Remote Sens Environ 154:358–367. https://doi.org/10.1016/j.rse.2013.09.032

Hubbell SP, He F, Condit R, Borda-de-Água L, Kellner J, ter Steege H (2008) How many tree species are there in the Amazon and how many of them will go extinct? Proc Natl Acad Sci 105:11498–11504

Jacquemoud S, Ustin SL (2001) Leaf Optical properties: a state of the art. In: 8th International Symposium on Physical Measurements & Signatures inRemote Sensing, Aussois, France, pp 223–232

Jacquemoud S, Verhoef W, Baret F, Bacour C, Zarco-Tejada PJ, Asner GP, Francois C, Ustin SL (2009) PROSPECT plus SAIL models: a review of use for vegetation characterization. Remote Sens Environ 113:S56–S66. https://doi.org/10.1016/j.rse.2008.01.026

Jetz W, Cavender-Bares J, Pavlick R, Schimel D, Davis FW, Asner GP, Guralnick R, Kattge J, Latimer AM, Moorcroft P, Schaepman ME, Schildhauer MP, Schneider FD, Schrodt F, Stahl U, Ustin SL (2016) Monitoring plant functional diversity from space. Nat Plants 2:16024. https://doi.org/10.1038/nplants.2016.24

Kampe TU, Johnson BR, Kuester M, McCorkel J (2011) Airborne remote sensing instrumentation for NEON: status and development. 2011 Aerospace conference

Kokaly RF, Asner GP, Ollinger SV, Martin ME, Wessman CA (2009) Characterizing canopy biochemistry from imaging spectroscopy and its application to ecosystem studies. Remote Sens Environ 113:S78–S91. https://doi.org/10.1016/j.rse.2008.10.018

Kress WJ, Erickson DL, Jones A, Swenson NG, Perez R, Sanjur O, Bermingham E (2009) Plant DNA barcodes and a community phylogeny of a tropical forest dynamics plot in Panama. Proc Natl Acad Sci 106:18621. https://doi.org/10.1073/pnas.0909820106

Martin RE, Asner GP, Sack L (2007) Genetic variation in leaf pigment, optical and photosynthetic function among diverse phenotypes of Metrosideros polymorpha grown in a common garden. Oecologia 151:387–400. https://doi.org/10.1007/s00442-006-0604-z

Martin RE, Chadwick K, Brodrick PG, Carranza-Jimenez L, Vaughn NR, Asner GP, Dana Chadwick K, Brodrick PG, Carranza-Jimenez L, Vaughn NR, Asner GP (2018) An approach for foliar trait retrieval from airborne imaging spectroscopy of tropical forests. Remote Sens 10:199. https://doi.org/10.3390/rs10020199

Marvin DC, Asner GP, Knapp DE, Anderson CB, Martin RE, Sinca F, Tupayachi R (2014) Amazonian landscapes and the bias in field studies of forest structure and biomass. Proc Natl Acad Sci 111:E5224–E5232. https://doi.org/10.1073/pnas.1412999111

Marvin DC, Koh LP, Lynam AJ, Wich S, Davies AB, Krishnamurthy R, Stokes E, Starkey R, Asner GP (2016) Integrating technologies for scalable ecology and conservation. Glob Ecol Conserv 7:262–275. https://doi.org/10.1016/J.GECCO.2016.07.002

McManus Chauvin K, Asner GP, Martin RE, Kress WJ, Wright SJ, Field CB (2018) Decoupled dimensions of leaf economic and anti-herbivore defense strategies in a tropical canopy tree community. Oecologia 186:765–782. https://doi.org/10.1007/s00442-017-4043-9

McManus KM, Asner GP, Martin RE, Dexter KG, Kress WJ, Field CB (2016) Phylogenetic structure of foliar spectral traits in tropical forest canopies. Remote Sens 8:196. https://doi.org/10.3390/rs8030196

Mooney HA, Cushman JH, Medina E, Sala OE, Schulze E-D (1996) Functional roles of biodiversity: a global perspective. p 493

Myers N, Mittermeier RA, Mittermeier CG, da Fonseca GAB, Kent J (2000) Biodiversity hotspots for conservation priorities. Nature 403:853–858. https://doi.org/10.1038/35002501

Papeş M, Tupayachi R, Martínez P, Peterson a TT, Powell GVNVN (2010) Using hyperspectral satellite imagery for regional inventories: a test with tropical emergent trees in the Amazon Basin. J Veg Sci 21:342–354. https://doi.org/10.1111/j.1654-1103.2009.01147.x

Reich PB, Wright IJ, Cavender-Bares J, Craine JM, Oleksyn J, Westoby M, Walters MB (2003) The evolution of plant functional variation: traits, spectra, and strategies. Int J Plant Sci 164:S143–S164. https://doi.org/10.1086/374368

Reichstein M, Bahn M, Mahecha MD, Kattge J, Baldocchi DD (2014) Linking plant and ecosystem functional biogeography. Proc Natl Acad Sci 111:13697–13702. https://doi.org/10.1073/pnas.1216065111

Running SW, Justice CO, Salomonson V, Hall D, Barker J, Kaufmann YJ, Strahler AH, Huete AR, Muller J-P, Vanderbilt V, Wan ZM, Teillet P, Carneggie D (1994) Terrestrial remote sensing science and algorithms planned for EOS/MODIS. Int J Remote Sens 15:3587–3620

Schimel DS, Asner GP, Moorcroft PR (2013) Observing changing ecological diversity in the Anthropocene. Front Ecol Environ 11:129

Schnitzer SA, Bongers F (2011) Increasing liana abundance and biomass in tropical forests: emerging patterns and putative mechanisms. Ecol Lett 14:397–406

Serbin SP, Singh A, McNeil BE, Kingdon CC, Townsend PA (2014) Spectroscopic determination of leaf morphological and biochemical traits for northern temperate and boreal tree species. Ecol Appl 24:1651–1669. https://doi.org/10.1890/13-2110.1

Tucker CJ, Townshend JRG (2000) Strategies for monitoring tropical deforestation using satellite data. Int J Remote Sens 21:1461–1471

Turner W (2014) Sensing biodiversity. Science 346:301–302. https://doi.org/10.1126/science.1256014

Turner W, Spector S, Gardiner N, Fladeland M, Sterling E, Steininger M (2003) Remote sensing for biodiversity science and conservation. Trends Ecol Evol 18:306–314. https://doi.org/10.1016/S0169-5347(03)00070-3

Ustin SL, Roberts DARA, Gamon JA, Gregory P, Green RO (2004) Using imaging spectroscopy to study ecosystem processes and properties. Bioscience 54:523–534

Ustin SL, Gitelson AA, Jacquemoud S, Schaepman M, Asner GP, Gamon JA, Zarco-Tejada P (2009) Retrieval of foliar information about plant pigment systems from high resolution spectroscopy. Remote Sens Environ 113(Suppl):S67–S77. https://doi.org/10.1016/j.rse.2008.10.019

Wold S, Sjöström M, Eriksson L (2001) PLS-regression: a basic tool of chemometrics. Chemom Intell Lab Syst 58:109–130. https://doi.org/10.1016/S0169-7439(01)00155-1

Wright IJ, Reich PB, Westoby M, Ackerly DD, Baruch Z, Bongers F, Cavender-Bares J, Chapin T, Cornelissen JHC, Diemer M, Flexas J, Garnier E, Groom PK, Gulias J, Hikosaka K, Lamont BB, Lee T, Lee W, Lusk C, Midgley JJ, Navas M-LL, Niinemets U, Oleksyn J, Osada N, Poorter H, Poot P, Prior L, Pyankov VI, Roumet C, Thomas SC, Tjoelker MG, Veneklaas EJ,

Villar R (2004) The worldwide leaf economics spectrum. Nature 428:821–827. https://doi. org/10.1038/nature02403
Wright SJ, Kitajima K, Kraft NJB, Reich PB, Wright IJ, Bunker DE, Condit R, Dalling JW, Davies SJ, DíAz S, Engelbrecht BMJ, Harms KE, Hubbell SP, Marks CO, Ruiz-Jaen MC, Salvador CM, Zanne AE (2010) Functional traits and the growth-mortality trade-off in tropical trees. Ecology 91:3664–3674. https://doi.org/10.1890/09-2335.1

Open Access This chapter is licensed under the terms of the Creative Commons Attribution 4.0 International License (http://creativecommons.org/licenses/by/4.0/), which permits use, sharing, adaptation, distribution and reproduction in any medium or format, as long as you give appropriate credit to the original author(s) and the source, provide a link to the Creative Commons license and indicate if changes were made.

The images or other third party material in this chapter are included in the chapter's Creative Commons license, unless indicated otherwise in a credit line to the material. If material is not included in the chapter's Creative Commons license and your intended use is not permitted by statutory regulation or exceeds the permitted use, you will need to obtain permission directly from the copyright holder.

Chapter 6
Remote Sensing for Early, Detailed, and Accurate Detection of Forest Disturbance and Decline for Protection of Biodiversity

Jennifer Pontius, Paul Schaberg, and Ryan Hanavan

6.1 Introduction

In many ways, biodiversity is a foundational component of healthy, productive forests and maintenance of the many ecosystem services that they provide (e.g., carbon sequestration, nutrient cycling, water filtration and provisioning, wildlife habitat). Forested landscapes are often characterized by a mosaic of species, age classes, and structural characteristics that results from natural patterns of disturbance. This diversity within stands and across forested landscapes increases resilience of larger forested ecosystems, enabling them to recover and maintain ecological function following disturbance (Thompson et al. 2009). But many pests and pathogens, particularly exotic invasive insects, as well as various abiotic stresses (e.g., pollution impacts or increases in climate extremes), have the potential to alter native populations, reduce biodiversity, and impact ecosystem function and service provisioning. This is particularly true for ecosystems dominated by keystone or foundational species, which exert a relatively large impact on community stability and ecosystem function (Ellison et al. 2010).

There are many examples of the impacts of pests and pathogens on biodiversity and ecological function in forested ecosystems. Dutch elm disease was introduced in the United States in the 1930s and the United Kingdom in the 1970s, with

J. Pontius (✉)
Rubenstein School of Environment and Natural Resources, University of Vermont, Burlington, VT, USA

USDA Forest Service, Northern Research Station, Burlington, VT, USA
e-mail: Jennifer.pontius@uvm.edu

P. Schaberg
USDA Forest Service, Northern Research Station, Burlington, VT, USA

R. Hanavan
USDA Forest Service, Forest Health Protection, Northeastern Area, Durham, NH, USA

© The Author(s) 2020
J. Cavender-Bares et al. (eds.), *Remote Sensing of Plant Biodiversity*,
https://doi.org/10.1007/978-3-030-33157-3_6

121

Fig. 6.1 Ancient
whitebark pines killed by
the recent mountain pine
beetle outbreak stand on a
windy ridge in Yellowstone
National Park. (Credit:
Adam Markham/CleanAir-
CoolPlanet.org, https://
www.fws.gov/cno/
newsroom/highlights/2017/
whitebark_pine/)

profound impacts on the biodiversity of rural landscapes (Harwood et al. 2011). The mountain pine beetle has impacted large swaths of coniferous and mixed forests in British Columbia, with severe impacts to avian biodiversity (Martin et al. 2006). In the western United States, pine blister rust has impacted biodiversity and ecological processes, particularly at high elevation sites where whitebark pine is a keystone species (Tomback and Achuff 2010, Fig. 6.1). Recent cases, such as the introduction of the Asian long-horned beetle and emerald ash borer to the United States, demonstrate the ongoing biosecurity challenges that currently face forested ecosystems.

Similarly, abiotic stresses can lead to declines that alter competition and biodiversity in the broader forest. For example, acid deposition that resulted from elevated inputs of sulfur and nitrogen pollution in the 1950s through 1980s led to declines in red spruce (*Picea rubens* Sarg.) (Schaberg et al. 2011) and sugar maple (*Acer saccharum* Marsh.) (Huggett et al. 2007) and increases in less sensitive species such as American beech (*Fagus grandifolia* Ehrh.) (Schaberg et al. 2001; Pontius et al. 2016). In another example, warming temperatures were associated with reductions in winter snowpacks, increased soil freezing, and root mortality that resulted in the broad-scale decline of yellow cedar (*Callitropsis nootkatensis*) but not sympatric species (Hennon et al. 2012). Warmer climates have also resulted in range expansion of native insects and disease with potential to further alter the landscape. For example, the southern pine beetle (*Dendroctonus frontalis*) continues to move north from the loblolly forests of the southern United States to pitch pine in the north.

Many resource managers cite the need for early detection of forest decline to minimize impacts of emergent stress agents (Genovesi et al. 2015; Sitzia et al. 2016). Research has shown that the earlier you can detect forest decline, the more successful management and control efforts will be (Epanchin-Niell and Hastings 2010). For invasive pests and pathogens, identifying the locations of incipient infestations is critical to minimizing spread, reducing ecosystem impacts, and targeting management and control (Mumford 2017).

But early detection also benefits the sustainable management of forested ecosystems responding to lower-level, chronic stress agents such as climate change and acid deposition. Such chronic stress agents often manifest in more subtle decline symptoms over many years. This slow and highly variable decline (some good years, some bad years) limits the ability to identify causal relationships, understand potential impacts to ecosystem function, and develop management strategies. As a result, we need to be able to *quantify* decline symptoms with greater detail and sensitivity to subtle changes, from the gradual loss of photosynthetic apparatus in response to initial stress, to reductions in canopy density, dieback, and ultimate mortality across the landscape.

Remote sensing (RS) has long been used to assess relative vegetation density, decline, and mortality. But landscape-scale assessment of small-scale or subtle decline symptoms has been more difficult. The spatial patial resolution of many sensors has limited our ability to detect small-scale decline in highly mixed pixels, while spectral resolution has limited our ability to detect early biogeochemical precursors to more severe decline symptoms. But as new sensors and modeling algorithms have come on board, there is a growing list of successful early decline detection efforts.

Here we present the science behind RS for the assessment of vegetation condition, with a focus on using these tools for more detailed and accurate monitoring of forest decline and disturbance. We also highlight the importance of this approach to inform the sustainable management of forested ecosystems and preservation of forest biodiversity.

6.2 The Basics of Forest Decline

In order to better understand how RS instruments can detect vegetation stress, and be used to quantify forest decline, it is important to understand the structural and physiological response of vegetation to stress. Any RS effort to detect or monitor decline is based on the sensor's ability to detect these biophysical changes that manifest following stress.

Trees adjust their physiology and form in response to environmental stimuli (e.g., light, temperature, moisture). Stress occurs when environmental conditions fall outside of the normal or optimal levels to which plants are adapted. As sessile organisms that cannot flee from the many stresses that they are routinely exposed to over their long life spans, trees have evolved enumerable mechanisms to avoid, mitigate, or rebound from stress. Some of these adaptations (e.g., protective pigments such as the yellow/orange carotenoids and red anthocyanins in leaves) can directly influence RS spectral measurements. Other stress adaptations (e.g., changes in carbohydrate storage and lipid and protein metabolism; Strimbeck et al. 2015) influence spectral characteristics indirectly through changes in leaf retention and life span. Here we walk through some of these physiological and structural changes relevant to RS efforts in more detail.

Leaf Size Small, emerging leaves can be difficult to detect via RS (e.g., White et al. 2014). Therefore, factors that delay or expedite bud break and leaf expansion, or lead to leaf wilting, curling, and folding can influence spectral signatures

Fig. 6.2 Leaf curl, wilt, and stunted expansion can result in decreased leaf area index that is commonly quantified in RS applications. (Credit: Eiku [CC BY-SA 4.0] from Wikimedia Commons)

Fig. 6.3 Many sensors can detect changes in leaf pigment concentration and function before chlorosis is visible to the human eye. (Credit: [CC0] https:// pxhere.com/en/ photo/575928)

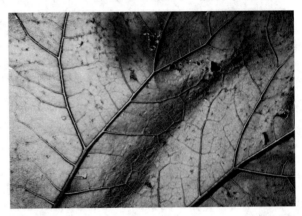

(Fig. 6.2). In addition, leaves that develop after episodic leaf mortality are often stunted, diminishing overall leaf area. Reduced leaf size can also result from carbohydrate losses associated with sucking insects, e.g., pear thrips (*Taeniothrips inconsequens*; Kolb and Teulon 1991), and insect herbivory can reduce the functional area of leaves through leaf consumption.

Leaf Chemistry and Physiology Plant pigments (chlorophylls essential in photosynthesis, xanthophylls that assist with light capture and protect leaves from photooxidation, and anthocyanins that have numerous protective capacities) are all spectrally responsive (Fig. 6.3). Therefore, environmental factors that influence their development and turnover (e.g., cold temperatures that can speed chlorophyll catabolism and trigger anthocyanin expression; Schaberg et al. 2017) can influence associated spectral signatures. Similarly, because leaf water content and chemistry have identifiable spectral features, environmental factors such as droughts, fertilization, and soil acidification can also influence spectral signatures.

Leaf Quantity and Longevity Despite remarkable and diverse capacities for stress response and protection, numerous biological and abiotic factors can reduce

Fig. 6.4 Peeling back the bark on green ash shows the girdling effect of the emerald ash borer (Agrilus planipennis). (Credit: USDA Forest Service)

leaf longevity or lead to significant defoliation. The most prominent factors causing foliar reductions vary across ecoregions (e.g., drought is a dominant factor in the western United States, whereas insect defoliation is prominent in the eastern states) and over time (e.g., episodic drought, cyclic insect outbreaks). However, numerous anthropogenic factors (e.g., ozone pollution, acid deposition, introduction of exotic pests and pathogens) have expanded the list of stress agents that can lead to significant defoliation. Some stress agents directly result in defoliation, but many stress agents impact other organs that crowns rely on, for example, insects such as bark beetles and the emerald ash borer (Fig. 6.4) and pathogens such as chestnut blight girdle stems. Invasive pests such as hemlock woolly adelgid extract photosynthate directly from phloem. Root freezing injury (e.g., yellow cedar decline; Hennon et al. 2012) can limit resource uptake. All of these stress agents can manifest as reduced leaf area index and canopy density.

Branch Dieback, Tree Decline, and Mortality   Repeated or severe direct damage to tree canopies or chronic imbalances in tree carbohydrate and/or stress response systems can lead to branch dieback. This dieback is typically first evident as mortality of the most distal portions of the crown (tip dieback) and can lead to significant carbon imbalances as the photosynthetic capacity of trees is outstripped by

Fig. 6.5 Dieback typically results in changes to spectral characteristics as pixels become dominated by understory or bark and soil surface features. (Credit: Joseph O'Brien, USDA Forest Service, Bugwood.org)

carbohydrate use associated with maintenance respiration as well as compensatory growth (e.g., epicormic branching), and seed production, which are often associated with decline. Significant crown loss exacerbates negative carbon balances, ultimately resulting in tree mortality. Temporary or partial crown dieback may be difficult to detect if it is not widespread, but protracted dieback, especially if it results in significant tree mortality, could dramatically alter spectral measurements in the near (during the decline event) and long terms if elevated mortality leads to significant changes in canopy density, gap fraction, species composition, or forest cover (Fig. 6.5).

6.3 RS Approaches to Forest Decline Detection

Aerial Sketch Mapping In the United States, federal and state forestry agencies have been conducting aerial detection surveys of forest decline for many decades (Fig. 6.6; Johnson and Wittwer 2008; Johnson and Ross 2008; McConnell 1999). This manual RS technique involves an observer mapping polygons by identifying host trees by crown shape and causal agent by damage signature from an aircraft. In the early decades (1950s–1980s), this was often deployed only in response to severe or widespread forest disturbance events, with limited flight lines and rough delineation of impacted stands onto paper maps. Now, organized by the national Forest Health Monitoring (FHM) program, many states are flown in their entirety each year to survey impacts from a suite of potential biotic and abiotic stressors and various disturbance types (e.g., defoliation, mortality, dieback), with mapping captured on digital, global positioning system (GPS)-enabled touchscreen tablets. Like other RS methods, ground validation adds confidence in the final map products. Aerial sketch mapping is currently the most widespread approach to forest condition mapping across the United States, and because of direct cooperation among federal and states agencies collecting and using the resulting maps, it also has the most direct link to land managers and decision-makers.

Fig. 6.6 Cessna 170-B survey plane mapping Douglas-fir beetle damage near Sutherlin, Oregon. (Credit: USDA Forest Service, Region 6, State and Private Forestry)

However, the mapping products generated vary based on differences in the base map scale used, observer bias, or agency emphasis (Kosiba et al. 2018). Products also vary year to year based on timing of flight and the visibility of different stress symptoms (e.g., early season vs. late season defoliators). Further, only decline symptoms that are severe enough, and in large enough patches to be visible to an observer in an aircraft traveling approximately 100 knots from an altitude of 1000–3000 feet above ground level, are mapped. As such, aerial sketch mapping can be highly subjective and should only be regarded as a coarse "snapshot" of landscape-level forest health.

Multispectral Sensors Terrestrial satellite RS began with the launch of the Landsat mission (then called the Earth Resources Technology Satellite (ERTS)) in 1972. Designed to supply regular images of Earth's surface, with multispectral bands designed to capture biospheric processes at medium-high spatial resolution, Landsat-1 enabled a revolution in terrestrial research (Williams et al. 2006). With continuous coverage since the 1972 launch, the family of Landsat sensors is particularly useful for studying forest change over time across regional to global scales (Fig. 6.7).

Initially, the broad, multispectral bands on the Landsat sensors were used to assess relative vegetative density, or "greenness." This was made possible by targeting the near-infrared (NIR) portions of the electromagnetic spectrum in addition to visible wavelengths. This "near-infrared plateau" is a region of strong reflectance in vegetation and is distinct from many other surface features such as soil, rock, and water, making it particularly useful for distinguishing vegetation from non-vegetative land cover types or assessing the relative amount of vegetation within mixed pixels. It is also highly responsive to common stress symptoms such as defoliation, chlorosis, and decreases in canopy density. Over the decades, scientists have developed a suite of vegetation indices to quantify vegetation condition and biophysical attributes (Table 6.1) that have been commonly used to assess changes in canopy cover (e.g., deforestation) and widespread defoliation or mortality.

July 19, 1984

August 12, 2010

Fig. 6.7 Landsat images from 1984 and 2010 show clear-cutting and forest regrowth in Washington State, highlighting the utility of multispectral sensors in detecting vegetation density and disturbance. (Credit: NASA image by Robert Simmon)

The use of multispectral sensors to identify more subtle or early decline symptoms is typically limited by the spectral resolution (few, broad-bands of spectral information to work with), spatial resolution (mixes of healthy and stressed vegetation in one pixel often mask the spectral stress signature of stressed individual trees), and temporal resolution (inability to acquire cloud-free images at intervals sufficient to detect change).

As the interest in RS products has grown, along with the range of applications, many commercial vendors have expanded access to multispectral products with both aerial and satellite platforms. We now have over 100 active satellite sensors with visible and NIR capabilities listed in the International Inst. for Aerospace Survey and Earth Sciences (Netherlands; formerly International Training Centre for

Table 6.1 A selection of broadband and narrowband vegetation indices useful for vegetation stress detection

Broadband indices	Abbreviation	Associated vegetation characteristics	Index formula (Landsat 7 ETM bands)	Reference
Difference vegetation index	DVI	Canopy cover, greenness	$B4 – B3$	Jordan (1969); Tucker (1979)
Enhanced vegetation index	EVI	Canopy cover, greenness	$2.5*((B4 – 33)/(B4 + (6 * 33) – (7.5 * B1) + 1))$	Huete et al. (1994); Huete et al. (2002)
Greenness index		Canopy cover, greenness	$B4/B3$	Sivanpillai et al. (2006)
	B7	Vegetation & soil water content	$B7$	Lillesand and Kiefer (1994)
	EL	Green biomass	$B5/B7$	Elvidge and Lyon (1985)
Modified soil adjusted vegetation index	MSAVI	Canopy cover, greenness	$0.5*(2*34 + 1 – (Sqr(((2 * B4 + 1) * (2 * B4 + 1))) – (8 * (B4 – B3))))$	Qi et al. (1994)
Moisture stress index	MSI	Canopy water content	$B5/B4$	Rock et al. (1986)
Normalized difference infrared index 5	NDII$_5$		$(B4 – B5)/(B4 + B5)$	Hardisky et al. (1983)
Normalized difference infrared index 7	NDII$_7$		$(B4–B7)/(34+ B7)$	Hunt and Rock (1989)
Normalized difference vegetation index	NDVI	Canopy greenness, condition, density	$(B4 – B3)/(B4 + B3)$	Deblonde and Cihlar (1993); Gamon et al. (1997); Myneni et al. (1995a, 1995b); Rouse et al. (1974)
Optomized soil adjusted vegetation index	OSAVI	Canopy density	$(B4 – b3)/(B4 + B3 + 0.16)$	Rondeaux et al. (1996)
Renormalized difference vegetation index	RDM	Leaf area index; chlorophyll	$Sqr((((B4 – B3)/(B4 + B3)) * (B4 – B3)))$	Roujean and Breon (1995)
Ratio vegetation index	RM	Chlorophyll	$B4/B3$	Pearson and Miller (1972)

(continued)

Table 6.1 (continued)

Broadband indices	Abbreviation	Associated vegetation characteristics	Index formula (Landsat 7 ETM bands)	Reference
Soil and atmospherically resistant vegetation index	SARM	Canopy density	$1.5 * ((B4 − (B3 − (B1 − B3))/(B4 − (B3 + (B1 − B3) + 0.5)))$	Kaufman and Tanre (1992); Huete et al. (1994)
Soil adjusted vegetation index	SAM	Soil adjusted canopy cover	$1.5 * ((B4 − B3)/(B4 + B3 + 0.5))$	Huete (1988)
	$SDRE_a$	General stress	$(B5 − B3) − (B4 − B2)$	Pontius et al. (2006)
	$SDRE_b$	General stress	$(B4 − B3) − (B3 − B2)$	Pontius et al. (2005a, 2005b)

Narrowband indices	Narrowband indices	Associated vegetation characteristics	Index formula narrowband formula broadband approximation	Landsat 7 ETM approximation	Reference
	Aoki	Chlorophyll	$R550/R800$	$B2/B4$	Aoki et al. (1981)
	BN_a	Chlorophyll	$R800 − R550$	$BB − B2$	Buschman and Nagel (1993)
	BN_b	Chlorophyll	$R800/R550$	$BB/B2$	Buschman and Nagel (1994)
Chlorophyll fluorescence ratio	CFR	Chlorophyll fluorescence	$F690/F735$	X	D'Ambrosio, et al. (1992)
	$Chap_A$	$Chlorophyll_a$	$R675/R700$	X	Chappelle et al. (1992)
	$Chap_B$	$Chlorophyll_a$	$R675/(R700 * R650)$	X	Chappelle et al. (1992)
Curvature index	CI	Chlorophyll and chlorophyll florescence	$(R683^2)/(R675 * R691)$	X	Zarco-Tejada et al. (2002)
	CS_a	Chlorophyll	$R694/R760$	$BB/B4$	Carter and Miller (1994)
	CS_b	General stress	$R694/R420$	$BB/B1$	Carter (1994)
	CS_c	$Chlorophyll_a$	$R605/R760$	X	Carter (1994); Carter and Miller (1994)

	CS_d	General stress	R710/R760	X	Carter (1994)
	CS_e	Chlorophyll	R695/R760	BB/B1	Carter (1994)
	$Datt_A$	Chlorophyll	R672 * (R550 * R708)	BB * (B2 *BB)	Datt (1998)
	$Datt_B$	Chlorophyll	FD754/FD704	X	Datt (1999)
	DattCS	Chlorophyll content	(R672/R550) * (R695/R 760)	X	Datt (1998)
Derivative chlorophyll index	DCI	Chlorophyll$_a$; chlorophyll$_b$, chlorophyll florescence	FD705/FD723	X	Zarco-Tejada et al. (2002)
	EZ	Chlorophyll	Sum FD 625 to 795	X ·	Elvidge and Chen (1995)
	FD_{703}	Chlorophyll$_a$	FD703	X	Boochs et al. (1990)
	FD_{720}	Chlorophyll$_b$	FD720	X	Boochs et al. (1990)
	Flo	Chlorophyll florescence; photo synthetic activity	FD690/FD735	X	Mohammed et al. (1995)
	FP	Leaf area index; chlorophyll	Sum FD 680 to 780	X	Filella and Penuelas (1994)
	Git_C	Chlorophyll	1/R700	1/B3	Gitelson et al. (1999); Gitelson et al. (2001)
	GM_a	Chlorophyll	R750/R550	B4/B2	Gitelson and Merzlyak (1994)
	GM_b	Chlorophyll	R750/R700	B4/B3	Gitelson and Merzlyak (1994)
	Mac	Chlorophyll	(R780 − R710)/(R780 − R680)	X	Maccioni et al. (2001)
Modified chlorophyll absorption in reflectance index	MCARI	Chlorophyll	((R700 − R670) − 0.2* (R700 − R550)) * (R700/ R670)	X	Daughtry et al. (2000)
Modified chlorophyll absorption in reflectance index 1	$MCARI_1$	Chlorophyll	1.2 * ((2.5 * (R800 − R670)) − (1.3 * (R800 − R550)))	1.2 * ((2.5 * (B3)) − (1.3 * (B4 − B2)))	Haboudane et al. (2004)
	McM	Chlorophyll	R700/R760	X	McMurtrey III et al. (1994)
	mND_{705}	Chlorophyll	(R750 − R705)/ (R750 + RR705 + 2R445)	(B4 − B3)/ (B4 + B3 + (2 * B1))	Sims and Gamon (2002)

(continued)

Table 6.1 (continued)

Narrowband indices	Narrowband indices	Associated vegetation characteristics	Index formula narrowband formula broadband approximation	Landsat 7 ETM approximation	Reference
Modified simple ratio	MSR	Leaf area index; fraction of photo synthetically active radiation	$((R800/R678) - 1)/ (Sqr((R800/ R670) + 1))$	$((B4/B3) - 1)/ (Sqr((B4/ B3) + 1))$	Chen (1996)
Modified simple ratio$_{705}$	MSR$_{705}$	Chlorophyll	$(R750 - R445)/(R705 - R445)$	$(B4 - B1)/ (B4 + B1)$	Sims and Gamon (2002)
Normalized difference index	NDI	Chlorophyll	$(R750 - R705)/ R750 + R705$	$(B4 - B3)/ (B4 - B3)$	Gitelson and Merzlyak (1994)
Normalized difference water index	NDWI	Canopy water content	$(R860 - R1240)/(R860 - R1240)$	$(B4 - B5)/ (B4 - B5)$	Gao (1996)
Modified triangular vegetation index	MTVI	Leaf area index	$1.2 * ((1.2 * (R800 - R550)) - (2.5 * (R670-R550)))$	$1.2 *((1.2 *(B4 - B2)) - (2.5 *(B3-B2)))$	Haboudane et al. (2004)
Normalized pigments chlorophyll ratio index	NPCI	Chlorophyll; general stress	$(R680 - R430)/ (R680 + R430)$	$(B3 - B1)/ (B3 + B1)$	Penuelas et al. (1994)
Normalized phae ophytinization index	NPQI	Chlorophyll; general stress	$(R415 - R435)/ (R415 + R435)$	X	Barnes (1992)
Photochemical reflectance index	PRI	Xanthophyll cycle activity	$(R531 - R570)/ (R531 + R570)$	X	Gamon et al. (1997); Rahman et al. (2001)
Pigment-specific normalized difference	PSND	Chlorophyll$_b$	$(R680 - R635)/ (R800 + R635)$	$(B4 - B3)/ (B4 + B3)$	Blackburn (1998, 1999)
Plant senescence reflectance index	PSRI	Xanthophyll cycle activity	$(R680 - R500))/R750$		Merzlyak et al. (1999)
Pigment specific simple ratio	PSSR	Chlorophyll	$R800/R635$	$B4/B3$	Blackburn (1998, 1999)

Index name	Abbreviation	Function	Equation	X	Reference
	R_{550}	Chlorophyll$_a$	R550	X	Carter (1993)
	R_{678}	General stress	R678	X	Pietrzykowski et al. (2006)
	R_{680}	Chlorophyll$_a$; chlorophyll florescence	R678	X	Mohammed et al. (1995)
	R_{687}	Chlorophyll florescence	R687	X	Meroni et al. (2008)
	R_{702}	Chlorophyll$_a$	R702	X	Carter (1993); Carter and Knapp (2001)
	R_{760}	Chlorophyll florescence	R760	X	Carter and Miller (1994)
	R_{950}	Canopy water content	R950	X	Williams and Norris (2001)
Red edge inflection point	REIP	Chlorophyll$_a$; vegetation greenness; canopy density	FD max near 700	X	Horler et al. (1983); Gitelson and Merzlyak (1996); Rock et al. (1988); Vogelmann et al. (1993)
Structure insensitive pigment index	SIPI	Carotenoid:chlorophyll	(R803 – R445)/(R800 – R680)	(B4 – B1)/(B4 – B3)	Penuelas et al. (1995)
Simple ratio pigment index	SRPI	Chlorophyll; general stress	R430/R680	B1/B3	Penuelas et al. (1993)
Triangular vegetation index	TVI	Chlorophyll	0.5 * (120 * (Ravg760to800 – Rcrvg530to570) – (200 * (Ravg650to680 – Ravg530to570))	0.5 * (120 * (B4 – B2) – 200 * (B3 – B2))	Broge and Leblanc (2001)
	Vog$_A$	Chlorophyll	R740/R720	X	Vogelmann et al. (1993)
	Vog$_B$	Chlorophyll	FD715/FD705	X	Vogelmann et al. (1993)
Water band index	WBI	Canopy water content	R900/R970		Penuelas et al. (1993)

Note that this is not all-inclusive but intended to demonstrate the wide range of algorithms available and how these have been linked to specific leaf pigment, canopy structure, or condition metrics

Where no name is specified for a vegetation index by the author, we assign an abbreviation based on index function or author initials. *R* indicates reflectance values from narrow spectral bands (~10 nm). *FD* indicates a spectral first derivative centered on the specified wavelength, *X* indicates that a broadband equivalent does not exist for sensors such as Landsat 7 ETM.

Aerial Survey) ITC Satellite and Sensor Database: https://webapps.itc.utwente.nl/sensor/default.aspx?view=allsensors

One particularly promising sensor for improved forest health detection includes Sentinel 2 (A and B), recently launched by the European Space Agency. This is the first civil Earth observation sensor to include three bands in the red edge, providing additional information to quantify vegetation condition. Its 5-day repeat time and 10 m pixels also improve its ability to detect more subtle decline symptoms. This temporal resolution has proven useful in identifying forest decline based on detecting changes in the spectra of declining trees relative to healthy ones over time (Zarco-Tejada et al. 2018). Geostationary sensors like the GOES-R series also provide a unique opportunity to monitor forest condition at rapid time intervals across large landscapes. With two visible and four infrared bands useful to inform vegetation condition, the Advanced Baseline Imager on GOES-16 can provide images every 5 minutes with a spatial resolution of 0.5–2 km.

Improvements in computing technologies and modeling techniques have also increased the utility of multispectral sensors in early vegetation decline detection (Lausch et al. 2017). For example, Pontius (2014) demonstrated that using a multi-temporal approach mimicking hyperspectral algorithms could successfully quantify a detailed decline scale using Landsat TM data. Over time, ongoing improvements in sensor resolution, computing capabilities, and modeling options will enable measurements of more subtle changes in reflectance associated with early decline detection.

Hyperspectral Sensors While multispectral sensors record electromagnetic radiation averaged over a broad "band" of wavelengths, a hyperspectral instrument records many adjacent narrow bands to image most of the spectrum within a set range. What makes these instruments so useful for vegetation assessment extends beyond the simple availability of *more* bands to work with. Typically, these bands record reflectance from much narrower regions of the electromagnetic spectrum. This narrowband design provides two key modeling capabilities that are not possible with broadband sensors: (1) narrow bands are able to target specific absorption features linked to specific physiological structures or processes that we can directly relate to plant stress response and (2) narrow, contiguous bands allow us to consider the overall shape of spectral signatures, including mathematical techniques (e.g., derivatives, area under the curve, slope of the line between key regions) that are not possible with broadband data.

Building off of the science of spectroscopy (the study of constituents and materials using specific wavelengths), RS analysts have used hyperspectral imagery to quantify specific vegetation constituents and processes. The best hyperspectral narrow bands to study vegetation are in the 400–2500 nm spectral range (Thenkabail et al. 2013; Fig. 6.8), enabling direct links to species composition, foliar chemistry, foliar function, and ecosystem characteristics (Smith et al. 2002; Williams and Hunt 2002; Kokaly et al. 2003; Asner and Heidebrecht 2003; Townsend et al. 2003; Carter et al. 2005; Cheng et al. 2006; Singh et al. 2015).

While it is generally believed that spectral changes in stressed vegetation are common across stress agents, the ability of hyperspectral sensors to target specific chemical, physiological, and morphological traits allows RS analysts to target and assess specific, early symptoms of decline and target detection efforts based on known physiological responses to a particular pest or pathogen. Lausch et al. (2013) targeted changes in chlorophyll absorption as an indicator of bark beetle-induced decline; Pontius et al. (2008) targeted chlorophyll fluoresce to map the invasive emerald ash borer (Pontius et al. 2008) and canopy density for detailed monitoring impacts of hemlock woolly adelgid (Pontius et al. 2005b).

Hyperspectral imagery has historically been limited in availability. NASA's Airborne Visible/Infrared Spectrometer (AVIRIS; Porter and Enmark 1987) hyperspectral sensor was the pioneer of airborne applications. But the launch of the NASA Hyperion Instrument (Pearlman et al. 2003) on the EO-1 satellite in 2000, and the addition of commercial vendors with aerial hyperspectral platforms (e.g., ITRES http://www.itres.com/; SPECIM http://www.specim.fi/hyperspectral-RS/), has increased the availability of hyperspectral imagery. The promise of new hyperspectral satellites such as the Environmental Mapping and Analysis Program (EnMAP http://www.enmap.org/mission.html) suggests there is potential for expanding applications in forest health monitoring and assessment. Recent examples include assessments of hemlock woolly adelgid-induced decline in the Catskills region of New York (Hanavan et al. 2015) and detection of drought-induced decline in the chaparral ecosystems of California (Coates et al. 2015). Fused hyperspectral and LiDAR imagery have also enabled the assessment of early decline at the canopy level in urban environments (e.g., Degerickx et al. 2018; Pontius et al. 2017).

6.4   Spectroscopy of Early Decline Detection

While different species have unique spectral signatures, there are similar changes in general spectral characteristics in response to stress (Buschmann and Nagel 1993). Many of these spectral features can be directly linked to the stress symptoms and physiological characteristics described above (Fig. 6.8). For example, changes in leaf chemistry and physiology are captured in the 480–520 nm (blue) and 600–680 nm (red) regions, where chlorophyll absorption is strong. But changes in this region are relatively small compared with the dramatic changes that can be seen with stress between 750 and 1300 nm. The sharp rise in reflectance between the red and NIR regions (red edge inflection point) can be used to quantify changes in both the slope of the spectral signature and the location of the inflection point of the slope in response to changes in leaf chemistry and canopy density. Spectral information at longer wavelengths (1650–2200 shortwave infrared) has also been useful in quantifying changes in leaf water content, often a key signal of early vegetation stress.

Often the most useful information about general canopy condition, density, and function is derived from combining bands from various regions in mathematical expressions referred to as vegetation indices (Elvidge and Chen 1995; Pinty et al. 1993). Sometimes these indices incorporate information from multiple wavelengths with known absorption features. But other times a nonresponsive "control" band

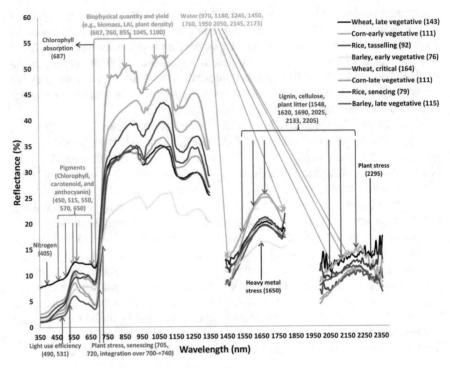

Fig. 6.8 Hyperspectral RS of vegetation condition is possible because of a suite of absorption and reflectance features across the visible and NIR spectra. (Credit: USGS by P. Thenkabail)

may be used to help account for differences in reflectance due to illumination or topography. Many vegetation indices have been designed for use with specific broadband sensors to assess general canopy characteristics such as relative "greenness," canopy density, or canopy condition (Table 6.1). But because of contributions in the field of spectroscopy, there is a wealth of literature that highlights specific regions of the electromagnetic spectrum (EMS) that are specifically associated with foliar chemistry, chlorophyll or carotenoid content, various metrics of photosynthetic activity, and other common stress markers (see Serbin et al. 2014, 2015; Singh et al. 2015).

Some of the vegetation indices listed in Table 6.1 are easily captured with widely available sensors. Others require reflectance information from narrow spectral regions that may only be accurately measured with hyperspectral sensors. Others may be located in regions that are outside of the EMS range of the imagery that is available. Thus, the number of available indices will depend on the imagery you have. Which index will prove most useful in detecting early canopy stress depends on the specific stress symptoms and the conditions of your study area. For example, in ecosystems with relatively sparse vegetation, a soil-adjusted vegetation index may work best to minimize the impact of background reflectance. Similarly, in ecosystems with very dense vegetation, you may need to select an index that does not saturate at high biomass levels. In most cases, you won't know which index, or set of indices and wavelengths, is best to use until you examine them as a part of your analyses. The best way to identify useful vegetation indices is detailed in the next section.

6.5   Techniques for Early Stress Detection

While mapping severe or widespread forest decline can be relatively straightforward using simple vegetation indices, it can be much more challenging to identify early or small-scale decline, particularly in mixed forests. For example, an insect outbreak may cause severe decline symptoms in the host tree species, but this signal may be washed out in a heterogeneous forest where reflectance from the larger canopy of other species dominates. Similarly, tree mortality is often accompanied by the release and ingrowth of understory vegetation. This can make detection of decline difficult as increased vegetation density from the understory masks the reduction in vegetation density in the upper canopy. Further, different species inherently have different chemical and structural characteristics, resulting in sometimes starkly different spectral signatures, even among healthy canopies. A healthy oak may be spectrally similar to a declining sugar maple. This underscores the importance of knowing the distribution of species across a landscape of interest and the characteristics of a "healthy" vs. "declining" spectral signature for a target forest type.

Because the identification of subtle stress characteristics relies on subtle changes in spectral characteristics, RS of early decline is very sensitive to anything that might alter spectral signatures. For example, an algorithm designed for early stress detection with one instrument may not be appropriate to apply to imagery from a different sensor. Even with a similar spectral, radiometric, and spatial configuration, differences in calibration may introduce differences that have nothing to do with the health of the canopy. Even when using the same instrument, atmospheric or illumination conditions may vary over time. For these reasons, it is important to calibrate each image to the specific conditions (atmospheric, illumination, canopy condition) at the time of acquisition.

There are several methodological approaches that can help to isolate and quantify decline symptoms, regardless of the sensor system (Pontius and Hallett 2014). Here we summarize the key components to identifying and quantifying early vegetation stress:

1. *Know the spectral characteristics of your baseline ecosystem.* While all vegetation has a common spectral curve, there are distinct differences in the spectral signature across different species and at different spatial resolutions. Because of inherent differences in foliar chemistry and canopy structure, a sugar maple has a spectral signature that is distinct from an eastern hemlock, even when both are in optimal health. Because of the spectral contribution from surrounding surface features, a healthy sugar maple in a heterogeneous forest will look different from

a healthy sugar maple grown in someone's front yard. Thus, it is important to know what the spectral signature for a pixel of your target ecosystem would look like in optimal condition.

There are many spectral libraries where "typical" spectra for a range of surface features can be downloaded and used for image calibration (e.g., ECOSTRESS Spectral Library https://speclib.jpl.nasa.gov/documents/jhu_desc or US Geological Survey (USGS) Spectral Library https://crustal.usgs.gov/speclab/QueryAll07a.php?quick_filter=vegetation). However, because of inherent differences between sensors, as well as atmospheric and illumination conditions at the time of image acquisition, it is best to also collect field spectra or identify homogeneous calibration pixels from across the imagery. Linking field data directly to the pixels will provide a spectral signature that is specific to the imagery you are using and ecosystem you are working in. This will serve as an important baseline and provide essential calibration data to model the species and stress condition of interest.

2. *Identify, quantify, and gather calibration data for the specific stress symptoms you expect to see.* While there are many common stress responses across vegetation types and stress agents, many symptoms can be species- or stress-specific. Of these, only some may be visible to the human eye. This is why it is important to identify the common stress symptoms you expect to see, from the earliest symptoms to the most obvious and severe decline, and design field data collection efforts that quantify each of those stress symptoms. Field calibration data should include measurements from locations across the imagery and cover the full range for each of these metrics that you would expect to manifest in the system you are studying and that you hope to quantify in your final product. These field data will provide valuable information as you analyze your imagery and model decline conditions across your study area.

For example, hemlock woolly adelgid feed on photosynthate stored within hemlock twigs, limiting the ability of trees to put on new growth. This may not be visible in a broad assessment of canopy vigor, but can be quantified in the field by collecting multiple branches from across the canopy and assessing the proportion of terminal branchlets that have put on new growth. This serves as a relatively quick and low-tech way to quantify foliar productivity and the reductions in new growth that are often the first sign of infestation. Similarly, the most obvious visible sign of emerald ash borer infestation in ash trees is often scarified bark that results from increased woodpecker activity. Woodpeckers strip bark as they feed on larvae, leaving obvious white markings. These telltale signs of early infestation can serve as a proxy for subtle biophysical changes in the canopy that are not yet visible to field crews.

Most often, decline manifests as many different concurrent stress symptoms (e.g., chlorosis and defoliation and dieback in various parts of the canopy) or a progression of decline symptoms that vary with the degree of impact (e.g., early decline manifests as chlorosis, later stages dominated by reductions in the live crown ratio, and ultimately mortality). In such cases, you may choose to develop an aggregate "field health" index that mathematically normalizes a suite of stress metrics into one summary metric (Pontius and Hallett 2014). This may be easier and more efficient than creating models to assess each of the various decline symptoms you expect to see in your target system or having to pick one decline metric to use.

3. *Calibrate imagery with field data.* In an ideal world, we would be able to develop one model that could be automated and applied to imagery over time and space regardless of sensor, acquisition condition, or location. Several automated RS tools currently available (see Sect. 6) have proven incredibly useful for monitoring large areas over time. But automated applications are limited in their ability to detect subtle, early decline, which requires careful calibration between the imagery acquired and ground conditions at the time that imagery was collected to make it possible to identify the targeted stress response while controlling for other sources of spectral variability. Ideally, field calibration data can be collected within several weeks of imagery acquisition (or at least before conditions on the ground change). GPS locations of field calibration sites link field data to the spectra of the associated pixel or pixels to calibrate the larger image.

Various proprietary software modules exist for spectral calibration, modeling, and analyses. These modules can range from simple classification techniques that match pixels to various stages of decline based on your calibration spectra, to more complex spectral unmixing algorithms that approximate the proportion of "healthy to declining" spectra contained within each pixel. Even without specialized RS software, simple statistics can be used to quantify relationships between spectral reflectance and derived vegetation indices using field calibration data. A common approach is to use correlations between individual vegetation indices and decline metrics to qualitatively assess canopy condition across the landscape. Another approach uses multivariate statistical models to identify the best combination of bands or vegetation indices to quantify the decline metric of interest. Regardless of the mathematical approach, accuracy and detail are ultimately determined by the quality and range field calibration data available for model development. This type of targeted calibration to match the timing, location, and sensor characteristics for each decline assessment maximizes accuracy and detail of the final products.

4. *Validate and assess accuracy to inform interpretation.* One of the dangers inherent in linking RS products with management applications is overconfidence in the RS products. There is error inherent in each component of the RS process, from incorrect sensor calibration, to the variability introduced by atmospheric, topographic, and georegistration errors. However, when presented with a RS product, many end users develop their plans without consideration of how accurate the product may be or how inaccuracies can be avoided.

Any RS product should include some measure of accuracy as well as any caveats that should be considered in its use. In some cases (e.g., the use of a vegetation index to qualitatively describe relative states of decline), it is sufficient to remind users that the scale presented is intended to be relative and does not necessarily identify stands in specific states of decline or resulting from specific stress agents. In other cases (e.g., the classification of pixels into levels of decline), we can use field data to present an accuracy assessment. Any accuracy assessment of classified image products should include overall accuracy as well as users' accuracy (percent of target pixels correctly classified; inverse = errors of omission) and producers' accuracy (percent of nontarget pixels that are *not* classified as the target class; inverse = errors of commission; Congalton 2001; Fassnacht et al. 2006). Splitting accuracy into users' and producers' values allows the end user to understand how false positives (saying a stand is dead when it is not) and false negatives (saying a stand is healthy when it is dead) can influence how the end product is used to inform management activities. For example, if overall accuracy in classifying forest mortality is 70% but almost all of the error results from false positives (many stands classified as dead when they are actually alive), end users may decide to limit management to locations with large clusters of predicted mortality or to clusters in higher-decline categories in order to avoid these common errors.

RS decline-detection products that result in ordinal classes of decline (e.g., healthy, degrees of decline, dead) can also be assessed for "fuzzy accuracy," which considers not only correct class assignments but also those within one ordinal class of the correct class. Products that provide a continuous decline metric can be used to produce more detailed accuracy metrics. Standard statistical regression techniques produce a coefficient of determination (r^2) to describe how well a statistical model fits the relationship between the input spectral variables and the output decline metric. Root mean square error, standard errors, and prediction errors can be used to place confidence bounds on predicted values. We can also examine how accuracy changes across the range of decline values predicted. For example, some models may be very good at quantifying severe decline but may not be able to detect early decline symptoms. Some models may overpredict early decline but underpredict severe decline. Standard statistical methods can be useful to examine how well your model works, which is critical to ensure that end users know how to best integrate your resulting RS products into their decision-making process.

A Nested Approach No one sensor, field methodology, or scale is appropriate for all applications. Different goals may require that you work at different scales (Fig. 6.9). The most detailed and accurate information about specific stress agents and response symptoms will always be obtained from on-the-ground field surveys (Tier 1). Such location-specific studies allow researchers to directly measure foliar chemistry, canopy structure, and spectral characteristics in situ. But these studies are limited in their utility to inform management across the broader landscape. Aerial sensors are often used to collect RS imagery at the local scale (Tier 2). Typically, this scale allows for the use of high spatial and spectral resolution imagery, ideally suited to detect forest stress conditions. However, such efforts may still be limited in geographic extent due to the high cost and computing needs. Most common is the use of broadband sensors at the regional-continental scale (Tier 3). Landsat sensors have been widely used for such applications, with sufficient spatial (30 m) and spectral resolution to prove useful in assessment of

Fig. 6.9 RS work occurs at a variety of scales, with benefits and limitations at each level. Sometimes the best approach includes nesting your analyses across multiple scales to gain a comprehensive understanding of the forest health dynamics on the ground

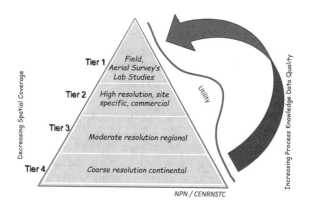

relative levels of forest decline. The recent addition of improved satellite sensors (e.g., Sentinel 2) is rapidly increasing the capability to cover broad landscapes at higher spatial resolutions. Global assessments (Tier 4) of forest condition typically require a reduction in spatial resolution in order to process information over vast geographic extents. The much larger, mixed pixels often mask subtle changes in vegetation condition but can be useful in time series analyses when focused on relative changes in vegetation indices on continental scales.

The best approach to mapping and modeling forest decline depends on the scale of the investigation, level of detail needed, resources available, and time frame. For example, a regional assessment may have to forgo spatial and spectral resolution (and predictive detail and accuracy) in order to achieve the spatial coverage desired. In contrast, a municipality concerned about the spread of a recently detected invasive insect pest may forgo widespread spatial coverage to maximize the spatial and spectral resolution necessary to identify individual, newly infested trees. Sometimes you are limited by what is available in terms of imagery, time, or financial resources. For example, it is impossible to go back in time to collect high-resolution imagery, but you may be able to make use of historical broadband satellite imagery for a general assessment of past conditions. In most cases RS products, even when not exactly matched to the user's needs, can still provide insight that is not available through traditional monitoring.

Perhaps the most comprehensive approach to detecting novel forest health issues is to combine approaches. For example, a broad landscape assessment can be useful to identify localized areas for more detailed image acquisition. Even better, examining the relationship between spectral characteristics from higher-resolution imagery could be used to train coarser resolution imagery for a larger-scale assessment. The key is to recognize that there is no *one* right approach and that perhaps there are several RS approaches that can be used to achieve your objectives.

Stakeholder Engagement For each of the steps suggested above, stakeholder engagement is critical to success. RS specialists typically are not experts in entomology, invasive species, tree physiology, or forest ecology, and may not be aware of the specific stress symptoms to target for a given application. Because we work in various locations, we are rarely experts on the ecological specifics of a new study area. Where can we find target species or stands in various stages of decline? What key landscape features or characteristics should be covered in our calibration to best inform management? We also may not be sure how the products we develop could be most useful to land managers and practitioners. Would a classification product be most useful, with simple "healthy vs. dead" groupings, or would a range of decline condition be better? Do we need to develop a species map first to better target the declining stands end users hope to find? Are they looking for potential healthy "refugia" areas for conservation, newly declining stands for intervention, or high-mortality stands for salvage? Knowing what they need will allow us to design our modeling outputs to best suit their needs.

To maximize the impact of the products you develop, we suggest engaging a range of stakeholders throughout the entire process, for example:

- Go beyond simply obtaining letters of support to include end users and other key stakeholders in proposal development and experimental design from the outset of a new project.
- Find practitioners in your study area to identify and visit potential field sites.
- Present at local and regional meetings with the specific intent to introduce the project and solicit feedback on product format and delivery (prior to obtaining results).
- Include stakeholders in fieldwork, training them in field methodologies and learning from their expertise. Creating a sense of ownership or investment in a project improves the chances that your final products will actually be used.
- Meet with potential users as products are developed to gauge if the format (metric scale/range, spatial resolution, file format, etc.) are useful and, if not, how you might modify products to meet their needs.
- In addition to presenting your results at scientific meetings, target professional meetings and workshops to reach end users.
- Make your data products easily discoverable and available. This may include posting final products in online databases or web portals. Be sure the format is not limiting. Google Earth provides a useful platform for users without ARC or other proprietary geocomputing resources.

Including stakeholders in this way not only helps maximize the utility and impact of your efforts but also builds bridges between scientific and management communities. Historically, there have been limited collaborations among land managers, practitioners, decision-makers, and the RS scientific community. In some cases, there has even been mistrust as products are promised but delivered on a scientific timeline rather than a management timeline. But there has been a recent push to include stakeholders in RS and modeling efforts, exemplified by the recent "Voices from the Land" project led by researchers at Harvard Forest (McBride et al. 2017). This stakeholder-driven approach used interviews with New Englanders to identify key outcomes and likely scenarios for modeling. Such steps can build relationships that can serve all communities interested in sustaining forested ecosystems.

6.6 Using RS to Inform Forest Management

The application of RS for vegetation stress detection has advanced rapidly, evolving from classical aerial survey and photointerpretation techniques to digital image processing, where manual interpretation has been replaced with machine learning to identify subtle signatures humans are incapable of seeing with the naked eye. This technological evolution has effectively transferred these tools to the sustainable management of forest resources, but limitations remain in their widespread use. Monitoring, detecting, and reporting on forest health threats has always been a priority of federal and state forestry agencies. Conversion of forest land and changes in land use; climate change, intensified storms, higher frequency and intensity of forest fires and concerns of host range recession; and the threat of introduction and establishment from invasive insects and diseases have created an even more urgent demand for improved *near-real-time* tools and products. The capabilities of most sensors and the applications on which they have been tested are impressive, and more promising techniques and approaches continue to build on field application.

Recently, several programs have been developed with the goal of advancing and improving RS applications for forest management, including online tools developed to bring RS products to the forest health management community in near real time. Here we present some examples of online resources developed to transfer RS products to end users on time scales useful to inform management and planning.

World Vegetation Health Index https://www.star.nesdis.noaa.gov/smcd/emb/vci/VH/vh_browse.php The National Oceanic and Atmospheric Administration (NOAA)-National Environmental Satellite, Data, and Information Service (NESDIS) has developed several RS products designed specifically to assess vegetation health across the globe. Their Center for Satellite Applications and Research (STAR) Vegetation Health Index (Fig. 6.10) uses Advanced Very High-Resolution Radiometer (AVHRR) imagery produced from the NOAA/NESDIS Global Area Coverage (GAC) data set from 1981 to the present, with 4 km spatial and 7-day composite temporal resolution. Common vegetation indices are used to estimate vegetation health, moisture, and temperature and serve as a proxy to monitor vegetation cover, density, productivity, and drought conditions, as well as phenological stages such as the start/end of the growing season. Outputs are scaled to a range (0 to 100), providing a relative assessment of vegetation condition rather than a prediction of actual decline symptoms or identification of stress agents. However, these products are useful for examining short-term changes in vegetation that can be used to identify widespread decline events such as drought, land degradation, or fire.

ForWarn Online Mapper http://forwarn.forestthreats.org/; https://forwarn.forest-threats.org/fcav2/ ForWarn Satellite-Based Change Recognition and Tracking (Fig. 6.11) is a near-real-time product from the US Forest Service that uses 250 m MODIS data to compare current NDVI to seasonally similar historic NDVI values to identify disturbance such as wildfires, windstorms, insects, disease outbreaks,

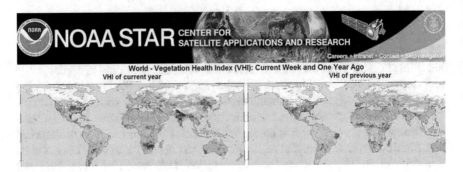

Fig. 6.10 The NOAA STAR World Vegetation Health Index visualization and data download portal

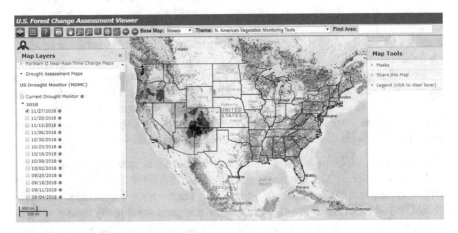

Fig. 6.11 The USFS ForWarn II online mapping portal provides weekly vegetation change and identification products dating back to 2003

logging, and land use change (Norman et al. 2013). Recent improvements in historical NDVI baseline data now provide the end user more tools to diagnose the severity and cause of changes in the mapping products.

Forest Disturbance Monitor (FDM) and Operational Remote Sensing (ORS) https://foresthealth.fs.usda.gov/FDM; http://foresthealth.fs.usda.gov/portal The US Forest Service Forest Disturbance Monitor (FDM; Fig. 6.12) is a forest disturbance web portal based on 16-day and 24-day MODIS composites that are updated every 8 days. FDM produces two forest disturbance products, 3-year Real-Time Forest Disturbance (RTFD) data and 5-year Trend Disturbance Data (TDD), providing near-real-time forest disturbance maps for land managers to target forest insect and disease events and complement aerial sketch mapping annual insect and disease surveys (IDSs; Chastain et al. 2015).

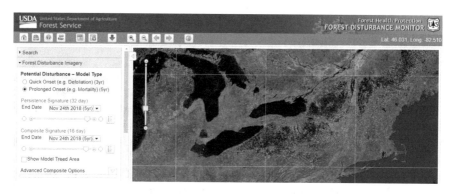

Fig. 6.12 The USFS Forest Disturbance Monitor online portal

To improve insect and disease surveys and facilitate the use of forest health information that RS products can provide, the USFS has recently initiated the Operational Remote Sensing (ORS) program. Similar to the FDM, ORS will use a phenology-based approach to intensifying surveys using 30 m Landsat and other moderate-resolution data.

Ecosystem Disturbance and Recovery Tracker (eDaRT) http://www.cstarsd3s. ucdavis.edu/systems/edart/ A collaboration among the University of California, Davis, Center for Southeastern Tropical Advanced Remote Sensing (CSTARS), and the US Forest Service, the Ecosystem Disturbance and Recovery Tracker (eDaRT; Koltunov et al. 2015) is an automated system that provides a suite of Landsat-derived products to identify and categorize changes in forest, shrubland, and herbaceous ecosystems. Currently, eDaRT products are not publicly available, but recent efforts are focused on expanding operations by the US Forest Service in California and elsewhere in the western United States in support of daily ecosystem management tasks.

Looking Ahead Because of the vast potential for RS to inform the sustainable management of terrestrial landscapes, there are several new Earth observation missions on the horizon. The European Space Agency (ESA) will launch Earth Explorer 7 in 2021 (https://www.esa.int/Our_Activities/Observing_the_Earth). This ecology mission, known as Biomass, is designed to characterize forests. The Biomass mission will be followed by the Earth Explorer 8 Fluorescence Explorer (FLEX) mission in 2022, with capabilities to quantify chlorophyll fluorescence in terrestrial vegetation. Landsat 9, part of the Earth observation data continuity mission from NASA (fast-tracked for December 2020 launch date), will maintain nearly 50 years of continuous Earth observation. This instrument is designed to simultaneously image 11 spectral bands, including a 15 m panchromatic band, with 12 bit radiometric resolution to increase sensitivity to small differences in reflectance. Such advances are critical to the early stress detection and detailed decline assessment that land managers need.

6.7   Management Applications: Limitations
and Opportunities

Thanks to continuing advances in computing and software technologies, we are poised to bring near-real-time RS products to more stakeholders. Applications like Google Earth Engine (https://earthengine.google.com/) now have the ability to automate image acquisition, preprocessing, and more complex modeling algorithms to provide critical forest health information across large landscapes at regular time intervals. Similarly, the ESA's Grid Processing on Demand (G-POD) provides an online environment where scientists can build and automate RS applications (https://gpod.eo.esa.int/). While several organizations (see ForWarn, FDM, and eDaRT above) are making final products from this type of rapid analysis and assessment operational for coarse forest health assessments and disturbance mapping efforts, higher-level products (higher spatial resolution, low-level stress detection) are not yet publicly available for use by broad stakeholder groups.

Currently, most RS efforts to detect incipient stress factors or detailed vegetation condition are conducted by the research community with scientific journals as their primary outputs. The more widespread use of more advanced RS techniques in forest management is primarily limited by:

- *The cost of image acquisition and expertise required to accurately calibrate sensors and validate products.* This is particularly true for hyperspectral efforts, which generate large amounts of data and require specialized expertise for pre-processing corrections, calibration, and data management. Computing advances and the growing commercial sector promise improved access, but for many land managers, cost is still a strong deterrent. Some organizations are hoping to make cutting-edge imagery more accessible. For example, NASA's Goddard's LiDAR, Hyperspectral, and Thermal Imager (G-LiHT) (https://gliht.gsfc.nasa.gov/) is a portable, airborne imaging system that simultaneously maps composition, structure, and function of terrestrial ecosystems using multispectral LiDARs (3-D information about the vertical and horizontal distribution of foliage and other canopy elements), hyperspectral imaging spectrometer to discern species composition and variations in biophysical variables (photosynthetic pigments and nutrient and water content), and a thermal camera to measure surface temperatures to detect heat and moisture stress (Cook et al. 2013). Owned and operated by NASA Goddard, this instrument has proven to be more affordable and accessible than comparable commercial vendors and may greatly expand access to cutting-edge sensor technologies for a variety of applications (Fig. 6.13).
- *The turnaround time required to deliver final mapping products.* Typically, the more irruptive forest health issues require immediate attention in the current growing season (e.g., pest outbreaks, extreme climate events, wildfires), while turnaround from RS projects doesn't always occur in the same year. This disparity between product delivery and product need is especially evident in studies

Fig. 6.13 NASA's G-LiHT online data portal

where method development is necessary and limits the adoption of more advanced RS efforts by the forest management community. However, the increased use of automated image processing scripts that make satellite image products available in near real time is expanding the use of traditional (vegetation index-based) relative assessments available for a variety of applications. The resulting online tools described above are being adopted by a range of state and federal agencies to inform management decisions.

- *Integration of mapping products into decision-making processes.* Even when RS products are available, there is no clear path on how to *use* the information they provide to inform decision-making. Land managers may reference mapping products to target specific locations, but more complete integration of spatial products into management plans can be challenging for those not used to working with spatial data. Foresters are typically trained in making decisions based on generalized inventories of forest stands or management units, not pixelated rasters across a landscape with a high degree of variability. End users may not be aware that mapping products should come with an accuracy assessment that informs how the information can best be used and how it impacts the overall confidence in the product. Many of these limitations can be resolved by scientists working more closely with end users as outlined in the Stakeholder Engagement section above. By working together, both scientists and land managers can learn from each other and so better use RS technologies to manage critical environmental resources.

6.8   Conclusions

While historically RS has been successfully used to assess and monitor vegetation condition on a coarse, relative scale, recent advances and new analysis techniques now enable us to also use RS to identify and track early decline, disturbance, and stress conditions in vegetative systems. Considering the environmental challenges currently facing terrestrial systems, this information is critical to inform management, policy, and planning in order to maintain the structure and function of these systems.

The challenge is for scientists to look beyond traditional approaches to vegetation assessment and target earlier or more subtle decline response resulting from incipient or chronic environmental stress agents (e.g., climate change, pollution). Key challenges include linking hyperspectral data to specific stress agents, extending the availability of higher-resolution imagery, and operationalizing near-real-time monitoring of the forest resource (Senf et al. 2017). Scientists must work closely with land managers to bring these new technologies to application in order to harness RS's full potential to inform the management of critical ecological resources.

References

Aoki M, Yabuki K, Totsuka T (1981) An evaluation of chlorophyll content of leaves based on the spectral reflectivity in several plants. Res Rep Nat Inst Environ Stud Jpn 66:125–130

Asner GP, Heidebrecht KB (2003) Imaging spectroscopy for desertification studies: comparing AVIRIS and EO-1 Hyperion in Argentina drylands. IEEE Trans Geosci Remote Sens 41(6):1283–1296

Barnes JD (1992) A reappraisal of the use of DMSO for the extraction and determination of chlorophylls a and b in lichens and higher plants. Environ Experim Bot 32(2):85–100

Blackburn GA (1998) Quantifying chlorophylls and carotenoids at leaf and canopy scales: an evaluation of some hyperspectral approaches. Remote Sens Environ 66(3):273–285

Blackburn GA (1999) Relationships between spectral reflectance and pigment concentrations in stacks of deciduous broadleaves. Remote Sens Environ 70:224–237

Boochs F, Kupfer G, Dockter K, Kuhbauh W (1990) Shape of the red edge as vitality indicator for plants. International Journal of Remote Sensing 11(10):1741–1753

Broge NH, Leblanc E (2001) Comparing prediction power and stability of broadband and hyperspectral vegetation indices for estimation of green leaf area index and canopy chlorophyll density. Remote Sens Environ 76(2):156–172

Buschmann C, Nagel E (1993) In vivo spectroscopy and internal optics of leaves as a basis for remote sensing of vegetation. Int J Remote Sens 14:711–722

Carter GA (1993) Responses of leaf spectral reflectance to plant stress. Am J Bot 80(3):239–243

Carter GA (1994) Ratios of leaf reflectances in narrow wavebands as indicators of plant stress. Int J Remote Sens 15(3):697–703

Carter GA, Knapp AK (2001) Leaf optical properties in higher plants: linking spectral characteristics to stress and chlorophyll concentration. Am J Bot 88(4):677–684

Carter GA, Miller RL (1994) Early detection of plant stress by digital imaging within narrow stress-sensitive wavebands. Remote Sens Environ 50(3):295–302

Carter GA, Knapp AK, Anderson JE, Hoch GA, Smith MD (2005) Indicators of plant species richness in AVIRIS spectra of a mesic grassland. Remote Sens Environ 98:304–316

Chappelle EW, Kim MS, McMurtrey JE III (1992) Ratio analysis of reflectance spectra (RARS): an algorithm for the remote estimation of the concentrations of chlorophyll a, chlorophyll b, and carotenoids in soybean leaves. Remote Sens Environ 39(3):239–247

Chastain RA, Fisk H, Ellenwood JR, Sapio FJ, Ruefenacht B, Finco MV, Thomas V (2015) Near-real time delivery of MODIS-based information on forest disturbances. In: Time-sensitive remote sensing. Springer, New York, pp 147–164

Chen J (1996) Evaluation of vegetation indices and a modified simple ratio for boreal applications. Can J Remote Sens 22(3):229–242

Cheng Y-B, Zarco-Tejada PJ, Riano D, Rueda CA, Ustin SL (2006) Estimating vegetation water content with hyperspectral data for different canopy scenarios: relationships between AVIRIS and MODIS indexes. Remote Sens Environ 105(2006):354–366

Coates AR, Dennison PE, Roberts DA, Roth KL (2015) Monitoring the impacts of severe drought on southern California chaparral species using hyperspectral and thermal infrared imagery. Remote Sens 7(11):14276–14291

Congalton RG (2001) Accuracy assessment and validation of remotely sensed and other spatial information. Int J Wildland Fire 10(3–4):321–328

Cook BD, Corp LW, Nelson RF, Middleton EM, Morton DC, McCorkel JT, Masek JG, Ranson KJ, Ly V, Montesano PM (2013) NASA Goddard's Lidar, Hyperspectral and Thermal (G-LiHT) airborne imager. Remote Sens Environ 5:4045–4066

D'ambrosio N, Szabo K, Lichtenthaler H (1992) Increase of the chlorophyll fluorescence ratio F690/F735 during the autumnal chlorophyll breakdown. Radiat Environ Biophys 31(1):51–62

Datt B (1998) Remote sensing of chlorophyll a, chlorophyll b, chlorophyll a + b, and total carotenoid content in eucalyptus leaves. Remote Sens Environ 66:111–121

Datt B (1999) Visible/near infrared reflectance and chlorophyll content in Eucalyptus leaves. Int J Remote Sens 20(14):2741–2759

Daughtry CST, Walthall CL, Kim MS, de Colstoun EB, McMurtrey JE (2000) Estimating corn leaf chlorophyll concentration from leaf and canopy reflectance. Remote Sens Environ 74(2):229–239. https://doi.org/10.1016/s0034-4257(00)00113-9

Deblonde G, Cihlar J (1993) A multiyear analysis of the relationship between surface environmental variables and NDVI over the Canadian landmass. Remote Sens Rev 7:151–177

Degerickx J, Roberts DA, McFadden JP, Hermy M, Somers B (2018) Urban tree health assessment using airborne hyperspectral and LiDAR imagery. Int J Appl Earth Obs Geoinf 73:26–38

Ellison AM, Barker-Plotkin AA, Foster DR, Orwig DA (2010) Experimentally testing the role of foundation species in forests: the Harvard Forest Hemlock Removal Experiment. Methods Ecol Evol 1(2):168–179. https://doi.org/10.1111/j.2041-210X.2010.00025.x

Elvidge CD, Chen Z (1995) Comparison of broad-band and narrow-band red and near-infrared vegetation indices. Remote Sens Environ 54(1):38–48

Elvidge CD, Lyon RJ (1985) Estimation of the vegetation contribution to the 1 65/2 22 μm ratio in airborne thematic-mapper imagery of the Virginia Range, Nevada. Int J Remote Sens 6(1):75–88

Epanchin-Niell RS, Hastings A (2010) Controlling established invaders: integrating economics and spread dynamics to determine optimal management. Ecol Lett 13(4):528–541

Fassnacht KS, Cohen WB, Spies TA (2006) Key issues in making and using satellite-based maps in ecology: a primer. For Ecol Manag 222:167–181

Filella I, Penuelas J (1994) The red edge position and shape as indicators of plant chlorophyll content, biomass, and hydric status. Int J Remote Sens 15(7):1459–1470

Gamon JA, Serrano L, Surfus JS (1997) The photochemical reflectance index: an optical indicator of photosynthetic radiation use efficiency across species, functional types, and nutrient levels. Oecologia 112(4):492–501

Gao BC (1996) Ndwi - a normalized difference water index for remote sensing of vegetation liquid water from space. Remote Sens Environ 58(3):257–266

Genovesi P, Carboneras C, Vila M, Walton P (2015) EU adopts innovative legislation on invasive species: a step towards a global response to biological invasions? Biol Invasions 17(5):1307–1311

Gitelson A, Merzlyak MN (1994) Quantitative estimation of chlorophyll-a using reflectance spectra: experiments with autumn chestnut and maple leaves. J Photochem Photobiol B Biol 22(3):247–252

Gitelson AA, Merzlyak MN (1996) Signature analysis of leaf reflectance spectra: algorithm development for remote sensing of chlorophyll. J Plant Physiol 148(3–4):494–500

Gitelson AA, Buschmann C, Lichtenthaler HK (1999) The chlorophyll fluorescence ratio F-735/F-700 as an accurate measure of the chlorophyll content in plants. Remote Sens Environ 69(3):296–302

Gitelson AA, Merzlyak MN, Chivkunova OB (2001) Optical properties and non-destructive estimation of anthocyanin content in plant leaves. Photochem Photobiol 74(1):38–45

Haboudane D, Miller JR, Pattey E, Zarco-Tejada PJ, Strachan IB (2004) Hyperspectral vegetation indices and novel algorithms for predicting green LAI of crop canopies: modeling and validation in the context of precision agriculture. Remote Sens Environ 90(3):337–352. https://doi.org/10.1016/j.rse.2003.12.013

Hanavan RP, Pontius J, Hallett R (2015) A 10-year assessment of hemlock decline in the Catskill Mountain region of New York State using hyperspectral remote sensing techniques. J Econ Entomol 108(1):339–349

Hardisky MA, Klemas V, Smart RM (1983) The influence of soil salinity, growth form, and leaf moisture on the spectral radiance of *Spartina alterniflora* canopies. Photogramm Eng Remote Sens 49(1):77–83

Harwood T, Tomlinson I, Potter C, Knight J (2011) Dutch elm disease revisited: past, present and future management in Great Britain. Plant Pathol 60(3):545–555

Hennon PE, D'Amore DV, Schaberg PG, Wittwer DT, Shanley CS (2012) Shifting climate, altered niche, and a dynamic conservation strategy for yellow-cedar in the North Pacific coastal rainforest. Bioscience 62(2):147–158

Horler D, DOCKRAY M, Barber J (1983) The red edge of plant leaf reflectance. Int J Remote Sens 4(2):273–288

Huete AR (1988) A soil adjusted vegetation index (SAVI). Remote Sens Environ 25(3):295–309. https://doi.org/10.1016/0034-4257(88)90106-x

Huete A, Justice C, Liu H (1994) Development of vegetation and soil indices for MODIS-EOS. Remote Sens Environ 49(3):224–234

Huete A, Didan K, Miura T, Rodriguez EP, Gao X, Ferreira LG (2002) Overview of the radiometric and biophysical performance of the MODIS vegetation indices. Remote Sensing of Environment 83(1–2):195–213

Huggett BA, Schaberg PG, Hawley GJ, Eagar C (2007) Long-term calcium addition increases growth release, wound closure, and health of sugar maple (Acer saccharum) trees at the Hubbard Brook Experimental Forest. Can J For Res 37(9):1692–1700

Hunt ER, Rock BN (1989) Detection of changes in leaf water content using near infrared and middle infrared reflectances. Remote Sens Environ 30(1):43–54. https://doi.org/10.1016/0034-4257(89)90046-1

Johnson EW, Ross J (2008) Quantifying error in aerial survey data. Aust For 71(3):216–222

Johnson E, Wittwer D (2008) Aerial detection surveys in the United States. Aust For 71(3):212–215

Jordan CF (1969) Derivation of leaf area index from quality of light on the forest floor. Ecology (Washington DC) 50(4):663–666. https://doi.org/10.2307/1936256

Kaufman YJ, Tanre D (1992) Atmospherically resistant vegetation index (ARVI) for EOS-MODIS. IEEE Trans Geosci Remote Sens 30(2):261–270

Kokaly RF, Despain DG, Clark RN, Livo KE (2003) Mapping vegetation in yellowstone national park using spectral feature analysis of AVIRIS data. Remote Sens Environ 84(3):437–456

Kolb T, Teulon D (1991) Relationship between sugar maple budburst phenology and pear thrips damage. Can J For Res 21(7):1043–1048

Koltunov A, Ramirez C, Ustin SL (2015) eDaRT: the ecosystem disturbance and recovery tracking system prototype supporting ecosystem management in California. NASA Carbon Cycle and Ecosystems Joint Science Workshop College Park, MD, April, 19–24

Kosiba AM, Meigs GW, Duncan J, Pontius J, Keeton WS, Tait E (2018) Spatiotemporal patterns of forest damage in the Northeastern United States: 2000–2016. Forest Ecology and Management 430:94–104.

Lausch A, Heurich M, Gordalla D, Dobner HJ, Gwillym-Margianto S, Salbach C (2013) Forecasting potential bark beetle outbreaks based on spruce forest vitality using hyperspectral remote-sensing techniques at different scales. For Ecol Manag 308:76–89

Lausch A, Erasmi S, King D, Magdon P, Heurich M (2017) Understanding forest health with remote sensing-part II—a review of approaches and data models. Remote Sens 9(2):129

Lillesand TM, Kiefer RW (1994) Remote sensing and photo interpretation, 3rd edn. John Wiley& Sons, New York

Maccioni A, Agati G, Mazzinghi P (2001) New vegetation indices for remote measurement of chlorophylls based on leaf directional reflectance spectra. J Photochem Photobiol 61(1,2):52–61

Martin K, Norris A, Drever M (2006) Effects of bark beetle outbreaks on avian biodiversity in the British Columbia interior: implications for critical habitat management. J Ecosyst Manag 7(3):10–24

McBride MF, Lambert KF, Huff ES, Theoharides KA, Field P, Thompson JR (2017) Increasing the effectiveness of participatory scenario development through codesign. Ecol Soc 22(3):16

McConnell TJ (1999) Aerial sketch mapping surveys, the past, present and future. In: Paper from the north American science symposium, toward a unified framework for inventorying and monitoring forest ecosystem resources, Guadalajara, Mexico

McMurtrey J III, Chappelle EW, Kim M, Meisinger J (1994) Distinguishing nitrogen fertilization levels in field corn (Zea mays L.) with actively induced fluorescence and passive reflectance measurements. Remote Sens Environ 47(1):36–44

Meroni M, Rossini M, Picchi V, Panigada C, Cogliati S, Nali C, Colombo R (2008) Assessing steady-state fluorescence and PRI from hyperspectral proximal sensing as early indicators of plant stress: the case of ozone exposure. Sensors 8(3):1740–1754. https://doi.org/10.3390/s8031740

Merzlyak MN, Gitelson AA, Chivkunova OB, Rakitin VY (1999) Non-destructive optical detection of pigment changes during leaf senescence and fruit ripening. Physiol Plant 106(1):135–141. https://doi.org/10.1034/j.1399-3054.1999.106119.x

Mohammed GH, Binder WD, Gillies SL (1995) Chlorophyll fluorescence - a review of its practical forestry applications and instrumentation. Scand J For Res 10(4):383–410

Mumford R (2017) New approaches for the early detection of tree health pests and pathogens. Impact 1(7):47–49

Myneni RB, Hall FG, Sellers PJ, Marshak AL (1995a) The interpretation of spectral vegetation indexes. IEEE Trans Geosci Remote Sens 33(2):481–486

Myneni RB, Maggion S, Iaquinta J, Privette JL, Gobron N, Pinty B, Kimes DS, Verstraete MM, Williams DL (1995b) Optical remote sensing of vegetation: modeling, caveats, and algorithms. Remote Sens Environ 51:169–188

Norman SP, Hargrove WW, Spruce JP, Christie WM, Schroeder SW (2013) Highlights of satellite-based forest change recognition and tracking using the ForWarn System. Gen Tech Rep SRS-GTR-180 Asheville, NC: USDA-Forest Service, Southern Research Station, 30 p, 180:1–30

Pearlman JS, Barry PS, Segal CC, Shepanski J, Beiso D, Carman SL (2003) Hyperion, a space-based imaging spectrometer. IEEE Trans Geosci Remote Sens 41:1160–1173

Pearson L, Miller LD (1972) Remote mapping of standing crop biomass for estimation of the productivity of the short-grass prairie, Pawnee National Grasslands, Colorado. In: Proceedings of the 8th international symposium on remote sensing of the environment, Ann Arbor, MI, 1972. ERIM, Ann Arbor, pp 1357–1381

Penuelas J, Filella I, Biel C, Serrano L, Save R (1993) The reflectance at the 950–970 nm region as an indicator of plant water status. Int J Remote Sens 14(10):1887–1905

Penuelas J, Gamon JA, Fredeen AL, Merino J, Field CB (1994) Reflectance indices associated with physiological changes in nitrogen-limited and water-limited sunflower leaves. Remote Sens Environ 48(2):135–146. https://doi.org/10.1016/0034-4257(94)90136-8

Penuelas J, Baret F, Filella I (1995) Semi-empirical indices to assess carotenoids/chlorophyll a ratio from leaf spectral reflectance. Photosynthetica 31:221–230

Pietrzykowski E, Stone C, Pinkard E, Mohammed C (2006) Effects of Mycosphaerella leaf disease on the spectral reflectance properties of juvenile Eucalyptus globulus foliage. For Pathog 36(5):334–348

Pinty B, Leprieur C, Verstraete MM (1993) Towards a quantitative interpretation of vegetation indices. Part I. Biophysical canopy properties and classical indices. Remote Sens Rev 7:127–150

Pontius J (2014) A new approach for forest decline assessments: maximizing detail and accuracy with multispectral imagery. Int J Remote Sens 35(9):3384–3402. https://doi.org/10.1080/014 31161.2014.903439

Pontius J, Hallett R (2014) Comprehensive methods for earlier detection and monitoring of forest decline. For Sci 60(6):1156–1163. https://doi.org/10.5849/forsci.13-121

Pontius J, Hallett R, Martin M (2005a) Assessing hemlock decline using visible and near-infrared spectroscopy: indices comparison and algorithm development. Appl Spectrosc 59(6):836–843. https://doi.org/10.1366/0003702054280595

Pontius J, Hallett R, Martin M (2005b) Using AVIRIS to assess hemlock abundance and early decline in the Catskills, New York. Remote Sens Environ 97(2):163–173. https://doi. org/10.1016/j.rse.2005.04.011

Pontius JA, Hallett RA, Jenkins JC (2006) Foliar chemistry linked to infestation and susceptibility to hemlock woolly adelgid (Homoptera: Adelgidae). Environ Entomol 35(1):112–120

Pontius J, Martin M, Plourde L, Hallett R (2008) Ash decline assessment in emerald ash borer-infested regions: a test of tree-level, hyperspectral technologies. Remote Sens Environ 112(5):2665–2676. https://doi.org/10.1016/j.rse.2007.12.011

Pontius J, Halman JM, Schaberg PG (2016) Seventy years of forest growth and community dynamics in an undisturbed northern hardwood forest. Can J For Res 46(7):959–967. https:// doi.org/10.1139/cjfr-2015-0304

Pontius J, Hanavan RP, Hallett RA, Cook BD, Corp LA (2017) High spatial resolution spectral unmixing for mapping ash species across a complex urban environment. Remote Sens Environ 199:360–369

Porter WM, Enmark HT (1987) A system overview of the airborne visible/infrared imaging spectrometer (AVIRIS). In: *Imaging Spectroscopy II*, vol 834. International Society for Optics and Photonics, Bellingham, WA. pp 22–32

Qi J, Chehbouni A, Huete AR, Kerr YH, Sorooshian S (1994) A modified soil adjusted vegetation index. Remote Sens Environ 48(2):119–126. https://doi.org/10.1016/0034-4257(94)90134-1

Rahman AF, Gamon JA, Fuentes DA, Roberts DA, Prentiss D (2001) Modeling spatially distributed ecosystem flux of boreal forest using hyperspectral indices from AVIRIS imagery. J Geophys Res-Atmos 106(D24):33579–33591

Rock BN, Vogelmann AF, Williams DL, Vogelmann DL, Hoshizaki T (1986) Remote detection of forest damage. Bioscience 36(7):439–445

Rock BN, Hoshizaki T, Miller JR (1988) Comparison of in situ and airborne spectral measurements of the blue shift associated with forest decline. Remote Sens Environ 24(1):109–127

Rondeaux G, Steven M, Baret F (1996) Optimization of soil-adjusted vegetation indices. Remote Sens Environ 55(2):95–107. https://doi.org/10.1016/0034-4257(95)00186-7

Roujean J-L, Breon F-M (1995) Estimating PAR absorbed by vegetation from bidirectional reflectance measurements. Remote Sens Environ 51(3):375–384

Rouse J, Hass R, Schell J, Deering D, Harlan J (1974) Monitoring the vernal advancement and retrogradation of natural vegetation. NASA Report, Greenbelt, MD

Schaberg PG, Dehayes DH, Hawley GJ (2001) Anthropogenic calcium depletion: a unique threat to forest ecosystem health? Ecosyst Health 7(4):214–228

Schaberg PG, Minocha R, Long S, Halman JM, Hawley GJ, Eagar C (2011) Calcium addition at the Hubbard Brook Experimental Forest increases the capacity for stress tolerance and carbon capture in red spruce (Picea rubens) trees during the cold season. Trees 25(6):1053–1061

Schaberg PG, Murakami PF, Butnor JR, Hawley GJ (2017) Experimental branch cooling increases foliar sugar and anthocyanin concentrations in sugar maple at the end of the growing season. Can J For Res 47(5):696–701

Senf C, Seidl R, Hostert P (2017) Remote sensing of forest insect disturbances: current state and future directions. Int J Appl Earth Obs Geoinf 60:49–60

Serbin SP, Singh A, McNeil BE, Kingdon CC, Townsend PA (2014) Spectroscopic determination of leaf morphological and biochemical traits for northern temperate and boreal tree species. Ecol Appl 24(7):1651–1669

Serbin SP, Singh A, Desai AR, Dubois SG, Jablonski AD, Kingdon CC et al (2015) Remotely estimating photosynthetic capacity, and its response to temperature, in vegetation canopies using imaging spectroscopy. Remote Sens Environ 167:78–87

Sims DA, Gamon JA (2002) Relationships between leaf pigment content and spectral reflectance across a wide range of species, leaf structures and developmental stages. Remote Sens Environ 81(2–3):337–354

Singh A, Serbin SP, McNeil BE, Kingdon CC, Townsend PA (2015) Imaging spectroscopy algorithms for mapping canopy foliar chemical and morphological traits and their uncertainties. Ecol Appl 25(8):2180–2197

Sitzia T, Campagnaro T, Kowarik I, Trentanovi G (2016) Using forest management to control invasive alien species: helping implement the new European regulation on invasive alien species. Biol Invasions 18(1):1–7

Sivanpillai R, Smith CT, Srinivasan R, Messina MG, Ben Wu X (2006) Estimation of managed loblolly pine stand age and density with Landsat ETM+ data. For Ecol Manag 223(1–3):247–254. https://doi.org/10.1016/j.foreco.2005.11.013

Smith ML, Ollinger SV, Martin ME, Aber JD, Hallett RA, Goodale CL (2002) Direct estimation of aboveground forest productivity through hyperspectral remote sensing of canopy nitrogen. Ecol Appl 12(5):1286–1302

Strimbeck GR, Schaberg PG, Fossdal CG, Schröder WP, Kjellsen TD (2015) Extreme low temperature tolerance in woody plants. Front Plant Sci 6:884

Thenkabail PS, Mariotto I, Gumma MK, Middleton EM, Landis DR, Huemmrich KF (2013) Selection of hyperspectral narrowbands (HNBs) and composition of hyperspectral twoband vegetation indices (HVIs) for biophysical characterization and discrimination of crop types using field reflectance and Hyperion/EO-1 data. IEEE J-STARS 6(2):427–439

Thompson I, Mackey B, McNulty S, Mosseler A (2009) Forest resilience, biodiversity, and climate change. In secretariat of the convention on biological diversity, Montreal. Technical Series No. 43. Vol 43, pp 1–67

Tomback DF, Achuff P (2010) Blister rust and western forest biodiversity: ecology, values and outlook for white pines. For Pathol 40(3–4):186–225

Townsend PA, Foster JR, Chastain RA, Currie WS (2003) Application of imaging spectroscopy to mapping canopy nitrogen in the forests of the central appalachian mountains using hyperion and aviris. IEEE Trans Geosci Remote Sens 41(6):1347–1354

Tucker CJ (1979) Red and photographic infrared linear combinations for monitoring vegetation. Remote Sens Environ 8:127–150

Vogelmann JE, Rock BN, Moss DM (1993) Red edge spectral measurements from sugar maple leaves. Int J Remote Sens 14(8):1563–1575

White K, Pontius J, Schaberg P (2014) Remote sensing of spring phenology in northeastern forests: a comparison of methods, field metrics and sources of uncertainty. Remote Sens Environ 148:97–107. https://doi.org/10.1016/j.rse.2014.03.017

Williams AP, Hunt ER (2002) Estimation of leafy spurge cover from hyperspectral imagery using mixture tuned matched filtering. Remote Sens Environ 82(2–3):446–456

Williams P, Norris K (2001) Near-infrared Technology in the Agricultural and Food Industries. American Association of Cereal Chemists, Inc., St. Paul

Williams DL, Goward S, Arvidson T (2006) Landsat. Photogramm Eng Remote Sens 72(10):1171–1178

Zarco-Tejada PJ, Miller JR, Mohammed GH, Noland TL, Sampson PH (2002) Vegetation stress detection through chlorophyll $a + b$ estimation and fluorescence effects on hyperspectral imagery. J Environ Qual 31(5):1433–1441

Zarco-Tejada PJ, Hornero A, Hernández-Clemente R, Beck PSA (2018) Understanding the temporal dimension of the red-edge spectral region for forest decline detection using high-resolution hyperspectral and sentinel-2a imagery. ISPRS J Photogramm Remote Sens 137:134

Open Access This chapter is licensed under the terms of the Creative Commons Attribution 4.0 International License (http://creativecommons.org/licenses/by/4.0/), which permits use, sharing, adaptation, distribution and reproduction in any medium or format, as long as you give appropriate credit to the original author(s) and the source, provide a link to the Creative Commons license and indicate if changes were made.

The images or other third party material in this chapter are included in the chapter's Creative Commons license, unless indicated otherwise in a credit line to the material. If material is not included in the chapter's Creative Commons license and your intended use is not permitted by statutory regulation or exceeds the permitted use, you will need to obtain permission directly from the copyright holder.

Chapter 7
Linking Leaf Spectra to the Plant Tree of Life

José Eduardo Meireles, Brian O'Meara, and Jeannine Cavender-Bares

7.1 Introduction

Evolution is the engine behind the diversity in leaf structure and chemistry that is captured in their spectral profiles, and, therefore, leaf spectra are inexorably linked to the tree of life. Our ability to distinguish species using spectra is a consequence of trait differences that arise and accumulate over evolutionary time. By the same token, the amount of variation that exists in different spectral regions is ultimately determined by the pace of evolution, convergence, and other evolutionary dynamics affecting the underlying leaf traits. There is an increasing interest in understanding leaf spectra through the lens of evolution and in the context of phylogenetic history (Cavender-Bares et al. 2016; McManus et al. 2016). Advances on this front will require, however, a good understanding of how evolutionary biologists leverage the tree of life to make inferences about evolution.

J. E. Meireles (✉)
Department of Ecology, Evolution and Behavior, University of Minnesota, Saint Paul, MN, USA

School of Biology & Ecology, University of Maine, Orono, ME, USA

B. O'Meara
Department of Ecology & Evolutionary Biology, University of Tennessee, Knoxville, TN, USA

J. Cavender-Bares
Department of Ecology, Evolution and Behavior, University of Minnesota, Saint Paul, MN, USA

© The Author(s) 2020
J. Cavender-Bares et al. (eds.), *Remote Sensing of Plant Biodiversity*,
https://doi.org/10.1007/978-3-030-33157-3_7

7.2   Evolutionary Trees

We refer to phylogenies in many different ways, and all of these terms appear in the literature. The terms *phylogeny*, *phylogenetic tree*, and *evolutionary tree* can be used interchangeably. We also use the term *tree of life* to refer to *the tree of life* (the evolutionary tree for all of life) or to the phylogeny of a really large group (or lineage) of organisms, such as the plant tree of life or the vertebrate tree of life.

7.2.1   How to Read Phylogenies

The idea that species descend from a common ancestor is at the very core of the theory of evolution. Evolutionary trees represent the branching structure of life and describe how species are related to each other similarly to how a genealogical tree recounts how people are related. A branch on a phylogenetic tree is a species; when it speciates, two (typically) descendant species arise. The two lineages coming from the same ancestor are known as sisters. These lineages can continue to branch, leading to more descendants. An ancestor and all its descendants are known as a clade: since these descendants all came from the same species, they share many inherited traits. Relatedness among organisms is encoded in the phylogeny's structure—its topology—which defines a series of lineages that are hierarchically nested. The branch lengths also usually convey information, such as the time since divergence, amount of molecular similarity, or number of generations (Fig. 7.1). Dated fossil information can be used to calibrate the age of some of the nodes in a phylogeny. This is one means by which branch lengths can be made to represent time fairly accurately. Typically, the spacing between tip nodes (the *y*-axis in Fig. 7.1a, b, and d) has no meaning, but it can sometimes be used to display information about the trait values of a species (Fig. 7.1c). Because no one has been taking notes of how lineages split over the last 4 billion years, phylogenetic trees must be estimated by analyzing current species data, generally DNA sequences, using models of evolution. This means that phylogenies are statistical inferences that have uncertainty about their topology and their branch lengths (Fig. 7.1d).

7.2.2   Why Care About Phylogenetic Accuracy?

An accurate phylogeny is key for understanding life. A phylogeny in which dandelions were more closely related to ferns than to roses would tell us a very different story about the evolution of flowering plants than would the true phylogeny, in which all flowering plants belong to a single lineage. In other words, the accuracy of the estimated tree topology—the structure of the relationships between species—matters to how we understand trait evolution. Accurately inferring divergence times

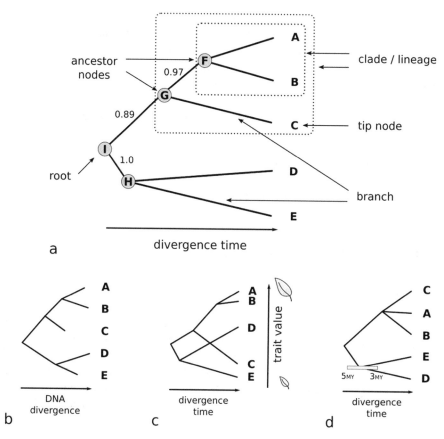

Fig. 7.1 Phylogenetic trees depict the inferred evolutionary relationships between species. (**a**) Clades (or lineages) are defined by a common ancestor and all of its descendants. Nodes are the branching points between descendants as well as tips, which are typically species. All nodes—tips and ancestors alike—share a common ancestor. The ancestral node from which all subsequent nodes of the tree descend is called the root. Confidence in the evolutionary relationships is shown above internal branches (maximum possible is 1). (**b**) Branch lengths (here shown along the *x*-axis) may represent divergence times, number of generations, or amount of molecular divergence. (**c**) In some cases the *y*-axis is used to display information about a quantitative trait—such as leaf size—in a tree known as a phenogram or a traitigram. (**d**) Unresolved relationships can be represented as three or more descendants stemming from the same ancestor, which is known as a polytomy. Uncertainty in divergence times are generally depicted with error bars at the internal nodes, if indicated at all

among species and lineages is also critical for making meaningful inferences about evolution. As we will discuss in the next section, estimates of the pace of trait evolution depends on the amount of change in a trait that occurs over a unit of time.

There are today several resources to help generate a good phylogenetic tree for a set of species. A common approach is to trim the whole plant tree of life—taken from the Open Tree of Life (Hinchliff et al. 2015) or Phylomatic (Webb and Donoghue 2005), for example—to the set of species of interest. A second option is to reconstruct the phylogeny from scratch using DNA sequences and then by time-calibrating the tree using fossil information and molecular clock models. Tree reconstruction is tricky and laborious, but there are many tools that can help (e.g., Antonelli et al. 2016; Pearse and Purvis 2013). Cobbling together a phylogenetic tree by manually assembling branches is not recommended for analysis of spectra or other traits.

Finally, as seen in the previous section, phylogenies are estimates, and systematists have means of assessing uncertainty in their topology and their branch lengths, which are together referred to as phylogenetic uncertainty. For example, the divergence between two lineages may have a mean of 20 million years and a confidence interval or 95% highest posterior density of 18–22 million years. That uncertainty can (and should) be carried over to downstream statistical analyses.

7.3 The Evolution of Quantitative Traits

The study of evolution is fundamentally concerned with describing how organisms change through time and with understanding the processes driving change. Evolutionary change, however, can be thought about at different phylogenetic and temporal scales. Because we are interested in understanding spectra in light of phylogenies, we will not discuss microevolutionary processes that occur at the population level such as genetic drift and natural selection. Instead, we will focus on describing macroevolution and how traits—such as leaf structure and chemical composition—change across entire lineages over long timescales (usually millions of years).

7.3.1 Macroevolutionary Models of Trait Evolution

Macroevolutionary models of trait evolution describe the long-term consequences of short timescale evolution. At any given time step, a trait value can increase or decrease due to mechanisms like selection, drift, and migration. For example, the reflectance in one spectral region may decrease due to selection for higher levels of a particular pigment, while reflectance in another spectral region may decrease due to a random change in leaf hair density. Many such changes occur over long evolutionary time in each lineage.

7.3.1.1 Brownian Motion

Most models for evolution of quantitative traits leverage the central limit theorem from statistics, which states that the sum of many random changes leads to a normal distribution. Because trait evolution at macroevolutionary scales integrates over many random changes in trait values (due to varied processes), it may be described by a normal distribution. This model of evolution is known as Brownian motion (Felsenstein 1985). The pace at which those changes accumulate is at the core of what we call the rate of evolution, and it is captured by the variance of the normal distribution (whose mean is the trait value at the root).

When lineages split, they start out with the same trait value and then diverge independently. It is easy then to see that the expected amount of trait variation between lineages depends on both the rate of evolution and on the divergence time. This leads to the expectation that trait values should be on average more similar among closely related taxa—which had little time to diverge—than among distantly related taxa. Such expectation is at the core of the concept of phylogenetic signal (see Sect. 7.3.2) and the idea that phylogenetic relatedness can be used as a proxy for functional similarity (Webb et al. 2002), particularly when integrating across a large number of traits (Cavender-Bares et al. 2009).

7.3.1.2   Ornstein–Uhlenbeck

With Brownian motion, an increase or decrease in a trait is equally likely, regardless of the current value of a trait (Fig. 7.2a). However, it could be more realistic to think of a trait as being pulled toward some optimum (or, similarly but not quite the same, away from extreme values). This force or "pull" could be due to many processes: it is often considered to be a pull toward some evolutionary optimum due to natural selection, but it could instead result from a bias in mutation toward a particular trait value, repulsion from extremes, or other factors that lead to a pattern that resembles a pull toward an optimum. The placement of the optimum, the strength of the pull, and the basic underlying rate of evolution are all parameters of this model, which is known as an Ornstein–Uhlenbeck process (Butler and King 2004). The degree of the pull toward the optimum is analogous to the strength of a rubber band linking

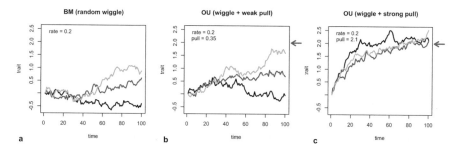

Fig. 7.2 Three independent realizations of the Brownian motion (BM) and Ornstein–Uhlenbeck (OU) processes. (**a**) In a BM model, trait values are equally likely to increase or decrease at each time step. (**b**, **c**) In contrast, traits in an OU model are more likely to move toward an optimum (represented by the red arrows). (**b**) When the evolutionary pull is weak, traits move slowly toward their optimum. (**c**) When the pull is strong, however, traits converge quickly toward their optimum

the evolving trait on one end and the optimum trait value on the other end. A weak rubber band will provide enough slack for the trait to wiggle around the optimum (Fig. 7.2b), whereas a strong rubber band will keep the evolving trait close to the optimum (Fig. 7.2c). The strength of the rubber band also affects how quickly the trait is pulled toward its optimum (Fig. 7.2b,c). The time a trait is expected to take to get halfway to the optimum is called the phylogenetic half-life, and this is an alternative way to think about the strength of the evolutionary pull.

Variation in traits within species, populations, and even individuals may result from responses to environmental conditions or have a genetic basis. Until recently, phylogenetic comparative methods largely ignored intraspecific variation and used species means instead. Ives et al. (2007) and Felsenstein (2008) devised methods to account for within species variation, which typically enters the model as the standard errors about the mean trait value of each species.

7.3.2 Phylogenetic Signal

Phylogenetic signal can be thought as the degree to which closely related species resemble each other. Two different metrics have been widely used to assess phylogenetic signal: Pagel's lambda (Pagel 1999) and Blomberg's K (Blomberg et al. 2003).

7.3.2.1 Pagel's Lambda

Pagel's lambda is a scalar for the correlation between the phylogenetic similarity matrix and the trait matrix. It has the effect of shrinking the internal branches (as opposed to the branches that lead to the tips) of a phylogeny, thereby reducing the expected species correlation due to shared evolutionary history (Fig. 7.3a–d). A lambda value of 0 indicates that trait correlations between species are independent from evolutionary history (Fig. 7.3d), whereas a lambda of 1 suggests that trait correlations are equal to the species correlation imposed by their shared evolutionary history (Fig. 7.3a), assuming a Brownian motion model of evolution.

7.3.2.2 Blomberg's K

Blomberg's K measures the degree to which trait variance lies within clades versus among clades. Brownian motion is used as an expectation. K values greater than 1 indicate that there is more variance among clades than expected by Brownian motion (Fig. 7.3e), while K values smaller than 1 imply that more variance is found within clades than expected under a Brownian motion model (Fig. 7.3f).

It is important to note that both Pagel's lambda and Blomberg's K are treewide metrics, meaning that they do not explicitly account for the heterogeneity in trait values among lineages. For example, an estimate of low phylogenetic signal in fruit

Fig. 7.3 This is how phylogenetic signal is inferred. (**a–d**) Pagel's lambda is equivalent to scaling the internal branches of the phylogeny, which reduces the expected covariance between species due to evolutionary history. (**e, f**) Blomberg's *K* measures phylogenetic signal by estimating the degree of variation between and within clades. (**e**) *K* value is high when most trait variation is found between clades instead of within them. (**f**) *K* values are low when trait variation is mostly within clades

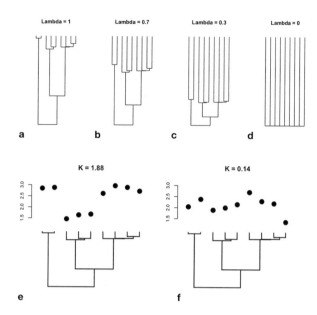

shape across all flowering plants does not imply the lack of phylogenetic signal in fruit shape within the oaks. Therefore, assessments of phylogenetic signal should be seen as indicators that are contingent on the scale of analysis and the particular species sampled instead of as general, hard truths.

It is also important to recognize that every calculation of the phylogenetic signal of a trait involves fitting an evolutionary model that comes with a series of assumptions. For example, most procedures to estimate phylogenetic signal using Blomberg's *K* are based on a single-rate Brownian motion model; using and reporting the measure of phylogenetic signal implicitly requires accepting the Brownian assumptions.

7.4 Evolution and Spectra

Chemical and structural leaf attributes that underlie plant spectra evolve through time. Because leaf spectra integrate over these evolved leaf attributes, they can carry information about phylogenetic relationships and leaf evolution. Given this, how would one go about analyzing spectra in a phylogenetic context?

One approach is to subject the spectra directly to an evolutionary analysis, essentially taking reflectance values at different bands across the spectrum to be a set of "traits." For example, McManus et al. (2016) estimated Pagel's lambda on spectra from Amazonian plants, assuming each band to be an independent trait. Cavender-Bares et al. (2016) used principal component analysis (PCA) to reduce the dimensionality of the spectral data before estimating phylogenetic signal on the resulting principal component axes using Blomberg's *K*.

a

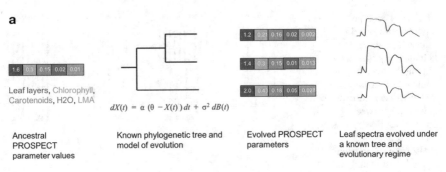

| Ancestral PROSPECT parameter values | Known phylogenetic tree and model of evolution | Evolved PROSPECT parameters | Leaf spectra evolved under a known tree and evolutionary regime |

Fig. 7.4 Integration of trait evolution and leaf spectral models enables estimation of evolutionary parameters from spectra and simulation of leaf spectra along a phylogeny. Ancestral leaf attributes evolve along a phylogenetic tree under a given evolutionary regime, generating the current leaf attributes that underlie spectra. From the evolved leaf attributes, radiative transfer models (RTMs) such as PROSPECT estimate spectra that carry the signature of the phylogeny

Fitting evolutionary models directly to spectra can be useful for identifying promising associations between phylogenetic history and plant spectral signatures. However, this approach is largely devoid of mechanism and does not allow us to verify that our inferences are biologically meaningful.

Another approach to integrate phylogenies and leaf spectra is to explicitly model the evolution of structural and chemical traits that underlie the spectrum. This approach matches more closely the reality of biology by acknowledging that any signal of evolution found in the spectra is an emerging property of the evolutionary dynamics of leaf traits (see Sect. 7.5.3). This idea can be implemented by coupling the models of trait evolution described in the previous section with leaf radiative transfer models (Fig. 7.4) that predict spectral profiles from a small set of leaf attributes (see Martin, Chap. 5; Ustin and Jacquemoud, Chap. 14).

This framework can be used in several ways. For example, we can simulate what leaf spectra would look like given a certain evolutionary model and phylogenetic tree (Sect. 7.4.1). Alternatively, given a phylogeny and a spectral data set, we can infer what ancestral spectra or ancestral traits were like if we assume a certain model of evolution. Finally, given a spectrum from an unknown plant, we could estimate how that plant is related to other plants (Sect. 7.4.3).

7.4.1 Simulating Leaf Spectra Under Different Evolutionary Regimes

A model that describes the evolution of leaf spectra mediated by the evolution of leaf traits enables us to simulate spectral data in a phylogenetically explicit way. This allows us to forecast how different evolutionary scenarios would affect the shape and diversity of spectral profiles we observe. For example, Fig. 7.5 shows how the different scenarios for the evolution of leaf structure—the number of layers parameter (N)

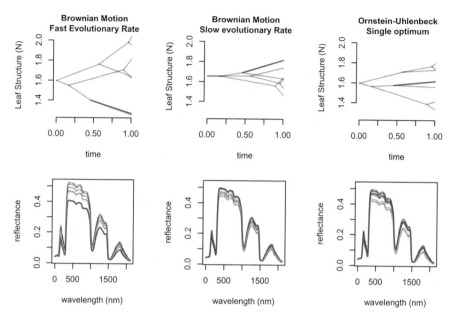

Fig. 7.5 Evolution of leaf structure under different evolutionary scenarios and consequences for leaf spectra. Top row depicts evolution according to an unbounded Brownian motion model at two different rates and according to an Ornstein–Uhlenbeck process. The bottom row shows spectra estimated with the PROSPECT5 model, where all leaf attributes evolved under the same model except for leaf structure, which evolved under the three scenarios outlined above

in PROSPECT5—result in different amounts of trait variability. A fast Brownian rate (top left, Fig. 7.5) results in higher trait variation than a slow Brownian rate (top center, Fig. 7.5). Evolution under an Ornstein–Uhlenbeck model also results in less variation than the fast Brownian model even though their rates of evolution are the same (top right, Fig. 7.5). The trait values shaped by evolution have a noticeable effect on the spectral profiles of those lineages (bottom panels, Fig. 7.5).

7.4.2 Making Evolutionary Inferences from Leaf Spectra

Integrating spectra and phylogenies raises the exciting prospect of leveraging spectra to estimate aspects of the evolutionary process and test hypotheses.

Some questions may be about evolutionary patterns in the spectra themselves. Those include investigations about phylogenetic signal or rates of evolution across the spectrum. For example, Cavender-Bares et al. (2016) and McManus et al. (2016) investigated how much phylogenetic signal is present in leaf spectra. Meireles et al. (in review) estimated how rates of evolution varied across the leaf spectrum of seed plants. Now, because we are interested in biology, evolutionary inference made at the spectral level will often need to be interpreted a posteriori.

Interpreting results correctly may pose some challenges, however. Who guarantees that the high rates of evolution in a particular spectral band really means that a certain trait is evolving at a fast pace? A potentially better approach is to infer traits from spectra first using either statistical (e.g., partial least squares regression) or RTM inversions (e.g., PROSPECT) and then study the evolution of those traits (see Serbin and Townsend, Chap. 3).

We can test hypotheses about how evolution affects leaf spectra because we can calculate the likelihood of spectral data being generated by different models of evolution, which can be compared to each other using a goodness of fit metric such as Akaike information criterion (AIC; Burnham and Anderson 2002). We foresee numerous interesting hypotheses being tested using this type of approach, especially related to evolutionary rates and convergent evolution.

Here is a hypothetical but realistic example: We could hypothesize that plant lineages that shift from sunny to shade habitats see an increase in their leaf chlorophyll content from 20 to 60 ug/cm^2, that is, they have a new chlorophyll content optimum, and that should be reflected in their spectra (Fig. 7.6). We used the predictive approach established in the previous subsection to simulate leaf spectra under that evolutionary scenario, which highlights the disparity in reflectance in the visible spectrum between sun and understory plants. We can then fit various models of evolution to the spectra (including one- and two-rate Brownian motion as well as a one-optimum Ornstein–Uhlenbeck and a two-optimum Ornstein–Uhlenbeck, the model under which the data were simulated), calculate their AIC, and compare models using AIC weights (Burnham and Anderson 2002), as shown in Fig. 7.6. In this simulated scenario, we find that indeed the best fit comes from having two different optima in the spectra correlating with chlorophyll content. However, in real data, we might find that there is a difference, but only in bands correlating with lignin content in leaves (which could reflect different herbivore or structural pressure); that there is a difference in optimum but that understory plants are much more constrained toward their optimum than plants from sunnier habitats; or that there is a change but it happens over longer time periods than we expect.

7.4.3 Leaf Spectra, Biodiversity Detection, and Evolution

As other chapters discuss, one approach to assessing biodiversity from plant spectra is to use spectral indices that correlate with species richness (Gamon et al., Chap. 16). Another approach is to use classification models (Clark et al. 2005; Asner and Martin 2011; Serbin and Townsend, Chap. 3). Using an empirical example within a single lineage (the oaks, genus *Quercus*), there is enough information in the spectra of leaves to significantly differentiate populations within a single species (*Quercus oleoides*) and assign them to the correct population most of the time. Different species can be correctly classified with even greater accuracy, and the four major oak clades in the example can be identified with very high accuracy (Cavender-Bares et al. 2016).

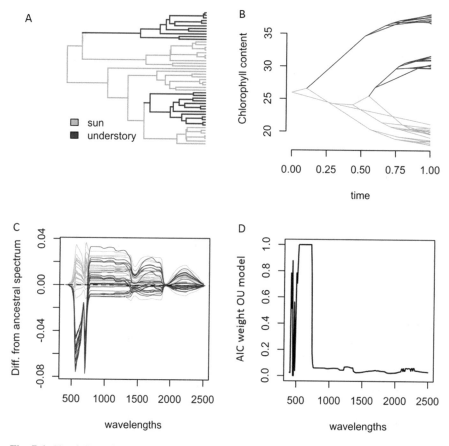

Fig. 7.6 Simulation of the evolution of chlorophyll content under a multiple optima Ornstein–Uhlenbeck model. (**a, b**) Macroevolutionary shifts from sun exposed to understory habitats (**a**) result in chlorophyll content being pulled toward different optima in different lineages (**b**). (**c**) Differences between the evolved spectra and the ancestral spectrum highlight the effect of chlorophyll evolution on the visible region of the spectrum. (**d**) We can use AIC to calculate how well various models of evolution, including the true multiple optima Ornstein–Uhlenbeck model, describe evolution across the spectrum. AIC weights suggest that the multiple optima Ornstein–Uhlenbeck model is preferred in the visible regions and nowhere else, which matches how the data were simulated

Evolutionarily-explicit diversity detection approaches could have enormous potential even when species cannot be identified. Biodiversity encompasses, among other things, which branches of the tree of life are found in an area how much evolutionary history that represents. Because plant spectral profiles can carry information about evolutionary history, they can be leveraged to assess the diversity of lineages instead of (or in addition to) the diversity in species or function. There are key conceptual advantages of taking this approach.

First, we can estimate lineage diversity at different phylogenetic scales when species-level detection performs poorly. As suggested in Fig. 7.7, leaf spectral

Fig. 7.7 Classification accuracy for different diversity levels of Quercus: (1) populations with Quercus oleoides, (2) 33 oak species, and (3) 4 clades of the genus Quercus. Accuracy was estimated from 300 independent PLS-DA iterations and summarized using Cohen's kappa. (Redrawn from Cavender-Bares et al. (2016))

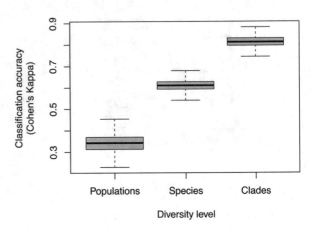

information can more accurately identify broad oak clades (kappa 0.81) than species (kappa 0.61) and than population within a species (kappa 0.34). It is thus possible that classification models can detect broad clades more accurately than they can detect very young clades or species.

Second, we know that species definitions change over time and that many species in hyperdiverse ecosystems are still unknown to science. How can we classify species that we do not yet know about? Using evolutionary models to estimate where an unknown spectrum belongs on the tree of life reduces the need for labeling, because it can reveal the taxa that the unknown sample is related to. This provides a means to estimate the phylogenetic diversity of a site even without species identities. The models of evolution described above should allow us to calculate the probability that the unknown spectrum belongs to different parts of the tree assuming that we know the correct evolutionary model and its parameter values. Developing the framework to achieve this would require filling in many gaps and detecting species at appropriate spatial resolutions. It also would require trusting many assumptions that go into evolutionary models, because we know that as we go deeper in phylogenetic time and evolutionary history, these models become increasingly complex (see Sect. 7.5.2).

7.4.4 Diversity Detection at Large Scales: Challenges and Ways Forward

The fact that spectra are tightly coupled with evolutionary history helps explain why hyperspectral data can be used for accurate taxonomic classification. It also provides a basis for using remotely sensed hyperspectral data for biodiversity composition monitoring.

Using RS hyperspectral data for biodiversity detection requires moving from the leaf level to the whole canopy level (Serbin and Townsend, Chap. 3; Martin, Chap. 5; Gamon et al., Chap. 16). We expect that canopy spectra, like leaf spectra, will show tight coupling to phylogenetic information. Branching architecture, leaf angles, and other structural traits of plants that contribute to spectral signals at the canopy level are themselves evolved traits and are potentially phylogenetically conserved. To the extent that remotely sensed hyperspectral data can capture the spectral profiles of individual canopies (Gamon et al., Chap. 16), hyperspectral data should be capable of detecting and identifying species (Morsdorf et al., Chap. 4) and lineages, following the logic presented above for leaves. Such an effort would require assembling vast libraries of spectral information across the plant tree of life for a given region of interest and comparing spectra obtained through remote sensing to those libraries.

Developing accurate classification models as the number of species and clades grow can be challenging. For example, Fig. 7.8 shows randomly assembled communities with different species diversity levels, where species spectra were simulated using PROSPECT5. As the number of species grows, the ability of a PLS-DA classification model to correctly classify species decreases.

This classification problem can be simplified by circumscribing the possible species pool. This could be done by estimating the potential pool of species or clades in a region based on other biodiversity monitoring and prediction approaches, including herbarium records, plant inventories or other types of in-situ data collection, and habitat suitability predictions (Pinto-Ledezma and Cavender-Bares, Chap. 9). Combining classification methods using hyperspectral data with prediction of species pools at regional scales across the globe could allow global plant compo-

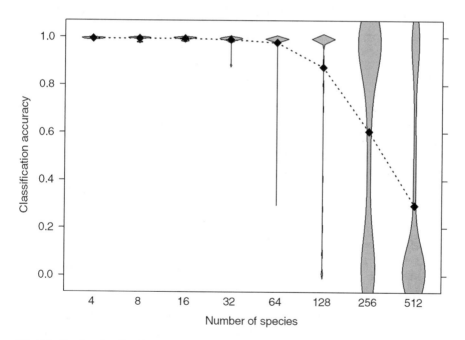

Fig. 7.8 Species classification accuracy of a PLS-DA model in simulated communities with different species richness. Spectra for each species were simulated using PROSPECT5, and communities with different diversity levels were randomly assembled

sition information to be detected when the spatial resolution of the data is sufficient to capture individuals.

7.5 Cautionary Notes

The integration of leaf spectra and phylogenies can provide breakthroughs in how we detect biodiversity, explain how spectral variation between species and lineages comes to be, and make inferences about the evolution of leaves. We should, nevertheless, be aware of the limitations inherent in making inferences about the deep past, be mindful of the sampling requirements and statistical assumptions of our analyses, and be careful to interpret our findings in a biologically meaningful way.

7.5.1 Is the Sampling Adequate for Making Evolutionary Inferences?

Inferences about the evolutionary process or that rely heavily on it—such as the degree of phylogenetic signal or the pace of evolution—are dependent on how well a lineage has been sampled. Evolutionary biologists usually target a particular lineage and strive to include in their analysis as many close relatives as possible regardless of their location. Ecologists, on the other hand, tend to focus on a specific geographic area of interest and end up sampling whatever species are there. This likely results in very severe undersampling of the total phylogenetic diversity represented by the particular species pool. For example, the 20 seed plants at a study site belong to a clade that has about 300,000 species, harbors incredible morphological and physiological diversity, and goes back 350 million years. Evolutionary analyses of undersampled will very likely yield poor estimates of the evolutionary parameters: The species in that area can tolerate a subset of all the climate conditions other seed plants can handle, for example.

In addition, ecological processes themselves can lead to bias in estimates of evolutionary parameters. For example, extremely arid conditions may act as an environmental filter that curbs colonization by species with low leaf water content, reducing the amount of variability in leaf succulence. As a consequence, estimates of the rate of evolution of leaf succulence based on species found in that hyper-arid community may be artificially low.

These caveats should be kept in mind when analyzing spectra in an evolutionary context. Finding that certain spectral regions have high phylogenetic signal in a large forest plot does not necessarily mean that those regions are truly phylogenetically conserved.

7.5.2   The More of the Tree of Life That Is Sampled, the More Complex Models Will (or Should) Be

Most of the models of evolution and phylogenetic signal statistics we saw here are actually rather simple. For example, a Brownian motion model has two parameters, the trait value at the root (mean) and the rate of evolution (variance). The single-rate Brownian motion model may reasonably describe the evolution of leaf water content in dogwoods (*Cornus*), but it would probably do a terrible job if you were analyzing all flowering plants because of the sheer heterogeneity and diversity that they possess (Felsenstein 2008; O'Meara 2012; Cornwell et al. 2014).

There is a trade-off: the most realistic model would have a different set of parameters at every time point on every branch but would have far more parameters to estimate than the data could support; a simple model of one set of parameters across all the time periods and species examined is clearly unrealistic. Most applications have used the simplest approach, but there are ways to allow for more complex models. Some of them test a priori hypotheses about heterogeneity in models of evolution: Biologists propose particular models linking sets of parameters on different parts of the tree (e.g., gymnosperms and angiosperms having different rates of evolution), and then the methods select between the possible models (Butler and King 2004; O'Meara et al. 2006). There are also methods that can automatically search across possible mappings to find the ones that fit best (Uyeda and Harmon 2014). In the case of multiple characters, such as reflectance at different wavelengths of light, there is also the question of whether different characters are evolving under the same or different models, and there are models to test that, as well (Adams and Otárola-Castillo 2013).

Early attempts to analyze spectra in an evolutionary context (Cavender-Bares et al. 2016; McManus et al. 2016; Meireles et al. in review) have used models that are maximally simple for each character (a single model applying for all taxa and times) and are nearly maximally complex between characters (each trait evolves independently of all others on the same common tree). Those approaches are computationally cheap but are at odds with our understanding of biology (i.e., models of evolution do vary among lineages) and physics (i.e., spectral bands do covary). Other ways of segregating complexity, such as models that incorporate heterogeneity among lineages and account for the covariance among spectral bands, remain potentially more fruitful ways of examining the diversity in leaf spectra.

7.5.3   Spectra Do not Evolve*, Leaves Do!

{*except when they do}

One could estimate the pace of evolution of the beaks of Darwin's finches from their photographs. But the photographs didn't evolve. Leaf spectra do capture many different aspects of the complex phenotype, and, we have seen in this chapter, each band of a spectrum can be analyzed as a trait in an evolutionary model. This does not necessarily mean that spectra themselves are traits nor that they themselves evolve. For example, there is no reason for evolution to favor lower reflectance at 660 nm. However, there may be biological reasons for natural selection to favor higher amounts of chlorophyll a in a leaf, which happens to absorb light at 660 nm. Terminology such as "evolution of spectra" or "spectral niches" may be efficient communication shortcuts but can also cause confusion. They may make it all too easy to lose sight of the biological mechanisms behind the observed phenomena.

Advances in analyzing spectral data in light of evolution will require keeping mechanisms in mind. That said, phylogenetic inference on spectra can be used as a discovery tool. Consistently finding high rates evolution in a spectral region not associated with a known function should trigger further investigation. Moreover, mechanistic thinking may end up proving us wrong and show that spectra in fact evolve (at least some regions). For example, increased leaf reflectance that prevents leaf overheating could be favored by evolution. In such a situation, high reflectance would result from "real" traits—such as bright hairs, cuticles, and waxes—but one can argue that there is biological meaning in the evolution of reflectance itself in this case.

7.5.4 *Ignore Phylogeny at Your Peril*

Phylogeny adds complexity to an analysis but has benefits in new insights (estimating ancestral leaf spectra, helping to go from observations to traits, and more). However, it can be tempting to analyze data on multiple species without accounting for shared evolutionary history. The problem with methods that ignore the phylogeny, such as partial least squares regression, is that they assume that species are independent data points. They are not! There is thus the risk of "overcounting" some parts of the tree of life: for example, if one wants to develop a model for all plants, and one has five oak species, a ginkgo, a pine, and a magnolia, the final model will essentially be an oak model with some deviations. However, the five oaks have shared much of their evolutionary history and so do not represent five independent instances of evolution. Phylogenies can be included into such analyses, and their importance appropriately scaled (in some cases, they will not affect results, but this is only knowable once the tree is used), and make results far more robust.

7.6 Moving Forward

The integration of leaf spectra and phylogenies using evolutionary models is still in its infancy. Phylogenetic models have the potential to unlock what drives evolution of the traits leading to different spectra. Spectra may have the potential, combined with phylogenies, to help identify species from afar, and even contain phylogenetic information themselves.

Acknowledgments We are grateful to Shan Kothari, Laura Williams, and Jesús Pinto-Ledezma for their feedback on an early version of this manuscript. We are also grateful to the National Institute for Mathematical and Biological Synthesis (NIMBioS) for funding the "Remotely Sensing Biodiversity" working group.

References

Adams DC, Otárola-Castillo E (2013) Geomorph: an R package for the collection and analysis of geometric morphometric shape data. Methods Ecol Evol 4(4):393–399

Antonelli A, Hettling H, Condamine FL, Vos K, Henrik Nilsson R, Sanderson MJ, Sauquet H et al (2016) Toward a self-updating platform for estimating rates of speciation and migration, ages, and relationships of taxa. Syst Biol 66(2):152–166. Oxford University Press: syw066

Asner G, Martin R (2011) Canopy phylogenetic, chemical and spectral assembly in a lowland Amazonian forest. New Phytol 189:999–1012

Blomberg SP, Garland T, Ives AR (2003) Testing for phylogenetic signal in comparative data: behavioral traits are more labile. Evolution 57(4):717–745

Burnham KP, Anderson DR (2002) Model selection and multimodel inference: a practical information-theoretic. Springer, New York, NY

Butler MA, King AA (2004) Phylogenetic comparative analysis: a modeling approach for adaptive evolution. Am Nat 164(6):683–695. The University of Chicago Press

Cavender-Bares J, Kozak KH, Fine PV, Kembel SW (2009) The merging of community ecology and phylogenetic biology. Ecol Lett 12:693–715

Cavender-Bares J, Meireles JE, Couture JJ, Kaproth MA, Kingdon CC, Singh A, Serbin SP et al (2016) Associations of leaf spectra with genetic and phylogenetic variation in oaks: prospects for remote detection of biodiversity. Remote Sens 8(3):221

Clark M, Roberts D, Clark D (2005) Hyperspectral discrimination of tropical rain 10 forest tree species at leaf to crown scales. Remote Sens Environ 96:375–398

Cornwell WK, Westoby M, Falster DS, FitzJohn RG, O'Meara BC, Pennell MW, McGlinn DJ et al (2014) Functional distinctiveness of major plant lineages. Edited by Amy Austin. J Ecol 102(2):345–356

Felsenstein J (1985) Phylogenies and the comparative method. Am Nat 125(1):1–15

Felsenstein J (2008) Comparative methods with sampling error and within-species variation: contrasts revisited and revised. Am Nat 171(6):713–725

Hinchliff CE, Smith SA, Allman JF, Gordon Burleigh J, Chaudhary R, Coghill LM, Crandall KA et al (2015) Synthesis of phylogeny and taxonomy into a comprehensive tree of life. Proc Natl Acad Sci U S A 112(41):12764–12769

Ives AR, Midford PE, Garland T (2007) Within-species variation and measurement error in phylogenetic comparative methods. Syst Biol 56(2):252–270

McManus K, Asner GP, Martin RE, Dexter KG, John Kress W, Field C (2016) Phylogenetic structure of foliar spectral traits in tropical forest canopies. Remote Sens 8(3):196

O'Meara BC (2012) Evolutionary inferences from phylogenies: a review of methods. Ann Rev Ecol Evol 43:267–285

O'Meara BC, Ané C, Sanderson MJ, Wainwright PC (2006) Testing for different rates of continuous trait evolution using likelihood. Evolution 60(5):922–933

Pagel M (1999) Inferring the historical patterns of biological evolution. Nature 401(6756):877–884

Pearse WD, Purvis A (2013) PhyloGenerator: an automated phylogeny generation tool for ecologists. Edited by Emmanuel Paradis. Methods Ecol Evol 4(7):692–698

Uyeda JC, Harmon LJ (2014) A novel Bayesian method for inferring and interpreting the dynamics of adaptive landscapes from phylogenetic comparative data. Syst Biol 63(6):902–918

Webb CO, Ackerly DD, McPeek MA, Donoghue MJ (2002) Phylogenies and Community Ecology. Annu Rev Ecol Syst 33:475–505

Webb CO, Donoghue MJ (2005) Phylomatic: tree assembly for applied phylogenetics. Mol Ecol Notes 5(1):181–183

Open Access   This chapter is licensed under the terms of the Creative Commons Attribution 4.0 International License (http://creativecommons.org/licenses/by/4.0/), which permits use, sharing, adaptation, distribution and reproduction in any medium or format, as long as you give appropriate credit to the original author(s) and the source, provide a link to the Creative Commons license and indicate if changes were made.

The images or other third party material in this chapter are included in the chapter's Creative Commons license, unless indicated otherwise in a credit line to the material. If material is not included in the chapter's Creative Commons license and your intended use is not permitted by statutory regulation or exceeds the permitted use, you will need to obtain permission directly from the copyright holder.

Chapter 8
Linking Foliar Traits to Belowground Processes

Michael Madritch, Jeannine Cavender-Bares, Sarah E. Hobbie, and Philip A. Townsend

8.1 Framework

Remote sensing (RS) of belowground processes via aboveground ecosystem properties and plant foliar traits depends upon (1) the ability to quantify ecosystem productivity and relevant plant attributes—including plant chemical composition and diversity—and (2) tight linkages between above- and belowground systems. These linkages can occur through the effects of aboveground inputs into belowground systems and/or through relationships between above- and belowground attributes and, in turn, between belowground relationships between plant roots and microbial communities and processes (i.e., fine-root turnover, mycorrhizal associations). The increasing ability of remotely sensed information to accurately measure productivity, ecologically important plant traits (Serbin and Townsend, Chap. 3, this volume; Wang et al. 2019), and plant taxonomic, functional, and phylogenetic diversity (Wang et al. 2019; Schweiger et al. 2018; Gholizadeh et al. 2019) creates new opportunities to observe terrestrial ecosystems. While the focus of RS tools is generally on aboveground vegetation characteristics, the tight linkage between above- and belowground systems through productivity and foliar chemistry means that many belowground processes can be inferred from remotely sensed information. Here, we focus on how the productivity and composition of foliar traits in plant communities influence belowground processes such as decomposition and nutrient cycling. We specifically consider foliar traits that are increasingly measurable via

M. Madritch (✉)
Department of Biology, Appalachian State University, Boone, NC, USA
e-mail: madritchmd@appstate.edu

J. Cavender-Bares · S. E. Hobbie
Department of Ecology, Evolution and Behavior, University of Minnesota, Saint Paul, MN, USA

P. A. Townsend
Department of Forest and Wildlife Ecology, University of Wisconsin, Madison, WI, USA

© The Author(s) 2020
J. Cavender-Bares et al. (eds.), *Remote Sensing of Plant Biodiversity*,
https://doi.org/10.1007/978-3-030-33157-3_8

airborne RS. Using two case studies, one in a clonal aspen (*Populus tremuloides*) forest system and one in a manipulated grassland biodiversity experiment, we demonstrate that plant foliar traits and vegetation cover, as measured via plant spectra (Wang et al. 2019), can provide critical information predictive of belowground processes.

8.2 How Are Belowground Processes and Microbial Communities Influenced by Aboveground Properties?

Belowground processes—including decomposition and nutrient cycling, which are mediated by microbial biomass, composition, and diversity—are heavily influenced by both the amount and chemistry of aboveground inputs. Quantifying the amount and quality of foliar components is a major aspect of trait-based ecology, which seeks to use functional traits, rather than taxonomic classification, to determine organisms' contributions to communities and ecosystems. Trait-based ecology has inherent strengths, including the ability to consider biological variation across both phylogenetic and spatial scales (Funk et al. 2017). While there is a range of accepted trait-based approaches in plant sciences (Funk et al. 2017), the emergence of the leaf economic spectrum (Wright et al. 2004) and the whole plant economic spectrum (Reich 2014) has clearly demonstrated that plant traits are important to ecosystem processes across multiple biological and spatial scales. Further, employing a trait-based approach to explore the relationships among plant function, biodiversity, and belowground processes allows us to take advantage of recent advances in RS to accurately measure plant traits across large spatial scales.

A. *Decomposition and Nutrient Cycling*—The productivity, composition, and diversity of aboveground communities influence belowground processes, in part through decomposition of leaf litter (Gartner and Cardon 2004; Hättenschwiler et al. 2005), root litter (Bardgett et al. 2014; Laliberté 2017), and root exudates (Hobbie 2015; Cline et al. 2018) and also through effects on soil organic matter (SOM) properties (Mueller et al. 2015) and soil physical structure (Gould et al. 2016). Several seminal reviews outlining the importance of biodiversity to ecosystem function (BEF) have focused specifically on the afterlife effects of litter diversity on decomposition (Hättenschwiler et al. 2005; Gessner et al. 2010).

B. *Microbial Community Composition*—Variation in the quantity and quality of organic inputs into belowground systems drives variation in belowground microbial communities and functioning (de Vries et al. 2012). Differences in aboveground communities are mirrored by those in belowground communities (Wardle et al. 2004; De Deyn and van der Putten 2005; Kardol and Wardle 2010). Across multiple spatial and taxonomic scales, variation in belowground microbial communities is driven by variation in plant traits associated with the leaf economic spectrum (de Vries et al. 2012). In general, fungi dominate decomposition of complex, low-quality substrates, while bacteria favor labile, high-quality substrates (Fig. 8.1, Bossuyt et al. 2001; Lauber et al. 2008).

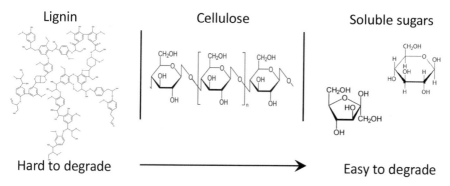

Fig. 8.1 Complex, recalcitrant compounds are typically degraded by fungi, while soluble, labile substrates are catabolized by bacteria

Microorganisms release extracellular enzymes, which degrade organic molecules outside of their cells, and likely differ among groups of microorganisms (Schneider et al. 2012). As a consequence, microbial composition and diversity are expected to influence decomposition and nutrient cycling. Most litter decomposition appears to be driven by fungal members, with *Ascomycota* dominating early degradation of cellulose and hemicellulose, followed by colonization by lignin-degrading *Basidiomycota* (Osono 2007; Schneider et al. 2012). Although lignin decomposition is dominated by fungal groups, some bacteria also degrade lignin (Kirby 2006; López-Mondéjar et al. 2016). Bacteria not directly involved with litter decomposition target the low molecular weight carbohydrates provided by fungal-derived extracellular enzymes (Allison 2005). The degradation of aromatic polyphenolics is largely limited to fungal member of the *Basidiomycota* phylum (Floudas et al. 2012). The wide structural variation among tannins (see section on carbon (polyphenols) results in a wide range of effects on specific microbial members (Kraus et al. 2003).

A challenge in predicting belowground processes such as decomposition and nutrient cycling from the diversity and quality of leaf litter inputs is that such an approach must also consider the diversity and function of belowground microbial communities. Belowground mycorrhizal communities can increase net primary production (NPP) and drive variation in plant communities (Wardle et al. 2004). Given the influence of plant traits on belowground processes, biodiversity may drive variation in decomposition through top-down (microbially driven) rather than bottom-up (substrate driven) forces (Srivasta et al. 2009). Several reviews have addressed the importance of belowground community diversity to ecosystem processes (e.g., Hättenschwiler et al. 2005; Gessner et al. 2010; Phillips et al. 2013; Bardgett and van der Putten 2014). Belowground diversity can influence aboveground factors such as NPP (Wardle et al. 2004; Eisenhauer et al. 2018) that then have important feedbacks to belowground processes. Decomposition is driven by a combination of both the microbial community and the quality and quantity of litter that those communities receive (e.g., Keiser et al. 2013; García-Palacios et al. 2016).

8.3 Mechanisms by Which Aboveground Vegetation Attributes Influence Belowground Processes

Aboveground community composition and vegetation chemistry are tightly linked with belowground communities through belowground inputs and subsequent decomposition and nutrient uptake (Hobbie 1992; Wardle et al. 2004). Plant biomass, structure, and chemical composition are all important drivers of belowground processes to such an extent that plant traits may be the dominant control on litter decomposition, outweighing the influence of climate even over large spatial scales (Cornwell et al. 2008).

8.3.1 Total Aboveground Inputs

Standing aboveground biomass and NPP are among the most important attributes of vegetation that impact belowground systems (Chapin et al. 2002) and are widely measured via RS techniques with increasing accuracy (Kokaly et al. 2009; Serbin et al., this issue). Belowground respiration is tightly linked with aboveground productivity (Högberg et al. 2001), and leaf litter can provide roughly half of organic inputs into some belowground systems (Coleman and Crossley 1996). The amount of aboveground biomass can be critical to litter decomposition (Lohbeck et al. 2015) and microbial community function and diversity (Fierer et al. 2009; Cline et al. 2018), and its influence may surpass the effects of plant quality, as measured by plant chemistry and functional traits (Lohbeck et al. 2015).

Plant traits related to biomass, such as leaf area index (LAI), are also linked to belowground processes, with belowground carbon (C) turnover peaking at intermediate LAI levels (Berryman et al. 2016; others). Importantly, LAI can be measured with RS products over large spatial scales (Serbin et al. 2014; Lausch et al., Chap. 13 this volume, Morsdorf et al. Chap. 4). While there have been few explicit links of remotely sensed LAI to soil respiration (but see Huang et al. 2015), the conceptual link has been recognized for decades (Landsberg and Waring 1997). Other remotely sensed variables tightly coupled with biomass, including vegetation cover (Wang et al. 2019), also predict soil respiration (Fig. 8.4).

The effects of biomass on belowground processes have been recognized by ecologists employing RS to estimate belowground C stocks (e.g., Bellassen et al. 2011). Across large scales, aboveground biomass is generally correlated with belowground root biomass (Cairns et al. 1997). While aboveground biomass is commonly measured, the calculation of belowground biomass is less common and is often limited to estimates of shoot biomass as a simple proportion of aboveground biomass (Mokany et al. 2006). Nonetheless, the belowground estimates based on aboveground measurements can be useful for estimating above- and belowground C stores via RS products over large spatial scales (Saatchi et al. 2011). Allocation of C to belowground systems varies among systems, with annual grassland systems differing from forested biomes in their allocation patterns of NPP (Litton et al. 2007). There are also large differences in above- and belowground linkages according to site fertility. In fertile sites the majority of NPP returned to the soil as labile fecal matter, whereas in infertile systems most NPP returned as recalcitrant plant litter (Wardle et al. 2004).

8.3.2 *Chemical Composition of Vegetation*

Beyond variation in total organic inputs to soil, variation in plant chemical composition is critical to belowground ecosystem processes. The physiological traits that comprise the plant economic spectrum developed by Wright et al. (2004) have important afterlife affects for belowground systems (Cornwell et al. 2008; Freschet et al. 2012; see review by Bardgett 2017). Variation in litter chemical quality can produce marked, long-term effects on litter decomposition rates and nutrient cycling in underlying soils, and litter quality has long been identified as key factor in determining decomposition rates (Tenney and Waksman 1929). Litter chemistry generally mirrors canopy chemistry (Hättenschwiler et al. 2008), making canopy chemistry a viable metric to estimate litter chemistry and subsequent belowground decomposition and nutrient cycling patterns. Aside from aboveground biomass, leaf nitrogen (N) and lignin content are often the dominant plant traits that drive variation in belowground process, particularly leaf litter decomposition (Aber and Mellilo 1982; Cadisch and Giller 1997), and both of these traits are readily derived from spectroscopy at multiple scales (Wessman et al. 1988; Serbin et al. 2014; Schweiger et al. 2018; Wang et al. 2019). RS of additional leaf traits important to belowground processes, such as plant secondary chemistry, is also increasingly measured via RS techniques (Kokaly et al. 2009; Asner et al. 2014; Serbin et al., this issue).

Nitrogen Foliar N is often the most important leaf trait driving variation in decomposition across biomes (Diaz et al. 2004; Cornwell et al. 2008; Handa et al. 2014). In some biomes leaf N is the only known leaf trait associated with leaf decomposition among wide ranges of species (Jo et al. 2016). Because canopy N has a tight correlation with plant carbon capture through photosynthesis, aboveground biomass, and belowground processes such as decomposition and N cycling rates, it is among the most common canopy traits measured via RS platforms (Martin and Aber 1997; Wessmen et al. 1998; Kokaly and Clark 1999; Martin et al. 1998, 2008; Ollinger et al. 2002; Townsend et al. 2003; Kokaly et al. 2009; Vitousek et al. 2009; Ollinger et al. 2013).

Leaf N is directly linked to plant productivity because most plant N is associated with metabolically active proteins, including RuBisCo. Leaf N content is driven by a trade-off between the benefits of increased photosynthetic potential and the costs associated with acquiring N along with the increased risk of herbivory (Diaz et al. 2016). In addition, leaf N can be indicative of plant growth strategies

(Wardle et al. 2004). Most short-term decomposition studies indicate that leaf N increases leaf litter decay (Cornwell et al. 2008). However, as decomposition progresses, leaf N may negatively affect the latter stages of decomposition, possibly due to interactions with lignified substrates (Berg 2014; discussed in brief below).

Carbon quality (lignin) The second most abundant natural polymer following cellulose is lignin, a complex phenolic polymer that wraps in and out of the structural polysaccharides in cell walls (Cadisch and Giller 1997). Due to its central roles in both aboveground biomass and belowground decomposition, lignin has been targeted as an important plant trait for RS techniques (Wessman et al. 1988; Serbin et al. 2014; Serbin and Townsend, Chap. 3). While lignin is a polyphenolic compound comprised of linked phenols (Horner et al. 1988), it is considered separate from other polyphenols because lignin is a primary structural component, whereas other polyphenols are a subset of secondary metabolites not directly involved with plant growth. The structure role of lignin and its low solubility also merit distinction from other polyphenolics when considering belowground processes (Hättenschwiler and Vitousek 2000). Lignin concentrations are negatively correlated with decomposition rates (Meentemeyer 1978; Melillo et al. 1982; Horner et al. 1988). The recalcitrant nature of lignin is due, in part, to its irregular structure and low energy yield, which largely limits its degradation to white-rot fungus members of *Basidiomycota* (Chapin et al. 2002).

The interaction of N and lignin during decomposition is not straightforward because N limits the early stages of decomposition, whereas lignin limits the latter stages of decomposition (Burns et al. 2013). Newly senesced leaves are composed largely of polysaccharides of holocellulose and lignin. High N availability will stimulate holocellulose decomposition in the early stages of decomposition but will then retard lignin decomposition in later stages of decomposition leading to lignified soil organic matter (SOM), potentially due to white-rot fungi favoring low N conditions (Berg 2014). The degradation of lignin is often a rate-limiting step during the later stages of decomposition because it protects cell wall polysaccharides physically and chemically (Talbot et al. 2012). Despite the changing roles that leaf N and lignin have over the course of decomposition, litter quality metrics such as C:N and lignin:N can explain variation in decomposition, with decomposition rates increasing with N in the early stages, but decreasing with N in the later stages, and decreasing with lignin (Fanin and Bertrand 2016).

While lignin almost universally retards decomposition, there is a large amount of variation within lignin compounds based on the proportion of specific monomers that varies across major plant groups (Thevenot et al. 2010). Angiosperm lignin tends to degrade more quickly than does gymnosperm lignin due to the specific identities of constituting moieties of lignin in each species (Higuchi 2006). The compact nature of gymnosperm lignin subunits is thought to protect them from enzymatic degradation (Hatakka and Hammel 2010). Functional measurements of lignin are often made via either acid digestion or thioglycolic acid methods that can then be used to calibrate spectroscopic methods (Brinkmann et al. 2002; Schweiger et al. 2018).

Carbon quality (polyphenols) In some ecosystems non-lignin carbon compounds (e.g., phenolics) explain more variation in decomposition than does either N or lignin (Hättenschwiler et al. 2011). Phenolics are the most widely distributed class of secondary plant metabolites and interact strongly with several aspects of nutrient cycling (Hättenschwiler and Vitousek 2000). Simple phenolics can prime (Fontaine et al. 2007), while large complex polyphenolics can retard (Coq et al. 2010) decomposition. Carbon quality—including the chemical composition of polyphenolics—can be more important to litter decomposition than is litter nutrient concentration (Hättenschwiler and Jørgensen 2010). Plant polyphenolics can be accurately measured via near-infrared spectroscopy (NIRS; Rupert-Nason et al. 2013), and by airborne imaging spectroscopy (Kokaly et al. 2009; Asner et al. 2014; Madritch et al. 2014; Serbin and Townsend, Chap. 3).

Though typically considered primarily for their aboveground defensive properties, phenolics in plant residues (leaf litter and roots) can have large influences on decomposition. Simple phenolics can increase soil respiration by providing a simple carbon source for microorganisms (Horner et al. 1988; Schimel et al. 1996; Madritch et al. 2007). Tannins are defined, in part, by their ability to bind to proteins (Bate-Smith 1975). The attributes of nonstructural polyphenolics that make them effective plant pathogen defenses also affect nonpathogenic fungi and microbes once litter enters the detrital food web; tannins do not discriminate between enzymes of plant pathogenic fungi or decomposing fungi. If tannins bind covalently with proteins to form polyphenolic-protein complexes, they become highly recalcitrant, and only basidiomycetes with polyphenol oxidase and earthworms can take advantage of these complex N sources (Hättenschwiler and Vitousek 2000). The inhibitory role of tannins on soil enzymes varies with specific tannin structure, which varies widely among species (Triebwasser et al. 2012). Tannins also have a limited ability to bind with carbohydrates and cellulose to form recalcitrant complexes (Horner et al. 1988; Kraus et al. 2003). The ability of polyphenolics to complex with proteins and other biochemicals is the primary method by which they influence soil respiration, litter decomposition, and soil N fluxes.

In addition to their influence on decomposition, nonstructural polyphenolics (which do not include lignin) influence N cycling by binding to and promoting retention of N-rich compounds including ammonium, amino acids, and proteins (Hättenschwiler and Vitousek 2000). Ayres (1997) suggested that condensed tannins may be more important to N cycling than to herbivore defense, since condensed tannins frequently have no anti-herbivory activity. Hättenschwiler et al. (2011) also proposed that polyphenolics, and tannins in particular, may be an important N conservation and recovery strategy for some species. This appears to be the case in *Populus tremuloides* systems, where high-tannin genotypes recovered more N than did low-tannin genotypes, especially when under severe herbivory (Madritch and Lindroth 2015). The high reactivity and branching structure of reactive hydroxyl sites also allow polyphenolics to complex with clay particles in soil and thereby influence several micronutrients in addition to N (Schnitzer et al. 1984).

Variation in plant phenolics is driven by several interacting factors. In general, polyphenolic concentrations in foliage are highest during the summer months (Feeny 1970). Summer coincides with both the onset of herbivory and the highest levels of photosynthetic activity. Herbivory-induced polyphenolic production is a well-documented aspect of plant-insect interactions (Herms and Mattson 1992; Baldwin 1994). The composition and quantity of phenolics vary among taxa at small and large phylogenetic scales. At large phylogenetic scales, condensed tannins are common in woody plants but almost absent in herbaceous species (Haslam 1989). At narrow phylogenetic scales, the concentration of polyphenolics is also under genetic control, and often there is considerable variation within the same species that can have important influences on belowground processes including litter decomposition and nutrient cycling (Lindroth et al. 2002; Schweitzer et al. 2005; Madritch et al. 2006, 2007).

8.3.3   Plant Diversity

Plant diversity, which can be accurately remotely sensed at some spatial scales (Wang et al. 2019; Gholizadeh et al. 2019), can influence belowground processes through its effects on productivity as well as on chemical diversity (Meier and Bowman 2008). Belowground diversity may be intrinsically linked to aboveground diversity because high plant diversity may provide a high diversity of litter quality and quantity to belowground systems that subsequently result in a high diversity of decomposers (Hooper et al. 2000). The specific relationship between aboveground plant communities and belowground microbial communities is context, system, and scale dependent (De Deyn and van der Putten 2005; Wu et al. 2011; Cline et al. 2018). For instance, Chen et al. (2018) found that plant diversity is coupled with soil beta diversity but not soil alpha diversity in grassland systems. Nonetheless, if aboveground diversity is indeed linked to belowground diversity, then aboveground estimates of plant diversity and plant traits could provide robust estimates of belowground processes.

Early work that focused on the influence of aboveground species diversity on litter decomposition yielded idiosyncratic results (Gartner and Cardon 2004; Hättenschwiler et al. 2005), with some studies reporting no effect of plant species diversity (e.g., Naeem et al. 1999; Wardle et al. 1999; Wardle et al. 2000; Knops et al. 2001), some reporting unpredictable results (Wardle and Nicholson 1996), and some reporting positive effects of plant species diversity on litter decomposition (Hector et al. 2000). Similar to aboveground processes, BEF studies that link aboveground diversity with belowground processes initially focused on aboveground species diversity (Scherer-Lorenzen et al. 2007; Ball et al. 2008; Gessner et al. 2010). The idiosyncratic relationship between species diversity and belowground processes led others to identify aboveground functional diversity and composition as more important to belowground processes than species diversity (Dawud et al. 2017).

Foliar chemistry is relevant to biodiversity and ecosystem functioning studies because plant chemistry varies widely among and within species and can influence belowground microbial communities and biogeochemical cycles (Cadisch and Giller 1997; Hättenschwiler and Vitousek 2000). It follows that variation in foliar traits important to decomposition (e.g., tannin concentration) will affect below-ground microbial communities and the basic biogeochemical cycles that sustain forested ecosystems. Some studies have supported a chemical diversity approach toward elucidating the belowground effects of aboveground diversity (Hoorens et al. 2003; Smith and Bradford 2003). Epps et al. (2007) demonstrated that account-ing for chemical variation was more informative regarding decomposition than was species diversity. While the usefulness of trait-based dissimilarity approaches remains somewhat equivocal (Frainer et al. 2015), there is increasing support for such trait-based approaches in explaining variation in leaf litter decomposition (Fortunel et al. 2009; Finerty et al. 2016; Jewell et al. 2017; Fujii et al. 2017). Handa et al. (2014) found that variation in leaf litter decomposition across widely different biomes was largely driven by commonly measured leaf traits such as N, lignin, and tannin content. At large scales, species traits rather than species diversity per se appears to at least partially drive variation in decomposition and belowground nutrient cycling.

In experimental systems, plant communities with high biodiversity result in high above- and belowground productivity (Tilman et al. 2001). The additional biomass that an ecosystem produces in diverse assemblages over what is expected from monocultures is called "overyielding" and has been documented in both grassland and forest experiments (Grossman et al. 2018; Weisser et al. 2017). The additional productivity results from several mechanisms acting simultaneously in more diverse communities, such as reduced pathogen attack, reduced seed limitation, and increased trait differences leading to "complementarity" in resource uptake (Weisser et al. 2017). Complementarity in resource use, particularly light harvesting, results in more efficient use of limiting resources and greater productivity (Williams et al. 2017). Similar patterns of greater productivity with higher diversity are observed in forest plots globally (Liang et al. 2016) although such patterns are scale dependent, and do not necessarily hold at large spatial extents (Chisholm et al. 2013). In natu-rally assembled grasslands, the relationship may not necessarily hold consistently (Adler et al. 2011). An open question, then, is the extent to which diversity and productivity are linked at large spatial scales in ecosystems globally. This is a ques-tion that can reasonably be addressed with remotely sensed measures of biodiversity and ecosystem productivity if scaling issues are appropriately considered (Gamon et al., Chap. 16). Plant diversity influences the quality of inputs and may allow for niche partitioning among functionally different microbes and may also influence productivity, the source of inputs of organic matter available to microbes, and microbial diversity. Through these linkages, foliar diversity has the potential to influence microbial diversity and function and hence belowground processes (Cline et al. 2018). The extent to which diversity and productivity, measured aboveground, can predict belowground microbial and soil processes is a question that is ready to be tackled at a range of scales across continents.

8.4   Case Studies

8.4.1   Remote Sensing of Belowground Processes via Canopy Chemistry Measurements

Plants act as aboveground signals for belowground systems. As such, RS of plant spectra can provide information about belowground systems. Plant spectra can provide a wealth of biological information important to plant physiology and community and ecosystem processes across multiple spatial scales (Cavender-Bares et al. 2017). Some researchers have used direct spectral measurements (e.g., NIRS) for direct measurements of soil characteristics (reviewed by Stenberg et al. 2010; Bellon-Maurel and McBratney 2011; Soriano-Disla et al. 2014), and there are limited examples of remotely sensed spectroscopic measurements of soils (reviewed by Ustin et al. 2004; Cecillon et al. 2009). Here we focus on remotely sensed spectral measurements of plant communities as a surrogate for belowground processes. The optical surrogacy hypothesis (sensu Gamon 2008) argues that plant spectra can serve as a surrogate for important belowground processes.

Direct spectral measurements have been used to assess belowground processes for decades. For instance, direct NIRS of leaf litter can be used to predict decomposition rates in a variety of systems (Gillon et al. 1993; Gillon et al. 1999; Shepherd et al. 2005; Fortunel et al. 2009; Parsons et al. 2011). RS of canopy traits to predict belowground processes is becoming increasingly useful. Spectroscopic measurement of $\delta^{15}N$ is of particular interest for ecosystem processes (Serbin et al. 2014) because stable N isotopes can provide important information regarding ecosystem N cycling (Robinson 2001; Hobbie and Hobbie 2006). RS of forest disturbance (e.g., fire severity) and subsequent belowground processes is relatively common (e.g., Holden et al. 2016). Sabetta et al. (2006) used hyperspectral imaging to predict leaf litter decomposition across four forest communities. Fisher et al. (2016) were able to distinguish between arbuscular and ectomycorrhizal tree-mycorrhizal associations using spectral information gleaned from Landsat data. While the above examples focus on remotely sensed spectral information, remotely sensed forest structural information developed from lidar data can also provide information about belowground systems, as Thers et al. (2017) were able to use remotely sensed lidar data to estimate belowground fungal diversity. The growing number of examples that employ remotely sensed data to provide information about belowground systems points to the potential of plant spectra to be used as surrogates for ecosystem processes.

8.4.2   Forest Systems: Aspen Clones Example

An example of optical surrogacy in practice is illustrated by work completed in trembling aspen (*Populus tremuloides*) systems across the Western and Midwestern USA. Trembling aspen is the most widespread native tree species in North America (Mitton and Grant 1996) and is an ecologically important foundation species across its native range (Lindroth and St. Clair 2013). Aspen is facing large and rapid declines in intraspecific biodiversity because concentrated patches of aspen are currently experiencing high mortality rates in North America (Frey et al. 2004; Worrall et al. 2008). This phenomenon, commonly referred to as sudden aspen decline (SAD), leads to the death of apparently healthy aspen stands in 3–6 years (Shields and Bockheim 1981; Frey et al. 2004). These natural history traits, combined with the ecological and economic significance of the species, make trembling aspen an ideal system to employ RS techniques to estimate genetic diversity and the consequences thereof for belowground processes.

Aspen typically reproduces clonally, often creating a patchwork of clones with many ramets (Fig. 8.2). Aspen clones vary widely in canopy chemistry traits that are important to belowground processes such as litter decomposition (Madritch et al. 2006). Several studies have highlighted the importance of plant genetic diversity to ecosystem processes (Madritch and Hunter 2002, 2003; Schweitzer et al. 2005; Crutsinger et al. 2006; Madritch et al. 2006, 2007) and community composition (Wimp et al. 2004, 2005; Johnson and Agrawal 2005). These recent advances demonstrate that genetic diversity affects fundamental ecosystem processes by influencing both above- and belowground communities (Hughes et al. 2008). The natural history traits of aspen, its clonal nature, genetically mediated variation in canopy chemistry, and the concomitant wide range of variation in foliar traits make it an ideal model system for RS of biodiversity.

Madritch et al. (2014) described how remotely sensed spectroscopic data from NASA's AVIRIS platform can be used to describe aboveground genetic and chemical variation in aspen forests across subcontinental spatial scales. This work built upon past work that demonstrated the ability of imaging spectroscopy to detect both aboveground chemistry (Townsend et al. 2003) and biodiversity (Clark et al. 2005) and employed imaging spectroscopy to discriminate intraspecific, genetic variation in aboveground chemistry and diversity. Because of the tight linkages between aboveground and belowground systems and because of the large variation in secondary chemistries important to belowground processes in aspen, this project also demonstrated the ability to predict belowground process via RS of forest canopy chemistry. Figure 8.3 illustrates both the

Fig. 8.2 Aerial photo showing color differentiation of genetically distinct aspen clones. Genotypes can be detected rapidly via remote sensing techniques

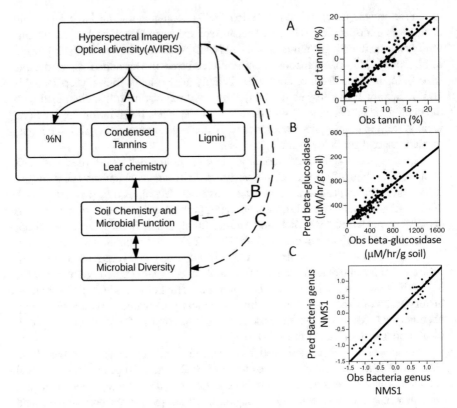

Fig. 8.3 Imaging spectroscopy links to several layers of ecological processes in aspen forests. (**a**) Partial least squares (PLS) prediction (pred) for condensed tannin concentration from AVIRIS data compared to observed (obs) tannin. (**b**) PLS prediction (pred) for soil b-glucosidase activity compared to observed (obs) b-glucosidase. (**c**) PLS prediction from AVIRIS spectra for bacterial diversity compared to observed bacterial diversity, where bacterial diversity is the first axis of an NMDS ordination of amplicon sequencing of rDNA (525f and 806r primers). (Tannin and soil enzyme data are from Madritch et al. (2014))

direct linkages between RS and canopy chemistry (A) and the subsequent indirect linkages to belowground function (B) and the microbial community (C). The indirect linkages represent the optical surrogacy hypothesis. Belowground attributes are not measured directly via RS, but rather RS of the forest canopy was able to provide detailed information regarding belowground process.

8.4.3 *Experiment Prairie Grassland System: Cedar Creek Example*

Vegetation differences between prairie and forested ecosystems have important consequences for above- and belowground linkages. Detrital inputs in forests are dominated by leaf litter, whereas they are dominated by root exudates and turnover in prairie systems that are frequently burned. We employed a parallel application of spectroscopic imagery to assess above- and belowground diversity and functioning at the grassland biodiversity experiment located at Cedar Creek Ecosystem Science Reserve (Tilman et al. 2001). Rather than a monospecific forest canopy, the grassland experiment consisted of replicated diversity treatments ranging from 1 to 16 perennial grassland species in 9 m × 9 m plots. This work had more technical challenges associated with it compared to the aspen forest project due to the inherent complexity of a mixed species system and the small spatial scale of the experimental plots.

The relationship between plant diversity and aboveground biomass in the Cedar Creek BioDIV experiment is well documented (Tilman et al. 2001, 2006). Schweiger et al. (2018) further demonstrate that both plant diversity and function are measurable via remotely sensed spectra within the experiment and that spectral diversity predicted productivity. Wang et al. (2019) used AVIRIS imagery to map functional traits across the experiment. Remotely sensed productivity and functional trait composition can thus be tested for linkages with belowground processes. In this system, the quantity of inputs had a large impact on fungal composition and diversity (Cline et al. 2018). Productivity, measured as annual aboveground biomass, given that it is annually burned, can be accurately detected as remotely sensed vegetation cover (Fig. 8.4a; Wang et al. 2019, following the method of Serbin et al. 2015). Remotely sensed vegetation cover, in turn, predicted fungal diversity, measured as operational taxonomic unit (OTU) richness (Fig. 8.4b), and cumulative soil respiration (Fig. 8.4c). In addition to the total organic matter inputs to the soil, chemical composition also influenced belowground microbial communities. For example, remotely sensed %N (Wang et al. 2019) was positively correlated with soil microbial biomass (Cavender-Bares et al., unpublished manuscript).

8.4.4 Challenges and Future Directions

Employing plant spectra to predict belowground processes has both caveats and advantages over traditional belowground sampling. One important caveat is that any prediction of belowground processes requires a solid understanding of the linkages between above- and belowground processes in any given system. Examples in the literature that link remotely sensed attributes of aboveground systems with belowground systems remain scarce, in part, because of the historic separation of the two disciplines. It is unclear how well remotely sensed plant attributes will predict microbial and soil processes across ecological systems. In the above forest example, aspen forests were generally uniform in canopy coverage. It was also a single-species system where leaf structure remained consistent across the study area, despite the large spatial sampling scheme. Consequently, most of the variation in aspen spectral signal was likely due to variation in canopy chemistry and biomass rather than leaf structure. Lastly, in this temperate forest system, leaf litter accounts for a large fraction of inputs into belowground systems, compared to systems such as Cedar Creek that are burned frequently and where fine-root turnover dominates belowground inputs.

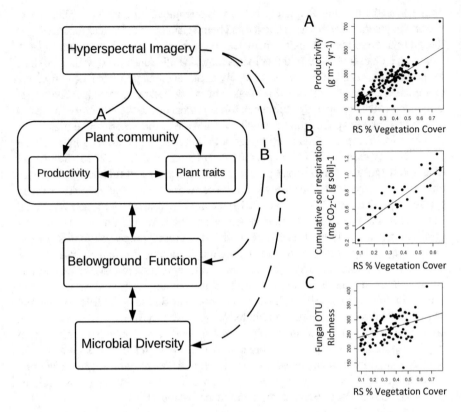

Fig. 8.4 Hyperspectral imagery links above- and belowground processes in a prairie ecosystem. (**a**) Remotely sensed vegetation cover significantly predicted aboveground plant productivity, $R^2 = 0.695$ (**a**); cumulative soil respiration (mg CO_2-C [g soil]$^{-1}$), $R^2 = 0.63$ (**b**); and fungal diversity, measured as OTU richness, $R^2 = 0.144$ (**c**). (Soil respiration and fungal diversity data are from Cline et al. (2018))

RS of aboveground properties poses further challenges that include separating the spectral signals important to canopy chemistry from those of physical properties of the forest canopy (Townsend et al. 2013). Larger challenges lie in the lack of accessibility of RS data and processing techniques to the broader ecological research community.

Several issues of scale present challenges to the application of RS to belowground systems. Large knowledge gaps remain in connecting the small spatial scale observations of traditional field studies with the large spatial scale observations of airborne or satellite RS platforms (Asner et al. 2015; Gamon et al., Chap. 16). In addition, there is a large mismatch in the spatial heterogeneity between above- and belowground systems, with belowground systems being notoriously heterogeneous across small spatial scales (Bardgett and van der Putten 2014). The majority of variation in belowground processes may be due to small, local-scale factors rather than large-scale factors such as climate (Bradford et al. 2016). In addition to issues associated with spatial scale, there are large spans in the scales of biodiversity and time. Speciose aboveground systems may contain upward of 600 species ha^{-1} (Lee et al. 2002), whereas soils contain many thousands of microbial "species" per gram of soil, with large numbers of endemics (Schloss and Handelsman 2006). Linking function to diversity remains a challenge in both systems and particularly in belowground systems where the functional role of the vast majority of species is unknown (Krause et al. 2014). Likewise, large differences in temporal scales exist between above- and belowground systems, with leaf responses to sunlight occurring on the order of seconds (Lambers et al. 1998), while the turnover of soil organic matter can take years to centuries (Bardgett and van der Putten 2014). Variation in temporal scales across systems is particularly important given that the importance of biodiversity to ecosystem processes increases with temporal scale (Cardinale et al. 2012; Reich et al. 2012).

Irrespective of RS, there are shortcomings associated with belowground measurement. For example, belowground measurements that use enzyme activity potentials as indicators of microbial function are widespread, but they are known to have numerous limitations (Nannipieri et al. 2018). Likewise, microbial diversity estimates based upon amplicon sequences of bacterial 16s rDNA have their own methodological and interpretive limitations (Schöler et al. 2017). Nonetheless, both enzyme activities and amplicon sequencing techniques provide useful information about belowground systems and are used widely enough to be compared across studies as long as protocols are consistent.

Advantages of using remotely sensed spectral properties of aboveground vegetation to predict belowground processes lie within the data-rich nature of imaging spectroscopy and the consequent ability to measure many more traits of the canopy than would otherwise be feasible with traditional benchtop methods. In Madritch et al. (2014), only four canopy traits were considered using traditional wet chemistry techniques (leaf tannin, N, C, lignin). These canopy foliar traits were expectedly well correlated with belowground processes. However, plant spectra themselves were better correlated with belowground processes than were plant leaf traits (Madritch et al. 2014). This strong relationship between plant spectra and belowground processing existed because the plant spectra provided quantitative information about many plant traits that were not measured via wet chemistry techniques. Potentially dozens of leaf traits important to belowground processes could be conveyed by plant spectra. The ability of plant spectra to capture many foliar attributes quickly and accurately is a large reason why plant spectra are useful for predicting belowground processes. In addition, identifying which regions of plant spectra are most variable and correlated with belowground process allows researchers to use spectra to identify plant traits important to soil processes. In short, the potential for RS products to link above- and belowground systems is promising but faces considerable obstacles.

References

Aber JD, Mellilo JM (1982) Nitrogen immobilization in decaying hardwood leaf litter as a function of initial nitrogen and lignin content. Can J Bot 60:2263–2269

Adler PB, Seabloom EW, Borer ET, Hillebrand H, Hautier Y, Hector A, Harpole WS, O'Halloran LR, Grace JB, Anderson TM, Bakker JD, Biederman LA, Brown CS, Buckley YM, Calabrese LB, Chu C-J, Cleland EE, Collins SL, Cottingham KL, Crawley MJ, Damschen EI, Davies KF, DeCrappeo NM, P a F, Firn J, Frater P, Gasarch EI, Gruner DS, Hagenah N, Hille Ris Lambers J, Humphries H, Jin VL, Kay AD, Kirkman KP, J a K, Knops JMH, La Pierre KJ, Lambrinos JG, Li W, MacDougall AS, McCulley RL, B a M, Mitchell CE, Moore JL, Morgan JW, Mortensen B, Orrock JL, Prober SM, D a P, Risch AC, Schuetz M, Smith MD, Stevens CJ, Sullivan LL, Wang G, Wragg PD, Wright JP, Yang LH (2011) Productivity is a poor predictor of plant species richness. Science 333:1750–1753

Allison SD (2005) Cheaters, diffusion, and nutrients constrain decomposition by microbial enzymes in spatially structured environments. Ecol Lett 8:626–635

Asner GP, Martin RE, Anderson CB, Knapp DE, (2015) Quantifying forest canopy traits: imaging spectroscopy versus field survey. Remote Sens Environ 158, 15–27

Asner GP, Martin RE, Carranza-Jiménez L, Sinca F, Tupayachi R, Anderson CB, Martinez P (2014) Functional and biological diversity of foliar spectra in tree canopies throughout the Andes to Amazon region. New Phytol 204:127–139

Ayres M, Clausen T, MacLean SF, Redman AM, Reichardt PB (1997) Diversity of structure and antiherbivore activity in condensed tannins. Ecology 78:1696–1712

Baldwin I (1994) Chemical changes rapidly induced by folivory. In: Bernays EA (ed) Insect-plant interactions. CRC Press, Tuscon, pp 1–23

Ball BA, Hunter MD, Kominoski JS, Swan CM, Bradford MA (2008) Consequences of non-random species loss for decomposition dynamics: experimental evidence for additive and non-additive effects. J Ecol 96:303–313

Bardgett RD (2017) Plant trait-based approaches for interrogating belowground function. Biol Environ Proc Royal Irish Acad 117:1–13

Bardgett RD, Van Der Putten WH (2014) Belowground biodiversity and ecosystem functioning. Nature 515:505–511

Bardgett RD, Mommer L, De Vries FT (2014) Going underground: root traits as drivers of ecosystem processes. Trends Ecol Evol 29:692–699

Bate-Smith E (1975) Phytochemistry of proanthocyanidins. Phytochemistry 14:1107–1113

Bellassen V, Delbart N, Le Maire G, Luyssaert S, Ciais P, Viovy N (2011) Potential knowledge gain in large-scale simulations of forest carbon fluxes from remotely sensed biomass and height. For Ecol Manag 261:515–530

Bellon-Maurel V, McBratney A (2011) Near-infrared (NIR) and mid-infrared (MIR) spectroscopic techniques for assessing the amount of carbon stock in soils - critical review and research perspectives. Soil Biol Biochem 43:1398–1410

Berg B (2014) Decomposition patterns for foliar litter – a theory for influencing factors. Soil Biol Biochem 78:222–232

Berryman E, Ryan MG, Bradford JB, Hawbaker TJ, Birdsey R (2016) Total belowground carbon flux in subalpine forests is related to leaf area index, soil nitrogen, and tree height. Ecosphere 7:e01418

Bossuyt H, Denef K, Six J, Frey SD, Merckx R, Paustian K (2001) Influence of microbial populations and residue quality on aggregate stability. Appl Soil Ecol 16:195–208

Bradford MA, Berg B, Maynard DS, Wieder WR, Wood SA (2016) Understanding the dominant controls on litter decomposition. J Ecol 104:229–238

Brinkmann K, Blaschke L, Polle A (2002) Comparison of different methods for lignin determination as a basis for calibration of near-infrared reflectance spectroscopy and implications of lignoproteins. J Chem Ecol 28:2483–2501

Burns RG, DeForest JL, Marxsen J, Sinsabaugh RL, Stromberger ME, Wallenstein MD, Weintraub MN, Zoppini A (2013) Soil enzymes in a changing environment: current knowledge and future directions. Soil Biol Biochem 58:216–234

Cadisch G, Giller K (1997) Driven by nature: plant litter quality and decomposition. CAB International, Wallingford

Cairns MA, Brown S, Helmer EH, Baumgardner GA (1997) Root biomass allocation in the world's upland forests. Oecologia 111:1–11

Cardinale BJ, Duffy JE, Gonzalez A, Hooper DU, Perrings C, Venail P, Narwani A, Mace GM, Tilman D, Wardle DA, Kinzig AP, Daily GC, Loreau M, Grace JB, Lariguaderie A, Srivasta DS, Naeem S (2012) Biodiversity loss and its impact on humanity. Nature 486:59–67

Cavender-Bares J, Gamon JA, Hobbie SE, Madritch MD, Meireles JE, Schweiger AK, Townsend PA (2017) Harnessing plant spectra to integrate the biodiversity sciences across biological and spatial scales. Am J Bot 104:966–969

Cecillon L, Barthes BG, Gomez C, Ertlen D, Genot V, Hedde M, Stevens A, Brun JJ (2009) Assessment and monitoring of soil quality using near infrared reflectance spectroscopy (NIRS). Eur J Soil Sci 60, 770–784

Chapin F, Matson P, Mooney H (2002) Principles of terrestrial ecosystem ecology. Springer, New York

Chen W, Xu R, Wu Y, Chen J, Zhang Y, Hu T, Yuan X, Zhou L, Tan T, Fan J (2018) Plant diversity is coupled with beta not alpha diversity of soil fungal communities following N enrichment in a semi-arid grassland. Soil Biol Biochem 116:388–398

Chisholm RA, Muller-Landau HC, Abdul Rahman K, Bebber DP, Bin Y, Bohlman SA, Bourg NA, Brinks J, Bunyavejchewin S, Butt N, Cao H, Cao M, Cárdenas D, Chang L-W, Chiang J-M, Chuyong G, Condit R, Dattaraja HS, Davies S, Duque A, Fletcher C, Gunatilleke N, Gunatilleke S, Hao Z, Harrison RD, Howe R, Hsieh C-F, Hubbell SP, Itoh A, Kenfack D, Kiratiprayoon S, Larson AJ, Lian J, Lin D, Liu H, Lutz JA, Ma K, Malhi Y, McMahon S, McShea W, Meegaskumbura M, Mohd. Razman S, Morecroft MD, Nytch CJ, Oliveira A, Parker GG, Pulla S, Punchi-Manage R, Romero-Saltos H, Sang W, Schurman J, Su S-H, Sukumar R, Sun I-F, Suresh HS, Tan S, Thomas D, Thomas S, Thompson J, Valencia R, Wolf A, Yap S, Ye W, Yuan Z, Zimmerman JK (2013) Scale-dependent relationships between tree species richness and ecosystem function in forests. J Ecol 101:1214–1224

Clark ML, Roberts DA, Clark DB (2005) Hyperspectral discrimination of tropical rain forest tree species at leaf to crown scales. Remote Sens Environ 96:375–398

Cline LC, Hobbie SE, Madritch MD, Buyarski CR, Tilman D, Cavender-Bares JM (2018) Resource availability underlies the plant-fungal diversity relationship in a grassland ecosystem. Ecology 99:204–216

Coleman DC, Crossley DA (1996) Fundamentals of soil ecology. Academic Press, San Diego

Coq S, Souquet J-M, Meudec E, Cheynier V, Hättenschwiler S (2010) Interspecific variation in leaf litter tannins drives decomposition in a tropical rain forest of French Guiana. Ecology 91:2080–2091

Cornwell WK, Cornelissen JHC, Amatangelo K, Dorrepaal E, Eviner VT, Godoy O, Hobbie SEHB, Kurokawa H, Pérez-Harguindeguy N, Quested HM, Santiago LS, Wardle DA, Wright IJ, Aerts R, Allison SD, van Bodegom P, Brovkin V, Chatain A, Callaghan TV, Diaz S, Garnier E, Gurvich DE, Kazakou E, Klein JA, Read J, Reich RB, Soudzilovskaia JA, Vaieretti MW, Westoby M (2008) Plant species traits are the predominant control on litter decomposition rates within biomes worldwide. Ecol Lett 11:1065–1071

Crutsinger GM, Collins MD, Fordyce JA, Gompert Z, Nice CC, Sanders NJ (2006) Plant genotypic diversity predicts community structure and governs an ecosystem process. Science 313:966–968

Dawud SM, Raulund-Rasmussen K, Ratcliffe S, Domisch T, Finér L, Joly FX, Hättenschwiler S, Vesterdal L (2017) Tree species functional group is a more important driver of soil properties than tree species diversity across major European forest types. Funct Ecol 31:1153–1162

De Deyn GB, van der Putten WH (2005) Linking aboveground and belowground diversity. Trends Ecol Evol 20:625–633

De Vries FT, Manning P, Tallowin JRB, Mortimer SR, Pilgrim ES, Harrison KA, Hobbs PJ, Quirk
    H, Shipley B, Cornelissen JHC, Kattge J, Bardgett RD (2012) Abiotic drivers and plant traits
    explain landscape-scale patterns in soil microbial communities. Ecol Lett 15:1230–1239
Díaz S, Hodgson JG, Thompson K, Cabido M, Cornelissen JHC, Jalili A, Montserrat-Martí G,
    Grime JP, Zarrinkamar F, Asri Y, Band SR, Basconcelo S, Castro- Díez P, Funes G, Hamzehee
    B, Khoshnevi M, Pérez-Harguindeguy N, Pérez-Rontomé MC, Shirvany FA, Vendramini F,
    Yazdani S, Abbas-Azimi R, Bogaard A, Boustani S, Charles M, Dehghan M, De Torres-Espuny
    L, Falczuk V, Guerrero-Campo J, Hynd A, Jones G, Kowsary E, Kazemi-Saeed F, Maestro-
    Martínez M, Romo-Díez A, Shaw S, Siavash B, Villar-Salvador P, Zak MR (2004) The plant
    traits that drive ecosystems: Evidence from three continents. Journal of Vegetation Science
    15:295–304
Díaz S, Kattge J, Cornelissen JHC, Wright IJ, Lavorel S, Dray S, Reu B, Kleyer M, Wirth C,
    Prentice IC, Garnier E, Bönisch G, Westoby M, Poorter H, Reich PB, Moles AT, Dickie J,
    Gillison AN, Zanne AE, Pierce S, Shipley B, Kirkup D, Casanoves F, Joswig JS, Günther A,
    Falczuk V (2016) The global spectrum of plant form and function. Nature 529:167–171
Eisenhauer N, Vogel A, Jensen B, Scheu S (2018) Decomposer diversity increases biomass pro-
    duction and shifts aboveground-belowground biomass allocation of common wheat. Sci Rep
    8:17894
Epps KY, Comerford NB, Reeves JB III, Cropper WP Jr, Araujo QR (2007) Chemical diversity—
    highlighting a species richness and ecosystem function disconnect. Oikos 116:1831–1840
Fanin N, Bertrand I (2016) Aboveground litter quality is a better predictor than belowground
    microbial communities when estimating carbon mineralization along a land-use gradient. Soil
    Biol Biochem 94:48–60
Feeny P (1970) Seasonal changes in oak leaf tannins an nutrients as a cause of spring feeding by
    winter moth caterpillars. Ecology 51:565–579
Fierer N, Strickland MS, Liptzin D, Bradford MA, Cleveland CC (2009) Global patterns in below-
    ground communities. Ecol Lett 12:1238–1249
Finerty GE, de Bello F, Bílá K, Berg MP, Dias ATC, Pezzatti GB, Moretti M (2016) Exotic or not,
    leaf trait dissimilarity modulates the effect of dominant species on mixed litter decomposition.
    J Ecol 104:1400–1409
Fisher JB, Sweeney S, Brzostek ER, Evans TP, Johnson DJ, Myers JA, Bourg NA, Wolf AT, Howe
    RW, Phillips RP (2016) Tree-mycorrhizal associations detected remotely from canopy spectral
    properties. Glob Chang Biol 22:2596–2607
Floudas D, Binder M, Riley R, Barry K, Blanchette RA, Henrissat B, Martínez AT, Otillar R,
    Spatafora JW, Yadav JS, Aerts A, Benoit I, Boyd A, Carlson A, Copeland A, Coutinho PM,
    de Vries RP, Ferreira P, Findley K, Foster B, Gaskell J, Glotzer D, Górecki P, Heitman J,
    Hesse C, Hori C, Igarashi K, Jurgens JA, Kallen N, Kersten P, Kohler A, Kües U, Kumar
    TKA, Kuo A, LaButti K, Larrondo LF, Lindquist E, Ling A, Lombard V, Lucas S, Lundell T,
    Martin R, McLaughlin DJ, Morgenstern I, Morin E, Murat C, Nagy LG, Nolan M, Ohm RA,
    Patyshakuliyeva A, Rokas A, Ruiz-Dueñas FJ, Sabat G, Salamov A, Samejima M, Schmutz
    J, Slot JC, St. John F, Stenlid J, Sun H, Sun S, Syed K, Tsang A, Wiebenga A, Young D,
    Pisabarro A, Eastwood DC, Martin F, Cullen D, Grigoriev IV, Hibbett DS (2012) The Paleozoic
    origin of enzymatic lignin decomposition reconstructed from 31 fungal genomes. Science
    336:1715–1719
Fontaine S, Barot S, Barré P, Bdioui N, Mary B, Rumpel C (2007) Stability of organic carbon in
    deep soil layers controlled by fresh carbon supply. Nature 450:277–280
Fortunel C, Garnier E, Joffre R, Kazakou E, Quested H, Grigulis K, Lavorel S, Ansquer P, Castro
    H, Cruz P, Doležal J, Eriksson O, Freitas H, Golodets C, Jouany C, Kigel J, Kleyer M, Lehsten
    V, Lepš J, Meier T, Pakeman R, Papadimitriou M, Papanastasis VP, Quétier F, Robson M,
    Sternberg M, Theau J-P, Thébault A, Zarovali M (2009) Leaf traits capture the effects of
    land use changes and climate on litter decomposability of grasslands across Europe. Ecology
    90:598–611
Frainer A, Moretti MS, Xu W, Gessner MO (2015) No evidence for leaf-trait dissimilarity effects
    on litter decomposition, fungal decomposers, and nutrient dynamics. Ecology 96:550–561

Freschet GT, Aerts R, Cornelissen JHC (2012) A plant economics spectrum of litter decompos-ability. Funct Ecol 16:56–65

Frey BR, Lieffers VJ, Hogg EH, Landhausser SM (2004) Predicting landscape patterns of aspen dieback: mechanisms and knowledge gaps. Can J For Res 34:1379–1390

Fujii S, Mori AS, Koide D, Makoto K, Matsuoka S, Osono T, Isbell F (2017) Disentangling rela-tionships between plant diversity and decomposition processes under forest restoration. J Appl Ecol 54:80–90

Funk JL, Larson JE, Ames GM, Butterfield BJ, Cavender-Bares J, Firn J, Laughlin DC, Sutton-Grier AE, Williams L, Wright J (2017) Revisiting the Holy Grail: using plant functional traits to understand ecological processes. Biol Rev 92:1156–1173

Gamon J (2008) Tropical remote sensing: opportunities and challenges. In: Kalacska M, Sanchez-Azofeifa GA (eds) Hyperspectral remote sensing of tropical and subtropical forests. CRC Press, New York, pp 297–305

García-Palacios P, Prieto I, Ourcival JM, Hättenschwiler S (2016) Disentangling the litter quality and soil microbial contribution to leaf and fine root litter decomposition responses to reduced rainfall. Ecosystems 19:490–503

Gartner TB, Cardon ZG (2004) Decomposition dynamics in mixed-species leaf litter. Oikos 104:230–246

Gessner MO, Swan CM, Dang CK, McKie BG, Bardgett RD, Wall DH, Hättenschwiler S (2010) Diversity meets decomposition. Trends Ecol Evol 25:372–380

Gholizadeh H, Gamon JA, Townsend PA, Zygielbaum AI, Helzer CJ, Hmimina GY, Yu R, Moore RM, Schweiger AK, Cavender-Bares J (2019) Detecting prairie biodiversity with airborne remote sensing. Remote Sens Environ 221:38–49

Gillon D, Joffre R, Dardenne P (1993) Predicting the stage of decay of decomposing leaves by near-infrared reflectance spectroscopy. Can J For Res-Revue Canadienne de Recherche Forestiere 23:2552–2559

Gillon D, Houssard C, Joffre R (1999) Using near- infrared reflectance spectroscopy to predict car-bon, nitrogen and phosphorus content in heterogeneous plant material. Oecologia 118:173–182

Gould IJ, Quinton JN, Weigelt A, De Deyn GB, Bardgett RD (2016) Plant diversity and root traits benefit physical properties key to soil function in grasslands. Ecol Lett 19:1140–1149

Grossman JJ, Vanhellemont M, Barsoum N, Bauhus J, Bruelheide H, Castagneyrol B, Cavender-bares J, Eisenhauer N, Ferlian O, Gravel D, Hector A, Jactel H, Kreft H, Mereu S, Messier C, Muys B, Nock C, Paquette A, Parker J, Perring MP, Ponette Q, Reich PB, Schuldt A, Staab M, Weih M, Clara D, Scherer-lorenzen M, Verheyen K (2018) Synthesis and future research direc-tions linking tree diversity to growth, survival, and damage in a global network of tree diversity experiments ☆. Environ Exp Bot 152:68–89

Handa IT, Aerts R, Berendse F, Berg MP, Bruder A, Butenschoen O, Chauvet E, Gessner MO, Jabiol J, Makkonen M, McKie BG, Malmqvist B, Peeters ETHM, Scheu S, Schmid B, Van Ruijven J, Vos VCA, Hättenschwiler S (2014) Consequences of biodiversity loss for litter decomposition across biomes. Nature 509:218–221

Haslam E (1989) Plant polyphenols: vegetable tannins revisited. Cambridge University Press, Cambridge

Hatakka A, Hammel KE (2010) Fungal biodegradation of lignocelluloses. In: Osiewacz HD (ed) The Mycota: a comprehensive treatise on fungi as experimental systems for basic and applied research. Springer, New York, pp 319–334

Hättenschwiler S, Jørgensen HB (2010) Carbon quality rather than stoichiometry controls litter decomposition in a tropical rain forest. J Ecol 98:754–763

Hättenschwiler S, Vitousek P (2000) The role of polyphenols in terrestrial ecosystem nutrient cycling. TREE 15:238–243

Hättenschwiler S, Tiunov AV, Scheu S (2005) Biodiversity and litter decomposition in terrestrial ecosystems. Ann Rev Ecol Evol Syst 36:191–218

Hättenschwiler S, Aeschlimann B, Coûteaux M-M, Roy J, Bonal D (2008) High variation in foli-age and leaf litter chemistry among 45 tree species of a neotropical rainforest community. New Phytol 179:165–75

Hättenschwiler S, Coq S, Barantal S, Handa IT (2011) Leaf traits and decomposition in tropical rainforests: revisiting some commonly held views and towards a new hypothesis. New Phytol 189:950–965

Hector A, Beale AJ, Minns A, Otway SJ, Lawton JH (2000) Consequences of the reduction of plant diversity for litter decomposition: effects through litter quality and microenvironment. Oikos 90:357–371

Herms DA, Mattson WJ (1992) The dilemma of plants: to grow or defend. Q Rev Biol 67:283–335

Higuchi T (2006) Look back over the studies of lignin biochemistry. J Wood Sci 52:2–8

Hobbie SE (1992) Effects of plant species on nutrient cycling. Trends Ecol Evol 7:336–339

Hobbie JE, Hobbie EA (2006) 15N in symbiotic fungi and plants estimates nitrogen and carbon flux rates in Arctic tundra. Ecology 87:816–822

Hobbie SE (2015) Plant species effects on nutrient cycling: revisiting litter feedbacks. Trends Ecol Evol 30:357–63

Högberg P, Nordgren A, Buchmann N, Taylor AFS, Ekblad A, Högberg MN, Nyberg G, Ottosson-Löfvenius M, Read DJ (2001) Large-scale forest girdling shows that current photosynthesis drives soil respiration. Nature 411:789–792

Holden SR, Rogers BM, Treseder KK, Randerson JT (2016) Fire severity influences the response of soil microbes to a boreal forest fire. Environ Res Lett 11:035004

Hooper DU, Bignell DE, Brown VK, Brussard L, Dangerfield JM, Wall DH, Wardle DA, Coleman DC, Giller KE, Lavelle P, van der Putten WH, de Ruiter PC, Rusek J, Silver WL, Tiedje JM, Wolters V (2000) Interactions between aboveground and belowground biodiversity in terrestrial ecosystems: patterns, mechanisms, and feedbacks. Bioscience 50:1049–1061

Hoorens B, Aerts R, Stroetenga M (2003) Does initial litter chemistry explain litter mixture effects on decomposition? Oecologia 137:578–586

Horner JD, Gosz JR, Cates RG (1988) The role of carbon-based plant secondary metabolites in decomposition in terrestrial ecosystems. Am Nat 132:869–883

Huang N, Gu L, Black TA, Wang L, Niu Z (2015) Remote sensing-based estimation of annual soil respiration at two contrasting forest sites 2306–2325

Hughes AR, Inouye BD, Johnson MTJ, Underwood N, Vellend M (2008) Ecological consequences of genetic diversity. Ecol Lett 11:609–623

Jewell MD, Shipley B, Low-Décarie E, Tobner CM, Paquette A, Messier C, Reich PB (2017) Partitioning the effect of composition and diversity of tree communities on leaf litter decomposition and soil respiration. Oikos 126:959–971

Jo I, Fridley JD, Frank DA (2016) More of the same? In situ leaf and root decomposition rates do not vary between 80 native and nonnative deciduous forest species. New Phytol 209:115–122

Johnson MTJ, Agrawal AA (2005) Plant genotype and environment interact to shape a diverse arthropod community on evening primrose (*Oenothera biennis*). Ecology 86:874–885

Kardol P, Wardle DA (2010) How understanding aboveground-belowground linkages can assist restoration ecology. Trends Ecol Evol 25:670–679

Keiser AD, Knoepp JD, Bradford MA (2013) Microbial communities may modify how litter quality affects potential decomposition rates as tree species migrate. Plant Soil 372:167–176

Kirby R (2006) Actinomycetes and lignin degradation. Adv Appl Microbiol 58:125–168

Knops JMH, Wedin D, Tilman D (2001) Biodiversity and decomposition in experimental grassland ecosystems. Oecologia 126:429–433

Kokaly RF, Clark RN (1999) Spectroscopic determination of leaf biochemistry using band-depth analysis of absorption features and stepwise multiple linear regression. Remote Sens Environ 67:267–287

Kokaly RF, Asner GP, Ollinger SV, Martin ME, Wessman CA (2009) Characterizing canopy biochemistry from imaging spectroscopy and its application to ecosystem studies. Remote Sens Environ 113:S78–S91

Kraus TEC, Dahlgren RA, Zasoski RJ (2003) Tannins in nutrient dynamics of forest ecosystems— a review. Plant Soil 256:41–66

Krause S et al (2014) Trait-based approaches for understanding microbial biodiversity and ecosystem functioning. Front Microbiol 5:251

Laliberté E (2017) Tansley insight - Below-ground frontiers in trait-based plant ecology. New Phytol 213:1597–1603

Lambers H, Chapin FS, Pons TL (1998) Plant physiological ecology. Springer-Verlag New York Inc., New York

Landsberg JJ, Waring RH (1997) A generalised model of forest productivity using simplified concepts of radiation-use efficiency, carbon balance and partitioning. For Ecol Manag 95:209–228

Lauber CL, Strickland MS, Bradford MA, Fierer N (2008) The influence of soil properties on the structure of bacterial and fungal communities across land-use types. Soil Biol Biochem 40:2407–2415

Lee HS, Davies SJ, Lafrankie J, Ashton PS (2002) Floristic and structural diversity of mixed dipterocarp forest in Lambir Hill National Park, Sarawak, Malaysia. J Trop For Sci 14:379–400

Liang J, Crowther TW, Picard N, Wiser S, Zhou M, Alberti G, Schulze E, Mcguire AD, Bozzato F, Pretzsch H, Paquette A, Hérault B, Scherer-lorenzen M, Barrett CB, Glick HB, Hengeveld GM, Nabuurs G, Pfautsch S, Viana H, Vibrans AC, Ammer C, Schall P, Verbyla D, Tchebakova N, Fischer M, Watson JV, Chen HYH, Lei X, Schelhaas M, Lu H, Gianelle D, Parfenova EI, Salas C, Lee E, Lee B, Kim HS, Bruelheide H, Coomes DA, Piotto D, Sunderland T, Schmid B, Gourlet-fleury S, Sonké B, Tavani R, Zhu J, Brandl S, Baraloto C, Frizzera L, Ba R, Oleksyn J, Peri PL, Gonmadje C, Marthy W, Brien TO, Martin EH, Marshall AR, Rovero F, Bitariho R, Niklaus PA, Alvarez-loayza P, Chamuya N, Valencia R, Mortier F, Wortel V, Engone-obiang NL, Ferreira LV (2016) Positive biodiversity-productivity relationship predominant in global forests. Science 354:aaf8957

Lindroth RL, St. Clair SB (2013) Adaptations of quaking aspen (Populus tremuloides Michx.) for defense against herbivores. For Ecol Manag 299:14–21

Lindroth RL, Osier TL, Barnhill HR, Wood SA (2002) Effects of genotype and nutrient availability on phytochemistry of trembling aspen (Populus tremuloides Michx.) during leaf senescence. Biochem Syst Ecol 30:297–307

Litton CM, Raich JW, Ryan MG (2007) Carbon allocation in forest ecosystems. Glob Chang Biol 13:2089–2109

Lohbeck M, Poorter L, Martínez-Ramos M, Bongers F (2015) Biomass is the main driver of changes in ecosystem process rates during tropical forest succession. Ecology 96:1242–1252

López-Mondéjar R, Zühlke D, Becher D, Riedel K, Baldrian P (2016) Cellulose and hemicellulose decomposition by forest soil bacteria proceeds by the action of structurally variable enzymatic systems. Sci Rep 6:25279

Madritch MD, Hunter MD (2002) Phenotypic diversity influences ecosystem functioning in an oak sandhills community. Ecology 83:2084–2090

Madritch MD, Hunter MD. 2003. Intraspecific litter diversity and nitrogen deposition affect nutrient dynamics and soil respiration. Oecologia 136:124–8.

Madritch MD, Kingdon CC, Singh A, Mock KE, Lindroth RL, Townsend PA, B PTRS, Madritch MD, Kingdon CC, Singh A, Mock KE, Lindroth RL, Townsend PA (2014) Imaging spectroscopy links aspen genotype with below-ground processes at landscape scales Imaging spectroscopy links aspen genotype with belowground processes at landscape scales

Madritch MD, Lindroth RL (2015) Condensed tannins increase nitrogen recovery by trees following insect defoliation. New Phytol 208:410–420

Madritch MD, Donaldson JR, Lindroth RL (2006) Genetic identity of Populus tremuloides litter influences decomposition and nutrient release in a mixed forest stand. Ecosystems 9:528–537

Madritch MD, Jordan LM, Lindroth RL (2007) Interactive effects of condensed tannin and cellulose additions on soil respiration. Can J For Res 37:2063–2067

Martin ME, Aber JD (1997) High spectral resolution remote sensing of forest canopy lignin, nitrogen, and ecosystem processes. Ecol Appl 7:431–443

Martin ME, Newman SD, Aber JD, Congalton RG (1998) Determining forest species composition using high spectral remote sensing data. Remote Sens Environ 65:249–254

Martin ME, Plourde LC, Ollinger SV, Smith M-L, McNeil BE (2008) A generalizable method for remote sensing of canopy nitrogen across a wide range of forest ecosystems. Remote Sens Environ 112:3511–3519

Meentemeyer V (1978) Macroclimate and lignin control of litter decomposition rates. Ecology 59(3):465–472

Meier CL, Bowman WD (2008) Links between plant litter chemistry, species diversity, and belowground ecosystem function. Proc Natl Acad Sci 105:19780–19785

Melillo JM, Aber JD, Muratone JF (1982) Nitrogen and lignin control of hardwood leaf litter decomposition dynamics. Ecology 63:621–626

Mitton JB, Grant MC (1996) Genetic variation and the natural history of quaking aspen. Bioscience 46:25–31

Mokany K, Raison RJ, Prokushkin AS (2006) Critical analysis of root:shoot ratios in terrestrial biomes. Glob Change Biol 12:84–96

Mueller KE, Hobbie SE, Chorover J, Reich PB, Eisenhauer N, Castellano MG, Chadwick OA, Dobies T, Hale CM, Jagodziński AM, Kalucka I, Kieliszewska-Rokicka B, Modrzyński J, Rożen A, Skorupski M, Sobczyk Ł, Stasińska M, Trocha LK, Weiner J, Wierzbicka A, Oleksyn J (2015) Effects of litter traits, soil biota, and soil chemistry on soil carbon stocks at a common garden with 14 tree species. Biogeochemistry 123:313–327

Naeem S, Tjossem SF, Byers D, Bristow C, Li S (1999) Plant neighborhood diversity and production. Ecoscience 6:355–365

Nannipieri P, Trasar-Cepeda C, Dick RP (2018) Soil enzyme activity: a brief history and biochemistry as a basis for appropriate interpretations and meta-analysis. Biol Fertil Soils. (2018 54:11–19

Ollinger SV, Smith ML, Martin ME, Hallett RA, Goodale CL, Aber JD (2002) Regional variation in foliar chemistry and N cycling among forests of diverse history and composition. Ecology 83:339–355

Ollinger SV, Reich PB, Frolking S, Lepine LC, Hollinger DY, Richardson AD (2013) Nitrogen cycling, forest canopy reflectance, and emergent properties of ecosystems. Proc Natl Acad Sci U S A 110:E2437

Osono T (2007) Ecology of ligninolytic fungi associated with leaf litter decomposition. Ecol Res 22:955–974

Parsons SA, Lawler IR, Congdon RA, Williams SE (2011) Rainforest litter quality and chemical controls on leaf decomposition with near-infrared spectrometry. J Plant Nutr Soil Sci 174:710–720

Phillips RP, Brzostek E, Midgley MG (2013) The mycorrhizal-associated nutrient economy: a new framework for predicting carbon-nutrient couplings in temperate forests. New Phytol 199:41–51

Reich PB, Tilman D, Isbell F, Mueller K, Hobbie SE, Flynn DFB, Eisenhauer N (2012) Impacts of biodiversity loss escalate through time as redundancy fades. Science 336:589–592

Reich PB (2014) The world-wide 'fast–slow' plant economics spectrum: a traits manifesto. J Ecol 102:275–301

Robinson D (2001) δ15N as an integrator of the nitrogen cycle. Trends Ecol Evol 16:153–162

Rubert-Nason KF, Holeski LM, Couture JJ, Gusse A, Undersander DJ, Lindroth RL (2013) Rapid phytochemical analysis of birch (Betula) and poplar (Populus) foliage by near-infrared reflectance spectroscopy. Anal Bioanal Chem 405:1333–44

Sabetta L, Zaccarelli N, Mancinelli G, Mandrone S, Salvatori R, Costantini ML, Zurlini G, Rossi L (2006) Mapping litter decomposition by remote-detected indicators. Ann Geophys 49:219–226

Saatchi SS, Harris NL, Brown S, Lefsky M, Mitchard ETA, Salas W (2011) Benchmark map of forest carbon stocks in tropical regions across three continents. 108

Scherer-Lorenzen M, Bonilla JL, Potvin C (2007) Tree species richness affects litter production and decomposition rates in a tropical bio-diversity experiment. Oikos 116:2108–2124

Schimel J, Van Cleve K, Cates R, Clausen T, Reichardt P (1996) Effects of balsam poplar (Populus balsamifera) tannins and low molecular weight phenolics on microbial activity in taiga floodplain soil: implications for changes in N cycling during succession. Can J Bot 74:84–90

Schloss PD, Handelsman J (2006) Toward a census of bacteria in soil. PLoS Comput Biol 2(7):e92

Schneider T, Keiblinger KM, Schmid E, Sterflinger-Gleixner K, Ellersdorfer G, Roschitzki B, Richter A, Eberl L, Zechmeister-Boltenstern S, Riedel K (2012) Who is who in litter decomposition? Metaproteomics reveals major microbial players and their biogeochemical functions. ISME J 6:1749–1762

Schnitzer M, Barr M, Hartenstein R (1984) Kinetics and characteristics of humic acids produced from simple phenols. Soil Biol Biochem 16:371–376

Schöler A, Jacquiod S, Vestergaard G, Schulz S, Schloter M (2017) Analysis of soil microbial communities based on amplicon sequencing of marker genes. 485–489

Schweiger AK, Cavender-Bares J, Townsend PA, Hobbie SE, Madritch MD, Wang R, Tilman D, Gamon JA (2018) Plant spectral diversity integrates functional and phylogenetic components of biodiversity and predicts ecosystem function. Nat Ecol Evol 2:976. https://doi.org/10.1038/s41559-018-0551-1

Schweitzer JA, Bailey JK, Hart SC, Wimp GM, Chapman SK, Whitham TG (2005) The interaction of plant genotype and herbivory deccelerate leaf litter decomposition and alter nutrient dynamics. Oikos 110:133–145

Serbin SP, Singh A, McNeil BE, Kingdon CC, Townsend PA (2014) Spectroscopic determination of leaf morphological and biochemical traits for northern temperate and boreal tree species. Ecol Appl 24:1651–1669

Serbin SP, Singh A, Desai AR, Dubois SG, Jablonski AD, Kingdon CC, Kruger EL, Townsend PA (2015) Remotely estimating photosynthetic capacity, and its response to temperature, in vegetation canopies using imaging spectroscopy. Remote Sens Environ 167, 78–87

Shepherd KD, Vanlauwe B, Gachengo CN, Palm CA (2005) Decomposition and mineralization of organic residues predicted using near infrared spectroscopy. Plant and Soil 277:315–333.

Shields WJ, Bockheim JG (1981) Deterioration of trembling aspen clones in the Great-Lakes region. Can J For Res 11:530–537

Smith VC, Bradford MA (2003) Do non-additive effects on decomposition in litter-mix experiments result from differences in resource quality between litters? Oikos 102:235–242

Soriano-Disla JM, Janik LJ, Viscarra Rossel RA, Macdonald LM, McLaughlin MJ (2014) The performance of visible, near-, and mid-infrared reflectance spectroscopy for prediction of soil physical, chemical, and biological properties. Appl Spectrosc Rev 49:139–186

Srivastava DS, Cardinale BJ, Downing AL, Duffy JE, Jouseau C, Sankaran M, Wright JP (2009) Diversity has stronger top-down than bottom-up effects on decomposition. Ecology 90:1073–1083

Stenberg B, Rossel RAV, Mouazen AM, Wetterlind J (2010) Visible and near infrared spectroscopy in soil science. In: Sparks DL (ed) Advances in agronomy, vol 107, pp 163–215

Talbot JM, Yelle DJ, Nowick J, Treseder KK (2012) Litter decay rates are determined by lignin chemistry. Biogeochemistry 108:279–295

Tenney FG, Waksman SA (1929) Composition of natural organic materials and their decomposition in the soil. IV. The nature and rapidity of decomposition of the various organic complexes in different plant materials, under aerobic conditions. Soil Sci 28:55–84

Thers H, Brunbjerg AK, Læssøe T, Ejrnæs R, Bøcher PK, Svenning J (2017) Lidar-derived variables as a proxy for fungal species richness and composition in temperate Northern Europe. Remote Sens Environ 200:102–113

Thevenot M, Dignac MF, Rumpel C (2010) Fate of lignins in soils: a review. Soil Biol Biochem 42:1200–1211

Tilman D, Reich PB, Knops J, Wedin D, Mielke T, Lehman C (2001) Diversity and productivity in a long-term grassland experiment. Science 294:843–845

Tilman D, Reich PB, Knops J (2006) Biodiversity and ecosystem stability in a decade-long grassland experiment. Nature 441:629–632

Townsend PA, Foster JR, Chastain RA, Currie WS (2003) Application of imaging spectroscopy to mapping canopy nitrogen in the forests of the central Appalachian Mountains using Hyperion and AVIRIS. IEEE Trans Geosci Remote Sens 41:1347–1354

Townsend PA, Serbin SP, Kruger EL, Gamon JA (2013) Disentangling the contribution of bio-logical and physical properties of leaves and canopies in imaging spectroscopy data. Proc Natl Acad Sci 110:E1074–E1074

Triebwasser DJ, Tharayil N, Preston CM, Gerard PD (2012) The susceptibility of soil enzymes to inhibition by leaf litter tannins is dependent on the tannin chemistry, enzyme class and vegeta-tion history. New Phytol 196:1122–1132

Ustin SL, Roberts DA, Gamon JA, Asner GP, Green RO (2004) Using imaging spectroscopy to study ecosystem processes and properties. Bioscience 54:523

Vitousek P, Asner GP, Chadwick OA, Hotchkiss S (2009) Landscape-level variation in for-est structure and biogeochemistry across a substrate age gradient in Hawaii. Ecology 90:3074–3086

Wang Z, Townsend PA, Schweiger AK, Couture JJ, Singh A, Hobbie SE, Cavender-Bares J (2019) Mapping foliar functional traits and their uncertainties across three years in a grassland experi-ment. Remote Sens Environ 221:405–416

Wardle DA (1999) Is "sampling effect" a problem for experiments investigating biodiversity-ecosystem function relationships? Oikos 87:403–407

Wardle DA, Nicholson KS (1996) Synergistic effects of grassland plant species on soil microbial biomass and activity: implications for ecosystem-level effects of enriched plant diversity. Funct Ecol 10:410–416

Wardle DA, Bonner KI, Barker GM (2000) Stability of ecosystem properties in response to above-ground functional group richness and composition. Oikos 89:11–23

Wardle DA, Bardgett RD, Klironomos JN, Setälä H, van der Putten WH, Wall DH (2004) Ecological linkages between aboveground and belowground biota. Science 304:1629–1633

Weisser WW, Roscher C, Meyer ST, Ebeling A, Luo G, Allan E, Beßler H, Barnard RL, Buchmann N, Buscot F, Engels C, Fischer C, Fischer M, Gessler A, Gleixner G, Halle S, Hildebrandt A, Hillebrand H, de Kroon H, Lange M, Leimer S, Le Roux X, Milcu A, Mommer L, Niklaus PA, Oelmann Y, Proulx R, Roy J, Scherber C, Scherer-Lorenzen M, Scheu S, Tscharntke T, Wachendorf M, Wagg C, Weigelt A, Wilcke W, Wirth C, Schulze E-D, Schmid B, Eisenhauer N (2017) Biodiversity effects on ecosystem functioning in a 15-year grassland experiment: patterns, mechanisms, and open questions. Basic Appl Ecol 23:1–73

Wessman CA, Aber JD, Peterson DL, Melillo JM (1988) Remote sensing of canopy chemistry and nitrogen cycling in temperate forest ecosystems. Nature 335:154–156

Williams LJ, Paquette A, Cavender-Bares J, Messier C, Reich PB (2017) Spatial complementarity in tree crowns explains overyielding in species mixtures. Nat Ecol Evol 1:0063

Wimp GM, Young WP, Woolbright SA, Martinsen GD, Keim P, Whitham TG (2004) Conserving plant genetic diversity for dependent animal communities. Ecol Lett 7:776–780

Wimp GM, Martinsen GD, Floate KD, Bangert RK, Whitham TG (2005) Plant genetic determi-nants of arthropod community structure and diversity. Evolution 59:61–69

Worrall JJ, Egeland L, Eager T, Mask RA, Johnson EW, Kemp PA, Shepperd WD (2008) Rapid mortality of Populus tremuloides in southwestern Colorado, USA. For Ecol Manag 255:686–696

Wright IJ, Reich PB, Westoby M, Ackerly DD, Baruch Z, Bongers F, Cavender- Bares J, Chapin T, Cornelissen JHC, Diemer M, Flexas J, Garnier E, Groom PK, Gulias J, Hikosaka K, Lamont BB, Lee T, Lee W, Lusk C, Midgley JJ, Navas M-L, Niinemets Ü, Oleksyn J, Osada N, Poorter H, Poot P, Prior L, Pyankov VI, Roumet C, Thomas SC, Tjoelker MG, Veneklaas EJ, Villar R (2004) The worldwide leaf economics spectrum. Nature 428:821–827

Wu T, Ayres E, Bardgett RD, Wall DH, Garey JR (2011) Molecular study of worldwide distribution and diversity of soil animals. PNAS 108:17720–17725

Open Access   This chapter is licensed under the terms of the Creative Commons Attribution 4.0 International License (http://creativecommons.org/licenses/by/4.0/), which permits use, sharing, adaptation, distribution and reproduction in any medium or format, as long as you give appropriate credit to the original author(s) and the source, provide a link to the Creative Commons license and indicate if changes were made.

The images or other third party material in this chapter are included in the chapter's Creative Commons license, unless indicated otherwise in a credit line to the material. If material is not included in the chapter's Creative Commons license and your intended use is not permitted by statutory regulation or exceeds the permitted use, you will need to obtain permission directly from the copyright holder.

Chapter 9
Using Remote Sensing for Modeling and Monitoring Species Distributions

Jesús N. Pinto-Ledezma and Jeannine Cavender-Bares

9.1 Introduction

What drives species distributions? This is one of the most fundamental questions in ecology, evolution, and biogeography, and it drew the attention of early naturalists (Gaston 2009; Guisan et al. 2017). Although the question is classic and its answers sometime seem obvious—for example, Alfred Russel Wallace recognized the effect of geographical and environmental features on species distributional ranges (Wallace 1860)—the answers are highly complex as a consequence of historical evolutionary and biogeographic processes and the spatial and temporal dynamics of abiotic and biotic factors (Soberón and Peterson 2005; Soberón 2007; Colwell and Rangel 2009).

Here we explore the potential of satellite remote sensing (S-RS) products to quantify species-environment relationships that predict species distributions. We propose several new metrics that take advantage of the high temporal resolution in Moderate Resolution Imaging Spectroradiometer (MODIS) leaf area index (LAI) and MODIS normalized difference vegetation index (NDVI) data products. Evaluating the potential of remotely sensed data in environmental niche modeling (ENM) and species distribution modeling (SDM) is an important step toward the long-term goal of improving our ability to monitor and predict changes in biodiversity globally. To achieve this, we first modeled the environmental/ecological niches for the American live oak species (*Quercus* section *Virentes*) using environmental variables derived from (1) interpolated climate surfaces data (i.e., WorldClim) and

J. N. Pinto-Ledezma (✉) · J. Cavender-Bares
Department of Ecology, Evolution and Behavior, University of Minnesota,
Saint Paul, MN, USA
e-mail: jpintole@umn.edu

© The Author(s) 2020
J. Cavender-Bares et al. (eds.), *Remote Sensing of Plant Biodiversity*,
https://doi.org/10.1007/978-3-030-33157-3_9

(2) S-RS products. Live oaks are a small lineage, descended from a common ancestor (for a discussion on phylogenetics, see Meireles et al., Chap. 7) that includes seven species that vary in geographic range size and climatic breadth and are distributed in both temperate and tropical climates from the southeastern United States, Mesoamerica, and the Caribbean (Cavender-Bares et al. 2015). Their variation in range size and climatic distributions, their distributions in both *highly* studied and *understudied* regions of the globe, and the second author's expert knowledge of their distributions make them an interesting case study for comparing SDM/ENMs that rely on classic data sources to those that use remotely sensed data sources, which have more consistent data accuracy and resolution. We used the live oaks as a test clade to evaluate the relationships among the modeled niches estimated from both sources of environmental data.

Given that the interpolated climate surfaces from WorldClim (Hijmans et al. 2005) are the most widely used data set for the study of species-environment relationships, we compare the performance of SDMs based on S-RS products to those based on WorldClim data. If there is a tight relationship between models from the two sources, this would indicate that the resultant models from S-RS products have similar performance to the resultant models from the WorldClim climatic predictors. Remotely sensed data products may provide an advantage in predicting species distributions in regions where climatic data is sparsely sampled. Although WorldClim provides interpolated climate surfaces for land areas across the world at multiple spatial resolutions, from 30 arc seconds (~1 km) to 10 arcmin (~18.5 km) (Hijmans et al. 2005), the spatial distribution of the base information (i.e., weather or climatic stations) used for interpolations is unevenly distributed across the world (Fig. 9.2c). This is not a small issue given the uncertainty associated with interpolated climatic variables when modeling species-environment relationships, especially in many tropical countries, where weather stations are frequently few and far apart (Soria-Auza et al. 2010). Given that tropical regions are precisely the regions where most species occur (Fig. 9.2c), finding alternative means to predict species is important for efforts to monitor and manage biodiversity globally. S-RS products, which provide quasi-global coverage of land and sea surfaces at high temporal and spatial resolution, represent promising alternatives that may be particularly important in the world's most biologically diverse regions. Our aim here is to provide an understanding of the potential of S-RS products to quantify species ecological niches and estimate species distributions rather than to develop a definitive ecological and geographical profile for the live oaks themselves. If the consistent accuracy and high spatial resolution of S-RS products can actually improve estimates of species distributions, they will represent an advance in our ability to predict where species are likely persist under changing environments. Ultimately, such predictions can be combined with other remote sensing (RS) means of detecting species and biodiversity (Meireles et al., Chap. 7; Bolch et al., Chap. 12; Record et al., Chap. 10) to enable global-scale biodiversity change detection.

9.2   Theoretical Background

9.2.1   The BAM Diagram

One way to explore the ubiquitous relationship between the spatial and temporal dynamics of abiotic and biotic factors is through the BAM framework (Fig. 9.1, Soberón and Peterson 2005; Soberón 2007), which formally describes the individual and joint effects of biotic factors (B; e.g., species interactions), abiotic factors (A; e.g., environmental conditions or abiotically suitable area), and movement (M; e.g., species dispersal capacity) in determining species distributions in a geographical space (G; e.g., the study region). Notice that all factors in the BAM framework are placed within a spatial context. Within the geographical space (G), three overlapping circles are shown, each of which represents suitable conditions for a given species. The intersection between all factors "B∩A∩M" represents the occupied distributional area (G0) or the "realized" or occupied niche. The intersection between biotic and abiotic factors "B∩A" represents the invadable distributional area (G1) or areas that can be colonized because suitable biotic and abiotic conditions and both present. The intersection between abiotic and movement factors "A∩M" represents the area where the species cannot be found. Finally, the union between the occupied and invadable areas "G0∪G1" represents the geographic potential distribution area (GP) or biotically reduced niche (see Soberón 2007; Peterson et al. 2011 for detailed explanations).

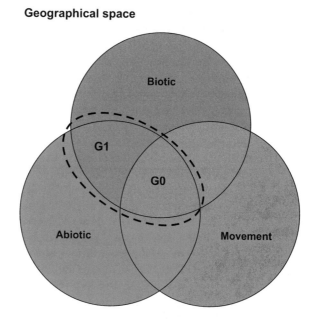

Fig. 9.1 The BAM diagram, where B biotic, A abiotically suitable area, and M movement or migration, illustrates the relationship among the three major determinants of species distributions. G1, the invadable distributional area, and G0, the occupied distributional area, represent the outcomes from the intersection between the major determinants

Geographical space

Biotic

G1

G0

Abiotic

Movement

9.2.2 Where Are We Now?

Although the BAM framework was developed to understand and quantify species-environment relationships (Soberón 2007; see also Soberón and Peterson 2005), the concept and investigation of species-environment relationships are long-standing, dating back to Wallace (Wallace 1860) and early ecologists (Grinnell 1904, 1917; Elton 1927; Holdridge 1947; Hutchinson 1957). These early naturalists originally established the theoretical principles to analyze and describe biogeographical distributions in relation to environmental patterns (Colwell and Rangel 2009). Interestingly, despite the large body of theoretical advances and empirical applications, the quantification of ecological niches and estimation of species distributions is still a challenging task (but see Sanín and Anderson 2018; Smith et al. 2018) and one of the most active areas in macroecological and biodiversity research (Franklin 2010; Peterson et al. 2011; Anderson 2013; Guisan et al. 2017).

In fact, since the first algorithm for modeling species-environment relationship was presented (BIOCLIM, Nix 1986), the number of publications has increased dramatically (Lobo et al. 2010; Booth et al. 2013). A simple search in Google Scholar for the terms "ecological niche model" and "species distribution model" (last accessed on December 30, 2018) returned 2,950 and 6,400 citations, respectively, for 1990–2018 (Fig. 9.2a). Interestingly, the number of publications on these topics increased markedly in the past 10 years (Fig. 9.2, see also Lobo et al. 2010) and continues to grow, particularly in studies that emphasize the application of ENMs and SDMs to environmental assessment, forecasting, and hindcasting species distributions (Anderson 2013; Elith and Franklin 2013; Guisan et al. 2017). Interestingly, although the number of publications increased in the last 10 years, most of the studies were performed in United States and Europe (Fig. 9.2b) in countries with a high density of weather stations (Fig. 9.2c), with much less emphasis on the most diverse regions of the globe. The increasing access to species occurrence data (e.g., Global Biodiversity Information Facility, GBIF) and environmental data (climatic and satellite derived) has created the opportunity not only to model species-environment relationships but to expand the theoretical and practical applications of ENM and SDM to different research programs and fields, including conservation biology, wildlife and ecosystem management, evolutionary biology, and public health (Franklin 2010; Peterson et al. 2011; Guisan et al. 2017), and to do so in remote regions where access is limited and predictions of species distributions have disproportionate importance.

Parallel to the development and evolution of ENM and SDM theory and applications, we have witnessed the growth of technological tools and S-RS products (Pettorelli et al. 2014a; Turner 2014). Many of these are particularly applicable for describing, quantifying, and mapping the spatial and temporal patterns of vegetation structure and function, the impacts of human activities, and environmental change (Turner et al. 2003; Pinto-Ledezma and Rivero 2014; Jetz et al. 2016; Cord et al. 2017) and more recently are used as predictors of broad patterns of biodiversity, including the associations between species co-occurrence patterns and ecosystem energy availability (Phillips et al. 2008; Pigot et al. 2016; Hobi et al. 2017). In addition, an unprecedented number of S-RS data and data products (S-RS) have

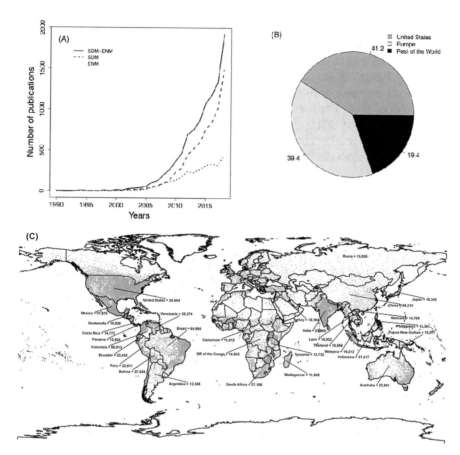

Fig. 9.2 (a) Number of publications containing the term "species distribution model" or "ecological niche model" between 1990 and 2018 (Google Scholar search December 31, 2018). The solid line represents the combination of both SDM and ENM, while dashed and dotted lines indicate the individual terms. (b) Percentage of ENM/SDM studies performed in the United States and Europe in relation to the total number of publications from (a). (c) Distribution of weather stations (green dots) used to create the interpolated climate surfaces (i.e., WorldClim) and the number of species for the 30 most diverse countries. The numbers correspond to the estimated number of species—vertebrates and vascular plants—for each country. Notice that for most countries the weather stations are sparse and have low coverage. (Source: WorldClim: Global weather stations, 2014 (http://databasin.org/dataset s/15a31dec689b4c958ee491ff30fcce75); biodiversity data: World Conservation Monitoring Centre of the United Nations Environment Programme (UNEP-WCMC), 2004)

been made freely available (Turner et al. 2003; Hobi et al. 2017) with the potential to track the spatial variation in the chemical composition of vegetation (Wang et al. 2019; Serbin and Townsend, Chap. 3), physiology, structure, and function (Lausch et al., Chap. 13; Serbin and Townsend, Chap. 3; Myneni et al. 2002; Saatchi et al. 2008; Jetz et al. 2016).

Despite the potential of S-RS products for measuring and modeling biodiversity (Gillespie et al. 2008; Pettorelli et al. 2014a, b; Turner 2014; Cord et al. 2013;

Zimmermann et al. 2007), attention has only recently turned to using these data in studies of species-environment relationships (Cord et al. 2013; West et al. 2016), and most studies use bioclimatic data such as WorldClim (but see Paz et al., Chap. 11; Record et al., Chap. 10). Although early attempts indicated that S-RS products do not seem to improve the accuracy in estimating species distributions (Pearson et al. 2004; Thuiller 2004; Zimmermann et al. 2007), more recent publications (Kissling et al. 2012; Cord et al. 2013) suggest that despite these apparent limitations, S-RS products provide better spatial resolution that allow the discrimination of habitat characteristics not captured when bioclimatic data are used (Saatchi et al. 2008; Cord et al. 2013), and they can be used as surrogates of biotic and/or functional predictors such as LAI that increase the performance of individual species models (Kissling et al. 2012; Cord et al. 2013).

9.3 Modeling Ecological Niches and Predicting Geographic Distributions

Although the terms ENM and SDM are often used synonymously in the literature, the two are not equivalent (Anderson 2012; Soberón et al. 2017). A comprehensive discussion of this topic is beyond the scope of this chapter but is provided elsewhere (see Peterson et al. 2011; Anderson 2012; Soberón et al. 2017). A crucial step in differentiating the two terms is to establish a distinction between environmental space and geographical space (Hutchinson's duality; Colwell and Rangel 2009). On the one hand, environmental space corresponds to a suite of environmental conditions at a given time (e.g., climate, topography); on the other hand, geographical space is the extent of a particular region or study area (Soberón and Nakamura 2009; Peterson et al. 2011) and includes important historical context. Thus, when modeling species ecological niches, we are modeling the existing abiotically suitable conditions for the species or the biotically reduced niche (Peterson et al. 2011; see also Fig. 9.1). However, when modeling species distributions, the intent is to project objects into geographical space (Fig. 9.1), and, depending on the factors considered, it is possible to estimate the occupied distributional area or the invadable distributional area (Soberón and Nakamura 2009; Peterson et al. 2011; Anderson 2012; Soberón et al. 2017).

9.3.1 Methods

9.3.1.1 Oak Species Data Sets

Occurrence data were downloaded from iDigBio between 20 and 24 July 2018, including localities collected by the authors, and cleaned for accuracy. Any botanical garden localities were discarded. All points were visually examined, and any localities that were outside the known range of the species, or in unrealistic locations (e.g., water bodies), were discarded.

9.3.1.2 Environmental Data Sets

For comparative purposes we obtained environmental data from two sources: (1) environmental variables derived from WorldClim and (2) S-RS data products (Fig. 9.3). Environmental variables derived from climatic data were obtained from the 10 to 2.5 arcmin WorldClim (Hijmans et al. 2005; spatial resolution of ~18.5 and 4.5 km at the equator, respectively) for annual mean temperature (BIO1), temperature seasonality (BIO4), minimum temperature of coldest month (BIO6), mean

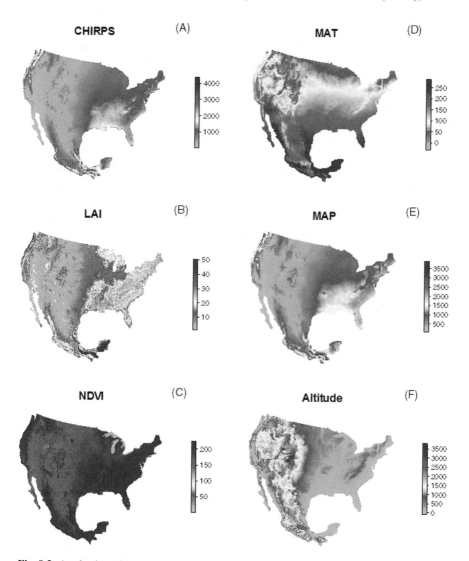

Fig. 9.3 A selection of S-RS products and climatic variables used in this study. The panels show (**a**) Climate Hazards group Infrared Precipitation with Stations (CHIRPS); (**b**) MODIS mean LAI; (**c**) MODIS mean NDVI; (**d**) mean annual temperature; (**e**) mean annual precipitation; and (**f**) altitude or mean elevation from Shuttle Radar Topography Mission (SRTM)

temperature of warmest quarter (BIO10), annual precipitation (BIO12), and precipitation seasonality (BIO15). These environmental variables were selected as critical for the distribution of oak species (Hipp et al. 2017) generally and were previously shown to be important in differentiating live oak (*Virentes*) species specifically (Cavender-Bares et al. 2011; Koehler et al. 2011; Cavender-Bares et al. 2015).

Environmental variables from S-RS products were obtained from MODIS over a 15-year period (2001–2015) from NASA using the interface EOSDIS Earthdata (https://earthdata.nasa.gov). Data include two MODIS Collection 5 land products: LAI (8-day temporal resolution) and NDVI (16-day temporal resolution). LAI and NDVI products (Fig. 9.3b, c) are derived from Terra/Aqua MOD15A2 and Terra MOD13A2, respectively (see Myneni et al. 2002 for a detailed explanation of MODIS products). We also obtained precipitation data from Climate Hazards group Infrared Precipitation with Stations (CHIRPS, Fig. 9.3a), an S-RS product designed for monitoring drought and global environmental land change (Funk et al. 2015). Notice that the original MODIS products present a spatial resolution of 1 km and CHIRPS, a spatial resolution of 3 arcmin or ~5.5 km at the equator. To standardize the spatial resolution of both MODIS products and CHIRPS, we upscaled the spatial resolution of MODIS products to that of CHIRPS.

Prior to following the ENM/SDM procedures (outlined below), we calculated five new metrics taking advantage of the high temporal resolution LAI and NDVI data by doing simple arithmetic calculations: LAI/NDVI cumulative, LAI/NDVI mean, LAI/NDVI max, LAI/NDVI min, and LAI/NDVI seasonality or the coefficient of variation (see Saatchi et al. 2008; Hobi et al. 2017 for details). These metrics represent the spatial variation in vegetation productivity over a year (Berry et al. 2007; Hobi et al. 2017) and allow detection of biodiversity changes, description of habitats of different species, and tracking of phenology within species geographical ranges (Fig. 9.4). We used these two S-RS products given LAI and NDVI

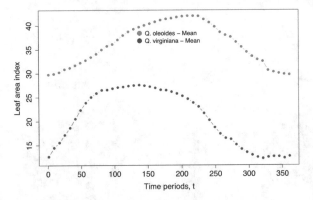

Fig. 9.4 Satellite remotely sensed vegetation phenology based on MODIS LAI product. The time periods (t) represent the 46 time intervals every 8 days within a year starting from 1 January. The curves for vegetation phenology represent the variation in LAI over a 1-year interval calculated as the mean LAI within a species geographical distribution at 8-day intervals averaged over 15 years (see Hobi et al. 2017 for details). Shown is seasonal variation in the temperate forest vegetation where *Q. virginiana* occurs in North America compared with seasonal variation in the tropical dry forest vegetation where *Q. oleoides* occurs in Mexico and Central America

provide information on net primary productivity, dynamics of the growing season, and vegetation seasonality, all potentially important variables for characterizing plant species ranges (Myneni et al. 2002; Saatchi et al. 2008). All data processing and metric calculations were performed in R v3.5 (R Core Team 2018) using customized scripts and core functions from the packages raster (Hijmans 2018), gdalUtils (Greenberg and Mattiuzzi 2018), and rgdal (Bivand et al. 2018). R scripts for data processing and metric calculations can be found at https://github.com/jesusNPL/RS-SDM_ENM.

9.3.1.3 Modeling Procedure

To model the ecological niche and distribution for oak species, we used an ensemble framework—prediction of a niche or a distributional area made by combining results of different modeling algorithms (Araújo and New 2007; Diniz-Filho et al. 2009). Within this framework we fit six species models and projected potential distributions for current environmental conditions for both environmental data sets (Table 9.1). The modeling algorithms included three statistical models (generalized linear models [GLM], generalized additive models [GAM], and adaptive regression splines [MARS]) and three machine learning models (MAXENT, support vector

Table 9.1 Combinations of environmental variables used for modeling live oak species-environment relationship under an ensemble framework

Source	Environmental predictors	Description
S-RS	CHIRPS	Climate Hazards group Infrared Precipitation with Stations
	LAI maximum	MODIS maximum leaf area index calculated over a year
	LAI mean	MODIS mean leaf area index calculated over a year
	LAI seasonality	MODIS seasonality of leaf area index calculated over a year
	LAI minimum	MODIS minimum leaf area index calculated over a year
	Altitude	Mean elevation from Shuttle Radar Topography Mission
S-RS2	CHIRPS	–
	LAI maximum	–
	LAI mean	–
	LAI seasonality	–
	LAI minimum	–
	NDVI maximum	MODIS maximum normalized difference vegetation index calculated over a year
	NDVI mean	MODIS mean normalized difference vegetation index calculated over a year
	NDVI seasonality	MODIS seasonality of normalized difference vegetation index calculated over a year
	NDVI minimum	MODIS minimum normalized difference vegetation index calculated over a year
	Altitude	–

(continued)

Table 9.1 continued

Source	Environmental predictors	Description
WorldClim	BIO 1	Mean annual temperature
	BIO 4	Temperature seasonality
	BIO 6	Minimum temperature of coldest month
	BIO 10	Mean temperature of warmest quarter
	BIO 12	Mean annual precipitation
	BIO 15	Precipitation seasonality
	Altitude	–

S-RS satellite remote sensing products. For comparative purposes, we used the same environmental variables from WorldClim at two spatial resolutions, 10 and 2.5 arcmin

machines [SVM], and Random Forest [RF]). A description for each algorithm is detailed in Franklin (2010) (see also Peterson et al. 2011). All algorithms were fit in R and used the packages dismo (Hijmans et al. 2017), kernlab (Karatzoglou et al. 2004), randomForest (Liaw and Wiener 2002), mgcv (Wood 2006), and earth (Milborrow 2016).

Within our ensemble framework, species' ecological niches are modeled using the six algorithms by fitting the occurrences of a single species and the predictors. The resulting six species models (one for each algorithm) are stacked into a single species model by averaging all models (Araújo and New 2007). We chose this approach because a major source of uncertainty in ENM/SDM arises from the algorithm used for modeling (Diniz-Filho et al. 2009; Qiao et al. 2015) and because the choice of the "best" modeling algorithm depends on the aims of the modeling applications (Peterson et al. 2011). Finally, using the stacked species models, we estimated macroecological patterns of species richness and the uncertainty associated with model parametrization. These patterns are less interesting in their own right for a small clade with only seven species, but they demonstrate an effective approach that can be applied to much larger groups of species.

We estimated live oak species richness by summing the projected potential species distributions; uncertainty was estimated as the variance attributable to the source of uncertainty (i.e., algorithms and their interactions) by performing a one-way analysis of variance (ANOVA) without replicates (Sokal and Rohlf 1995). The resulting uncertainty map shows regions with low and high uncertainty associated with the source of uncertainty (i.e., algorithm).

Statistical Analyses

To explore the performance of environmental data derived from RS for ENM/SDM compared to traditionally used environmental data from climatic variables (e.g., WorldClim), we evaluated the relationship between the modeled ecological niches from: (1) S-RS products; and (2) environmental variables from WorldClim. In doing so, we used correlation analyses corrected according Clifford's method to obtain the effective degrees of freedom for Pearson's coefficients while controlling for spatial autocorrelation (Clifford et al. 1989). Statistical analyses were performed in R using the package SpatialPack (Vallejos and Osorio 2014).

9.3.2   Results

Live oak models calibrated using different sources and combinations of environmental predictors (Table 9.1) within the ensemble framework generally provided similar suitability distributions (Fig. 9.5). Interestingly, increasing the number of predictors or increasing model complexity (S-SR2 in Table 9.1) affected model performance as measured by the Cohen's Kappa coefficient and AUC (area under the receiver operating characteristic curve) indices (Table 9.2), and thus affected the geographic predictions: Complex models tended to have higher statistical performance but to underestimate the distributions of live oak species when compared with simpler models (Fig. 9.5). Individual live oak species models made from S-RS products and WorldClim differed somewhat in their performances (see Table 9.2 and Fig. 9.5). Models from WorldClim tended to have slightly better statistical performance in inferring species distributions based on the AUC and Kappa criteria. However, these metrics do not capture differences in the precision and spatial resolution of the approaches. In several species, the WorldClim models predicted low precision locations compared to the S-RS data. In particular, the IUCN (International Union for Conservation of Nature) red-listed narrow endemic Brandegee Oak (*Quercus brandegeei*) in southern Baja California is very imprecisely predicted compared with the S-RS data. Using high-resolution interpolated climatic predictors did not improve the performance of individual models (WC25 in Table 9.2) and returned similar suitability predictions to those estimated under lower spatial resolution climatic predictors (Fig. 9.5). Although WorldClim models seems to have better statistical performance as shown in Table 9.2, we can at most discriminate the accuracy of interpolating continuous surface-derived models, only when we are inferring habitat suitability models (ENM) and not the projected species geographical distribution (SDM). Using S-RS data as predictors not only helps to identify the species habitat suitability but also incorporates local ecological conditions necessary to predict local species distributions and co-occurrence (Radeloff et al. 2019). This is because S-RS data have the potential to get at biological mechanisms, for example, through the detection of species phenological variation over space and time (Figs. 9.3c and 9.4).

When macroecological patterns of species richness and uncertainty maps were constructed, we observed similar patterns of species richness between maps made from the simpler combination of S-RS and WorldClim models (Fig. 9.6a, c, and d; see Table 9.1 for a description of the environmental combinations of S-RS and WorldClim). Notably, species richness estimation from the complex S-RS tends to restrict live oak assemblages to southeastern North America (Fig. 9.6b), which is the

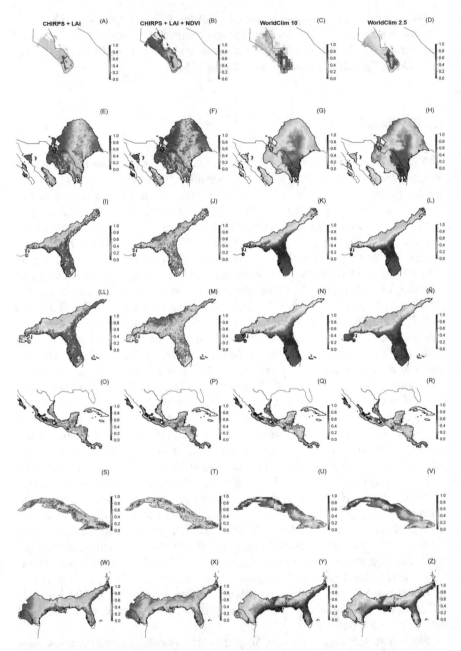

Fig. 9.5 Maps of predicted distributions for seven live oak species under three combinations of environmental variables. Legend colors represent values of suitability, where 1 and 0 represent maximum and minimum suitability, respectively. (**a–d**) *Quercus brandegeei*; (**e–h**) *Quercus fusi-formis*; (**i–l**) *Quercus geminata*; (**LL–Ñ**) *Quercus minima*; (**o–r**) *Quercus oleoides*; (**s–v**) *Quercus sagraena*; and (**w–z**) *Quercus virginiana*. Notice that each species model was estimated using an ensemble framework such that each species model represents the average of six algorithms weighted by their AUC. For representation purposes we cropped the predicted distributions using the species geographical ranges obtained from BIEN database (Enquist et al. 2016)

Table 9.2 Model accuracy assessment for the three combinations of environmental predictors

Species	Environmental combinations	Threshold	AUC	Kappa
Quercus brandegeei	S-RS	0.5221	0.9901	0.7989
	S-RS2	0.4808	0.9983	0.9044
	WC10	0.5865	0.9992	0.9244
	WC25	0.5816	0.9996	0.9321
Quercus fusiformis	S-RS	0.4764	0.9224	0.6192
	S-RS2	0.5026	0.9432	0.7115
	WC10	0.4137	0.9205	0.6073
	WC25	0.4153	0.9020	0.5814
Quercus geminata	S-RS	0.4730	0.9470	0.6439
	S-RS2	0.4196	0.9737	0.7365
	WC10	0.4958	0.9853	0.7965
	WC25	0.5829	0.9621	0.7650
Quercus minima	S-RS	0.4311	0.9587	0.6241
	S-RS2	0.4601	0.9848	0.7311
	WC10	0.4985	0.9813	0.7429
	WC25	0.5232	0.9708	0.7158
Quercus oleoides	S-RS	0.4917	0.8408	0.5678
	S-RS2	0.4947	0.8660	0.6113
	WC10	0.5527	0.9320	0.7788
	WC25	0.4891	0.9128	0.7121
Quercus sagraena	S-RS	0.4975	0.9788	0.7185
	S-RS2	0.5196	0.9945	0.8311
	WC10	0.5423	0.9809	0.7522
	WC25	0.5758	0.9818	0.7666
Quercus virginiana	S-RS	0.3712	0.9088	0.6292
	S-RS2	0.3613	0.9396	0.7064
	WC10	0.4204	0.9424	0.7263
	WC25	0.4314	0.9231	0.7215

AUC area under the ROC curve, Kappa Cohen's Kappa coefficient, *S-RS* CHIRPS + LAI + Altitude, *S-RS2* CHIRPS + LAI + NDVI + Altitude, *WC10 and WC25* WorldClim + Altitude at spatial resolution of 10 and 2.5 arcmin, respectively

only region where three live oak species co-occur. Uncertainty maps (Fig. 9.6e–h) show that uncertainty values for the simpler S-RS models (Fig. 9.6e) are lower when compared with the other models and WorldClim models tend to show high uncertainty values across south-central North America (e.g., Texas).

The geographical relationship between the individual species models made for the four combinations of environmental predictors varied depending on the species evaluated (Table 9.3), although they show positive relationships in all cases. In general, spatial relationships between species models from S-RS products (i.e., S-RS vs S-RS2) and WorldClim (i.e., WC10 vs WC25) were strongly correlated (see also Paz et al., Chap. 11), while the relationship between models made from S-RS products

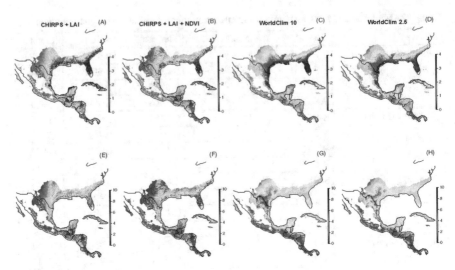

Fig. 9.6 Macroecological patterns of species richness (top panel) and uncertainty (bottom panel) for live oak species quantified under four combinations of environmental variables. WorldClim data were used at two spatial resolutions, 10 and 2.5 arcmin. See Table 9.1 for a description of the environmental combinations. Numbers on legends for the top panel represent the number of species within each pixel, where 4 means that four live oak are co-occurring in those pixels. Numbers on legends on the bottom panel represent the percentage of uncertainty or variance between algorithms, where higher values represent higher uncertainty

and WorldClim data varied slightly; the simpler models made from the S-RS data showed stronger spatial relationship to the WorldClim models (Table 9.3). Interestingly, the spatial relationships increased as a function of increasing the species potential distributions. For example, weaker spatial relationships ($r = 0.5384$ for S-RS/WC10 and $r = 0.6360$ for S-RS/WC25) were found for *Quercus brandegeei*, and stronger spatial relationships were found for dwarf live oak (*Quercus minima*; $r = 0.8150$ for S-RS/WC10 and $r = 0.8332$) and southern live oak (*Quercus virginiana*; $r = 0.8872$ for S-RS/WC10 and $r = 0.8523$ for S-RS/WC25) that are distributed across the southeastern United States (Fig. 9.5LL–Ñ and Fig. 9.5W–Z, respectively).

Finally, we found that macroecological patterns of species richness derived from the four sets of environmental predictors were strongly correlated (Table 9.4). Although the algorithms used for modeling have been emphasized as a major source of uncertainty (Diniz-Filho et al. 2009; Qiao et al. 2015), by applying the ensemble framework, we found that uncertainties due to the modeling algorithm were low (<10%) for the four sets of environmental predictors and showed similar distribution estimates (Fig. 9.6e–h). These results suggest that our results are not biased by applying a particular algorithm. Interestingly, we found low correlations between uncertainty predictions under S-RS and WC comparisons (Table 9.4), which potentially could suggest an associated error due to the predictors used to build the models.

Table 9.3 Spatial correlations between live oak ENMs estimated under three combinations of environmental variables

Species	Correlation	r	F	d.f.	P
Quercus brandegeei	RS/RS2	0.8441	40.8471	16.4851	0.0000
	RS/WC10	0.5384	5.7733	14.1398	0.0306
	RS/WC25	0.6360	10.3226	15.2004	0.0057
	RS2/WC10	0.6315	4.4031	6.6384	0.0762
	RS2/WC25	0.7043	7.0943	7.2068	0.0315
	WC10/WC25	0.8844	20.2487	5.6377	0.0048
Quercus fusiformis	RS/RS2	0.9105	102.5446	21.1581	0.0000
	RS/WC10	0.5735	8.1106	16.5511	0.0113
	RS/WC25	0.6117	8.7692	14.6656	0.0099
	RS2/WC10	0.5251	8.2619	21.6973	0.0089
	RS2/WC25	0.5652	8.9715	19.1088	0.0074
	WC10/WC25	0.9588	128.5055	11.2890	0.0000
Quercus geminata	RS/RS2	0.8764	62.1065	18.7489	0.0000
	RS/WC10	0.7099	14.3529	14.1270	0.0020
	RS/WC25	0.6951	12.1338	12.9824	0.0040
	RS2/WC10	0.6615	9.5259	12.2447	0.0092
	RS2/WC25	0.6589	8.6050	11.2139	0.0134
	WC10/WC25	0.9868	260.9438	7.0118	0.0000
Quercus minima	RS/RS2	0.8758	35.2577	10.7082	0.0001
	RS/WC10	0.8150	15.2887	7.7296	0.0048
	RS/WC25	0.8332	17.4921	7.7073	0.0033
	RS2/WC10	0.8051	13.1280	7.1263	0.0082
	RS2/WC25	0.8240	15.0107	7.0984	0.0059
	WC10/WC25	0.9878	195.5017	4.8702	0.0000
Quercus oleoides	RS/RS2	0.8945	930.5677	232.4530	0.0000
	RS/WC10	0.3946	29.0474	157.5251	0.0000
	RS/WC25	0.4130	35.6662	173.4187	0.0000
	RS2/WC10	0.3913	22.1329	122.4110	0.0000
	RS2/WC25	0.4059	26.5171	134.4395	0.0000
	WC10/WC25	0.9290	423.5265	67.2515	0.0000
Quercus sagraena	RS/RS2	0.7268	76.7276	68.5351	0.0000
	RS/WC10	0.6090	33.5887	56.9651	0.0000
	RS/WC25	0.6626	42.6603	54.5144	0.0000
	RS2/WC10	0.4627	17.5238	64.3187	0.0001
	RS2/WC25	0.5028	20.8935	61.7460	0.0000
	WC10/WC25	0.9333	83.3701	12.3397	0.0000
Quercus virginiana	RS/RS2	0.9064	47.3462	10.2891	0.0000
	RS/WC10	0.8872	42.5539	11.5129	0.0000
	RS/WC25	0.8523	44.9094	16.9107	0.0000
	RS2/WC10	0.7865	24.3299	14.9996	0.0002
	RS2/WC25	0.7748	32.6135	21.7171	0.0000
	WC10/WC25	0.9662	279.5733	19.9292	0.0000

S-RS CHIRPS + LAI + Altitude, *S-RS2* CHIRPS + LAI + NDVI + Altitude, WC10 and WC25 WorldClim + Altitude, at spatial resolution of 10 and 2.5 arc-min, respectively. *r* Pearson's correlation coefficient, *F* F-statistic, *d.f.* degrees of freedom, *P* associated p-value

Table 9.4 Spatial correlation between estimations of live oak species richness and uncertainty quantified under three combinations of environmental variables

Component	Correlation	r	F	d.f.	P
Species richness	RS/RS2	0.8991	98.5384	23.3534	0.0000
	RS/WC10	0.7420	27.0162	22.0477	0.0000
	RS/WC25	0.7224	25.6607	23.5094	0.0000
	RS2/WC10	0.7269	29.3608	26.2069	0.0000
	RS2/WC25	0.7120	28.6270	27.8452	0.0000
	WC10/WC25	0.9858	791.2799	22.9570	0.0000
Uncertainty	RS/RS2	0.8163	47.6775	23.8806	0.0000
	RS/WC10	0.3549	2.5399	17.6208	0.1288
	RS/WC25	0.4561	5.4356	20.6974	0.0299
	RS2/WC10	0.2315	1.2401	21.9056	0.2775
	RS2/WC25	0.3164	2.8755	25.8479	0.1019
	WC10/WC25	0.9425	136.3926	17.1534	0.0000

S-RS CHIRPS + LAI + Altitude, *S-RS2* CHIRPS + LAI + NDVI + Altitude, *WC10 and WC25* WorldClim + Altitude at spatial resolution of 10 and 2.5 arcmin, respectively

9.4 Perspectives

The field of ecological niche and species distribution modeling contributes significantly to our capacity to evaluate and describe the effect of geographical and environmental features on species distributions and has become in one of the most widely applied tools for the assessment of the impact of climate change and human activities on species and communities, biological invasions, epidemiology, and conservation biology (Peterson et al. 2011; Guisan et al. 2017). However, despite important advances in theory (Soberón 2007; Colwell and Rangel 2009; Soberón and Nakamura 2009; Peterson et al. 2011), methods, and algorithms (reviewed in Duarte et al. 2019; see also Warren et al. 2018) and practical applications (Guillera-Arroita et al. 2015; Cord et al. 2017; Sanín and Anderson 2018), most studies still rely on the use of interpolated climate data as environmental predictors (Saatchi et al. 2008; Waltari et al. 2014). In this article we compare the performance of environmental data derived from interpolated climate surfaces data (i.e., WorldClim) and S-RS products data (i.e., LAI and NDVI). Specifically, using live oaks as a case study, we show the advances and potential caveats in using S-RS data in describing and predicting species-environment relationships. Overall, our analyses show that S-RS products perform, as well as products from interpolated climate surfaces as environmental predictors (Tables 9.2, 9.3, and 9.4), and indeed present quite similar results for both species environmental suitability and macroecological patterns (Figs. 9.5 and 9.6), similar to Paz et al. (Chap. 11). However, they have the potential to provide more precise estimates of species distributions at higher spatial resolution.

In our example, we used different grain sizes for both data sets: WorldClim (10 and 2.5 arcmin or ~18.5 and ~ 4.5 km at the equator, respectively) and S-RS products (3 arcmin or ~5.5 km at the equator). Although changing grain size in the predictors might be anticipated to affect model performance, different lines of evidence indicate that model performance is not affected by grain resolution, but rather by species response to the environmental conditions in the study region (Guisan et al. 2007, see also Fig. 9.5). Our results show that enhancing spatial resolution of interpolated climatic data does not improve the spatial resolution at which species distributions can be accurately predicted. The quality of interpolated climate surfaces such as WorldClim, which depends on climatic stations as data sources, has been ignored as a source of uncertainty in studies of species-environment relationships—for example, Hijmans et al. (2005) used a variable number of weather stations for their interpolations, 47.554, 24.542, and 14.930 for precipitation, mean temperature, and maximum and minimum temperature, respectively—especially in the tropics (Fig. 9.2c), where weather stations are sparse (Hijmans et al. 2005; Soria-Auza et al. 2010). This source of uncertainty can be avoided using S-RS products, which have continuous (from daily to monthly) and quasi-global environmental information, including precipitation, temperature, and biophysical variables that represent different components of vegetation and ecosystems (Funk et al. 2015; Cord et al. 2017; Radeloff et al. 2019).

Fine and broad spatial and temporal scale data derived from S-RS, which have only been available in the last ~20 years (Turner 2014), can be used to improve the evaluation of species-environment relationships. A number of research avenues remain to be pursued to better understand the potential of S-RS data and their products in quantifying species ecological niches and estimating species distributions. For example, applying the same framework presented here to other species or clades (including vertebrates and invertebrates) or applying more complex frameworks (e.g., Peterson and Nakazawa 2008; Waltari et al. 2014) may shed light on the potential of S-RS products as predictors for the analysis of species-environment relationships. This is important because ENMs/SDMs are used as predictive models that can be extrapolated across space and time to forecast and monitor biodiversity under a changing global climate (Peterson and Nakazawa 2008; Warren 2012).

9.4.1 Should We Use S-RS Data for ENM/SDM?

Whether S-RS data should replace other environmental data in modeling niches and projecting species distribution depends on the modeling purposes (Peterson et al. 2011). In fact, modeling species niches and projecting distributions involves relating a set of species occurrences to relevant environmental predictors. In essence, ENM/SDM based only on climatic variables would tend to return broad predictions (Coudun et al. 2006, see also right panel in Fig. 9.5), particularly because climatic data are useful in describing macroecological patterns of species distributions and communities (Lin and Wiens 2017; Manzoor et al. 2018), while ENM/SDM based on S-RS data alone allows the discrimination of local features not captured by climatic information (Coudun et al. 2006; Saatchi et al. 2008; Radeloff et al. 2019).

Nevertheless, it seems that overall statistical model accuracy in this example is not improved (Pearson et al. 2004; Thuiller 2004; see also Table 9.2). Given that climatic and S-RS data provide information at different spatial and temporal scales, a promising option would be to use both sources of environmental predictors to model species distributions to achieve "the best of both worlds" (Saatchi et al. 2008; Pradervand et al. 2013).

More accurate predictions of species distributions are critical for the development of conservation and management actions if we are to meet the challenges posed by global change (Coudun et al. 2006; Zimmermann et al. 2007; Cord et al. 2013). Our point here is to facilitate and demonstrate the potential for the use of S-RS data for predicting species distributions and modeling environmental niches. The results we show here and those of others (e.g., Saatchi et al. 2008; Waltari et al. 2014) indicate that S-RS data provide a valuable complement to other environmental variables for ENM/SDM.

Another potential and important research direction is the use of S-RS products that have high temporal resolution, such as LAI (Fig. 9.3b), as biophysical variables that represent ecosystem functions (Cord et al. 2013, 2017). These products allow the exploration of dynamics of vegetation growth and seasonality in vegetation function, fundamental features that characterize vegetation form and function (Myneni et al. 2002; Hobi et al. 2017). Here, using metrics derived from MODIS LAI in combination with other S-RS products (Table 9.1), we show that, using relevant biophysical variables, it is possible to predict distributions similar to those predicted from climate data alone (Fig. 9.5, Tables 9.2 and 9.3). In fact, a recent study (Simões and Peterson 2018) found that including biotic predictors can improve ENMs even while increasing model complexity, such that the combination of abiotic and biotic predictors improves model performance (Simões and Peterson 2018). To confirm this conclusion, substantial effort would be needed, including new methodological and conceptual approaches, to disentangle the real contribution of S-RS products—spatial and temporal features of S-RS products that improve statistical model performance—as predictors of species distributions. Nonetheless, our results highlight an advance on the use of relevant predictors for modeling species-environment relationships.

In addition, recent macroecological studies have used these products to relate annual vegetation productivity to continental and global patterns of species richness (Pigot et al. 2016; Hobi et al. 2017; Coops et al. 2018), providing spatially explicit support for the use of satellite data products in predicting biodiversity. These advances point to an exciting avenue for the study of the distribution and assembly of biological communities (Ferrier and Guisan 2006). For example, S-RS products can be used for the development of stacked species distribution models (S-SDM, see Fig. 9.6) that can be integrated into novel biodiversity modeling frameworks, such as Spatially Explicit Species Assemblage Modelling (SESAM, Guisan and Rahbek 2011) or the Hierarchical Modelling of Species Communities (HMSC, Ovaskainen et al. 2017), aimed at predicting composition and distribution of species and communities (Mateo et al. 2017).

Finally, the example presented here is meant to spur further theoretical, methodological, and empirical research aimed at developing a Global Biodiversity Observatory (Geller et al., Chap. 20). Explicit incorporation of biotic information into species-environment modeling may turn our focus away from the use only of climatic information toward the "complete" evaluation of the drivers that determine the species distributions (Fig. 9.1).

9.4.2   Enabling Large-Scale Biodiversity Change Detection

Since the last millennium, rising human population and activity have been major drivers of environmental change on Earth, with consequences for the distribution and abundance of biodiversity and associated ecosystem functioning (Tilman 1997; Tylianakis et al. 2008). Thus, improving large-scale biodiversity change detection is crucial to the development of effective policies that advance conservation and management of species and communities.

Such efforts are critical to enhancing efforts to develop a Global Biodiversity Observatory (Geller et al., Chap. 20; Jetz et al. 2016). Research interest in using S-RS has increased in recent years given its high potential for monitoring global biodiversity and detecting change (Turner 2014; Jetz et al. 2016). For example, it is possible to identify shifts in vegetation structure or to monitor the dynamics of the growing season of an entire region, or within a specific species geographical range (Fig. 9.4) using time series S-RS products such as LAI—half of the total green leaf area per unit of horizontal ground surface area (Xiao et al. 2014)—which has a temporal resolution of 8 days. This is particularly important given that ENM/SDM theory assumes that species' niches are stable across time and space and that species and their environments are at pseudo-equilibrium, suggesting that species are occupying all suitable areas (Guisan and Thuiller 2005). However, the environment is dynamic and can change even at small scales; species ranges can thus expand and retract across time, varying within species lifetimes as well as over evolutionary timescales encompassing many generations. Long-term series of S-RS data products (i.e., spatial and temporal) supply remarkable opportunities for assessing and monitoring the state of the Earth's surface and, combining with species-environment relationship modeling, provide new frontiers for the prediction of species distributions and species monitoring across time and space (Randin et al. 2020). Indeed, using biophysical variables derived from high-resolution S-RS products (i.e., LAI) allows the identification of geographic areas where species actually occur (Fig. 9.7) and thus has the potential for enhancing the predictions of a set of species that could occur in an area—species pool—that is used for species assignments from direct RS detection using hyperspectral data (see simulation in Fig. 7.8, Section 9.4.2 in Meireles et al., Chap. 7).

In addition, enhancing predictive models of the species expected to be present in a given geographic region can be coupled with other means of detecting which species are present based on spectroscopic imaging (Serbin et al. 2015; Bolch et al.,

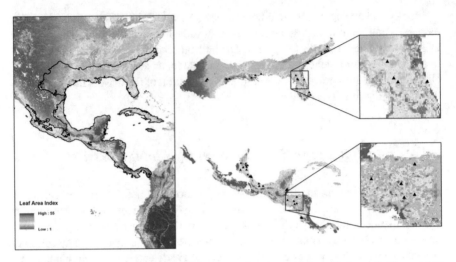

Fig. 9.7 Mean LAI estimated at 8-day intervals averaged over 15 years (left panel). The overlaid continuous lines correspond to the geographical ranges of *Q. virginiana* and *Q. oleoides* obtained from BIEN database. Predicted distributions for *Q. virginiana* (top right panel) and *Q. oleoides* (bottom right panel) are based on S-RS products, which include the temporal variation in LAI shown in A. The triangles over the maps represent occurrence points used for calibration (where the authors have collected specimens), and the boxes represent a zoom over a specific area of the predicted species distributions. Note that high values of the predicted distributions coincide with the occurrence points

Chap. 12; Meireles et al., Chap. 7), reducing the complexity of species identification algorithms. For example, imaging spectroscopy allows mapping of functional traits, by estimating vegetation traits for each pixel in an image (Wang et al. 2019; Asner et al. 2017; Martin, Chap. 5). Plant spectra obtained from imaging spectroscopy at different spatial resolutions can in turn be used to detect different aspects and traits—within- and between-species differences in morphology, foliar chemistry, life history strategies—of plant species (Ustin and Gamon 2010; Cavender-Bares et al. 2017; Schweiger et al. 2018) and the correct identification of different taxonomic levels from populations to species to clades (Cavender-Bares et al. 2016). Thus, the integration of spectral approaches with techniques for modeling species ecological niches has the potential to produce reliable information of species distributions and co-occurrence, filling current gaps about species-environment relationships at a range of spatial scales and levels of organization—from species to communities—increasing the accuracy of direct detection assignments, and enabling monitoring of changes in biodiversity, one of the premises for the sustainable management of the biosphere (Pinto-Ledezma and Rivero 2014; Fernández et al., Chap. 18).

Acknowledgments We thank Julián Velasco for his advice with ENM/SDM analyses. J.N.P-L. was supported by the University of Minnesota College of Biological Sciences' Grand Challenges in Biology Postdoctoral Program. We acknowledge the bioDISCOVERY community at the University of Zürich for important discussions and the NSF Research Coordination Network "Biodiversity Across Scales."

References

Anderson RP (2012) Harnessing the world's biodiversity data: promise and peril in ecological niche modeling of species distributions. Ann N Y Acad Sci 1260(1):66–80. https://doi.org/10.1111/j.1749-6632.2011.06440.x

Anderson RP (2013) A framework for using niche models to estimate impacts of climate change on species distributions. Ann N Y Acad Sci 1297(1):8–28. https://doi.org/10.1111/nyas.12264

Araújo MB, New M (2007) Ensemble forecasting of species distributions. Trends Ecol Evol 22(1):42–47. https://doi.org/10.1016/j.tree.2006.09.010

Asner GP, Martin RE, Knapp DE et al (2017) Airborne laser-guided imaging spectroscopy to map forest trait diversity and guide conservation. Science 355(6323):385–389. https://doi.org/10.1126/science.aaj1987

Berry S, Mackey B, Brown T (2007) Potential applications of remotely sensed vegetation greenness to habitat analysis and the conservation of dispersive fauna. Pacific Cons Bio 13(2):120. https://doi.org/10.1071/pc070120

Bivand R, Keitt T, Rowlingson B (2018) rgdal: bindings for the 'geospatial' data abstraction library. R package version 1.3-4. https://CRAN.R-project.org/package=rgdal.

Booth TH, Nix HA, Busby JR et al (2013) Bioclim: the first species studies. Divers Distrib 20(1):1–9. https://doi.org/10.1111/ddi.12144

Cavender-Bares J, Gonzalez-Rodriguez A, Pahlich A et al (2011) Phylogeography and climatic niche evolution in live oaks (Quercus series Virentes) from the tropics to the temperate zone. J Biogeogr 38(5):962–981. https://doi.org/10.1111/j.1365-2699.2010.02451.x

Cavender-Bares J, Gonzalez-Rodriguez A, Eaton DA et al (2015) Phylogeny and biogeography of the American live oaks (Quercus subsection Virentes): a genomic and population genetics approach. Mol Ecol 24(14):3668–3687. https://doi.org/10.1111/mec.13269

Cavender-Bares J, Meireles JE, Couture J et al (2016) Associations of leaf spectra with genetic and phylogenetic variation in oaks: prospects for remote detection of biodiversity. Remote Sens 8(3):221. https://doi.org/10.3390/rs8030221

Cavender-Bares J, Gamon JA, Hobbie SE et al (2017) Harnessing plant spectra to integrate the biodiversity sciences across biological and spatial scales. Am J Bot 104(7):966–969. https://doi.org/10.3732/ajb.1700061

Clifford P, Richardson S, Hemon D (1989) Assessing the significance of the correlation between two spatial processes. Biometrics 45(1):123–134. https://doi.org/10.2307/2532039

Colwell RK, Rangel TF (2009) Hutchinson's duality: the once and future niche. Proc Natl Acad Sci U S A 106(Supplement_2):19651–19658. https://doi.org/10.1073/pnas.0901650106

Coops NC, Kearney SP, Bolton DK et al (2018) Remotely-sensed productivity clusters capture global biodiversity patterns. Sci Rep 8(1):16261. https://doi.org/10.1038/s41598-018-34162-8

Cord AF, Meentemeyer RK, Leitão PJ et al (2013) Modelling species distributions with remote sensing data: bridging disciplinary perspectives. J Biogeogr 40(12):2226–2227. https://doi.org/10.1111/jbi.12199

Cord AF, Brauman KA, Chaplin-Kramer R et al (2017) Priorities to advance monitoring of ecosystem services using earth observation. Trends Ecol Evol 32(6):416–428. https://doi.org/10.1016/j.tree.2017.03.003

Coudun C, Gégout JC, Piedallu C et al (2006) Soil nutritional factors improve models of plant species distribution: an illustration with Acer campestre (L.) in France. J Biogeogr 33(10):1750–1763. https://doi.org/10.1111/j.1365-2699.2005.01443.x

Diniz-Filho JAF, Bini L, Rangel TF et al (2009) Partitioning and mapping uncertainties in ensembles of forecasts of species turnover under climate change. Ecography 32(6):897–906. https://doi.org/10.1111/j.1600-0587.2009.06196.x

Duarte A, Whitlock SL, Peterson JT (2019) Species distribution modeling. In: Encyclopedia of Ecology, pp 189–198. https://doi.org/10.1016/b978-0-12-409548-9.10572-x

Elith J, Franklin J (2013) Species distribution modeling. In: Encyclopedia of biodiversity, pp 692–705. https://doi.org/10.1016/b978-0-12-384719-5.00318-x

Elton C (1927) Animal Ecology. Sedgwick and Jackson, London

Enquist BJ, Condit R, Peet RK et al (2016) Cyberinfrastructure for an integrated botanical information network to investigate the ecological impacts of global climate change on plant biodiversity. PeerJ Preprints 4:e2615v2. https://doi.org/10.7287/peerj.preprints.2615v2

Ferrier S, Guisan A (2006) Spatial modelling of biodiversity at the community level. J Appl Ecol 43(3):393–404. https://doi.org/10.1111/j.1365-2664.2006.01149.x

Franklin J (2010) Mapping species distribution: spatial inference and prediction. Cambridge University Press, Cambridge

Funk C, Peterson P, Landsfeld M et al (2015) The climate hazards infrared precipitation with stations—a new environmental record for monitoring extremes. Sci Data 2:150066. https://doi.org/10.1038/sdata.2015.66

Gaston KJ (2009) Geographic range limits: achieving synthesis. Proc R Soc Lond B Biol Sci 276(1661):1395–1406. https://doi.org/10.1098/rspb.2008.1480

Gillespie TW, Foody GM, Rocchini D et al (2008) Measuring and modelling biodiversity from space. Prog Phys Geogr 32(2):203–221. https://doi.org/10.1177/0309133308093606

Greenberg JA, Mattiuzzi M (2018) gdalUtils: wrappers for the geospatial data abstraction library (GDAL) utilities. R package version 2.0.1.14. https://CRAN.R-project.org/package=gdalUtils.

Grinnell J (1904) The Origin and Distribution of the Chest-Nut-Backed Chickadee. The Auk, 21(3), 364–382. https://doi.org/10.2307/4070199

Grinnell J (1917) The Niche-Relationships of the California Thrasher. The Auk, 34(4), 427–433. https://doi.org/10.2307/4072271

Guillera-Arroita G, Lahoz-Monfort JJ, Elith J et al (2015) Is my species distribution model fit for purpose? Matching data and models to applications. Glob Ecol Biogeogr 24(3):276–292. https://doi.org/10.1111/geb.12268

Guisan A, Thuiller W (2005) Predicting species distribution: offering more than simple habitat models. Ecol Lett 8(9):993–1009. https://doi.org/10.1111/j.1461-0248.2005.00792.x

Guisan A, Graham CH, Elith J et al (2007) Sensitivity of predictive species distribution models to change in grain size. Divers Distrib 13(3):332–340. https://doi.org/10.1111/j.1472-4642.2007.00342.x

Guisan A, Rahbek C (2011) SESAM - a new framework integrating macroecological and species distribution models for predicting spatio-temporal patterns of species assemblages. Journal of Biogeography, 38(8), 1433–1444. https://doi.org/10.1111/j.1365-2699.2011.02550.x

Guisan A, Thuiller W, Zimmermann NE (2017) Habitat suitability and distribution models. Cambridge University Press, Cambridge. https://doi.org/10.1017/9781139028271

Hijmans RJ, Phillips S, Leathwick J et al (2017) dismo: species distribution modeling. R package version 1.1-4. https://CRAN.R-project.org/package=dismo.

Hijmans RJ (2018) raster: Geographic Data Analysis and Modeling. R package version 3.0-7. https://CRAN.R-project.org/package=raster

Hijmans RJ, Cameron SE, Parra JL et al (2005) Very high resolution interpolated climate surfaces for global land areas. Int J Climatol 25(15):1965–1978. https://doi.org/10.1002/joc.1276

Hipp AL, Manos PS, González-Rodríguez A et al (2017) Sympatric parallel diversification of major oak clades in the Americas and the origins of Mexican species diversity. New Phytol 217(1):439–452. https://doi.org/10.1111/nph.14773

Hobi ML, Dubinin M, Graham CH et al (2017) A comparison of dynamic habitat indices derived from different MODIS products as predictors of avian species richness. Remote Sens Environ 195:142–152. https://doi.org/10.1016/j.rse.2017.04.018

Holdridge LR (1947) Determination of world plant formations from simple climatic data. Science 105(2727):367–368. https://doi.org/10.1126/science.105.2727.367

Hutchinson GE (1957) Concluding Remarks. Cold Spring Harbor Symposia on Quantitative Biology, 22(0), 415–427. https://doi.org/10.1101/sqb.1957.022.01.03

Jetz W, Cavender-Bares J, Pavlick R et al (2016) Monitoring plant functional diversity from space. Nat Plants 2(3):16024. https://doi.org/10.1038/nplants.2016.24

Karatzoglou A, Smola A, Hornik K et al (2004) Kernlab – an S4 package for Kernel methods in R. J Stat Softw 11(9):1–20. http://www.jstatsoft.org/v11/i09/.

Kissling WD, Dormann CF, Groeneveld J et al (2012) Towards novel approaches to modelling biotic interactions in multispecies assemblages at large spatial extents. J Biogeogr 39(12): 2163–2178. https://doi.org/10.1111/j.1365-2699.2011.02663.x

Koehler K, Center A, Cavender-Bares J (2011) Evidence for a freezing tolerance-growth rate trade-off in the live oaks (Quercus series Virentes) across the tropical-temperate divide. New Phytol 193(3):730–744. https://doi.org/10.1111/j.1469-8137.2011.03992.x

Liaw A, Wiener M (2002) Classification and regression by randomForest. R News 2:18–22

Lin LH, Wiens JJ (2017) Comparing macroecological patterns across continents: evolution of climatic niche breadth in varanid lizards. Ecography 40(8):960–970. https://doi.org/10.1111/ecog.02343

Lobo JM, Jiménez-Valverde A, Hortal J (2010) The uncertain nature of absences and their importance in species distribution modelling. Ecography 33(1):103–114. https://doi.org/10.1111/j.1600-0587.2009.06039.x

Manzoor SA, Griffiths G, Lukac M (2018) Species distribution model transferability and model grain size – finer may not always be better. Sci Rep 8(1):7168. https://doi.org/10.1038/s41598-018-25437-1

Mateo RG, Mokany K, Guisan A (2017) Biodiversity models: what if unsaturation is the rule? Trends Ecol Evol 32(8):556–566. https://doi.org/10.1016/j.tree.2017.05.003

Milborrow S (2016) Earth: multivariate adaptive regression splines. R package version 4.4.4. Retrieved from https://CRAN.R-project.org/package=earth.

Myneni R, Hoffman S, Knyazikhin Y et al (2002) Global products of vegetation leaf area and fraction absorbed PAR from year one of MODIS data. Remote Sens Environ 83(1–2):214–231. https://doi.org/10.1016/s0034-4257(02)00074-3

Nix HA (1986) A biogeographic analysis of Australian elapid snakes. In: Longmore R (ed) Atlas of elapid snakes of Australia: Australian Flora and Fauna Series, vol 7. Bureau of Flora and Fauna, Canberra, pp 4–15

Ovaskainen O, Tikhonov G, Norberg A et al (2017) How to make more out of community data? A conceptual framework and its implementation as models and software. Ecol Lett 20(5):561–576. https://doi.org/10.1111/ele.12757

Pearson RG, Dawson TP, Liu C (2004) Modelling species distributions in Britain: a hierarchical integration of climate and land-cover data. Ecography 27(3):285–298. https://doi.org/10.1111/j.0906-7590.2004.03740.x

Peterson AT, Martínez-Meyer E, Soberón J et al (2011) Ecological niches and geographic distributions. Monographs in population biology, vol 49. Princeton University Press, Princeton

Peterson AT, Nakazawa Y (2008) Environmental data sets matter in ecological niche modelling: an example with Solenopsis invicta and Solenopsis richteri. Glob Ecol Biogeogr 17(1):135–144. https://doi.org/10.1111/j.1466-8238.2007.00347.x

Pettorelli N, Safi K, Turner W (2014a) Satellite remote sensing, biodiversity research and conservation of the future. Philos Trans R Soc Lond Ser B Biol Sci 369(1643):20130190–20130190. https://doi.org/10.1098/rstb.2013.0190

Pettorelli N, Laurance WF, O'Brien TG et al (2014b) Satellite remote sensing for applied ecologists: opportunities and challenges. J Appl Ecol 51(4):839–848. https://doi.org/10.1111/1365-2664.12261

Phillips LB, Hansen AJ, Flather CH (2008) Evaluating the species energy relationship with the newest measures of ecosystem energy: NDVI versus MODIS primary production. Remote Sens Environ 112(9):3538–3549. https://doi.org/10.1016/j.rse.2008.04.012

Pigot AL, Tobias JA, Jetz W (2016) Energetic constraints on species coexistence in birds. PLoS Biol 14(3):e1002407. https://doi.org/10.1371/journal.pbio.1002407

Pinto-Ledezma JN, Rivero ML (2014) Temporal patterns of deforestation and fragmentation in lowland Bolivia: implications for climate change. Clim Chang 127(1):43–54. https://doi.org/10.1007/s10584-013-0817-1

Pradervand JN, Dubuis A, Pellissier L et al (2013) Very high resolution environmental predictors in species distribution models. Prog Phys Geogr 38(1):79–96. https://doi.org/10.1177/0309133313512667

Qiao H, Soberón J, Peterson AT (2015) No silver bullets in correlative ecological niche modelling: insights from testing among many potential algorithms for niche estimation. Methods Ecol Evol 6(10):1126–1136. https://doi.org/10.1111/2041-210x.12397

R Core Team (2018) R: a language and environment for statistical computing. R Foundation for Statistical Computing, Vienna, Austria. https://www.R-project.org/

Radeloff VC, Dubinin M, Coops NC et al (2019) The dynamic habitat indices (DHIs) from MODIS and global biodiversity. Remote Sens Environ 222:204–214. https://doi.org/10.1016/j.rse.2018.12.009

Randin CF, Ashcroft MB, Bolliger J et al. (2020) Monitoring biodiversity in the Anthropocene using remote sensing in species distribution models. Remote Sens Environ 239:111626 https://doi.org/10.1016/j.rse.2019.111626

Smith AB, Godsoe W, Rodríguez-Sánchez F et al (2018) Niche estimation above and below the species level. Trends Ecol Evol 34:260. https://doi.org/10.1016/j.tree.2018.10.012

Saatchi S, Buermann W, ter Steege H et al (2008) Modeling distribution of Amazonian tree species and diversity using remote sensing measurements. Remote Sens Environ 112(5):2000–2017. https://doi.org/10.1016/j.rse.2008.01.008

Sanín C, Anderson RP (2018) A framework for simultaneous tests of abiotic, biotic, and historical drivers of species distributions: empirical tests for north American Wood warblers based on climate and pollen. Am Nat 192(2):E48–E61. https://doi.org/10.1086/697537

Serbin SP, Singh A, Desai AR et al (2015) Remotely estimating photosynthetic capacity, and its response to temperature, in vegetation canopies using imaging spectroscopy. Remote Sens Environ 167:78–87. https://doi.org/10.1016/j.rse.2015.05.024

Simões MVP, Peterson AT (2018) Importance of biotic predictors in estimation of potential invasive areas: the example of the tortoise beetle *Eurypedus nigrosignatus*, in Hispaniola. PeerJ 6:e6052. https://doi.org/10.7717/peerj.6052

Soberón J (2007) Grinnellian and Eltonian niches and geographic distributions of species. Ecol Lett 10(12):1115–1123. https://doi.org/10.1111/j.1461-0248.2007.01107.x

Soberón J, Nakamura M (2009) Niches and distributional areas: concepts, methods, and assumptions. Proc Natl Acad Sci U S A 106(Supplement_2):19644–19650. https://doi.org/10.1073/pnas.0901637106

Soberón J, Peterson AT (2005) Interpretation of models of fundamental ecological niches and species' distributional areas. Biodivers Inform 2. https://doi.org/10.17161/bi.v2i0.4

Soberón J, Osorio-Olvera L, Peterson AT (2017) Diferencias conceptuales entre modelación de nichos y modelación de áreas de distribución. Rev Mex Biodivers 88(2):437–441. https://doi.org/10.1016/j.rmb.2017.03.011

Sokal RR, Rohlf FJ (1995) Biometry: the principles and practice of statistics in biological research, 3rd edn. W. H. Freeman and Co., New York

Soria-Auza RW, Kessler M, Bach K et al (2010) Impact of the quality of climate models for modelling species occurrences in countries with poor climatic documentation: a case study from Bolivia. Ecol Model 221(8):1221–1229. https://doi.org/10.1016/j.ecolmodel.2010.01.004

Schweiger AK, Cavender-Bares J, Townsend PA et al (2018) Plant spectral diversity integrates functional and phylogenetic components of biodiversity and predicts ecosystem function. Nat Ecol Evol 2(6):976–982. https://doi.org/10.1038/s41559-018-0551-1

Thuiller W (2004) Patterns and uncertainties of species' range shifts under climate change. Glob Chang Biol 10(12):2020–2027. https://doi.org/10.1111/j.1365-2486.2004.00859.x

Tilman D (1997) The influence of functional diversity and composition on ecosystem processes. Science 277(5330):1300–1302. https://doi.org/10.1126/science.277.5330.1300

Turner W (2014) Sensing biodiversity. Science 346(6207):301–302. https://doi.org/10.1126/science.1256014

Turner W, Spector S, Gardiner N et al (2003) Remote sensing for biodiversity science and conservation. Trends Ecol Evol 18(6):306–314. https://doi.org/10.1016/s0169-5347(03)00070-3

Tylianakis JM, Didham RK, Bascompte J et al (2008) Global change and species interactions in terrestrial ecosystems. Ecol Lett 11(12):1351–1363. https://doi.org/10.1111/j.1461-0248.2008.01250.x

Ustin SL, Gamon JA (2010) Remote sensing of plant functional types. New Phytol 186(4):795–816. https://doi.org/10.1111/j.1469-8137.2010.03284.x

Vallejos R, Osorio F (2014) Effective sample size of spatial process models. Spat Stat 9:66–92. https://doi.org/10.1016/j.spasta.2014.03.003

Wallace AR (1860) On the zoological geography of the Malay archipelago. Zool J Linnean Soc 4(16):172–184. https://doi.org/10.1111/j.1096-3642.1860.tb00090.x

Waltari E, Schroeder R, McDonald K et al (2014) Bioclimatic variables derived from remote sensing: assessment and application for species distribution modelling. Methods Ecol Evol 5(10):1033–1042. https://doi.org/10.1111/2041-210x.12264

Wang Z, Townsend PA, Schweiger AK et al (2019) Mapping foliar functional traits and their uncertainties across three years in a grassland experiment. Remote Sens Environ 221:405–416. https://doi.org/10.1016/j.rse.2018.11.016

Warren DL (2012) In defense of "niche modeling". Trends Ecol Evol 27(9):497–500. https://doi.org/10.1016/j.tree.2012.03.010

Warren DL, Beaumont LJ, Dinnage R et al (2018) New methods for measuring ENM breadth and overlap in environmental space. Ecography 42:444. https://doi.org/10.1111/ecog.03900

West AM, Evangelista PH, Jarnevich CS et al (2016) Integrating remote sensing with species distribution models; mapping tamarisk invasions using the software for assisted habitat modeling (SAHM). J Vis Exp (116). https://doi.org/10.3791/54578

Wood SN (2006) Generalized additive models. Chapman and Hall/CRC, Boca Raton

Xiao Z, Liang S, Wang J et al (2014) Use of general regression neural networks for generating the GLASS leaf area index product from time-series MODIS surface reflectance. IEEE Trans Geosci Remote Sens 52(1):209–223. https://doi.org/10.1109/tgrs.2013.2237780

Zimmermann NE, Edwards TC, Moisen GG et al (2007) Remote sensing-based predictors improve distribution models of rare, early successional and broadleaf tree species in Utah. J Appl Ecol 44(5):1057–1067. https://doi.org/10.1111/j.1365-2664.2007.01348.x

Open Access   This chapter is licensed under the terms of the Creative Commons Attribution 4.0 International License (http://creativecommons.org/licenses/by/4.0/), which permits use, sharing, adaptation, distribution and reproduction in any medium or format, as long as you give appropriate credit to the original author(s) and the source, provide a link to the Creative Commons license and indicate if changes were made.

The images or other third party material in this chapter are included in the chapter's Creative Commons license, unless indicated otherwise in a credit line to the material. If material is not included in the chapter's Creative Commons license and your intended use is not permitted by statutory regulation or exceeds the permitted use, you will need to obtain permission directly from the copyright holder.

Chapter 10
Remote Sensing of Geodiversity as a Link to Biodiversity

Sydne Record, Kyla M. Dahlin, Phoebe L. Zarnetske, Quentin D. Read,
Sparkle L. Malone, Keith D. Gaddis, John M. Grady, Jennifer Costanza,
Martina L. Hobi, Andrew M. Latimer, Stephanie Pau, Adam M. Wilson,
Scott V. Ollinger, Andrew O. Finley, and Erin Hestir

10.1 Conserving Nature's Stage

Biodiversity is essential for ecosystem functioning and ecosystem services (Chapin et al. 1997; Yachi and Loreau 1999). Yet rapid global change is altering biodiversity and endangering its vital functions, with human-caused habitat deterioration being the number one cause of biodiversity loss (Sala et al. 2000). In addition, climate change is directly affecting individual species abundances and distributions and indirectly affecting species via biotic interactions (Walther et al. 2002). When combined, these effects lead to novel ecological communities for which there are no modern analogs (Williams and Jackson 2007). Although species have continually experienced shifts in climate, the recent rate of temperature change is more rapid than in any other timeframe in the past 10,000 years (Marcott et al. 2013), and temperatures are expected to rise even faster in the near future (Smith et al. 2015). In light of these rapid global changes, a major challenge for biodiversity scientists is to generate robust statistical models that describe and predict biodiversity in space and

S. Record (✉)
Department of Biology, Bryn Mawr College, Bryn Mawr, PA, USA
e-mail: srecord@brynmawr.edu

K. M. Dahlin
Department of Geography, Environment, & Spatial Sciences, Michigan State University,
East Lansing, MI, USA

Ecology, Evolutionary Biology, and Behavior Program, Michigan State University,
East Lansing, MI, USA

P. L. Zarnetske · Q. D. Read
Ecology, Evolutionary Biology, and Behavior Program, Michigan State University,
East Lansing, MI, USA

Department of Integrative Biology, Michigan State University, East Lansing, MI, USA

© The Author(s) 2020
J. Cavender-Bares et al. (eds.), *Remote Sensing of Plant Biodiversity*,
https://doi.org/10.1007/978-3-030-33157-3_10

time, from which changes in hot spots (highs) and cold spots (lows) of biodiversity may indicate shifts in ecosystem functions and services.

Contemporary strategies for addressing and managing biodiversity loss align with a metaphor developed by G. Evelyn Hutchinson in his book *The Ecological Theater and the Evolutionary Play* from Shakespeare's *As You Like It* (Hutchinson 1965). In Act II, Scene VII, of *As You Like It*, Shakespeare wrote, "All the world's a stage, and all the men and women merely players. They have their exits and their entrances." In Hutchinson's metaphor, the world's biota comprises the players, and the script is an evolutionary play. More recently, the metaphor has been extended to consider the Earth's abiotic setting as the stage (Beier et al. 2015).

Conservation efforts often emphasize management plans for the actors [e.g., Essential Biodiversity Variables (EBVs)] (Fernandez and Pereira, Chap. 18).

S. L. Malone
Department of Biological Sciences, Florida International University, Miami, FL, USA

K. D. Gaddis
National Aeronautics and Space Administration, Washington, D.C., USA

J. M. Grady
Department of Biology, Bryn Mawr College, Bryn Mawr, PA, USA

Ecology, Evolutionary Biology, and Behavior Program, Michigan State University, East Lansing, MI, USA

Department of Integrative Biology, Michigan State University, East Lansing, MI, USA

J. Costanza
Department of Forestry and Environmental Resources, NC State University, Research Triangle Park, NC, USA

M. L. Hobi
Swiss Federal Research Institute WSL, Birmensdorf, Switzerland

SILVIS Lab, Department of Forest and Wildlife Ecology, University of Wisconsin-Madison, Madison, WI, USA

A. M. Latimer
Department of Plant Sciences, UC Davis, Davis, CA, USA

S. Pau
Department of Geography, Florida State University, Tallahassee, FL, USA

A. M. Wilson
Geography Department, University at Buffalo, Buffalo, NY, USA

S. V. Ollinger
Department of Natural Resources and the Environment, University of New Hampshire, Durham, NH, USA

A. O. Finley
Department of Geography, Environment, & Spatial Sciences, Michigan State University, East Lansing, MI, USA

Department of Forestry, Michigan State University, East Lansing, MI, USA

E. Hestir
School of Engineering, University of California, Merced, CA, USA

For instance, the US Endangered Species Act and International Union for Conservation of Nature (IUCN) Red List focus on individual species (ESA 1973; IUCN 2001). However, an inherent challenge to managing species is that, during the course of a play, the actors move across the stage. Geo-referenced fossils from the paleoecological record provide evidence of how species' geographic ranges shifted in the past as Earth's climate fluctuated (Williams and Jackson 2007; Veloz et al. 2012). For instance, in terms of estimating EBVs (Fernandez and Pereira, Chap. 18), species distribution models (SDMs) are one of the most common tools for understanding how species ranges might shift over time and space (Elith and Leathwick 2009; Record and Charney 2016), but they are fraught with statistical (Record et al. 2013) and biological shortcomings (Belmaker et al. 2015; Charney et al. 2016; Evans et al. 2016) that hamper their ability to reliably inform management. Given the challenges of managing species whose ranges might be shifting in response to climate change (Veloz et al. 2012), there is interest in focusing conservation efforts on areas that are likely to support biodiversity and on the processes that generate it (Pressey et al. 2007; Anderson and Ferree 2010; Beier and Brost 2010). Indeed, The Nature Conservancy, one of the world's leading nonprofit conservation organizations, has adopted the rallying cry of "conserving nature's stage" (Beier et al. 2015). Conserving nature's stage entails identifying parcels of Earth that are valuable for their geodiversity and for their capacity to support diverse life forms today and into the future.

Geodiversity has been defined in several ways (see Table 1.2 in Gray 2013). Some definitions of geodiversity refer to variability in soil, geological, and geomorphological features and the processes that give rise to them (Gray 2013 and references therein). Other definitions tend to have a wider scope and also include topography, hydrology, and climate (Benito-Calvo et al. 2009; Parks and Mulligan 2010). These more inclusive definitions of geodiversity capture variability in the entire geosphere (Hjort et al. 2012) that link to important drivers of biodiversity (e.g., energy, water, and nutrients (Richerson and Lum 1980; Kerr and Packer 1997)). The geosphere includes the lithosphere, atmosphere, hydrosphere, and cryosphere (Williams 2012) and processes within and among them and encompasses the abiotic components of Earth's "Critical Zone," or the portion of Earth where biotic and abiotic processes support life on Earth's surface (NRC 2001). Just as the Critical Zone arises from interactions among abiotic and biotic processes, geodiversity is not separated from biotic influences and biodiversity. A key step in the prioritization of conservation areas using this approach is to understand the relationships between biodiversity and geodiversity. Remotely sensed biodiversity and geodiversity data have the potential to answer questions of scale to better inform conservation decisions because they can provide coverage at nearly continuous large spatial extents (i.e., regional to global) and at fine spatial and temporal resolutions (Fig. 10.1 for a spatial example). Here, we provide an overview of remotely sensed data sources that can be used to measure geodiversity and biodiversity to better understand biodiversity-geodiversity relationships, which is a key step in conserving nature's stage.

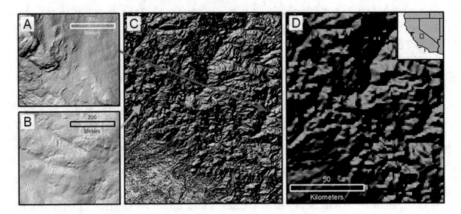

Fig. 10.1 Topography at different spatial grains. Hillshade maps calculated from digital elevation models (DEMs) at 1 m resolution (**a**) and (**b**), 90 m resolution (**c**), and 1 km resolution (**d**). The inset map in (**d**) shows the locations of panels (**c**) and (**d**) in California, which have the same extent. Data for panels (**a**) and (**b**) are from the National Ecological Observatory Network's (NEON) Airborne Observation Platform Light Detection and Ranging (LiDAR) system (Kampe et al. 2010). Data for panels (**c**) and (**d**) are from the Shuttle Radar Topography Mission (SRTM) via earthenv.org (Robinson et al. 2014)

10.2   Geodiversity Indices

Geodiversity represents an opportunity for habitat differentiation (Radford 1981) and available niche space (Dufour et al. 2006) that is thought to support biodiversity (Gray 2008). The continuous nature of remote sensing (RS) data enables exploration of novel measures of geodiversity. In this section we focus our discussion on metrics of variability, although absolute values (e.g., minimum and maximum thresholds) of some geographical features are also informative for understanding species' limits and ultimately species diversity. Studies have used two aspects of variability: the absolute range of conditions and the spatial configuration of these conditions (Spehn and Körner 2005; Dufour et al. 2006; Jackova and Romportl 2008; Serrano et al. 2009; Hjort and Luoto 2010; Hjort and Luoto 2012). The range in conditions is an estimate of the different elements in the area of interest. Given sampling units larger than the minimum pixel resolution, the proportional area covered by distinct geographical features could be used to calculate an evenness index of geodiversity. Categorical features have also treated geodiversity variables similarly to species with measured presences or abundances in various geodiversity metrics (Serrano et al. 2009; Tuanmu and Jetz 2015).

Alternatively, geodiversity could be quantified as variability in continuous observations such as elevation or climate. A focus on variability allows for different geological contexts (past and present) to be taken into account. One of the most common measures of environmental heterogeneity is elevational range (Stein et al. 2014), simply the absolute difference between elevation at two sites or sample units (i.e., among or within sites, respectively). Using elevation as an example, the average difference, squared, between the elevation in a focal cell and all other cells in a sample unit could be used as a measure of topographic heterogeneity. The coefficient of variation is a similar measure of heterogeneity, though it is standardized to the mean elevation of the sample unit. Pairwise site differences in multiple geographical features can be used as predictors in matrix regression such as generalized dissimilarity models (Ferrier et al. 2007) or more generally a Mantel test (Tuomisto et al. 2003; Legendre et al. 2005), though mechanistic interpretation is limited when geographical features are combined in this way.

Additional approaches include a geodiversity atlas that classifies areas as having very high, high, moderate, low, and very low geodiversity (Kozlowski 1999), quantifying geodiversity in terms of total component resource potential (i.e., energy, water, space, and nutrients; Parks and Mulligan 2010), and the geodiversity index (Gd) that relates the variety of physical elements (i.e., geomorphological, hydrological, soils) with the roughness and surface of the previously established geomorphological units according to the formula:

$$Gd = \frac{EgR}{lnS} \tag{10.1}$$

where Eg is the number of different physical elements, R is the coefficient of roughness of the unit, and S is the surface of the unit (km^2). The Gd is a semiquantitative scale that permits the establishment of five values of geodiversity, from very low to very high for each homogeneous unit. It is argued that use of Gd would allow easier comparison of units and aid suitable management of protected areas (Serrano et al. 2009; Hjort and Luoto 2010; Tukiainen et al. 2017).

With continuously measured remotely sensed geographical features, the sample unit (i.e., grain size) can be modified to examine within site and total site (and thus between sites) geodiversity. Additionally, RS data can uniquely address how relationships between geodiversity and biodiversity change across scales. Various combinations of changing grain and extent (change grain maintain extent, change extent maintain grain, change grain and extent) could be examined to explore scaling relationships (Barton et al. 2013).

10.3 Remote Sensing of Geodiversity

In the following sections, we describe the different components of geodiversity (Table 10.1), some of the ways they can be quantified, and the current state of technologies available to measure them remotely via airborne or satellite observations (Table 10.2). To match current interests in global biodiversity databases (e.g., the Global Biodiversity Information Facility, gbif.org), and because of the importance of scaling from local to much larger extents, we focus here on globally available data; however, we also mention some local scale RS applications. In particular, given that more and more remotely sensed data have been made publically available, we highlight open access remotely sensed geodiversity data.

Table 10.1 Elements of geodiversity

Lithosphere	Geology	Minerals	
		Rocks	
		Unconsolidated solids	
		Fossils	
	Geomorphology	Tectonics	
	Soils	Soil chemical properties	
		Soil physical properties	
	Topography	Elevation	
		Landforms (e.g., ridges, spurs)	
		Slope	
		Aspect	
		Energy	
		Roughness	
Atmosphere	Climate and weather	Temperature	Extreme events
		Precipitation	
		Wind	
Hydrosphere	Surface water		
	Groundwater		
Cryosphere	Ice		
	Snow		

Adapted from Serrano et al. (2009)

Table 10.2 Examples of remotely sensed geodiversity elements

Geosphere	Geodiversity element	RS data set
Lithosphere	Geology	Ground-penetrating radar (GPR)
	Topography	Advanced Spaceborne Thermal Emission and Reflection Radiometer (ASTER)
		Shuttle Radar Topography Mission (SRTM)
		Sentinel-2
Atmosphere	Surface temperature	MODIS (Moderate Resolution Imaging Spectroradiometer) surface temperature
		AVHRR (Advanced Very High Resolution Radiometer) surface temperature
		Sentinel-3
	Rainfall	Tropical Rainfall Measurement Mission (TRMM)
		Global Precipitation Measurement (GPM) mission
	Wind direction and speed	Quick Scatterometer (QuickSCAT)
		Rapid Scatterometer (RapidScat)
Hydrosphere	Soil moisture	ESA's Soil Moisture and Ocean Salinity (SMOS)
		NASA's Soil Moisture Active Passive (SMAP) observatory
	Gravity anomalies	Gravity Recovery and Climate Experiment (GRACE)
Cryosphere	Ice sheet mass balance	Geoscience Laser Altimeter System (GLAS) sensor onboard the Ice, Cloud, and land Elevation Satellite (ICESat)

10.3.1   Lithosphere

10.3.1.1   Lithosphere: Topography

Topographic barriers can influence geographic patterns of biodiversity by physically isolating populations of plants and animals (Janzen 1967). Topography also can be used as an indirect measure of microclimate, as topographic position can influence temperature and precipitation (e.g., Ollinger et al. 1995). Topography of the lithosphere crust is often represented by elevation (the height above sea level of a given point on the ground) or bathymetry (the depth to the bottom of a water body). In February 2000 the SRTM radar system flew on the US Space Shuttle Endeavour for 11 days collecting radar-derived elevation data from 60°N to 56°S. These data were originally released at 90 m resolution; however, in 2015, 30 m data (1 arc second) were released for the entire SRTM extent. There are many other sources of elevation data including NASA Advanced Spaceborne Thermal Emission and Reflection Radiometer ASTER (Fig. 10.2), *active* radar satellites designed for ice measurement (see the Cryosphere section), and more. NASA is currently working to develop a best available digital elevation model (DEM) for the planet, NASADEM. For this the entire SRTM data set will be reprocessed, Geoscience Laser Altimeter System (GLAS) data will be incorporated to remove artifacts, and the Advanced Spaceborne Thermal Emission and Reflection Radiometer Global Digital Elevation Model version 2 (ASTER) and Global Digital Elevation Map (GDEM) V2 DEMs will be used for refinement.

Elevation is only one of many products under the umbrella of topography. Slope (the angle between two elevation points) and aspect (the direction a slope is facing) are two of the many indices that can be derived from elevation data. Importantly, most of these indices are kernel-dependent, meaning they rely on data not just from an individual point but from surrounding points as well. For example, ArcGIS 10.3 (ESRI; Redlands, California) calculates the slope of a given pixel (elevation value) as the maximum slope between that center pixel and the eight surrounding pixels. The "terrain" function in the raster package in R statistical software (Hijmans and van Etten 2019) permits several different methods for calculating slope based on either a 4- or an 8-cell kernel, and these calculations differ slightly from those in the Geospatial Data Abstraction Library (GDAL; gdal.org). Environment for Visualizing Images (ENVI) software (Harris Geospatial Solutions, Broomfield, Colorado) allows the user to select any kernel size then fits a quadratic surface to the entire kernel, calculating slope and other parameters based on that surface (Wood 1996). These different methods could lead to somewhat different results; in particular, the selection of a small versus a large kernel could change the slope estimated. Imagine, for example, with fine-grained data, the inside of a tip-up pit on the side of a north-facing slope. The local aspect could be south facing, while a larger kernel could reveal that the landscape is north facing.

Beyond slope and aspect, there are many other kernel-dependent topographic measures. For instance, Topographic Position Index (TPI) is defined as the difference between a central pixel and the mean of its surrounding pixels. Terrain

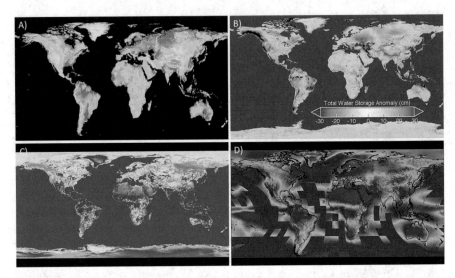

Fig. 10.2 Four examples of geodiversity variables derived from National Aeronautics and Space Administration (NASA) data products. (**a**) Earth's elevation, from which topographic diversity can be calculated, from 2009 imagery from the Advanced Spaceborne Thermal Emission and Reflection Radiometer (ASTER) instrument aboard NASA's Terra satellite (30 m spatial resolution). Image courtesy of NASA/JPL/METI/ASTER Team, NASA's Goddard Space Flight Center. https://svs.gsfc.nasa.gov/11734. Elevation in meters shown with yellows being lower in elevation than greens or reds. (**b**) Gravity Recovery and Climate Experiment's (GRACE) Terrestrial Water Storage Anomaly as of April 2015 relative to a 2002–2015 mean. Image courtesy of NASA's Scientific Visualization Studio (1° spatial resolution). (**c**) Soil Moisture Active Passive (SMAP) global radiometer map. Image courtesy of NASA (9 km spatial resolution). H-polarized brightness temperatures are shown in degrees Kelvin with warmer colors (reds and oranges) showing warmer temperatures and cooler colors (blues and yellows) showing cooler temperatures. (**d**) Mean annual cloud frequency (%; reds indicate higher cloud frequency than blues) over 2000–2014 derived from NASA's Moderate Resolution Imaging Spectroradiometer (MODIS) satellites (Wilson and Jetz 2016; 1 km spatial resolution)

Ruggedness Index (TRI), in contrast, is the mean of the difference between the central pixel and its surrounding pixels (Riley et al. 1999). Wood (1996) describes a number of convexity and curvature metrics based on the first and second derivatives of the quadratic surface described above. DEMs can also be classified into topographic features or landforms like peaks, ridges, channels, and pits, though these definitions depend on specific threshold values that may either be prescribed by software or defined by the user. Incident solar radiation can also be calculated for a given day or aggregated for a year based on a given point's elevation and latitude and the elevations of surrounding pixels. Although this section mainly describes DEM-derived morphometric landforms, it is also important to acknowledge that the genesis of landforms interacts with the ecology of a system. For instance, two hills with similar shapes may have very different associated vegetation if one is sandy (e.g., a dune) and the other is made of tills (e.g., end moraine).

While SRTM-derived products are typically used to produce "best available" topographic information, a challenge with SRTM is that the mission occurred only once. In geologically and tectonically active areas and areas where humans are influencing geology, satellite-derived data can be used to detect even very small changes over time. For example, Ge et al. (2014) used synthetic aperture radar (SAR) interferometry to detect subsidence in the Bandung Basin (Indonesia) likely due to groundwater extraction. Yun et al. (2015) used SAR data to map areas of change and potential damage after the 2015 Gorkha earthquake in central Nepal. Using SAR instruments in concert with LiDAR instruments on airborne flights has allowed for greater than 30 cm vertical accuracy (Corbley 2010). The launch of the Global Ecosystem Dynamics Investigation (GEDI) mission onboard the International Space Station has the potential to allow for improved global topographic data (Stavros et al. 2017).

10.3.1.2 Lithosphere: Geology and Soils

Geology consists of several subdisciplines, including lithology, tectonics, volcanology, and seismology. A modern geologic "map" in a geographic information system (GIS) framework may include polygons outlining the different substrate types and their ages, lines showing faults, and points identifying small outcrops or places where cores were collected. These static (unchanging through time) representations are developed through the painstaking work of geologists who gather in-situ records of rock type and estimates of geologic feature extents. Geologic maps vary in quality and access due largely to the density and biases of field technicians. When considering long-term evolutionary histories that generate deeper phylogenetic patterns, geological processes of uplift and erosion can become important (Cowling et al. 2009). Nevertheless, for more historically proximate species, community assembly, the available minerals, substrate structure, and topography are likely to play a more important role, especially in plants. For example, although all locations across the Mauna Loa environmental matrix in Hawai'i share a common parent material, differences in age, texture, and nutrient availability (due to variation in climate and weathering) lead to dramatically different vegetation patterns (Vitousek et al. 1992).

Similar to geologic maps, soil maps are typically developed through fieldwork and image interpretation for a single time period. Nevertheless, soils have higher spatial variability than bedrock and may change rapidly in response to natural or man-made disturbance. Recently there have been calls to improve the quality and dynamism of soil maps (Grunwald et al. 2011). The SoilGrids1km data product (Hengl et al. 2014) is one such example. It is a modeled product that relies on indirect remotely sensed variables, such as Moderate Resolution Imaging Spectroradiometer (MODIS), leaf area index (LAI), land surface temperature (LST), and topography from the SRTM to produce estimates at six depths of soil organic carbon, soil pH, sand, silt, and clay fractions, bulk density, cation-exchange capacity, coarse fragments, and depth to bedrock.

Imaging spectroscopy has been broadly applied for geologic mapping (Goetz et al. 1985; Gupta 2013). Multispectral imagery, like NASA's ASTER instrument and the European Space Agency's (ESA's) Sentinel-2 satellite, that is part of the

Copernicus program has long been used for mapping lithography in exposed surface environments (Rowan and Mars 2003; Hewson et al. 2005; Massironi et al. 2008; van der Werff and van der Meer 2016). Hyperspectral imagery has been used successfully to map minerals in many low-vegetation landscapes. For example, the Hyperion sensor, aboard the now decommissioned EO-1 satellite, was used to map mineralogy in Australia (Cudahy et al. 2001). The Airborne Visible/Infrared Imaging Spectrometer (AVIRIS) and new AVIRIS-Next Generation missions continue to push the boundary of imaging spectroscopy used in mineral mapping (Krause et al. 1993; Crowley 1993; Green et al. 1998). These instruments can also provide information on soil nutrient availability in areas dominated by vegetation cover via the influence of soils on foliar chemistry (e.g., Ollinger et al. 2002).

Ground-based RS has also provided insights for subsurface geologic mapping. For instance, ground-penetrating radar (GPR) uses radar pulses to map the relative densities of materials belowground and effectively maps soil and bedrock in layers (Davis and Annan 1989). Airborne GPR can greatly enhance the temporal and spatial resolution of geologic maps (Catapano et al. 2014; Campbell et al. 2018).

10.3.2 Atmosphere: Climate and Weather

Climate is an important control on mineral weathering, soil formation, and landforms (Jenny 1941). Surface temperature and cloud cover are readily observed with RS. The Advanced Very High Resolution Radiometers [AVHRR; National Oceanographic and Atmospheric Administration (NOAA)] have been collecting surface radiation data in the visible, infrared, and thermal spectra with twice-daily global coverage since 1981 that currently gathers data at ~1 km spatial resolution. AVHRR data can be used to map cloud cover and land and water surface temperatures; however, changes in satellite technology and the lack of onboard calibration in the AVHRR sensors have made the use of these data challenging due to a need for standardization of data across satellite technologies (Cao et al. 2008). The launch of the MODIS sensors on NASA's Terra (launched in 1999) and Aqua (launched in 2002) satellites and ESA's Sentinel-3 satellite as part of the Copernicus program (3-A launched in 2016 and 3-B launched in 2018) significantly improved global mapping capabilities. The two MODIS sensors map most of the planet twice a day with 36 bands ranging from the visible to the thermal infrared. The MODIS bands were selected to capture properties of the land surface but also ocean properties, atmospheric water vapor, surface temperature, and clouds (Fig. 10.2). Products from MODIS, such as surface temperature and cloud presence, have been used either to directly map climate variables for use in ecological research (e.g., Cord and Rödder 2011; Wilson and Jetz 2016) or to inform modeled climate products like Worldclim-2 (Fick and Hijmans 2017). Furthermore, surface temperature can better characterize plant ecological differences (Still et al. 2014) because it more accurately captures canopy temperature, which is not the same as air temperature, and because many air temperature products (such as Worldclim-2) are interpolated (see Pinto-Ledézma and Cavender-Bares, Chap. 9).

Satellite-derived rainfall products are estimated through a combination of measurements, including surface reflectance of clouds (i.e., cloud coverage, type, and top temperature), passive microwave (i.e., column precipitation content, cloud water and ice, rain intensity and type), and lightning sensors. The Tropical Rainfall Measurement Mission (TRMM) operated from 1997 to 2015, providing information on rainfall amount and intensity and lightning activity globally every 3 hours at 5 km resolution from 38°N to 38°S. As a follow-up to TRMM, the Global Precipitation Measurement (GPM) mission relies on a constellation of satellites, including a core GPM observatory, to produce 0.1° resolution data every 30 minutes from 60°N to 60°S. Initiated in 2014, GPM allows new explorations of extreme weather events. Like MODIS temperature measurements, TRMM and GPM precipitation measures have been directly incorporated into ecological research (e.g., Deblauwe et al. 2016) and used to inform modeled climate products like the Climate Hazards Group Infrared Precipitation with Station data product (CHIRPS; Funk et al. 2015).

There is also a broad set of efforts to generate reanalysis products that combine the history of Earth observations to develop temporally and spatially consistent global models of climatic and environmental variables. For instance, the NASA Modern-Era Retrospective Analysis for Research and Applications (MERRA) models close to 800 radiative and physical properties of the Earth's atmosphere at 3- to 6-hour time steps from 1979 to present at ~50 km spatial resolution (Rienecker et al. 2011). While this obviously sacrifices spatial resolution, these efforts open the door for longer-term analysis of climatic influence on biologic phenomena.

One commonly overlooked source of geologic substrate lies in the atmosphere. Airborne dust particles provide an essential source of nutrients in many environments and can originate from sources hundreds to thousands of miles away (Chadwick et al. 1999). Aeolian transport of phosphorus from North Africa to South America is thought to be an important driver of Amazonian productivity (e.g., Okin et al. 2004). Studies have mapped dust sources and rates using MODIS products (Ginoux et al. 2012) and produced 3-D models of dust transportation using LiDAR on the Cloud-Aerosol Lidar and Infrared Pathfinder Satellite Observation (CALIPSO) satellite (Yu et al. 2015). Furthermore, the SeaWinds instrument on the Quick Scatterometer (QuickSCAT) satellite and the subsequent Rapid Scatterometer (RapidSCAT) aboard the International Space Station measures wind speed and direction over the ocean's surface.

10.3.3 Hydrosphere

The hydrosphere consists of the water on, in, and above Earth's surface and is known to have a large influence in structuring riparian and aquatic communities of organisms (reviewed by Atkinson et al. 2017). The hydrosphere interacts with other types of geodiversity in the lithosphere, cryosphere, and atmosphere. Topography alone can be used to indirectly provide a crude estimate of many hydrological variables, including watershed size, soil water content (Moore et al. 1991), flow paths, and surface water. In addition, two types of satellite data can be used to estimate soil moisture and groundwater, which in some systems are important drivers of plant diversity because drought sensitivity may shape plant distributions (e.g., Engelbrecht et al. 2007). The ESA's Soil Moisture and Ocean Salinity (SMOS; launched 2009) and NASA's Soil Moisture Active Passive observatory (SMAP; launched 2015; Fig. 10.2) both use microwave radiometers to detect surface soil moisture globally in areas with low topographic variation and low-vegetation cover. The Gravity Recovery and Climate Experiment (GRACE; launched in 2002; Fig. 10.2) is a pair of satellites that measure gravity anomalies around the world, allowing researchers to estimate available groundwater reserves and their change over time.

Water quality is a critical driver of aquatic biodiversity across taxa, from plants to animals (Stendera et al. 2012). Watershed disturbance, sediment runoff, and nutrient pollution are major aquatic biodiversity stressors, affecting phytoplankton and aquatic and wetland vegetation abundance and diversity (Lacoul and Freedman 2006; Mouillot et al. 2013) and higher trophic levels (e.g., zooplankton, shrimps, larval fish, and birds (Thackeray et al. 2010). Optical RS can be used to retrieve a limited but important set of water quality variables, including particulate and dissolved organic and inorganic matter, chlorophyll-a, as well as other phytoplankton pigments like the phycocyanins common in potentially harmful cyanobacteria blooms. Surface or "skin" water temperature is measured from instruments with thermal bands (Giardino et al. 2018; Alcântara et al. 2010). The major limitation in RS of water quality is in sensor resolution. Sensors must have a fine enough pixel size to resolve water bodies, with high enough radiometric sensitivity to detect small changes in a dark target (10% or less of the total signal received by the sensor, Muller-Karger et al. 2018; Hestir et al. 2015). While some water quality products are publically distributed with limited spatial coverage [e.g., United Nations Educational, Scientific and Cultural Organization (UNESCO) regions], free data processors distributed by NASA (Sea-viewing Data Analysis System [SeaDAS]) and the ESA (Sentinel Application Platform) enable users to compute their own water quality products.

10.3.4   Cryosphere

Earth's fossil record illustrates how changes in glacial cover over time have governed the distribution of biodiversity (e.g., Veloz et al. 2012), and many aspects of the globe's biodiversity are influenced by snow, ice, and permafrost (reviewed by Vincent et al. 2011). The frozen parts of the Earth system, the cryosphere, can be detected with a number of different RS tools. The cryosphere can be divided into several different components—seasonally snow-covered land, permafrost, glaciers and ice sheets, lake ice, and sea ice. Because cloud cover is a frequent problem at high latitudes, cryosphere RS often relies on longwave techniques that can pass through clouds. A recent book, *Remote Sensing of the Cryosphere* (Tedesco 2014), describes these tools and methods in great detail; here we review some of the major techniques. In all of the discussion below, the importance of change over time is paramount; inter- and intra-annual variation in snow and ice cover are important drivers of physical and biological processes.

The 3-D extent of snow and ice can easily be mapped using optical techniques; snow reflects strongly in the visible and near-infrared (NIR) range but absorbs in the shortwave infrared (SWIR), making it spectrally distinct from other white objects such as rooftops and clouds. These distinctions may still be challenging with multi-spectral sensors, but hyperspectral sensors permit mapping of snow versus clouds and even some estimation of snow particle size (e.g., Burakowski et al. 2015). Passive microwave sensors can be used to estimate snow depth and snow water equivalent, while active microwave sensors can map liquid water content. Tools and techniques for mapping snow are reviewed by Dietz et al. (2011).

Ice and permafrost features can be mapped with many of the tools and methods described in preceding sections. Snow cover can be mapped using optical sensors and methods; subsidence of the cryosphere can be mapped with SRTM (near global extent, 30–90 m spatial resolution, single snapshot in time) and SAR (airborne, 2 m spatial resolution); and passive microwave radiometers such as SMOS (global extent, 50 km spatial resolution, 3-day temporal resolution) and SMAP (near global extent for low-vegetation areas, 9–36 km spatial resolution, 8-day temporal resolution) can be used to map frozen versus thawed ground surfaces (Entekhabi et al. 2014). Because glaciers and ice sheets are fundamentally a combination of snow, ice, and liquid water, many of the techniques described above, such as optical sensors and passive microwave radiometers, can be used to map their extent and status. In addition, the GLAS sensor onboard the Ice, Cloud, and land Elevation Satellite (ICESat; near global spatial extent, 70 m spatial resolution, 91-day temporal resolution from 2003 to 2009) permitted the mapping of ice sheet mass balance (Zwally et al. 2011). ICESat-2 is scheduled for launch in 2018 (global spatial extent, 14 km spatial resolution, 91-day temporal resolution). SAR has also been used to map ice flow on Antarctica (Rignot et al. 2011).

Sea, lake, and river ice cover can be mapped using optical techniques (Jeffries et al. 2005), while thickness has been measured using ICESat and passive microwave sensors (e.g., Kwok and Rothrock 2009). The difference between first-year sea ice and older sea ice can be identified by changes in salinity using multichannel passive microwave sensors like the Advanced Microwave Scanning Radiometer for Earth Observing System (AMSR-E) onboard NASA's Aqua satellite (global spatial extent, 474 km spatial resolution, 12-hour temporal resolution, operational 2002–2015). River ice mapping is critical for monitoring and predicting river habitat quality and duration for a variety of organisms (e.g., Charney and Record 2016; Pavelsky and Zarnetske 2017). The extent and duration of river icing types have been mapped with different polarizations of passive microwave data from Canada's RADARSAT-1 (1995–2013) and RADARSAT-2 (launched 2007) (Weber et al. 2003; Jeffries et al. 2005; Yoshikawa et al. 2007 for aufeis features) and with MODIS Terra (Pavelsky and Zarnetske 2017).

10.4 Remote Sensing of Biodiversity

Approaches for using RS to track biodiversity are reviewed in several chapters in this book (Fernandez and Pereira, Chap. 18; Serbin and Townsend, Chap. 4; Meireles et al., Chap. 7). Biodiversity has many forms—including taxonomic, functional, genetic, and phylogenetic diversity (Serbin and Townsend, Chap. 4; Meireles et al., Chap. 7). Each form may exhibit different relationships with both geophysical and biological drivers, owing to a variety of mechanisms (Gaston 2000; Lomolino et al. 2010). For example, reorganization of organisms in response to changing environments leads to species assemblages becoming more or less similar through biotic homogenization or differentiation (Baiser et al. 2012). Such biotic homogenization/ differentiation is usually characterized taxonomically (Olden and Rooney 2006). However, functional traits (i.e., traits representing the interface between species and their environment) possessed by species are often more important to ecosystem functions valued by society (Baiser and Lockwood 2011) and may be more appropriate to use in assessing biodiversity-ecosystem function relationships (Flynn et al. 2011). Many functional traits may also exhibit a phylogenetic signal (Srivastava et al. 2012), so it is important to consider multiple measures of diversity (i.e., taxonomic, functional, and phylogenetic) when assessing patterns of biodiversity (Serbin and Townsend, Chap. 4; Meireles et al., Chap. 7; Lausch et al. 2016; Lausch et al. 2018).

One caveat to measures of biodiversity generated from high-resolution RS data is that as the spatial resolution of data increases, the spatial extent typically decreases (Turner 2014; Gamon et al., Chap. 16). This limitation hinders our ability to understand how biodiversity relates to different drivers (e.g., geodiversity) at different spatial scales to better inform conservation decisions. There have been recent calls from scientists for new satellite missions and data integration efforts to address this issue (Schimel et al., Chap. 19). For instance, Jetz et al. (2016) call for a Global Biodiversity Observatory to generate worldwide remotely sensed data on several plant functional traits. Petorelli et al. (2016) and Fernández and Pereira (Chap. 18) identify satellite RS data that, given technological and algorithmic developments in the near future, could be capable of meeting the criteria of EBVs for conservation outlined by the international Group on Earth Observations—Biodiversity Observation Network (GEO BON) at a global spatial extent.

Until finer resolution, remotely sensed biodiversity data exist at large spatial extents, data available from in-situ measurements of organisms can inform the relationships between biodiversity and geodiversity. Publically available biodiversity data with geographic locations include expert range maps of individual species from IUCN (IUCN 2017), occurrence data [e.g., Global Biodiversity Information Facility (GBIF, GBIF 2016); Botanical Information and Ecology Network (BIEN, Enquist et al. 2016)], citizen science networks [e.g., Invasive Plant Atlas of New England, IPANE, Bois et al. 2011], and national [e.g., US Forest Service Forest Inventory and Analysis (FIA), Bechtold and Paterson 2005)] and international inventory networks [e.g., the Amazon Forest Inventory Network (RAINFOR), Peacock et al. 2007].

Each of these data sets comes with its own uncertainties (e.g., observation errors) and user challenges. For instance, citizen science data require detailed metadata on the sampling process to ensure that citizen scientists are able to reduce error and bias as they collect data and to enable those analyzing the data to model potential uncertainty (Bird et al. 2014). Despite these sources of uncertainty and logistical hurdles, these data provide a useful starting point for understanding the relationship between biodiversity and geodiversity.

10.5 A Case Study Linking RS of Geodiversity to Tree Diversity in the Eastern United States

To motivate explorations of the relationship between biodiversity and geodiversity with remotely sensed data, we provide an example using biodiversity data from the FIA program of the US Forest Service (O'Connell et al. 2017) and geodiversity data on elevation from SRTM. We selected elevation as a covariate because patterns of tree diversity often vary with elevation (Körner 2012). While some studies promote the use of many geodiversity components (Serrano et al. 2009; Hjort and Luoto 2010; Bailey et al. 2017; Tukiainen et al. 2017), a great deal of the variation in geodiversity is captured by the standard deviation in elevation (Hjort and Luoto 2012), which is used in this analysis.

The FIA program uses a two-phase protocol to characterize the nation's forest resources. In phase one, all land in the United States is categorized as either "forested" or "not forested" using remotely sensed data. In phase two, in every 2428 ha of land classified as forested, one permanent FIA plot is placed for in-situ sampling. Each FIA plot consists of four 7.2-m-fixed-radius subplots wherein all trees >12.7 cm diameter at breast height are measured. FIA plot measurements began in the 1940s, but a consistent nationwide sampling protocol was not implemented until 2001. In the analysis presented, we used data from the most recent full plot FIA inventory from 2012–2016; the SRTM data were collected in 2009. Although there is not perfect temporal overlap in the geodiversity and biodiversity data used in this example, we do not expect that topography at a spatial resolution of 50 km would have changed much over the time period encompassed by both data sets for this part of the world.

We fixed the spatial extent of the analysis to the contiguous United States east of 100°W longitude (n = 90,250 plots total) and selected a grain size of 50 km for calculating alpha (within site), beta (turnover between sites), and gamma (total across all sites) diversities within a radius centered on each FIA plot. All biodiversity metrics were based on species abundances as quantified by the total basal area of each tree species in each plot. Alpha diversity was calculated as the median abundance-weighted effective species number of all plots falling within a 50 km radius of the focal FIA plot, including the focal plot. Beta diversity was calculated as the mean abundance-weighted pairwise Sørensen dissimilarity of all pairs of plots within a 50 km radius of the focal plot, including the focal plot. Gamma

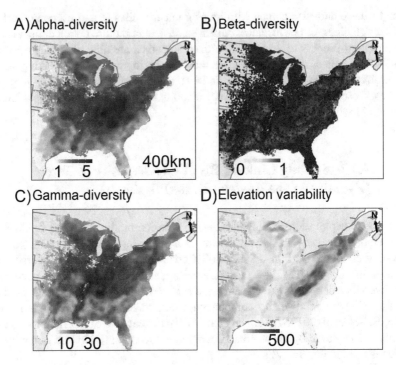

Fig. 10.3 Mapped variation in tree diversity calculated within 50 km radii. Tree data come from the Forest Inventory and Analysis of the US Forest Service (FIA, O'Connell et al. 2017. (**a**) Taxonomic alpha diversity. (**b**) Taxonomic beta diversity. (**c**) Taxonomic gamma diversity. (**d**) The standard deviation of all elevation pixels within the radius from 30 m Shuttle Radar Topography Mission (SRTM) data

diversity was calculated as the aggregated effective species number of all plots within a 50 km radius of the focal plot, including the focal plot. For each 50 km radius centered on a focal plot, we computed the standard deviation of elevation across pixels within the radius from 30 m SRTM data (Fig. 10.3). To avoid edge effects, all plots within 100 km of the political borders of the United States were excluded, retaining 80,411 plots. To avoid pseudo-replication, we generated 999 subsamples of plots separated by at least 100 km, yielding ~370 plots per subsample. Because of the saturating relationship between biodiversity and geodiversity, we fit natural splines with 3 degrees of freedom to relate all focal plots' univariate diversity to elevation standard deviation (SD) (linear regression for alpha and gamma diversity; beta regression for beta diversity), and goodness of fits of the models were assessed with r-squared (Fig. 10.4).

This example shows how the relationships between biodiversity and geodiversity for a subset of different biodiversity metrics vary depending on the metric of biodiversity calculated. Here beta and gamma diversity do not show a strong relationship ($r^2 = 0.03$ and $r^2 = 0.07$, respectively; Figs. 10.3 and 10.4) with geodiversity, but alpha diversity shows a stronger, positive relationship with elevation variability

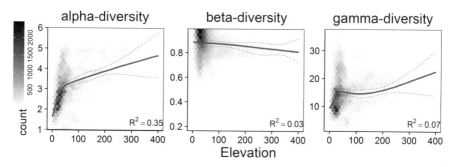

Fig. 10.4 The relationships between three measures of tree taxonomic diversity (alpha, beta, and gamma) and geodiversity (i.e., elevation standard deviation) at a spatial resolution of 50 km. Points indicate the aggregated plots, and the red line indicates the natural spline relationship fitted with a linear regression model for alpha and gamma diversity and a beta regression model for beta diversity. Dotted lines represent the 2.5% and 97.5% quantiles of predicted values across 999 spatially stratified random subsamples of the data, and the given r-squared value is the mean across all the subsamples

($r^2 = 0.35$; Figs. 10.3 and 10.4). Interestingly, in a sister study, Zarnetske et al. (2019) found that FIA tree diversity with a different spatial extent—in California, Washington, and Oregon—showed a different relationship with elevation variability. Furthermore, beta and gamma diversity showed a strong increasing relationship with elevation variability, whereas alpha diversity did not. This comparison between the case study illustrated in this chapter and the results of Zarnetske et al. (2019) highlights the importance of considering how the relationship between geodiversity and biodiversity may change with different spatial scales (Gamon et al., Chap. 16). Bailey et al. (2017) also showed that landforms detected with airborne RS at smaller spatial resolutions explained more of the variation in alpha diversity of alien vascular plants in Great Britain than did climate measured at larger spatial resolutions. While these examples do not provide an exhaustive exploration of the ways in which tree diversity responds to geodiversity, they clearly show how remotely sensed data may help us understand the relationships between geodiversity and biodiversity and how these relationships may be different in different geographic areas.

This example shows how the relationship between taxonomic biodiversity and geodiversity depends on the biodiversity metric chosen. There are various methodologies for calculating biodiversity metrics and various facets of biodiversity (e.g., functional, taxonomic, phylogenetic), and the theoretical pros and cons of each remain controversial (e.g., Jost 2007; Clark 2016), so it may not be obvious which metric is the best. Furthermore, different conclusions may be drawn depending on the types of taxa used in the analysis.

In a similar vein, the choice of an appropriate geodiversity metric may not be obvious. Here we use a single measure of geodiversity, standard deviation of elevation. However, different definitions of the term *geodiversity* include different components of geology, topography, and, in some instances, climate (Parks and Mulligan 2010; Gray 2013). The amalgamation of these different variables to characterize geodiversity as a whole is an area in need of development.

10.5.1 Challenges and Opportunities

10.5.1.1 The Interplay Between Biodiversity and Geodiversity over Time

Although we have focused thus far on the effects of geodiversity on biodiversity, biodiversity can also affect geodiversity. Ecosystem engineers (Jones et al. 1994) and foundation species (Record et al. 2018) can influence biodiversity through habitat formation (Hastings et al. 2007). Geodiversity can be modified by species impacting the structure and function of landscape features. For example, elephants dig, form trails, and trample (Haynes 2012), and vegetation and sediment interact to form streams and coastal dunes (Zarnetske et al. 2012; Atkinson et al. 2017). In turn, these species-modified features can feed back to mediate the strength and direction of biotic interactions among species and ultimately influence patterns of biodiversity (Zarnetske et al. 2017). Even climate can be influenced by biodiversity and biogeographic patterns. Forests directly affect Earth's climate through atmospheric exchange (Bonan 2008). If shrubs expand by 20% and continue to dominate in areas north of 60°N latitude, for example, Arctic annual temperature could increase by 0.66°C–1.84°C, via decreased albedo and increased evapotranspiration (Bonfills et al. 2012).

Many of these feedbacks between biodiversity and geodiversity are not detectable given a single snapshot in time and require longer time series. RS with repeat samples taken as satellites orbit the Earth provide data with high spatial and deep temporal coverage that can be used to assess changes in the dominance of a species within a community (Pau and Dee 2016). Changes in the dominance structure of communities (or its counterpart, evenness) should be early indicators of global change because these changes occur before the complete loss or replacement of species (Hillebrand et al. 2008). Furthermore, tracking dominant species should be especially important for quantifying biomass or abundance-driven ecosystem functions and services (Pau and Dee 2016). For instance, Cavanaugh et al. (2013) used 28 years of Landsat imagery to map the poleward expansion of mangroves, which are important in preventing coastal erosion, in the eastern United States. Furthermore, the 45-year time series of Landsat data provide an excellent opportunity for detecting changes in habitat due to species, which may have extreme impacts on the abiotic stage.

10.5.1.2 Scale and Expertise Mismatches

The relationships between geodiversity and biodiversity are likely to change across spatial and temporal scales. For instance, a focused spotlight shining down on one part of the stage (e.g., the tip of a mountaintop) might exhibit different covariation between geodiversity and biodiversity than a broad swath of light on another portion of the stage (e.g., an expansive low-lying valley). Spatial patterns of biodiversity and geodiversity are each scale dependent (Rahbek 2005; Bailey et al. 2017; Cavender-Bares et al., Chap. 2; Gamon et al., Chap. 16), and it is well established that ecological processes influencing the assembly of communities of organisms are scale dependent (Levin 1992; McGill 2010). A spatially explicit framework for conceptualizing community assembly describes external filters (e.g., climate or soils) that sort species from a regional pool at a spatial scale larger than the community and internal filters that sort species into a community from a subset of the species that make it through the external filter (e.g., microenvironmental heterogeneity, biotic interactions; Violle et al. 2012; Fig. 10.5). These "assembly rules" about how communities form remain a controversial paradigm with uncertainty about which processes operate at which scales (McGill 2010; Belmaker et al. 2015). Observing and quantifying relationships between geodiversity and biodiversity and how these relationships change with scale, however, are essential for moving forward regardless of one's position on these controversies. To most effectively use geodiversity to help explain and predict patterns of biodiversity, we need a framework that addresses the scaling relationship between biodiversity and geodiversity.

Furthermore, there are important disconnects in both scale and expertise between biodiversity science and RS (Petorelli et al. 2014) that once addressed will aid in the development of such a framework. Whereas the availability of remotely sensed geodiversity data products has increased, many of the scales are too coarse to reflect the environmental and biological conditions that often drive more fine-scaled spatially heterogeneous biodiversity patterns (Nadeau et al. 2017) and thus may require complex post-processing techniques unfamiliar to most biodiversity scientists before they can be used appropriately in biodiversity models. Also, there are likely many important aspects of geodiversity that at this time can only be derived through in-situ measurements and cannot be remotely sensed. Determining how physical and biological drivers influence biodiversity across spatial and temporal scales is a central focus of ecology. However, most models predicting future patterns of biodiversity assume broad-scale climatic drivers—temperature and precipitation—are sole drivers and leave out important biological drivers (Zarnetske et al. 2012; Record et al. 2013). Biological drivers such as dispersal ability and biotic interactions (e.g., competition) are often mediated by the structure of the landscape, including geophysical feature configuration, topographic complexity, and habitat patch arrangement (Zarnetske et al. 2017). Yet a significant knowledge gap remains about how the relationships between biodiversity and its geophysical and biological drivers change with respect to space and time—perhaps owing to the scale mismatch between fine-scale point-level biodiversity data and many coarse scale remotely sensed data products.

Many ecological questions are addressed at scales much finer than the grain size of MODIS or GPM, which makes statistical downscaling a necessity for remotely sensed products to be used. Yet the landscape of options for statistical downscaling is vast and complex (Pourmokhtarian et al. 2016). In addition, the increasing availability of airborne topographic data like LiDAR makes the possibility of finer-grain analysis even more viable, yet these data also bring another dimension of complexity and a lack of standardization across platforms and methods.

Open access analytical tools and training will provide ways forward given data downloading and processing challenges. The Application for Extracting and

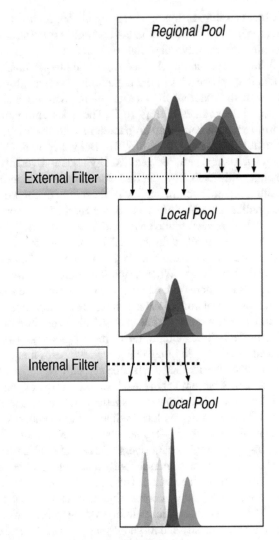

Fig. 10.5 Conceptual diagram adapted from Violle et al. (2012) showing how spatially explicit (i.e., local versus regional) filters influence the assembly of traits in an observed community (bottom schematic). The regional species pool (top schematic) contains all of the species capable of seeding into the local community. However, the observed local community may only contain a subset of the species in the regional pool after species have passed through a series of filters. Both internal and external filters encompass different aspects of the stage, whereas internal filters may also include the actors. Examples of external filters include broad-scale climate or soil types for which some species may not have physiological tolerances. Internal filters include microenvironmental heterogeneity and/or biotic interactions. In this schematic, the traits that passed through the external and internal filters partition in the observed local community, perhaps due to competitive effects between species for resources

Exploring Analysis Ready Samples (AppEEARS) offered by NASA and US Geological Survey (USGS) is a user-friendly tool that enables simple and efficient downloads and transformations of geospatial data from a number of federal data archives from the United States [e.g., the Land Process Distributed Active Archive Center (LP DAAC)]. Additionally, Geomorphons provides a user-friendly interface that automates the calculation of complex geodiversity features from topography data (Jasiewicz and Stepinski 2013). Training the next generation of ecologists and conservation biologists in RS will be integral to overcoming some of these hurdles and bridging the gaps between RS and ecology and conservation.

10.6 Conclusion

Cross-scale studies of relationships between geodiversity and biodiversity using RS and large field-based data sets hold promise for evaluating processes underlying biodiversity and identifying scales and methods for its monitoring and management. Realizing this potential will require more interaction among biodiversity scientists, geoscientists, RS experts, and statisticians to reconcile the challenges associated with differences in scales, available data products, disciplinary barriers, and available methods for connecting geodiversity to biodiversity. These challenges are far from trivial, but overcoming them has the potential to result in key ecological insights that will help us to be better stewards of the entire ecological theater.

Acknowledgments We thank J. Cavender-Bares, J. Gamon, A. Lausch, M. Madritch, F. Schrodt, P. Townsend, and S. Ustin for comments on the chapter while at NIMBioS. Funding for the NASA bioXgeo working group was provided by the National Aeronautics and Space Administration (NASA) Ecological Forecasting Program, Earth Science Division, Grant #NNX16AQ44G. Additional support for PLZ, KD, QR, JG, and AF came from Michigan State University. SR was supported by Bryn Mawr College KG Fund. AMW acknowledges support from NASA's Ecological Forecasting Program, Earth Science Division, Grant #NNX16AQ45G. MH was supported by the NASA Biodiversity Program and MODIS Science Program, Grant #NNX14AP07G. SO acknowledges support from the NSF Macrosystems Biology Program, Grant #638688. We thank the National Center for Ecological Analysis and Synthesis (NCEAS) for housing the working group meetings. KDG was supported by an AAAS Science and Technology Policy Fellowship served at NASA. The views expressed in this paper do not necessarily reflect those of NASA, the US Government, or the American Association for the Advancement of Science.

References

Alcântara EH, Stech JL, Lorenzzetti JA, Bonnet MP, Casamitjana X, Assireu AT, de Moraes L, Novo EM (2010) Remote sensing of water surface temperature and heat flux over a tropical hydroelectric reservoir. Remote Sens Environ 114:2651–2665

Anderson MG, Ferree CE (2010) Conserving the stage: climate change and the geophysical underpinnings of species diversity. PLoS One 5:e11554

Atkinson CL, Allen DG, Davis L, Nickerson ZL (2017) Incorporating ecogeomorphic feedbacks to better understand resiliency in streams: a review and directions forward. Geomorphology 305:123–140

Baiser B, Lockwood JL (2011) The relationship between functional and taxonomic homogenization. Glob Ecol Biogeogr 20:134–144

Baiser B, Olden JD, Record S, Lockwood JL, McKinney ML (2012) Pattern and process of biotic homogenization in the New Pangaea. P R Soc B 279:4772–4777

Barton PS, Cunningham SA, Manning AD, Gibb H, Lindenmayer DB, Didham RK (2013) The spatial scaling of beta diversity. Glob Ecol Biogeogr 22:639–647

Bechtold WA, Paterson PL (eds) (2005) The enhanced forest inventory and analysis program: national sampling design and estimation procedure. General technical report SRS-80. USA Department of Agriculture, Forest Service Southern Research Station, Asheville, NC, USA

Beier P, Brost B (2010) Use of land facets to plan for climate change: conserving the arenas, not the actors. Cons Biol 24:701–710

Beier P, Hunter ML, Anderson M (2015) Special section: conserving nature's stage. Cons Biol 29:613–617

Belmaker J, Zarnetske P, Tuanmu M-N, Zonneveld S, Record S, Strecker A, Beaudrot L (2015) Empirical evidence for the scale dependence of biotic interactions. Glob Ecol Biogeogr 24:750–761

Benito-Calvo A, Pérez-González A, Magri O, Meza P (2009) Assessing regional geodiversity: the Iberian Peninsula. Earth Surf Proc Land 34:1433–1445

Bird TJ, Bates AE, Lefcheck JS, Hill NA, Thomson RJ, Edgar GJ, Stuart-Smith RD, Wotherspoon S, Krkosek M, Stuart-Smith JF, Pecl GT (2014) Statistical solutions for error and bias in global citizen science datasets. Biol Conserv 173:144–154

Bois ST, Silander JA, Mehrhoff LJ (2011) Invasive plant atlas of New England: the role of citizens in the science of invasive alien species detection. Bioscience 61:763–770

Bonan GB (2008) Forests and climate change: forcings, feedbacks, and the climate benefits of forests. Science 320:1444–1449

Bonfills CJW, Phillips TJ, Lawrence DM, Cameron-Smith P, Riley WJ, Subin ZM (2012) On the influence of shrub height and expansion on northern high latitude climate. Environ Res Lett 7:015503

Burakowski EA, Ollinger SV, Lepine LC, Schaaf CB, Wang Z, Dibb JE, Hollinger DY, Kim J, Erb A, Martin ME (2015) Spatial scaling of reflectance and surface albedo over a mixed-use, temperate forest landscape during snow-covered periods. Remote Sens Environ 158:465–477

Campbell S, Affleck RT, Sinclair S (2018) Ground-penetrating radar studies of permafrost, periglacial, and near-surface geology at McMurdo Station, Antarctica. Cold Reg Sci Technol 148:38–49

Cao C, Xiong S, Wu A, Wu X (2008) Assessing the consistency of AVHRR and MODIS L1B reflectance for generating fundamental climate data records. J Geophys Res-Atmos 113:1–10

Catapano I, Antonio Affinito, Gianluca Gennarelli, Francesco di Maio, Antonio Loperte, Francesco Soldovieri, (2014) Full three-dimensional imaging via ground penetrating radar: assessment in controlled conditions and on field for archaeological prospecting. Applied Physics A 115 (4):1415–1422

Cavanaugh KC, Kellner JR, Forde AJ, Gruner DS, Parker JD, Rodriguez W, Feller IC (2013) Poleward expansion of mangroves is a threshold response to decreased frequency of extreme cold events. P Natl Acad Sci USA 111:723–727

Chadwick OA, Derry LA, Vitousek PM, Huebert BJ, Hedin LO (1999) Changing sources of nutrients during four million years of ecosystem development. Nature 397:491–497

Chapin FS, Walker BH, Hobbs RJ, Hooper DU, Lawton JH, Sala OE, Tilman D (1997) Biotic control over the functioning of ecosystems. Science 277:500–504

Charney ND, Record S (2016) Combining incidence and demographic modelling approaches to evaluate metapopulation parameters for an endangered riparian plant. AOB Plants 8:1–11

Charney ND, Babst F, Poulter B, Record S, Trouet VM, Frank D, Enquist BJ, Evans MEK (2016) Observed forest sensitivity to climate implies large changes in 21st century North American forest growth. Ecol Lett 19:1119–1128

Clark JS (2016) Why species tell more about traits than traits about species: predictive analysis. Ecology 97:1979–1993

Corbley K (2010) GEOSAR-making mapping the 'impossible' possible. GEO Inf 13:43–47

Cord A, Rödder D (2011) Inclusion of habitat availability in species distribution models through multi-temporal remote sensing data? Ecol Appl 21:3285–3298

Cowling RM, Proches S, Partridge TC (2009) Explaining the uniqueness of the Cape flora: incorporating geomorphic evolution as a factory for explaining its diversification. Mol Phylogenet Evol 51:64–74

Crowley JK (1993) Mapping playa evaporite minerals with AVIRIS data: a first report from Death Valley, California. Remote Sens Environ 44:337–356

Cudahy TJ, Hewson R, Huntington JF, Quigley MA, Barry PS (2001) The performance of the satellite-borne Hyperion hyperspectral VNIR-SWIR imaging system for mineral mapping at Mount Fitton, South Australia. In: Geoscience and remote sensing symposium (IGARSS 2001), pp 314–316

Davis JL, Annan AP (1989) Ground-penetrating radar for high-resolution mapping of soil and rock stratigraphy. Geophys Prospect 37:531–551

Deblauwe V, Droissart V, Bose R, Sonké B, Blach-Overgaard A, Svenning JC, Wieringa JJ, Ramesh BR, Stévart T, Couvreur TLP (2016) Remotely sensed temperature and precipitation data improve species distribution modelling in the tropics. Glob Ecol Biogeogr 25:443–454

Dietz AJ, Kuenzer C, Gessner U (2011) Remote sensing of snow—a review of available methods. Int J Remote Sens 33:4049–4134

Dufour A, Gadallah F, Wagner HH, Guisan A, Buttler A (2006) Plant species richness and environmental heterogeneity in a mountain landscape: effects of variability and spatial configuration. Ecography 29:573–584

Elith J, Leathwick JR (2009) Species distribution models: ecological explanation and prediction across space and time. Ann Rev Ecol Evol Syst 40:677–697

Engelbrecht BMJ, Comita LS, Condit R, Kursar TA, Tyree MT, Turner BL, Hubbell SP (2007) Drought sensitivity shapes species distribution patterns in tropical forests. Nature 447:80–83

Enquist BJ, Condit R, Peet RK, Schildhauer M, Thiers BM (2016) Cyberinfrastructure for an integrated botanical information network to investigate the ecological impacts of global climate change on plant biodiversity. Peer J 4:e2615v2

Entekhabi D, Yueh S, O'Neill PE, Kellogg KH, Allen A, Bindlish R, Brown M, Chan S, Colliander A, Crow WT (2014) SMAP handbook: soil moisture active passive

ESA. Endangered Species Act 1973 16 U.S.C. § 1531 et seq.

Evans MEK, Merow C, Record S, McMahon SM, Enquist BJ (2016) Towards process-based range modeling of many species. Trends Ecol Evol 31:860–871

Ferrier S, Manion G, Elith J, Richardson K (2007) Using generalized dissimilarity modelling to analyse and predict patterns of beta diversity in regional biodiversity assessment. Div Dist 13:252–264

Fick SE, Hijmans RJ (2017) WorldClim 2: new 1-km spatial resolution climate surfaces for global land areas. Int J Climatol 37:4302–4315

Flynn DFB, Mirotchnick N, Jain M, Palmer MI, Naeem S (2011) Functional and phylogenetic diversity as predictors of biodiversity-ecosystem function relationships. Ecology 92:1573–1581

Funk CP, Peterson P, Landsfeld M, Pedreros D, Verdin J, Shukla S, Husak G, Rowland J, Harrison L, Hoell A, Michaelsen J (2015) The climate hazards infrared precipitation with stations - a new environmental record for monitoring extremes. Sci Data 2:150066

Gaston KJ (2000) Global patterns in biodiversity. Nature 405:220–227

GBIF (2016) Global biodiversity information facility. Available at: https://gbif.org

Giardino C, Brando VE, Gege P, Pinnel N, Hochberg E, Knaeps E, Reusen I, Doerffer R, Bresciani M, Braga F, Foerster S, Champollion N, Dekker A (2018) Imaging spectrometry of inland and coastal waters: state of the art, achievements and perspectives. Surv Geophys 40:1–29

Ginoux P, Prospero JM, Gill TE, Hsu NC, Zhao M (2012) Global-scale attribution of anthropogenic and natural dust sources and their emission rates based on Modis deep blue aerosol products. Rev Geophys 50:1–36

Goetz AFH, Vane G, Solomon JE, Rock BN (1985) Imaging spectrometry for earth remote sensing. Science 228:1147–1153

Gray M (2008) Geoheritage 1. Geodiversity: a new paradigm for valuing and conserving geoheritage. Geosci Can 35:2

Gray M (2013) Geodiversity: valuing and conserving abiotic nature, 2nd edn. Wiley-Blackwell, London

Green RO, Eastwood ML, Sarture CM, Chrien TG, Aronsson M, Chippendale BJ, Faust JA, Pavri BE, Chovit CJ, Solis M, Olah MR (1998) Imaging spectroscopy and the airborne visible/infrared imaging spectrometer (AVIRIS). Remote Sens Environ 65:227–248

Grunwald S, Thompson JA, Boettinger JL (2011) Digital soil mapping and modeling at continental scales: finding solutions for global issues. Soil Sci Soc Am J 75:1201

Gupta RP (2013) Remote sensing geology, 2nd edn. Springer, New York

Hastings A, Byers JE, Crooks JA, Cuddington K, Jones CG, Lambrinos JG, Talley TS, Wilson WG (2007) Ecosystem engineering in space and time. Ecol Lett 10:153–164

Haynes G (2012) Elephants (and extinct relatives) as earth-movers and ecosystem engineers. Geomorphology 157:99–107

Hengl T, De Jesus JM, MacMillan RA, Batjes NH, Heuvelink GBM, Ribeiro E, Samuel-Rosa A, Kempen B, Leenaars JGB, Walsh MG, Gonzalez MR (2014) SoilGrids1km - global soil information based on automated mapping. PLoS One 9:e105992

Hestir EL, Brando VE, Bresciani M, Giardino C, Matta E, Villa P, Dekker AG (2015) Measuring freshwater aquatic ecosystems: the need for a hyperspectral global mapping satellite mission. Remote Sens Environ 167:181–195

Hewson RD, Cudahy TJ, Mizuhiko S, Ueda K, Mauger AJ (2005) Seamless geological map generation using ASTER in the Broken Hill-Curnamona Province of Australia. Remote Sens Environ 99:159–172

Hijmans RJ, Cameron SE, Parra JL, Jones PG, Jarvis A (2005) Very high resolution interpolated climate surfaces for global land areas. Int J Climatol 25:1965–1978

Hijmans R, van Etten J (2019) raster: Geographic data analysis and modeling. R Package version 3.0–7

Hillebrand H, Bennet DM, Cadotte M (2008) Consequences of dominance: a review of evenness effects on local and regional ecosystem processes. Ecology 89:1510–1520

Hjort J, Luoto M (2010) Geodiversity of high-latitude landscapes in northern Finland. Geomorphology 115:109–116

Hjort J, Luoto M (2012) Can geodiversity be predicted from space? Geomorphology 153:74–80

Hjort J, Heikkinen RK, Luoto M (2012) Inclusion of explicit measures of geodiversity improve biodiversity models in a boreal landscape. Biodivers Conserv 21:3487–3506

Hutchinson GE (1965) The ecological theater and the evolutionary play. Yale University Press, Connecticut

International Union for Conservation of Nature and Natural Resources (2001) IUCN red list categories and criteria (Int. Union Conserv. Nat. Nat. Resour. Species Survival Commission, Cambridge, U.K.) Version 3.1

IUCN (2017) The IUCN red list of threatened species. Version 2017–2. Available at: http://www.iucnredlist.org

Jackova K, Romportl D (2008) The relationship between geodiversity and habitat richness in Sumava National Park and Krivoklatsko PLA (Czech Republic): a qualitative analysis approach. J Landsc Ecol 1:23–28

Janzen DH (1967) Why mountain passes are higher in the tropics. Am Nat 101:233–249

Jasiewicz J, Stepinski TF (2013) Geomorphons—a pattern recognition approach to classification and mapping of landforms. Geomorphology 182:147–156

Jeffries MO, Morris K, Kozlenko N (2005) Ice characteristics and processes, and remote sensing of frozen rivers and lakes. In: Duguay CR, Pietroniro A (eds) Remote sensing in northern hydrology: measuring environmental change. American Geophysical Union, Washington D.C., pp 63–90

Jenny H (1941) Factors of soil formation. McGraw-Hill, New York

Jetz W, Cavender-Bares J, Pavlick R, Schimel D, Davis FW, Asner GP, Guralnick R, Kattge J, Latimer AM, Moorcroft P, Scaepman ME, Schildhauer MP, Schneider FD, Schrodt F, Stahl U, Ustin SL (2016) Monitoring plant functional diversity from space. Nat Plants 2:e3

John W. Williams, Stephen T. Jackson, (2007) Novel climates, no-analog communities, and ecological surprises. Frontiers in Ecology and the Environment 5 (9):475-482

Jones CG, Lawton JH, Shachak M (1994) Organisms as ecosystem engineers. In: Samson FB, Knopf FL (eds) Ecosystem management. Springer, New York, pp 130–147

Joseph J. Bailey, Doreen S. Boyd, Jan Hjort, Chris P. Lavers, Richard Field, (2017) Modelling native and alien vascular plant species richness: At which scales is geodiversity most relevant?. Global Ecology and Biogeography 26(7):763–776

Jost L (2007) Partitioning diversity into independent alpha and beta components. Ecology 88:2427–2439

Kampe TU, Johnson BR, Kuester M, Keller M (2010) NEON: the first continental-scale ecological observatory with airborne remote sensing of vegetation canopy biochemistry and structure. J Appl Remote Sens 4:043510

Kerr JT, Packer L (1997) Habitat heterogeneity as a determinant of mammal species richness in high-energy regions. Nature 385:252–254

Körner C (2012) Alpine treelines: functional ecology of the global high elevation tree limits. Springer, New York, pp 21–30

Kozlowski S (1999) Programme of geodiversity conservation in Poland. Pol Geol Inst Spec Papers 2:15–19

Krause FA, Lefkoff AB, Dietz JB (1993) Expert system-based mineral mapping in northern Death Valley, California/Nevada, using the airborne visible/infrared imaging spectrometer (AVIRIS). Remote Sens Environ 44:309–336

Kwok R, Rothrock DA (2009) Decline in Arctic Sea ice thickness from submarine and ICESat records. Geophys Res Lett 36:1958–2008

Lacoul P, Freedman B (2006) Environmental influences on aquatic plants in freshwater ecosystems. Environ Rev 14:89–136

Lausch A, Bannehr L, Beckmann M, Boehm C, Feilhauer H, Hacker JM, Heurich M, Jung A, Klenke R, Neumann C, Pause M, Rocchini D, Schaepman ME, Schmidtlein S, Schulz K, Selsam P, Settele J, Skidmore AK, Cord AF (2016) Linking earth observation and taxonomic, structural, and functional biodiversity: local to ecosystem perspectives. Ecol Indic 70:317–339

Lausch A, Borg E, Bumberger J, Dietrich P, Heurich M, Huth A, Jung A, Klenke R, Knapp S, Mollenhauer H, Paasche H, Paulheim H, Pause M, Schweitzer C, Schmulius C, Settele J, Skidmore A, Wegmann M, Zacharias S, Kirsten T, Schapeman M (2018) Understanding forest health with remote sensing, part III: requirements for a scalable multi-source forest health monitoring network based on data science approaches. Remote Sens 10:1120

Lawler JJ, Ackerly DD, Albano CM, Anderson MG, Dobrowski SZ, Gill JL, Heller NE, Pressey RL, Leidner AK, Böhm M, He KS, Nagendra H, Dubois G, Fatoyinbo T, Hansen MC, Paganini M, de Klerk HM, Asner GP, Kerr JT, Estes AB, Schmeller DS, Heiden U, Rocchini D, Pereira HM, Turak E, Fernandez N, Lausch A, Cho MA, Alcaraz-Segura D, McGeoch MA, Turner W, Mueller A, St-Louis V, Penner J, Petteri V, Belward A, Reyers B, Geller GN (2016) Framing the concept of satellite remote sensing essential biodiversity variables: challenges and future directions. Remote Sens Ecol Cons 2:122–131

Legendre P, Borcard D, Peres-Neto PR (2005) Analyzing beta diversity: partitioning the spatial variation of community composition data. Ecol Monogr 75:435–450

Levin SA (1992) The problem of pattern and scale in ecology. Ecology 73:1943–1967

Linlin Ge, Alex Hay-Man Ng, Xiaojing Li, Hasanuddin Z. Abidin, Irwan Gumilar, (2014) Land subsidence characteristics of Bandung Basin as revealed by ENVISAT ASAR and ALOS PALSAR interferometry. Remote Sensing of Environment 154:46-60

Lomolino MV, Brown JH, Sax DF (2010) Reticulations and reintegration of "A biogeography of the species" in Island Biogeography Theory Eds. Losos JB, Ricklefs RE. Princeton University Press, Princeton, NJ

Marcott KC, Shakun JD, Clark PU, Mix AC (2013) A reconstruction of regional and global temperature for the past 11,300 years. Science 339:1198–1201

Massironi M, Bertoldi L, Calafa P, Visona D, Bistacchi A, Giardino C, Schiavo A (2008) Interpretation and processing of ASTER data for geological mapping and granitoids detection in the *Saghro massif* (eastern anti-atlas, Morocco). Geosphere 4:736–759

McGill BJ (2010) Matters of scale. Science 328:575–576

Moore I. D., R. B. Grayson, A. R. Ladson, (1991) Digital terrain modelling: A review of hydrological, geomorphological, and biological applications. Hydrological Processes 5(1):3–30

Mouillot D, Graham NAJ, Villéger S, Mason NWH, Bellwood DR (2013) A functional approach reveals community responses to disturbances. Trends Ecol Evol 28:167–177

Muller-Karger FE, Hestir E, Ade C, Turpie K, Roberts DA, Siegel D, Miller RJ, Humm D, Izenberg N, Morgan F et al (2018) Satellite sensor requirements for monitoring essential biodiversity variables of coastal ecosystems. Ecol Appl 28:749–760

Nadeau CP, Urban M, Bridle JR (2017) Coarse climate change projections for species living in a fine-scaled world. Glob Change Biol 23:12–24

Nathalie Pettorelli, Martin Wegmann, Andrew Skidmore, Sander Mücher, Terence P. Dawson, Miguel Fernandez, Richard Lucas, Michael E. Schaepman, Tiejun Wang, Brian O'Connor, Robert H.G. Jongman, Pieter Kempeneers, Ruth Sonnenschein, Allison K. Leidner, Monika Böhm, Kate S. He, Harini Nagendra, Grégoire Dubois, Temilola Fatoyinbo, Matthew C. Hansen, Marc Paganini, Helen M. de Klerk, Gregory P. Asner, Jeremy T. Kerr, Anna B. Estes, Dirk S. Schmeller, Uta Heiden, Duccio Rocchini, Henrique M. Pereira, Eren Turak, Nestor Fernandez, Angela Lausch, Moses A. Cho, Domingo Alcaraz-Segura, Mélodie A. McGeoch, Woody Turner, Andreas Mueller, Véronique St-Louis, Johannes Penner, Petteri Vihervaara, Alan Belward, Belinda Reyers, Gary N. Geller, Doreen Boyd, (2016) Framing the concept of satellite remote sensing essential biodiversity variables: challenges and future directions. Remote Sensing in Ecology and Conservation 2(3):122–131

NRC. National Research Council (2001) Basic research opportunities in earth science, pp 14–15 The National Academies Press, Washington, DC, USA. https://doi.org/10.17226/9981

O'Connell BM, Conkling BL, Wilson AM, Burrill EA, Turner JA, Pugh SA, Christensen G, Ridley T, Menlove J (2017) The forest inventory and analysis database: database description and user's manual version 7.0 for phase 2. U.S. Department of Agriculture, Forest Service, p 830, Washington, DC, USA

Okin GS, Mahowald N, Chadwick OA, Artaxo P (2004) Impact of desert dust on the biogeochemistry of phosphorus in terrestrial ecosystems. Global Biogeochem Cy 18:2

Olden JD, Rooney TP (2006) On defining and quantifying biotic homogenization. Glob Ecol Biogeogr 15:113–120

Ollinger SV, Aber JD, Federer CA, Lovett GM, Ellis J (1995) Modeling physical and chemical climatic variables across the northeastern U.S. for a geographic information system. USDA Forest Service General Technical Report NE-191

Ollinger SV, Smith ML, Martin ME, Hallett RA, Goodale CL, Aber JD (2002) Regional variation in foliar chemistry and soil nitrogen status among forests of diverse history and composition. Ecology 83:339–355

Parks KE, Mulligan M (2010) On the relationship between a resource based measure of geodiversity and broad scale biodiversity patterns. Biodivers Conserv 19:2751–2766

Pau S, Dee LE (2016) Remote sensing of species dominance and the value for quantifying ecosystem services. Remote Sense Ecol Conserv 2:141–151

Pavelsky TM, Zarnetske JP (2017) Rapid decline in river icings detected in arctic Alaska: implications for a changing hydrologic cycle and river ecosystems. Geophys Res Lett 44(7):2016GL072397. https://doi.org/10.1002/2016GL072397

Peacock J, Baker TR, Lewis SL, Lopez-Gonzalez G, Phillips OL (2007) The RAINFOR database: monitoring forest biomass and dynamics. J Veg Sci 18:535–542

Petorelli N, Laurance WF, O'Brien TG, Wegmann M, Nagendra H, Turner W (2014) Satellite remote sensing for applied ecologists: opportunities and challenges. J Appl Ecol 51:839–848

Pourmokhtarian A, Driscoll CT, Campbell JL, Hayhoe K, Stoner AMK (2016) The effects of climate downscaling technique and observation dataset on modeled ecological responses. Ecol Appl 26:1321–1337

Radford AE (1981) Natural heritage: classification, inventory and information. University of North Carolina Press, Chapel Hill

Rahbek C (2005) The role of spatial scale and the perception of large-scale species-richness patterns. Ecol Lett 8:224–239

Record S, Charney ND (2016) Modeling species ranges. Chance 29:31–37

Record S, Fitzpatrick MC, Finley AO, Veloz S, Ellison AM (2013) Should species distribution models account for spatial autocorrelation? A test of model projections across eight millennia of climate change. Glob Ecol Biogeogr 22:760–771

Record S, McCabe T, Baiser B, Ellison AME (2018) Identifying foundation species in North American forests using long-term data on ant assemblage structure. Ecosphere 9:e01239

Richerson PJ, Lum K (1980) Patterns of plant species and diversity in California: relation to weather and topography. Am Nat 116:504–536

Rienecker MM, Suarez MJ, Gelaro R, Todling R, Bacmeister J, Liu E, Bosilovich MG, Schubert SD, Takacs L, Kim GK, Bloom S (2011) MERRA: NASA's modern-era retrospective analysis for research applications. J Clim 24:3624–3648

Rignot E, Mouginot J, Scheuchl B (2011) Ice flow of the Antarctic ice sheet. Science 333:1427–1430

Riley SJ, DeGloria SD, Elliot R (1999) A terrain ruggedness index that quantifies topographic heterogeneity. Intermountain J Sci 5:23–27

Robert L. Pressey, Mar Cabeza, Matthew E. Watts, Richard M. Cowling, Kerrie A. Wilson, (2007) Conservation planning in a changing world. Trends in Ecology & Evolution 22(11):583–592

Robinson N, Regetz J, Guralnick RP (2014) EarthEnv-DEM90: a nearly global, void-free, multiscale smoothed, 90m digital elevation model from fused ASTER and SRTM data. ISPRS J Photogramm 87:57–67

Rowan LC, Mars JC (2003) Lithologic mapping in the Mountain Pass, California area using advanced spaceborne thermal emission and reflection radiometer (ASTER) data. Remote Sens Environ 84:350–366

Sala OE, Chapin FS, Armesto JJ, Berlow E, Bloomfield J, Dirzo R, Huber-Sanwald E, Huenneke LF, Jackson RB, Kinzig A, Leemans R, Lodge DM, Mooney HA, Oesterheld M, Poff NL, Sykes MT, Walker BH, Walker M, Wall DH (2000) Global biodiversity scenarios for the year 2100. Science 287:1770–1774

Sang-Ho Yun, Kenneth Hudnut, Susan Owen, Frank Webb, Mark Simons, Patrizia Sacco, Eric Gurrola, Gerald Manipon, Cunren Liang, Eric Fielding, Pietro Milillo, Hook Hua, Alessandro Coletta, (2015) Rapid Damage Mapping for the 2015 7.8 Gorkha Earthquake Using Synthetic Aperture Radar Data from COSMO–SkyMed and ALOS-2 Satellites. Seismological Research Letters 86(6):1549–1556

Serrano E, Ruiz-Flaño P, Arroyo P (2009) Geodiversity assessment in a rural landscape: Tiermes-Caracena area (Soria, Spain). Memorie desrittive della carta geologica d'Italia 87:173–180

Smith SJ, Edmonds J, Hartin CA, Mundra A, Calvin K (2015) Near-term acceleration in the rate of temperature change. Nat Clim Chang 5:333–336

Spehn EM, Körner C (2005) A global assessment of mountain biodiversity and its function. In: Huber UM, BugmannMel KM, Reasoner A (eds) Global change and mountain regions. Springer, Dordrecht, pp 393–400

Srivastava DS, Cadotte MW, MacDonald AAM, Marushia RG, Mirotchnick N (2012) Phylogenetic diversity and the functioning of ecosystems. Ecol Lett 15:637–648

Stavros EN, Schimel D, Pavlick R, Serbin S, Swann A, Ducanson L, Fisher JB, Fassnacht F, Ustin S, Dubayah R, Schweiger A (2017) ISS observations offer insights into plant function. Nature Ecol Evol 1:0194

Stein A, Gerstner K, Kreft H (2014) Environmental heterogeneity as a universal driver of species richness across taxa, biomes, and spatial scales. Ecol Lett 17:866–880

Stendera S, Adrian R, Bonada N, Cañedo-Argüelles M, Hugueny B, Januschke K, Pletterbauer F, Hering D (2012) Hydrobiologia 696:1–28

Still CJ, Pau S, Edwards EJ (2014) Land surface temperature captures thermal environments of C3 and C4 grasses. Glob Ecol Biogeogr 23:286–296

Tedesco M (2014) Remote sensing of the cryosphere. Wiley-Blackwell, Hoboken

Thackeray SJ, Sparks TH, Fredericksen M, Burthe S, Bacon PJ, Bell JR, Botham MS, Brereton TM, Bright PW, Carvalho L, Clutton-Brock T, Dawson A, Edwards M, Elliott JM, Harrington R, Johns D, Jones ID, Jones JT, Leech DI, Roy DB, Scott WA, Smith M, Smithers RJ, Winfield IJ, Wanless S (2010) Trophic level asynchrony in rates of phenological change for marine, freshwater and terrestrial environments. Glob Change Biol 16:3304–3313

Tuanmu M-N, Jetz W (2015) A global, remote sensing-based characterization of terrestrial habitat heterogeneity for biodiversity and ecosystem modeling. Glob Ecol Biogeogr 24:1329–1339

Tukiainen H, Bailey JJ, Field R, Kangas K, Hjort J (2017) Combining geodiversity with climate and topography to account for threatened species richness. Cons Biol 31:364–375

Tuomisto H, Ruokolainen K, Aguilar M, Sarmiento A (2003) Floristic patterns along a 43-km long transect in an Amazonian rainforest. J Ecol 91:743–756

Turner W (2014) Sensing biodiversity. Science 346:301–302

van der Werff H, van der Meer F (2016) Sentinel-2A MSI and Landsat 8 OLI provide data continuity for geological remote sensing. Remote Sens 8(11):883

Veloz SD, Williams JW, Blois JL, Feng H, Otto-Bliesner B, Liu Z (2012) No-analog climates and shifting realized niches during the late quaternary: implications for 21st-century predictions by species distribution models. Glob Change Biol 18:1698–1713

Vincent WF, Callaghan TV, Dahl-Jensen D, Johansson M, Kovacs KM, Michel C, Prowse T, Reist JD, Sharp M (2011) Ecological implications of changes in the Arctic cryosphere. Ambio 40:87–99

Violle C, Enquist BJ, McGill BJ, Jiang L, Albert CH, Hulshof C, Jung V, Messier J (2012) The return of the variance: intraspecific variability in community ecology. Trends Ecol Evol 27:244–252

Vitousek PM, Aplet G, Turner D, Lockwood JJ (1992) The Mauna Loa environmental matrix: foliar and soil nutrients. Oecologia 89:372–382

Walther G-R, Post E, Convey P, Menzel A, Parmesan C, Beebee TJC, Fromentin J-M, Hoegh Guldberg O, Bairlein F (2002) Ecological responses to recent climate change. Nature 416:389–395

Weber F, Nixon D, Hurley J (2003) Semi-automated classification of river ice types on the Peace River using RADARSAT-1 synthetic aperture radar (SAR) imagery. Can J Civil Eng 30:11–27

Williams RS (2012) Introduction—changes in the Earth's cryosphere and global environmental change in the earth system. In: Williams RS, Ferrigno JG (eds) State of the earth's cryosphere at the beginning of the 21st century—glaciers, global snow cover, floating ice, and permafrost and periglacial environments: U.S. Geological Survey Professional Paper 1386-A. U.S. Geological Survey, Reston, p 23

Williams JW, Jackson ST (2007) Novel climates, no-analog communities, and ecological surprises. Front Ecol Environ 5:475–482

Wilson AM, Jetz W (2016) Remotely sensed high-resolution global cloud dynamics for predicting ecosystem and biodiversity distributions. PLoS Biol 14:1–20

Wood J (1996) Scale-based characterisation of digital elevation models. Innovat GIS 3:163-175

Yachi S, Loreau M (1999) Biodiversity and ecosystem productivity in a fluctuating environment: the insurance hypothesis. P Natl Acad Sci USA 96:1463–1468

Yoshikawa K, Hinzman LD, Kane DL (2007) Spring and Aufeis (icing) hydrology in Brooks Range, Alaska. J Geophys Res-Biogeosci 112:G04S43

Yu H, Chin M, Bian H, Yuan T, Prospero JM, Omar AH, Remer LA, Winker DM, Yang Y, Zhang Y, Zhang Z (2015) Quantification of trans-Atlantic dust transport from seven-year (2007-2013) record of CALIPSO LIDAR measurements. Remote Sens Environ 15:232–249

Zarnetske PL, Skelly DK, Urban MC (2012) Biotic multipliers of climate change. Science 336:15–30

Zarnetske PL, Baiser B, Strecker A, Record S, Belmaker J, Tuanmu M-N (2017) The interplay between landscape structure and biotic interactions. Curr Landscape Ecol Rep 2:12–29

Zarnetske PL, Record S, Dahlin K, Read Q, Grady JM, Costanza J, Finley AO, Gaddis K, Hobbi M, Latimer A, Malone S, Ollinger S, Pau S, Turner W, Wilson A (2019) Connecting biodiversity and geodiversity with remote sensing across scales. Glob Ecol Biogeog 28:548-556.

Zwally HJ, Li J, Brenner AC, Beckley M, Cornejo HG, Di Marzio J, Giovinetto MB, Neumann TA, Robbins J, Saba JL, Yi D, Wang W (2011) Greenland ice sheet mass balance: distribution of increased mass loss with climate warming; 2003-07 versus 1992-2002. J Glaciol 57:88–102

Open Access This chapter is licensed under the terms of the Creative Commons Attribution 4.0 International License (http://creativecommons.org/licenses/by/4.0/), which permits use, sharing, adaptation, distribution and reproduction in any medium or format, as long as you give appropriate credit to the original author(s) and the source, provide a link to the Creative Commons license and indicate if changes were made.

The images or other third party material in this chapter are included in the chapter's Creative Commons license, unless indicated otherwise in a credit line to the material. If material is not included in the chapter's Creative Commons license and your intended use is not permitted by statutory regulation or exceeds the permitted use, you will need to obtain permission directly from the copyright holder.

Chapter 11
Predicting Patterns of Plant Diversity and Endemism in the Tropics Using Remote Sensing Data: A Study Case from the Brazilian Atlantic Forest

Andrea Paz, Marcelo Reginato, Fabián A. Michelangeli, Renato Goldenberg, Mayara K. Caddah, Julián Aguirre-Santoro, Miriam Kaehler, Lúcia G. Lohmann, and Ana Carnaval

11.1 Introduction

The spatial distribution of species is unquestionably tied to environments, particularly temperature and precipitation (Hutchinson 1957). By exploring this correlation, multiple studies have demonstrated that environmental descriptors are able to predict geographic patterns of biological diversity reasonably well (Peters et al. 2016;

A. Paz (✉) · A. Carnaval
Department of Biology, City College of New York, New York, NY, USA

Biology Program, The Graduate Center, City University of New York, New York, NY, USA

M. Reginato
Departamento de Botânica, Instituto de Biociências, Universidade Federal do Rio Grande do Sul, Porto Alegre, RS, Brazil

F. A. Michelangeli
Biology Program, The Graduate Center, City University of New York, New York, NY, USA

Institute of Systematic Botany, The New York Botanical Garden, The Bronx, NY, USA

R. Goldenberg
Universidade Federal do Paraná, Curitiba, PR, Brazil

M. K. Caddah
Universidade Federal de Santa Catarina, Florianopolis, SC, Brazil

J. Aguirre-Santoro
Biology Program, The Graduate Center, City University of New York, New York, NY, USA

Instituto de Ciencias Naturales, Facultad de Ciencias, Universidad Nacional de Colombia, Bogotá, Colombia

M. Kaehler · L. G. Lohmann
Departamento de Botânica, Instituto de Biociências, Universidade de São Paulo, São Paulo, SP, Brazil

© The Author(s) 2020
J. Cavender-Bares et al. (eds.), *Remote Sensing of Plant Biodiversity*,
https://doi.org/10.1007/978-3-030-33157-3_11

Zellweger et al. 2016). Temperature, for example, has been repeatedly shown to be a good predictor of the species that inhabit a given area (the taxonomic dimension of biodiversity, e.g., Peters et al. 2016). However, the power to predict the distinct dimensions of biodiversity varies within and across groups of organisms. For instance, the contribution of different measures of temperature and precipitation appears to be idiosyncratic when multiple taxa are compared (Rompré et al. 2007; Laurencio and Fitzgerald 2010; Peters et al. 2016; Zellweger et al. 2016). Moreover, and in contrast to species richness (SR), the relationships between climate and the geographic distribution of evolutionary diversity in a region (i.e., the phylogenetic dimension of biodiversity), as well as the relationships between climate and endemism, have been less explored. Still, those relationships appear weaker due to the relatively larger contribution of history, biogeography, and contingency in the spatial distribution of lineages (da Silva et al. 2012; Barratt et al. 2017).

Most of those advances have relied on the use of climatic data sets that are interpolated from weather station data (Hijmans et al. 2005), summarizing spatial patterns of temperature and precipitation. These include the widely used WorldClim data set (Hijmans et al. 2005), country-specific data sets (e.g., Cuervo-Robayo et al. 2014), and the hybrid CHELSA database (Karger et al. 2017). The ease by which biodiversity scientists can access and download these databases, and the fact that they provide global-scale climatic information at biologically relevant scales (up to 1 km), have resulted in a sharp increase in the number of studies that explore the correlations between climate and biodiversity patterns. Yet the accuracy and the effectiveness of these global climatic descriptors have been questioned (Soria-Auza et al. 2010). Because the distribution of weather stations around the world is unequal, the confidence in those data sets is reduced in undersampled areas, which frequently correspond to the most biodiverse areas on Earth (see Pinto-Ledézma and Cavender-Bares, Chap. 9).

In this chapter, we explore the use of bioclimatic variables built from long-term climatologies derived from remote sensing (RS) as predictors of biodiversity patterns. We focus in a megadiverse region, with high topographic complexity: the Brazilian Atlantic Forest hotspot. We evaluate whether climate, inferred from RS sources, predicts which areas accumulate the highest diversity of species, evolutionary lineages, and endemism. For that, we use distribution and phylogenetic data from three plant clades representing different life forms, that are commonly found in the Brazilian Atlantic Forest: melastomes (178 species of shrubs and trees), bromeliads (43 species of epiphytes), and bignones (131 species of lianas). We also evaluate what (if any) gains emerge from the use of climatic descriptors based on RS, rather than weather stations, for this area. Given the sharp altitudinal changes observed in the Brazilian Atlantic Forest hotspot, it has been proposed that interpolated weather station data may perform more poorly than variables derived from RS (Waltari et al. 2014).

11.2 Study System

The Brazilian Atlantic Forest harbors one of the highest levels of endemism and threat globally, representing one the world's hotspots of biodiversity (Myers et al. 2000). Although only about 16% of the original forest persists (Ribeiro et al. 2009), the Atlantic Forest is topographically and environmentally complex, spanning more than 1,700 m in altitude and about 25° of latitude (Ribeiro et al. 2009). Climatic analyses of the forest, along with molecular studies of its biota, suggest that it encompasses multiple environmental spaces and associated species pools. More specifically, the northern (mostly lowland) and southern (mostly montane) elements are largely different in species composition and have responded differently to past climatic changes (Carnaval et al. 2014; Leite et al. 2016).

Melastomes represent the first clade selected for our study. The tribe Miconieae (Melastomataceae) is exclusively Neotropical, with ca. 1,900 species, mostly shrubs and small trees, but also herbs, lianas, epiphytes, and large trees (Michelangeli et al. 2004, 2008; Goldenberg et al. 2008). In the Atlantic Forest, the tribe is represented by ca. 310 species, 70% of which are endemic ("Flora do Brasil 2020"; Goldenberg et al. 2009). These species are largely grouped into three clades: the *Leandra* clade with ca. 215 species (Caddah 2013; Reginato and Michelangeli 2016), the *Miconia* section *Chaenanthera* clade (Goldenberg et al. 2018), and the *Miconia* sect. *discolor* clade. Most of these species are small trees and shrubs (although the *Pleiochiton* clade contains 12 species of shrubby epiphytes; Reginato et al. 2010, 2013), and the great majority are bee pollinated and have berry fruits that are dispersed by birds. In the Atlantic Forest, species of Miconieae are found throughout most environments and at all elevations, with species ranges varying from widely distributed within the domain and beyond, to microendemics found in a single mountain top (Michelangeli et al. 2008).

Bromeliads represent the second clade included in this investigation. The Bromeliaceae is an almost exclusively Neotropical family, with ca. 3,300 species of terrestrial or epiphytic rosette-forming herbs. In the Atlantic Forest, the Bromeliaceae is represented by 816 species, over 75% of which are endemic (Martinelli et al. 2009). The data set used here represents a clade of 70 species belonging to the *Ronnbergia-Wittmackia* alliance (Aguirre-Santoro et al. 2016; Aguirre-Santoro 2017). With the exception of one species, the basal grade of 26 species of *Wittmackia* is composed of species restricted to the Atlantic Forest. All of them are tank-forming epiphytes found in forested environments, many with very restricted distributions (Aguirre-Santoro 2017). In the Atlantic Forest, *Wittmackia* is found predominantly in the central and northern states.

Bignones are the third plant clade used in this analysis. The tribe Bignonieae (Bignoniaceae) originated at around 50 million years ago (mya) in the Brazilian Atlantic Forest and subsequently occupied Amazonia and the dry areas of Central Brazil (Lohmann et al. 2013). The group is very diverse ecologically, including species pollinated by hummingbirds, butterflies, bees, and bats (Gentry 1974; Alcantara and Lohmann 2010). Ant-plant interactions are extremely common and play an important role in herbivore defense (Nogueira et al. 2015). Most species in the family are dispersed by wind or water (Lohmann 2004).

Former drying of Neotropical climates, and the Andean orogeny, seems to have represented key diversification drivers for tribe Bignonieae (Lohmann et al. 2013). Today, it includes 383 species and 21 genera (Lohmann and Taylor 2014), representing the most diverse and abundant clade of lianas in Neotropical forests (Lohmann 2006). All species of the tribe are distributed among three main clades: (i) the "multiples of four clade" (referring to the multiples of four phloem wedges), with ca. 135 species (Lohmann 2006); (ii) the "*Fridericia* and Allies clade," with around 132 species (Kaehler et al. 2019); and (iii) the "*Adenocalymma-Neojobertia*" clade, with ca. 75 species (Fonseca and Lohmann 2018). The remaining species of the Tribe are distributed among eight small genera (Lohmann 2006).

11.3   Methods

To investigate the relationships between climate and biodiversity patterns in the Atlantic rainforest of Brazil, we selected three clades of angiosperms with different life forms, i.e., shrubs and small trees (tribe Miconieae, Melastomataceae), epiphytic herbs (the *Ronnbergia/Wittmackia* alliance, Bromeliaceae), and lianas (the "*Fridericia* and Allies" clade of tribe Bignonieae, Bignoniaceae).

For each group, we combined geo-referenced occurrence data from each species with information about its evolutionary relationships. Using personal field data, published records, and geo-referenced herbarium information, we gathered locality information for 352 species and 22,338 unique locality points vetted by experts for spatial and taxonomic accuracy as follows: (i) melastomes, 178 species and 10,253 records of members of tribe Miconieae; (ii) bromeliads, 43 species and 4,606 records of members of the *Ronnbergia/Wittmackia* alliance; and (iii) Bignones, 132 species and 7,480 records of members of the "*Fridericia* and Allies" clade of tribe Bignonieae (Lohmann, unpublished data; see Meyer et. al 2008 for further details on this data set).

For each species, we used the locality data to generate a multiple convex polygon representing its range, which was then converted into a gridded map (~5 km resolution). Maps of the individual species were then stacked, allowing us to compute the total number of species per pixel. Information about the species composition at each grid cell was then combined with published and novel data on the phylogenetic relationships among species of melastomes (Caddah 2013; Reginato and Michelangeli 2016; Goldenberg et al. 2018), bromeliads (Aguirre-Santoro et al. 2016), and bignones (Kaehler et al. 2019), to provide a measurement of phylogenetic diversity (PD) per pixel, using Faith's phylogenetic diversity index (Faith 1992). This metric quantifies the evolutionary history included in every community by adding the branch lengths leading to each taxon present in the community (Faith 1992).

We also identified pixels holding high or low levels of phylogenetic endemism (PE) by including information about the range of each species' sister taxon

(PE, Rosauer et al. 2009). This metric takes into account the evolutionary history (as branch lengths) and spatial restriction (here as range estimates; Rosauer et al. 2009). To allow for comparisons across plant groups with distinct life histories and environmental envelopes, we performed these analyses separately for each clade (melastomes, bromeliads, and bignones). We used the Biodiverse software (Laffan et al. 2010) to map the geographical patterns of SR, the PD, and the phylogenetic endemism of each group.

We then gathered climatic descriptors for the Atlantic Forest region using two sources of climatic data, both at a 0.05° resolution (~5 km): one derived from RS instruments (Deblauwe et al. 2016) and another derived from interpolated weather station data (WorldClim database; Hijmans et al. 2005). Both databases describe environmental variation in the form of 19 bioclimatic variables that reflect spatial and temporal differences in precipitation and temperature (Bio1-19, as defined in the WorldClim database). These data were estimated with the same formulae, across data sets. While the WorldClim data reflect bioclimatic conditions estimated from interpolated weather station information, the database of Deblauwe et al. (2016) was built based on temperature information from NASA's Moderate Resolution Imaging Spectroradiometer (MODIS) and precipitation from the Climate Hazards Group InfraRed Precipitation with Station (CHIRPS) data. To reduce collinearity between the 19 bioclimatic variables, we employed a variance inflation factor (VIF), retaining only those variables with VIF < 5 in both datasets in all analyses. This left us with seven variables from each source, in both cases bio 3, 8, 9 13,18, and 19, plus bio 2 for the dataset based on weather station data (WorldClim), and bio 7 for the RS-based (Deblauwe et al. 2016) dataset (see Table 11.1 for bioclimatic variable descriptions).

To investigate how much of the spatial patterns of SR, PD, and phylogenetic endemism can be explained by each set of climatic descriptors, we ran conditional autoregressive (CAR) models on the pooled data from each group. CAR models

Table 11.1 Bioclimatic variables used as predictors for analyses, after removing variables with high variance inflation factor (VIF)

Variable	Description
Bio 2	Mean diurnal range [mean of monthly (max temp–min temp)]
Bio 3	**Isothermality (Bio 2/Bio 7) ∗100**
Bio 7	Temperature annual range
Bio 8	**Mean temperature of the wettest quarter**
Bio 9	**Mean temperature of the driest quarter**
Bio 13	**Precipitation of the wettest month**
Bio 18	**Precipitation of the warmest quarter**
Bio 19	**Precipitation of the coldest quarter**

In bold, variables used in both the RS- and weather station-derived data sets

Bio 2 was used only in the weather station-based analysis; bio 7 was used only in the RS-based analysis implement a multiple spatial regression in which the covariance among residuals considers the neighborhood of each evaluated cell (Rangel et al. 2006). To evaluate model fit, we computed pseudo-R^2 for our models, including both the full model and the predictor-only effect (i.e., removing the effect of space and spatial autocorrelation). Also, to detect areas of potential concentration or overdispersion of the regression residuals, we generated residual maps for models of SR and PD.

11.4 Results and Discussion

The spatial patterns of SR, PD, and phylogenetic endemism vary across groups. While melastomes show higher PD and SR along the east coast of Brazil, bromeliads accumulate PD and SR in the northern region of the Atlantic Forest, and bignones in the central portion of the domain. Phylogenetic endemism concentrates in the coastal mountains for melastomes, in the northern coast for bromeliads, and toward the west and northwest for bignones (Fig. 11.1).

Bioclimatic variables derived from RS products have excellent predictive power for both SR and PD. The full CAR models built from RS sources performed well, irrespective of plant group ($R^2 > 0.89$, Table 11.2). Prediction of phylogenetic ende-

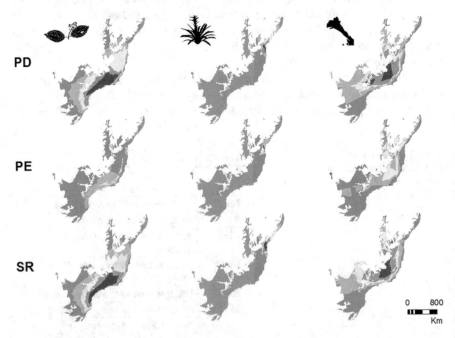

Fig. 11.1 Patterns of biodiversity for three plant clades in the Brazilian Atlantic Forest, from left to right: melastomes, bromeliads, and bignones. Each row corresponds to one biodiversity index, top to bottom: phylogenetic diversity (PD), phylogenetic endemism (PE), and species richness (SR). Warmer colors represent higher values of each of the indices

Table 11.2 Predictive power of models using either RS-based variables (RS) or weather station-derived variables (WC) as predictors of phylogenetic diversity (PD), phylogenetic endemism (PE), and species richness (SR) in three plant clades from the Brazilian Atlantic Forest: melastomes, bromeliads, and bignones

Clade	Predictors	Predicted	Full model R^2	Non-space R^2
Melastomes	RS	PD	0.96	**0.61**
		PE	0.62	0.01
		SR	**0.97**	**0.61**
	WC	PD	0.96	0.57
		PE	0.62	0.02
		SR	0.96	0.55
Bromeliads	RS	PD	**0.91**	**0.37**
		PE	**0.75**	**0.03**
		SR	**0.92**	**0.37**
	WC	PD	0.88	0.19
		PE	0.74	0.02
		SR	0.89	0.19
Bignones	RS	PD	0.94	0.58
		PE	0.57	0.10
		SR	0.96	0.72
	WC	PD	0.94	**0.59**
		PE	0.57	0.10
		SR	0.96	**0.74**

Numbers in bold have higher predictive power when comparing RS and WC for a single group

mism based on RS sources was not as successful: the performance of the CAR models was fair to good, irrespective of plant clade (R^2 0.57–0.75, Table 11.2). These observed differences in predictive power are not surprising and stand in agreement with the expectation that phylogenetic endemism may be more strongly impacted by historical processes and former climates (e.g., Late Quaternary) than by contemporary descriptors (Rosauer and Jetz 2015).

The predictive power of the models built with weather station data (WorldClim; Hijmans et al. 2005) was comparable to that of models based on satellite information, similar to Pinto-Ledézma and Cavender-Bares (Chap. 9), showing only slightly lower R^2 values overall (within 0.01, Table 11.2). This difference tended to increase (i.e., with models based on RS data performing better than those built with weather station data) when spatial autocorrelation effects were removed from the analyses (Table 11.2).

Climatic descriptors derived from both RS information (Deblauwe et al. 2016) and weather station data (Hijmans et al. 2005) failed to predict spatial patterns of phylogenetic endemism (PE) when decoupled from space (R^2 0.01–0.1, Table 11.2). Geography is naturally expected to impact maps of endemism because this analysis of geographical restriction of evolutionary history explicitly incorporates space in its calculations (Rosauer et al. 2009). Still, when this spatial imprint is removed from the data, we notice that contemporary climates are unable to predict the distribution of lineage restriction—in agreement with previous suggestions that historical climates, or the stochasticity of the processes associated with colonization or extinction, may have an important role in determining phylogenetic endemism (Carnaval et al. 2014; Rosauer and Jetz 2015).

Although the performance of the CAR models varied across diversity measures and taxa, the analyses decoupled from space recovered consistently lower predictive power in bromeliads, irrespective of diversity measures (Table 11.2). Unlike the other two groups, this clade is composed mainly of microendemics and represented only in a relatively small region of the Atlantic Forest (Fig. 11.1). We hypothesize that the larger influence of space, history, and chance events (particularly related to local extinctions) may be responsible for the lower correspondence between spatial patterns of biodiversity and climatic descriptors in groups of species that are narrowly distributed. This being true, it is expected that the predictive power of correlative models of biodiversity such as those presented here—including the use of RS data—will perform best in tropical groups in which most species have relatively large ranges.

The spatial distribution of the residuals of the correlation between biodiversity metrics and climate data differed across clades. In melastomes, they were homogeneously distributed across the forest, while for both bromeliads and bignones large residuals of SR and PD were observed in areas with low overall diversity. In bignones, residuals were especially concentrated in the south—where large geographic extensions showed more or less PD than expected (Fig. 11.2). Particularly the southern portion of the forest shows higher PD of bignones than expected, given the models based on climate data. This may be related to these plants' sensitivity to altitude, which limits their growth (Lohmann, pers. obs.). Bignoniaceae species are sensitive to temperature and precipitation, with abundance and species richness responding positively in warmer climates, and strongly negative in wetter climates. Thus, the species richness, or phylogenetic diversity, is increased in warmer and drier dry-season habitats (Punyasena et al. 2008). The south of the Atlantic forest is montane and the areas with larger residuals in South Brazil have dry, though cool winters. Given that no topographic variables were included in our models, this overprediction appears reasonable.

11.5   Conclusions and Future Directions

Community-level data from three representative tropical plant groups that include lianas, shrubs, and trees demonstrate that the use of RS data describing temperature and precipitation accurately predicts the spatial distribution of two essential biodiversity metrics (SR and PD) in a biodiversity hotspot. This predictive power is reduced when the approach is applied to a clade of spatially restricted (narrow endemic) species, such as bromeliads. Across all plant groups, predictive power is lower for diversity indices highly influenced by historical contingency and spatial configuration, such as phylogenetic endemism. For predictive purposes, and at the

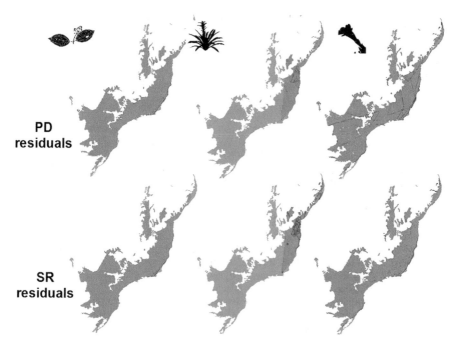

Fig. 11.2 Residuals of the CAR models using remote sensing variables as predictors of phylogenetic diversity (PD) and species richness (SR) for three groups of plants from the Brazilian Atlantic Forest, from left to right: melastomes, bromeliads, and bignones. Blue denotes areas with negative residuals (sites where less diversity is observed than expected under the model). Areas with positive residuals are shown in red. Darker shades of blue and red represent larger residuals

spatial scale of the Atlantic Forest, the performance of RS-based climate descriptors is comparable to, or slightly better than, that of weather station-based databases. These results show promise for predicting different dimensions of diversity in the tropics, based on RS data, especially for widely distributed groups. This approach may be particularly relevant in groups or regions for which direct or indirect species identification through RS (e.g., hyperspectral images) is feasible or available. It also may be extended to other groups of plants, and to animals. Future directions of this work include testing whether RS-based predictions of biodiversity work similarly well in other biological groups, biomes, and geographical areas, while also potentially including additional variables of interest, such as topography and historical climates.

References

Aguirre-Santoro J (2017) Taxonomy of the *Ronnbergia* Alliance (Bromeliaceae: Bromelioideae): new combinations, synopsis, and new circumscriptions of Ronnbergia and the resurrected genus Wittmackia. Plant Syst Evol 303:615–640

Aguirre-Santoro J, Stevenson D, Michelangeli F (2016) Molecular phylogenetics of the *Ronnbergia* Alliance (Bromeliaceae, Bromelioideae) and insights about its morphological evolution. Mol Phylogenet Evol 100:1–20

Alcantara S, Lohmann LG (2010) Evolution of floral morphology and pollination system in Bignonieae (Bignoniaceae). Am J Bot 97:782–796

Barratt CD, Bwong BA, Onstein RE et al (2017) Environmental correlates of phylogenetic endemism in amphibians and the conservation of refugia in the Coastal Forests of Eastern Africa. Divers Distrib 23:875–887

Caddah M (2013) Estudos Taxonomicos e filogenéticos em *Miconia* sect. *Discolor* (Melastomataceae, Miconieae). PhD thesis. UNICAMP, Campinas

Carnaval AC, Waltari E, Rodrigues MT et al (2014) Prediction of phylogeographic endemism in an environmentally complex biome. Proc R Soc B Biol Sci 281:20141461

Cuervo-Robayo AP, Téllez-Valdés O, Gómez-Albores MA et al (2014) An update of high-resolution monthly climate surfaces for Mexico. Int J Climatol 34:2427–2437

da Silva FR, Almeida-Neto M, do Prado VHM, Haddad CFB, de Cerqueira Rossa-Feres D (2012) Humidity levels drive reproductive modes and phylogenetic diversity of amphibians in the Brazilian Atlantic Forest. J Biogeogr 39:1720–1732

Deblauwe V, Droissart V, Sonke B et al (2016) Remotely sensed temperature and precipitation data improve species distribution modelling in the tropics. Glob Chang Biol 25(4):443–454

Faith DP (1992) Conservation evaluation and phylogenetic diversity. Biol Conserv 61:1–10

Flora do Brasil 2020 Jardim Botânico do Rio de Janeiro

Fonseca L, Lohmann LG (2018) Combining high-throughput sequencing and targeted loci data to infer the phylogeny of the "*Adenocalymma-Neojobertia*" clade. Mol Phylogenet Evol 123:1–15

Gentry AH (1974) Coevolutionary patterns in Central American Bignoniaceae. Ann Mo Bot Gard 61:728–759

Goldenberg R, Penneys DS, Almeda F, Judd WS, Michelangeli F (2008) Phylogeny of *Miconia* (Melastomataceae): initial insights into broad patterns of diversification in a megadiverse neotropical genus. Int J Plant Sci 169:963–979

Goldenberg R, Guimarães P, Kriebel R, Romero R (2009) Melastomataceae. In: Stehmann JR, Forzza R, Salino A et al (eds) Plantas da Floresta Atlantica. Instituto de Pesquisas Jardim Botânico do Rio de Janeiro, Rio de Janeiro, pp 330–343

Goldenberg R, Reginato M, Michelangeli F (2018) Disentangling the infrageneric classification of megadiverse taxa from Mata Atlantica: phylogeny of *Miconia* section *Chaenanthera* (Melastomataceae. Miconieae). Taxon 67:537–551

Hijmans RJ, Cameron SE, Parra JL, Jones PG, Jarvis A (2005) Very high resolution interpolated climate surfaces for global land areas. Int J Climatol 25:1965–1978

Hutchinson (1957) Concluding remarks. Cold Spring Harb Symp Quant Biol 22:415–427

Kaehler M, Michelangeli FA, Lohmann LG (2019) Fine-tuning the circumscription of Fridericia (Bignonieae, Bignoniaceae). Taxon 68(3). https://doi.org/10.1002/tax.12121

Karger DN, Conrad O, Böhner J et al (2017) Data descriptor: climatologies at high resolution for the earth's land surface areas. Scientific Data 4:1–20

Laffan SW, Lubarsky E, Rosauer DF (2010) Biodiverse, a tool for the spatial analysis of biological and related diversity. Ecography 33:643–647

Laurencio D, Fitzgerald LA (2010) Environmental correlates of herpetofaunal diversity in Costa Rica. J Trop Ecol 26:521–531

Leite YLR, Costa LP, Loss AC et al (2016) Neotropical forest expansion during the last glacial period challenges refuge hypothesis. Proc Natl Acad Sci 113:1008–1013

Lohmann LG (2004) Bignoniaceae. In: Smith N, Mori S, Henderson A, Stevenson SH (eds) Flowering plants of the neotropics. Princeton University Press, Princeton, pp 51–53

Lohmann LG (2006) Untangling the phylogeny of neotropical lianas (Bignonieae, Bignoniaceae). Am J Bot 93:304–318

Lohmann LG, Taylor CM (2014) A new generic classification of tribe Bignonieae (Bignoniaceae). Ann Mo Bot Gard 99:348–489

Lohmann LG, Bell CD, Calió MF, Winkworth RC (2013) Pattern and timing of biogeographical history in the Neotropical tribe Bignonieae (Bignoniaceae). Bot J Linn Soc 171:154–170

Martinelli G, Vieira C, Leitman P, Costa A, Forzza R (2009) Bromeliaceae. In: Stehmann JR, Forzza R, Salino A et al (eds) Plantas da Floresta Atlantica. Instituto de Pesquisas Jardim Botânico do Rio de Janeiro, Rio de Janeiro, pp 186–204

Meyer L, Diniz-Filho JAF, Lohmann LG (2008) A comparison of hull methods for estimating species ranges and richness maps. Plant Ecol Diversity 10:389–401

Michelangeli F, Penneys D, Giza J et al (2004) A preliminary phylogeny of the tribe Miconieae (Melastomataceae) based on nrlITS sequence data and its implications on inflorescence position. Taxon 53:279–290

Michelangeli F, Judd WS, Penneys DS et al (2008) Multiple events of dispersal and radiation of the tribe Miconieae (Melastomataceae) in the Caribbean. Bot Rev 74:53–77

Myers N, Mittermeier RA, Mittermeier CG, da Fonseca GAB, Kent J (2000) Biodiversity hotspots for conservation priorities. Nature 403:853–858

Nogueira A, Rey P, Alcántara J, Feitosa R, Lohmann LG (2015) Geographic mosaic of plant evolution: Extrafloral nectary variation mediated by ant and herbivore assemblages. PLoS One 10:e0123806

Peters MK, Hemp A, Appelhans T et al (2016) Predictors of elevational biodiversity gradients change from single taxa to the multi-taxa community level. Nat Commun 7:13736

Punyasena SW, Eshel G, McElwain JC (2008) The influence of climate on the spatial patterning of Neotropical plant families. J Biogeogr 35:117–130

Rangel TFLVB, Diniz-Filho JAF, Bini LM (2006) Towards an integrated computational tool for spatial analysis in macroecology and biogeography. Glob Ecol Biogeogr 15:321–327

Reginato M, Michelangeli F (2016) Untangling the phylogeny of Leandra sensu str. (Melastomataceae, Miconieae). Mol Phylogenet Evol 96:17–32

Reginato M, Michelangeli F, Goldenberg R (2010) Phylogeny of Pleiochiton A. Gray (Melastomataceae, Miconieae): total evidence. Bot J Linn Soc 162:423–434

Reginato M, Baumgratz J, Goldenberg R (2013) A taxonomic revision of Pleiochiton (Melastomataceae, Miconieae). Brittonia 65:16–41

Ribeiro MC, Metzger JP, Martensen AC, Ponzoni FJ, Hirota MM (2009) The Brazilian Atlantic Forest: how much is left, and how is the remaining forest distributed? Implications for conservation. Biol Conserv 142:1141–1153

Rompré G, Robinson WD, Desrochers A, Angehr G (2007) Environmental correlates of avian diversity in lowland Panama rain forests. J Biogeogr 34:802–815

Rosauer DF, Jetz W (2015) Phylogenetic endemism in terrestrial mammals. Glob Ecol Biogeogr 24:168–179

Rosauer D, Laffan SW, Crisp MD, Donnellan SC, Cook LG (2009) Phylogenetic endemism: a new approach for identifying geographical concentrations of evolutionary history. Mol Ecol 18:4061–4072

Soria-Auza RW, Kessler M, Bach K et al (2010) Impact of the quality of climate models for modelling species occurrences in countries with poor climatic documentation: a case study from Bolivia. Ecol Model 221:1221–1229

Waltari E, Schroeder R, McDonald K, Anderson RP, Carnaval A (2014) Bioclimatic variables derived from remote sensing: assessment and application for species distribution modelling. Methods Ecol Evol 5:1033–1042

Zellweger F, Baltensweiler A, Ginzler C et al (2016) Environmental predictors of species richness in forest landscapes: abiotic factors versus vegetation structure. J Biogeogr 43:1080–1090

Open Access This chapter is licensed under the terms of the Creative Commons Attribution 4.0 International License (http://creativecommons.org/licenses/by/4.0/), which permits use, sharing, adaptation, distribution and reproduction in any medium or format, as long as you give appropriate credit to the original author(s) and the source, provide a link to the Creative Commons license and indicate if changes were made.

The images or other third party material in this chapter are included in the chapter's Creative Commons license, unless indicated otherwise in a credit line to the material. If material is not included in the chapter's Creative Commons license and your intended use is not permitted by statutory regulation or exceeds the permitted use, you will need to obtain permission directly from the copyright holder.

Chapter 12
Remote Detection of Invasive Alien Species

Erik A. Bolch, Maria J. Santos, Christiana Ade, Shruti Khanna, Nicholas T. Basinger, Martin O. Reader, and Erin L. Hestir

12.1 Introduction

Invasive alien species (IAS) are non-native species with a rapid spread potential that can have negative ecological, environmental, and economic effects on the environments where they have been introduced (Masters and Norgrove 2010). The current rate and variety of species invasions is unprecedented in the fossil record (Ricciardi 2007). Global rates of invasion increased from around 8 records per year in 1800 to 1.5 per day in 1996. Although this rate may be partly the result of better record keeping, the rate is consistent across most taxa and shows little sign of slowing down (Seebens et al. 2017). Driven by climate change, invasion is expected to continue apace as global temperatures continue to rise and human societies and economies become increasingly connected around the world (Penk et al. 2016; van Kleunen et al. 2015).

E. A. Bolch (✉) · C. Ade · E. L. Hestir
University of California Merced, School of Engineering, Environmental Systems, Merced, CA, USA
e-mail: ebolch@ucmerced.edu; cade@ucmerced.edu; ehestir@ucmerced.edu

M. J. Santos · M. O. Reader
University of Zürich, Department of Geography and University Research Priority Program in Global Change and Biodiversity, Zürich, Switzerland
e-mail: maria.j.santos@geo.uzh.ch; martin.reader@geo.uzh.ch

S. Khanna
University of California Davis, Department of Land, Air, and Water Resources, Davis, CA, USA
e-mail: shrkhanna@ucdavis.edu

N. T. Basinger
University of Georgia, Department of Crop and Soil Sciences, Athens, GA, USA
e-mail: nicholas.basinger@uga.edu

© The Author(s) 2020
J. Cavender-Bares et al. (eds.), *Remote Sensing of Plant Biodiversity*,
https://doi.org/10.1007/978-3-030-33157-3_12

12.1.1 Invasive Alien Species and Global Environmental Change

Human-mediated IAS introductions, deliberate or unintentional, tend to be much faster than natural processes (e.g., wind, animal; Theoharides and Dukes 2007; Hulme 2009; Pyšek et al. 2009; Seebens et al. 2017). Invasion pathways differ between taxa; intentional transport (escape and release) is most important for plants and vertebrates, while unintentional transport is more significant for invertebrates, algae, and microorganisms (Saul et al. 2017). Roads, tracks, and waterways create natural and artificial corridors for invasion, exposing ecosystems to invasion, particularly in emerging economies where development is rapid (Mortensen et al. 2009; Masters and Norgrove 2010). Globally, the continued expansion of tourism, air transport, and trade is dramatically heightening propagule pressure and subsequent invasion (Hulme 2015).

Global environmental changes, particularly changes in climate and weather patterns, nutrient cycles, and land use, generally drive increasing invasions while also making invasion prevalence, impacts, and feedbacks to the Earth system less predictable (Bradley et al. 2010; Dukes and Mooney 1999). These same change processes can also alter IAS transport and introduction mechanisms, hindering monitoring and control (Hellmann et al. 2008; Walther et al. 2009) and making it more challenging to predict future spread. Moreover, these changes stress ecosystems and increase invasion success (Simberloff 2000). Climate and land use changes drive species range shifts, potentially creating new invasion hotspots (Bellard et al. 2013; Bradley et al. 2010) while decreasing invasion risk and increasing recovery potential in other regions (Allen and Bradley 2016). Thus, observing the geographic patterns of the spread of IAS is critical to understand their origins, pathways, and invasion processes on a changing planet.

12.1.2 Biodiversity Impacts and Global Relevance

Biodiversity provides ecosystems with the capacity to respond to biotic and abiotic conditions and stress, often used as an indicator of ecosystem resilience. IAS threaten biodiversity through competition, hybridization, population reduction, and extinction of native species and modification of habitat. It has been estimated that 42% of all threatened or endangered species are at risk primarily because of IAS (Pimentel et al. 2005). IAS are able to thrive because they arrive in new ecosystems without coevolved local competitors, parasites, and pathogens to regulate their numbers (Keane and Crawley 2002) and are potentially able to exploit resources and niche spaces that natives cannot (Byers and Noonburg 2003; Levine 2000). Hybridization with local organisms reduces genetic diversity and further increases extinction risk (Mooney and Cleland 2001). For example, cheatgrass (*Bromus tectorum*) introduction to the Great Basin in North America resulted in decreases in biodiversity and dramatic changes in the ecosystem, as cheatgrass eliminated native competing shrubs (and thus species dependent on them) and increased fire frequency in the region (Pimentel et al. 2005). Ecosystem services losses and subsequent economic impacts of IAS are also high, from agriculture, forestry, and fisheries production losses to decreased recreation and tourism revenues (Pimentel et al. 2005). As of 2005, direct costs of invasive species and their management in the United States alone reach around $120 billion per year, excluding the degradation of invaluable ecosystem services (Pimentel et al. 2005). Globally, costs of invasions and IAS management exceed those of natural disasters by an order of magnitude (Ricciardi et al. 2011).

The increasing economic and ecosystem impacts of IAS require international cooperation given the transboundary nature of IAS transport, spread, and impacts (Fig. 12.1). In recognition of the global threat IAS pose to biodiversity, ecosystems, economies, and livelihoods, the Convention on Biological Diversity (CBD) Aichi Target #9 specifically addresses IAS: "By 2020, invasive alien species and pathways are identified and prioritized, priority species are controlled or eradicated and measures are in place to manage pathways to prevent their introduction and establishment." The International Union for Conservation of Nature supports Aichi Target #9 through its global network of scientific and policy experts in the Invasive Species Specialist Group (ISSG), maintaining several databases including the Global Invasive Species Database (GISD) and the Global Register of Introduced and Invasive Species (GRIIS). The Intergovernmental Science-Policy Platform on Biodiversity and Ecosystem Services (IPBES) administered by the UN Environment Programme (UNEP) includes Deliverable 3(b)(ii): "Thematic assessment on invasive alien species and their control." This indicates that IPBES will be assessing IAS status and producing a deliverable directly to policy-makers to assist in preservation of biodiversity and ecosystem services.

IAS-driven disturbances disproportionately affect developing countries, where livelihoods often depend on local natural resources that are threatened if IAS

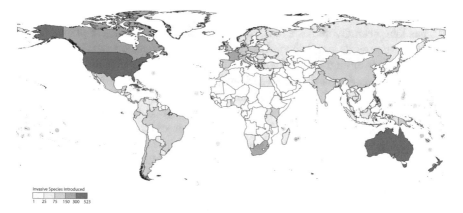

Fig. 12.1 2016 Estimates of global IAS introductions by country from the Global Invasive Species Database (GISD). (Data acquired from Turbelin et al. (2017) for reproduction)

become prevalent (Masters and Norgrove 2010). Therefore, minimizing IAS spread is necessary to meet the targets in UN Sustainable Development Goal 15, *Life on Land*, which has a target focusing specifically on preventing introduction, controlling, and eradicating IAS. The European Environmental Agency (EEA 2012) has also developed an "invasive alien species in Europe" indicator summarizing the trends of invasions since 1900 and the greatest biodiversity threats. Meanwhile the US National Invasive Species Council coordinates and facilitates data interoperability across data providers and users, including defining data standards, formats, and protocols and facilitating cooperation across sectors and governments (National Invasive Species Council 2016).

In order to reduce the pressure of IAS on biodiversity and ecosystems, globally integrated approaches to IAS prioritization, management, and control are needed. Fundamental to international cooperation is cross-border policy and cooperation and transboundary assessments that are implemented within a global monitoring framework (Latombe et al. 2017). Following the Essential Biodiversity Variable (EBV) concept (see Fernández et al., Chap. 18), essential variables for invasion monitoring have recently been proposed by Latombe et al. (2017) to underpin a global monitoring system for IAS. Essential variables for IAS include occurrence, alien status, and alien species impact. Remote sensing (RS) is a valuable observation tool in this new EBV framework because it can be used to identify locations, cover, abundance, biomass, and other traits of IAS. Because it provides synoptic spatial, routine monitoring with fine scale, high-resolution RS can be used to identify sources of IAS and pathways for spread. RS-enabled IAS location data can inform control decisions and, with routine monitoring, can be used to quantify trends and predict invasion processes into the future to support policy decisions and management actions aimed at preventing undesired spread.

12.1.3   Remote Sensing for Detection of Plant Invasions

RS has long been favored as a tool for IAS mapping, specifically for plants, due to its ability to provide synoptic views over large geographical extents. This provides an advantage over field surveys, which are often limited to a small areas and may be in difficult to access locations. Historically, RS has been crucial in IAS detection. As far back as the 1970s, color infrared (IR) photos captured from airplanes were used to target herbicide applications to control water hyacinth (*Eichhornia crassipes*) infestations (Rouse et al. 1975). Over time, the state of the science has progressed substantially. Current technologies such as hyperspectral imaging spectroscopy and light detection and ranging (lidar) make it possible to detect and differentiate plant species within the same functional groups. Coupled with advances in image processing algorithms, these technologies have enabled accurate, repeatable RS measurements over time, providing consistent monitoring records to support control efforts.

Three factors make mapping IAS using RS most viable (He et al. 2015). First, when the IAS is the dominant growth form or has large homogeneous patches, it is easier to train a classifier to recognize it. For example, water hyacinth is sometimes the only IAS in lakes, so mapping it is as easy as separating bright green vegetation from spectrally dark water (Venugopal 2002). This is feasible with simple color IR aerial photography (Rouse et al. 1975) or multispectral satellite data such as that from Satellite Pour l'Observation de la Terre (SPOT) or Landsat. Second, when the target IAS has a unique phenology, it is easier to distinguish from native plants during some parts of the year. For example, Andrew and Ustin (2008) identified perennial pepperweed (*Lepidium latifolium*) during its flowering period, when it was spectrally most distinct from the surrounding marsh due to its unique white flowers. Temporally rich imagery can be used to identify the ideal time period for differentiation along with high spectral resolution to distinguish phenological differences. Third, the target IAS has a unique chemistry or biophysiology. For example, Khanna et al. (2011) differentiated water hyacinth from other co-occurring floating aquatic macrophytes using differences in canopy water content, since water hyacinth is a succulent with a higher plant-water content than co-occurring species water primrose (*Ludwigia peploides*) and water pennywort (*Hydrocotyle ranunculoides*). This requires a spectrally rich data set that is capable of quantifying canopy biochemistry. These three requirements are well matched with the three domains of RS data: spatial, temporal, and spectral.

Invasion detection often involves species mapping, which requires much more data than functional-type or general biodiversity mapping. Often hyperspectral imagery uses phenology to time the image capture and additional ancillary data such as altitude are necessary. As mentioned, sensors collect information in three primary domains: spectral, spatial, and temporal (an additional fourth domain, radiometric resolution, is critical for aquatic and marine applications – see Sect. 12.2.3 for more details). As a rule of thumb, hyperspectral imagery is rich in data in the spectral domain, aerial imagery from piloted and unpiloted aircraft in the spatial domain, and satellite imagery in the time domain. Each of these platforms and sensor types has trade-offs between the three domains and is typically only strong in one. Selecting the best platform/sensor and fusing the collected imagery with appropriate supplementary data results in the best classification maps. Each species and habitat presents unique challenges for identifying and mapping IAS using RS, which we elaborate upon further in the chapter. Regardless of habitat, the general process of detecting and mapping IAS remains the same and consists of the following steps (see also outlined Fig. 12.2):

1. *Identify the target species and/or area.* What IAS is affecting biodiversity, ecosystem services, or other economic functions in your area (e.g., transportation)? What do you know about your target IAS (e.g., spectral characteristics, phenology, ecosystem function, habitat requirements)? Do you know, or can you hypothesize, the IAS extent and community composition of other species in the area?

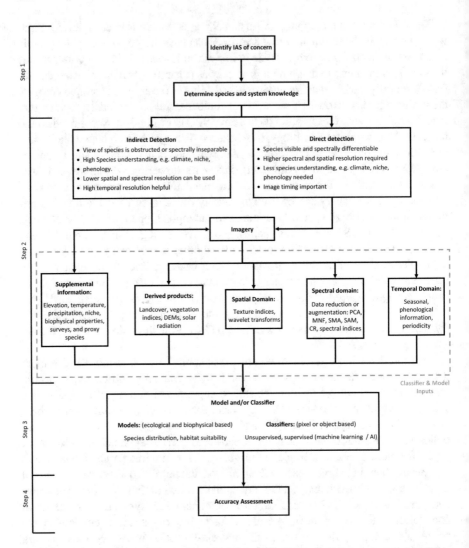

Fig. 12.2 General workflow for detecting IAS using RS. DEM, digital elevation models; PCA, principal component analysis; MNF, minimum noise fraction; SMA, spectral mixture analysis; SAM, spectral angle mapping; CR, continuum removal

2. *Determine the appropriate platform/sensor and identify/collect supplementary data based on species and habitat knowledge.* Target species can be detected using direct or indirect methods. Direct detection uses spectral data and derived products from imagery. Indirect detection utilizes the ecological relationships between species and their environment to predict distribution.

Each species and habitat discussed in this chapter has specific characteristics that can be exploited to detect IAS. Exploitable differences can exist in the temporal, spatial, or spectral domains. The temporal domain consists of data collection timing and revisit timing. For example, if an IAS flowers at an earlier or later time than its surroundings, this information can be used to time image acquisition for when the target species appears most spectrally different. The spatial domain consists of pixel size and overall geographic coverage, and the spectral domain consists of the number of wavelengths, the position and bandwidth of wavelengths measured, and the spectral range of the sensor at which radiance can be measured reliably. Sensors typically have trade-offs among these domains based upon sensor design, size limitations, and data volume limitations. For example, in the spatial domain, there is a trade-off between overall coverage area or swath width and pixel size; both can be forced to increase, but at the expense of sensor size, which limits the platform it can be mounted on. There are also trade-offs between domains, mostly related to platforms. Most satellite platforms have larger pixel sizes than other platforms (20–100 s of meters) but have quick revisit time (days to weeks) and greater geographic coverage. Airborne platforms have a longer revisit time due to costs and logistics and smaller spatial coverage but offer smaller pixel size (centimeters to meters) and often support hyperspectral sensors. Unmanned aircraft systems (UAS) offer quick revisit time, on-demand deployment, and small pixel size but have very limited spatial coverage and limited spectral resolution due to size restrictions.

When direct detection is not possible due to canopy cover or other factors, indirect methods can be used to predict species locations. Species knowledge regarding habitat constraints or coexisting species can be used to govern a model using other data products. These data include things like digital elevation models (DEMs), climate layers, soil moisture, and any factor restricting species location. In some situations, these data can also be combined with direct detection methods to improve results.

3. *Enhance data and model/classify.* A model or classifier can be thought of as a set of rules or a mathematical function that uses pixel data to assign or predict class membership. This can either be supervised, where training data (pixels or spectra that have been identified previously) are used to define classes, or unsupervised, where classes are formed based upon pixel spectral/statistical similarity. Usually, atmospherically corrected surface reflectance data are provided to the classifier. Often, image enhancement is conducted to increase the information content of the input data. In addition to reflectance data, enhanced products can also be supplied to the classifier. Methods to enhance spectral data include spectral indices, principal component analysis (PCA), and minimum noise fraction (MNF). Spectral indices are combinations of spectral reflectance from two or more wavelengths that highlight a given reflectance or absorption feature and often indicate relative abundance of features of interest; for example, the Normalized Difference Vegetation Index (NDVI) is a normalized difference ratio of red and near-infrared (NIR) bands commonly used as an indicator of vegetation vigor.

With hyperspectral data, many narrowband indices are available that provide additional data about plant traits, including light use efficiency from the photochemical reflectance index (PRI; Gamon et al. 1997), canopy nitrogen from the normalized difference nitrogen index (NDNI; Serrano et al. 2002), canopy water content from the normalized difference water index (NDWI; Gao 1995), and a large number of leaf pigment indices [see Sims and Gamon (2002) for an overview]. Continuum removal is another technique used to target absorption features. For each pixel reflectance, a convex hull is fit over the top of the spectrum, absorption features are normalized to that hull, and the depth of a specific absorption feature (e.g., leaf water content) can be quantified. PCA is a linear transformation method that maximizes the variance of the data. When applied to a hyperspectral image, it produces a series of components that correspond to linear combinations of the original bands aligned to represent the variation within the original data set, with the first component being the plane responsible for the most variation. This allows for determining the most significant characteristics within an image that relate to classes. Minimum noise fraction transformation (MNF) rescales the noise in the data (a process called noise whitening), enabling the analyst to eliminate bands containing too much sensor noise and leaving only coherent image data.

Commonly used classification techniques include random forest, a supervised machine learning algorithm that constructs many decision trees and utilizes their outputs to get an accurate class prediction based upon training data, and maximum likelihood estimation (MLE), a supervised classification method in which parameter values of a statistical model are determined that maximize the chance that the process described by the model was actually observed. All of these data enhancement and classification methods can be performed using open-source software, such as R (https://www.r-project.org/) and Python (https://www.python.org/), where many packages are available to use, or in commercial software, such as ENVI (https://www.harrisgeospatial.com/).

4. *Assess accuracy.* One of the most important considerations is accuracy assessment following mapping. Depending on the objectives of the study, some types of error may be acceptable, while some may not. Typical accuracy metrics for image classification include overall accuracy, user's accuracy, producer's accuracy, and Kappa coefficient. Overall accuracy is the probability that an image classifier will correctly classify a pixel. This metric does not account for the number of validation pixels per class and may be misleading if a similar number is not used for each class. User's accuracy and producer's accuracy may be better metrics for assessing the classification. User's accuracy (error of commission) is the fraction of correctly classified pixels with regard to all pixels classified. Producer's accuracy (errors of omission) is the fraction of correctly classified pixels with regard to all ground reference validation pixels. In some situations, such as automated weed management in agriculture, overall accuracy and producer's accuracy may not be as much of a concern as user's accuracy because identifying small amounts of weeds (IAS) as crops may be okay, but spraying crops misidentified as IAS could be more damaging to crop yields than the IAS themselves. An example where maximizing producer's accuracy may be more important would be in mapping IAS to understand species spread and the invasion process; any omitted species data as changes are monitored over time could affect process understanding and spread predictions. The last metric, the Kappa coefficient, can be useful for comparing multiple classification methods within the same data set. The Kappa coefficient is a measure of how closely the resulting

overall accuracy of a classifier compares with expected accuracy, a random classification of pixels from the data set. One final consideration for accuracy assessment is the importance of having independent validation data that were not used in the mapping procedure. If accuracy is assessed with training data, it only measures how good the classifier is for those specific data that it was trained on, but the classifier may not be as accurate with other non-training pixels within the image.

12.2   Invasive Plants in Natural and Agroecosystems

Each ecosystem and IAS combination presents unique challenges for identification and mapping using RS. This is due to different landscape configurations, community composition, canopy structures, climates, habitat characteristics, and plant phenology. Each of these characteristics can be used to inform the optimal instrumentation for IAS detection and mapping. For this reason, we have separated IAS detection methods by biome and then split into more specific ecosystems and case studies.

12.2.1   Forests

Around one-third of Earth's land surface is covered by forests. Forests are critical ecosystems, holding a very large proportion of global biodiversity. They are responsible for a large fraction of the global carbon storage and fluxes, strongly influence local and global water cycle processes, and provide fundamental goods and services to humanity (Foley et al. 2007). Globally, there are 26 types of forests, from taiga to tropical, all characterized by the unique ecological adaptations of trees to local climate, geology, and ecological conditions.

Forests invasions come in two types: (i) tree invasions (13 trees are in the top 100 world's most invasive alien species, Lowe et al. 2000); and (ii) when other plants, such as vines and shrubs, or animals invade (Resasco et al. 2007; Cheng et al. 2007; Santos and Whitham 2010). Detection of invasion by tree species requires the direct detection of tree canopies (e.g., Asner et al. 2008a, b). Invasion of forests by other plants or animals can be detected directly, for example, when the IAS covers the canopy (Cheng et al. 2007), or indirectly, by measuring canopy leaf-off (Resasco et al. 2007; Wilfong et al. 2009), or through detection of pest impacts (Näsi et al. 2015; Ortiz et al. 2013).

Several studies have used optical RS data to directly detect invasion by tree species. One of the earliest approaches performed texture analysis on simulated satellite panchromatic imagery from historical 2 m aerial photography to map the invasive acacia (*Acacia mearnsii*) in South Africa (Hudak and Wessman 1998). Ramsey III et al. (2002) used 0.5 and 1.0 m color-infrared aerial photographs to map Chinese tallow (*Sapium sebiferum*) in Louisiana and Texas. They used a k-means classifier to discriminate IAS with relative success, attributed to the differences in senescence colors between the IAS and the native vegetation. A subsequent study scaled this approach to satellites, using a combination of Hyperion, Landsat 5, and aerial photos to define characteristic spectral signatures from 400 to 950 nm for Chinese tallow (Ramsey III et al. 2005). Pearlstine et al. (2005) also used aerial photos with larger spatial resolution (37 × 25 m) to map Brazilian pepper tree (*Schinus terebinthifolius*) using texture analysis on red, green, and NIR bands to identify the IAS relatively well.

Multispectral satellite data have been used to map tree IAS with varying levels of success. Fuller (2005) performed a supervised classification of IKONOS (2 m) and Landsat ETM+ (30 m) data to detect broad-leafed paperbark (*Melaleuca quinque-nervia*) in Florida; the timing of imagery was chosen to enhance IAS separability. Cuneo et al. (2009) also used Landsat Enhanced Thematic Mapper (ETM) data to map African olive (*Olea europaea cuspidata*) in Australia based on spectral dissimi-larity with the native *Eucalyptus* spp. with an accuracy of 85% and very low confu-sion between the species. More recently, decadal-scale time series afforded by sustained land imaging have enabled increased accuracy in cases where phenologi-cal cycles can distinguish IAS. Diao and Wang (2016) used a long time series to use the phenological changes in tamarisk for high-accuracy classification. Hoyos et al. (2010) mapped glossy privet (*Ligustrum lucidum*) in Argentina using a time series of Landsat TM data and machine learning (support vector machines, SVM), achiev-ing classification accuracies of 89%.

Several studies used imaging spectroscopy to map tree IAS (He et al. 2011; Bradley 2014), e.g., tamarisk (Hamada et al. 2007; Carter et al. 2009), black cherry (*Prunus serotina*), black locust (*Robinia pseudoacacia*) and northern red oak (*Quercus rubra*) (Boschetti et al. 2007), Brazilian pepper (Lass and Prather 2004), and fire tree (*Myrica faya*) (Asner et al. 2008a, b). The studies determined charac-teristic IAS spectral profiles (sensu Ramsey III et al. 2005), compared spectral pro-files across species using techniques such as SAM (e.g., Lass and Prather 2004), and correlated them with ground measurements (e.g., Asner et al. 2008a, b).

Lidar in combination with imaging spectroscopy has been found useful for assessments of tree IAS (Huang and Asner 2009). For example, Asner et al. (2008a, b) combined imaging spectroscopy and lidar to detect fire tree in Hawaii and mea-sure impacts on forest canopy biochemistry (Fig. 12.3). Hantson et al. (2012) mapped black cherry and beach rose (*Rosa rugosa*) in the Netherlands, finding that the additional height information from lidar improved classification accuracy by 12% over imaging spectroscopy data alone.

Direct detection of IAS on the tree canopy has also been studied. For example, Cheng et al. (2007) used imaging spectroscopy to detect kudzu (*Pueraria montana*) in a pine forest in Western Georgia, United States. They used an MNF transform and SAM to differentiate the spectral profile of the IAS from the native forest. Wu et al. (2006) mapped the invasive climbing fern (*Lygodium microphyllum*) in the Florida Everglades with a supervised classification of IKONOS imagery to show how it established in different parts of the forest. Although successful, their results under-estimated fern extent in the understory.

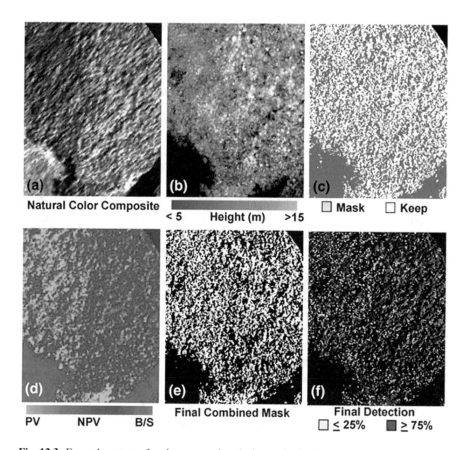

Fig. 12.3 Example output of each automated analysis step in the hyperspectral-lidar data fusion and invasive species detection process from Asner et al. (2008a, b). This 53 ha example of the study site in Hawaii shows (**a**) basic reflectance imagery that demonstrates the prescreening of the spectrometer image data by (**b**) minimum vegetation height modeling from lidar data (ground, black; shorter canopies, red/dark blue; taller canopies, yellow/white); (**c**) shadow masking based on 3-D structure of the canopies with respect to solar angle and sensor geometry (shadow, gray; sunlit, white); (**d**) live/dead fractional cover masking from AutoMCU (a spectral mixture analysis) modeling (PV, green; NPV, blue; bare/shade, pink); and (**e**) the final detection of an invasive tree based on spectral endmember bundles and AutoMCU-S algorithm (invader, yellow/red; native, green)

Indeed, invasion of the forest understory is relatively understudied. Dense canopies mask understory contribution to the RS signal. To address this, researchers have turned to leveraging forest phenology to directly detect the understory when it is most visible. Resasco et al. (2007) mapped the historical spread of Amur honeysuckle (*Lonicera maackii*) during leaf-off conditions of the native forest using the Soil Adjusted Atmospheric Resistant Vegetation Index calculated from Landsat TM and ETM+ from 1999 to 2006. Wilfong et al. (2009) found that using a difference image measuring the difference between leaf-on and leaf-off conditions better predicted Amur honeysuckle cover than a single image. Evangelista et al. (2009) used a species distribution model to predict tamarisk (*Tamarisk ramosissima*) distribution over time based on vegetation indices derived from Landsat ETM+ data, with a 90% classification accuracy. Kimothi et al. (2010) used Indian RS satellite data to map another understory IAS, the West Indian lantana (*Lantana camara*), using texture analysis of images from September, February, and April. The dense leaf canopy meant separation of IAS was not possible in September, but classification accuracies were >90% in the other images after leaf fall. Barbosa et al. (2016) mapped subcanopy strawberry guava (*Psidium cattleianum*) outbreak with imaging spectroscopy and lidar and tested the accuracy of a machine learner, biased-SVM (BSVM), and mixture-tuned matched filtering (MTMF; a partial unmixing classification algorithm similar in principle to MNF) across canopy layers. While both methods allowed the estimation of the fraction of canopy layers that were invaded, the BSVM used information across the entire spectrum, while the MTMF did not, which may limit the applicability of MTMF when spectra of IAS are similar to "background" native species.

Indirect methods are another alternative to study understory IAS. Joshi et al. (2006) mapped Siam weed (*Chromolaena odorata*) in the understory using Landsat ETM+ and an artificial neural network to predict forest density and canopy light penetration and then subsequently predict Siam weed seed production. They found that 93% of the IAS seed production was predicted by the light intensity reaching the understory and concluded that this method worked relatively well to detect the IAS, despite the spatial resolution limiting detection to well-established IAS patches.

In summary, the most common method to detect tree IAS and map their distribution are to use their characteristic spectral signatures and dissimilarity with that of the native vegetation (Lass and Prather 2004). Tree IAS likely affect both the forest's spatial structure as reflected in texture metrics (Pearlstine et al. 2005) and its 3-D structure, as shown with lidar (Asner et al. 2008a, b). To maximize the ability to detect invasive tree species, the use of the full visible (VIS) to shortwave infrared (SWIR) spectrum with imaging spectroscopy has shown clear advantages (Martin, Chap. 5), for example, in detecting the fire tree (Asner et al. 2008a, b) and for detecting bamboo (*Dendrocalamus* sp.) and slash pine (*Pinus elliottii*; Amaral et al. 2015). Alternatively, other studies selected specific bands that maximized discrimination and eliminated potential noise from nondiscriminating parts of the spectrum (Boschetti et al. 2007). While the advantages of imaging spectroscopy are obvious, data are not yet readily available to detect and map many tree IAS, especially in early stage invasion stages, although the upcoming launch of several hyperspectral satellite sensors will soon change this. Many tree IAS have different phenology than the native forest, either staying green longer, greening earlier, or flowering or budding later (Landmann et al. 2015); or they may be evergreen in a deciduous forest (Diao and Wang 2016). Timing imagery acquisition to maximize phenological differences has resulted in good classification accuracy (Ramsey III et al. 2002). Finally, using pixel sizes that match a tree canopy allows the detection of single invading trees (Bradley 2014). However, this can be very time-consuming, costly, and perhaps less systematic and viable across large areas or for early detection.

There are several analysis considerations for mapping tree IAS. A first, and perhaps most important, aspect is that tree IAS detection is prone to higher classification error (Bradley 2014) than other classifications, given the similarity in the spectral characteristics of trees to each other relative to other plant functional types. Spectral similarity between invasive and native trees may influence accuracy (Lass and Prather 2004), so ensemble classifications are recommended as well as other approaches that maximize spectral differences such as taking into account phenology. The examples cited above illustrate the value of a good field sampling design (Ramsey III et al. 2002) that covers the diversity of canopy structures (Hudak and Wessman 1998) and community compositions within the area of interest, since heterogeneity affects overall classification accuracy. In all of the studies discussed here, we observed a trade-off between omission and commission errors, where classification accuracy seems to be positively correlated with commission errors. Thus, we recommend that several accuracy metrics should be reported rather than just overall accuracy to give a better understanding of which species contribute to commission errors and which areas are more uncertain in IAS distribution maps.

12.2.2 Rangelands and Grasslands

Grasslands cover approximately one-third of the Earth's surface (Latham et al. 2014), account for at least 30% of primary production by terrestrial vegetation (Grace et al. 2006), and, after forests, are the largest terrestrial carbon sinks (Anderson 1991; Derner and Schuman 2007; Grace et al. 2006). There are two main classes of grasslands, tropical/subtropical (also known as savanna) and temperate, which can further be described by three different subclasses: human generated, highly managed natural, and rangelands (Ali et al. 2016). Regardless of classification, these regions serve as a major source of animal feed and are heavily influenced by changes in climate and fire dynamics (D'Antonio and Vitousek 1992; Brooks et al. 2004). Contrary to popular belief, grasslands and rangelands harbor large amounts of biodiversity (Murphy et al. 2016); however, they are under threat as IAS continue to invade. This threatens biodiversity not only through direct losses by IAS replacing native grasses but also through indirect impacts to ecosystems by changing fire regimes (D'Antonio and Vitousek 1992; Balch et al. 2013), supporting wind erosion (Weisberg et al. 2017), and serving as a facilitator for plant viruses (Ingwell and Bosque-Pérez 2015).

IAS in grasslands may be monitored directly or indirectly because not all species or all grassland ecosystems are good candidates for RS measurements. IAS in grasslands can be difficult to monitor. They are often indistinguishable from native plants due to spectral similarities or the nature in which they grow—in small patches, mixed with native vegetation (Shafii et al. 2004). Often indirect methods are most appropriate because they do not rely solely on discrimination between similar vegetation functional types. Indirect methods include multisource data for inferring IAS distributions and coupled RS observations and modeling. For example, the National Land Cover Database (NLCD), which is derived from Landsat data, has been used in combination with EROS Moderate Resolution Imaging Spectroradiometer (eMODIS) vegetation products (Jenkerson et al. 2010) to create a cheatgrass index based on phenology (Fig. 12.4; Boyte et al. 2015). Climate variable models such as Daymet (Thornton et al. 2018) that use DEMs created from Shuttle Radar Topography Mission (SRTM) data have been used in combination with eMODIS vegetation products to monitor the spread of cheatgrass (Downs et al. 2016).

Phenological differences are helpful for distinguishing native from non-native grasses. Given their frequent temporal resolution and global coverage, satellite optical sensors, such as Landsat TM/ETM+/OLI, SPOT, Sentinel-2, or, in some cases, Moderate Resolution Imaging Spectroradiometer (MODIS), have been used in several studies to map invaded grasslands. Cheatgrass, one of the top invaders in North America, greens up in early spring and senesces before native grasses, making it a suitable target species for RS approaches that leverage phenology differences (Fig. 12.4). Various studies across the United States have paired field data with multi-seasonal imagery selected during the green up (April–May) and senescent period to successfully map cheatgrass spread (Peterson 2005; Singh and Glenn 2009;

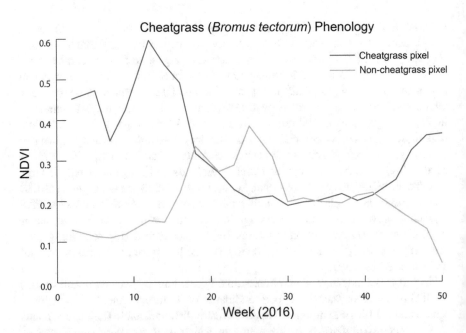

Fig. 12.4 Cheatgrass phenological differences from native sagebrush (*Artemisia* spp.) shown using eMODIS NDVI. Note that sagebrush (non-cheatgrass) greens up later in the year, allowing for development of the cheatgrass index (Boyte et al. 2015; Boyte and Wylie 2017)

West et al. 2017) and die-off (Boyte et al. 2015; Weisberg et al. 2017). Rather than just using images selected during green up and senescence periods, extracting phenology metrics from vegetation indices to refine cheatgrass classifications has also been successful (Bradley and Mustard 2008); however, in arid and semiarid environments, these indices can be highly influenced by rock and soil and should be used with caution (Singh and Glenn 2009). Huang and Geiger (2008) showed that a multi-date imaging approach can be successful even when natural phenologies of natives and nonnatives are similar. For example, native grasses and Lehmann lovegrass (*Eragrostis lehmanniana*) responded differently to unusual amounts of cool season precipitation, which allowed new tissues in invasive species to grow, making the two grasses distinguishable with multi-date imaging (Huang and Geiger 2008).

Imaging spectroscopy to map grassland IAS also often depends on differences in phenology, but the higher spectral resolution and typically higher spatial resolution afforded by airborne platforms often allow for more detailed and early detection maps. Image acquisition timing is important for species that exhibit differences in coloration throughout the year, such as flowering species or deciduous shrubs. In the case of leafy spurge (*Euphorbia esula*), hyperspectral instruments are better equipped to detect changes in flowering and thus have a higher success rate when compared to multispectral instruments (Mitchell and Glenn 2009). Leafy spurge has characteristic yellow flowers that bloom in early summer, and tamarisk leaves turn from yellow-orange to orange-brown in autumn before leaf drop. This distinct pigmentation enables remote detection using both imaging spectroscopy (Williams and Hunt Jr 2002; Glenn et al. 2005) and multispectral data (Anderson et al. 1993; Everitt et al. 1995; Evangelista et al. 2009). The blue-green color of new stems and the red-brown color of older stems help detection of spotted knapweed (*Centaurea maculosa*) from imaging spectroscopy (Lass et al. 2002; Lawrence et al. 2006). For early detection of goldenrod (*Solidago altissima*), an invasive moist tall grass in Japan, hyperspectral images acquired during early spring before full development of the grass canopy make it easier to map the exposed understory (Ishii and Washitani 2013).

Differences in canopy architecture or plant morphological traits, such as plant height and pubescence, can also be exploited when plants share similar phenologies or imagery is unavailable when growth cycles show key differences. Broom snakeweed (*Gutierrezia sarothrae*), for example, has an erect leaf canopy structure that results in a dark image response (Everitt et al. 1987; Yang and Everitt 2010). Spotted knapweed tends to inhibit the growth of other vegetation; the resulting increase in visible bare soil can help identify places where spotted knapweed grows (Lass et al. 2002; Lawrence et al. 2006).

In summary, multispectral sensors that provide free and open access to global imagery are used regularly for IAS detection in grasslands because their predefined temporal resolution offers recurring overpasses and at the very least provides seasonal imagery. This supports time series analyses and multi-date classification techniques. Looking to the future, changes in grassland species composition are anticipated to have the largest impact on Africa because it is home to the largest savannas, which cover roughly 50% of the continent (Campbell 1996; Grace et al.

2006). In addition, savannas in South American and Central Asia and temperate grasslands in the Western United States will also be heavily impacted. Thus, free, open-access, global mapping satellite RS data sets are especially important for grassland IAS detection and monitoring. These sensors lack the fine spatial resolution and full spectrum afforded by airborne imaging spectroscopy, which may be necessary to separate native and non-native species of the same functional type. However, multispectral imagery is often used for viewing widespread and abundant invasives, which is key for monitoring overall ecosystem invasion onset and die-off, but offers little help in terms of real-time or early IAS detection. In both cases, the minimum percent cover required for mapping can vary across similar ecosystems (Bradley 2014) and depends on sensor resolution and on how distinguishable the invader is from the background. Even when a non-native grass is spectrally distinguishable, an acceptable detection rate is not always possible when patch sizes are small relative to pixel resolution (Mladinich et al. 2006). Therefore, to ensure successful mapping, IAS targets must differ from the native community spectrally, phenologically, texture/morphologically, or architecturally (Bradley 2014). Analysis considerations must include a careful evaluation of the relationship between vegetation characteristics and sensor resolutions, particularly in the spatial, spectral, and temporal domains.

12.2.3 Aquatic Ecosystems

Although they cover a small portion of the Earth's surface, aquatic ecosystems are disproportionately important to global diversity. They are among the most diverse and productive ecosystems on Earth and provide vital ecosystem services (Tabacchi et al. 1998; Barbier et al. 2011). Aquatic ecosystems encompass multiple gradients, such as water intermittency, microtopography, and salinity, leading to complex environmental heterogeneity (Junk et al. 1989; Mitsch and Gosselink 2007). This mosaic of diverse environmental conditions supports high biodiversity through multiple niches (Tockner et al. 2000; Ward et al. 2002).

Biodiversity losses in coastal and freshwater aquatic ecosystems are among the highest in the world (Dudgeon et al. 2015; Waycott et al. 2009; Vörösmarty et al. 2010). At least 30%–50% of the world's wetlands have been lost (Finlayson 2012; Hu et al. 2017), and up to 35% of the extent of critical habitats like seagrasses and mangroves have been destroyed just in the twentieth century (UNESCO 2018). These ecosystems are among the most vulnerable to invasion because they are highly connected, are used extensively by humans, and often are geographically close to invasion foci such as ports or urban areas (Gherardi 2007; Williams and Grosholz 2008).

Plants in aquatic ecosystems can be broadly classified into five functional types or sets of species that occupy distinct spatial niches along the gradient from water to land and often have similar characteristics. The five functional types considered here are from land to water: riparian forests with shrubs and trees, emergent reeds and sedges, floating macrophytes, submerged macrophytes and macroalgae, and phytoplankton. Differentiating among these functional types with RS is achievable, but species-level detection within each community is more difficult due to similar survival strategies. Each functional type has its own challenges regarding species detection.

Many studies have successfully mapped IAS in aquatic environments using direct detection. Depending on the objectives of the study and the functional type being examined, spectral, spatial, and temporal requirements vary. In simple systems, high spatial resolution aerial photos can often be used to map species functional types as well as single species by taking advantage of unique attributes or phenology (Marshall and Lee 1994; Everitt et al. 1999, 2003). Mapping multiple species within the same functional types has been less successful using aerial photos. In these situations, more spectral information is needed to differentiate at the species level due to varying community complexity and species attributes (e.g., Khanna et al. 2011). Multispectral data have also been used with varying levels of success to map IAS in simple systems such as lakes invaded by just one species (essentially a two-class system; Venugopal 2002) or lakes with floating and some submerged vegetation (a three-class system; Everitt et al. 2003; Verma et al. 2003; Albright et al. 2004). Many classification methods have been used within aquatic ecosystems with varying degrees of success, including unsupervised classifiers, such as k-means and ISODATA (Ackleson and Klemas 1987; Dogan et al. 2009) and simple supervised classifiers, such as maximum likelihood and minimum distance (Malthus and George 1997; Vis et al. 2003; Nelson et al. 2006; Jollineau and Howarth 2008; Phinn et al. 2008; Yuan and Zhang 2008; Dogan et al. 2009), as well as more advanced machine learning methods (Malthus and George 1997; Nelson et al. 2006; Hestir et al. 2008, 2012; Everitt et al. 2011; Santos et al. 2012, 2016). While some studies have been successful and have even been operationalized into routine monitoring for invasive species management and reporting (sensu Santos et al. 2009; Santos et al. 2016), in many studies it is difficult to judge classification efficacy because accuracy assessment is missing or unusual, often not having independent validation data. Overall, machine learning algorithms seemed to have performed best. Within functional types, some specific strategies seem to work best as well. We highlight these below.

12.2.3.1 Riparian

Riparian plants are often more difficult to differentiate at the species level than emergent and floating plants due to higher number of species and life forms, and a complex canopy structure, similar to forest IAS detection. Riparian IAS sometimes grow in monocultures, which may be easier to detect (e.g., giant reed, *Arundo donax*). Other IAS can grow embedded in the native community similar to grasslands, making them harder to map using RS (e.g., yellow star-thistle, *Centaurea solstitialis*).

From an RS perspective, the layered canopy, many species, and mixed pixels make it hard to map target IAS within this complex community mosaic.

Community complexity can often be overcome by taking advantage of differences in phenology. Acquiring imagery during flowering or senescence when the target IAS is most distinct from its surrounding vegetation may allow for detection at the species level. For example, Landsat ETM+ and QuickBird have been used to take advantage of correct timing and fine spatial resolution, respectively, to distinguish riparian IAS (Laba et al. 2008; West et al. 2017). Frequently, increasing spectral data further has been necessary to detect riparian IAS. (Ustin et al. 2002; Laba et al. 2005; Hamada et al. 2007; Andrew and Ustin 2008).

Another concept used to map riparian plants is adding contextual information such as distance from channel and elevation (Fig. 12.5; Andrew and Ustin 2009). Contextual information can also help in improving accuracy of detection across various techniques (Maheu-Giroux and de Blois 2005; Andrew and Ustin 2008) or

Fig. 12.5 A sample vertical cross-section of the lidar returns on a transect perpendicular to a given channel shows the relationships among ground cover, elevation, and distance to a channel at Rush Ranch, California, USA (top). Current and predicted distribution (3 m window topography model) of perennial pepperweed at Rush Ranch, California, USA, overlain on a true color mosaic of airborne hyperspectral imagery (HyMap). Potential distribution was mapped as the majority rule of 25 individual classification tree models (bottom). (Derived from Andrew and Ustin (2009))

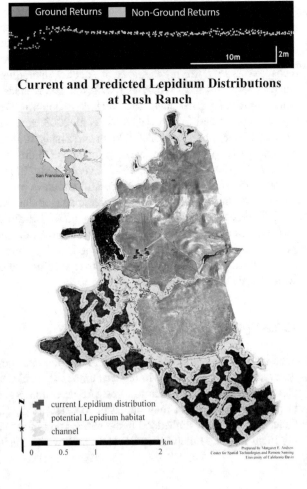

Current and Predicted Lepidium Distributions at Rush Ranch

in building species occupancy models based upon species ecological knowledge to predict future invasions or direction of spread (Andrew and Ustin 2009; Rocchini et al. 2015).

12.2.3.2   Emergent

Within the emergent functional type, the canopy is relatively uniform, composed generally of only grasses, sedges, and reeds. These species are often mixed, and patch sizes remain small even among species tending to grow as monocultures. The canopy structure is typically erectrophilic, and spectral mixing with water is common, even with fine spatial scale imagery. In addition to spectral information and temporal information, the texture of invasive and native species patches can be leveraged in mapping IAS and may be used to improve success. Samiappan et al. (2017) used four methods to calculate texture indices as inputs into a SVM algorithm to map common reed (*Phragmites australis*). They took advantage of the 5 m spatial resolution afforded by airborne (in this case UAS) imagery, though they cautioned such an approach is unlikely to work if patches of IAS are smaller than a few pixels or more mixed. However, texture has been shown to be advantageous even with moderate spatial resolution imagery. For example, Arzandeh and Wang (2003) successfully differentiated common reed and cattail (*Typha angustifolia*) using Landsat TM by adding texture indices to increase pixel spectral information content. For these reasons, hyperspectral aerial surveys have offered the best data source for classifications for emergent communities. Using sensors such as CASI, AVIRIS, and HyMap, many studies have mapped the emergent community, differentiating species within submerged and floating functional types (Hestir et al. 2008; Jollineau and Howarth 2008; Hunter et al. 2010; Khanna et al. 2011; Hestir et al. 2012; Zhao et al. 2012). Occasionally, both spectrally rich and temporally strategic data have been used together to map IAS (Laba et al. 2005; Hamada et al. 2007; Pu et al. 2008).

12.2.3.3   Floating Macrophytes

Floating macrophytes have a simple canopy structure with vegetation growing close to the water surface. They can spread over large areas and often grow as monocultures, so mapping them using RS has been relatively easy, except when two or more floating species co-occur in a single ecosystem (Khanna et al. 2011; Cavalli et al. 2009). Floating macrophyte mats often appear very similar spectrally, for example, water hyacinth, water primrose, and pennywort (*Centella asiatica*) (Khanna et al. 2011). Cavalli et al. (2009) separated three floating species with Landsat ETM+ data using spectral linear mixture modeling trained by high-quality spectral libraries developed from field spectroscopy. However, without detailed spectral libraries for a location, hyperspectral data are needed to differentiate between similar, bright green uniform mats of floating species (Yang 2007; Khanna et al. 2011).

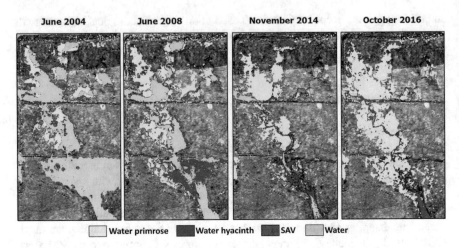

Fig. 12.6 Water primrose expansion into open water and submerged vegetation habitat (June 2008 and November 2014) and finally into emergent marsh habitat (October 2016). (Reproduced from Khanna et al. 2018)

Using hyperspectral data from HyMap and AVIRIS, Khanna et al. (2018) monitored how water primrose spread over a 12-year period (Fig. 12.6) and showed how it fundamentally changed biophysical and ecological characteristics of the ecosystem, including successional pathways.

12.2.3.4 Submerged Macrophytes

Mapping submerged macrophytes and macroalgae presents additional challenges due to the presence of the water column. Detection of these plants is complicated by the combined effects of inherent optical properties (IOPs) of the water column, which are influenced by the diffuse attenuation of the water column itself and the absorbing and scattering properties of its dissolved and suspended matter, and the apparent optical properties (AOPs), which are controlled by weather, sun, and sensor view angles (which can lead to sun glint or insufficient signal returns) as well as the influence of the air-water interface (Mertes et al. 1993; Bostater Jr. et al. 2004; Morel and Bélanger 2006; Hestir et al. 2008). IOPs are difficult to account for because water quality and depth can vary spatially and temporally with runoff, geomorphological gradients, meteorological conditions, flow conditions, land use practices, tidal stage, and phytoplankton phenology and community changes (Vis et al. 2003; Nelson et al. 2006; Hestir et al. 2008). Radiative transfer approaches are useful for classifying submerged species. Typically, they use either use model inversion or look-up tables to solve the radiative transfer model in the water column to distinguish different optically active constituents (e.g., phytoplankton and other pigments, suspended non-algal particulates, colored dissolved organic matter) and differentiate bathymetry and bottom type (see Odermatt et al. 2012; Giardino et al. 2018 for comprehensive reviews of the approach). These approaches are often preferred because, being physics-based, they are in principle generalizable and transferable across sensors and systems (Giardino et al. 2010, 2012; Malthus et al. 2012; Hestir et al. 2015). However, such approaches require detailed spectral information on specific water body IOPs, which are difficult to collect and not generally available (Matthews 2011; Lymburner et al. 2016). In these approaches, bottom type is typically mapped to just a few broad classes (e.g., sand/sediment, rock, submerged plants, coral), so species-level detections are not common in the literature (Dörnhöfer and Oppelt 2016). However, Santos et al. (2012) were able to show species-level discrimination of submerged macrophytes at the leaf level and could differentiate native from non-native submerged macrophytes at the canopy level from HyMap airborne imaging spectroscopy in a turbid estuary in California.

Often the dominant species is invasive, so even community-level maps can still reveal important processes about IAS spread and persistence and the effects of invasion on ecosystem function. Santos et al. (2016) successfully mapped submerged macrophyte spread and persistence over several years using the airborne imaging spectrometer HyMap, highlighting invasion pathways (Fig. 12.7) in the upper San Francisco estuary in California, USA. Hestir et al. (2008, 2012) mapped submerged aquatic vegetation using the same airborne imaging spectrometer and used those maps to show that increased vegetation cover significantly contributed to the increased water clarity of the system (Hestir et al. 2016).

To circumvent some of the confounding factors of the air-water interface and water column for mapping submerged macrophytes, hydroacoustics are often used for bed delineation and height and density quantification (Winfield et al. 2007). These require intensive boat surveys (which limit access), do not provide species-level discrimination, and can provide significantly different results for the same system due to lack of standardization in signal processing approaches (Radomski and Holbrook 2015). Recently it has been argued that RS imagery approaches are, despite several limitations, overall more efficacious than hydroacoustic surveys (McIntyre et al. 2018).

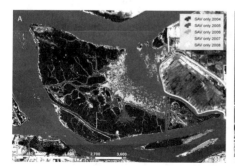

Fig. 12.7 (**a**) Map of submerged aquatic vegetation (SAV) spread near Sherman Island, CA, from 2004 to 2008. (**b**) Map of SAV persistence from 2004 to 2008 at Sherman Island, CA

12.2.3.5 Phytoplankton

Commercial shipping and the exchange of ballast water is one of the main pathways of IAS spread in marine and aquatic environments around the world. It is difficult to characterize phytoplankton species as native or non-native due to limited inventories, varying morphology and complex synonymy based on regional environmental differences, and the spontaneous "appearance" of new species (Olenina et al. 2010). Nonetheless, many phytoplankton species have been documented to have spread via ballast water (Subba Rao et al. 1994; Olenin et al. 2000), and species recorded in ships' ballast water are increasing in abundance (Olenina et al. 2010). Rapid shifts in species composition and large harmful algal blooms in coastal and inland waters have cascading effects on community structure for waterfowl, marine mammals, fish, shellfish, and benthic communities and are a constant concern for biodiversity conservation and ecosystem managers (Anderson et al. 2002).

In the water column, different phytoplankton pigments have key spectral absorption features that can be resolved in order to make inferences about their functional type. Chlorophyll a, the key diagnostic pigment for many diatoms, absorbs strongly at 435–438 and 660 nm. Cyanobacteria, the common culprit of large-scale harmful "blue-green" algal blooms, show absorption features at 490–625 nm. Floating algae have spectral features in the 550–900 nm range. *Mesodinium rubrum*, the photosynthetic ciliate that causes red tides, contains the pigment phycoerythrin, which fluoresces in the yellow peak (565–570 nm; Dierssen et al. 2015).

With the exception of key diagnostic pigments that allow direct estimation of the concentration of certain species (e.g., coccolithophores, Mesodinium), RS of phytoplankton species is typically limited to detection of phytoplankton functional types or groups (based on taxonomic criteria or biogeochemical function) or phytoplankton size class (based on size range) (Bracher et al. 2017). Most detection algorithms rely on radiative transfer models that account for bio-optical properties (e.g., pigment composition, absorption, and backscattering), empirical relationships that relate chlorophyll a concentrations measured via satellite with in-situ measurements of diagnostic marker pigments determined from high-performance liquid chromatography (HPLC) or ecological models that predict phytoplankton functional type presence based on different abiotic and biotic parameters. Moisan et al. (2012) and Bracher et al. (2017) provide an overview on the state of the science for RS phytoplankton species detection. Sathyendranath et al. (2014) and Mouw et al. (2017) provide details on most of the current algorithms and procedures for phytoplankton functional type mapping from RS.

Mapping phytoplankton functional types in coastal and inland waters is still challenging, however. Current land missions lack the temporal resolution to make frequent, repeated observations at the scale of tidal, riverine, meteorological, and biotic processes (e.g., growth, grazing, senescence) that drive phytoplankton variability (Muller-Karger et al. 2018). Phytoplankton and water quality change on the scale of hours to days due to runoff, advection, and mixing. Kudela et al. (2015) used time series of field hyperspectral observations to show that phytoplankton blooms can be displaced by cyanobacteria in a few days. Hestir et al. (2015)

documented similar rapid changes in cyanobacteria from hyperspectral measurements. Chen et al. (2010) observed phytoplankton blooms that evolve over 2–3 days in Tampa Bay. After 13 years of observations in Long Island Sound, Dierssen et al. (2015) concluded that monthly measurements are insufficient to quantify episodic plankton blooms. While they documented a bloom of a ciliate that could only be detected with hyperspectral measurements, of yellow fluorescence, only one such image has ever been collected of this area and this was with the Hyperspectral Imager for the Coastal Ocean (HICO) that ceased operations in 2014.

Mapping submerged phytoplankton, macrophytes, and macroalgae is one of the most challenging aspects of IAS detection in aquatic systems. Well-calibrated hyperspectral data with good radiometric quality is crucial when mapping submerged phytoplankton, macrophytes, and macroalgae to the species level. Due to the low reflectance, noise can severely affect data. Because of signal attenuation within the water column, typically less than 10% of the signal measured at the top of the atmosphere comes from the water column and the submerged community. The reduction in signal as water depth increases above submerged species can be seen in Fig. 12.8. Thus, atmospheric correction, sensor performance, accuracy, and radiometric quality are especially important for the water column and submerged aquatic macrophytes (Muller-Karger et al. 2018). Space-based sensors designed to meet such requirements are targeted at oceans, with pixels on the order of 250–1000 m, far exceeding the spatial resolution needed for macrophyte mapping. Recent land-observing sensors such as Sentinel 2A/2B, SPOT 6/7, and Landsat 8 OLI have higher signal-to-noise ratios and improved calibration algorithms. Hence, mapping submerged macrophytes could become more feasible, although mapping individual species is likely still a continuing challenge without high spectral resolution data.

In summary, RS of aquatic IAS requires moderate to fine spatial resolution, high spectral resolution, and, for submerged IAS, high radiometric resolution. We are optimistic that future global mapping missions with climate-relevant mission durations can improve riparian and aquatic IAS mapping by enabling time-based

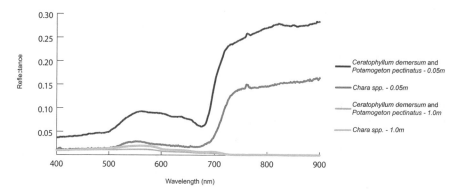

Fig. 12.8 Water column effects on reflectance of the submerged aquatic vegetation species hornwort (*Ceratophyllum demersum*), sago pondweed (*Potamogeton pectinatus*), and green algae (*Chara spp.*) from 5 cm water column height to 1 m water column height

approaches such as phenology signatures. Even without high spectral resolution data, RS of aquatic macrophytes is progressing. For example, through radiative transfer modeling, it has been shown to be robust for mapping aquatic macrophyte morphological traits in temperate systems (e.g., leaf area index, fractional cover, and biomass) across floating, emergent, and submerged macrophytes, which can be used to better quantify nutrient uptake, community dynamics, and invasion hotspots (Villa et al. 2014, 2015, 2017). The rapidly developing science of drone and UAS imagery also raises the potential to map IAS using differences in texture or using segmentation tools to do object-based mapping, especially when the area being mapped is small.

12.2.4   Agroecosystems

Agroecosystems are unique ecosystems due to the extraordinarily high anthropogenic interventions and pressures placed on them. Unlike other ecological systems, agricultural systems have more controlled environmental conditions with limited plant biodiversity. Crops are often grown as a monoculture, in uniform rows with highly regulated demography. Though crop species are often robust and herbicide resistant, many IAS are also developing resistance to herbicide, making them more invasive with increasing impacts on crops. With a rising global population, there is increased pressure on agricultural systems to increase productivity. IAS consume resources meant for crops and reduce yield, productivity, and income for farmers. In corn and soybean, two of the major crops grown in the United States losses due to IAS have been estimated at $17 billion in soybean and $27 billion in corn annually, approximately 50% of the yield of each of these crops (Soltani et al. 2016, 2017). IAS can become established in agroecosystems as in any other system, through both natural (wind, water, animals, forceful dehiscence) and artificial (machinery, crop seed, livestock feed, spreading of crop, and livestock waste) means. The application of water and nutrients also complicates the system by enhancing IAS' ability to compete with crops and reproduce. Often the effects of IAS depend on the crops present. Certain IAS may be problematic in some crops but not others due to crop management practices (time of planting, tillage, irrigation, mulch, registered herbicides, rotation).

To effectively detect IAS in agricultural systems, RS must meet the challenge of detecting IAS before they become competitive with crops. Field spectroscopy has been shown to be effective for discrimination of IAS from crops (Basinger 2018; Koger et al. 2004a, b; Gray et al. 2009), but it is not the most efficient due to the short duration of such field campaigns, since detection must then occur within a small window during one growing cycle. Research has long been published on the use of satellites or other airborne sensors for IAS detection in agriculture (Hunt et al. 2007; Menges et al. 1985), but these methods often lack the spatial and/or temporal resolution needed to detect IAS intermixed with a crop species.

One factor that aids in detection is that IAS tend to emerge in patches or patterns associated with farm management practices. For example, plants growing outside of the uniform row formations are often IAS and can be treated. Most studies typically investigate a single IAS. However, IAS are often intermixed, making them hard to distinguish from each other. Additionally, Basinger (2018) found that using field spectroscopy, IAS detection is not uniform across cropping systems and suggested that improved IAS detection may require crop-specific parameters for accurate IAS detection and control.

Hyperspectral data, as seen in Fig. 12.9, have also been demonstrated to enable detection of IAS density within the crop and determination of when in the planting cycle IAS are most readily detectable, especially during early growth stages (Basinger 2018). If only a few spectral bands are available, it can be very difficult to differentiate between species during the first few weeks after planting. So far, the most promising platform for IAS detection appears to be UAS. They have the necessary spatial resolution to locate IAS at early stages in the growing cycle, before they can spread or be obscured by the crop canopy, and UAS can be launched whenever necessary to collect imagery.

The main challenges of using RS in agroecosystems are associated with data latency (which impedes rapid IAS management on the part of producers) and the necessity of early growth cycle detection (where many species appear similar). Current market solutions tend to focus on active sensors or the use of artificial lighting rather than passive sensors. Commercial early IAS management systems used active proximal sensors to spot and spray IAS with herbicides. However, while

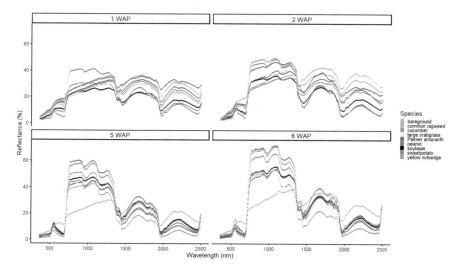

Fig. 12.9 Spectra of four crop species, cucumber (*Cucumis sativus*), peanut (*Arachis hypogaea*), soybean (*Glycine max*), and sweet potato (*Ipomoea batatas*), and four IAS, common ragweed (*Ambrosia artemisiifolia*), large crabgrass (*Digitaria sanguinalis*), Palmer amaranth (*Amaranthus palmeri*), and yellow nutsedge (*Cyperus esculentus*) over the first 10 weeks after being planted in 2016. (Data from Basinger (2018))

these systems can detect vegetation, they are not able to detect small IAS (Blackshaw et al. 1998) or distinguish between the crop and IAS. They thus rely solely on a priori assumptions about timing of emergence of IAS relative to crop species.

In summary, using RS in agroecosystems is only useful to growers within the timeline of crop cycles. IAS control is most effective when plants are small, but this is when they are also most difficult to detect and differentiate from the crop. IAS detection often requires high spatial and temporal resolution due to synchronous and asynchronous IAS emergence with the crop and sometimes high spectral resolution to deal with similar appearances during early growth stages. Implementing data-based management decisions is difficult if monitoring is not near constant due to the necessity of rapid responses. Thus, the use RS for the control of IAS has seen limited adoption in agriculture, despite a long history of research. However, UAS have become more common because the technology now meets several of the requirements for RS of IAS in agricultural settings.

12.2.5   Urban Ecosystems

More than half of all people live in urban areas, and this proportion is expected to increase substantially during this century. Urban ecosystems differ from agricultural or natural systems in terms of structural properties related to the built/natural ratio of the landscape; built area includes impervious and permeable built environments and the connecting infrastructure. Urban ecosystems have been colonized by increasing numbers of IAS (Paap et al. 2017; Hui et al. 2017). These ecosystems are unique because trees and other ornamental species in private and public city gardens are often non-native and can be sources of IAS to surrounding areas (Paap et al. 2017; Mayer-Pinto et al. 2017). IAS richness in urban areas is positively correlated with housing density (Gavier-Pizarro et al. 2010), urban wastelands (Bonthoux et al. 2014; Maurel et al. 2010), green infrastructure (Hostetler et al. 2011), and roads (Rupprecht et al. 2015). By harboring IAS, cities may unwittingly act as sources of IAS to surrounding agroecosystems and natural ecosystems (Paap et al. 2017; McLean et al. 2017).

Use of RS for IAS detection and mapping in urban environments is essential to gauge the affect of urban plants, which are often non-native, on the surrounding ecosystems. Detection has been successful with many forms of RS. For example, Shouse et al. (2012) used a combination of 0.3 m color aerial photographs and multispectral Landsat data to map bush honeysuckle (*Lonicera maackii*) under the forest canopy in an urban park in Louisville, Kentucky, USA. They conducted an object-based classification, a supervised classification, and constructed a species distribution model, with accuracies above 75%, especially for the object-based classification. This high accuracy can be attributed to extended greened-up seasons and high spatial resolution. Hyperspectral data has been used to detect Himalayan blackberry (*Rubus armeniacus*) and English ivy (*Hedera helix*) in nonforested areas of Surrey, British Columbia, Canada (Chance et al. 2016a). Classification accuracies were higher than 75% for both; the potential for spectral separability was maximized by the choice of wavelength regions, and the researchers were able to increase accuracy using a random forest classifier, due to higher capability of under-canopy detection (Chance et al. 2016b). Because urban ecosystems are smaller than other ecosystems and more complex, high spatial resolution is necessary to detect IAS within the mostly non-vegetative ground cover.

Lidar with spectral data also has proven effective for mapping vegetation within urban areas. By combining lidar with hyperspectral imaging and a random forest classifier to map tree species including honey locust (*Gleditsia triacanthos*) in Surrey, British Columbia, Canada. Liu et al. (2017) further improved classification accuracy, showing the power of data fusion. Other studies have combined lidar data with IKONOS multispectral data to detect whether Chinese privet (*Ligustrum sinense*) invasion changed urban forest structure in Charlotte, North Carolina, USA (Singh et al. 2015). These researchers also found that a random forest built with lidar-derived metrics produced the best results.

RS of urban IAS, however, has some unique challenges. Because most of the ground is covered by manmade features, it is difficult to detect green areas and map and identify individual species (Alonzo et al. 2014). With sufficient spatial resolution, these challenges can be overcome. The most successful approach to date is to use a combination of hyperspectral and lidar, which yields spectral, structural, and height information.

In summary, detection of IAS in urban environments requires high spatial resolution to differentiate natural from built environments, high spectral resolution to identify species, and sufficient temporal resolution to detect IAS at different stages of invasion. While this is an emerging field with a growing literature, relatively few studies of IAS in urban environments have used RS data, and further research is needed in different geographical settings, invasion process phases, and urban density conditions.

12.3 Summary, Conclusions, and Prospectus

Invasive species are a major direct driver of biodiversity loss because they outcompete native species for local resources, eventually replacing or displacing them. They also cause indirect losses because they do not assume all of the ecological roles of the replaced native species. As they spread, IAS modify nutrient availability, nutrient cycling, soil chemistry, water quality, hydrology, food webs, habitats, and other ecosystem functions (Gordon 1998; Scheffer et al. 2003; Dukes and Mooney 2004; Hestir et al. 2016; Khanna et al. 2018), impairing ecosystem function. In addition to causing functional changes, IAS also modify ecosystem structure by physically changing canopy structures in forests and water quality in aquatic ecosystems. Increasing global changes related to climate, nutrient cycles, and land use will potentially change transport and introduction mechanisms of IAS in a way that provides a competitive advantage for new IAS, likely reducing effectiveness of control strategies. The acceleration in global change and biodiversity loss degrades ecosystem resilience, threatening valuable ecosystem services. To preserve these services will require global cooperation on IAS monitoring and control with RS is a critical tool.

Each biome discussed in this chapter contains a unique complement of species. As a result, a different method of RS and data fusion works best for each. However, some methodologies can be valuable in all circumstances, such as increasing spectral information content. If only one IAS is of interest and it differs from its surroundings, multispectral data or use of photographs and texture analysis may be enough to identify and map it. However, in most cases there are multiple IAS competing with one another and with native vegetation, with varying canopy complexity and functional types. In such scenarios, difference in phenological characteristics can be exploited for identification. For example, an IAS might be identified through differing flowering times, flower colors, or earlier or later periods of senescence relative to surrounding vegetation. This requires temporally dense data. In cases where the invasion scenario is not simple or the data are not temporally sufficient, fusion between RS and other data sources (e.g., habitat models, DEMs, climate models) can be used to improve accuracy.

Data collection in the three domains of RS (spectral, spatial, and temporal) can be optimized for a species based on the ecosystem type and image analysis approach. For forests, lidar data are often a good addition to spectral information because they can provide information on height and physical crown structure. For species below the forest canopy, indirect methods such as models based on ecological knowledge of the species may be necessary, or imagery may simply be collected during a leaf-off period. IAS in grasslands often have similar spectral properties to natives, requiring hyperspectral data, strategic image timing, or indirect modeling methods. Aquatic ecosystems introduce many confounding factors due to presence of water and its associated processes, necessitating high radiometric quality and good calibration. Because this biome is so complex, hyperspectral information and customized image timing are a must for differentiating IAS. Additionally, radiative transfer modeling is often necessary to detect submerged and water column IAS. Agroecosystems have minimal diversity, so fewer spectral data are required. However, frequent assessment is necessary to allow a timely response to minimize crop loss. RS detection of IAS in urban ecosystems requires varying methods and unique adaptations because of the high potential for introductions and unusual landscape features, such as impervious surfaces.

These factors underscore the importance of mission design for two key data collection platforms. First, airborne platforms (piloted and unpiloted), which are vital to rapid, local-scale assessments, must acquire data at key times relevant to IAS phenology. As temperatures and biodiversity losses continue to increase, plant phenology is expected to continue to change (Primack et al. 2015; Wolf et al. 2017) and airborne acquisition strategies must adjust accordingly. Second, satellite platforms are critical to providing global-scale systematic monitoring of IAS. Current and future missions must include high spectral resolution sensors with the capability to create climate-relevant time series (a duration on the order of approximately a

Fig. 12.10 Accumulation of RS data over time makes RS a powerful tool for monitoring and understanding the spread of IAS, as well as filling an important role in IAS management. (Image credit: Vanessa Tobias, California Department of Fish and Wildlife)

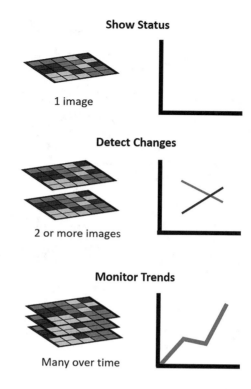

decade) to characterize phenology for widespread IAS detection, particularly for grasslands and forests (Fig. 12.10).

To date, most IAS management has been reactive. RS can help land managers see where IAS occur, target removal, monitor rates of growth and expansion, and evaluate treatment effectiveness. The future of the field is in prevention. Novel research is starting to focus on invasion processes, impacts, and management assessments (e.g., Santos et al. 2009; Hestir et al. 2016; Santos et al. 2016). Further research is needed to understand how RS can be fully integrated into understanding the invasion process, from arrival to establishment and spread. Freely available time series data alongside increasing amounts of field data related to early detection of IAS may allow the achievement of such a goal. For this reason, it is important to cultivate a cross-disciplinary understanding of the invasion process and the effects IAS on ecosystems and biodiversity. Two promising developments that will increase support for IAS mapping and monitoring are the upcoming Surface Biology and Geology (SBG) global mapping hyperspectral satellite (Schimel, Chap. 19) and the use of UAS imaging spectroscopy. The new satellite offers potential to improve mapping of IAS on a global scale; though limited by spatial resolution, it will still be capable of species level identification in many situations. UAS offer high spatial resolution mapping on demand, providing flexibility and simplification of RS missions, reducing costs compared with manned flights, and improving safety. These two developments will drastically improve the volume of data being collected and, with scientific innovation, help minimize economic and environmental impacts of IAS (Fig. 12.11).

Fig. 12.11 New off the shelf UAS systems offer hyperspectral image collection, providing new capabilities, and presenting new challenges. This image shows a preliminary classification of floating IAS and emergent vegetation overlayed on top of and RGB flightline mosaic. The flight took approximately 20 min to cover a 250 m × 200 m area with 5.5 cm spatial resolution. Raw data size for all of the flightlines is roughly 40 gb

References

Ackleson SG, Klemas V (1987) Remote sensing of submerged aquatic vegetation in lower Chesapeake Bay: a comparison of Landsat MSS to TM imagery. Remote Sens Environ 22:235–248. https://doi.org/10.1016/0034-4257(87)90060-5

Albright TP, Moorhouse TG, Mcnabb TJ (2004) The rise and fall of water hyacinth in Lake Victoria and the Kagera river basin. J Aquat Plant Manage 42:73–84

Ali I, Cawkwell F, Dwyer E et al (2016) Satellite remote sensing of grasslands: from observation to management. J Plant Ecol 9:649–671. https://doi.org/10.1093/jpe/rtw005

Allen JM, Bradley BA (2016) Out of the weeds? Reduced plant invasion risk with climate change in the continental United States. Biol Conserv 203:306–312. https://doi.org/10.1016/j.biocon.2016.09.015

Alonzo M, Bookhagen B, Roberts DA (2014) Urban tree species mapping using hyperspectral and lidar data fusion. Remote Sens Environ 148:70–83. https://doi.org/10.1016/j.rse.2014.03.018

Amaral CH, Roberts DA, Almeida TIR, Souza Filho CR (2015) Mapping invasive species and spectral mixture relationships with neotropical woody formations in southeastern Brazil. ISPRS J Photogramm Remote Sens 108:80–93. https://doi.org/10.1016/j.isprsjprs.2015.06.009

Anderson JM (1991) The effects of climate change on decomposition processes in grassland and coniferous forests. Ecol Appl 1:326–347. https://doi.org/10.2307/1941761

Anderson GL, Hanson JD, Haas RH (1993) Evaluating landsat thematic mapper derived vegetation indices for estimating above-ground biomass on semiarid rangelands. Remote Sens Environ 45:165–175. https://doi.org/10.1016/0034-4257(93)90040-5

Anderson DM, Glibert PM, Burkholder JM (2002) Harmful algal blooms and eutrophication: nutrient sources, composition, and consequences. Estuaries 25:704–726. https://doi.org/10.1007/bf02804901

Andrew ME, Ustin SL (2008) The role of environmental context in mapping invasive plants with hyperspectral image data. Remote Sens Environ 112:4301–4317. https://doi.org/10.1016/j. rse.2008.07.016

Andrew ME, Ustin SL (2009) Habitat suitability modelling of an invasive plant with advanced remote sensing data. Divers Distrib 15:627–640. https://doi.org/10.1111/j.1472-4642.2009.00568.x

Arzandeh S, Wang J (2003) Monitoring the change of phragmites distribution using satellite data. Can J Remote Sens 29:24–35. https://doi.org/10.5589/m02-077

Asner GP, Jones MO, Martin RE et al (2008a) Remote sensing of native and invasive species in Hawaiian forests. Remote Sens Environ 112:1912–1926. https://doi.org/10.1016/j. rse.2007.02.043

Asner GP, Knapp DE, Kennedy-Bowdoin T et al (2008b) Invasive species detection in Hawaiian rainforests using airborne imaging spectroscopy and LiDAR. Remote Sens Environ 112:1942–1955. https://doi.org/10.1016/j.rse.2007.11.016

Balch JK, Bradley BA, D'Antonio CM, Gómez-Dans J (2013) Introduced annual grass increases regional fire activity across the arid western USA (1980-2009). Glob Chang Biol 19:173–183. https://doi.org/10.1111/gcb.12046

Barbier EB, Hacker SD, Kennedy C et al (2011) The value of estuarine and coastal ecosystem services. Ecol Monogr 81:169–193. https://doi.org/10.1890/10-1510.1

Barbosa JM, Asner GP, Martin RE et al (2016) Determining subcanopy *Psidium cattleianum* invasion in Hawaiian forests using imaging spectroscopy. Remote Sens 8:33. https://doi. org/10.3390/rs8010033

Basinger NT (2018) Interference and spectral changes of Palmer Amaranth (*Amaranthus palmeri S. Wats.*) and large crabgrass (*Digitaria sanguinalis (L.) Scop.*) density in sweet potato and soybean and use of spectroscopy for discrimination of crop and weed species. Dissertation, North Carolina State University

Bellard C, Thuiller W, Leroy B et al (2013) Will climate change promote future invasions? Glob Chang Biol 19:3740–3748. https://doi.org/10.1111/gcb.12344

Blackshaw RE, Molnar LJ, Lindwall CW (1998) Merits of a weed-sensing sprayer to control weeds in conservation fallow and cropping systems. Weed Sci 46:120–126

Bonthoux S, Brun M, Di Pietro F et al (2014) How can wastelands promote biodiversity in cities? A review. Landsc Urban Plan 132:79–88. https://doi.org/10.1016/j.landurbplan.2014.08.010

Boschetti M, Boschetti L, Oliveri S et al (2007) Tree species mapping with airborne hyperspectral MIVIS data: the Ticino Park study case. Int J Remote Sens 28:1251–1261. https://doi. org/10.1080/01431160600928542

Bostater CR Jr, Ghir T, Bassetti L et al (2004) Hyperspectral remote sensing protocol development for submerged aquatic vegetation in shallow waters. In: Remote sensing of the ocean and sea ice 2003. International Society for Optics and Photonics, pp 199–216

Boyte SP, Wylie BK (2017) A time series of herbaceous annual cover in the sagebrush ecosystem. U.S. Geological Survey data release. https://doi.org/10.5066/F71J98QK

Boyte SP, Wylie BK, Major DJ (2015) Mapping and monitoring cheatgrass dieoff in rangelands of the northern Great Basin, USA. Rangel Ecol Manag 68:18–28. https://doi.org/10.1016/j. rama.2014.12.005

Bracher A, Bouman HA, Brewin RJW et al (2017) Obtaining phytoplankton diversity from ocean color: a scientific roadmap for future development. Front Mar Sci 4:55. https://doi.org/10.3389/ fmars.2017.00055

Bradley BA (2014) Remote detection of invasive plants: a review of spectral, textural and phenological approaches. Biol Invasions 16:1411–1425. https://doi.org/10.1007/s10530-013-0578-9

Bradley BA, Mustard JF (2008) Comparison of phenology trends by land cover class: a case study in the Great Basin, USA. Glob Chang Biol 14:334–346. https://doi. org/10.1111/j.1365-2486.2007.01479.x

Bradley BA, Blumenthal DM, Wilcove DS, Ziska LH (2010) Predicting plant invasions in an era of global change. Trends Ecol Evol 25:310–318. https://doi.org/10.1016/j.tree.2009.12.003

Brooks ML, D'Antonio CM, Richardson DM et al (2004) Effects of invasive alien plants on fire regimes. Bioscience 54:677. https://doi.org/10.1641/0006-3568(2004)054[0677:EOIAPO]2.0 .CO;2

Brown SD, Dooling RJ (1993) Perception of conspecific faces by budgerigars (Melopsittacus undulatus): II. Synthetic models. J Comp Psychol 107:48–60. https://doi.org/10.1037/0735-7036.107.1.48

Byers JE, Noonburg EG (2003) Scale dependent effects of biotic resistance to biological invasion. Ecology 84:1428–1433. https://doi.org/10.1890/02-3131

Campbell BM (ed) (1996) The Miombo in transition: woodlands and welfare in Africa. CIFOR. Bogor, Indonesia

Carter GA, Lucas KL, Blossom GA et al (2009) Remote sensing and mapping of tamarisk along the Colorado River, USA: a comparative use of summer-acquired Hyperion, Thematic Mapper and QuickBird data. Remote Sens 1:318–329. https://doi.org/10.3390/rs1030318

Cavalli RM, Laneve G, Fusilli L et al (2009) Remote sensing water observation for supporting Lake Victoria weed management. J Environ Manag 90:2199–2211. https://doi.org/10.1016/j.jenvman.2007.07.036

Chance CM, Coops NC, Crosby K, Aven N (2016a) Spectral wavelength selection and detection of two invasive plant species in an urban area. Can J Remote Sens 42:27–40. https://doi.org/1 0.1080/07038992.2016.1143330

Chance CM, Coops NC, Plowright AA et al (2016b) Invasive shrub mapping in an urban environment from hyperspectral and LiDAR-derived attributes. Front Plant Sci 07:1528

Chen Z, Hu C, Muller-Karger FE, Luther ME (2010) Short-term variability of suspended sediment and phytoplankton in Tampa Bay, Florida: observations from a coastal oceanographic tower and ocean color satellites. Estuar Coast Shelf Sci 89:62–72. https://doi.org/10.1016/j.ecss.2010.05.014

Cheng Y-B, Tom E, Ustin SL (2007) Mapping an invasive species, kudzu (Pueraria montana), using hyperspectral imagery in western Georgia. J Appl Remote Sens 1:013514. https://doi.org/10.1117/1.2749266

Convention on Biological Diversity (CBD) (2010) Year in review. Secretariat of the Convention on Biological Diversity, Montreal, p 2009

Cuneo P, Jacobson CR, Leishman MR (2009) Landscape-scale detection and mapping of invasive African Olive (Olea europaea L. ssp. cuspidata Wall ex G. Don Ciferri) in SW Sydney, Australia using satellite remote sensing. Appl Veg Sci 12:145–154. https://doi.org/10.1111/j.1654-109X.2009.01010.x

D'Antonio CM, Vitousek PM (1992) Biological invasions by exotic grasses, the grass/fire cycle, and global change. Annu Rev Ecol Syst 23:63–87. https://doi.org/10.1146/annurev.es.23.110192.000431

Derner JD, Schuman GE (2007) Carbon sequestration and rangelands: a synthesis of land management and precipitation effects. J Soil Water Conserv 62:77–85

Diao C, Wang L (2016) Incorporating plant phenological trajectory in exotic saltcedar detection with monthly time series of Landsat imagery. Remote Sens Environ 182:60–71. https://doi.org/10.1016/j.rse.2016.04.029

Dierssen H, McManus GB, Chlus A et al (2015) Space station image captures a red tide ciliate bloom at high spectral and spatial resolution. Proc Natl Acad Sci 112:14783–14787

Dogan OK, Akyurek Z, Beklioglu M (2009) Identification and mapping of submerged plants in a shallow lake using QuickBird satellite data. J Environ Manag 90:2138–2143. https://doi.org/10.1016/j.jenvman.2007.06.022

Dörnhöfer K, Oppelt N (2016) Remote sensing for lake research and monitoring - recent advances. Ecol Indic 64:105–122. https://doi.org/10.1016/j.ecolind.2015.12.009

Downs J, Larson K, Cullinan V (2016) Mapping cheatgrass across the range of the greater sage-grouse: linking biophysical, climate and remote sensing data to predict cheatgrass occurrence. Pacific Northwest National Laboratory. https://static1.squarespace.com/static/5016c7a324ac93bdfdfb930c/t/582b58e746c3c44fc1e27c19/1479235820777/2016_9MappingCheatgrassSageGrouseRange.pdf. Accessed 30 Nov 2018

Dudgeon D, Arthington AH, Gessner MO et al (2015) Freshwater biodiversity: importance, threats, status and conservation challenges. Society 81:163–182. https://doi.org/10.1017/S1464793105006950

Dukes JS, Mooney HA (1999) Does global change increase the success of biological invaders? Trends Ecol Evol 14:135–139. https://doi.org/10.1016/S0169-5347(98)01554-7

Dukes JS, Mooney HA (2004) Disruption of ecosystem processes in western North America by invasive species. Rev Chil Hist Nat 77:411–437. https://doi.org/10.4067/S0716-078X2004000300003

European Environmental Agency (EEA) (2012) Invasive alien species indicators in Europe – a review of streamlining European biodiversity (SEBI) indicator 10. EEA Technical Report No 15/2012. European Environment Agency, Copenhagen

Evangelista PH, Stohlgren TJ, Morisette JT, Kumar S (2009) Mapping invasive tamarisk (*Tamarix*): a comparison of single-scene and time-series analyses of remotely sensed data. Remote Sens 1:519–533. https://doi.org/10.3390/rs1030519

Everitt JH, Pettit RD, Alaniz MA (1987) Remote sensing of broom snakeweed (*Gutierrezia sarothrae*) and spiny aster (*Aster spinosus*). Weed Sci 35:295–302

Everitt JH, Anderson GL, Escobar DE et al (1995) Use of remote-sensing for detecting and mapping leafy spurge (*Euphorbia-Esula*). Weed Technol 9:599–609. https://doi.org/10.1017/s0890037x00023915

Everitt JH, Yang C, Escobar DE et al (1999) Using remote sensing and spatial information technologies to detect and map two aquatic macrophytes. J Aquat Plant Manag 37:71–80

Everitt JH, Yang C, Flores DG (2003) Light reflectance characteristics and remote sensing of waterlettuce. J Aquat Plant Manag 41:39–44. https://doi.org/10.2307/4003041

Everitt JH, Yang C, Summy KR et al (2011) Evaluation of hyperspectral reflectance data for discriminating six aquatic weeds. J Aquat Plant Manag 49:94–100

Finlayson MC (2012) Forty years of wetland conservation and wise use. Aquat Conserv Mar Freshw Ecosyst 22:139–143. https://doi.org/10.1002/aqc.2233

Foley JA, Asner GP, Costa MH et al (2007) Amazonia revealed: forest degradation and loss of ecosystem goods and services in the Amazon Basin. Front Ecol Environ 5:25–32. https://doi.org/10.1890/1540-9295(2007)5[25:arfdal]2.0.co;2

Fuller DO (2005) Remote detection of invasive melaleuca trees (*Melaleuca quinquenervia*) in South Florida with multispectral IKONOS imagery. Int J Remote Sens 26:1057–1063. https://doi.org/10.1080/01430060512331314119

Gamon JA, Serrano L, Surfus JS (1997) International association for ecology the photochemical reflectance index: an optical indicator of photosynthetic radiation use efficiency across species, functional types, and nutrient levels. Oecologia 112:492–501

Gao B-C (1995) Normalized difference water index for remote sensing of vegetation liquid water from space. In: Imaging spectrometry, Orlando, June 1995, vol 2480, SPIE (International Society for Optics and Photonics), pp 225–237. https://doi.org/10.1117/12.210877

Gavier-Pizarro GI, Radeloff VC, Stewart SI et al (2010) Housing is positively associated with invasive exotic plant species richness in New England, USA. Ecol Appl 20:1913–1925. https://doi.org/10.1890/09-2168.1

Gherardi F (2007) Biological invasions in inland waters: an overview. In: Biological invaders in inland waters: profiles, distribution, and threats. Springer Netherlands, Dordrecht, pp 3–25

Giardino C, Bresciani M, Villa P, Martinelli A (2010) Application of remote sensing in water resource management: the case study of Lake Trasimeno, Italy. Water Resour Manag 24:3885–3899. https://doi.org/10.1007/s11269-010-9639-3

Giardino C, Candiani G, Bresciani M et al (2012) BOMBER: a tool for estimating water quality and bottom properties from remote sensing images. Comput Geosci 45:313–318. https://doi.org/10.1016/j.cageo.2011.11.022

Giardino C, Brando VE, Gege P et al (2018) Imaging spectrometry of inland and coastal waters: state of the art, achievements and perspectives. Surv Geophys 40:1–29. https://doi.org/10.1007/s10712-018-9476-0

Glenn NF, Mundt JT, Weber KT et al (2005) Hyperspectral data processing for repeat detection of small infestations of leafy spurge. Remote Sens Environ 95:399–412. https://doi.org/10.1016/j.rse.2005.01.003

Gordon DR (1998) Effects of invasive, non-indigenous plant species on ecosystem processes: lessons from Florida. Ecol Appl 8:975. https://doi.org/10.2307/2640955

Grace J, José JS, Meir P et al (2006) Productivity and carbon fluxes of tropical savannas. J Biogeogr 33:387–400. https://doi.org/10.1111/j.1365-2699.2005.01448.x

Gray CJ, Shaw DR, Bruce LM (2009) Utility of hyperspectral reflectance for differentiating soybean (*Glycine max*) and six weed species. Weed Technol 23:108–119. https://doi.org/10.1614/WT-07-117.1

Hamada Y, Stow DA, Coulter LL et al (2007) Detecting tamarisk species (*Tamarix spp.*) in riparian habitats of southern California using high spatial resolution hyperspectral imagery. Remote Sens Environ 109:237–248. https://doi.org/10.1016/j.rse.2007.01.003

Hantson W, Kooistra L, Slim PA (2012) Mapping invasive woody species in coastal dunes in the Netherlands: a remote sensing approach using LIDAR and high-resolution aerial photographs. Appl Veg Sci 15:536–547. https://doi.org/10.1111/j.1654-109X.2012.01194.x

He KS, Rocchini D, Neteler M, Nagendra H (2011) Benefits of hyperspectral remote sensing for tracking plant invasions. Divers Distrib 17:381–392. https://doi.org/10.1111/j.1472-4642.2011.00761.x

He KS, Bradley BA, Cord AF et al (2015) Will remote sensing shape the next generation of species distribution models? Remote Sens Ecol Conserv 1:4–18. https://doi.org/10.1002/rse2.7

Hellmann JJ, Byers JE, Bierwagen BG, Dukes JS (2008) Five potential consequences of climate change for invasive species. Conserv Biol 22:534–543. https://doi.org/10.1111/j.1523-1739.2008.00951.x

Hestir EL, Khanna S, Andrew ME et al (2008) Identification of invasive vegetation using hyperspectral remote sensing in the California Delta ecosystem. Remote Sens Environ 112:4034–4047. https://doi.org/10.1016/j.rse.2008.01.022

Hestir EL, Greenberg JA, Ustin SL (2012) Classification trees for aquatic vegetation community prediction from imaging spectroscopy. IEEE J Sel Top Appl Earth Obs Remote Sens 5:1572–1584. https://doi.org/10.1109/jstars.2012.2200878

Hestir EL, Brando VE, Bresciani M et al (2015) Measuring freshwater aquatic ecosystems: the need for a hyperspectral global mapping satellite mission. Remote Sens Environ 167:181–195. https://doi.org/10.1016/j.rse.2015.05.023

Hestir EL, Schoellhamer DH, Greenberg J et al (2016) The effect of submerged aquatic vegetation expansion on a declining turbidity trend in the Sacramento-San Joaquin River Delta. Estuar Coasts 39:1100–1112. https://doi.org/10.1007/s12237-015-0055-z

Hostetler M, Allen W, Meurk C (2011) Conserving urban biodiversity? Creating green infrastructure is only the first step. Landsc Urban Plan 100:369–371. https://doi.org/10.1016/j.landurbplan.2011.01.011

Hoyos LE, Gavier-Pizarro GI, Kuemmerle T et al (2010) Invasion of glossy privet (*Ligustrum lucidum*) and native forest loss in the Sierras Chicas of Córdoba, Argentina. Biol Invasions 12:3261–3275. https://doi.org/10.1007/s10530-010-9720-0

Hu S, Niu Z, Chen Y et al (2017) Global wetlands: potential distribution, wetland loss, and status. Sci Total Environ 586:319–327. https://doi.org/10.1016/j.scitotenv.2017.02.001

Huang CY, Asner GP (2009) Applications of remote sensing to alien invasive plant studies. Sensors (Switzerland) 9:4869–4889. https://doi.org/10.3390/s90604869

Huang CY, Geiger EL (2008) Climate anomalies provide opportunities for large-scale mapping of non-native plant abundance in desert grasslands. Divers Distrib 14:875–884. https://doi.org/10.1111/j.1472-4642.2008.00500.x

Hudak AT, Wessman CA (1998) Textural analysis of historical aerial photography to characterize woody plant encroachment in South African Savanna. Remote Sens Environ 66:317–330. https://doi.org/10.1016/S0034-4257(98)00078-9

Hui C, Richardson DM, Visser V (2017) Ranking of invasive spread through urban green areas in the world's 100 most populous cities. Biol Invasions 19:3527–3539. https://doi.org/10.1007/s10530-017-1584-0

Hulme PE (2009) Trade, transport and trouble: managing invasive species pathways in an era of globalization. J Appl Ecol 46:10–18. https://doi.org/10.1111/j.1365-2664.2008.01600.x

Hulme PE (2015) Invasion pathways at a crossroad: policy and research challenges for managing alien species introductions. J Appl Ecol 52:1418–1424. https://doi. org/10.1111/1365-2664.12470

Hunt ER, Daughtry CST, Kim MS, Williams AEP (2007) Using canopy reflectance models and spectral angles to assess potential of remote sensing to detect invasive weeds. J Appl Remote Sens 1:013506. https://doi.org/10.1117/1.2536275

Hunter PD, Gilvear DJ, Tyler AN et al (2010) Mapping macrophytic vegetation in shallow lakes using the compact airborne spectrographic imager (CASI). Aquat Conserv Mar Freshw Ecosyst 20:717–727. https://doi.org/10.1002/aqc.1144

Ingwell LL, Bosque-Pérez NA (2015) The invasive weed *Ventenata dubia* is a host of *Barley yellow dwarf virus* with implications for an endangered grassland habitat. Weed Res 55:62–70. https://doi.org/10.1111/wre.12110

Ishii J, Washitani I (2013) Early detection of the invasive alien plant *Solidago altissima* in moist tall grassland using hyperspectral imagery. Int J Remote Sens 34:5926–5936. https://doi.org/1 0.1080/01431161.2013.799790

Jenkerson CB, Maiersperger T, Schmidt G (2010) eMODIS: a user-friendly data source. US Geological Survey. https://doi.org/10.3133/ofr20101055

Jollineau MY, Howarth PJ (2008) Mapping an inland wetland complex using hyperspectral imagery. Int J Remote Sens 29:3609–3631. https://doi.org/10.1080/01431160701469099

Joshi C, De Leeuw J, Van Andel J et al (2006) Indirect remote sensing of a cryptic forest understorey invasive species. For Ecol Manag 225:245–256. https://doi.org/10.1016/j.foreco.2006.01.013

Junk WJ, Bayley PB, Sparks RE (1989) The flood-pulse concept in river-floodplain systems. Proc Int Large River Symp Can Spec Publ Fish Aquat Sci 106:110–127

Keane RM, Crawley MJ (2002) Exotic plant invasions and the enemy release hypothesis. Trends Ecol Evol 17:164–170. https://doi.org/10.1016/S0169-5347(02)02499-0

Khanna S, Santos MJ, Ustin SL, Haverkamp PJ (2011) An integrated approach to a biophysiologically based classification of floating aquatic macrophytes. Int J Remote Sens 32:1067–1094. https://doi.org/10.1080/01431160903505328

Khanna S, Santos MJ, Boyer JD et al (2018) Water primrose invasion changes successional pathways in an estuarine ecosystem. Ecosphere 9(9). https://doi.org/10.1002/ecs2.2418

Kimothi MM, Anitha D, Vasistha HB et al (2010) Remote sensing to map the invasive weed, *Lantana camara* in forests. Trop Ecol 51:67–74

Koger CH, Shaw DR, Reddy KN, Bruce LM (2004a) Detection of pitted morningglory (*Ipomoea lacunosa*) with hyperspectral remote sensing. I. Effects of vegetation ground cover and reflectance properties. Weed Sci 52:230–235. https://doi.org/10.1614/ws-03-083r1

Koger CH, Shaw DR, Reddy KN, Bruce LM (2004b) Detection of pitted morningglory (*Ipomoea lacunosa*) with hyperspectral remote sensing. II. Effects of vegetation ground cover and reflectance properties. Weed Sci 52:230–235. https://doi.org/10.1614/ws-03-083r1

Kudela RM, Palacios SL, Austerberry DC et al (2015) Application of hyperspectral remote sensing to cyanobacterial blooms in inland waters. Remote Sens Environ 167:196–205. https://doi. org/10.1016/j.rse.2015.01.025

Laba M, Tsai F, Ogurcak D et al (2005) Field determination of optimal dates for the discrimination of invasive wetland plant species using derivative spectral analysis. Photogramm Eng Remote Sensing 71:603–611. https://doi.org/10.14358/pers.71.5.603

Laba M, Downs R, Smith S et al (2008) Mapping invasive wetland plants in the Hudson River National Estuarine Research Reserve using QuickBird satellite imagery. Remote Sens Environ 112:286–300. https://doi.org/10.1016/j.rse.2007.05.003

Landmann T, Piiroinen R, Makori DM et al (2015) Application of hyperspectral remote sensing for flower mapping in African savannas. Remote Sens Environ 166:50–60. https://doi. org/10.1016/j.rse.2015.06.006

Lass LW, Prather TS (2004) Detecting the locations of Brazilian pepper trees in the everglades with a hyperspectral sensor. Weed Technol 18:437–442. https://doi.org/10.1614/wt-03-174r

Lass LW, Thill DC, Shafii B, Prather TS (2002) Detecting spotted knapweed (*Centaurea maculosa*) with hyperspectral remote sensing technology. Weed Technol 16:426–432. https://doi. org/10.1614/0890-037x(2002)016[0426:dskcmw]2.0.co;2

Lass LW, Prather TS, Glenn NF et al (2005) A review of remote sensing of invasive weeds and example of the early detection of spotted knapweed (*Centaurea maculosa*) and babysbreath (*Gypsophila paniculata*) with a hyperspectral sensor. Weed Sci 53:242–251. https://doi.org/10.1614/ws-04-044r2

Latham J, Cumani R, Rosati I, Bloise M (2014) Global Land Cover SHARE (GLC-SHARE) database Beta-Release Version 1.0

Latombe G, Pyšek P, Jeschke JM et al (2017) A vision for global monitoring of biological invasions. Biol Conserv 213:295–308. https://doi.org/10.1016/j.biocon.2016.06.013

Lawrence RL, Wood SD, Sheley RL (2006) Mapping invasive plants using hyperspectral imagery and Breiman Cutler classifications (randomForest). Remote Sens Environ 100:356–362. https://doi.org/10.1016/j.rse.2005.10.014

Levine JM (2000) Species diversity and biological invasions: relating local process to community pattern. Science 288:852–854. https://doi.org/10.1126/science.288.5467.852

Liu L, Coops NC, Aven NW, Pang Y (2017) Mapping urban tree species using integrated airborne hyperspectral and LiDAR remote sensing data. Remote Sens Environ 200:170–182. https://doi.org/10.1016/j.rse.2017.08.010

Lowe S, Browne M, Boudjelas S, De Poorter M (2000) 100 of the world's worst invasive alien species: a selection from the global invasive species database. Invasive Species Specialist Group, Auckland

Lymburner L, Botha E, Hestir E et al (2016) Landsat 8: providing continuity and increased precision for measuring multi-decadal time series of total suspended matter. Remote Sens Environ 185:108–118. https://doi.org/10.1016/j.rse.2016.04.011

Maheu-Giroux M, De Blois S (2005) Mapping the invasive species *Phragmites australis* in linear wetland corridors. Aquat Bot 83:310–320. https://doi.org/10.1016/j.aquabot.2005.07.002

Malthus TJ, George DG (1997) Airborne remote sensing of macrophytes in Cefni Reservoir, Anglesey, UK. Aquat Bot 58:317–332. https://doi.org/10.1016/S0304-3770(97)00043-0

Malthus TJ, Hestir EL, Dekker AG, Brando VE (2012) The case for a global inland water quality product. In: 2012 IEEE international geoscience and remote sensing symposium (IGARSS), Munich, July 2012. pp 5234–5237. https://doi.org/10.1109/igarss.2012.6352429

Marshall TR, Lee PF (1994). Mapping aquatic macrophytes through digital image analysis of aerial photographs: an assessment. J Aquat Plant Manag, 32:61–66

Masters G, Norgrove L (2010) Climate change and invasive alien species. UK CABI Work Pap 1. https://www.cabi.org/Uploads/CABI/expertise/invasive-alien-species-working-paper.pdf. Accessed 3 Dec 2018

Matthews MW (2011) A current review of empirical procedures of remote sensing in inland and near-coastal transitional waters. Int J Remote Sens 32:6855–6899. https://doi.org/10.1080/01431161.2010.512947

Maurel N, Salmon S, Ponge JF et al (2010) Does the invasive species *Reynoutria japonica* have an impact on soil and flora in urban wastelands? Biol Invasions 12:1709–1719. https://doi.org/10.1007/s10530-009-9583-4

Mayer-Pinto M, Johnston EL, Bugnot AB et al (2017) Building 'blue': an eco-engineering framework for foreshore developments. J Environ Manag 189:109–114. https://doi.org/10.1016/j.jenvman.2016.12.039

McIntyre K, McLaren K, Prospere K (2018) Mapping shallow nearshore benthic features in a Caribbean marine-protected area: assessing the efficacy of using different data types (hydroacoustic versus satellite images) and classification techniques. Int J Remote Sens 39:1117–1150. https://doi.org/10.1080/01431161.2017.1395924

McLean P, Gallien L, Wilson JRU et al (2017) Small urban centres as launching sites for plant invasions in natural areas: insights from South Africa. Biol Invasions 19:3541–3555. https://doi.org/10.1007/s10530-017-1600-4

Menges RM, Nixon PR, Richardson AJ (1985) Light reflectance and remote sensing of weeds in agronomic and horticultural crops. Weed Sci 33:569–581. https://doi.org/10.2307/4044150

Mertes LAK, Smith MO, Adams JB (1993) Estimating suspended sediment concentrations in surface waters of the Amazon River wetlands from Landsat images. Remote Sens Environ 43:281–301. https://doi.org/10.1016/0034-4257(93)90071-5

Mitchell JJ, Glenn NF (2009) Leafy spurge (*Euphorbia esula*) classification performance using hyperspectral and multispectral sensors. Rangel Ecol Manag 62:16–27. https://doi.org/10.2111/08-100

Mitsch WJ, Gosselink JG (2007) Wetlands. Wiley, Hoboken

Mladinich CS, Bustos MR, Stitt S et al (2006) The use of Landsat 7 enhanced thematic mapper plus for mapping leafy spurge. Rangel Ecol Manag 59:500–506. https://doi.org/10.2111/06-027R1.1

Moisan TAH, Sathyendranath S, Bouman HA (2012) Ocean color remote sensing of phytoplankton functional types. In: Fatoyinbo L (ed) Remote sensing of biomass-principles and applications. InTech, Croatia. https://doi.org/10.5772/17174

Mooney HA, Cleland EE (2001) The evolutionary impact of invasive species. Proc Natl Acad Sci 98:5446–5451. https://doi.org/10.1073/pnas.091093398

Morel A, Bélanger S (2006) Improved detection of turbid waters from ocean color sensors information. Remote Sens Environ 102:237–249. https://doi.org/10.1016/j.rse.2006.01.022

Mortensen DA, Rauschert ESJ, Nord AN, Jones BP (2009) Forest roads facilitate the spread of invasive plants. Invasive Plant Sci Manag 2:191–199. https://doi.org/10.1614/ipsm-08-125.1

Mouw CB, Hardman-Mountford NJ, Alvain S et al (2017) A consumer's guide to satellite remote sensing of multiple phytoplankton groups in the global ocean. Front Mar Sci 4(41). https://doi.org/10.3389/fmars.2017.00041

Muller-Karger FE, Hestir E, Ade C et al (2018) Satellite sensor requirements for monitoring essential biodiversity variables of coastal ecosystems. Ecol Appl 28:749–760. https://doi.org/10.1002/eap.1682

Murphy BP, Andersen AN, Parr CL (2016) The underestimated biodiversity of tropical grassy biomes. Philos Trans R Soc B Biol Sci 371:20150319. https://doi.org/10.1098/rstb.2015.0319

Näsi R, Honkavaara E, Lyytikäinen-Saarenmaa P et al (2015) Using UAV-based photogrammetry and hyperspectral imaging for mapping bark beetle damage at tree-level. Remote Sens 7:15467–15493. https://doi.org/10.3390/rs71115467

National Invasive Species Council (2016) Management plan: 2016–2018. NISC, Washington, D.C.

Nelson SAC, Cheruvelil KS, Soranno PA (2006) Satellite remote sensing of freshwater macrophytes and the influence of water clarity. Aquat Bot 85:289–298. https://doi.org/10.1016/j.aquabot.2006.06.003

Odermatt D, Gitelson A, Brando VE, Schaepman M (2012) Review of constituent retrieval in optically deep and complex waters from satellite imagery. Remote Sens Environ 118:116–126. https://doi.org/10.1016/j.rse.2011.11.013

Olenin S, Gollasch S, Jonušas S, Rimkutė I (2000) En-route investigations of plankton in ballast water on a ship's voyage from the Baltic Sea to the open Atlantic coast of Europe. Int Rev Hydrobiol, 85:577–596. https://doi.org/10.1002/1522-2632(200011)85:5/6<577::aid-iroh577>3.0.co;2-c

Olenina I, Wasmund N, Hajdu S et al (2010) Assessing impacts of invasive phytoplankton: the Baltic Sea case. Mar Pollut Bull 60:1691–1700. https://doi.org/10.1016/j.marpolbul.2010.06.046

Ortiz SM, Breidenbach J, Kändler G (2013) Early detection of bark beetle green attack using TerraSAR-X and RapidEye data. Remote Sens 5:1912–1931. https://doi.org/10.3390/rs5041912

Paap T, Burgess TI, Wingfield MJ (2017) Urban trees: bridge-heads for forest pest invasions and sentinels for early detection. Biol Invasions 19:3515–3526. https://doi.org/10.1007/s10530-017-1595-x

Pearlstine L, Portier KM, Smith SE (2005) Textural discrimination of an invasive plant, *Schinus terebinthifolius*, from low altitude aerial digital imagery. Photogramm Eng Remote Sens 71:289–298. https://doi.org/10.14358/PERS.71.3.289

Penk MR, Jeschke JM, Minchin D, Donohue I (2016) Warming can enhance invasion success through asymmetries in energetic performance. J Anim Ecol 85:419–426. https://doi.org/10.1111/1365-2656.12480

Peterson EB (2005) Estimating cover of an invasive grass (*Bromus tectorum*) using tobit regression and phenology derived from two dates of Landsat ETM + data. Int J Remote Sens 26:2491–2507. https://doi.org/10.1080/01431160500127815

Phinn S, Roelfsema C, Dekker A et al (2008) Mapping seagrass species, cover and biomass in shallow waters: an assessment of satellite multi-spectral and airborne hyper-spectral imaging systems

in Moreton Bay (Australia). Remote Sens Environ 112:3413–3425. https://doi.org/10.1016/j.
rse.2007.09.017

Pimentel D, Zuniga R, Morrison D (2005) Update on the environmental and economic costs associated with alien-invasive species in the United States. Ecol Econ 52:273–288. https://doi.
org/10.1016/j.ecolecon.2004.10.002

Primack RB, Laube J, Gallinat AS, Menzel A (2015) From observations to experiments in phenology research: investigating climate change impacts on trees and shrubs using dormant twigs.
Ann Bot 116:889–897. https://doi.org/10.1093/aob/mcv032

Pu R, Gong P, Tian Y et al (2008) Using classification and NDVI differencing methods for monitoring sparse vegetation coverage: a case study of saltcedar in Nevada, USA. Int J Remote Sens
29:3987–4011. https://doi.org/10.1080/01431160801908095

Pyšek P, Jarošík V, Pergl J et al (2009) The global invasion success of Central European plants
is related to distribution characteristics in their native range and species traits. Divers Distrib
15:891–903. https://doi.org/10.1111/j.1472-4642.2009.00602.x

Radomski P, Holbrook BV (2015) A comparison of two hydroacoustic methods for estimating
submerged macrophyte distribution and abundance: a cautionary note. J Aquat Plant Manag
53:151–159

Ramsey III EW, Nelson GA, Sapkota SK et al (2002) Mapping Chinese tallow with color-infrared
photography. Photogramm Eng Remote Sens 68:251–255

Ramsey III E, Rangoonwala A, Nelson G, Ehrlich R (2005) Mapping the invasive species, Chinese
tallow, with EO1 satellite Hyperion hyperspectral image data and relating tallow occurrences
to a classified Landsat Thematic Mapper land cover map. Int J Remote Sens 26:1637–1657.
https://doi.org/10.1080/01431160512331326701

Resasco J, Hale AN, Henry MC, Gorchov DL (2007) Detecting an invasive shrub in a deciduous
forest understory using late-fall Landsat sensor imagery. Int J Remote Sens 28:3739–3745.
https://doi.org/10.1080/01431160701373721

Ricciardi A (2007). Are modern biological invasions an unprecedented form of global change?.
Conserv Biol 21(2):329–336. https://doi.org/10.1111/j.1523-1739.2006.00615.x

Ricciardi A, Palmer ME, Yan ND (2011) Should biological invasions be managed as natural disas-
ters? Bioscience 61:312–317. https://doi.org/10.1525/bio.2011.61.4.11

Rocchini D, Andreo V, Förster M et al (2015) Potential of remote sensing to predict species invasions: a modelling perspective. Prog Phys Geogr 39:283–309. https://doi.
org/10.1177/0309133315574659

Rouse JW, Benton AR, Toler RW, Haas RH (1975) Three examples of applied remote sensing of vegetation. In: NASA earth resources survey symposium, vol 1-C. NASA, Houston,
pp 1797–1810

Rupprecht CDD, Byrne JA, Garden JG, Hero J-M (2015) Informal urban green space: a trilingual
systematic review of its role for biodiversity and trends in the literature. Urban For Urban
Green 14:883–908. https://doi.org/10.1016/j.ufug.2015.08.009

Samiappan S, Turnage G, Hathcock L et al (2017) Using unmanned aerial vehicles for high-
resolution remote sensing to map invasive Phragmites australis in coastal wetlands. Int J
Remote Sens 38:2199–2217. https://doi.org/10.1080/01431161.2016.1239288

Santos MJ, Whitham TG (2010) Predictors of Ips confusus outbreaks during a record drought in
Southwestern USA: implications for monitoring and management. Environ Manag 45:239–
249. https://doi.org/10.1007/s00267-009-9413-6

Santos MJ, Khanna S, Hestir EL et al (2009) Use of hyperspectral remote sensing to evaluate efficacy of aquatic plant management. Invasive Plant Sci Manag 2:216–229. https://doi.
org/10.1614/IPSM-08-115.1

Santos MJ, Hestir EL, Khanna S, Ustin SL (2012) Image spectroscopy and stable isotopes elucidate functional dissimilarity between native and nonnative plant species in the aquatic environment. New Phytol 193:683–695. https://doi.org/10.1111/j.1469-8137.2011.03955.x

Santos MJ, Khanna S, Hestir EL et al (2016) Measuring landscape-scale spread and persistence
of an invaded submerged plant community from airborne remote sensing. Ecol Appl 26:1733–
1744. https://doi.org/10.1890/15-0615

Sathyendranath S, Aiken J, Alvain S (2014) Phytoplankton functional types from space. In: Reports and monographs of the international ocean-colour coordinating group. International Ocean-Colour Coordinating Group, p 163

Saul WC, Roy HE, Booy O et al (2017) Assessing patterns in introduction pathways of alien species by linking major invasion data bases. J Appl Ecol 54:657–669. https://doi.org/10.1111/1365-2664.12819

Scheffer M, Szabo S, Gragnani A et al (2003) Floating plant dominance as a stable state. Proc Natl Acad Sci U S A 100:4040–4045. https://doi.org/10.1073/pnas.0737918100

Seebens H, Blackburn TM, Dyer EE et al (2017) No saturation in the accumulation of alien species worldwide. Nat Commun 8:14435. https://doi.org/10.1038/ncomms14435

Serrano L, Peñuelas J, Ustin SL (2002) Remote sensing of nitrogen and lignin in Mediterranean vegetation from AVIRIS data: decomposing biochemical from structural signals. Remote Sens Environ 81:355–364. https://doi.org/10.1016/S0034-4257(02)00011-1

Shafii B, Price WJ, Prather TS et al (2004) Using landscape characteristics as prior information for Bayesian classification of yellow starthistle. Weed Sci 52:948–953. https://doi.org/10.1614/WS-04-042R1

Shouse M, Liang L, Fei S (2012) Identification of understory invasive exotic plants with remote sensing: in urban forests. Int J Appl Earth Obs Geoinf 21:525–534. https://doi.org/10.1016/j.jag.2012.07.010

Simberloff D (2000) Global climate change and introduced species in United States forests. Sci Total Environ 262:253–261. https://doi.org/10.1016/S0048-9697(00)00527-1

Sims DA, Gamon JA (2002) Relationships between leaf pigment content and spectral reflectance across a wide range of species, leaf structures and developmental stages. Remote Sens Environ 81:337–354. https://doi.org/10.1016/S0034-4257(02)00010-X

Singh N, Glenn NF (2009) Multitemporal spectral analysis for cheatgrass (*Bromus tectorum*) classification. Int J Remote Sens 30:3441–3462. https://doi.org/10.1080/01431160802562222

Singh KK, Davis AJ, Meentemeyer RK (2015) Detecting understory plant invasion in urban forests using LiDAR. Int J Appl Earth Obs Geoinf 38:267–279. https://doi.org/10.1016/j.jag.2015.01.012

Soltani N, Dille JA, Burke IC et al (2016) Potential corn yield losses from weeds in North America. Weed Technol 30:979–984. https://doi.org/10.1614/WT-D-16-00046.1

Soltani N, Dille JA, Burke IC et al (2017) Perspectives on potential soybean yield losses from weeds in North America. Weed Technol 31:148–154. https://doi.org/10.1017/wet.2016.2

Subba Rao DV, Sprules WG, Locke A, Carlton JT (1994) Exotic phytoplankton species from ships' ballast waters: risk of potential spread to mariculture sites on Canada's east coast. Can data Rep Fish Aquat Sci 937:1–51

Tabacchi E, Correll DL, Hauer R et al (1998) Development, maintenance and role of riparian vegetation in the river landscape. Freshw Biol 40:497–516. https://doi.org/10.1046/j.1365-2427.1998.00381.x

Theoharides KA, Dukes JS (2007) Plant invasion across space and time: factors affecting nonindigenous species success during four stages of invasion. New Phytol 176:256–273. https://doi.org/10.1111/j.1469-8137.2007.02207.x

Thornton PE, Thornton MM, Mayer BW et al (2018) Daymet: daily surface weather data on a 1-km grid for North America, Version 3. ORNL DAAC, Oak Ridge. https://doi.org/10.3334/ORNLDAAC/1328

Tockner K, Malard F, Ward JV (2000) An extension of the food pulse concept. Hydrol Process 2883:2861–2883

Turbelin AJ, Malamud BD, Francis RA (2017). Mapping the global state of invasive alien species: patterns of invasion and policy responses. Global Ecol Biogeogr 26(1):78–92. https://doi.org/10.1111/geb.12517

UNESCO (2018). UNESCO'S commitment to biodiversity. In M. Bouamrane (ed.). Paris: United Nations educational, scientific and cultural organization, https://www.unesdoc.unesco.org/images/0026/002652/265200e.pdf

Ustin SL, DiPietro D, Olmstead K et al (2002) Hyperspectral remote sensing for invasive species detection and mapping. IEEE Int Geosci Remote Sens Symp 3:1658–1660. https://doi.org/10.1109/IGARSS.2002.1026212

van Kleunen M, Dawson W, Essl F et al (2015) Global exchange and accumulation of non-native plants. Nature 525:100–103. https://doi.org/10.1038/nature14910

Venugopal G (2002) Monitoring the effects of biological control of water hyacinths using remotely sensed data: a case study of Bangalore, India. Singap J Trop Geogr 19:91–105. https://doi.org/10.1111/1467-9493.00027

Verma R, Singh SP, Ganesha Raj K (2003) Assessment of changes in water-hyacinth coverage of water bodies in northern part of Bangalore city using temporal remote sensing data. Curr Sci 84:795–804

Villa P, Mousivand A, Bresciani M (2014) Aquatic vegetation indices assessment through radiative transfer modeling and linear mixture simulation. Int J Appl Earth Obs Geoinf 30:113–127. https://doi.org/10.1016/j.jag.2014.01.017

Villa P, Bresciani M, Bolpagni R et al (2015) A rule-based approach for mapping macrophyte communities using multi-temporal aquatic vegetation indices. Remote Sens Environ 171:218–233. https://doi.org/10.1016/j.rse.2015.10.020

Villa P, Pinardi M, Tóth VR et al (2017) Remote sensing of macrophyte morphological traits: implications for the management of shallow lakes. J Limnol 76:109–126. https://doi.org/10.4081/jlimnol.2017.1629

Vis C, Hudon C, Carignan R (2003) An evaluation of approaches used to determine the distribution and biomass of emergent and submerged aquatic macrophytes over large spatial scales. Aquat Bot 77:187–201. https://doi.org/10.1016/S0304-3770(03)00105-0

Vörösmarty C, McIntyre P, Gessner M et al (2010) Global threats to human water security and river biodiversity. Nat Commun 467:555–561. https://doi.org/10.1038/nature09440

Walther GR, Roques A, Hulme PE et al (2009) Alien species in a warmer world: risks and opportunities. Trends Ecol Evol 24:686–693. https://doi.org/10.1016/j.tree.2009.06.008

Ward JV, Tockner K, Arscott DB, Claret C (2002) Riverine landscape diversity. Freshw Biol 47:517–539. https://doi.org/10.1046/j.1365-2427.2002.00893.x

Waycott M, Duarte CM, Carruthers TJB et al (2009) Accelerating loss of seagrasses across the globe threatens coastal ecosystems. Proc Natl Acad Sci 106:12377–12381. https://doi.org/10.1073/pnas.0905620106

Weisberg PJ, Dilts TE, Baughman OW et al (2017) Development of remote sensing indicators for mapping episodic die-off of an invasive annual grass (Bromus tectorum) from the Landsat archive. Ecol Indic 79:173–181. https://doi.org/10.1016/j.ecolind.2017.04.024

West AM, Evangelista PH, Jarnevich CS et al (2017) Using multi-date satellite imagery to monitor invasive grass species distribution in post-wildfire landscapes: an iterative, adaptable approach that employs open-source data and software. Int J Appl Earth Obs Geoinf 59:135–146. https://doi.org/10.1016/j.jag.2017.03.009

Wilfong BN, Gorchov DL, Henry MC (2009) Detecting an invasive shrub in deciduous forest understories using remote sensing. Weed Sci 57:512–520. https://doi.org/10.1614/WS-09-012.1

Williams SL, Grosholz ED (2008) The invasive species challenge in estuarine and coastal environments: marrying management and science. Estuar Coasts 31:3–20. https://doi.org/10.1007/s12237-007-9031-6

Williams AP, Hunt ER Jr (2002) Estimation of leafy spurge cover from hyperspectral imagery using mixture tuned matched filtering. Remote Sens Environ 82:446–456. https://doi.org/10.1016/S0034-4257(02)00061-5

Winfield IJ, Onoufriou C, O'Connell MJ et al (2007) Assessment in two shallow lakes of a hydroacoustic system for surveying aquatic macrophytes. In: Gulati RD, Lammens E, De Pauw N, Van Donk E (eds) Hydrobiologia. Springer Netherlands, Dordrecht, pp 111–119

Wolf AA, Zavaleta ES, Selmants PC (2017) Flowering phenology shifts in response to biodiversity loss. Proc Natl Acad Sci 114:3463–3468. https://doi.org/10.1073/pnas.1608357114

Wu Y, Rutchey K, Wang N, Godin J (2006) The spatial pattern and dispersion of Lygodium microphyllum in the Everglades wetland ecosystem. Biol Invasions 8:1483–1493. https://doi.org/10.1007/s10530-005-5840-3

Yang C (2007) Evaluating airborne hyperspectral imagery for mapping waterhyacinth infestations. J Appl Remote Sens 1:013546. https://doi.org/10.1117/1.2821827

Yang C, Everitt JH (2010) Comparison of hyperspectral imagery with aerial photography and multispectral imagery for mapping broom snakeweed. Int J Remote Sens 31:5423–5438. https://doi.org/10.1080/01431160903369626

Yuan L, Zhang LQ (2008) Mapping large-scale distribution of submerged aquatic vegetation coverage using remote sensing. Ecol Inform 3:245–251. https://doi.org/10.1016/j.ecoinf.2008.01.004

Zhao D, Jiang H, Yang T et al (2012) Remote sensing of aquatic vegetation distribution in Taihu Lake using an improved classification tree with modified thresholds. J Environ Manag 95:98–107. https://doi.org/10.1016/j.jenvman.2011.10.007

Open Access This chapter is licensed under the terms of the Creative Commons Attribution 4.0 International License (http://creativecommons.org/licenses/by/4.0/), which permits use, sharing, adaptation, distribution and reproduction in any medium or format, as long as you give appropriate credit to the original author(s) and the source, provide a link to the Creative Commons license and indicate if changes were made.

The images or other third party material in this chapter are included in the chapter's Creative Commons license, unless indicated otherwise in a credit line to the material. If material is not included in the chapter's Creative Commons license and your intended use is not permitted by statutory regulation or exceeds the permitted use, you will need to obtain permission directly from the copyright holder.

Chapter 13
A Range of Earth Observation Techniques for Assessing Plant Diversity

Angela Lausch, Marco Heurich, Paul Magdon, Duccio Rocchini, Karsten Schulz, Jan Bumberger, and Doug J. King

13.1 Understanding Plant Diversity with Remote Sensing

Stress, disturbance, and resource limitations such as anthropogenic changes to ecosystems all lead to changes in biodiversity and vegetation diversity (Cardinale et al. 2012) on different scales of biological organization as well as disturbances in the interactions between trophic levels and ecosystem functions, impairing ecosystem services such as pollination or soil fertility (Cord et al. 2017). Vegetation diversity is multidimensional, multifactorial, and tremendously complex in time and space (Lausch et al. 2018a). This level of complexity can only be fully understood when

---

The original version of this chapter was revised. The correction to this chapter is available at https://doi.org/10.1007/978-3-030-33157-3_21

A. Lausch (✉)
Department Computational Landscape Ecology, Helmholtz Centre for Environmental Research–UFZ, Leipzig, Germany

Geography Department, Humboldt University Berlin, Berlin, Germany
e-mail: angela.lausch@ufz.de

M. Heurich
Bavarian Forest National Park, Department of Conservation and Research, Grafenau, Germany

Chair of Wildlife Ecology and Wildlife Management, University of Freiburg, Freiburg, Germany
e-mail: marco.heurich@npv-bw.bayern.de

P. Magdon
Chair of Forest Inventory and Remote Sensing, Georg-August-University Göttingen, Göttingen, Germany
e-mail: pmagdon@gwdg.de

D. Rocchini
Department of Biological, Geological, and Environmental Sciences, Alma Mater Studiorum University of Bologna, Italy
e-mail: duccio.rocchini@unibo.it

© The Author(s) 2020, Corrected Publication 2020
J. Cavender-Bares et al. (eds.), *Remote Sensing of Plant Biodiversity*,
https://doi.org/10.1007/978-3-030-33157-3_13

monitoring approaches are applied to record different characteristics of vegetation (i.e., phylo-diversity; taxonomic, structural, functional, and trait diversity on different levels of biotic organization—molecular, genetic, individual, species, population, community, biome, ecosystem, and landscape). Different processes and drivers influence the resilience of vegetation diversity (Fig. 13.1).

To record the status, stress, disturbances, and resource limitations in vegetation diversity, we have to differentiate between two monitoring approaches: (i) in-situ approaches, whereby the most important monitoring concepts are the phylogenetic species concept (PSC, Eldredge and Cracraft 1980), the biological species concept (BSC, Mayr 1942) and the morphological species concept (MSC, Mayr 1969) and (ii) physically based approaches of remote sensing (RS) (Lausch et al. 2018b). Unlike in-situ approaches, RS records the biochemical, biophysical, physiognomic, morphological, structural, phenological, and functional characteristics of vegetation diversity at all scales, from the molecular and individual plant levels to communities and the entire ecosystem, based on the principles of image spectroscopy across the electromagnetic spectrum from the visible to the microwave (Ustin and Gamon 2010). When compared with the traits approach of the MSC used by taxonomists, RS approaches are not able to record the same number and characteristics of traits or trait variations as the in-situ approaches (Homolová et al. 2013; Lausch et al. 2016a).

Traits and trait variations that can be recorded using RS techniques are hereafter referred to as *spectral traits* (ST), and the changes to their spectral characteristics are referred to as *spectral trait variations* (STV). The overall approach is referred to as the *remote sensing-spectral trait/spectral trait variations* (RS-ST/STV) concept for monitoring biodiversity (Lausch et al. 2016b) as well as geodiversity (Lausch et al. 2019) (Fig. 13.7).

Traits bridge the gap between in-situ and RS monitoring approaches. Species traits have allowed us to take a completely new direction and to gain a better understanding of fundamental questions of status, stress, disturbances, resource limitations, and resilience in biodiversity—i.e., "why organisms live where they do and how they will respond to environmental change" (Green et al. 2008). Therefore, ecologists are increasingly focusing on traits rather than species to better understand

K. Schulz
University of Natural Resources and Life Sciences (BOKU), Institute of Hydrology and Water Management, Vienna, Austria
e-mail: karsten.schulz@boku.ac.at

J. Bumberger
Department of Monitoring and Explorations Technologies, Helmholtz Centre for Environmental Research – UFZ, Leipzig, Germany
e-mail: jan.bumberger@ufz.de

D. J. King
Departement of Geography and Environmental Studies, Geomatics and Landscape Ecology Lab, Carleton University, Ottawa, ON, Canada
e-mail: doug.king@carleton.ca

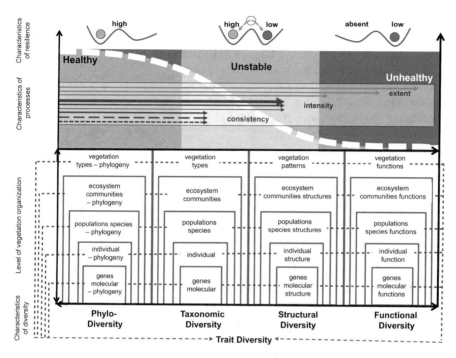

Fig. 13.1 Schematic diagram of the different levels of vegetation organization from genes up to vegetation types showing characteristics of phylo-diversity, taxonomic diversity, structural diversity, functional diversity, and trait diversity. This also shows how the different characteristics of the processes (the extent, process intensity, process consistency, resilience, and their characteristics) all influence the resilience and health of vegetation. (From Lausch et al. 2018a)

the status, changes, health, and resilience of ecosystems (Cernansky 2017) and the internal patterns and heterogeneity of communities and landscapes (Lausch et al. 2015a). To understand the complexity of ecosystems, no one monitoring approach, no single model, scale, or RS platform on its own is sufficient to discern the effects of processes and different drivers of vegetation diversity (Lausch et al. 2018a, b).

This chapter introduces the different ranges of EO techniques for assessing vegetation diversity. The focus here is to give an overview of existing close-range RS platforms as well as air- and spaceborne RS platforms for assessing plant diversity.

13.2 Range of EO Platforms to Assess Plant Diversity

RS sensors are mounted on different platforms such as camera traps or handheld, or they may have fixed supports (e.g., a tripod) or towers for field-based spectral measurements. Drones, aircraft (airborne RS), or satellites (spaceborne RS) are also used in field-based studies, depending on the spatial scale of the study (Gamon et al., Chap. 16). In laboratories, cameras and sensors may be mounted in plant phenomics facilities or ecotrons (Fig. 13.2; Lausch et al. 2017). The characteristics of all RS approaches are the same, irrespective of the platform. Vegetation stress, disturbance, and diversity result in variations in spectral radiance or reflectance that are recorded using RS in a nondestructive manner. The RS sensor on the platform records the spectral radiance at a distance of just a few millimeters up to thousands of kilometers to the object of interest.

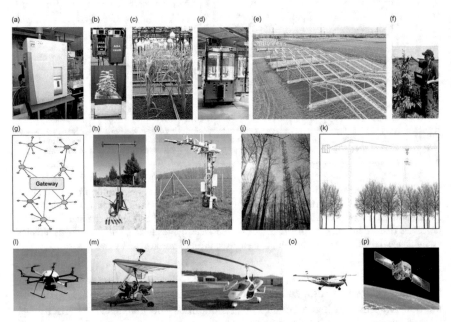

Fig. 13.2 Overview of different close-range, air-, and spaceborne RS platforms for assessing plant and vegetation diversity and vegetation health. (**a**) Laboratory spectrometer; (**b**) ash trees monitored in a close-range RS spectral laboratory (manual) with imaging hyperspectral sensors AISA-EAGLE/HAWK (Modified after Brosinsky et al. 2013); (**c**) automated plant phenomics facilities; (**d**) ecotrons (Modified after Türke et al. 2017); (**e**) Global Change Experimental Facility (GCEF)/Helmholtz-Zentrum für Umweltforschung (UFZ), Germany as platforms with different RS sensors (photo: A. Künzelmann/UFZ); (**f**) manual measuring with field spectrometer; (**g**) WSNs; (**h**) one sensor node of the WSN (Graphic, photo **g**, **h** by J. Bumberger and H. Mollenhauer/UFZ); (**i**) flux tower with different RS instruments, test area grassland/UFZ; (**j**) flux tower with different RS instruments, test area Hohe Holz/UFZ (Photo **i**, **j** by C. Rebmann/UFZ); (**k**) mobile crane with RS sensors; (**l**) unmanned aerial systems (UAS)—drone with different RS sensors; (**m**) microlight of the UFZ with different RS sensors like the AISA-EAGLE (hyperspectral 400–970 nm); (**n**) gyrocopter of the Institute for Geoinformation and Surveying, Dessau, Germany, with different RS sensors (Photo by L. Bannehr); (**o**) Cessna; (**p**) Spaceborne RS platforms. (Modified after Lausch et al. 2017)

13.2.1   Close-Range EO Approaches

13.2.1.1   Spectral Laboratory

The reactions of plants to stress phenomena depend on the plant species (Müller 2009; Teodoro et al. 2016). Teodoro et al. (2016) analyzed the different strategies of Brazilian tree species like *Campomanesia pubescens* (Myrtaceae), *Eremanthus seidellii*, and *Lessingianthus warmingianus* (Asteraceae) to cope with drought stress. The results showed different reactions and trade-offs to maintain plant functioning under drought stress conditions. Moreover, the ability of different tree species to adapt to climate change is still not well understood (Beck and Müller 2007). Reactions of woody plants to stress factors such as drought can often only be observed years later in the form of biochemical, physiological, or geometric changes to woody plant traits (Buddenbaum et al. 2015b). Therefore, specific in-situ investigations need to be conducted on the stress reactions of different taxonomic plants to determine the spectral responses to different drivers.

With the help of close-range laboratory spectroscopy (see Fig. 13.2a, b), extensive long-term stress monitoring can be carried out that takes into account entire vegetation periods as well as investigations over several years. Scenarios specifically targeted at investigating different stress factors such as stress from drought, ozone levels, fungal infestations, pesticide deposits, or temperature increases or decreases are conducted under comparable settings and environmental conditions, enabling good inputs for models and eliminating confounding factors. In addition to imaging and nonimaging spectrometer measurements, a broad range of parameters for vegetation traits, soil, and climate can be measured with in-situ approaches. Brosinsky et al. (2013) investigated the spectral response from the impacts of flooding on the physiological stress reactions of ash trees *Fraxinus excelsior L.* over a 3-month period, whereas Buddenbaum et al. (2015b) modeled the photosynthesis rate of young European beech trees under drought stress using hyperspectral visible infrared and hyperspectral thermal sensors. They created high spatial resolution (cm) maps of photosynthetic activity using the photochemical reflectance index (PRI), fluorescence, and temperature. Other approaches have derived the different phenology indicators of barley with imaging hyperspectral RS over its entire development period (Lausch et al. 2015b).

13.2.1.2   Plant Phenomics Facilities

One of the most important challenges in plant biology and vegetation stress physiology is the qualitative, quantitative, and spectroscopic recording of plant species phenotypes to gain a better understanding of interactions between the genotype and the phenotype. The genotype of a plant species comprises its genetic information, while the phenotype represents the physiological, morphological, anatomical, and development characteristics as well as interactions with the environment, resource limitations, and stress factors (see Cavender-Bares et al., Chap. 2). The interaction of the genotype with its environment give rise to the functional and structural traits of plants and their specific phenotype (Großkinsky et al. 2015a, b; Pieruschka and Lawson 2015). Insights into the role of genotype and phenotype in plant stress physiology can be gained not only from recording individual plant spectral trait-stress factor interactions but also by including the entire genotype-epigenetic-phenotype-environment matrix (Mittler and Blumwald 2010). This can be achieved by recording phenotypical plant traits in plant phenomics facilities (Furbank 2009; Großkinsky et al. 2015a, b) (see Fig. 13.2c). Due to the high number of plant species, plant phenomics facilities have been established all over the world that collaborate as part of the International Plant Phenotyping Network (IPPN, http://www.plant-phenotyping.org/), where in an automated and often robotic manner, noninvasive measurement methods such as RS techniques are implemented, enabling a holistic and quantitative recording of the phenotype of a plant over its entire development period at a reasonable cost (Ehrhardt and Frommer 2012; Fiorani and Schurr 2013).

Plant phenomics facilities thus include comparable analyses of genotype-phenotype interactions under experimental as well as natural growth conditions. The goal of plant phenomics facilities is to implement and develop innovative noninvasive measurement methods and RS techniques such as stereo hyperspectral, RGB, thermal, and fluorescence cameras, laser scanning instruments, or x-ray tomographs (Fiorani and Schurr 2013). Data from such facilities are then saved in databases (Krajewski et al. 2015) to make such information available for future research with airborne and spaceborne RS applications.

With plant phenomics facilities, crucial investigations have been carried out on the effects of different plant stresses on photosynthetic performance (Jansen et al. 2009; Konishi et al. 2009; Rascher 2007). This research on chlorophyll fluorescence and its acquisition using spectroscopic techniques forms the basis for developing the Fluorescence Explorer (FLEX) sensors (Kraft et al. 2012; Rascher 2007; Rascher et al. 2015). On the basis of its very high spectral resolution of 0.3–3.0 µm, FLEX will be the first satellite that is able to directly measure the solar-induced chlorophyll fluorescence and thus the stress levels in plants and other types of vegetation using RS.

13.2.1.3 Ecotrons

Ecotrons are controlled environmental facilities (see Fig. 13.2d) for the investigation of plant and animal populations and ecosystem processes under near-natural conditions using noninvasive methods (Lawton et al. 1993; Türke et al. 2017). They differ from greenhouse experiments because not only plant populations, but interactions between plant and animal populations, can be investigated. Furthermore, ecotrons enable investigations of aboveground and belowground interactions, which drive the relationship between plant diversity and ecosystem function (Eisenhauer 2018).

Ecological processes and material flows can be measured by ecotrons with non-invasive methods, while at the same time, the environmental conditions are controlled and regulated. Ecosystems that are investigated by ecotrons are thus closed systems. There is no undesired input or outflow of water, nutrients, resources, organisms, or gases, or input from undesired disturbance variables or stress factors. All changes taking place in such ecosystem processes are documented and can be compared with one another and with different scenarios in a standardized manner. In ecotrons, biodiversity is manipulated at different trophic levels at the same time. In this manner, the responses of different species and genotypes and species to stress, disturbances, or resource limitations and their effects on ecosystem functions can be examined. This approach enables a much better recording and understanding of aboveground and belowground interactions between different plant and animal species, microorganisms, and abiotic factors, as well as material and energy flows. The integration of close-range RS sensors in ecotrons is still very new and in need of further development if we are to understand the complete system of soil-vegetation-climate-biotic interactions with spectral response.

13.2.1.4 WSNs, Sensorboxes

WSNs can be used to record complex vegetation processes both extensively and continually in a noninvasive, cost-effective, and automated manner (Hart and Martinez 2006).

The implementation of wireless mobile and stationary sensor networks in terrestrial environmental systems (Fig. 13.2g, h) enables high-frequency in-situ information to be recorded using various sensor types (e.g., thermal, multispectral, hyperspectral, soil moisture, air condition). Another advantage of mobile wireless ad hoc sensor networks is their self-organizing infrastructure, leading to significant reduction of cost and time consumption for installation, maintenance, and operation.

WSNs are being implemented more frequently in environmental and vegetation monitoring (Hwang et al. 2010; Mollenhauer et al. 2016) in agriculture and the food industry (Mafuta et al. 2013; Ruiz-Garcia et al. 2009), for monitoring terrestrial and underground conditions such as soils, and for aquatic applications (Yick et al. 2008). They have also been used for experimental platforms such as greenhouses or the GCEF (Mollenhauer et al. 2016). In the context of vegetation health, WSNs are implemented to detect and verify forest fires in real time (Liyang Yu et al. 2005; Lloret et al. 2009) or to demonstrate the effects of the 2015 El Niño extreme drought on the sap flow of trees in eastern Amazonia (Mauro et al. 2016).

WSNs have also been used to record how important processes of soil-plant-atmosphere interactions; vegetation processes such as transpiration, carbon uptake and storage, and water stripping from clouds are affected by climatic variation and the temporal and spatial structure of the vegetation interior in whole ecosystems (Oliveira et al. 2016). Teodoro et al. (2016) used WSN to demonstrate the interplay between hydraulic traits, growth performance, and stomata regulation capacity in three shrub species in a tropical montane scrubland of Brazil under contrasting water availability. The results showed that these plant species employ different strategies in the regulation of hydraulic and stomatal conductivity during drought stress and thus substantiate the need for setting up WSN for different plant species and communities (Teodoro et al. 2016).

Grassland ecology experiments in remote locations requiring quantitative analysis of biomass, which is a key ecosystem variable, are becoming increasingly widespread but are still limited by manual sampling methodologies. To provide a cost-effective automated solution for biomass determination, several photogrammetric techniques have been examined to generate 3-D point cloud representations of plots, which are used to estimate aboveground biomass. Methods investigated include structure from motion (SfM) techniques (Kröhnert et al. 2018; see Fig. 13.3).

13.2.1.5   Towers

Flux towers involve an integrated sampling approach (see Fig. 13.2i, j, k) that supports the acquisition of different ecosystem parameters such as carbon dioxide, water vapor, and energy fluxes as they cycle through the atmosphere, as well as vegetation and soil parameters. FLUX towers are often coupled with sensor technologies such as airborne RS or soil sensors. Towers acquire individual point and local area information and are of particular importance in terms of long-term in-situ measurement for the calibration and validation of air- and spaceborne RS data. By linking flux towers to an international network (FLUXNET, Baldocchi et al. 2001), greater understanding of ecological processes and changes to vegetation health has been achieved using RS (Chen 2016; Yang et al. 2016). Towers and mobile in-situ stations are often combined as global sensor networks. Furthermore, the physiological reactions of plant species and communities depend on the taxonomy and phylogeny of plant species characteristics and numerous abiotic ecosystem variables as well as the intensity of land use (Garnier et al. 2007). Simple drones are also avail-

Fig. 13.3 Generated 3-D representations of *Onobrychis viciifolia* and *Daucus carota* using structure from motion (SfM) techniques as well as the use of a time-of-flight (TOF) 3-D camera, a laser light sheet triangulation system, and a coded light projection system. (From Kröhnert et al. 2018)

able, enabling a mapping of the distribution of plant species in mixed grassland communities using close-range imaging spectroscopy (Lopatin et al. 2017). The special value of both fixed and mobile towers or simple close-range RS platforms is that vegetation diversity can be monitored more frequently and with a higher spatial resolution. Table 13.1 lists the advantages and disadvantages of close-range EO approaches to monitor and assess vegetation diversity.

13.2.2 Air- and Spaceborne RS Platforms and Sensors

13.2.2.1 Unmanned Aerial Systems (UAS)

In recent years, UAS has become an important RS technology in spatial ecology. Nowadays a plethora of platforms, including fixed-wing and rotor-based systems, can carry multispectral, hyperspectral, thermal, LiDAR, and radar sensors and can navigate autonomously on predefined routes using global navigation satellite system (GNSS). With the increased availability and simplicity, such platforms are being used more and more in ecological research and monitoring (Anderson and Gaston 2013). In this context two essential characteristics of UAS are relevant:

(i) *High flexibility and low cost of operation:* UASs offer high flexibility in terms of payloads, flight time, and flight specifications such as altitude, time of day, and weather condition. When compared with manned aircraft or satellites, it is much easier to plan and conduct an image acquisition campaign once a UAS and a trained pilot are available. Due to low fixed costs, UAS can be cheaper than manned planes and helicopters.

(ii) *High spatial and temporal resolutions:* Within the technical and legal limitations, flight heights of UAS can be freely set and typically range from a couple of meters to hundreds of meters. Depending on the sensor system, images with very high spatial resolution (<5 cm) can be acquired when flown at low altitudes. The high flexibility of operation and the low image acquisition costs enable users to efficiently create multitemporal image series.

In the context of biodiversity monitoring, UASs are used in vegetated ecosystems to obtain optical images with high spatial and spectral resolution and 3-D point clouds of the Earth's surface and vegetation structures.

In grassland ecosystems, high-resolution UAS images are used to map habitat types (Cruzan et al. 2016) or single target species such as weeds (Hardin and Jackson 2005). In recent studies, proximal RS using scaffolds has been used to link species and functional diversity to spectral traits (Schweiger et al. 2018). Here the high spatial resolution is of utmost importance because grassland plants are typically small and highly mixed (Lu et al. 2016). Very high spatial resolution imagery offers the potential for both community- and plant-based analysis (Lopatin et al. 2017). However, even with spatial resolutions <1 cm, species identification of individuals is challenging and might only work under favorable conditions such as low structural

Table 13.1 Close-range Remote Sensing (RS) approaches and their advantages and disadvantages for monitoring and assessing plant diversity

Close-range RS approaches	Advantages	Disadvantages
Spectral analyses of plant species (Asner et al. 2015; Asner and Martin 2009)	I. Basis for conducting research on the spectral characteristics of specific biochemical, biophysical, and morphological traits in various organs of plants, including leaves and flowers II. Storage in spectral databases for validation and calibration III. Basis for the spectral fingerprints (SFP) of the vegetation IV. Basis for conducting research on taxonomic, phylogenetic, genetic, epigenetic, or morphological-functional features	Analysis on molecular level Geometric, structural, distribution, population, and community effects are not measurable
Spectral laboratory (manual operation) (Brosinsky et al. 2013; Buddenbaum et al. 2015a, b; Buddenbaum and Hill 2015; Doktor et al. 2014; Lausch et al. 2013) *Plant phenomics facilities and ecotrons* (fully automatic operation) (Ehrhardt and Frommer 2012; Fiorani and Schurr 2013; Furbank 2009; Großkinsky et al. 2015a, b; Li et al. 2014; Pieruschka and Lawson 2015; Virlet et al. 2015)	1. Long-term monitoring is possible (entire vegetation period, over several years, specific investigations of impact phases of stressors on plant plants) 2. Experimental stress analyses are possible (drought stress, heavy metals, tropospheric ozone, flooding, flood stress, nitrogen loads, etc.) 3. Extensive measurement program is possible for biotic, abiotic, and climate conditions within the spectral laboratory 4. Storage in spectral databases for validation and calibration 5. Comparative analyses can be conducted under natural or artificial conditions to investigate the influence of artificial light, geometry effects, or additional effects on the spectral signal 6. Multisensor recording at specific plant development stages a possible	Development of the measuring boxes for the sensors (automated) Age and development stages of trees are a limiting factor (often only trees up to age 5 can be recorded)

(continued)

Table 13.1 (continued)

Close-range RS approaches	Advantages	Disadvantages
Tower (flux tower) with different noninvasive measuring technologies as well as RS technology (mobile, permanently installed) http://www.fluxnet.ornl.gov/ *Phenocams* (Brown et al. 2016)	Advantages II, IV, 1, 3, 4, 6 above also apply (a) Links with international networks are possible (b) Important ground-truth RS information for plant health under natural growth conditions, with certain variables	Local results for a particular site, which do not enable results for extensive areas, but are limited to the forest stand under investigation Primarily nonimaging sensor technology can be implemented
WSNs (WSN) (Hwang et al. 2010; Liyang Yu et al. 2005; Lloret et al. 2009; Mafuta et al. 2013; Mauro et al. 2016; Mollenhauer et al. 2015, 2016; Oliveira et al. 2016; Ruiz-Garcia et al. 2009; Teodoro et al. 2016)	Advantages II, IV, 1, 3, 4, 6, a, b above also apply Long-term monitoring with high time frequencies WSN enables results over more extensive areas from the network distribution Terrestrial sensor networks as well as aquatic WSNs are possible	The number of wireless sensor nodes determines the accuracy of information over extensive areas Primarily nonimaging sensor technology can be implemented
Field measurements (manual operation) Long-term vegetation monitoring experiments (Bruelheide et al. 2014; Hantsch et al. 2013; Hector et al. 2011; Scherer-Lorenzen et al. 2007)	Spectral measurements directly on trees Investigation of geometric effects (different heights, recording angle) Measuring various biochemical, biophysical, and structural variables in organs (roots, leaf, stem) of a tree Recording microclimate information about soil, water, climate of a tree	Not applicable: IV, 1, 2, 3, 4, 5, 6

complexity, low spatial overlap, and low number of species (Lopatin et al. 2017). Schweiger et al. (2018) showed a strong relationship between functional diversity of grassland species and spectral traits collected using a hyperspectral sensor mounted on a scaffolding. Following this new approach, detection of individuals is no longer needed to monitor functional aspects of biodiversity. If models are to be developed that link the spectral signals to properties of plant diversity, it is important that both the field data collection and the image campaign are synchronized, particular in highly dynamic ecosystems such as grasslands with high land-use intensities. The high flexibility of UAS is a major advantage in such situations.

In vegetated ecosystems, vegetation structure is a key characteristic that is strongly related to the diversity of many taxa. UAS high-resolution images are used to characterize different aspects of vegetation structure: Getzin et al. (2014) used high-resolution RGB images to create canopy gap maps. They showed strong relations between spatial gap metrics and herbal plant species diversity in temperate forests. 3-D point clouds derived from UAS images can be used to characterize the

3-D vegetation structure, for example, by deriving canopy height models to characterize the structural complexity (Saarinen et al. 2017) or by describing vegetation
structures directly based on the vertical profiles of the 3-D point clouds (Wallace
et al. 2016).

Currently, the efficient use of UASs is limited to areas of less than a couple of
square kilometers. Therefore, their main application in the context of biodiversity
assessments is for sample-based observations (e.g., plots, transects) where relationships between spectral and structural traits and components of vegetation diversity
can be established. Given the option to create dense time series at user-defined frequencies with low effort, UASs offer scientists new opportunities for scale-
appropriate measurement of ecological phenomena (Anderson and Gaston 2013)
such as phenological and other seasonal effects on canopy reflectance. Thus, they
can be used to bridge the gap between scale of observation and the scale of the ecological phenomena that long existed in the temporal and spatial domain when using
air- or spaceborne platforms. Therefore, UAS technology needs to be considered as
an important intermediate-scale technology for biodiversity monitoring systems
and upscaling from field-based measurements and models to larger area estimation.

13.2.2.2 Optical RS

The relationship between optical spectral variability over space or time and species
diversity can be used to optimize the inventory of species diversity, so priority may
be given to sites that are spectrally more different and hence more diverse in species
composition (Rocchini et al. 2005). Such analyses can be conducted at different
spatial extents and resolutions, from a few meters [e.g., using high-resolution
(~1–3 m multispectral) satellite data such as Worldview or GeoEye] to 10–30 m
(e.g., Sentinel, Landsat) up to large spatial grain and extent [e.g., Moderate
Resolution Imaging Spectroradiometer (MODIS) data from 250 m to 1000 m].

Alpha Diversity

Alpha diversity is the number of species living within a given local area and is a
measure of within-ecosystem species richness. Most research dealing with RS-based
estimates of alpha diversity has focused on mapping localized biodiversity hot
spots, based on the spectral variation hypothesis (SVH, (Palmer et al. 2002)). The
SVH states that the spatial variability in the remotely sensed signal, i.e., the spectral
heterogeneity, is expected to be positively related to environmental heterogeneity
and could therefore be used as a powerful proxy of species diversity. In other terms,
the greater the habitat heterogeneity, the greater the local species diversity within it,
regardless of the taxonomic group under consideration. Besides random variation in
species distribution, higher heterogeneity habitats will host a higher number of species each occupying a particular niche (niche difference model, Nekola and
White 1999).

Different modeling techniques have been used to model the local species diversity-spectral heterogeneity relationship, ranging from simple univariate models (Gould 2000), to multivariate statistics (Feilhauer and Schmidtlein 2009), to neural networks (Foody and Cutler 2003) and generalized additive models (GAMs, Parviainen et al. 2009). A number of different measures of spectral heterogeneity have been proposed and used to assess ecological heterogeneity and thus species diversity (Cavender-Bares et al., Chap. 2). Many of these are related to the variability in a spectral space of different pixel values, such as the variance or texture in a neighborhood of the spectral response (Gillespie 2005) or the distance from the spectral centroid, which may be represented as the mean of spectral values in a multidimensional system whose axes are represented by each image band or by principal components where noise related to band collinearity has been removed (Rocchini 2007). Moreover, in addition to the use of common vegetation indices such as the Normalized Difference Vegetation Index (NDVI), some studies have demonstrated an increase in the strength of the relationship when using additional spectral information (e.g., Landsat bands 5 and 7 in the shortwave infrared (SWIR) (Rocchini 2007) and (Nagendra et al. 2010)).

Beta Diversity

While alpha diversity is related to local variability, species turnover (beta diversity) is a crucial parameter when trying to identify high-biodiversity areas (Baselga 2013). In fact, for a given level of local species richness, high beta diversity leads to high global diversity of the area. This is one of the basic rules underpinning the concept of irreplaceability of protected areas (e.g., Wegmann et al. 2014).

In some cases spatial distance/dispersal ability might not be the only driver of species turnover, which seems to be more strictly related to environmental conditions. Hence, models have been built to relate species and spectral turnover to explain their potential relationship and its causes (Rocchini et al. 2018b). In some cases, spatial distance accounted only for a small fraction of variance in species similarity, while environmental variation is expected to account for a much larger one. When using spatial distances, distance decay does not necessarily account for environmental heterogeneity (Palmer and Michael 2005), especially in heavily fragmented landscapes. Thus, the use of spectral distances for summarizing beta diversity patterns may be more reliable because this method explicitly takes environmental heterogeneity into account instead of mere spatial distances among sites. Therefore, it is expected that the higher the spectral distance among sites, the higher their difference in terms of environmental niches, potentially leading to higher beta diversity.

A straightforward method for measuring beta diversity is to calculate the differences between pairs of plots in terms of their species composition using one of the many (dis)similarity coefficients proposed in the ecological literature (e.g., Legendre and Legendre 1998) and assess the spectral turnover variability derived remotely from the variation in species composition among sites. This has been mainly related to spectral distance decay models in which species similarity decays once spectral distance increases, using all pairwise distances among N plots, based on an a priori defined statistical sampling design.

Another powerful method to estimate beta diversity is related to the so-called spectral species concept (Féret and Asner 2014). This approach is based on the preliminary unsupervised clustering of spectral data, assigning each pixel to a "spectral species." After spectral clustering, the image is divided into homogeneous elementary surface units, and a dissimilarity metric is then used to compute pairwise dissimilarity between each pair of surface units. Finally, the resulting dissimilarity matrix is processed using nonmetric multidimensional scaling to project elementary units in a 3-D Euclidean space, allowing the creation of a map in the standard red-green-blue (RGB) color system. Such a map expresses changes in species composition with changes in color or color intensity.

While the previously described methods are powerful in describing and estimating diversity from space, they are mainly related to spectral heterogeneity measurement, with no direct relationship with drivers of diversity, such as climate drivers, which might be better estimated by thermal RS.

A very important milestone in biodiversity research was the development of plant functional types (PFTs) such as the Ellenberg indicator values (Schmidtlein 2005) or the CSR-strategy types (C, competitive species; S, stress-tolerant species; R, ruderal species), which altered their functional traits as a consequence of the adaptation to changes in abiotic conditions and/or human pressures such as land-use intensity or management practices. Schmidtlein et al. (2012) developed the foundations for linking RS with this biodiversity concept. Rocchini et al. (2018a) used this research as a basis for calculating a global biodiversity index, namely, "Rao's Q." The Rao's Q is calculated on a set of CSR score maps (derived from Schmidtlein et al. 2012) to estimate the diversity of functional-type probability in space (Rocchini et al. 2018a, see Fig. 13.4).

Rao's Index
■ <0.01
  0.101
  0.195
■ 0.289
■ 0.384

Fig. 13.4 Rao's quadratic diversity metric applied to a MODIS-derived 250 m pixel NDVI map of the world NDVI (date 2016-06-06, http://land.copernicus.eu/global/products/ndvi), resampled at 2 km resolution with a moving window of 5 pixels. (Copyright: License number: 4466960473531. From Rocchini et al. (2018a)). Courtesy: Matteo Marcantonio

13.2.2.3   Thermal RS

Thermal RS detects the energy emitted from Earth's surface as electromagnetic radiation in the thermal infrared spectral range (TIR, 3–15 μm). This energy can be radiated by all bodies with a temperature above absolute zero and is dependent on the surface temperature and the thermal properties (emissivity) of the observed target (Kuenzer et al. 2013; Künzer and Dech 2013).

Land surface temperature (LST) is one of the most important state variables representing the coupled interaction of the surface energy and water balance from local to global scale (e.g., Kustas et al. 2003). LST is highly influenced by the radiative, thermal, and hydraulic properties of the soil-plant-atmosphere system and has therefore been recognized as one of the high-priority parameters of the International Geosphere and Biosphere Program (IGBP, Townshend et al. 1994).

Various RS platforms and sensors currently provide TIR data at different spatial, spectral, and temporal resolutions. The most common include the Advanced Very High Resolution Radiometer (AVHRR) onboard the Polar Orbiting Environmental Satellites (POES); Landsat 5, 7, and 8; the MODIS sensor on board the NASA Terra and Aqua satellite, the Advanced Spaceborne Thermal Emission and Reflection Radiometer (ASTER) on the Terra Earth observing satellite platform; and Sea and Land Surface Temperature Radiometer (SLSTR) onboard the Sentinel-3 mission.

Although LST is rarely used by ecologists (Wang et al. 2010), a number of applications are closely linked to understanding landscape and biodiversity characteristics. Most often, LST is taken as source to estimate evapotranspiration (see Krajewski et al. 2006 for a review). LST is highly controlled by atmospheric conditions, but also by stomata conductance and plant-available soil moisture (Bonan 2008). In this sense, monitoring of LST with sufficiently high spatial and temporal resolution is able to provide valuable information about the water and energy exchange between the soil-plant-atmosphere continuum and related photosynthetic activities of the vegetation (see Fig. 13.5). Differences in the spatiotemporal behavior of LST can therefore be related to different plant/species distributions and/or to differences

(a)   (b)

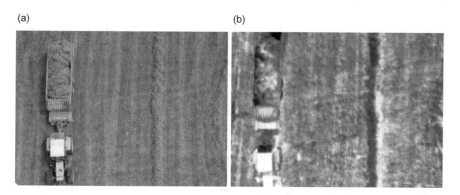

Fig. 13.5 Optical (**a**) and TIR (**b**) image of a ScaleX field campaign test site in July 2016 at the TERENO pre-alpine grassland site, Fendt, Germany. Elevated land surface temperatures (yellow) are detected, especially for the rows of hay mounds facing the sun

related to local energy, water, or nutrient conditions. Examples will be briefly described of methods to disentangle the effects of water, energy, and nutrients on plants in the context of vegetation.

Müller et al. (2014, 2016) applied principal component analysis (PCA) to extract dominant LST patterns from time series (28 scenes covering 12 years) of ASTER TIR images of the mesoscale Attert catchment in midwestern Luxembourg. The PCA-component values for each pixel were related to land use/vegetation data and to geological and soil texture data, indicating a strong information signal in the temporal dynamics of LST data with regard to plant diversity.

Environmental disturbances have been investigated (e.g., by Duro et al. 2007) making use of the negative relationship between vegetation density and LST. Mildrexler et al. (2007) proposed a disturbance detection index based on this principle that uses the 16-day MODIS Enhanced Vegetation Index (EVI) and 8-day LST. They were able to successfully detect disturbance events such as wildfire, irrigated vegetation, precipitation variability, and the recovery of disturbed landscapes at the continental scale.

Sun and Schulz (2015) could demonstrate that an integration of TIR data from Landsat 5 and 8 was able to significantly enhance the classification results for different aggregation levels of land-use and land cover categories for a mesoscale catchment in Luxembourg. This indicates the high potential of TIR data to support more specific and selective plant species monitoring as relevant for biodiversity research.

Environmental stress induced by long-term heat waves and/or a limited availability of water is likely to reduce stomata conductance, limit transpiration, and thereby increase leaf surface temperature (Stoll and Jones 2007). The difference between air temperature and leaf temperature combined with information on vegetation density can serve as an indicator of plant stress. Hoffmann et al. (2016) used a spectral vegetation index and LST data from cameras mounted on UAVs to develop a water deficit index (WDI). The WDI was highly correlated to eddy covariance measurements of latent heat fluxes over a growing season, and that was used to map spatially distributed water demands of various crops.

Environmental stress may also cause changes in leaves and the structure of plants, dependent on their biophysiological characteristics. Buitrago et al. (2016) found that two plant species [European beech, (*Fagus sylvatica*) and rhododendron (*Rhododendron catawbiense*)], when exposed to either water or temperature stress, experience significant changes in TIR radiance. The changes in TIR in response to stress were similar within a species, regardless of the stress. However, changes in TIR spectra differed between species, and these differences could be explained by changes in the microstructure and biochemistry of leaves (e.g., cuticula).

Overall, the potential for exploiting LST information data in plant biodiversity research is manifold. While LST is easily measured by thermometers at the point scale, satellite RS TIR data are needed in order to derive LST routinely at high temporal and spatial resolutions over large spatial extents. However, the derivation of LST from TIR data is a difficult task because such radiance measurements depend not only on LST but also on surface emissivity and atmospheric conditions (Li and Becker 1993). Therefore, besides cloud detection and radiometric calibration, corrections for emissivity and atmospheric effects have to be carried out. A large number of studies have addressed these issues in the past. It is beyond the scope of this section to summarize these studies, but an excellent review is provided by Dash et al. (2002).

13.2.2.4   Light Detection and Ranging (LiDAR)

LiDAR is an active RS technique in which short pulses of laser light emitted from a scanning device are distributed across a wide area and their reflections from objects are subsequently recorded by a sensor. The distance to the objects can be calculated from the elapsed time and the speed of light. The absolute position of the reflection can be reconstructed using the position recorded by the Global Positioning System (GPS) and the orientation of the sensor determined by the inertial navigation system (INS). The result is a set of 3-D points that represents the scanned surface from which the pulses were reflected. More detailed descriptions of LiDAR technology can be found in Popescu (2011) and Wehr and Lohr (1999).

The primary characteristic that makes LiDAR well suited for monitoring plant biodiversity, vegetation structure, and landscape diversity is the penetration of light beams below the forest canopy. When a LiDAR beam hits the top of the canopy, the beam is reflected by leaves, needles, and branches, and the reflection is recorded by the receiver. If the energy of the beam is still high when it hits the first reflective surface, the beam will split and can penetrate farther through openings in the canopy until it hits additional vegetation, which can again cause reflections. This process continues until a massive reflector, such as a tree trunk or the ground, reflects the beam or until the signal becomes too weak. These properties of LiDAR beams allow a detailed reconstruction of 3-D vegetation structures below the forest canopy, which cannot be provided by passive RS techniques (Koch et al. 2014). Hence, LiDAR RS is a valuable technique for monitoring plant diversity and vegetation structure, and it adds a further dimension to the properties of optical RS.

LiDAR systems can be classified as discrete-return systems or full-waveform systems, based on the capabilities of data recording. At the onset of LiDAR development, sensors were only able to record either the first or the last reflection of the LiDAR beam, which is generally the top of trees and the terrain, respectively. As LiDAR evolved, discrete-return systems were developed, which were able to record a fixed number of range measurements per LiDAR beam, usually up to four to five. The returns were based on thresholds, which were integrated into the proprietary detection method (Thiel and Wehr 2004).

With the more recently developed full-waveform systems, the entire pathway of the LiDAR beam through the canopy can be detected and recorded (Wagner and Ullrich 2004). The post-processing of this data can be applied to theoretically extract an unlimited number of echoes. Moreover, with Gaussian decomposition—

the standard procedure for waveform decomposition—additional echo attributes, such as amplitude and intensity of the return signal, can be provided, which can support the classification process. As a result, full-waveform data provide a much more detailed characterization of the vertical vegetation structure. In this way, important indicators for vegetation structure and biodiversity, e.g., vegetation height and cover of the different vegetation layers, can be estimated with a lower bias and higher consistency (Reitberger et al. 2008).

Nowadays, laser-based instruments are mounted on all kinds of RS platforms, including stationary or mobile scanners and terrestrial-, drone-, and aircraft-based platforms (e.g., the well-established airborne LiDAR scanning). To record a variety of structural parameters, it is possible to combine information from LiDAR sensors with optical, thermal, or radar RS sensors (Joshi et al. 2015, 2016). Li et al. (2014) provide an overview of 3-D imaging techniques for describing plant phenotyping of vegetation. Rosell and Sanz (2012) review methods and applications of 3-D imaging techniques for the geometric characterization of tree crops in agricultural systems. Wulder et al. (2012) provide a review of LiDAR sampling for characterizing landscapes.

The LiDAR systems used for ecological applications generally have a beam footprint of less than 1 m diameter on the ground. These so-called small-footprint systems are preferred because they provide a good link between the LiDAR beam and the structural vegetation attributes that could subtly change as a consequence of stress or damage, sometimes within individual trees. By comparison, large-footprint systems have beam diameters of up to scores of meters on the ground; e.g., the Geoscience Laser Altimeter System (GLAS) instrument mounted on the Ice, Cloud, and land Elevation Satellite (ICESat) platform has a footprint of 38 m (Schutz et al. 2005). Such systems can be used to model and map broad vegetation structural attributes and are well suited for detecting structural vegetation characteristics across large areas.

The most important environmental application of LiDAR is the precise mapping of terrain and surface elevations. Such digital terrain models (DTMs) or digital surface models (DSMs) can be useful in determining topographic information important for plant growth and monitoring of vegetation structure and biodiversity, e.g., changes in vegetation height or density resulting from succession or natural disturbance (Heurich 2008). Many filtering methods have been developed to extract terrain elevation from point clouds, which produces DTMs with high spatial resolution and root mean square errors (RMSEs) of 0.15–0.35 m (Andersen et al. 2005; Heurich 2008; Sithole and Vosselman 2004). No other RS technique has the ability to deliver DTMs of similar quality within dense vegetation. Recent studies show that it is even possible for LiDAR to detect objects located on the ground surface. Coarse woody debris, as an example, is an important indicator of past disturbances that might influence biodiversity because it provides habitat to a multitude of plant and animal species and plays an important role in the forest carbon cycle.

Because of its characteristics, LiDAR is well suited for measuring biophysical parameters of vegetation, such as tree dimensions and canopy properties. Two main approaches have been developed over recent years. The area-based approach is a straightforward methodology in which the height distribution of the LiDAR beam reflections is analyzed for a given area. In the first step, plenty of different "LiDAR metrics," e.g., maximum height or fractional cover, are calculated for each area. The second step is model calibration, where these metrics are compared to on-the-ground survey data such as plant species richness, aboveground biomass (AGB), or vertical and horizontal vegetation structure. In the final step, the models are used to estimate the selected biodiversity indicators for large areas using square grid cells. Such an analysis is generally conducted using a priori stratification of structural vegetation types and plant species. In the years that followed, this methodology was proven to be able to determine key biophysical vegetation variables on a larger scale. To date, this method has been shown to deliver a precision of 4–8% for height, 6–12% for mean diameter, 9–12% for basal area, 17–22% for stem number, and 11–14% for volume estimations of boreal forests (Maltamo et al. 2006; Næsset 2002, 2007). Because of the highly accurate estimation of important vegetation structural parameters, the area-based approach was further developed and adapted to operational forest inventories in boreal forests of Scandinavia. Similar accuracies have also been achieved for the temperate zone, although the more complex vegetation structures in this zone, especially the higher number of tree and plant species and higher amount of biomass, led to less accurate estimations and more effort in stratification and ground measurement to obtain species-specific results (Heurich and Thoma 2008; Latifi et al. 2010, 2015).

The second methodology is the individual-tree approach, which has the objective of extracting data on single trees and modeling the tree properties. The procedure consists of four steps. In the first step, individual trees are delineated by dividing each crown into segments with techniques originally used for raster analysis, such as watershed analysis and local maxima detection (Heurich 2008; Persson et al. 2002). However, these techniques do not take advantage of the full information of the 3-D point cloud, and therefore, trees beneath the crown surface cannot be detected. For this reason, new methods based on 3-D point clouds have been developed over recent years (Tang et al. 2013; Yao et al. 2012). When these novel techniques are employed, more than 80% of the trees of the upper canopy level can be detected. Moreover, tree detection in the lower canopy is much improved compared to 2-D techniques. In the second step, parameters of each tree (e.g., height, species, and crown parameters) are derived. Tree height can be determined by measuring the distance of the highest reflection of a LiDAR beam within the tree segment and the DTM, with an accuracy of less than 2 m and a slight underestimation (Heurich 2008). The third step is the model calibration of the biophysical parameters of the tree, namely, diameter at breast height (DBH), volume, and biomass, using trees measured on the ground as a reference. The tree crown can be modeled using convex hulls and alpha shapes. The fourth step involves the application of these models to predict DBH, volume, and biomass of all trees delineated by LiDAR. Based on these crown representations, basic attributes reflecting tree health can be derived, e.g., total volume, crown length, crown area, and crown base height (Yao et al. 2012). The extracted parameters of individual trees also form the basis for identifying the tree species by calculating point cloud and waveform features within the 2-D or 3-D representation of the tree and with the help of classification techniques (Reitberger et al. 2008). While differentiation between deciduous and coniferous trees is highly accurate (>80%, up to ca. 97%), differentiation within these classes is more difficult and leads to a higher classification error. Moreover, it is possible to distinguish between living trees, standing dead trees, and snags (Yao et al. 2012) and to map dead trees at the plot or stand level. However, 3-D LiDAR has its limitations in differentiating between trees species and dead trees when not combined with multispectral optical data. One drawback of the individual-tree approach is that the LiDAR beam loses some transmission on its way through the canopy and is therefore not always suitable for smaller understory trees, which results in their underestimation. To overcome this problem, methods have been developed to predict diameter distributions of forest stands based on detectable trees in the upper canopy and LiDAR-derived information on the vertical forest structure and density (Lefsky et al. 2002).

In addition to the traditional parameters related to forestry, a multitude of traits that describe the ecological conditions of the forest can be estimated with LiDAR sensors. One key element for assessing plant diversity and vegetation structure is canopy cover, which is defined as the projection of the tree crowns onto the ground divided by ground surface area. This parameter can be easily obtained from LiDAR data by dividing the number of returns measured above a certain height threshold by the total number of returns. Many studies have proven the strong ($R^2 > 0.7$) relationship between this LiDAR metric and ground measurements. By using hemispherical images or other ground-based instruments for calibration, leaf area index (LAI) and solar radiation can also be derived from LiDAR data with a high precision over large areas (Moeser et al. 2014). Because canopy metrics are affected by sensor and flight characteristics, it is recommended that each campaign be calibrated to obtain high-quality results. However, it has been shown that even without calibration, fairly reliable results can be obtained.

Vertical vegetation structure is highly relevant for the description of forest and vegetation heterogeneity and highly important for biodiversity studies. A widely used LiDAR metric for representing vertical canopy complexity is the coefficient of variation. High coefficient values correspond to more diverse multilayer stands, whereas low values represent single-layer stands. The coefficient of variation can be applied at point clouds, the digital crown model, or individual trees. Zimble et al. (2003) applied this principle and classified vegetation types according to stand structure with an overall accuracy of 97%.

Another approach is the partitioning of the vertical structure into different height layers in relation to ecological importance. Latifi et al. (2015) divided the canopy into height layers according to phytosociological mapping standards and found a strong relationship to various LiDAR metrics in regression models. Similar approaches were used by Ewald et al. (2014) to represent understory offering protection for birds and deer and to detect forest regeneration. A more recent study applied a 3-D segmentation algorithm to estimate regeneration cover with an accuracy of 70%. LiDAR-derived information about the vertical structure is also used for the assessment of forest fuels and their vertical distribution, which are important input variables in forest fire models used in fire management.

In summary, LiDAR RS is a powerful tool for monitoring vegetation structure and plant diversity. It delivers detailed and accurate information about forest properties down to the scale of the individual tree and is therefore regarded as the gold standard for determining vegetation structure. Nowadays, LiDAR is widely applied in RS research as a reference to test the accuracy of other methods and is used in practical forest management in the boreal zone of Scandinavia (Næsset 2007). The development of new sensors will lead to multi- or hyperspectral LiDAR technology, which will combine the advantages of today's LiDAR and optical sensors. These systems will be able to collect accurate 3-D information and calibrated spectral information without facing the problems of varying illumination in the tree crowns. Furthermore, the resolution of the data will increase, thereby enabling parameter extraction at the branch level.

13.2.2.5 Radar

Several reviews have been conducted on radar alone or radar and optical sensors for vegetation applications relevant to habitat and biodiversity. They typically include classification of vegetation or land cover types, biophysical modeling of parameters such as biomass or tree height, and ecosystem disturbance detection and mapping (e.g., Balzter 2001; Treuhaft et al. 2004; Lu 2006, which includes summaries of four previous reviews; Lutz et al. 2008; Bergen et al. 2009; Lowry et al. 2009; Koch 2010; Nagendra et al. 2013; Tiner et al. 2014; White et al. 2015; Timothy et al. 2016; Baltzer 2017).

Systems and Techniques

Active radar is the focus of this section because the resolution of passive sensors is generally too coarse for all but large extent studies. In active radar, transmitted pulses interact with scattering elements of the surface in terms of their dielectric properties, size, and arrangement. In vegetation, moisture (increasing dielectric constant) and more complex stem-branch-leaf arrangements result in increased backscatter intensity. Much research has been conducted using physically based models to characterize and understand backscatter effects in vegetated canopies (e.g., Sun and Ranson 1995; Ningthoujam et al. 2016). Spatial and temporal variations in these properties associated with different vegetation types, age distribution, health, and management provide information or indicators of potential habitat and biodiversity. Radar data are available at different frequencies/wavelengths; X-, C-, and L-bands (2.5–3.75 cm, 3.75–7.5 cm, and 15–30 cm wavelengths, respectively) are the most common on satellite platforms. S-band (7.5–15 cm) has been deployed on a couple of satellites, and new S- and P-band (30–100 cm) satellite sensors are planned for the near future (e.g., NISAR L- and S-bands; NovaSAR S-band;

BIOMASS P-band). Backscatter of shorter wavelengths is generally from the upper canopy, while longer wavelengths penetrate farther into vegetation. Combining shorter and longer wavelength data can be advantageous to detect contributions of multiple scattering from different vertical portions of the canopy as well as ground surface scattering and double bounce ground-stem scattering. Transmitted and received radar signal polarizations [commonly horizontal (H) and vertical (V)] can provide information on vegetation composition and structure. Co-polarized signals (e.g., HH) respond more to trunk-ground configurations (double bounce scattering), particularly at longer wavelengths, while cross-polarizations (e.g., HV) respond more to canopy woody biomass (Ningthoujam et al. 2017). Combinations such as ratios (e.g., HH/HV) can therefore enhance spatial differences in scattering types and magnitude due to varying vegetation density (Mitchard et al. 2011). Similarly, steeper incidence angles generally penetrate farther into the canopy, particularly in leaf-on conditions, but multiple angles may enhance differences in vertical structure (Henderson and Lewis 1998). All of the above characteristics related to the degree of canopy penetration provide opportunity for analysis of vertical structural complexity and composition diversity by using multiple bands, polarizations, and incidence angles. Radar image texture information has also been found to be useful in classification (Simard et al. 2000) and biophysical modeling (Kuplich et al. 2005).

Polarimetric data, where phase information is preserved (Ulaby et al. 1987), allows additional analysis of polarization parameters and, through decomposition analysis, the relative contributions from the various scattering mechanisms (e.g., surface, volume, double bounce) that are associated with canopy structural characteristics. Commonly applied decomposition techniques include (van Zyl 1989; Cloude et al. 1996; Freeman and Durden 1998; Yamaguchi et al. 2005; Touzi 2007). Several others were designed to build on or correct issues with previous techniques (Hong and Wdowinski 2014). The Interferometric SAR (InSAR, e.g., Balzter (2001)), provides a plain language description) incorporates transmission from two different angles either simultaneously or in repeat passes (less preferable given potential decorrelation of the response signals between passes). The phase differences between the radar response signals can be used to estimate the scattering phase height center, which is associated with canopy density and arrangement. They can also be used to generate digital elevation models (DEMs), with longer wavelengths that penetrate the canopy being more suitable. Using multiple parallel baselines, SAR tomography has been used to construct a 3-D representation of a given volume such as a forest (e.g., Reigber and Moreira 2000).

Classification and Biophysical Modeling Applications

As with other RS technologies, landscape or vegetation diversity as an indicator of biodiversity can be mapped through thematic classification. This can be accomplished across a broad gradient from nonvegetated to dense vegetation classes, to map landscape cover types (e.g., Devaney et al. 2015) that may be associated with the spatial distribution of biodiversity. For example, within the context of developing understanding of the spatial distribution of sensitive arctic shore habitat and biodiversity areas in the event of oil spills, Banks et al. (2014a) used decomposition parameters in a comparison of three unsupervised polarimetric classifiers for their potential in mapping multiple classes of substrate (nonvegetated), tundra vegetation, and wetland type. Similarly, Varghese et al. (2016) classified water, settlements, agriculture, shrub/scrub, and three forest density classes in a comparison of parameters derived from six decomposition techniques. In an analogous manner, specific classes within an ecotype can be mapped. For example, mapping the diversity of given classes within a wetland complex (e.g., Touzi and Deschamps 2007; Gosselin et al. 2013; Dingle Robertson et al. 2015; Hong et al. 2015; Dubeau et al. 2017) can aid identification of the variety of habitat conditions available and potential biodiversity. In general, radar data have not been found to provide consistently better overall classification accuracy than optical data. However, since radar data are typically complementary and not highly correlated with optical data, they can provide additional information for certain vegetation classes that can be distinguished by structure in cases where optically derived spectral reflectance and vegetation indices are similar. Thus, combining radar and optical imagery has often been shown to improve the accuracy of such classes over either data type alone (e.g., Bergen et al. 2007; Wang et al. 2009; Bwangoy et al. 2010; Banks et al. 2014b).

An alternative approach to thematic classification is estimation of vegetation structure parameters that can serve as indicators of potential habitat diversity or biodiversity, for example, the average or spatial heterogeneity of AGB, LAI, vegetation height, and stem and branch parameters. Luckman et al. (1997), Lucas et al. (2006), and Le Toan et al. (2004), among others, have reported that the backscatter-AGB relationship typically saturates in the range of 100–150 t/ha. However, AGB spatial variability can be mapped in environments with lower vegetation density (e.g., Häme et al. 2013), and efforts to produce suitable models with a higher saturation threshold by improving data information content are common. For example, use of the following has proven beneficial: cross-polarized (HV) data rather than co-polarized; longer wavelengths that penetrate deeper into the canopy (Santos et al. 2003); ratios such as VV/HH (Manninen et al. 2009 for LAI); shorter-to-longer wavelength ratios such as C−/L-bands (Foody et al. 1997); averaging of multitemporal data sets to reduce moisture/rain effects (Englhart et al. 2011); and integrating optical and radar data (Vaglio et al. 2017). Imhoff et al. (1997) modeled canopy parameters using steep incidence angle C-, L-, and P-band airborne radar and found strong correlations for C-HV and LAI, L-VV and branch surface area or volume, and P-VV with bole surface area or volume; these relationships were then used to map broad avian habitat classes. Bergen et al. (2009) combined biomass estimates from C- and L-band backscatter with Landsat vegetation classification, thereby improving habitat mapping for three bird species over use of vegetation type alone.

InSAR has been used to estimate canopy height and height variance, which can be an indicator of vegetation type, structural complexity, and age diversity. Canopy height is most commonly estimated from the difference between scattering phase height center estimates derived from short wavelength (X- or C-band) InSAR and a ground DEM derived from longer wavelength InSAR (L- or P-bands, Neeff et al. 2005; Balzter et al. 2007a) or from another source such as LiDAR (e.g., Kellndorfer et al. 2004; Andersen et al. 2008; Tighe 2012). Tighe et al. (2009) applied this approach with the addition of correction factors for various forest ecotypes in US and Canadian environments from semiarid to boreal. Integrating polarimetric response with InSAR (i.e., PolInSAR; Cloude and Papathanassiou 1998) can improve InSAR estimates of tree height (e.g., Balzter et al. 2007b). Tomography has also shown promise in modeling the vertical distribution of canopy biomass, but multiple acquisitions must be conducted within a short time to minimize temporal decorrelation. The DEMs produced from InSAR can also be used to generate topographic indices for analysis of topographic complexity or roughness related to habitat diversity and biodiversity (Turner et al. 2003; Kuenzer et al. 2014). Fusion of optical imagery and/or LiDAR with InSAR (e.g., as reviewed in Treuhaft et al. 2004) can also improve vertical canopy and topographic information.

Mapping of disturbance or environmental change can serve as an indicator of potential impacts on habitat diversity and biodiversity. Many studies have been conducted in diverse applications that cannot be fully reviewed here. Most early applications were in mapping of deforestation, particularly in tropical regions where deforestation had become a major issue (e.g., Rignot et al. 1997; van der Sanden and Hoekman 1999). Temporal data have become widely used in land cover change analysis using classification approaches (e.g., Thapa et al. 2013), analysis of backscatter change (e.g., Whittle et al. 2012; Mermoz and Le Toan 2016), and biophysical modeling for biomass loss (e.g., Mitchard et al. 2011). Other major radar applications of environmental change with implications for biodiversity impacts are burn and inundation detection and mapping. Early fire impact studies focused on backscatter variations related to burn intensity classes (e.g., Kasischke et al. 1992). More recent work has included temporal backscatter data in pre−/postburn analysis (e.g., Tanase et al. 2015) and polarimetric analysis and decomposition in modeling biomass changes due to fire (Martins et al. 2016). Radar is particularly useful in detecting inundation under vegetated canopies due to specular reflection off the water surface; in the case of inundated forests, penetration of the canopy by longer wavelengths occurs with double bounce scattering off the water surface and tree trunks (e.g., Kim et al. 2009) and phase differences between different polarizations (e.g., Rignot et al. 1997).

Overall, use of radar for biodiversity and landscape diversity analysis, modeling, mapping, and monitoring follows similar approaches to optical RS. Diversity of land cover types or specific classes within a given ecotype may be directly mapped using classification or modeling, while estimation of biophysical variables can serve as indicators of spatial heterogeneity and potential habitat diversity or biodiversity. The main contributions of radar are in its unique response to vegetation structure that complements spectral reflectance characteristics of vegetation in the optical regions. With the multitude of wavelengths, incidence angles, and polarizations available, as well as the capability to acquire and process InSAR and polarimetric data, much promise has been shown for mapping and monitoring land cover diversity- and biodiversity-related vegetation metrics. New satellite systems are being developed, and particularly the 2021 BIOMASS P-band InSAR mission (Le Toan et al. 2011) will provide consistent means for global biomass mapping and monitoring.

13.3 Conclusion and Further Work

Traits, drivers, and effects on biodiversity exist on all spatiotemporal scales. Air- and spaceborne RS data capture processes and patterns in ecosystems, but often without the knowledge of the cause of the phenomenon and real high-frequency ground information. Therefore, close-range RS platforms that record RS information at high frequency must be coupled with air- and spaceborne RS platforms (see Fig. 13.6).

No monitoring approach alone is sufficient, comprehensive, cost-effective, and flexible enough to perform vegetation health monitoring from local to global scales and for short- to long-term processes as well as to monitor changes in phylo-, taxonomic, functional, and trait diversity and to assess the resilience of ecosystems. Therefore, the development and application of a multisource vegetation diversity and health monitoring network (MUSO-VDH-MN) is important where multisource data (close-range, air-, and spaceborne RS data) as well as different in-situ monitoring approaches can be linked in an effort to compensate for the shortcomings of one approach with the advantages of another and to achieve additional benefits for VH monitoring. A future MUSO-VDH-MN should therefore contain the following elements (see Fig. 13.7):

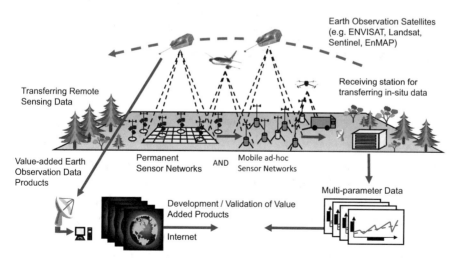

Fig. 13.6 Linking different approaches (high-frequency WSNs up to spaceborne satellites) with relative frequency monitoring, sensors, and different platforms of RS to better describe, explain, predict, and understand vegetation diversity with RS techniques as well as improve the calibration and validation of RS data. (From Lausch et al. 2018a)

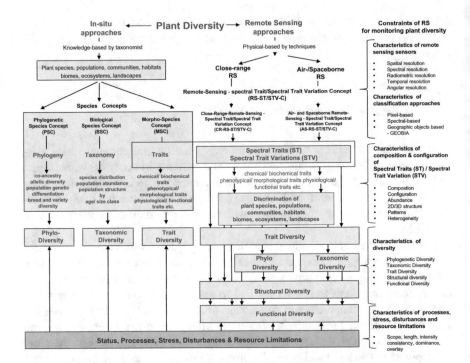

Fig. 13.7 Overview of in-situ approaches—the phylogenetic species concept (PSC), the biological species concept (BSC), the morphological species concept (MSC), and the RS-spectral trait/spectral trait variation (RS-ST/STV) concept, which integrates the close-range RS approaches and the air−/spaceborne RS approach. The different in-situ and RS approaches are crucial for determining phylo-, taxonomic, structural, trait diversity as well as functional diversity, in order to be able to monitor and assess status, stress, shifts, disturbances, or resource limitations at different levels of vegetation organization. Components that need to be included for a future multisource vegetation diversity and health monitoring network (MUSO-VDH-MN): (I) linking of existing monitoring approaches; (II) integration of existing data, networks, and platforms; and (III) the use of data science as a bridge for handling and coupling big vegetation diversity and health data. (Modified after Lausch et al. 2018a)

(i) *Coupling of the different monitoring approaches (in-situ and RS) for plant diversity.*

(ii) *The integration and linking of multisource data and RS platforms.* MUSO-VDH-MN should integrate the following data and site survey platforms:

Species/habitats: Data from site surveys for species, species lists, metabarcoding, microgenomics (Bush et al. 2017), and phenotyping (Deans et al. 2015) and data from museums, lysimeters, plant phenomic facilities (Furbank 2009), controlled environmental facilities (ecotrons, Lawton et al. 1993), long-term ecological research (Mueller et al. 2010), spectral laboratory experiments, and biodiversity ecosystem functioning experiments (Bruelheide et al. 2014)

RS: optical (multispectral, hyperspectral), thermal, radar, LiDAR data, laboratory, tower, camera traps, WSNs, drones, and close-range, air- and space-borne RS platforms. Additionally, it should link monitoring databases, networks, citizen science information, abiotic (soil, water, air) information, and social and economic information.

(iii) *Data science, linked open data, and semantic web as a bridge for understanding and monitoring vegetation diversity.* For further information see also Lausch et al. (2015c, 2018a, b) (Fig. 13.7).

Acknowledgments We particularly thank the researchers for the Hyperspectral Equipment of the Helmholtz Centre for Environmental Research—UFZ and TERENO funded by the Helmholtz Association and the Federal Ministry of Education and Research, Germany. At the same time, we truly appreciate the support that we received from the project "GEOEssential: Essential Variables workflows for resource efficiency and environmental management" (ERA-NET Cofund Grant, Grant Agreement No. 689443). Finally, we thank the NIMBioS working group on Remote Sensing of Biodiversity.

References

Andersen HE, McGaughey RJ, Reutebuch SE (2005) Estimating forest canopy fuel parameters using LIDAR data. Remote Sens Environ 94:441–449. https://doi.org/10.1016/j.rse.2004.10.013

Andersen HE, McGaughey RJ, Reutebuch SE, Andersen H-E, McGaughey RJ, Reutebuch SE, Andersen HE, McGaughey RJ, Reutebuch SE (2008) Assessing the influence of flight parameters, interferometric processing, slope and canopy density on the accuracy of X-band IFSAR-derived forest canopy height models. Int J Remote Sens 29:1495–1510. https://doi.org/10.1080/01431160701736430

Anderson K, Gaston KJ (2013) Lightweight unmanned aerial vehicles will revolutionize spatial ecology. Front Ecol Environ 11:138–146. https://doi.org/10.1890/120150

Asner GP, Martin RE (2009) Airborne spectranomics: mapping canopy chemical and taxonomic diversity in tropical forests. Front Ecol Environ 7:269–276. https://doi.org/10.1890/070152

Asner GP, Anderson CB, Martin RE, Tupayachi R, Knapp DE, Sinca F (2015) Landscape biogeochemistry reflected in shifting distributions of chemical traits in the Amazon forest canopy. Nat Geosci 8:567–573. https://doi.org/10.1038/ngeo2443

Baldocchi D, Falge E, Lianhong G, Olson R, Hollinger D, Running S, Anthoni P, Bernhofer C, Davis K, Evans R, Gu LH, Olson R, Hollinger D, Running S, Anthoni P, Bernhofer C, Davis K, Evans R, Fuentes J, Goldstein A, Katul G, Law B, Lee XH, Malhi Y, Meyers T, Munger W, Oechel W, Paw UKT, Pilegaard K, Schmid HP, Valentini R, Verma S, Vesala T, Wilson K, Wofsy S, Paw UKT, Pilegaard K, Schmid HP, Valentini R, Verma S, Vesala T, Wilson K, Wofsy S (2001) FLUXNET: a new tool to study the temporal and spatial variability of ecosystem-scale carbon dioxide, water vapor, and energy flux densities. Bull Am Meteorol Soc 82:2415–2434. https://doi.org/10.1175/1520-0477(2001)082<2415:FANTTS>2.3.CO;2

Balzter H (2001) Forest mapping and monitoring with interferometric synthetic aperture radar (InSAR). Prog Phys Geogr 25:159–177. https://doi.org/10.1177/030913330102500201

Balzter H (2017) Earth observation for land and emergency monitoring. University of Leicester Leicester

Balzter H, Luckman A, Skinner L, Rowland C, Dawson T (2007a) Observations of for-est stand top height and mean height from interferometric SAR and LiDAR over a coni-fer plantation at Thetford Forest, UK. Int J Remote Sens 28:1173–1197. https://doi.org/10.1080/01431160600904998

Balzter H, Rowland CS, Saich P (2007b) Forest canopy height and carbon estimation at Monks Wood National Nature Reserve, UK, using dual-wavelength SAR interferometry. Remote Sens Environ 108:224–239. https://doi.org/10.1016/j.rse.2006.11.014

Banks SN, King DJ, Merzouki A, Duffe J (2014a) Characterizing scattering behaviour and assess-ing potential for classification of Arctic shore and near-shore land covers with fine quad-pol RADARSAT-2 data. Can J Remote Sens 40:291–314. https://doi.org/10.1080/07038992.2014.979487

Banks SN, King DJ, Merzouki A, Duffe J (2014b) Assessing RADARSAT-2 for mapping shoreline cleanup and assessment technique (SCAT) classes in the Canadian Arctic. Can J Remote Sens 40:243–267. https://doi.org/10.1080/07038992.2014.968276

Baselga A (2013) Multiple site dissimilarity quantifies compositional heterogeneity among several sites, while average pairwise dissimilarity may be misleading. Ecography (Cop) 36:124–128. https://doi.org/10.1111/j.1600-0587.2012.00124.x

Beck W, Müller J (2007) Impact of heat and drought on tree and stand vitality – dendroecologi-cal methods and first results from level 2-plots in southern Germany. Schr Forstl Fak Univ Göttingen Nord Forstl Versuchsanst 142:120–128

Bergen KM, Gilboy AM, Brown DG (2007) Multi-dimensional vegetation structure in modeling avian habitat. Ecol Inform 2:9–22. https://doi.org/10.1016/j.ecoinf.2007.01.001

Bergen KM, Goetz SJ, Dubayah RO, Henebry GM, Hunsaker CT, Imhoff ML, Nelson RF, Parker GG, Radeloff VC (2009) Remote sensing of vegetation 3-D structure for biodiversity and habitat: review and implications for lidar and radar spaceborne missions. J Geophys Res Biogeosciences 114. https://doi.org/10.1029/2008JG000883

Bonan GB (2008) Ecological climatology: concepts and applications, 2nd edn. *Geogr Res* 48:221–222. https://doi.org/10.1111/j.1745-5871.2009.00640.x

Brosinsky A, Lausch A, Doktor D, Salbach C, Merbach I, Gwillym-Marginato S, Pause M (2013) Analysis of spectral vegetation signal characteristics as a function of soil moisture condi-tions using hyperspectral remote sensing. J Indian Soc Remote Sens 42:311–324. https://doi.org/10.1007/s12524-013-0298-8

Brown TB, Hultine KR, Steltzer H, Denny EG, Denslow MW, Granados J, Henderson S, Moore D, Nagai S, Sanclements M, Sánchez-Azofeifa A, Sonnentag O, Tazik D, Richardson AD (2016) Using phenocams to monitor our changing Earth: toward a global phenocam network. Front Ecol Environ 14:84–93. https://doi.org/10.1002/fee.1222

Bruelheide H, Nadrowski K, Assmann T, Bauhus J, Both S, Buscot F, Chen XY, Ding B, Durka W, Erfmeier A, Gutknecht JLM, Guo D, Guo LD, Härdtle W, He JS, Klein AM, Kühn P, Liang Y, Liu X, Michalski S, Niklaus PA, Pei K, Scherer-Lorenzen M, Scholten T, Schuldt A, Seidler G, Trogisch S, von Oheimb G, Welk E, Wirth C, Wubet T, Yang X, Yu M, Zhang S, Zhou H, Fischer M, Ma K, Schmid B (2014) Designing forest biodiversity experiments: general con-siderations illustrated by a new large experiment in subtropical China. Methods Ecol Evol 5:74–89. https://doi.org/10.1111/2041-210X.12126

Buddenbaum H, Hill J (2015) PROSPECT inversions of leaf laboratory imaging spectros-copy – a comparison of spectral range and inversion technique influences. Photogramm – Fernerkundung – Geoinf 2015:231–240. https://doi.org/10.1127/pfg/2015/0264

Buddenbaum H, Rock G, Hill J, Werner W (2015a) European journal of remote sensing measuring stress reactions of beech seedlings with PRI, fluorescence, temperatures and emissivity from VNIR and thermal field imaging spectroscopy. Eur J Remote Sens 48:263–282. https://doi.org/10.5721/EuJRS20154815

Buddenbaum H, Stern O, Paschmionka B, Hass E, Gattung T, Stoffels J, Hill J, Werner W (2015b) Using VNIR and SWIR field imaging spectroscopy for drought stress monitoring of beech seedlings. Int J Remote Sens 36:4590–4605. https://doi.org/10.1080/01431161.2015.1084435

Buitrago MF, Groen TA, Hecker CA, Skidmore AK (2016) Changes in thermal infrared spectra of plants caused by temperature and water stress. ISPRS J Photogramm Remote Sens 111:22–31. https://doi.org/10.1016/j.isprsjprs.2015.11.003

Bush A, Sollmann R, Wilting A, Bohmann K, Cole B, Balzter H, Martius C, Zlinszky A, Calvignac-Spencer S, Cobbold CA, Dawson TP, Emerson BC, Ferrier S, Gilbert MTP, Herold M, Jones L, Leendertz FH, Matthews L, Millington JDAA, Olson JR, Ovaskainen O, Raffaelli D, Reeve R, Rödel M-OO, Rodgers TW, Snape S, Visseren-Hamakers I, Vogler AP, White PCLL, Wooster MJ, Yu DW (2017) Connecting Earth observation to high-throughput biodiversity data. Nat Ecol Evol 1:0176. https://doi.org/10.1038/s41559-017-0176

Bwangoy JRB, Hansen MC, Roy DP, De Grandi G, Justice CO (2010) Wetland mapping in the Congo Basin using optical and radar remotely sensed data and derived topographical indices. Remote Sens Environ 114:73–86. https://doi.org/10.1016/j.rse.2009.08.004

Cardinale BJ, Duffy JE, Gonzalez A, Hooper DU, Perrings C, Venail P, Narwani A, Mace GM, Tilman D, Wardle DA, Kinzig AP, Daily GC, Loreau M, Grace JB, Larigauderie A, Srivastava DS, Naeem S (2012) Biodiversity loss and its impact on humanity. Nature 486:59–67. https://doi.org/10.1038/Nature11148

Cernansky R (2017) Biodiversity moves beyond counting species. Nature 546:22–24. https://doi.org/10.1038/546022a

Chen X (2016) A case study using remote sensing data to compare biophysical properties of a forest and an urban area in Northern Alabama, USA. J Sustain For 35:261–279. https://doi.org/10.1080/10549811.2016.1166969

Cloude SR, Papathanassiou KP (1998) Polarimetric SAR interferometry. IEEE Trans Geosci Remote Sens 36:1551–1565. https://doi.org/10.1109/36.718859

Cloude SR, Pettier E, Claude SR, Pottier E (1996) A review of target decomposition theorems in radar polarimetry. IEEE Trans Geosci Remote Sens 34:498–518. https://doi.org/10.1109/36.485127

Cord AF, Brauman KA, Chaplin-Kramer R, Huth A, Ziv G, Seppelt R (2017) Priorities to advance monitoring of ecosystem services using Earth observation. Trends Ecol Evol 32:1–13. https://doi.org/10.1016/j.tree.2017.03.003

Cruzan MB, Weinstein BG, Grasty MR, Kohrn BF, Hendrickson EC, Arredondo TM, Thompson PG (2016) Small unmanned aerial vehicles (micro-UAVs, drones) in plant ecology. Appl Plant Sci 4:1600041. https://doi.org/10.3732/apps.1600041

Dash P, Göttsche F-M, Olesen F-S, Fischer H (2002) Land surface temperature and emissivity estimation from passive sensor data: theory and practice-current trends. Int J Remote Sens 23:2563–2594. https://doi.org/10.1080/01431160110115041

Deans AR, Lewis SE, Huala E, Anzaldo SS, Ashburner M, Balhoff JP, Blackburn DC, Blake JA, Burleigh JG, Chanet B, Cooper LD, Courtot M, Csösz S, Cui H, Dahdul W, Das S, Dececchi TA, Dettai A, Diogo R, Druzinsky RE, Dumontier M, Franz NM, Friedrich F, Gkoutos GV, Haendel M, Harmon LJ, Hayamizu TF, He Y, Hines HM, Ibrahim N, Jackson LM, Jaiswal P, James-Zorn C, Köhler S, Lecointre G, Lapp H, Lawrence CJ, Le Novére N, Lundberg JG, Macklin J, Mast AR, Midford PE, Mikö I, Mungall CJ, Oellrich A, Osumi-Sutherland D, Parkinson H, Ramirez MJ, Richter S, Robinson PN, Ruttenberg A, Schulz KS, Segerdell E, Seltmann KC, Sharkey MJ, Smith AD, Smith B, Specht CD, Squires RB, Thacker RW, Thessen A, Fernandez-Triana J, Vihinen M, Vize PD, Vogt L, Wall CE, Walls RL, Westerfeld M, Wharton RA, Wirkner CS, Woolley JB, Yoder MJ, Zorn AM, Mabee P (2015) Finding our way through phenotypes. PLoS Biol 13. https://doi.org/10.1371/journal.pbio.1002033

Devaney J, Barrett B, Barrett F, Redmond J, O'Halloran J (2015) Forest cover estimation in Ireland using radar remote sensing: a comparative analysis of forest cover assessment methodologies. PLoS One 10:1–27. https://doi.org/10.1371/journal.pone.0133583

Dingle Robertson L, King DJ, Davies C (2015) Object-based image analysis of optical and radar variables for wetland evaluation. Int J Remote Sens 36:5811–5841

Doktor D, Lausch A, Spengler D, Thurner M (2014) Extraction of plant physiological status from hyperspectral signatures using machine learning methods. Remote Sens 6:12247–12274. https://doi.org/10.3390/rs61212247

Dubeau P, King DJ, Unbushe DG, Rebelo LM (2017) Mapping the Dabus Wetlands, Ethiopia, using random forest classification of Landsat, PALSAR and topographic data. Remote Sens 9:1–23. https://doi.org/10.3390/rs9101056

Duro DC, Coops NC, Wulder MA, Han T (2007) Development of a large area biodiversity monitoring system driven by remote sensing. Prog Phys Geogr 31:235–260. https://doi.org/10.1177/0309133307079054

Ehrhardt DW, Frommer WB (2012) New technologies for 21st century plant science. Plant Cell 24:374–394

Eisenhauer N (2018) Aboveground-belowground interactions drive the relationship between plant diversity and ecosystem function. Res Ideas Outcomes 4:e23688. https://doi.org/10.3897/rio.4.e23688

Eldredge N, Cracraft J (1980) Phylogenetic patterns and the evolutionary process. Columbia University Press, New York

Englhart S, Keuck V, Siegert F (2011) Aboveground biomass retrieval in tropical forests – the potential of combined X- and L-band SAR data use. Remote Sens Environ 115:1260–1271. https://doi.org/10.1016/j.rse.2011.01.008

Ewald M, Dupke C, Heurich M, Müller J, Reineking B (2014) LiDAR remote sensing of forest structure and GPS telemetry data provide insights on winter habitat selection of european roe deer. Forests 5:1374–1390. https://doi.org/10.3390/f5061374

Feilhauer H, Schmidtlein S (2009) Mapping continuous fields of forest alpha and beta diversity. Appl Veg Sci 12:429–439. https://doi.org/10.1111/j.1654-109X.2009.01037.x

Féret J-B, Asner GP (2014) Mapping tropical forest canopy diversity using high-fidelity imaging spectroscopy. Ecol Appl 24(6):1289–1296. Published by: Wiley on behal. Ecol. Appl 24, 1289–1296

Fiorani F, Schurr U (2013) Future scenarios for plant phenotyping. Annu Rev Plant Biol 64:267–291. https://doi.org/10.1146/annurev-arplant-050312-120137

Foody GM, Cutler MEJ (2003) Tree biodiversity in protected and logged Bornean tropical rain forests and its measurement by satellite remote sensing. J Biogeogr 30:1053–1066

Foody GM, Green RM, Lucas RM, Curran PJ, Honzak M, Do Amaral I (1997) Observations on the relationship between SIR-C radar backscatter and the biomass of regenerating tropical forests. Int J Remote Sens 18:687–694. https://doi.org/10.1080/014311697219024

Freeman A, Durden SL (1998) A three-component scattering model for polarimetric SAR data. IEEE Trans Geosci Remote Sens 36:963–973. https://doi.org/10.1109/36.673687

Furbank RT (2009) Foreword: plant phenomics: from gene to form and function. Funct Plant Biol 36:v. https://doi.org/10.1071/FPv36n11_FO

Garnier E, Lavorel S, Ansquer P, Castro H, Cruz P, Dolezal J, Eriksson O, Fortunel C, Freitas H, Golodets C, Grigulis K, Jouany C, Kazakou E, Kigel J, Kleyer M, Lehsten V, Lepš J, Meier T, Pakeman R, Papadimitriou M, Papanastasis VP, Quested H, Quétier F, Robson M, Roumet C, Rusch G, Skarpe C, Sternberg M, Theau JP, Thébault A, Vile D, Zarovali MP (2007) Assessing the effects of land-use change on plant traits, communities and ecosystem functioning in grasslands: a standardized methodology and lessons from an application to 11 European sites. Ann Bot 99:967–985. https://doi.org/10.1093/aob/mcl215

Getzin S, Nuske RS, Wiegand K (2014) Using unmanned aerial vehicles (UAV) to quantify spatial gap patterns in forests. Remote Sens 6:6988–7004. https://doi.org/10.3390/rs6086988

Gillespie TW (2005) Predicting woody-plant species richness in tropical dry forests: a case study from South Florida, USA. Ecol Appl 15:27–37. Published by: Wiley on behalf of the Ecological Society of America Stable. http://www.jstor.org/stable/4543333

Gosselin G, Touzi R, Cavayas F (2013) Radarsat-2 wetland classification using the Touzi decomposition: case of the Lac Saint-Pierre RAMSAR wetland. Can J Remote Sens 36:491–506

Gould W (2000) Remote sensing of vegetation, plant species richness, and regional biodiversity hotspots. Ecological Ecol Appl 10:1861–1870. Published by: Wiley on behalf of the Ecological Society of America Stable. http://www.jstor.org/stable/2641244

Green JL, Bohannan JM, Whitaker RJ, Bohannan BJM, Whitaker RJ (2008) Microbial biogeography: from taxonomy to traits. Science 320:1039–1043. https://doi.org/10.1126/science.1153475

Großkinsky DK, Pieruschka R, Svensgaard J, Rascher U, Christensen S, Schurr U, Roitsch T (2015a) Phenotyping in the fields: dissecting the genetics of quantitative traits and digital farming. New Phytol 207:950–952. https://doi.org/10.1111/nph.13529

Großkinsky DK, Svensgaard J, Christensen SRT (2015b) Plant phenomics and the need for physiological phenotyping across scales to narrow the genotype-to-phenotype knowledge gap. J Exp Bot 66:5429–5440

Häme T, Rauste Y, Antropov O, Ahola H a, Kilpi J (2013) Improved mapping of tropical forests with optical estimation. Sel Top Appl Earth Obs Remote Sens IEEE J 6:92–101. https://doi.org/10.1109/JSTARS.2013.2241020

Hantsch L, Braun U, Scherer-Lorenzen M, Bruelheide H (2013) Species richness and species identity effects on occurrence of foliar fungal pathogens in a tree diversity experiment. Ecosphere 4:art 81. https://doi.org/10.1890/es13-00103.1

Hardin PJ, Jackson MW (2005) An unmanned aerial vehicle for rangeland photography. Rangel Ecol Manag. https://doi.org/10.2111/1551-5028(2005)058[0439:AUAVFR]2.0.CO;2

Hart JK, Martinez K (2006) Environmental sensor networks: a revolution in the Earth system science? Earth-Sci Rev 78:177–191. https://doi.org/10.1016/j.earscirev.2006.05.001

Hector A, Philipson C, Saner P, Chamagne J, Dzulkifli D, O'Brien M, Snaddon JL, Ulok P, Weilenmann M, Reynolds G, Godfray HCJ (2011) The Sabah biodiversity experiment: a long-term test of the role of tree diversity in restoring tropical forest structure and functioning. Philos Trans R Soc B Biol Sci 366:3303–3315. https://doi.org/10.1098/rstb.2011.0094

Henderson FM, Lewis AJ (1998) Principles and applications of imaging radar. Manual of remote sensing, vol 2. Wiley, New York

Heurich M (2008) Automatic recognition and measurement of single trees based on data from airborne laser scanning over the richly structured natural forests of the Bavarian Forest National Park. For Ecol Manag 255:2416–2433. https://doi.org/10.1016/j.foreco.2008.01.022

Heurich M, Thoma F (2008) Estimation of forestry stand parameters using laser scanning data in temperate, structurally rich natural European beech (Fagus sylvatica) and Norway spruce (Picea abies) forests. Forestry 81:645–661. https://doi.org/10.1093/forestry/cpn038

Hoffmann H, Nieto H, Jensen R, Guzinski R, Zarco-Tejada P, Friborg T (2016) Estimating evaporation with thermal UAV data and two-source energy balance models. Hydrol Earth Syst Sci 20:697–713. https://doi.org/10.5194/hess-20-697-2016

Homolová L, Maenovsky Z, Clevers JGPW, Garcia-Santos G, Schaepman ME (2013) Review of optical-based remote sensing for plant trait mapping. Ecol Complex 15:1–16. https://doi.org/10.1016/j.ecocom.2013.06.003

Hong SH, Wdowinski S (2014) Double-bounce component in cross-polarimetric SAR from a new scattering target decomposition. IEEE Trans Geosci Remote Sens 52:3039–3051. https://doi.org/10.1109/TGRS.2013.2268853

Hong S, Kim H, Wdowinski S, Feliciano E (2015) Evaluation of polarimetric SAR decomposition for classifying wetland vegetation types. Remote Sens 7:8563–8585. https://doi.org/10.3390/rs70708563

Hwang J, Shin C, Yoe H (2010) Study on an agricultural environment monitoring server system using wireless sensor networks. Sensors 10:11189–11211. https://doi.org/10.3390/s101211189

Imhoff ML, Sisk TD, Milne A, Morgan G, Orr T (1997) Remotely sensed indicators of habitat heterogeneity: use of synthetic aperture radar in mapping vegetation structure and bird habitat. Remote Sens Environ 60:217–227. https://doi.org/10.1016/S0034-4257(96)00116-2

Jansen M, Gilmer F, Biskup B, Nagel KA, Rascher U, Fischbach A, Briem S, Dreissen G, Tittmann S, Braun S, De Jaeger I, Metzlaff M, Schurr U, Scharr H, Walter A (2009) Simultaneous phenotyping of leaf growth and chlorophyll fluorescence via Growscreen Fluoro allows detection of stress tolerance in Arabidopsis thaliana and other rosette plants. Funct Plant Biol 36:902–914. https://doi.org/10.1071/FP09095

Joshi N, Mitchard ETA, Woo N, Torres J, Moll-rocek J, Ehammer A (2015) Mapping dynamics of deforestation and forest degradation in tropical forests using radar satellite data. Environ Res Lett 10:34014. https://doi.org/10.1088/1748-9326/10/3/034014

Joshi N, Baumann M, Ehammer A, Fensholt R, Grogan K, Hostert P, Jepsen MR, Kuemmerle T, Meyfroidt P, Mitchard ETA, Reiche J, Ryan CM, Waske B (2016) A review of the application of optical and radar remote sensing data fusion to land use mapping and monitoring. Remote Sens 8:1–23. https://doi.org/10.3390/rs8010070

Kasischke ES, Bourgeauchavez LL, French NHF, Harrell P, Christensen NL (1992) Initial observations on using SAR to monitor wildfire scars in boreal forests. Int J Remote Sens 13:3495–3501. https://doi.org/10.1080/01431169208904137

Kellndorfer J, Walker W, Pierce L, Dobson C, Fites JA, Hunsaker C, Vona J, Clutter M (2004) Vegetation height estimation from Shuttle Radar Topography Mission and National Elevation Datasets. Remote Sens Environ 93:339–358. https://doi.org/10.1016/j.rse.2004.07.017

Kim JW, Lu Z, Lee H, Shum CK, Swarzenski CM, Doyle TW, Baek SH (2009) Integrated analysis of PALSAR/Radarsat-1 InSAR and ENVISAT altimeter data for mapping of absolute water level changes in Louisiana wetlands. Remote Sens Environ 113:2356–2365. https://doi.org/10.1016/j.rse.2009.06.014

Koch B (2010) Status and future of laser scanning, synthetic aperture radar and hyperspectral remote sensing data for forest biomass assessment. ISPRS J Photogramm Remote Sens 65:581–590. https://doi.org/10.1016/j.isprsjprs.2010.09.001

Koch B, Kattenborn T, Straub C, Vauhkonen J (2014) Segmentation of forest to tree objects. In: Maltamo M, Næsset E, Vauhkonen J (eds) Forestry applications of airborne laser scanning, vol 27. Springer, Dordrecht, pp 89–112

Konishi A, Eguchi A, Hosoi F, Omasa K (2009) 3D monitoring spatio–temporal effects of herbicide on a whole plant using combined range and chlorophyll a fluorescence imaging. Funct Plant Biol 36:874. https://doi.org/10.1071/FP09108

Kraft S, Del Bello U, Bouvet M, Drusch M, Moreno J (2012) FLEX: ESA's Earth explorer 8 candidate mission. Int Geosci Remote Sens Symp:7125–7128. https://doi.org/10.1109/IGARSS.2012.6352020

Krajewski WF, Anderson MC, Eichinger WE, Entekhabi D, Hornbuckle BK, Houser PR, Katul GG, Kustas WP, Norman JM, Peters-Lidard C, Wood EF (2006) A remote sensing observatory for hydrologic sciences: a genesis for scaling to continental hydrology. Water Resour Res 42:1–13. https://doi.org/10.1029/2005WR004435

Krajewski P, Chen D, Cwiek H, Van Dijk ADJ, Fiorani F, Kersey P, Klukas C, Lange M, Markiewicz A, Nap JP, Van Oeveren J, Pommier C, Scholz U, Van Schriek M, Usadel B, Weise S (2015) Towards recommendations for metadata and data handling in plant phenotyping. J Exp Bot 66:5417–5427. https://doi.org/10.1093/jxb/erv271

Kröhnert M, Anderson R, Bumberger J, Dietrich P, Harpole WS, Maas HG (2018) Watching grass grow – a pilot study on the suitability of photogrammetric techniques for quantifying change in aboveground biomass in grassland experiments. Int Arch Photogramm Remote Sens Spat Inf Sci – ISPRS Arch 42:539–542. https://doi.org/10.5194/isprs-archives-XLII-2-539-2018

Kuenzer C, Guo H, Ottinger M, Zhang J, Dech S (2013) Spaceborne thermal infrared observation – an overview of most frequently used sensors for applied research. Remote Sens Digit Image Process:131–148. https://doi.org/10.1007/978-94-007-6639-6_7

Kuenzer C, Ottinger M, Wegmann M, Guo H (2014) Earth observation satellite sensors for biodiversity monitoring: potentials and bottlenecks. Int J Remote Sens 35:6599–6647. https://doi.org/10.1080/01431161.2014.964349

Künzer C, Dech S (2013) Infrared remote sensing – sensors, methods, applications. Springer, Dordrecht

Kuplich TM, Curran PJ, Atkinson PM (2005) Relating SAR image texture to the biomass of regenerating tropical forests. Int J Remote Sens 26:4829–4854. https://doi.org/10.1080/01431160500239107

Kustas WP, French AN, Hatfield JL, Jackson TJ, Susan Moran M, Rango A, Ritchie JC, Schmugge TJ, Moran MS, Rango A, Ritchie JC, Schmugge TJ (2003) Remote sensing research in hydrometeorology. Photogramm Eng Remote Sens 69:631–646. https://doi.org/10.14358/PERS.69.6.631

Latifi H, Nothdurft A, Koch B (2010) Non-parametric prediction and mapping of standing timber volume and biomass in a temperate forest: application of multiple optical/LiDAR-derived predictors. Forestry 83:395–407. https://doi.org/10.1093/forestry/cpq022

Latifi H, Fassnacht FE, Muller J, Tharani A, Dech S, Heurich M (2015) Forest inventories by LiDAR data: a comparison of single tree segmentation and metric-based methods for inventories of a heterogeneous temperate forest. Int J Appl Earth Obs Geoinf 42:162–174. https://doi.org/10.1016/j.jag.2015.06.008

Lausch A, Pause M, Schmidt A, Salbach C, Gwillym-Margianto S, Merbach I (2013) Temporal hyperspectral monitoring of chlorophyll, LAI, and water content of barley during a growing season. Can J Remote Sens 39:191–207. https://doi.org/10.5589/m13-028

Lausch A, Blaschke T, Haase D, Herzog F, Syrbe R-U, Tischendorf L, Walz U (2015a) Understanding and quantifying landscape structure – a review on relevant process characteristics, data models and landscape metrics. Ecol Model 295:31–41. https://doi.org/10.1016/j.ecolmodel.2014.08.018

Lausch A, Salbach C, Schmidt A, Doktor D, Merbach I, Pause M (2015b) Deriving phenology of barley with imaging hyperspectral remote sensing. Ecol Model 295:123–135. https://doi.org/10.1016/j.ecolmodel.2014.10.001

Lausch A, Schmidt A, Tischendorf L (2015c) Data mining and linked open data – new perspectives for data analysis in environmental research. Ecol Model 295:5–17. https://doi.org/10.1016/j.ecolmodel.2014.09.018

Lausch A, Bannehr L, Beckmann M, Boehm C, Feilhauer H, Hacker JM, Heurich M, Jung A, Klenke R, Neumann C, Pause M, Rocchini D, Schaepman ME, Schmidtlein S, Schulz K, Selsam P, Settele J, Skidmore AK, Cord AF (2016a) Linking Earth observation and taxonomic, structural and functional biodiversity: local to ecosystem perspectives. Ecol Indic 70:317–339. https://doi.org/10.1016/j.ecolind.2016.06.022

Lausch A, Erasmi S, King DJ, Magdon P, Heurich M (2016b) Understanding forest health with remote sensing -part I – a review of spectral traits, processes and remote-sensing characteristics. Remote Sens 8:1029. https://doi.org/10.3390/RS8121029

Lausch A, Erasmi S, King D, Magdon P, Heurich M (2017) Understanding forest health with remote sensing-part II – a review of approaches and data models. Remote Sens 9:129. https://doi.org/10.3390/rs9020129

Lausch A, Borg E, Bumberger J, Dietrich P, Heurich M, Huth A, Jung A, Klenke R, Knapp S, Mollenhauer H, Paasche H, Paulheim H, Pause M, Schweitzer C, Schmulius C, Settele J, Skidmore A, Wegmann M, Zacharias S, Kirsten T, Schaepman M (2018a) Understanding forest health with remote sensing, part III: requirements for a scalable multi-source forest health monitoring network based on data science approaches. Remote Sens 10:1120. https://doi.org/10.3390/rs10071120

Lausch A, Olaf B, Stefan K, Leitao P, Jung A, Rocchini D, Schaepman ME, Skidmore AK, Tischendorf L, Knapp S (2018b) Understanding and assessing vegetation health by in-situ species and remote sensing approaches. Methods Ecol Evol 9:1799 1809. https://doi.org/10.1111/2041-210X.13025

Lausch A, Baade J, Bannehr L, Borg E, Bumberger J, Chabrilliat S, Dietrich P, Gerighausen H, Glässer C, Hacker JM, et al (2019) Linking remote sensing and geodiversity and their traits relevant to biodiversity—part I: soil characteristics. Remote Sens 11:2356. https://doi.org/10.3390/rs11202356

Lawton JH, Naeem S, Woodfin RM, Brown VK, Gange A, Godfray HCJ, Heads PA, Lawler S, Magda D, Thomas CD, Tompson LJ, Young S (1993) The Ecotron: a controlled environmental facility for the investigation of population and ecosystem processes. Philos Trans Biol Sci 341:181–194. https://doi.org/10.1098/rstb.1993.0102

Le Toan T, Quegan S, Woodward I, Lomas M, Delbart N, Picard G (2004) Relating radar remote sensing of biomass to modelling of forest carbon budgets. Clim Chang 67:379–402. https://doi.org/10.1007/s10584-004-3155-5

Le Toan T, Quegan S, Davidson MWJ, Balzter H, Paillou P, Papathanassiou K, Plummer S, Rocca F, Saatchi S, Shugart H, Ulander L (2011) The BIOMASS mission: mapping global forest biomass to better understand the terrestrial carbon cycle. Remote Sens Environ 115:2850–2860. https://doi.org/10.1016/j.rse.2011.03.020

Lefsky MA, Cohen WB, Parker GG, Harding DJ (2002) Lidar remote sensing for ecosystem studies. Bioscience 52:19. https://doi.org/10.1641/0006-3568(2002)052[0019:LRSFES]2.0.CO;2

Legendre P, Legendre L 1998 Numerical ecology, 2nd English edn. Amsterdam, Elsevier. https://doi.org/10.1017/CBO9781107415324.004

Li Z-L, Becker F (1993) Feasibility of land surface temerature and emissivity determination from AVHRR data. Remote Sens Environ 43:67–85

Li L, Zhang Q, Huang D (2014) A review of imaging techniques for plant phenotyping. Sensors (Switzerland) 14:20078–20111. https://doi.org/10.3390/s141120078

Lloret J, Garcia M, Bri D, Sendra S (2009) A wireless sensor network deployment for rural and forest fire detection and verification. Sensors 9:8722–8747. https://doi.org/10.3390/s91108722

Lopatin J, Fassnacht FE, Kattenborn T, Schmidtlein S (2017) Mapping plant species in mixed grassland communities using close range imaging spectroscopy. Remote Sens Environ 201:12–23. https://doi.org/10.1016/j.rse.2017.08.031

Lowry J, Hess L, Rosenqvist A (2009) Mapping and monitoring wetlands around the world using ALOS PALSAR: the ALOS Kyoto and carbon initiative wetlands products. Innov Remote Sens Photogramm:105–120. https://doi.org/10.1007/978-3-540-93962-7

Lu D (2006) The potential and challenge of remote sensing-based biomass estimation. Int J Remote Sens 27:1297–1328. https://doi.org/10.1080/01431160500486732

Lu B, He Y, Liu H (2016) Investigating species composition in a temperate grassland using unmanned aerial vehicle-acquired imagery. In: 4th international workshop on Earth observation and remote sensing applications, EORSA 2016 – proceedings. IEEE pp 107–111. https://doi.org/10.1109/EORSA.2016.7552776

Lucas RM, Cronin N, Lee A, Moghaddam M, Witte C, Tickle P (2006) Empirical relationships between AIRSAR backscatter and LiDAR-derived forest biomass, Queensland, Australia. Remote Sens Environ 100:407–425. https://doi.org/10.1016/j.rse.2005.10.019

Luckman A, Baker J, Kuplich TM, Corina da Costa FY, Alejandro CF (1997) A study of the relationship between radar backscatter and regenerating tropical forest biomass for spaceborne SAR instruments. Remote Sens Environ 60:1–13. https://doi.org/10.1016/S0034-4257(96)00121-6

Lutz DA, Washington-Allen RA, Shugart HH (2008) Remote sensing of boreal forest biophysical and inventory parameters: a review. Can J Remote Sens 34:S286–S313. https://doi.org/10.5589/m08-057

Mafuta M, Zennaro M, Bagula A, Ault G, Gombachika H, Chadza T (2013) Successful deployment of a wireless sensor network for precision agriculture in Malawi. Int J Distrib Sens Netw 2013:1–13. https://doi.org/10.1155/2013/150703

Maltamo M, Eerikäinen K, Packalén P, Hyyppä J (2006) Estimation of stem volume using laser scanning-based canopy height metrics. Forestry 79:217–229. https://doi.org/10.1093/forestry/cpl007

Manninen T, Korhonen L, Voipio P, Lahtinen P, Stenberg P (2009) Leaf area index (LAI) estimation of boreal forest using wide optics airborne winter photos. Remote Sens 1:1380–1394. https://doi.org/10.3390/rs1041380

Martins F d SRV, dos Santos JR, Galvão LS, Xaud HAM (2016) Sensitivity of ALOS/PALSAR imagery to forest degradation by fire in northern Amazon. Int J Appl Earth Obs Geoinf 49:163–174. https://doi.org/10.1016/j.jag.2016.02.009

Mauro Brum J, Oliveira RS, Gutierrez J, Licata JPTG (2016) Effects of the 2015 El-Niño extreme drought on the sapflow of trees in eastern Amazonia. In: Proceedings of the soil-plant-atmosphere interactions in a tropical montane cloud forest. Emerging issues in tropical ecohydrology, 2016 – AGU CHAPMAN conference, Cuenca, Ecuador, pp 5–9

Mayr E (1942) Systematics and the origin of species: from the viewpoint of a zoologist. Nature. https://doi.org/10.1038/151347a0

Mayr E (1969) The biological meaning of species. Biol J Linn Soc 1:311–320. https://doi.org/10.1111/j.1095-8312.1969.tb00123.x

Mermoz S, Le Toan T (2016) Forest disturbances and regrowth assessment using ALOS PALSAR data from 2007 to 2010 in Vietnam, Cambodia and Lao PDR. Remote Sens 8:1–22. https://doi.org/10.3390/rs8030217

Mildrexler DJ, Zhao M, Heinsch FA, Running SW (2007) A new satellite-based methodology for continental-scale disturbance detection. Ecol Appl 17:235–250. https://doi.org/10.1890/1051-0761(2007)017[0235:ANSMFC]2.0.CO;2

Mitchard ETA, Saatchi SS, Lewis SL, Feldpausch TR, Woodhouse IH, Sonké B, Rowland C, Meir P (2011) Measuring biomass changes due to woody encroachment and deforestation/degradation in a forest-savanna boundary region of central Africa using multi-temporal L-band radar backscatter. Remote Sens Environ 115:2861–2873. https://doi.org/10.1016/j.rse.2010.02.022

Mittler R, Blumwald E (2010) Genetic engineering for modern agriculture: challenges and perspectives. Annu Rev Plant Biol 61:443–462. https://doi.org/10.1146/annurev-arplant-042809-112116

Moeser D, Roubinek J, Schleppi P, Morsdorf F, Jonas T (2014) Canopy closure, LAI and radiation transfer from airborne LiDAR synthetic images. Agric For Meteorol 197:158–168. https://doi.org/10.1016/j.agrformet.2014.06.008

Mollenhauer H, Schima R, Assing M, Mollenhauer O, Dietrich P, Bumberger J (2015) Development of innovative and inexpensive optical sensors in wireless ad-hoc sensor networks for environmental monitoring. In: 12th EGU general assembly, in Wien, Austria, 12–17 April, 2015

Mollenhauer H, Remmler P, Schuhmann G, Lausch A, Merbach I, Assing M, Mollenhauer O, Dietrich P, Bumberger J (2016) Adaptive multichannel radiation sensors for plant parameter monitoring. In: Geophysical research abstracts vol. 18, EGU (European Geosciences Union General Assembly) 2016–7238, 2016 EGU General Assembly 2016, Austria, Vienna, 17–22 April 2016

Mueller F, Baessler C, Schubert H, Klotz S (2010) Term ecological research Between theory and application. Springer Science + Business Media B.V. https://doi.org/10.1007/978/-90-481-8782-9_1

Müller J (2009) Forestry and water budget of the lowlands in northeast Germany – consequences for the choice of tree species and for forest management. J Water L Dev 13:133–148. https://doi.org/10.2478/v10025-010-0024-7

Müller B, Bernhardt M, Schulz K (2014) Identification of catchment functional units by time series of thermal remote sensing images. Hydrol Earth Syst Sci 18:5345–5359. https://doi.org/10.5194/hess-18-5345-2014

Müller B, Bernhardt M, Jackisch C, Schulz K (2016) Estimating spatially distributed soil texture using time series of thermal remote sensing – a case study in central Europe. Hydrol Earth Syst Sci 20:3765–3775. https://doi.org/10.5194/hess-20-3765-2016

Næsset E (2002) Predicting forest stand characteristics with airborne scanning laser using a practical two-stage procedure and field data. Remote Sens Environ 80:88–99. https://doi.org/10.1016/S0034-4257(01)00290-5

Næsset E (2007) Airborne laser scanning as a method in operational forest inventory: status of accuracy assessments accomplished in Scandinavia. Scand J For Res 22:433–442. https://doi.org/10.1080/02827580701672147

Nagendra H, Rocchini D, Ghate R, Sharma B, Pareeth S (2010) Assessing plant diversity in a dry tropical forest: comparing the utility of landsat and ikonos satellite images. Remote Sens 2:478–496. https://doi.org/10.3390/rs2020478

Nagendra H, Lucas R, Honrado JP, Jongman RHG, Tarantino C, Adamo M, Mairota P (2013) Remote sensing for conservation monitoring: assessing protected areas, habitat extent, habitat condition, species diversity, and threats. Ecol Indic 33:45–59. https://doi.org/10.1016/j.ecolind.2012.09.014

Neeff T, Dutra LV, Dos Santos JR, Da Costa Freitas C, Araujo LS (2005) Tropical forest measurement by interferometric height modeling and P-band radar backscatter. For Sci 51:585–594

Nekola JC, White PS (1999) Special paper: the distance decay of similarity in biogeography and ecology. J Biogeogr 26:867–878. https://doi.org/10.2307/2656184

Ningthoujam RK, Tansey K, Balzter H, Morrison K, Johnson SCM, Gerard F, George C, Burbidge G, Doody S, Veck N, Llewellyn GM, Blythe T (2016) Mapping forest cover and forest cover change with airborne S-band radar. Remote Sens 8. https://doi.org/10.3390/rs8070577

Ningthoujam RK, Balzter H, Tansey K, Feldpausch TR, Mitchard ETA, Wani AA, Joshi PK (2017) Relationships of S-band radar backscatter and forest aboveground biomass in different forest types. Remote Sens 9:1–17. https://doi.org/10.3390/rs9111116

Oliveira RS, Eller CB, Burgess S, de Barros FV, Muller C, Bittencourt P (2016) Soil-plant-atmosphere interactions in a tropical montane cloud forest. In: Proceedings of the soil-plant-atmosphere interactions in a tropical montane cloud forest. Emerging issues in tropical Ecohydrology, 2016 – AGU CHAPMAN conference, Cuenca, Ecuador, 5–9 June 2016

Palmer MW, Michael W (2005) Distance decay in an old-growth neotropical forest. J Veg Sci 16:161–166

Palmer MW, Earls PG, Hoagland BW, White PS, Wohlgemuth T (2002) Quantitative tools for perfecting species lists. Environmetrics 13:121–137. https://doi.org/10.1002/env.516

Parviainen M, Luoto M, Heikkinen RK (2009) The role of local and landscape level measures of greenness in modelling boreal plant species richness. Ecol Model 220:2690–2701. https://doi.org/10.1016/j.ecolmodel.2009.07.017

Persson Å, Holmgren J, Söderman U (2002) Detecting and measuring individual trees using an airborne laser scanner. Photogramm Eng Remote Sens 68:925–932. 0099-1112/02/6809-925$3.0

Pieruschka R, Lawson T (2015) Preface. J Exp Bot 66:5385–5387. https://doi.org/10.1093/jxb/erv395

Popescu SCC (2011) Lidar remote sensing. Advances in environmental remote sensing: sensors, algorithms, and applications. CRC Press, Taylor & Francis Group, Boca Raton London New York

Rascher U (2007) FLEX – fluorescence explorer: a remote sensing approach to quantify spatio-temporal variations of photosynthetic efficiency from space. Photosynth Res 91:PS234

Rascher U, Alonso L, Burkart A, Cilia C, Cogliati S, Colombo R, Damm A, Drusch M, Guanter L, Hanus J, Hyvärinen T, Julitta T, Jussila J, Kataja K, Kokkalis P, Kraft S, Kraska T, Matveeva M, Moreno J, Muller O, Panigada C, Pikl M, Pinto F, Prey L, Pude R, Rossini M, Schickling A, Schurr U, Schüttemeyer D, Verrelst J, Zemek F (2015) Sun-induced fluorescence – a new probe of photosynthesis: first maps from the imaging spectrometer HyPlant. Glob Chang Biol 21:4673–4684. https://doi.org/10.1111/gcb.13017

Reigber A, Moreira A (2000) First demonstration of airborne SAR tomography using multibaseline L-band data. IEEE Trans Geosci Remote Sens 38:2142–2152. https://doi.org/10.1109/36.868873

Reitberger J, Krzystek P, Stilla U (2008) Analysis of full waveform LIDAR data for the classification of deciduous and coniferous trees. Int J Remote Sens 29:1407–1431. https://doi.org/10.1080/01431160701736448

Rignot E, Salas WA, Skole DL (1997) Mapping deforestation and secondary growth in Rondonia, Brazil, using imaging radar and thematic mapper data. Remote Sens Environ 59:167–179. https://doi.org/10.1016/S0034-4257(96)00150-2

Rocchini D (2007) Effects of spatial and spectral resolution in estimating ecosystem α-diversity by satellite imagery. Remote Sens Environ 111:423–434. https://doi.org/10.1016/j.rse.2007.03.018

Rocchini D, Butini SA, Chiarucci A (2005) Maximizing plant species inventory efficiency by means of remotely sensed spectral distances. Glob Ecol Biogeogr 14:431–437. https://doi.org/10.1111/j.1466-822x.2005.00169.x

Rocchini D, Bacaro G, Chirici G, Da Re D, Feilhauer H, Foody GM, Galluzzi M, Garzon-Lopez CX, Gillespie TW, He KS, Lenoir J, Marcantonio M, Nagendra H, Ricotta C, Rommel E, Schmidtlein S, Skidmore AK, Van De Kerchove R, Wegmann M, Rugani B (2018a) Remotely sensed spatial heterogeneity as an exploratory tool for taxonomic and functional diversity study. Ecol Indic 85:983–990. https://doi.org/10.1016/j.ecolind.2017.09.055

Rocchini D, Luque S, Pettorelli N, Bastin L, Doktor D, Faedi N, Feilhauer H, Feret J-B, Foody GM, Gavish Y, Godinho S, Kunin WE, Lausch A, Leitão P, Marcantonio M, Neteler M, Ricotta C, Schmidtlein S, Vihervaara P, Nagendra H (2018b) Measuring β-diversity by remote sensing: a challenge for biodiversity monitoring. Methods Ecol Evol 9:1787–1798. https://doi.org/10.1111/2041-210X.12941

Rosell JRR, Sanz R (2012) A review of methods and applications of the geometric characterization of tree crops in agricultural activities. Comput Electron Agric 81:124–141. https://doi.org/10.1016/j.compag.2011.09.007

Ruiz-Garcia L, Lunadei L, Barreiro P, Robla JI (2009) A review of wireless sensor technologies and applications in agriculture and food industry: state of the art and current trends. Sensors (Basel) 9:4728–4750. https://doi.org/10.3390/s90604728

Saarinen N, Vastaranta M, Rosnell T, Hakala T, Honkavaara E, Wulder MA, Luoma V, Imai NN, Ribeiro EAW, Holopainen M, Survey NL, Centre PF, Canada NR, Columbia B, Sensing R, Ecology F, Mensuration F, Inventory F (2017) UAV-based photogrammetric point clouds and hyperspectral imaging for mapping biodiversity indicators in boreal forests. XLII: 25–27. https://doi.org/10.5194/isprs-archives-XLII-3-W3-171-2017

Santos JR, Freitas CC, Araujo LS, Dutra LV, Mura JC, Gama FF, Soler LS, Sant'Anna SJS, Sant'Anna SJS (2003) Airborne P-band SAR applied to the aboveground biomass studies in the Brazilian tropical rainforest. Remote Sens Environ 87:482–493. https://doi.org/10.1016/j.rse.2002.12.001

Scherer-Lorenzen M, Schulze ED, Don A, Schumacher J, Weller E (2007) Exploring the functional significance of forest diversity: a new long-term experiment with temperate tree species (BIOTREE). Perspect Plant Ecol Evol Syst 9:53–70. https://doi.org/10.1016/j.ppees.2007.08.002

Schmidtlein S (2005) Imaging spectroscopy as a tool for mapping Ellenberg indicator values. J Appl Ecol 42:966–974. https://doi.org/10.1111/j.1365-2664.2005.01064.x

Schmidtlein S, Feilhauer H, Bruelheide H (2012) Mapping plant strategy types using remote sensing. J Veg Sci 23:395–405. https://doi.org/10.1111/j.1654-1103.2011.01370.x

Schutz BE, Zwally HJ, Shuman CA, Hancock D, DiMarzio JP (2005) Overview of the ICESat mission. Geophys Res Lett 32:1–4. https://doi.org/10.1029/2005GL024009

Schweiger AK, Cavender-Bares J, Townsend PA, Hobbie SE, Madritch MD, Wang R, Tilman D, Gamon JA (2018) Plant spectral diversity integrates functional and phylogenetic components of biodiversity and predicts ecosystem function. Nat Ecol Evol 2:976–982. https://doi.org/10.1038/s41559-018-0551-1

Simard M, Saatchi SS, De Grandi G (2000) The use of decision tree and multiscale texture for classification of JERS-1 SAR data over tropical forest. IEEE Trans Geosci Remote Sens 38:2310–2321. https://doi.org/10.1109/36.868888

Sithole G, Vosselman G (2004) Experimental comparison of filter algorithms for bare-Earth extraction from airborne laser scanning point clouds. ISPRS J Photogramm Remote Sens 59:85–101. https://doi.org/10.1016/j.isprsjprs.2004.05.004

Stoll M, Jones HG (2007) Thermal imaging as a viable tool for monitoring plant stress. J Int des Sci la Vigne du Vin 41:77–84

Sun G, Ranson KJ (1995) A three-dimensional radar backscatter model of forest canopies. IEEE Trans Geosci Remote Sens 33(2):372

Sun L, Schulz K (2015) The improvement of land cover classification by thermal remote sensing. Remote Sens 7:8368–8390. https://doi.org/10.3390/rs70708368

Tanase MA, Kennedy R, Aponte C (2015) Radar burn ratio for fire severity estimation at canopy level: an example for temperate forests. Remote Sens Environ 170:14–31. https://doi.org/10.1016/j.rse.2015.08.025

Tang SJ, Dong PL, Buckles BP (2013) Three-dimensional surface reconstruction of tree canopy from lidar point clouds using a region-based level set method. Int J Remote Sens 34:1373–1385. https://doi.org/10.1080/01431161.2012.720046

Teodoro GS, Eller CB, Pereira L, Brum Jr M, Oliveira RS (2016) Interplay between stomatal regulation capacity, hydraulic traits and growth performance in three shrub species in a tropical montane scrubland under contrasting water availability. In: Proceedings of the soil-plant-atmosphere interactions in a tropical montane cloud forest. Emerging issues in tropical ecohydrology, 2016 – AGU CHAPMAN conference, Cuenca, Ecuador, 5–9. June 2016, contrasting water availability

Thapa RB, Shimada M, Watanabe M, Motohka T, Shiraishi T (2013) The tropical forest in south East Asia: monitoring and scenario modeling using synthetic aperture radar data. Appl Geogr 41:168–178. https://doi.org/10.1016/j.apgeog.2013.04.009

Thiel K, Wehr A (2004) An overview and measurement principle analysis. Int Arch Photogramm Remote Sens Spat Inf Sci 36:14–18

Tighe ML (2012) Empirical assessment of multi-wavelength synthetic aperture radar for land cover and canopy height estimation. Carleton University, Ottawa

Tighe ML, King D, Balzter H, McNairn H, Tighe ML, Balzter H, McNairn H (2009) Comparison of X/C-HH InSAR and L-PolInSAR for canopy height estimation in a lodgepole pine forest. In: Proceedings of 4th international workshop on science and applications of SAR polarimetry and polarimetric interferometry, (4) pp 26–30

Timothy D, Onisimo M, Cletah S, Adelabu S, Tsitsi B (2016) Remote sensing of aboveground forest biomass: a review. Trop Ecol 57:125–132

Tiner RW, Lang MW, Klemas VV (2014) Remote sensing of wetlands: applications and advances. Geosci Remote Sens, IEEE Trans. https://doi.org/10.1109/TGRS.1983.350471

Touzi R (2007) Target scattering decomposition in terms of roll-invariant target parameters. IEEE Trans Geosci Remote Sens 45:73–84. https://doi.org/10.1109/TGRS.2006.886176

Touzi R, Deschamps ARG (2007) Wetland characterization using polarimetric RADARSAT-2 capability. Can J Remote Sens 33:56–67

Townshend JR, Justice CO, Skole D, Malingreau JP, Cihlar J, Teillet P, Sadowski F, Ruttenberg S (1994) The 1 km resolution global data set: needs of the international geosphere biosphere programme! Int J Remote Sens 15:3417–3441. https://doi.org/10.1080/01431169408954338

Treuhaft RN, Law BE, Asner GP (2004) Forest attributes from radar interferometric structure and its fusion with optical remote sensing. Bioscience 54:561. https://doi.org/10.1641/0006-3568(2004)054[0561:FAFRIS]2.0.CO;2

Türke M, Feldmann R, Fürst B, Hartmann H, Herrmann M, Klotz S, Mathias G, Meldau S, Ottenbreit M, Reth S, Schäder M, Trogisch S, Buscot F, Eisenhauer N (2017) Multitrophische Biodiversitätsmanipulation unter kontrollierten Umweltbedingungen im iDiv Ecotron. In: Multitrophische Biodiversitätsmanipulation Unter Kontrollierten Umweltbedingungen Im IDiv Ecotron pp 107–114

Turner W, Spector S, Gardiner N, Fladeland M, Sterling E, Steininger M (2003) Remote sensing for biodiversity science and conservation. Trends Ecol Evol. https://doi.org/10.1016/S0169-5347(03)00070-3

Ulaby F, Held D, Donson M, McDonald KA, Senior T (1987) Relating polaization phase difference of SAR signals to scene properties. IEEE Trans Geosci Remote Sens GE-25:83–92. https://doi.org/10.1109/TGRS.1987.289784

Ustin SL, Gamon JA (2010) Remote sensing of plant functional types. New Phytol 186:795–816. https://doi.org/10.1111/j.1469-8137.2010.03284.x

Vaglio GL, Pirotti F, Callegari M, Chen Q, Cuozzo G, Lingua E, Notarnicola C, Papale D (2017) Potential of ALOS2 and NDVI to estimate forest above-ground biomass, and comparison with lidar-derived estimates. Remote Sens 9. https://doi.org/10.3390/rs9010018

van der Sanden JJ, Hoekman DH (1999) Potential of airborne radar to support the assessment of land cover in a tropical rain forest environment. Remote Sens Environ 68:26–40. https://doi.org/10.1016/S0034-4257(98)00099-6

van Zyl JJ (1989) Unsupervised classification of scattering behavior using radar polarimetry data. IEEE Trans Geosci Remote Sens 27:36–45. https://doi.org/10.1109/36.20273

Varghese AO, Suryavanshi A, Joshi AK, Varghese AO, Suryavanshi A, Joshi AK (2016) Analysis of different polarimetric target decomposition methods in forest density classification using C band SAR data. Int J Remote Sens 37:694–709. https://doi.org/10.1080/01431161.2015.1136448

Virlet N, Costes E, Martinez S, Kelner JJ, Regnard JL (2015) Multispectral airborne imagery in the field reveals genetic determinisms of morphological and transpiration traits of an apple tree hybrid population in response to water deficit. J Exp Bot 66:5453–5465. https://doi.org/10.1093/jxb/erv355

Wagner W, Ullrich ATM (2004) From single-pulse to full-waveform airborne laser scanners: potential and practical challenges. Int Arch Photogramme-try Remote Sens Spat InfSci.:201–206. https://doi.org/10.1007/s10044-005-0018-2

Wallace L, Lucieer A, Malenovsky Z, Turner D, Vopenka P (2016) Assessment of forest structure using two UAV techniques: a comparison of airborne laser scanning and structure from motion (SfM) point clouds. Forests 7:1–16. https://doi.org/10.3390/f7030062

Wang K, Franklin SE, Guo X, He Y, McDermid GJ (2009) Problems in remote sensing of landscapes and habitats. Prog Phys Geogr 33:747–768. https://doi.org/10.1177/0309133309350121

Wang K, Franklin SE, Guo X, Cattet M (2010) Remote sensing of ecology, biodiversity and conservation: a review from the perspective of remote sensing specialists. Sensors 10:9647–9667. https://doi.org/10.3390/s101109647

Wegmann M, Santini L, Leutner B, Safi K, Rocchini D, Bevanda M, Latifi H, Dech S, Rondinini C (2014) Role of African protected areas in maintaining connectivity for large mammals. Philos Trans R Soc B Biol Sci. https://doi.org/10.1098/rstb.2013.0193

Wehr A, Lohr U (1999) Airborne laser scanning – an introduction and overview. ISPRS J Photogramm Remote Sens 54:68–82. https://doi.org/10.1016/s0924-2716(99)00011-8

White L, Brisco B, Dabboor M, Schmitt A, Pratt A (2015) A collection of SAR methodologies for monitoring wetlands. Remote Sens. https://doi.org/10.3390/rs70607615

Whittle M, Quegan S, Uryu Y, Stüewe M, Yulianto K (2012) Detection of tropical deforestation using ALOS-PALSAR: a Sumatran case study. Remote Sens Environ 124:83–98. https://doi.org/10.1016/j.rse.2012.04.027

Wulder MA, White JC, Nelson RF, Naesset E, Ørka HO, Coops NC, Hilker T, Bater CW, Gobakken T (2012) Lidar sampling for large-area forest characterization: a review. Remote Sens Environ 121:196–209. https://doi.org/10.1016/j.rse.2012.02.001

Yamaguchi Y, Moriyama T, Ishido M, Yamada H (2005) Four-component scattering model for polarimetric SAR image decomposition. IEEE Trans Geosci Remote Sens 43:1699–1706. https://doi.org/10.1109/TGRS.2005.852084

Yang Y, Guan H, Batelaan O, McVicar TR, Long D, Piao S, Liang W, Liu B, Jin Z, Simmons CT (2016) Contrasting responses of water use efficiency to drought across global terrestrial ecosystems. Sci Rep 6:23284. https://doi.org/10.1038/srep23284

Yao W, Krzystek P, Heurich M (2012) Tree species classification and estimation of stem volume and DBH based on single tree extraction by exploiting airborne full-waveform LiDAR data. Remote Sens Environ 123:368–380. https://doi.org/10.1016/j.rse.2012.03.027

Yick J, Mukherjee B, Ghosal D (2008) Wireless sensor network survey. Comput Netw 52:2292–2330. https://doi.org/10.1016/j.comnet.2008.04.002

Yu L, Wang N, Meng X (2005) Real-time forest fire detection with wireless sensor networks. In: Proceedings 2005 international conference on wireless communications, networking and mobile computing. IEEE, pp 1214–1217. https://doi.org/10.1109/WCNM.2005.1544272

Zimble DA, Evans DL, Carlson GC, Parker RC, Grado SC, Gerard PD (2003) Characterizing vertical forest structure using small-footprint airborne LiDAR. Remote Sens Environ 87:171–182. https://doi.org/10.1016/S0034-4257(03)00139-1

Open Access This chapter is licensed under the terms of the Creative Commons Attribution 4.0 International License (http://creativecommons.org/licenses/by/4.0/), which permits use, sharing, adaptation, distribution and reproduction in any medium or format, as long as you give appropriate credit to the original author(s) and the source, provide a link to the Creative Commons license and indicate if changes were made.

The images or other third party material in this chapter are included in the chapter's Creative Commons license, unless indicated otherwise in a credit line to the material. If material is not included in the chapter's Creative Commons license and your intended use is not permitted by statutory regulation or exceeds the permitted use, you will need to obtain permission directly from the copyright holder.

Chapter 14
How the Optical Properties of Leaves Modify the Absorption and Scattering of Energy and Enhance Leaf Functionality

Susan L. Ustin and Stéphane Jacquemoud

14.1 Introduction

Leaves interact with light in ways that create a spectral footprint of the terrestrial environment of our planet. Most of the visible light penetrating the Earth's atmosphere is absorbed by leaves, and at wavelengths around 700 nm, just beyond the red visible bands, this pattern abruptly reverses, to reflect about half of the incoming light from 700 to 1000 nm. This region of rapid change in reflectance is termed the "red edge" and produces a distinct pattern in which the Earth's albedo is brighter for wavelengths longer than the red edge than for shorter wavelengths (Arnold et al. 2002; Montañés-Rodríguez et al. 2006). This pattern is a result of the abundance of green leaves in the terrestrial environment absorbing light for photosynthesis, with the red edge the manifestation of the long wavelength edge of chlorophyll pigment absorption. This is one example of how leaf optical properties (the absorption and scattering of different wavelengths of sunlight) permit detection of leaf functional properties.

Because the primary physiological processes and functional properties of seed plants are homologous, evolutionary constraints have optimized physiological properties (Jacquemoud and Ustin 2008; Ustin and Gamon 2010). This means that among higher plant species, concentrations of individual biochemicals in the suite of major biochemicals may vary. So considering overall biochemical traits, different species

S. L. Ustin (✉)
Department of Land, Air and Water Resources, University of California Davis, Davis, CA, USA
e-mail: slustin@ucdavis.edu

S. Jacquemoud
Institut de Physique du Globe de Paris – Sorbonne Paris Cité, 8 Université of Paris Diderot, Paris, France
e-mail: jacquemoud@ipgp.jussieu.fr

© The Author(s) 2020
J. Cavender-Bares et al. (eds.), *Remote Sensing of Plant Biodiversity*,
https://doi.org/10.1007/978-3-030-33157-3_14

may have different concentrations of biochemicals that together produce a distinct pattern. Minor biochemicals, such as specific defensive compounds, may reveal more about biodiversity by their presence or absence than by their concentrations.

In their paper on the worldwide leaf economics spectrum, Wright et al. (2004) showed that plant investments in leaf traits represent long-term adaptations to climate characteristics such as length of the growing season, air temperature, and precipitation. Global research on the concept of relating leaf traits to ecosystem functionality (e.g., Wright et al. 2005; Ordoñez et al. 2009; Kattge et al. 2011) and the use of these properties to better understand functional adaptations has rapidly expanded.

In recent years, there has been strong interest in using optical properties to elucidate biodiversity patterns, and identification of functional traits with phylogenetic associations has become a key new objective of spectroscopy and remote sensing (RS) (Martin, Chap. 5; Meireles et al. Chap. 7. With rapid losses of biodiversity, there is a need to improve understanding, identify hot spots, and predict how patterns of biodiversity may change in the future. This has led to renewed interest in whether the optical properties of plants can be understood in a phylogenetic context as well as their functional processes.

14.2   On the Optical Spectrum of Seed Plants

There are many definitions of what constitutes the optical spectrum. Most narrowly, it is visible light—the part of the electromagnetic spectrum that can be seen by the human eye (wavelengths 400–700 nm). The full solar spectrum includes all the wavelengths of electromagnetic energy from the sun that reach the Earth's surface. These wavelengths start in the zone of ultraviolet (UV) A (generally longer than 370 nm) and include visible light, near-infrared (NIR, 700–1000 nm), and shortwave-infrared (SWIR, 1000–3000 nm, also termed middle-infrared in some disciplines). Solar energy interacts with a leaf across the full range of wavelengths to produce the leaf's optical properties, which are determined by its biochemical and biophysical characteristics. In recent years, with improved detector technology, it has become possible to measure reflected sunlight with satellite and airborne imagers that have sufficient spectral resolution to access the absorption patterns of an increasing number of chemical compounds. It is the variation in the full suite of chemistry and scattering properties that allows identification of plant species from their leaf spectra— the patterns of absorption and reflection across all wavelengths that can be measured in the solar spectrum.

Seed plants have three basic types of leaves. Monocot and dicot leaves of angiosperms typically have a wide blade, and conifers have needle-shaped leaves. Figure 14.1 shows examples of typical spectra of evergreen (*Quercus wislizeni*) and deciduous (*Quercus douglasii*) dicot leaves, deciduous leaves of a monocot (*Zea mays*), and evergreen conifer needles (*Pinus ponderosa*). The overall shape of the leaf spectra is similar, with low reflectance across visible wavelengths due to absorption by photosynthetic pigments (Gates et al. 1965). High reflectance is

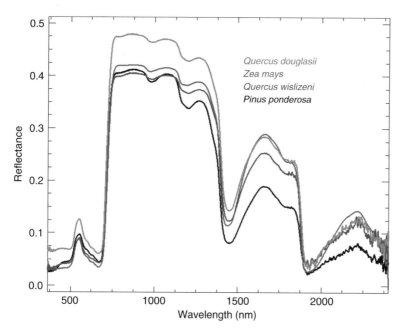

Fig. 14.1 Typical leaf spectra of evergreen and deciduous species, measured on field-grown leaves in June. *Quercus douglasii* (blue oak) is a deciduous dicot species, *Zea mays* (corn) is a deciduous monocot, *Q. wislizeni* (black oak) is an evergreen dicot species, and *Pinus ponderosa* (ponderosa pine) is a needle-leaf conifer

characteristic of the NIR, where generally 10% or less of radiation is absorbed (Jacquemoud and Ustin 2008); intermediate reflectance is characteristic of the SWIR region (1500–2500 nm), where energy is primarily absorbed by water in leaves (Carter 1991) or by plant residues when leaves are dry. Cell wall compounds such as cellulose and lignin or other biochemicals found in the cytoplasm such as proteins and sugars account for many overlapping absorption features in the SWIR.

Differences between spectra in Fig. 14.1 relate to differences in biochemical composition and concentration of pigments, water, and cell wall structural materials or are due to the sources of scattering at either the leaf surface or from internal cellular structures. Despite the similarity of overall shape, Fig. 14.1 clearly shows taxa-specific differences in reflectance across the spectrum. These differences at the leaf scale contribute to our ability to map land cover types and distinguish plant communities, genera, and species, as discussed in other chapters in this book.

Regardless of the diversity in leaf anatomy and morphology, biochemical and biophysical properties exhibit consistent absorption features in the optical spectrum and have been detected in modern high-fidelity airborne imaging spectrometers (e.g., Kokaly et al. 2009; Ustin et al. 2009; Féret et al. 2011; Asner and Martin 2016; Asner et al. 2015). The principles of spectroscopy indicate that many of the chemical compounds with absorption features in the solar spectrum (280–3000 nm) can be detected and identified based on their spectral reflectance features. Both differences in the amount of energy absorbed and the anatomical and morphological differences that enhance scattering of light at different wavelengths allow us to differentiate related taxa. While this chapter is focused on the leaf scale, complications arise at larger scales; even when just scaling to the canopy, it can be difficult to detect leaf optical properties due to the presence of additional materials adding to the measured spectrum (e.g., live and dead leaves, flowers, fruit, bark, and understory, both vegetation and soil). However, in some cases, if the absorptivity of the material is weak, the absorption at the canopy [pixel] scale can be enhanced. An example of this is observed for foliar water content when measured on a single leaf or a spectrum from multiple leaves (e.g., Roberts et al. 2004; Kokaly et al. 2009). Our ability to measure leaf optical properties from airborne and satellite sensors varies with spatial scales, as discussed in Chap. 16.

14.3 Leaf Reflectance Patterns

Reflectance is the fraction of light reflected from the leaf surface (R_s). It is composed of two parts: specular reflectance, which reflects directly off the surface in the forward direction, and diffuse reflectance, which scatters light in all directions from the surface. Light can be specularly scattered at some wavelengths and diffusely scattered at others, depending on the scale of the roughness of the surface. Specular reflectance happens when light intersects a surface that is smooth, i.e., one with particles much smaller than the wavelengths contacting it. If the surface is rough, that is, composed of particles about the size of the wavelengths of light or larger, it will scatter light diffusely. Specular reflection is a leaf property that is determined by the structure and chemical composition of the cuticle; thus, differences among species are potentially related to biodiversity questions.

The fraction of light that is reflected from the interior of the leaf, R_i, is the diffuse or multiply scattered component. Some fraction of the beam of light that enters the leaf's interior will be absorbed, some transmitted through the leaf, and some will be scattered back out of the upper surface of the leaf. Only the fraction of the incident light that is reflected from the interior of the leaf carries information about the biochemical and structural properties of the leaf.

Reflectance patterns in the visible spectrum are primarily due to photosynthetic pigments that absorb about 90% or more of the incoming light (Gates et al. 1965). Water is the second strongest absorbing molecule in leaves; it absorbs strongly in the SWIR region, with several smaller vibrational overtone absorptions in the NIR (Carter 1991). Because there are no strongly absorbing molecules in the NIR, plants reflect or transmit all but about 10% of the incoming radiation in this region (Jacquemoud and Ustin 2008).

While the spectral shapes of leaves are generally consistent across all green plants, they differ between species and plant functional types. Evergreen leaves usually have thicker cell walls and smaller cells and are more compact than deciduous leaves. Consequently, their spectral signatures generally have lower reflectance in the NIR and SWIR (Fig. 14.1). Phenological changes over the growing season are expressed in leaf reflectance by declining pigment and water contents and increases in the mass of secondary cell wall components as the growing season nears its end. Similar patterns are also observed when comparing leaves from mesic to arid habitats.

In late spring, when the leaves shown in Fig. 14.1 were measured, phenological differences are generally minimized because leaves are near their growth peak. The leaves in Fig. 14.1 show they have high water content as the water absorption features in the NIR, around 970 nm and 1240 nm, are relatively deep for non-succulent leaves. Water causes absorption at all wavelengths longer than 1400 nm, decreasing reflectance across the spectrum. The small absorption feature observed near 1800 nm is from cellulose and other related structural C (C) compounds. In addition to these few large absorption features, many small absorptions exist, only some of which are identified with a specific biochemical. The wide variety of secondary biochemicals that exist in plant leaves and their possible range of concentrations provides a spectral palette that can be used to identify individual species in optical data.

Clearly, the environmental conditions that a plant is exposed to, including soil properties, weather, and its phenological age, alter the leaf's optical properties. The genetic heritage modulates the types of responses of a species to environmental conditions. The reflection of light from the leaf and transmission through the leaf are determined by what wavelengths of light are absorbed by the various biochemical compounds in leaves (chlorophylls, carotenoids, water, cellulose and lignin, proteins, etc.) and the relative strength of the absorptions.

The scattering of light at the leaf surface depends on the structure of the epidermis, the waxes, cutin, and protrusions such as leaf hairs (Ehleringer et al. 1976) and on the orientation of the leaf to the beam of light (Comstock and Mahall 1985; James and Bell 2000). The variety of leaf properties are expressions of different adaptive strategies among species and are related to their functional traits (Serbin and Townsend, Chap. 3). For example, differences in epidermal structure cause leaves of one species to have a bluish powder coating, those of a different species to appear white, and those of another to have a shiny smooth green surface.

Within the leaf, scattering occurs between cells and between organelles within cells (Vogelmann 1993). Vogelmann et al. (1996a) used fiber optics to study scattering processes within cells and tissues to show how leaf anatomy modifies the internal light environment to optimize photosynthetic performance under different habitat conditions.

14.4 Leaf Transmittance Patterns

Transmittance is the fraction of light that enters a leaf and is eventually scattered out the opposite surface. The transmittance spectrum approximates the reflectance spectrum, but they are not exact copies of each other. Transmittance can be greater or lesser than reflectance (Knipling 1970; Wooley 1971; Jacquemoud and Ustin 2008), depending on leaf thickness, number of leaves light passes through, and their optical properties. Thus, transmittance spectra carry information about leaf traits and biodiversity but are seldom used in remote sensing (RS) except in field-based or laboratory studies.

14.5   Leaf Absorptance Patterns

The absorption of light is determined by the absorbing molecules in the leaf balanced by the structural properties that scatter light, such as air spaces and water–air interfaces. In the visible spectrum, it is photosynthetic pigments, primarily chlorophylls and carotenoids that strongly absorb light. Other non-photosynthetic pigments also absorb in this wavelength region, such as anthocyanins (a large and diverse group of flavonoids that are involved with leaf color but also colors of flowers and fruit). Anthocyanins (and more generally flavonoids) provide photoprotection from UV light (Stapleton 1992; Steyn et al. 2002)), such as in alpine environments or during early leaf development (Chalker-Scott 1999; Karageorgou and Manetas 2006) when the photosynthetic machinery is not fully developed. Figure 14.2 shows two adjacent evergreen shrubs in early spring, both widely planted cultivars, one (*Photinia x fraseri*) with red expanding leaves and the other (*Xylosma congesta*) with orange colored expanding leaves. Such differences could indicate pH differences in the vacuoles or different combinations of anthocyanin pigments, or combinations of anthocyanin and carotenoid pigments. Lee and Collins

Fig. 14.2 Red and orange expanding leaves on adjacent shrubs. Color differences could represent different anthocyanin molecules, different pH environments in the vacuoles, or a mixture of carotenoid and anthocyanin pigments

(2001) evaluated leaf ontogeny and found that spring leaf expansion was usually correlated with high anthocyanin pigments in the mesophyll tissue. Along with phenolic compounds, anthocyanins provide protection of the pigment molecules during senescence (Matile 2000) and defensive functions against herbivory (Hamilton and Brown 2001) and during leaf senescence. Some molecules absorb in the UV and blue wavelengths, ranging from the simplest phenol to complex polyphenols like tannic acid. Kokaly and Skidmore (2015) recently demonstrated detection of a phenolic absorption in plants at 1660 nm. Phenolic compounds often provide regulatory and defensive functions. Because of the great diversity of non-photosynthetic pigments and photosynthetic accessory pigments (carotenoids), pigments provide a basis for discriminating among plant species in optical data. Suites of pigments often occur in specific families or clades in agreement with molecular phylogeny (Lee and Collins 2001).

In the NIR, there are no strongly absorbing compounds, so a high proportion of light is reflected or transmitted. Light is often scattered multiple times, increasing the probability of absorption before being reflected or transmitted out of the leaf. For example, Wooley (1971) reported 4% NIR absorptance in soybean (*Glycine max*), and Everitt et al. (1985) found only 5% NIR absorptance in buffalo grass (*Bouteloua dactyloides*) leaves. Scattering within the leaf is related to the internal cellular structure, especially at cell membrane and air interfaces, where light can be reflected and refracted. Allen et al. (1970) showed the volume of intercellular air spaces was highly correlated with NIR reflectance. Multiple scattering of photons causes the NIR reflectance to be much higher than reflectance of visible or SWIR wavelengths, where absorptions by pigments and water result in single scattering processes (light is absorbed or scattered on its first interaction).

Secondary water absorption features in leaves are found around 980 nm and 1240 nm (Carter 1991). The diversity of plant adaptations to different water regimes results in a wide range of leaf water contents among species. The percent water content is generally positively correlated with increasing leaf thickness, but the opposite may occur. Thus, a sclerophyllous leaf species like *Adenostoma fasciculatum* (chamise) can have higher leaf mass area (LMA = 1 divided by specific leaf area) and low water content, but a succulent species may have high water content and a high leaf mass per area (Ackerly et al. 2002; Vendramaini et al. 2002).

The SWIR part of the leaf spectrum is dominated by water absorption when the leaves are living and by leaf chemical constituents when dry. Many plant compounds have absorptions in the SWIR, including cellulose, lignin, nitrogen (N), sugars, starch, and waxes. Interpretation of these absorption features in dry leaves is complicated because many molecules have absorptions at overlapping wavelengths and we lack the specific absorption coefficients to identify them in the data. The cell wall C compounds comprise the largest fraction of the dry biomass of a leaf, and the absorption feature around 1750 nm is generally attributed to these materials (Kokaly et al. 2009). Kokaly (2001) also identified two absorptions at 2054 nm and 2172 nm that cause broadening of the 2100 nm absorption feature due to N compounds. The complexity of relationships among species adaptations for water, structural carbohydrates, and nutrients provides a strong basis for detecting species diversity in RS imagery, at least in local to regional studies (Asner and Martin 2016).

14.6   Physical Processes Underlying Leaf Optical Properties

Leaf anatomy and phylogeny determine how leaf cells (epidermis, mesophyll) and specialized tissues (xylem and phloem conducting tissues and the stomatal complex) are arranged (Al-Edany and Al-Saadi 2012) and how this affects reflectance. There are many modifications, but the standard arrangement of cells and tissues in dicot plants is strongly asymmetric, commonly flattened in the dorsiventral orientation with the adaxial side of the leaf oriented upward (Fig. 14.3). This places the chloroplast-rich palisade parenchyma near the upper surface to intercept incoming light and the spongy mesophyll and large air spaces on the abaxial side of the leaf near the lower surface where all or most of the stomata are located. Clearly, this bifacial arrangement is adapted to facilitate gas exchange and photosynthesis (Parkhurst 1986). Leaves on plants grown under high light tend to have more compact parenchyma than leaves in low light environments, which have more air spaces.

In another common arrangement, grass leaves are isobilateral with the interior of the leaf filled with generalized mesophyll parenchyma cells and stomata on both surfaces. Conifer needles are more cylindrical with the interior filled with generalized parenchyma cells and a centralized vascular bundle, separated from the mesophyll by an endodermal cell layer. Conifer needles are compact, with little air space between cells. This reduces the interior scattering of light within leaves and contributes to the low NIR reflectance of conifer needles compared with broadleaf plants.

Another leaf type has symmetrical palisade parenchyma on both the top and bottom of the leaf. This is common in dicot species with extreme erectophile leaf orientation, such as the hanging adult leaves of eucalyptus species such as *Eucalyptus*

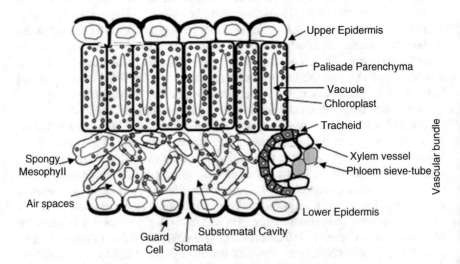

Fig. 14.3 Dorsiventral cross section of a dicot leaf with the adaxial surface at the top. This is the most common structure of dicot leaves, with a distinct asymmetry oriented toward incoming light from the adaxial surface

nitens. The more rounded juvenile leaves of this species have larger air spaces than the adult leaves, suggesting they may be adapted to more rapid growth but have less tolerance to drought (Gras et al. 2005). This type of adult leaf anatomy increases light interception at lower solar zenith angles closer to sunrise and sunset and enhances potential for photosynthesis.

The asymmetric leaf structure of typical dicots enhances capture of energy for photosynthesis by orienting the leaf upright, facilitated by branching angle and petiole orientation. It makes a difference whether light enters the leaf from the upper or lower surfaces because the anatomical structures and biochemistry across the leaf are different. The effect of the asymmetric distribution on reflectance and absorptance patterns is shown in Fig. 14.4 for photosynthetic light at two solar zenith angles entering the leaf from the upper or lower side of the leaf.

The downward flux constitutes most of the net flux when the solar zenith angle is higher (left side); when the solar zenith angle is lower (right side), the upward flux is much larger. The panel (14.4. #1 and 14.4. #2) with the typical dorsiventral orientation has higher net flux through the epidermis than panels (14.4. #3 and 14.4. #4) with the abaxial side receiving the incident flux. About 20% of the net flux enters the palisade parenchyma in 14.4. #1 and 14.4. #2, while 14.4.#3 and 14.4. #4 show very low net flux. Because the chlorophyll concentration is highest in the palisade parenchyma, comparatively little photosynthesis occurs in the spongy mesophyll, with higher net flux of 675 nm light (suitable for chlorophyll a absorption in photosystem II).

Panel 14.4. #1, with the typical dorsiventral orientation (adaxial side up) at the higher solar zenith angle, has higher net flux downward (about 20%) through the epidermis than Panel 14.4. #2 with the lower solar zenith angle, where the net flux is much lower in both palisade and spongy parenchyma and the upward flux is much larger. Panel 14.4. #3, with the abaxial side up, shows little difference in upward flux, but most of the net flux continues to be located in the palisade parenchyma. Panel 14.4. #4, with the abaxial side receiving the light at the lower zenith angle, has virtually no net flux into the palisade parenchyma and high upward flux in the abaxial epidermis. Because the chlorophyll concentration is highest in the palisade parenchyma, there is little photosynthesis elsewhere, even with high net flux of 675 nm light (red wavelength region, suitable for chlorophyll *a* absorption).

14.7 The Epidermis

The epidermis is the outermost layer of leaf cells and is generally one cell layer thick, but some species have several cell layers. The epidermis lacks pigments and is generally transparent to light. The outer cuticle surface is covered by wax to limit uptake and loss of gases, except at the stomatal complexes, which are generally located on the abaxial (lower) side of the leaf. Stomata are composed of two kidney-shaped stomatal cells and two to four guard cells at the ends. The stomata are generally located above an open space in the mesophyll where gases can collect, termed

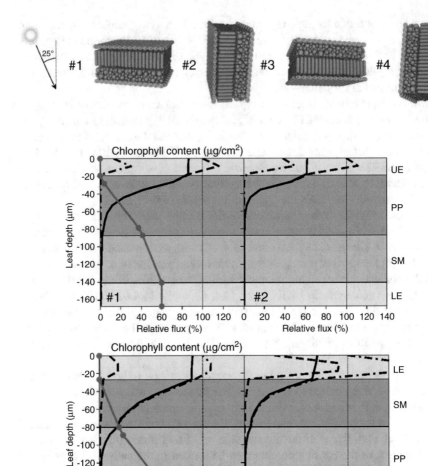

Fig. 14.4 Top panel illustrates the four orientations selected to simulate the absorption profiles of a dicot leaf. Lower panels show the distribution of relative downward flux $\Phi_\downarrow(z)$, dashed line; relative upward flux $\Phi_\uparrow(z)$, dash-dot line; and relative net flux $\Phi_n(z)$, solid line for light at 675 nm wavelength inside a horizontal dorsiventral leaf with light incident on the palisade parenchyma (#1 and #2) and light incident on the spongy mesophyll (#3 and #4), for illumination zenith angles of 25° (#1 and #3) and 65° (#2 and #4). Gray lines (filled circles) indicate the cumulative chlorophyll concentration, expressed in μg cm^{-2}. Symbols are UE upper epidermis, PP palisade parenchyma, SM spongy mesophyll, and LE lower epidermis. (Panels 1 and 2 were modified from Ustin et al. (2001) with permission from Wiley and Sons. Panels 3 and 4 were not previously published)

the stomatal (or substomatal) cavity. The stomata open and close, regulated by turgor pressure, to allow gases, including water vapor, C dioxide, and oxygen, to exchange between the outside air and the stomatal cavity. Because the anatomical structures of species are different across the interior of the leaf, their biochemistry is also specialized, causing the asymmetric distributions of reflectance and absorption patterns as light passes through the leaf, shown in Fig. 14.4, for photosynthetic light at two solar zenith angles entering from the lower side of the leaf.

Xerophytes and hydrophytes illustrate the extremes of leaf adaptations for cuticle characteristics. Aquatic species that float on the surface like water hyacinth (*Eichhornia crassipes*) and water lily (*Nymphaea* sp.) have stomata on the upper, adaxial side that is open to the atmosphere. Submerged aquatic species, like Brazilian waterweed (*Egeria densa*), hydrilla (*Hydrilla verticillata*), and Eurasian watermilfoil (*Myriophyllum spicatum*), lack stomata or they are nonfunctional, and their cuticle is thin and reduced to allow direct gas exchange with the water. These species typically have other morphological traits that support adaptation to the aquatic habitat, including very small leaves. In contrast, xerophytes like semiarid grasses, e.g., the beach grass *Ammophila breviligulata*, and conifers or succulent species like agaves and cacti, have stomata in deep pits that reduce transpiration by retaining high vapor pressure in the cavity. In addition, other traits typically present in xerophytes include thick cuticles, pubescence, and a reduced stomatal complex. Leaf traits such as these tend to be clustered, representing a suite of adaptive traits; thus, to identify a taxon, we expect several traits to be present in particular configurations; thus, potential identification is enhanced by patterns of traits (see Serbin and Townsend, Chap. 3; Morsdorf et al., Chap. 4; Bolch et al., Chap. 12).

14.7.1 Surface Characteristics of Epidermal Cells

Epidermal cells can be coated with smooth wax, which enhances specular scattering off the surface of leaves. This type of surface scattering occurs when the leaf is oriented to cause forward scattering of incoming beam. Light specularly scattered from a leaf surface has the same wavelength composition and intensity as the incoming beam, so it does not provide any information about the interior of the leaf. Smooth, waxy leaves having shiny, glabrous surfaces are often found in young leaves of broadleaf shrubs, trees, or herbaceous understory species. In woody plants, this is generally a mechanism to avoid absorbing excess photosynthetically active radiation under high light conditions. Species that have this trait include members of *Cinnamomum* in the laurel family (e.g., *Cinnamomum camphora* and *C. parthenoxylon*) and *Magnolia grandiflora*, which are found in warm subtropical habitats, often in the understory.

The waxy cuticle or outer layer of the epidermis produces a 3-D structure of waxes and cutin of variable thickness and cell types that create a diverse range of textured surfaces and colors. Leaf traits like thick cuticles are common to a wide range of plants such as columnar cacti (e.g., *Pilosocereus leucocephalus*) and conifer species like the blue Colorado spruce, *Picea pungens*. These plants are often found in high light or drought-stressed environments, or at high elevations, where they may be exposed to high UV radiation.

14.7.2 *Epidermal Cell Shape and Function*

The outer (adaxial) surfaces of epidermal cells form convex shapes. This shape has been shown to focus light into the palisade parenchyma cells (Fig. 14.5), increasing light capture for photosynthesis (Haberlandt 1914; Martin et al. 1989; Bone et al. 1989; Poulson and Vogelmann 1990). Light focusing is a widespread property in seed plants and is common in prairies, deserts, and deciduous forests (Vogelmann et al. 1996b), especially in plants from understory species (e.g., *Medicago sativa*, *Impatiens* spp.) and other species growing in low light conditions. Light focusing increases the photon density (up to 10×) inside the palisade parenchyma cells, and given the mobility of chloroplasts in the palisade parenchyma cells, they assist in optimizing the light environment within the leaf (Poulson and Vogelmann 1990).

The shape of the epidermal cells affects the focal length. By changing the turgor pressure in the epidermal cells, a plant can decrease focal length to increase the absorbing area in shaded habitats or increase it to penetrate deeper into the palisade parenchyma. This ability increases the probability of light absorption and thus is beneficial for leaves in low light environments (Vogelmann et al. 1996a, b; Smith et al. 1997).

Fig. 14.5 Refraction of rays at two different solar zenith angles of 0° (perpendicular to surface) and 15° off vertical in epidermal cells of *Anthurium warocqueanum*. Light is focused on the palisade parenchyma below the single layer of epidermal cells. (Reproduced from Poulson et al. (1989) with permission from the Optical Society of America)

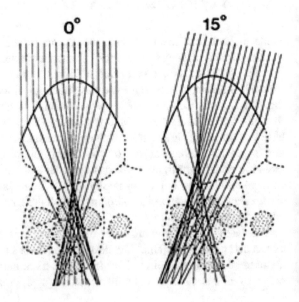

14.7.3 Epidermal Cell Index of Refraction

As light passes into and through the epidermal cell walls and into mesophyll cells, it is scattered in new directions, based on differences in the velocity with which different wavelengths move between the cell solution and the air spaces. The refractive index determines how much the light is bent between different media, as described by the Snell–Descartes' law, which says that the biochemical constituents of the leaf determine the speed of light passing through it relative to its speed in a vacuum. The effective refractive index of leaves varies with the biochemical composition of different species. It also varies with the wavelength of light, as seen when white light is split into its individual colors by a prism. Each wavelength is bent at a different angle, from 40° to 42°, causing separation of the colors. The angles of incidence and refraction at the interfaces between the mesophyll and the cell walls influence the leaf's optical properties by affecting the probability that light is multiply scattered through the cell interior, escapes directly out of the leaf after the first interaction with a surface, or is absorbed.

The refractive index is a complex number in which the real part is the refractive index and the imaginary part is related to the extinction coefficient (also called the mass absorption coefficient), which accounts for light attenuation when photons pass through a medium. These values change across the optical spectrum, and it has not been easy to determine the refractive index for most plant compounds; only pure liquid water has been fully characterized. Thus, in most cases, the refractive index and extinction coefficient cannot be measured directly and must be estimated from measurements of properties that depend on them, such as reflectance and transmittance. Since the 1950s, a long list of investigators have improved knowledge of the real and imaginary parts of the refractive index for water in different regions of the electromagnetic spectrum (e.g., Segelstein 1981; Hale and Querry 1973; Wieliczka et al. 1989) as shown in Fig. 14.6. Because these coefficients change with the phase of water (vapor, liquid, or solid) and its temperature, the phase needs to be specified. Curcio and Petty (1951) were among the first to accurately measure absorption coefficients for liquid water between 700 nm and 2500 nm at 20 °C; they identified absorptions at 760, 970, 1190, 1450, and 1940 nm. Except for the water absorption feature at 760 nm, the liquid water bands at 970 and 1190 nm are readily observed in canopy and leaf spectra. In the laboratory, we observe the much stronger liquid water absorptions at 1450 and 1940 nm; however, in satellite and airborne imagery, these wavelengths are usually saturated with atmospheric water vapor, which has significantly higher concentration over the full atmospheric column than does liquid water in the leaf. As with other molecular absorptions (pigments, liquid water, dry biomass, etc.), as the amount of water vapor increases, the wavelength bands on the shoulders of the absorption maximum also absorb energy, and the feature expands over more wavelengths. Generally, acquisition of airborne imagery is avoided under rainfall or high water vapor conditions because clouds obscure the ground in optical imagery; hence, archives have little data acquired under high liquid water atmospheric conditions.

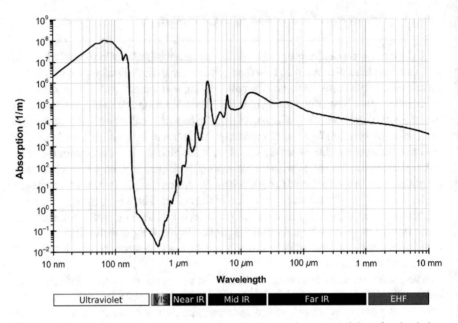

Fig. 14.6 Liquid water absorption spectrum showing the imaginary part of the refractive index, the absorption coefficient. Created 1 July 2008, compiled by Kebes (https://commons.wikimedia. org/wiki/File:Absorption_spectrum_of_liquid_water.png). Accessed 30 Nov 2018) at English Wikipedia CC by SA 3.0 (http://creativecommons.org/licenses/by-sa/3.0). Based on Segelstein (1981), Hale and Querry (1973), and Wieliczka et al. (1989); other references at http://omlc.ogi. edu/spectra/water/abs/index.html

14.8   The Mesophyll

The leaf mesophyll includes all cells between the two epidermal layers and consists of one or more types of ground parenchyma, depending on the cellular specialization. Because these tissues are arranged differently in dicots, monocots, and conifers, spectral responses frequently differ among the three groups. Other taxa, including mosses and ferns, also have different characteristic foliar anatomy and biochemistry that allows them to be distinguished from higher plants by their spectral characteristics (Vogelmann and Moss 1993: Bubier et al. 2007; Van Gaalen et al. 2007). Monocots and conifers generally have one type of ground tissue, called chlorenchyma, which are chloroplast-containing parenchyma cells, while dicot leaves typically have two types of parenchyma cells: elongated densely packed cells that are oriented perpendicular to the upper leaf surface and termed palisade parenchyma and irregular-shaped cells with a high proportion of air spaces in the tissue, called spongy mesophyll (Fig. 14.3). The mesophyll tissue also includes the vascular tissue that transports water and nutrients into the leaf and carbohydrates out of the leaf. These cells are generally within a larger tissue termed the vascular bundle, which includes xylem and phloem cells that conduct fluids and fiber cells and tracheids that provide structural support for the leaf and the conducting tissue. The conducting cells are connected throughout the leaf in a network of veins.

14.8.1 Mesophyll Index of Refraction

The refractive indices of cell walls in the visible wavelengths have been empirically estimated to be 1.4 (Knipling 1970), 1.425 (Gausman 1974), and 1.415 in living cells and 1.53 in dried cell walls (Woolley 1975). It is very difficult to characterize the refractive index for pigments, cellulose, cell walls, proteins, etc. given our inability to measure them in situ. One reason the refractive index is known for so few plant biochemicals is that in their functional state, these cell constituents are bound to membranes in complexes with proteins and other molecules, so when they are isolated, their 3-D structure has been lost along with their bond structures and other molecular interactions. The function of proteins, pigments, enzymes, (including ribulase-1,5-bisphosphate carboxylase, RUBP-Case, RuBisCO), and amino acids (all of which contain N) depends on their structure, and empirical methods have not worked well to obtain their absorption coefficients in vivo, so they cannot be accurately predicted in radiative transfer models. It has long been known that N is generally allocated proportional to optimal photosynthesis (Field and Mooney 1986). More than half of all leaf N is allocated to photosynthetic proteins (Makino and Osmond 1991; Hikosaka and Terashima 1996). Because N forms many types of bonds, it has been necessary to estimate the total concentration of foliar N from training data in empirical models, based on either photosynthesis models or statistical models like partial least squares and other self-learning methods (see Serbin and Townsend, Chap. 3). In recent years, statistical methods have become the preferred approach to estimate leaf N concentration. These can be accurately applied as long as the new data have the same statistical structure and ranges as the original test data. The total concentration of foliar N is often predicted directly from empirical models using methods like partial least squares regression (Smith et al. 2002; Ollinger and Smith 2005; Singh et al. 2015) or, more recently, by self-learning methods, such as wavelets, used by Cheng et al. (2014) to estimate leaf mass area (dry weight · area^{-1}). These methods can easily be overfitted, so care is needed to produce a realistic result (Féret et al. 2011). Nonetheless, these models can be accurate for the vegetation types and concentrations they are trained against, and they are being used to estimate functional properties and biological diversity (Asner et al. 2014a, b; Asner and Martin 2016; Féret et al. 2014a, b).

14.8.2 Molecular Absorption Processes

For a wavelength of light to be absorbed, the amount of energy in a photon must equal the specific energy difference between the resting (ground) state of the electron (S_0) and its excited state in an allowable unoccupied higher energy level (Fig. 14.7). It is the separation of water into an H+ and an OH– ion that provides the electron that is transferred through the electron transport chain in photosynthesis. The magnesium ion in the tetrapyrrolic ring of the chlorophyll molecule helps initially stabilize the charged state long enough to transfer it to a phaeophytin in the

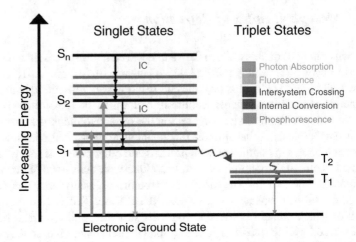

Fig. 14.7 Jablonski diagram showing energy levels for a chlorophyll α molecule excited by light absorption. The excited electron states S_1 and S_2 are excited singlet states at wavelengths effective for electron transfer for photosystem II. Gray lines represent spin multiplicity states (usually vibrational or rotational) within the main electronic energy states (bold lines). The Soret bands (S_n) excite electrons for photosystem I at the shorter wavelength excitation peak for chlorophyll β, overlapping with chlorophyll α and facilitating energy transfer. As electrons return to the ground state, energy can undergo internal conversion (IC) to the S_2 or S_1 states or be released by emission as fluorescence. Under slow energy transfer pathway (termed *intersystem crossing*), the T_1 and T_2 excited triplet states can release energy as phosphorescence. Derived from multiple sources

excited P680* complex and from there to plastoquinol, the first step in the electron transport chain. The energy is passed to the cytochrome b6f complex and to plastocyanin in the path to the reaction center at photosystem I (PS I) and, from there, ultimately to NADP+ and ADP. In their reduced state, they provide energy to reduce CO_2 to a 3-C sugar in the Calvin cycle. The enzyme catalyzing the CO_2 reduction reaction is RuBisCO, which is the most abundant protein on Earth (Ellis 1979; Cooper 2000; Raven 2013).

14.8.3 Leaf Biochemistry and Energy Absorption in the Solar Spectrum

The chemistry of plant species may be highly variable in terms of secondary compounds, but essential physiological functions are similar. Because seed plants share a common ancestry and face common requirements to survive, younger species retain structures based on their genetic heritage to capture and use sunlight. For a species to survive in a particular environment, the suite of leaf traits must be consistent with long-term patterns of environmental resources, and investments in plant tissues and biochemical composition are determined by metabolic activities, as illustrated by correlations among environmental conditions, leaf chemistry, and physiological functions and as shown in the TRY leaf trait database (Kattge et al. 2011). Studies show loose stoichiometric relationships between different biochemical elements such as C, N, and phosphorus and growth rates and biomass accumulation. Although the relationships vary with conditions and species, typically, C shows relationships with X (biochemical) where the C/X ratio increases under nutrient-limited growth, increasing light intensity, and partial pressure of CO_2 and as a function the species (Sterner et al. 2002).

We expect that to maintain high levels of productivity, leaf chemistry associated with light harvesting and reduction of CO_2 must be correlated, such as chlorophyll a and b and N concentrations, and correlations with other nutrients and water content are all available at appropriate levels for leaves to achieve high productivity. This is formulated in concepts of plant functional morphology (Tilman 1985; Chapin et al. 1993), which emphasize trade-offs in allocation of resources. For example, allocation should shift to shoots under low light and high nutrient conditions and to roots under high light and low nutrient conditions. Ackerly (1999) show that under conditions of high fertility, high rates of growth produce self-shading in older leaves that are lower in the canopy. This limits available light and results in declining rates of assimilation for older leaves. Field (1983) reported that older leaves had reduced N concentrations and photosynthetic potential. Other studies have shown that canopies with steep light gradients exhibit greater declines in photosynthetic capacity than those with small gradients (Mooney et al. 1981). These relationships result in high nutrient environments being favorable to deciduous trees with high growth rates, low C/N ratios, and high rates of leaf and root turnover. Under low nutrient conditions and high light environments, plants exhibit slower growth rates and higher root/shoot ratios. Competition would favor slow growth and low stature species with high nutrient retention and higher C/N ratios. The ratio of leaf dry mass per unit leaf area seems to be highly correlated with growth potential, and species with low mass per unit area have high potential growth rates and high rates of C uptake, while species with high leaf mass area have low growth potential but are generally more stress tolerant (Wright et al. 2004). Of course, all species fall somewhere on this range, but annuals are expected to be at the higher growth end, deciduous woody species are expected to have higher values than evergreen species, and those with thick leathery evergreen leaves are expected to score among the lowest values.

Today, with the development of imaging spectrometry, there is active research aimed at understanding the significance of different assemblages of leaf traits (and their associated chemistry), the roles they play in adaptation for specific habitats, and how they can be detected from spectral patterns measured with optical sensors (Féret and Asner 2014a; Asner et al. 2014b, 2015; Serbin et al. 2015; Singh et al. 2015; Couture et al. 2016). The concept of detecting plant traits has jumped from RS and ecological research to rapid testing of crop breeding of new genotypes through high-throughput phenotyping (Araus and Cairns 2014; Li et al. 2014). The current procedures are derived from precision farming but involve high spatial resolution (using differential GPS) multiple RS inputs, commonly including lidar, thermal infrared, imaging spectrometers, fluorescence imagers, and multiband imagery to rapidly characterize phenotypic variability in relation to desired crop attributes like resistance to disease or other stressors while retaining high growth potential (Bai et al. 2016; Tanger et al. 2017). Such combined methods would provide better resolution of biodiversity in natural systems than use of a single instrument data type.

14.8.4   Photosynthetic Pigments

Light absorption by pigments in the chloroplast produces a unique absorption pattern in the visible spectrum, with higher absorption in the blue and red wavelengths than in the green wavelengths. All higher plants have chlorophyll a and b in their photosynthetic tissues. Absorption features in the visible to NIR part of the spectrum are predominantly caused by excitation of electrons in a process call electronic transitions, in contrast with bending and stretching of molecules in the infrared bands.

Chlorophyll b is nearly identical to chlorophyll a (Fig. 14.8) except that an aldehyde replaces the methyl on the chlorin ring, opposite the phytol tail. This difference affects which wavelengths are absorbed; chlorophyll b has peak absorptions at 450 nm and 642 nm) blue–green), whereas chlorophyll a absorbs primarily at 590–720 nm (orange–red). Because its absorption peaks in these bands are at longer (455 vs. 429 nm) and shorter (642 vs 659 nm) wavelengths than chlorophyll a, chlorophyll b can transfer its excited electron to the reaction center of chlorophyll a (P680). The chemical composition and structure of accessory pigments increase the range of wavelengths that can capture energy for photosynthesis. While all seed plants all share chlorophyll a and b, they differ in the concentrations of the suite of carotenoids found in chloroplasts. The composition of chlorophylls, carotenoids, and anthocyanin pigments in different species provide a basis for using remote sensing data to differentiate species and perhaps phylogenetic relationships among related species and could contribute to biodiversity monitoring.

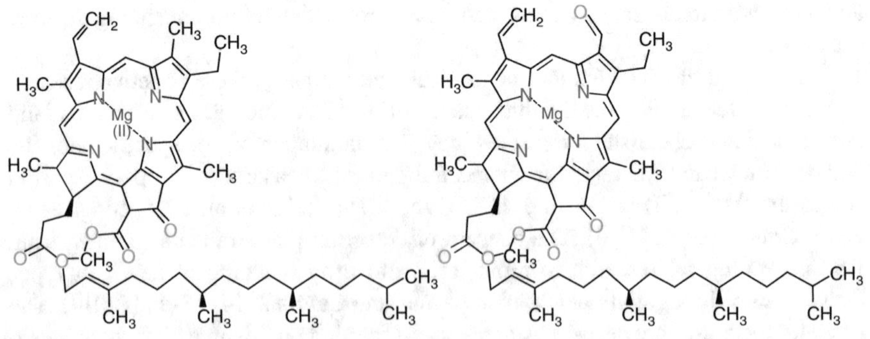

Fig. 14.8 Chlorophyll α and β molecules. (Modified from ChemSpider http://www.chemspider. com and reproduced with permission of the Royal Society of Chemistry)

The optical properties of pigments have been characterized after extraction from chloroplasts (Fig. 14.9). However, these properties are not the same as they are in the intact chloroplast and leaf because extraction alters the chemical environment and destroys the bond structure in their functional state of the pigment–protein complexes. The light-harvesting pigment–protein complexes are associated with other molecules in the chloroplast that affect their three-dimensional configuration and, hence, their absorption patterns. The polarity and water content of the solvents used to extract chlorophylls also shift their peak absorption wavelengths (Lichtenthaler 1987, Fig. 14.9).

One alternative to using extractive chemistry to determine the absorption coefficients of pigments is to use inversion of radiative transfer models. The PROSPECT family of models are the most widely used leaf optical properties models; the recent

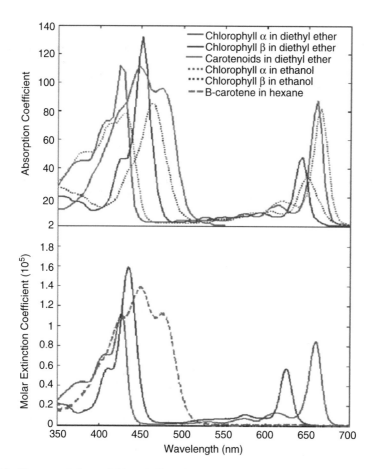

Fig. 14.9 The spectral shape of chlorophylls and carotenoids extracted in different solvents. (Data redrawn from Lichtenthaler (1987) and Du et al. (1998); reproduced from Ustin et al. (2009), with permission from Elsevier)

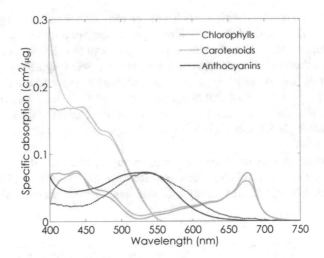

Fig. 14.10 Specific absorption coefficients for total chlorophyll and carotenoids derived from PROSPECT-D (Féret et al. 2017) are shown in solid lines. Dashed lines are the specific absorption coefficients from PROSPECT 5 (Féret et al. 2008). The specific absorption coefficients of anthocyanins were measured by Peters and Nobel (2014). (From Féret et al. (2017), reprinted with permission from Elsevier)

version, PROSPECT-D (Féret et al. 2017), used a large database of leaf spectra and chemistry to predict the in-situ absorption coefficients for chlorophyll, carotenoid, and anthocyanin pigments. Figure 14.10 shows that the modeled absorption coefficients of in-situ pigments are broadened and compressed relative to the extracted pigments.

14.8.5 Accessory Pigments

In the light-harvesting complex, chlorophyll b and carotenoids enhance capture of photons and pass them to the reaction centers of the two photosystems. Carotenoids are produced in plastids in all plant organs (Beisel et al. 2010); however, those in leaf chloroplasts are critical to photosynthetic functioning. Carotenoid species are under constant synthesis and degradation (Beisel et al. 2010), although total concentrations remain approximately equal to the concentration of chlorophyll *a*. Carotenoid molecules are composed of 40-C polyene backbone structures with different side chains. Common chloroplast carotenes include α-carotene and β-carotene, which have oxygen-free structures and are considered primary carotenoids due to their photosynthetic function. Lutein and other forms of carotenoids that have oxygen in their structure are known as xanthophyll pigments; they help regulate energy in the chloroplast (Lichtenthaler 1987). Concentrations of specific carotenoids and pool size vary between species (Thayer and Björkman 1990) and with environmental conditions such as sun vs. shade, cold temperature, and others (Thayer and Björkman 1990, Demmig-Adams and Adams 1992, Hannoufa and Houssain 2012). For example, some species with shade-grown leaves may have high concentrations of α-carotene, and some with sun-grown leaves have only trace concentrations (Thayer and Björkman 1990). Comparing among species with shade-grown leaves, concentrations of lutein or neoxanthin can vary as much as factors of two (Thayer and Björkman 1990). Such differences in concentrations of specific carotenoids and/or their pool sizes provide potential for use in identifying biodiversity patterns if remote sensing instruments and analytics have power to resolve some of these differences.

14.8.6 Xanthophyll Pigments

Oxygen-containing carotenoids contribute to regulation of photosynthetic functioning. For example, a reversible bond changes between violaxanthin and zeaxanthin as light environments change from low light to high light conditions, causing a small increase in reflectance around 530 nm that protects the photosynthetic reaction center from the additional light. Gamon et al. (1992) provided the first experimental evidence that this signal could be measured with spectrometers. The photochemical reflectance index (PRI) by Gamon et al. (1992) has had extensive use and is assumed to follow short-term (minutes to hours) xanthophyll cycle changes; over longer periods, the PRI more likely represents changes in chlorophyll/carotenoid ratios (Gamon et al. 2015).

14.8.6.1 Apparent Concentration vs. Actual Concentration

In some cases, the apparent concentration of a chemical is overestimated or underestimated due to the probability of light interacting with weakly or strongly absorbing chemicals in nonhomogeneous media. An in vivo phenomenon termed the *detour effect* increases the probability of absorption for nonhomogeneous distributions of weakly absorbing molecules (Fukshansky et al. 1993; Terashima et al. 2009). This effect lengthens the optical path length within the leaf and increases photon scattering, enhancing the potential for a weakly absorbing molecule to interact with the photon (Terashima et al. 2009). A different phenomenon termed the *sieve effect* has, generally speaking, an opposite effect. The sieve effect decreases the expected absorption by concentrating strongly absorbing molecules (pigments, water, and other compounds) in a small area of the cell volume (e.g., in organelles). Consequently, the molecules have more limited opportunities to interact with a photon and be absorbed. It effectively reduces the path length, and its effect is most noticeable at wavelengths where light is strongly absorbed [e.g., chlorophyll pigments in the blue and red wavelengths (Evans et al. 2004) or water in the SWIR (Baranoski and Eng 2007)]. Terashima et al. (2009) show that for weakly absorbed light such as in green wavelengths, the loss of absorption due to the sieve effect is minimal, while the gain in absorption by the detour effect is large. The detour effect can increase absorption at 550 nm sufficient to rival absorption in blue and red bands for photosynthesis. Such interactions make quantifying the concentrations of specific compounds and correctly attributing their impact on physiological processes subject to potentially significant errors. Despite difficulties in separating the effect of each pigment in the leaf, the in vivo spectral shapes of different species are fairly conservative within related taxa and are often distinctive of the taxonomic group over much of the wavelength region between 450 nm and 700 nm; these patterns have been used to identify genera, species, and even phenotypes (Asner and Martin 2016; Junker and Ensminger 2016).

14.8.7 Non-photosynthetic Pigments

Anthocyanins are a diverse group (more than 540 pigments identified in nature; Anderson and Francis 2004) of colored, water-soluble flavonoids found in the vacuoles of many seed plants (Hrazdina et al. 1978). Pigment colors range from blue to red, depending on pH. Anthocyanins have been associated with many benefits to plants (Lee and Gould 2002). They attract pollinators and animals that disperse seeds and fruits, protect plants growing at high elevations from UV light, and protect against cold (Chalker-Scott 1999; Lee 2000). Anthocyanins are common in understory plants (Lee 2002). The protective mechanism of anthocyanins in leaves develops during early stages of leaf expansion (e.g., young leaves in Fig. 14.2) before leaves are photosynthetically functional in some species (Landi et al. 2015). Lee and Collins (2001) showed within-family phylogenetic relationships of anthocyanins for 399 woody tropical taxa. Other flavonoids also contribute to physiological regulation, providing chemical signals to other parts of the plant or to other plants, or promoting or inhibiting interactions with other organisms. Because of their strong absorption in the red wavelengths, anthocyanins can be confused with chlorophylls, resulting in overestimation of photosynthetic capacity. Gitelson et al. (Gitelson 2012) developed semi-analytic three-band models to separately estimate the concentrations of chlorophylls, carotenoids, and anthocyanin foliar pigments. In collaboration with Gitelson, Féret et al. (2017) updated most recent PROSPECT-D model (Féret et al. 2017) fusing a linear relationship among six independent data sets between predicted and measured anthocyanin. Féret et al. (2017) improved the specific absorption coefficients for chlorophylls, carotenoids, and anthocyanins as shown in Fig. 14.10.

14.8.8 Brown Pigments

When pigments and their protein complexes degrade, they form colored chemical residues. These are poorly described but constitute the "brown" pigments or colored pigment residues in the leaf when it senesces and dies. These result from oxidation and polymerization of cell constituents. Without pigment residues, cellulose and other cell wall materials would be a nearly colorless white. Regardless of plant species, there is a general trajectory of color change over time. Brown pigment residues produce the shades of light tan colors found in recently dead leaves. The residues further degrade and become darker over winter. Eventually, they become dark brown and chemically and spectrally indistinguishable from humus (Ziechmann 1964). The visible near-infrared shape of the absorption spectrum for recently dead plant residues is generally monotonic, lowest at 400 nm and increasing in a polynomial curve to about 900 nm.

14.9   Leaf Water Content

The optical properties of water are better known than those of any other plant biochemical. In recent years, there have been several attempts to improve the absorption coefficients for liquid water; but since there have been no major changes, we assume this property is mostly resolved for liquid water.

The optical properties of water are known with much greater precision than are those of pigments or other molecules. Water absorptions are due to vibration of the molecules, with a vibrational frequency that approximates simple harmonic motion when excited by absorbing a quantum of energy. Because hydrogen atoms are small, vibrations have large amplitude. Water vapor has three fundamental vibrational modes with the dipole moments changing in the direction of the vibration (Fig. 14.11). The first mode for liquid water at 25 °C is symmetric stretching; in this case, both hydrogen ions vibrate simultaneously, which is at mode ν_1 with a 3050 nm absorption feature. The second vibrational mode is bending the covalent bonds, which occurs when the two hydrogen atoms vibrate by moving toward and away from each other. This bending mode is at mode ν_2 and it causes a strong absorption at 6080 nm. The third mode is asymmetric stretching that results from one hydrogen ion being attracted toward the oxygen while the other moves away. The absorption wavelength for this ν_3 mode is 2870 nm. Vibrational modes for liquid water at shorter wavelengths (401–1900 nm) are combination modes. Vibrational modes are restricted in liquid water and ice by hydrogen bonds. *Libration* describes the back-and-forth rotation of the hydrogen ions in liquid water when its motion is restricted.

The infrared spectrum of water results from vibrational overtones and combinations with librations. Palmer and Williams (1974) conducted a detailed study, including verification of prior authors (e.g., Hale and Querry 1973) and reported 94 optical constants between the UV and the NIR. They report liquid water values related to those in Table 14.1 for measurements taken at 27 °C to be 769, 847, 973, 980, 1205, 1443, 1786, and 1927 nm. Kou et al. (1993) established absorption coefficients for ice between 1440 and 2500 nm, showing that these peaks are at longer wavelengths than those of liquid water when measured at warmer temperatures. Wieliczka et al. (1989) developed an improved measuring device, measured the real part of the refractive index, and used it with Kramers–Kronig methods to compute

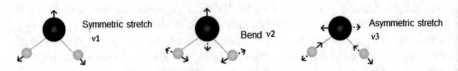

Fig. 14.11 The vibration modes in liquid water from overtone and combined overtone bands in the near-infrared (NIR)

Table 14.1 Vibrational–rotational transitions for water vapor, liquid water (near 100 °C), and ice

Gas	Liquid (near 100 °C)	Liquid (near 0 °C)	Ice	Vibrational assignment
Wavelength (nm)	Wavelength (nm)	Wavelength (nm)	Wavelength (nm)	
723	740	770	800	$3\nu_1 + \nu_3$
823	840	847	909	$2\nu_1 + \nu_2 + \nu_3$
1025	967	979	942	$2\nu_1 + \nu_3$
1250	1160	1200	1135	$\nu_1 + \nu_2 + \nu_3$
1492	1425	1453	1380	$\nu_1 + \nu_3$
1780	1786	1780	–	$\nu_2 + \nu_{3+}\, \nu_L$
1988	1916	1938	1875	$\nu_2 + \nu_3$

From Workman and Weyer (2008), modified with permission from Taylor and Francis
ν_1 is the symmetric stretch, ν_2 is the bending mode, ν_3 is the asymmetric stretch, and ν_L is the undefined "intermolecular mode"

the imaginary part. Buiteveld et al. (1994) conducted detailed measurements of pure water over temperatures from 25 to 40.5 °C and improved absorption coefficients over the range of 300–800 nm. Féret et al. (2017) combined these in developing the PROSPECT-D model and tested the inversion on leaf data from six data sets with a total of 521 leaves. From this, they derived the absorption coefficients for pure water, shown in Fig. 14.12 covering the 400–2500 nm region.

The atmosphere is nearly always saturated with water vapor at 1450 nm and 1940 nm (Fig. 14.13). Published imaging spectrometer data are typically shown with these wavelengths deleted since only in the driest deserts is there a possibility of observing the ground surface from airborne or satellite sensors (see also Segelstein's (1981) absorption coefficients, Fig. 14.6). A series of increasingly strong water vapor absorptions are observed from the visible region, around 512 nm to the SWIR at–2500 nm. Because the absorption maximum of each phase of water at 940–1020 nm is offset by 30–40 nm, it is possible to identify and quantify whether water is present in vapor, liquid, or solid phases in imaging spectroscopy data having bandwidths of 5–10 nm. The second absorption region for detecting the three phases of water is located at 1100–1300 nm. Absorptions are too strong beyond

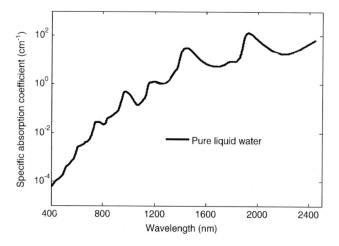

Fig. 14.12 Féret et al. (2008) developed a new absorption coefficient for water for the PROSPECT 4 and 5 model built by combining data from Buiteveld et al. (1994) at 400–800 nm, Kou et al. (1993) at 800–1232 nm, and Wieliczka et al. (1989) at 1232–2500 nm. The model was evaluated by testing on six independent leaf data sets. (Reproduced by permission from Elsevier)

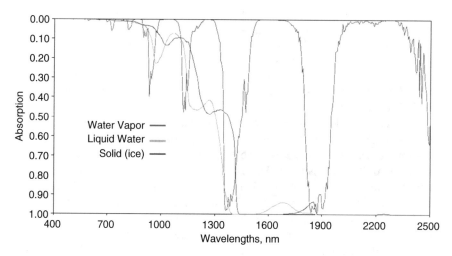

Fig. 14.13 The absorption spectrum at 400–2500 nm for vapor, liquid, and ice phases of water. For each absorption feature, water vapor (shown in red) is at the shortest wavelengths, and ice (shown in blue) is at the longest wavelengths. (Figure from Robert O. Green, NASA Jet Propulsion Laboratory)

these wavelengths to easily measure them in the optical region. Airborne imaging spectrometer data use one or both of the two wavelength regions (940–1020 nm and 1100–1300 nm) for calibrating an RT model for absorptions and scattering in the atmosphere to fit to the measured spectrum. The water vapor, liquid water, and ice concentrations are then estimated from the best fit of the RT model.

14.10   Cell Wall Constituents

The organic compounds that make up the cell wall comprise about 75–90% of the dry weight of the leaf. They vary with the location in the leaf, by phenology (season), with environmental stressors, and by species. They also vary depending on whether the walls are primary or secondary structures or part of the vascular bundles. Common constituents include cellulose, hemicellulose, lignin, proteins, and pectins. They are large molecules of variable molecular composition; their –OH units form extensive hydrogen bonds that link the molecules together, providing both flexibility and strength. They form solid structures and are mostly insoluble. This makes them poor candidates for spectroscopy, so they are generally measured with other techniques.

Cellulose is the most abundant organic polymer on Earth (Klemm et al. 2005). It is a polysaccharide forming a straight chain of many to thousands linked D-glucose units (Fig. 14.14a) with a nonreducing end (shown on left) linked to repeating glucose units by oxygen and ending with a reducing oxygen ion. The –OH groups form hydrogen bonds across chains, holding them together and providing structure. Species can be identified based on the relative abundance of these molecules.

Hemicellulose refers to any of several heteropolymers that form the matrix polysaccharides in most plant cell walls (Fig. 14.14b). Hemicellulose monomers include hexoses and fructoses such as xyloglucans, xylans, and glucans. Figure 14.15 illustrates some differences in hemicellulose subunit structures. Their chemical composition is similar to that of cellulose, so they are spectrally similar and are not generally separately identified in leaf spectra. The structure of these molecules, however, varies between species, suggesting that they could contribute to identifying biodiversity from RS data and possibility different species or related species.

Lignins are a widely distributed class of complex amorphous organic, branched, and cross-linked polymers (Fig. 14.14c) of the approximate composition $C_{31}H_{34}O_{11}$. They form the structural materials of cell walls and other support tissues in leaves, such as the walls of tracheids in the vascular system. Lignin is unusually heterogeneous in composition and lacks a defined primary molecular structure. Figure 14.14c illustrates the cross-linked phenol units that make it very slow to decompose. Eventually, the recalcitrant residues become a major fraction of the soil humus. Humus is important in the global C cycle for sequestering soil C and nutrients and retaining soil moisture. Lignins interact with soil organic matter and N turnover rates, and this affects lignin stabilization (Thevenot et al. 2010), so they are important in biogeochemical cycling. The variety of lignin molecules found in different plant species makes quantification difficult, but some forms are known to occur in specific plant clades, providing a basis for estimating diversity between lower vascular plants, conifers and angiosperms (Weng and Chapple 2010) (Fig. 14.15).

Although fundamental frequencies are known for C–H, C=O, C–N, O–H, and N–H bonds, dried and powdered leaf materials have overlapping spectral properties of various cell wall molecules, proteins, enzymes, amino acids, sugars, starches, waxes, and other biomolecules that make it difficult to isolate specific molecules

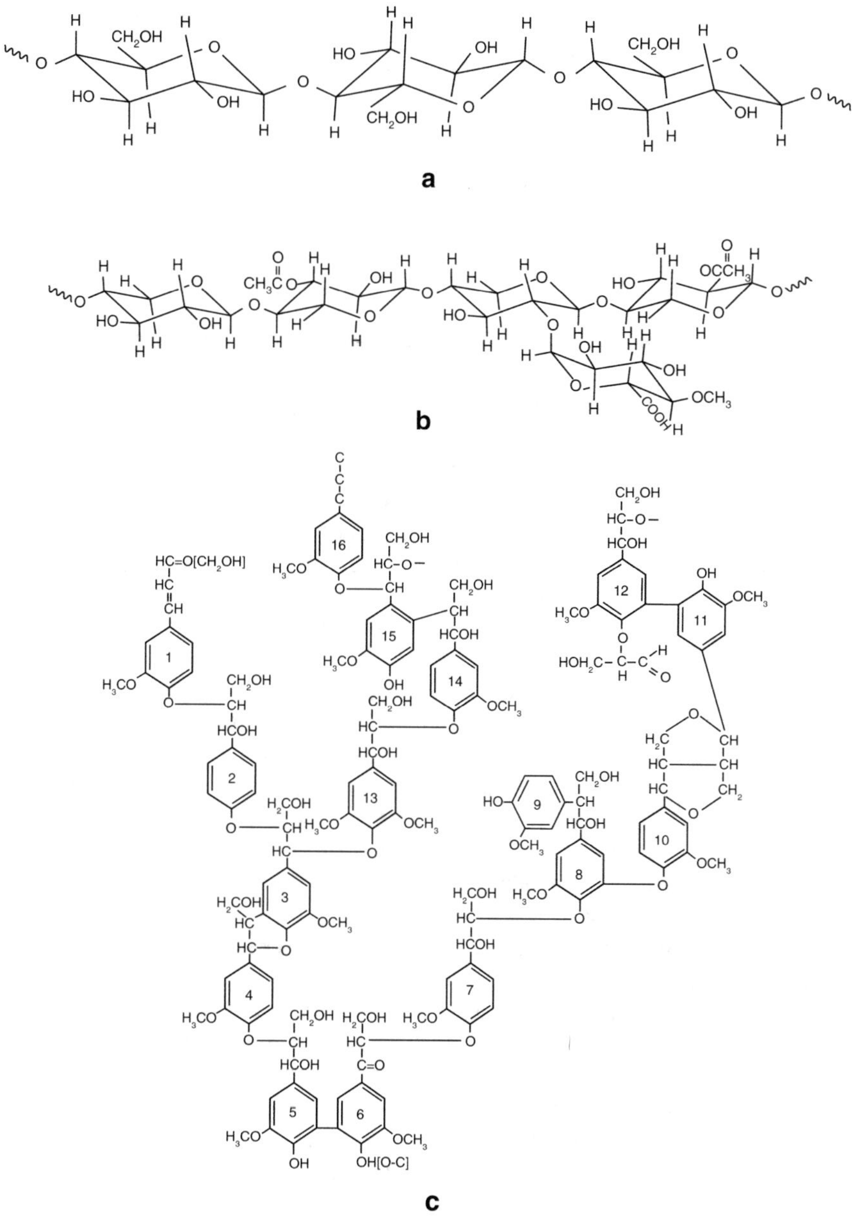

Fig. 14.14 Repeating structures of cellulose (**a**), hemicellulose (**b**), and lignin (**c**). (From K.K. Pandy, Journal of Applied Polymer Science 1999, with permission of John Wiley and Sons. Lignin structure was reproduced from E. Alder (1977) with permission of Springer)

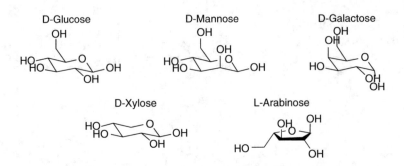

Fig. 14.15 Examples of hemicellulose structures From Pierson et al. (2013), open access

(Curran 1989) and nearly impossible to untangle their spectral properties in intact leaves, at least with the spectral resolution and models available today. However, the variability in these properties makes it possible to use various regression, classification, and self-learning techniques to identify genera or species without knowing exactly the biochemical composition that identifies them. Features in the NIR and SWIR are relatively weak and spectrally broad, having originated as harmonics (integer multiples of the fundamental frequency) and overtones (frequencies higher than the fundamental frequency) from wavelengths in the UV and middle-infrared ranges (2500–6000 nm).

While the absorption spectra are not identical, cellulose, starch, and sugar have strong similarities due to their hydrocarbon chain chemical structures. The PROSPECT models continue to consider these compounds together as "dry matter content," given the uncertainty in their absorption coefficients. Retrieval of dry matter in the models permits estimation of dry biomass; when expressed on a leaf area basis, this yields leaf mass area (dry biomass/leaf area), a measure shown to be highly correlated with photosynthetic production (Poorter et al. 2009). The presence or absence of various lignins, humic acids, and aromatic polyphenols can be determined from absorptions at 1420 and 1920 nm that are related to O–H bonds and C=O vibrations, with shoulders at 1700 and 2100 nm that are related to aromatic C–H bonds (Ziechmann 1964). Kokaly and Skidmore (2015) recently reported a narrow feature for aromatic C–H bonds in phenolic compounds of various woody species and non-hydroxylated aromatics at 1660 nm. Phenolic compounds are generally considered important in plant defense.

Curran (1989) noted that absorption features in leaves are broadened by multiple scattering and often interfere with one another. He cites an example where the first overtones of the N-H and O-H stretch overlap for most of their width. Thus, most studies have opted to analyze spectral data and relationships by identifying taxa or identifying leaf traits using various statistical methods such as multiple stepwise regression (e.g., Serrano et al. 2002), partial least squares regression (PLSR; e.g., Smith et al. 2002; Ollinger et al. 2008), discriminant function analysis (Filella et al. 1994), continuum removal (Kokaly and Clark 1999; Kokaly 2001), wavelets (Cheng et al. 2011, 2012, 2014; Kalacska et al. 2015), or a combination of PLSR, nested

Fig. 14.16 Composite absorption coefficient for dry leaf chemistry used in the PROSPECT 4 and 5 model that was modified from earlier PROSPECT versions. Jacquemoud et al. (2000) and Féret et al. (2008), with permission from Elsevier

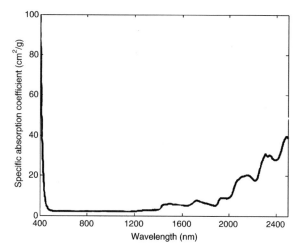

random effects–analysis of variance, and discriminant function analysis (Asner and Martin 2011; Asner et al. 2014b). Newer methods include Gaussian methods and various types of deep learning (Eches et al. 2011; Verrelst et al. 2012; Bazi et al. 2014; Sun et al. 2015) Some authors also combine leaf and canopy radiative transfer models with empirical models (e.g., Asner et al. 2014a, b; Gatellu-Etchegorry and Bruniquel-Pinel 2001). Féret et al. (2008), in developing PROSPECT 5, modified the specific absorption coefficient shown in Fig. 14.16 for dry matter (a composite representing the chemistry of dry leaves) based on a spectrum from Frederic Baret (CNES, unpublished) that Jacquemoud et al. (2009) tested against 245 dry leaves in the LOPEX93 database (Hosgood et al. 1995).

14.11   Conclusion

Researchers have sought to understand the optical properties of plant leaves for the past century or more. Great strides have been made in recent years, but much remains to be resolved before the full chemistry of leaves can routinely be determined from spectroscopy. This is a period of dynamic growth, at least partially because of the interdisciplinary research community— spanning physics, engineering, botany, ecology, RS, statistics, computer science, and modeling— that is interested in solving these problems. Much of the current interest originates from a desire to accurately quantify climate change impacts on global ecosystems, better understand global biodiversity patterns and their functional traits, find ways to monitor changes in global biodiversity, understand sustainability of production of natural and agricultural ecosystems, and better understand global biogeochemical cycles, specifically the C, N, and water budgets. Based on the number of papers estimating biodiversity and the methods applied, it is clear this is an active area of research but also that we are still at the early stages of using optical features to determine alpha, beta, and gamma biodiversity and to develop robust measures for more completely understanding species mixtures and biodiversity.

References

Ackerly D (1999) Self-shading, carbon gain and leaf dynamics: a test of alternative optimality models. Oecologia 119:300–310

Ackerly DD, Knight CA, Weiss SB, Barton K, Starmer KT (2002) Leaf size, specific leaf area and microhabitat distribution of chaparral woody plants: contrasting patterns in species level and community level analysis. Oecologia 130:449–457

Adler E (1977) Lignin chemistry—past, present and future. Wood Sci Technol 11:169–218

Al-Edany TY, Al-Saadi SAAM (2012) Taxonomic significance of anatomical characteristics in some species of the Family Myrtaceae. Am J Plant Sci 3:19475

Allen WA, Gausman HW, Richards AJ, Wiegand CL (1970) Mean effective optical constants of 13 kinds of plant leaves. Appl Opt 9:2573. https://doi.org/10.1364/AO.9.002573

Anderson ØM, Francis GW (2004) Techniques of pigment identification. Plant pigments and their manipulation. Annu Plant Rev 14:293–341

Araus JL, Cairns JE (2014) Field high-throughput phenotyping: the new crop breeding frontier. Trends Plant Sci 19:52–61

Arnold L, Gillet S, Lardiere O, Riaud P, Schneider J (2002) A test for the search for life on extrasolar planets. Looking for the terrestrial vegetation signature in the Earthshine spectrum. Astron Astrophys 392:231–237

Asner GP, Martin RE (2011) Canopy phylogenetic, chemical and spectral assembly in a lowland Amazonian forest. New Phytol 189:999–1012

Asner GP, Martin RE (2016) Spectranomics: emerging science and conservation opportunities at the interface of biodiversity and remote sensing. Global Ecol Cons 8:212–219

Asner GP, Martin RE, Carranza-Jiménez L, Sinca F, Tupayachi R et al (2014a) Functional and biological diversity of foliar spectra in tree canopies throughout the Andes to Amazon region. New Phytol 204:127–139

Asner GP, Martin RE, Tupayachi R, Anderson CB, Sinca F et al (2014b) Amazonian functional diversity from forest canopy chemical assembly. Proc Natl Acad Sci U S A 111:5604–5609

Asner GP, Anderson CB, Martin RE, Tupayachi R, Knapp DE et al (2015) Landscape biogeochemistry reflected in shifting distributions of chemical traits in the Amazon forest canopy. Nat Geosci 8:567–573

Bai G, Ge Y, Hussain W, Baenziger PS, Graef G (2016) A multi-sensor system for high throughput field phenotyping in soybean and wheat breeding. Comput Electron Agr 128:181–192

Baranoski GVG, Eng D (2007) An investigation on sieve and detour effects affecting the interaction of collimated and diffuse infrared radiation (750–2500m) with plant leaves. IEEE Trans Geosci Remote S 45:2593–2599

Bazi Y, Alajlan N, Melgani F, Al Hichri H, Malek S (2014) Differential evolution extreme learning machine for the classification of hyperspectral images. IEEE GeoSci Remote S 11:1066–1070

Beisel KG, Jahnke S, Hofmann D, Köppchen S, Schurr U, Matsubara S (2010) Continuous turnover of carotenes and chlorophyll a in mature leaves of Arabidopsis revealed by $^{14}CO_2$ pulse-chase labeling[OA]. Plant Physiol 152:2188–2199

Blankenship RE (2010) Early evolution of photosynthesis. Plant Physiol 154:434–438

Bone RA, Lee DW, Norman JM (1989) Epidermal cells functioning as lenses in leaves of tropical rain-forest shade plants. Appl Opt 24:1408–1412

Bubier JL, Moore TR, Bledzki LA (2007) Effects of nutrient addition on vegetation and carbon cycling in an ombrotrophic bog. Glob Chang Biol 13:1168–1186

Buiteveld H, Hakvoort JMH, Donze M (1994) The optical properties of pure water. In: Jaffe JS (ed) SPIE, the International Society for Optical Engineering, Proc Ocean Opt XVII, vol 2258, pp 174–183

Carter GA (1991) Primary and secondary effects of water content on the spectral reflectance of leaves. Am J Bot 78:916–924

Cazzonelli C (2011) Carotenoids in nature: insights from plants and beyond. Funct Plant Biol 38(11):833–847

Chalker-Scott L (1999) Environmental significance of anthocyanins in plant stress responses. Phytochem Photobiol 70:1–9

Chapin FS, Autumn K, Pugnaire F (1993) Evolution of suites of traits in response to environmental-stress. Am Nat 142(S):S78–S92

Cheng T, Rivard B, Sanchez-Azofeifa AG (2011) Spectroscopic determination of leaf water content using continuous wavelet analysis. Remote Sens Environ 115:659–670

Cheng T, Rivard B, Sanchez-Azofeifa AG, Féret JB, Jacquemoud S, Ustin SL (2012) Predicting leaf gravimetric water content from foliar reflectance across a range of plant species using continuous wavelet analysis. J Plant Phys 169:1134–1142

Cheng T, Rivard B, Sanchez-Azofeifa AG, Féret JB, Jacquemoud S, Ustin SL (2014) Deriving leaf mass per area (LMA) from foliar reflectance across a variety of plant species using continuous wavelet analysis. ISPRS J Photogramm Rem S 87:28–38

Comstock JP, Mahall BE (1985) Drought and changes in leaf orientation for two California shrubs: *Ceanothus megacarpus* and *Ceanothus crassifolius*. Oecologia 65:531–535

Cooper GM, Hausman RE (2004) The chloroplast genome. In: Cooper GM, Hausman RE (eds) The cell: a molecular approach, 3rd edn. ASM Press, Washington, D.C., pp 417–418. ISBN 0878932143

Couture JJ, Singh A, Rubert-Nason KF, Serbin SP, Lindroth RL, Townsend PA (2016) Spectroscopic determination of ecologically relevant plant secondary metabolites. Methods Ecol Evol 7:1402–1412

Curcio JA, Petty CC (1951) The near infrared absorption spectrum of liquid water. J Optical Soc Am 41:302–304

Curran PJ (1989) Remote sensing of foliar chemistry. Remote Sens Environ 30:271–278

Demmig-Adams B, Gilmore AM, Adams WW III (1996) In vivo functions of carotenoids in higher plants. FASEB J 10:403–412

Du H, Fuh RCA, Li J, Corkan LA, Lindsey JS (1998) PhotochemCAD: a computer-aided design and research tool in photochemistry. Photochem Photobiol 68:141–142

Eches O, Dobigeon N, Tourneret JY (2011) Enhancing hyperspectral image unmixing with spatial correlations. IEEE Trans Geosci Remote Sens 49:4239–4247

Ehleringer J, Björkman O, Mooney HA (1976) Leaf pubescence: effects on absorptance and photosynthesis in a desert shrub. Science 192:376–377

Ellis RJ (1979) The most abundant protein in the world. Trends Biochem Sci 4:241–244

Evans JR, Vogelmann TC, Williams WE, Gorton HL (2004) Chloroplast to leaf. In: Smith WC, Vogelmann TC, Critchley C (eds) Photosynthetic adaptation: chloroplast to landscape. Springer, New York, pp 15–41

Everitt JH, Richardson AJ, Gausman HW (1985) Leaf reflectance-nitrogen-chlorophyll relations in buffelgrass. Photogramm Eng Rem S 5:463–466

Féret J-B, Asner GP (2014a) Microtopographic controls on lowland Amazonian canopy diversity from imaging spectroscopy. Ecol Appl 24:1297–1310

Féret J-B, Asner GP (2014b) Mapping tropical forest diversity using high-fidelity imaging spectroscopy. IEEE Trans Geosci Remote S 51:73–84

Féret J-B, Françios C, Asner GP, Gitelson AA, Martin RE et al (2008) PROSPECT 4 and 5: advances in the leaf optical properties model separating photosynthetic pigments. Remote Sens Environ 112:3030–3043

Féret J-B, François C, Gitelson A, Asner GP, Barry KM, Panigada C, Richardson AD, Jacquemoud S (2011) Optimizing spectral indices and chemometric analysis of leaf chemical properties using radiative transfer modeling. Remote Sens Environ 115:2742–2750

Féret J-B, Gitelson AA, Noble SD, Jacquemoud S (2017) PROSPECT-D: towards modeling leaf optical properties through a complete lifecycle. Remote Sens Environ 193:204–215

Field CB (1983) Allocating leaf nitrogen for the maximization of carbon gain – leaf age as a control on the allocation program. Oecologia 56:341–347

Field CB, Mooney H (1986) The photosynthesis – nitrogen relationship in wild plants. In: Givinsh TJ (ed) On the economy of form and function. Cambridge University Press, Cambridge, pp 25–55

Filella I, Serrano L, Serra J, Peñuelas J (1994) Evaluating wheat nitrogen status with canopy reflectance indices and discriminant analysis. Crop Sci 35:1400–1405

Fukshansky L, Martinez V, Remisowsky A, McClendon J, Ritterbusch A, Richter T et al (1993) Absorption spectra of leaves corrected for scattering and distributional error: a radiative transfer and absorption statistics treatment. Photochem Photobiol 57:538–555

Gamon JA, Peñuelas J, Field CB (1992) A narrow-waveband spectral index that tracks diurnal changes in photosynthetic efficiency. Remote Sens Environ 41:35–44

Gamon JA, Kovalchuck O, Wong CYS, Harris A, Garrity SR (2015) Monitoring seasonal and diurnal changes in photosynthetic pigments with automated PRI and NDVI sensors. Biogeosciences 12:4149–4159

Gatellu-Etchegorry JP, Bruniquel-Pinel V (2001) A modeling approach to assess the robustness of spectrometric predictive equations for canopy chemistry. Remote Sens Environ 76:1–15

Gates DM, Keegan HJ, Schleter JC, Weidner VR (1965) Spectral properties of plants. Appl Opt 4:11–20

Gausman HW (1974) Leaf reflectance of near-infrared. Photogramm Eng 40:183–191

Gitelson AA (2012) Nondestructive estimation of foliar pigment (chlorophylls, carotenoids, and anthocyanins) contents: evaluating a semi analytical three-band model. In Thenkabail PS, Lyon JG, Huete A (eds) Hyperspectral remote sensing of vegetation. CRC Press-Taylor and Francis Group, Boca Raton; p 141–165

Gras EK, Read J, Mach CT, Sanson GD, Clissold FJ (2005) Herbivore damage, resource richness and putative defenses in juvenile versus adult Eucalyptus leaves. Aust J Bot 53:33–44

Haberlandt G (1914) Physiological plant anatomy. Perception of light. Macmillan and Co., London

Hale GM, Querry MR (1973) Optical constants of water in the 200nm to 200μm wavelength region. Appl Opt 12:555–563

Hamilton WD, Brown SP (2001) Autumn tree colours as a handicap signal. Proc Royal Soc B-Biol Sci 286(1475):1489–1493

Hannoufa A, Houssain Z (2012) Regulation of carotenoid accumulation in plants. Biocatal Agric Biotechnol 1(3):198–202

Hosgood B, Jacquemoud S, Andreoli G, Verdebout J, Pedrini G et al (1995) Leaf Optical properties EXperiment 93 (LOPEX93) report EUR-16095-EN. European Commission, Joint Research Centre, Institute for Remote Sensing Applications, Ispra

Hrazdina G, Wagner GJ, Siegelman HW (1978) Subcellular localization of enzymes of anthocyanin biosynthesis in protoplasts. Phytochemistry 17:53–56

Jacquemoud S, Ustin S (2008) Modeling leaf optical properties. Photobiological Sciences Online. Environmental Photobiology. http://www.photobiology.info/#Environ. Accessed 27 Nov 2018

Jacquemoud S, Bacour C, Poilvé H, Frangi J-P (2000) Comparison of four radiative transfer models to simulate plant canopies reflectance: direct and inverse models. Remote Sens Environ 74:471–481

Jacquemoud S, Verhoef W, Baret F, Bacour C, Zarco-Tejada PJ et al (2009) PROSPECT + SAIL models: a review of use for vegetation characterization. Remote Sens Environ 113(S1):S56–S66

James SA, Bell DT (2000) Leaf orientation, light interception and stomatal conductance of Eucalyptus globulus ssp. globulus leaves. Tree Physiol 20:815–823

Junker LV, Ensminger I (2016) Fast detection of leaf pigments and isoprenoids for ecophysiological studies, plant phenotyping and validating remote–sensing of vegetation. Physiol Planta 158:369–381

Kalacska M, Lalonde M, Moore TR (2015) Estimation of foliar chlorophyll and nitrogen content in an ombrotrophic bog from hyperspectral data: scaling from leaf to image. Remote Sens Environ 169:270–279

Karageorgou P, Manetas Y (2006) The importance of being red when young: anthocyanins and the protection of young leaves of *Quercus coccifera* from insect herbivory and excess light. Tree Physiol 26:613–621

Kattge J, Diaz S, Lavorel S, Prentice IC, Leadley P et al (2011) TRY – a global database of plant traits. Glob Chang Biol 17:2905–2935

Kebes (2008) Liquid water absorption and refractive index. Compiled by Krebes (Created 1 July 2008), at English Wikipedia CC by SA 3.0 (http://creativecommons.org/licenses/by-sa/3.0). Based largely on publications by Segelstein (1981), Hale & Quarry (1973), and Wieliczka et al. (1989), but Krebes provides other references at: http://omlc.ogi.edu/spectra/water/abs/index.html. Accessed 18 Nov 2018

Kiang NY, Siefert J, Govindjee, Blankenship RE (2007) Spectral signatures of photosynthesis. I. Review of earth organisms. Astrobiology 7:222–251

Klemm D, Heublein B, Fink H-P, Bohn A (2005) Cellulose: fascinating biopolymer and sustainable raw material. Angew Chem Int Edit 44(22):3358–3393. https://doi.org/10.1002/anie.200460587

Knipling EB (1970) Physical and physiological basis for the reflectance of visible and near-infrared radiation from vegetation. Remote Sens Environ 1:155–159

Kokaly RF (2001) Investigating a physical basis for spectroscopic estimates of leaf nitrogen concentration. Remote Sens Environ 75:153–161

Kokaly RF, Clark RN (1999) Spectroscopic determination of leaf biochemistry using band-depth analysis of absorption features and stepwise multiple linear regression. Remote Sens Environ 67:267–287

Kokaly RF, Skidmore AK (2015) Plant phenolics and absorption features in vegetation reflectance spectra near 1.66μm. Int J Appl Earth Observ Geoinform 43:55–83

Kokaly RF, Asner GP, Ollinger SV, Martin ME, Wessman CA (2009) Characterizing canopy biochemistry from imaging spectroscopy and its application to ecosystem studies. Remote Sens Environ 113:S78–S91

Kou L, Labrie D, Chylek P (1993) Refractive indices of water and ice in the 0.65 to 2.5-μm spectral range. Appl Opt 32:3531–3540

Landi M, Tattini M, Gould KS (2015) Multiple functional roles of anthocyanins in plant-environment interactions. Environ Exp Bot 119:4–17

Lee DW (2002) Anthocyanins in leaves: distribution, phylogeny and development. Adv Bot Res 37:37–53

Lee DW, Collins TM (2001) Phylogenetic and ontogenetic influences on the distribution of anthocyanins and betacyanins in leaves of tropical plants. Int J Plant Sci 162:1141–1153

Lee DW, Gould KS (2002) Anthocyanins in leaves and other vegetative organs: an introduction. Adv Bot Res Incorporating Adv Plant Path 37:1–16

Li L, Zhang Q, Huang D (2014) A review of imaging techniques for plant phenotyping. Sensors 14:20078–20111. https://doi.org/10.3390/s141120078

Lichtenthaler HK (1987) Chlorophylls and carotenoids: pigments of photosynthetic biomembranes. Method Enzymol 148:350–382

Martin GM, Josserand SA, Bornman JF, Vogelmann TC (1989) Epidermal focusing and the light environment within leaves of *Medicago sativa*. Physiol Plant 76:485–492

Matile P (2000) Biochemistry of Indian summer: physiology of autumnal leaf coloration. Exp Gerontol 35:145–158

Milborrow BV (2001) The pathway of biosynthesis of abscisic acid in vascular plants: a review of the present state of knowledge of ABA synthesis. J Exp Bot 52(359):1145–1164

Montañés-Rodríguez P, Pallé E, Goode PR, Martin Tores FJ (2006) Vegetation signature in the observed globally integrated spectrum of earth considering simultaneous cloud data: applications for extrasolar planets. Astrophys J 651:544–552

Mooney HA, Field C, Gulmon SL, Bazzaz FA (1981) Photosynthetic capacity in relation to leaf position in desert versus old-field annuals. Oecologia 50:109–112

Ollinger S, Smith ML (2005) Net primary production and canopy nitrogen in a temperate forest landscape: an analysis using imaging spectroscopy, modeling and field data. Ecosystems 8:760–778

Ollinger SV, Richardson AD, Martin ME, Hollinger DY, Frolking SE et al (2008) Canopy nitrogen, C assimilation, and albedo in temperature and boreal forests: functional relations and potential climate feedbacks. Proc Natl Acad Sci U S A 105:19336–19341

Ordoñez JC, Van Bodegom PM, Witte J-P, Wright IJ, Reich PB et al (2009) A global study of relationships between leaf traits, climate and soil measures of nutrient fertility. Global Ecol Biodiver 18:137–149

Owen NL, Thomas DW (1989) Infrared studies of "hard" and "soft" woods. Appl Spectrosc 43:451–455

Palmer KF, Williams D (1974) Optical properties in the near infrared. J Optical Soc Am 64:1107–1110

Pandy KK et al (1999) A study of chemical structure of soft and hardwood and wood polymers by FTIR spectroscopy. J Appl Polym Sci 71:1969–1975

Parkhurst DF (1986) Internal leaf structure: a three-dimensional perspective. In: Givnish TJ (ed) On the economy of plant form and function. Cambridge University Press, Cambridge, pp 215–249

Peters RD, Noble SD (2014) Spectrographic measurement of plant pigments from 300 to 800 nm. Remote Sens Environ 148:119–123

Pierson Y, Bobbink F, Yan N (2013) Alcohol mediated liquefaction of lignocellulosic materials: a mini review. Chem Eng Process Tech 1(2):1014

Poorter H, Niinemets Ü, Poorter L, Wright IJ, Villar R (2009) Causes and consequence of variation in leaf mass per area (LMA): a meta-analysis. New Phytol 182:565–588

Poulson ME, Vogelmann TC (1990) Epidermal focusing and effects upon photosynthetic light-harvesting in leaves of *Oxalis*. Plant Cell Environ 13:803–811

Raven JA (2013) Rubisco: still the most abundant protein on earth? New Phytol 198:1–3

Roberts DA, Ustin SL, Ogunjemiyo S, Greenberg J, Dobrowski SZ, Chen J, Hinckley TM (2004) Spectral and structural measures of northwest forest vegetation at leaf to canopy scales. Ecosystems 7:545–562

Schaepman-Strub G, Schaepman ME, Painter TH, Dangel S, Martonchik JV (2006) Reflectance quantities in optical remote sensing-definitions and case studies. Remote Sens Environ 103:28–42

Segelstein D (1981) The complex refractive index of water. M.S. thesis, University of Missouri-Kansas City. Figure derived from Philip Laven www.philiplaven.com/refractive index of water. Accessed 7 Dec 2016

Serbin SP, Singh A, Desai AR, Dubois SG, Jablonski AD et al (2015) Remotely estimating photosynthetic capacity, and its response to temperature, in vegetation canopies using imaging spectroscopy. Remote Sens Environ 167:78–87

Serrano L, Peñeulas J, Ustin SL (2002) Remote sensing of nitrogen and lignin in Mediterranean vegetation from AVIRIS data: decomposing biochemical from structural signals. Remote Sens Environ 81:355–364

Singh A, Serbin SP, McNeil BE, Kingdon CC, Townsend PA (2015) Imaging spectroscopy algorithms for mapping canopy foliar chemical and morphological traits and their uncertainties. Ecol Appl 25:2180–2197

Smith WK, Vogelmann TC, DeLucia EH, Bell DT, Shepherd KA (1997) Leaf form and photosynthesis. BioSci 47:785–793

Smith M-L, Ollinger SV, Martin ME, Aber JD, Hallett RA et al (2002) Direct estimation of aboveground forest productivity through hyperspectral remote sensing of canopy nitrogen. Ecol Appl 12:1286–1302

Stapleton AE (1992) Ultraviolet radiation and plants: burning questions. Plant Cell 4:1353–1358

Sterner RW, Elser JJ, Gorokhova E, Fagan WF, Markow TA et al (2002) Ecological stoichiometry: the biology of elements from molecules to the biosphere. Princeton University Press, Princeton, p 441

Steyn WJ, Wand SJE, Holcroft DM, Jacobsm G (2002) Anthocyanins in vegetative tissues: a proposed unified function in photoprotection. New Phytol 155:349–361

Sun S, Zhong P, Xiao HT, Wang RS (2015) Active learning with Gaussian process classifier for hyperspectral image classification. IEEE Trans Geosci Remote Sens 53:1746–1760

Tanger P, Klassen S, Mojica JP, Lovell JT, Moyers BT et al (2017) Field-based high throughput phenotyping rapidly identifies genomic regions controlling yield components in rice. Nature Sci Rep 7:43839. https://doi.org/10.1038/srep42839

Terashima I, Fujita T, Inoue T, Chow WS, Oguchi R (2009) Green light drives leaf photosynthesis more efficiently than red light in strong white light: revisiting the enigmatic question of why leaves are green. Plant Cell Phys 50:684–697

Thayer SS, Björkman O (1990) Leaf xanthophyll content and composition in sun and shade determined by HPLC. Photosyn Res 23:331–343

Thevenot M, Dignac M-F, Rumpel C (2010) Fate of lignins in soil: a review. Soil Biol Biochem 42:1200–1211

Tilman D (1985) The resource-ratio hypothesis of plant succession. Am Nat 125:827–852

Ustin SL, Gamon JA (2010) Remote sensing of plant functional types. New Phytol 186:795–816

Ustin SL, Jacquemoud S, Govaerts Y (2001) Simulation of photon transport in a three-dimensional leaf: implications for photosynthesis. Plant Cell Environ 24:1095–1103

Ustin SL, Gitelson AA, Jacquemoud S, Schaepman M, Asner GP, Gamon JA, Zarco-Tejada P (2009) Retrieval of foliar information about plant pigment systems from high resolution spectroscopy. Remote Sens Environ 113:S67–S77

Ustin SL, Riaño D, Hunt ER Jr (2012) Estimating canopy water content from spectroscopy. Israel J Plant Sci 60:9–23

Van Gaalen KE, Flagan LB, Peddle DR (2007) Photosynthesis, chlorophyll fluorescence and spectral reflectance in *Sphagnum* moss at varying water contents. Oecologia 153:19–28

Vendramaini F, Díaz S, Gurvich DE, Wilson PJ, Thompson K et al (2002) Leaf traits as indicators of resource-use strategy in floras with succulent species. New Phytol 154:147–157

Verrelst J, Camps-Valls G, Delegido J, Moreno J (2012) Retrieval of vegetation biophysical parameters using Gaussian process techniques. IEEE Trans GeoSci Remote S 50:1832–1843

Vogelmann TC (1993) Plant tissue optics. Annu Rev Plant Phys 44:231–251

Vogelmann JR, Moss DM (1993) Spectral reflectance measurements in the genus Sphagnum. Remote Sens Environ 45:273–279

Vogelmann TC, Nishio JN, Smith WK (1996a) Leaves and light capture: light propagation and gradients of carbon fixation within leaves. Trends Plant Sci 1:65–71

Vogelmann TC, Bornman JF, Yates DJ (1996b) Focusing of light by leaf epidermal cells. Physiol Plant 98:48–56

Weng J-K, Chapple C (2010) The origin and evolution of lignin biosynthesis. New Phytol 1878:273–285

Wieliczka DM, Weng SS, Querry MR (1989) Wedge shaped cell for highly absorbent liquids – infrared optical –constants of water. Appl Opt 28:1714–1719

Wooley JT (1971) Reflectance and transmittance of light by leaves. Plant Physiol 47:656–662

Wooley JT (1975) Refractive index of soybean leaf cell walls. Plant Physiol 55:172–174

Workman J Jr, Weier L (2012) Practical guide and spectral atlas for interpretative infrared spectroscopy. Francis and Taylor Group, Boca Raton

Wright IJ, Reich PB, Westerby M, Ackerly DD, Baruch Z et al (2004) The worldwide leaf economics spectrum. Nature 428:821–827

Wright IJ, Reich PB, Cornelissen JHC, Falster DS, Groom PK et al (2005) Modulation of leaf economic traits and trait relationships by climate. Global Ecol Biogeo 14:411–421

Ziechmann W (1964) Spectroscopic investigations of lignin, humic substances and peat. Geochim Cosmochim Acta 28:1555–1566

Open Access This chapter is licensed under the terms of the Creative Commons Attribution 4.0 International License (http://creativecommons.org/licenses/by/4.0/), which permits use, sharing, adaptation, distribution and reproduction in any medium or format, as long as you give appropriate credit to the original author(s) and the source, provide a link to the Creative Commons license and indicate if changes were made.

The images or other third party material in this chapter are included in the chapter's Creative Commons license, unless indicated otherwise in a credit line to the material. If material is not included in the chapter's Creative Commons license and your intended use is not permitted by statutory regulation or exceeds the permitted use, you will need to obtain permission directly from the copyright holder.

Chapter 15
Spectral Field Campaigns: Planning and Data Collection

Anna K. Schweiger

15.1 Introduction

Field spectrometry is the measurement of spectral properties of (Earth) surface features in the environment (McCoy 2005) and is particularly relevant to plant biodiversity detection. Generally, molecular composition and arrangement and scattering properties of the measured media influence the spectral response (Goetz et al. 1985). Spectral characteristics of plants depend on chemical, structural, anatomical, and morphological leaf characteristics and whole plant traits, including plant height and shape, canopy architecture, branching structure, and the distribution of foliage within canopies (Cavender-Bares et al. 2017; Ustin and Gamon 2010; Serbin and Townsend, Chap. 3; Ustin and Jacquemoud, Chap. 14). Plant spectra provide a wealth of information about how plants use nutrients, light, and water; how these resources are shared within plant communities; and how patterns of resource use influence ecosystem functions and processes, including nutrient and water cycling and the provisioning of resources and habitat for other trophic levels. Spectroscopy of vegetation and plant biodiversity is part of the larger field of biophysics, which uses theories and methods of physics, such as optics, to understand biological systems.

For a long time, field campaigns were regarded as relatively unimportant for remote sensing (RS) studies. Using the term "ground data" or, more generally, "surface reference data" instead of "ground truth" has been suggested, since the latter implies that field data can be easily collected and are relatively "error-free"

A. K. Schweiger (✉)
Department of Ecology, Evolution and Behavior, University of Minnesota, Saint Paul, MN, USA

Institut de recherche en biologie végétale, Université de Montréal, Montréal, QC, Canada

© The Author(s) 2020
J. Cavender-Bares et al. (eds.), *Remote Sensing of Plant Biodiversity*,
https://doi.org/10.1007/978-3-030-33157-3_15

(Justice and Townshend 1981). In RS studies, ground data still are mainly collected for accuracy assessments or the validation of map products. Most often, land cover or vegetation maps are produced through image interpretation (Bartholomé and Belward 2005; Bicheron et al. 2008). Although the importance of accuracy assessments has been pointed out in the literature (Stehman 2001; Justice and Townshend 1981; Johannsen and Daughtry 2009), validation of map products usually means using high-resolution RS images to validate coarser resolution maps (Congalton et al. 2014). Map validations are often based on agreement among random points (Stehman 2001)—i.e., the extent to which land cover classes that random points fall into match the investigator's interpretation of land cover visible in an image. While this procedure makes sense for global map products distinguishing few vegetation classes, local map products clearly benefit from the collection of ground data for validation. However, the importance of ground reference data goes far beyond map accuracy assessments. In fact, ground reference data are essential for remote sensing of plant biodiversity. Collecting ground reference data during spectral field campaigns provides a great opportunity to bridge the gap between RS science and ecology, two fields that are uniquely positioned to together develop methods to assess biodiversity across large spatial scales, continuously and in a detailed way. These assessments are needed to provide information about the current status of ecosystems; to predict the distribution of biodiversity, ecosystem function, and ecosystem processes into the future; and to counteract detrimental changes in ecosystems associated with global change. One reason to advocate for field campaigns is that remotely sensed images, which provide information pixel by pixel, always obscure part of the information on the ground, with the amount of hidden information depending on pixel size (Atkinson 1999). In order to understand the information provided by remotely sensed images of vegetation, it is critical to study the spectral characteristics of plants, their links to plant traits, and their influence on ecosystem properties at the sub-pixel level, because spectral variation is progressively lost when spectra of individual plants and non-vegetation features blend together at increasing spatial resolutions (Atkinson 1999).

This chapter deals mainly with planning field work and the collection of vegetation spectra with field spectrometers on the ground, which can subsequently be linked to other ecological data and/or RS data to investigate biological phenomena. Data collection for airborne spectroscopy is discussed as well, while other RS methods such as unmanned aerial systems (UASs), towers, and trams are covered in more detail in Gamon et al. (Chap. 16). Focus is also placed on data organization and management, particularly because these aspects of planning tend to receive less attention than, e.g., planning of sample collection, yet they are critical to a successful field campaign.

This chapter was written in full awareness that "good practices" are ever-evolving. The relative importance of, and acquisition methods for, ground data, including ecological data, depends on the research question, on the project goals, as well as on study scale, spectroscopic methods and RS data used, budget, time, site accessibility, and the personnel and their training (Justice and Townshend 1981). Likewise, spectral processing standards evolve and software goes out of date quickly. The examples included in this chapter are intended to illustrate what worked in particular situations and to point out pitfalls to avoid; many other approaches are as valuable. Flexibility (knowledge about different techniques and tools, a plan B, etc.) is important for adjusting to particular circumstances and challenges. The "best practice" is probably to learn about several "good practices"; read protocols; talk to field ecologists, data administrators, geographic information system (GIS) professionals, programmers, and communication experts; and get some hands-on experience. A selection of excellent protocols is available from Australia's Terrestrial Ecosystem Research Network (TERN, http://www.auscover.org.au/wp-content/uploads/AusCover-Good-Practice-Guidelines_web.pdf), the Field Spectroscopy Facility at UK's Natural Environment Research Council (NERC, http://fsf.nerc.ac.uk/resources/guides/), the US National Ecological Observatory Network (NEON, http://data.neonscience.org/documents), the Global Airborne Observatory (GAO, https://gao.asu.edu/spectranomics), and the Canadian Airborne Biodiversity Observatory (CABO, http://www.caboscience.org), among others. For more in-depth coverage of particular topics, see texts on the general principles of RS (e.g., Warner et al. 2009) and RS of vegetation (e.g., Jones and Vaughan 2010; Thenkabail et al. 2012), field methods in RS (e.g., McCoy 2005), spatial statistics (e.g., Stein et al. 2002), vegetation sampling (e.g., Bonham 2013), and plant trait measurements (e.g., Perez-Harguindeguy et al. 2013).

15.1.1 Why Plan? The Data Life Cycle

Central to every field campaign are the research questions and proposed explanations outlined in the form of testable hypotheses. It seems natural that planning the science (What data do we need to tackle our questions? What methods are available?) and planning the logistics (Where do we collect data and when? What resources do we need?) often rank above planning data organization and communication. However, starting a project with a data management plan (DMP) has a series of advantages. A DMP integrates several planning aspects in a structured way; it ensures the long-term sustainability of a project and its data, which is important not only because sustainability furthers scientific advancement (e.g., through data sharing and the reuse of data in meta-analysis) but also because it provides accountability for spending resources on research. DMPs are usually required in research proposals and make, through self-defined standards on data acquisition, data formats, documentation, and archiving, scientific work, including collaborations, more effective.

Funding sources often have their own guidelines about the structure and content of a DMP. Although only some of them might be required or relevant for a particular project, common components include:

- *Data collection and documentation*: description of the types, formats, and volumes of data and samples and other materials collected, observed, or generated during a project, including existing data sources; description of the methods for data collection, observation, and generation, including derivative data; standards for ensuring data quality, including repeated measurements, sampling design, naming conventions, version control, and folder structure; description of the documentation standards for data and metadata format and content; and the software used for analyses

- *Ethical, legal, and security issues:* details regarding the protection of privacy, confidentiality, security, and intellectual property rights, including information about access, use, reuse, and distribution rights; the time of data storage; possible changes to these rights over time; and strategies for settling disagreements
- *Archiving:* description of storage needs for data, samples, and other research products; plans for long-term preservation, access, and security, including details on the parties and organizations involved; backup strategies; selection criteria for long-term storage; community standards for documentation; and file formats

A DMP covers all aspects of the data life cycle (Corti et al. 2014), including the following phases:

- *Discovery and planning*: designing the research project and planning data management; planning data collection and consent for data sharing; outlining processing protocols and templates; and developing strategies for discovering existing data sources
- *Data collection*: collecting data, including observations, measurements, recordings, experimentations, and simulations; capturing and creating metadata; and acquiring existing third-party data
- *Data processing and analysis*: entering, digitizing, transcribing, and translating data and metadata; checking, validating, cleaning, and anonymizing data, where necessary; deriving, describing, and documenting data and metadata; analyzing and interpreting data; producing research outputs; authoring publications; citing data sources; and managing and storing data
- *Publishing and sharing*: establishing copyright of data; creating discoverable metadata and user documentations; publishing, sharing, and distributing data and metadata; managing access to data; and archiving
- *Long-term management*: migrating data to best format and suitable media; backing up and storing data; gathering and producing metadata and documentation; and preserving and curating data
- *Reusing data*: conducting secondary analysis; undertaking follow-up research and conducting research reviews; scrutinizing findings; and using data for teaching and learning

Compiling a DMP, establishing guidelines for data and metadata collection and documentation, and outlining data use policies early in the planning phase is good practice. Starting discussions about how to organize data during or after data collection is a difficult task; reorganizing file structures, renaming files, and explaining and setting up new data structures will rarely be a top priority once data collection has started, and new data sets are ready to work with. Many organizations and agencies have their own standards (e.g., https://ngee-arctic.ornl.gov/data-policies) and can provide a good starting point when thinking about one's own.

Communication strategies are another planning aspect that should not be overlooked (Sect. 15.2.2). Timely communication with site administrators is not only central to receiving permits and critical information; it also brings opportunities for public engagement during fieldwork, which is one of the most publicly visible parts of the scientific process. Even if site visits last for only a couple of hours, planning is important. Interactions with the public can happen at any time, so being prepared to answer questions and give a brief project overview in plain language, and perhaps having a flyer ready to hand out, can provide valuable opportunities for science communication. The support of stakeholders, such as site managers, local communities, and authorities, not only is important for a successful research project but also plays a critical role in determining the degree to which ecological research enters in public discourse and ultimately results in broader impact. Moreover, fieldwork brings opportunities for connecting researchers from different disciplines, which can aid in developing a common language, lead to new collaborations, and make projects more effective. Good research plans and communication strategies increase the chances for fruitful exchanges.

From the perspective of a project's feasibility in terms of time, personnel, and budget, proper planning allows field campaigns to stick to their schedule (which is important because ecological processes change over time) and to the collection of data that are relevant for answering particular questions (it is easy to keep bolting on new measurements that slow down and jeopardize the main focus of a study). Moreover, adjusting to particular situations and handling challenges becomes easier when a detailed plan and the reasons behind it are clearly communicated to the research team. Clarity on the daily responsibilities and the project aims also help to keep research teams motivated.

15.1.2 Spectral Models and Scales of Measurement

Models are simplified descriptions of some aspect of the world and usually how it works (Fleishman and Seto 2009; Horning et al. 2010). Modeling is a multistep iterative process to formulate, by abstraction and idealization, a representation of reality (conceptual model), specify it mathematically (mathematical model), and "solve it," which usually involves translating the math into computer code (computational model; Dahabreh et al. 2017). Models are used to test hypotheses, to assess relationships between response variables and factors that influence them, to investigate interactions between parts of a system, to make predictions about how a system will likely behave in the future, and to test how well models calibrated with data from the past fit current conditions (also known as hindcasting). Ideally, a model describes the full extent of the phenomenon of interest, but in practice, there are limits to the variables that can be determined in any given study. These limits can be formally described by model boundaries, which are as any ordering/bordering system an attempt to say something about (or attain power over) what is incompletely understood (or under-controlled; Jones 2009; Szary 2015). For the purpose of this chapter, model boundaries illustrate the situations under which a model is likely valid to some degree; ideally, likelihood and degree of validity are mathematically established. Model boundaries can be biological (e.g., specific to ecosystems, species, life stages), physical (e.g., specific to latitudinal, geological, hydrological, and topographical extents, or specific in time and place), or political (e.g., specific to regions or countries). Further, models can be classified as reductionist or system-based, quantitative or conceptual, correlative or mechanistic, static or dynamic, and hybrids thereof (Horning et al. 2010). Model boundaries and modeling approaches should be determined early in the planning phase and reported. Some model limitations are likely beyond the researcher's control, such as instances in which data can only be acquired within certain political units. However, the choice of model to describe a particular system should be made deliberately; and the modeling approach should determine data collection, and not vice versa.

The data needed to investigate a phenomenon of interest with spectroscopy depend not only on the research question, the modeling approach, and model boundaries but also on the aim of the analysis (e.g., model calibration, validation, interpretation) and the level of spectral data acquisition (leaf-level spectroscopy, proximal, airborne, satellite RS; see Sect. 15.3., Gamon et al. Chap. 16). Data for model calibration and validation should match the conceptual model's boundaries (e.g., the model's temporal and spatial scale) and the modeling approach. For example, while quantitative and correlative models are ideally based on relatively uncorrelated or orthogonal variables, conceptual and mechanistic models ideally include all variables relevant for a particular study system. Drawing inferences from models and applying them to make predictions are only justifiable when model accuracy has been assessed (Horning et al. 2010); a model can give a very accurate description of a particular system, but one would not know until its accuracy is assessed.

Model calibration describes the process of determining the values of parameters so that model outputs fit the observed data. Internal validation refers to testing a model's ability to explain the data used to populate the model. One common method for this is cross-validation. During cross-validation, the data set is split into calibration and (internal) validation data, the calibration data are used to fit the model, the model coefficients are applied to the validation data, and predicted and measured values from the validation data are statistically compared to assess model fit. Usually the data are split repeatedly, and model statistics (and often model parameters) are averaged across the number of splits. In k-fold cross-validation, k indicates the number of random data subsets or splits. One subset is omitted from model calibration and used for validation, and the process is repeated until all subsets have been left out once. For small data sets, leave-one-out cross-validation is particularly useful; here, only one sample is omitted from model calibration and used for validation, and the process is repeated until all samples have been left out once. In contrast to internal validation, external validation refers to a model's ability to predict observations not used for model development (Dahabreh et al. 2017), which is critical for evaluating model performance and transferability. External validation involves either leaving out samples from the internal calibration-validation process or collecting additional independent data, followed by the evaluation of agreement between model output and observed data without any attempt to modify model parameters to improve fit.

Spectral models can be categorized into empirical-statistical and physical models and combinations between the two (see Verrelst et al. 2015). Empirical-statistical models are based on the relationship between the spectral behavior of certain spectral bands or the entire spectrum (the predictors or independent variables) and the vegetation characteristic(s) of interest (the dependent variable(s)). Empirical-statistical models generally use regression or clustering algorithms and aim to predict vegetation characteristics or class membership from a population of spectral data that were not included in the modeling process. For calibrating empirical-statistical models, it is essential to collect representative field data (see Sect. 15.2.3). Generally, this means that for the ecosystem, time of year, and area of interest, data should cover the range of values for which predictions are intended to be made or the number of vegetation classes with suitable replication, as well as the range of environmental conditions present in that area. Sample size should be large enough and samples distributed evenly across the expected range of values, classes, and environmental gradients to allow samples to be left out from model development and enable external validation.

Regression techniques are generally used for modeling and predicting continuous vegetation characteristics, such as biomass and chemical or structural composition (Ustin et al. 2009; Serbin et al. 2014; Schweiger et al. 2015a, b; Couture et al. 2016), or relative proportions of vegetation properties, such as the abundances of species, plant functional types, or vegetation types (Schmidtlein et al. 2012; Lopatin et al. 2017; Fassnacht et al. 2016; Féret and Asner 2014; Schweiger et al. 2017). Univariate, multivariate, linear, and nonlinear regressions are common for modeling and predicting vegetation characteristics from few spectral bands or from spectral indices. Spectral indices are used to infer vegetation status, including plant stress, and ecosystem parameters, inducing productivity, from empirical or physical relationships between spectra and plant traits. Many spectral indices have been published (see, e.g., https://cubert-gmbh.com/applications/vegetation-indices/). The most widely used indices include the normalized difference vegetation index (NDVI; Rouse Jr et al. 1974; Tucker 1979), an indicator of vegetation greenness, and modified versions such as the soil-adjusted vegetation index (SAVI; Huete 1988) and the photochemical reflectance index (PRI; Gamon et al. 1992). The NDVI and its variants have been shown to correlate well with biomass, LAI, and the photosynthetic capacity of canopies. The PRI estimates light-use efficiency and can be used to estimate gross primary productivity (GPP) and assess environmental stress (Sims and Gamon 2002). Spectral indices can also be used directly to estimate certain environmental characteristics (Anderson et al. 2010; Pettorelli et al. 2011; Wang et al. 2016). For example, the NDVI has been used for predicting and mapping taxonomic diversity (e.g., Gould 2000), the rationale being the expected increase in ecological niches with increasing energy or resources in ecosystems (Brown 1981; Wright 1983; Bonn et al. 2004). In addition, advances in sensor technology allow capturing aboveground productivity in ecologically meaningful units, such as the annual amount, variation, and minimum of photosynthetically active radiation (fPAR), which has been found to explain global patterns of mammalian, amphibian, and avian diversity to a substantial degree (Coops et al. 2018). The idea behind spectral indices is that they are generally applicable and transferable; however, ground reference data are still needed to assess their accuracy. In addition, site-specific data are also needed to recalibrate spectral indices, for example, by selecting the optimal wavelengths for the sensor used and by estimating site-specific model coefficients, because index responses often vary with the particular context.

Latent variable methods, such as partial least squares regression (PLSR) and partial least squares discriminant analysis (PLSDA), were developed for chemometrics and specifically deal with the high degree of autocorrelation inherent in data with high spectral resolution (Wold et al. 1983; Martens 2001). PLSR is a standard method for modeling and predicting continuous vegetation characteristics and PLSDA for determining class membership from spectral data. Clustering methods, including principal component analysis (PCA), principal coordinate analysis (PCoA), and linear discriminant analysis (LDA), are frequently used to explore patterns, such as the degree to which species or plant functional types cluster separately from each other in spectral space. In addition, machine learning algorithms [e.g., random forest (RF) or support vector machines (SVNs)] and deep learning methods [e.g., convolutional neural networks (CNNs)] can be used for classification problems, including the identification of vegetation types based on physiognomic attributes (e.g., forest, shrubland, grassland, cropland), for the detection of plant pathogens and other stresses (Pontius et al. 2005; Herrmann et al. 2018), and for species detection (Clark et al. 2005; Fassnacht et al. 2016; Kattenborn et al. 2019).

Physical models are based on causal physical relationships between electromagnetic radiation and vegetation properties; in spectroscopy, these models are called radiative transfer models (RTMs). For leaf optical properties, the RTM PROSPECT (Jacquemoud and Baret 1990; Jacquemoud et al. 2009) models leaf reflectance and transmittance based on leaf chlorophyll a and b content, "brown pigment" content, equivalent water thickness, leaf dry matter content, and a leaf structure parameter indicating mesophyll thickness and density. For modeling optical properties of canopies, leaf-level RTMs can be combined with canopy RTMs (Jacquemoud et al. 2009), which incorporate structural canopy parameters, including leaf area index (LAI, the ratio of leaf area to ground surface area), leaf inclination, a hot spot parameter (a function of the ratio of leaf size to canopy height), as well as soil reflectance and measurement characteristics, including sun and viewing angle. Frequently used canopy RTMs include SAIL (Verhoef 1984), GeoSAIL (Huemmrich 2001), and DART (Gastellu-Etchegorry et al. 2004). Although RTMs do not model all interactions between plants and light, because they cannot incorporate all characteristics of leaves and vegetation canopies that influence the spectral response, they are useful for simulating spectra and retrieving estimates about plant characteristics. In "forward mode," RTMs simulate spectra from the vegetation parameters incorporated into the model, e.g., by using the expected range of values for chlorophyll content, SLA, and other parameters, in certain ecosystems or for certain species. In "backward mode," RTMs estimate vegetation characteristics from spectra (Vohland and Jarmer 2008; Weiss et al. 2000). Model inversion is usually done by running an RTM in forward mode and systematically varying the input parameters using lookup tables (LUT) with different trait combinations, until the measured spectral signal is sufficiently well approximated. The most plausible input parameter combinations can then be averaged to provide estimates of vegetation traits. In principle, RTM inversion can be conducted without ground reference data collected on-site (e.g., when input parameters can be sourced from plant trait databases, such as TRY; Kattge et al. 2011) or determined based on expert knowledge. However, inversion of RTMs is generally an ill-posed problem in the sense that there is not a single solution but rather multiple solutions to model inversions (i.e., multiple input parameters can yield the same output spectra). Ground reference data are important for RTMs because they can be used to limit the ranges of possible input values (Combal et al. 2003) and are essential for model validation.

15.2 Planning Field Campaigns

This section includes thoughts about data organization (Sect. 15.2.1) and communication (Sect. 15.2.2), before covering the planning of data collection in more detail (Sect. 15.2.3).

15.2.1 Data Organization

Data organization schemes help define and implement guidelines to make project management and collaborations more efficient and ensure long-term project sustainability and the reproducibility of research (https://ropensci.github.io/reproducibility-guide/). Guidelines for folder structure, file names, documentation, file formats, data sharing and archiving, version control, and data backups are all part of data organization. When archiving is handled by a third party, researchers need to consider how to structure data and metadata to match external requirements. Generally, it is good practice to work backward and start with identifying where project data will be stored long term and which data and metadata standards will make long-term storage possible and data sets discoverable and reusable later. For instance, it is important to use community standards for taxon names, units, and keywords and to store data in file formats that are nonproprietary (open), unencrypted, and in common use by the research community. The US Library of Congress has released a recommended file format statement (http://www.loc.gov/

preservation/resources/rfs/) and provides detailed format description documents for different categories of digital data, including data sets, images, and geospatial data. Once data characteristics for long-term storage are known, one can define short-term storage structures and the workflow leading from raw data collection, to cleaned-up data, to preliminary and final results and products. Data backups should match the original data structure so recovery requires only a few minutes. Ideally, backups are done automatically, continuously, and incrementally, and data history is preserved. Data recovery should be tested on a regular basis. Additionally, it is important to check how long data backups are being sustained. When storage space is limited, it makes sense to use a time-dependent structure, such as keeping daily backups for a year, biweekly backups for 3 years, and monthly backups thereafter. Several resources provide details about good data management practices, including the Oak Ridge National Laboratory Distributed Active Archive Center (https://daac. ornl.gov/datamanagement/) and the rOpenSci initiative (https://ropensci.github.io/ reproducibility-guide/).

Fundamentals of data organization include (see, e.g., Cook et al. 2018):

- *Definition of file/folder content*: Keep similar measurements in one data file or folder (e.g., if the documentation/metadata for data are the same, then the data products should all be part of one data set).
- *Variable definition*: Describe the variable name and explicitly state the units and formats in the metadata; use commonly accepted names, units, and formats and provide details on the standards used (e.g., SI units, ISO standards, nomenclature standards); use format consistently throughout the file; use a consistent code (e.g., −9999) for missing values; and use only one variable per measurement (e.g., avoid reporting coordinates in more than one coordinate system or time in several time zones).
- *Consistent data organization*: Do not change or rearrange columns in the original data; include header rows (first row should contain file name, data set title, author, date, and companion file names); use column headings to describe the content of each column; include one row for variable names and one for variable units; and make sure either each row in a file represents a complete record (with columns representing all variables that make up the record) or each variable is placed in an individual row (e.g., for relational databases).
- *Stability of file formats*: Avoid proprietary formats, and prefer formats encoding information with a lossless algorithm (e.g., text, comma/tab-separated values, SQL, XML, HTML, TIFF, PNG, GIF, WAV, postscript formats).
- *Descriptive file names*: Use descriptive, unique file names; use ASCII characters only and avoid spaces (e.g., start with ISO date, followed by descriptive file name: 20180430_siteA_plotB_vegSurvey); remember that file names are not a replacement for metadata; explain naming structure of files in metadata; organize files logically; and make sure directory structure and file names are both human- and machine-readable (check operating or database system limitations on file name length and allowed characters).

- *Processing information*: Consider including information on software or programming language and version; provide well-documented code and information about data transformation.
- *Quality checks*: Ensure that data are delimited and lined up in proper columns; check that there are no missing values (blank cells); scan for impossible and anomalous values; perform and review statistical summaries; and map location data.
- *Documentation*: Document content of data set; reason for data collection; investigator; current contact person; time, location, and frequency of data collection; spatial resolution of data; sampling design; measurement protocol and methods used, including references; processing information; uncertainty, precision, accuracy, and known problems with the data set; processing information; assumptions regarding spatial and temporal representativeness; data use and distribution policy; ancestors and offspring of data set (including references to publications).
- Data protection: Create backup copies often and without user interference (automatically, continuously, incrementally); three copies (original, on-site external, off-site) are ideal; test restoring information; and use checksums to ensure that copies are identical.
- *Data preservation:* Preserve well-structured data files with variables, units, and values defined; documentation and metadata records; materials from project wiki/websites; files describing the project, protocols, and field sites (including photos); and project proposal (at least parts) and publications in open-access archives. Check platform standards for data archiving beforehand.

The best way to organize data depends on the project, size of the team, and degree of interaction among team members, among other things. It is good practice to think about ways to organize data early in the project-planning phase and to include at least the core project team in these discussions. However, differences in personal work styles can be a challenge for reaching agreements; the larger the team, the more difficult this becomes. In such cases, top-down approaches to data organization can be a good option, especially ones that have been tested before. Laying out data organization schemes at an early project stage and inviting people's feedback at this stage is good practice. Clearly, there should be room for discussion during later project stages as well (particularly when the existing organization scheme is not working as expected), but generally adjustments become more complicated the longer projects are running. Data organization schemes are intended to make daily workflows, data exchange, and data archiving easier; they should not cause an extra workload. Research teams are much more likely to adapt a particular organizing scheme when it is simple and intuitive, and everyone is much more likely to stick to a system when its benefits are obvious. In the end, even the best organization structure fails when no one is following it.

Box 15.1 An Example of Folder Structure

It can be advantageous for research groups and institutions to implement common data standards and a file structure that forms the backbone of a data organization scheme and does not have to be discussed for every new project. Data standards also promote reproducible research. One option is to set up file directories separated by content. The example below structures a higher-level directory (e.g., the project directory) into *docu_work*, *docu_pub*, *orig_data*, *data_work*, *data_pub*, *gis_work*, *gis_pub*, *maps*, and *printout* folders. The key feature of this structure is the distinction between *work* folders, containing work in progress and intermediate results, and *pub* folders, containing the final versions and results. After the completion of a project, the *orig_data*, *maps*, *printout*, and all *pub* folders are archived (publicly and internally), while all *work* folders get deleted from a public or shared drive. Backups are kept at a certain frequency for a certain amount of time (e.g., daily backups for a year, biweekly backups for 3 years, monthly backups for 10 years), and personal copies can of course be kept as long as necessary. A folder structure like this makes archiving data easy because at each project stage it is clear which data sets, documents, and products will be preserved. The *pub*, *orig_data*, and *maps* folders should contain everything a person without knowledge about the project would need to repeat the analysis, including basic project background and workflow descriptions in the *docu_pub* folder, but nothing unnecessary, such as intermediate results. One testable reproducibility goal could be that a person with appropriate analytical skills but without any information about the project would be able to re-create and explain a main result, including the rationale behind the analysis, after 1 workday without any external help. For GIS heavy projects, it may make sense to separate analysis, results, and products based on geospatial data from those using other data sources. In this example, map products are reproducible based on data from *gis_pub* and the layouts found in the *maps* folder.

Short Description of Contents and Management of Example Folder Structure

Folder type	Content description and management
docu folders	These are for the proposal, project descriptions, documentation of workflows, planning documents, minutes of meetings, manuscripts, photos, etc. File names could, for example, start with the ISO date followed by a descriptive name. Files are usually organized into subfolders. Typical file formats include .doc, .txt., .pdf, and .tiff. The *docu_work* folder is deleted after the project is finished. The *docu_pub* folder contains the final versions in a non-editable format, such as .pdf. It should contain the essentials of the project background and all workflows needed to repeat the analysis; detailed descriptions should either be left out or clearly flagged, e.g., as "additional information." The *docu_pub* folder also includes information about the use of corporate or project identity styles (use of logos, colors, fonts, etc.). This folder gets archived after the project is finished

Folder type	Content description and management
orig_ data folder	This folder includes raw data acquired during the project. These data never get changed. "Read only" permission is advisable; metadata files describing the data sets are critical. It is good practice to check backup copies when new data sets are added. From here data can be copied to the *data_work* folder, e.g., if the format needs to be changed or different data sets are being combined into a master data set. This folder can contain proprietary file formats, in which case it is important to include details about the software and version used to access files. This folder gets archived after the project is finished
data folders	These are where analyses happen. The folders usually contain subfolders, e.g., for code, data input, and data output. Work copies of data copied from *orig_data* are saved here. Metadata describing any changes to the original data are important, including variable transformations, references for methodology, software, and version used. Typical file formats are .csv and .txt. The *data_work* folder includes preliminary results and is deleted after the project is finished. All analysis steps are being documented in the *docu* folders. The *data_pub* folder contains final scripts, final results, compiled master data sets, etc. This folder gets archived after the project is finished
gis folders	These are similar to *data* folders but for geospatial data. This folder contains geo data that have been modified from the original data. Processing details are included in the metadata; original data remain in *orig_data*. Typical file formats include geotiff, .tiff, and .shp. The *gis_work* folder is for intermediate steps and is deleted after the project is finished. The *gis_pub* folder is for final results and gets archived
maps folder	This folder is for data associated with maps, layouts, and styles. It gets archived after the project is finished. All paths in maps should refer to data in the *data_pub* or *orig_data* folders when the project is finished
printout folder	This folder contains final products including publications, maps, posters, and presentations, usually in a non-editable format. This folder gets archived after the project is finished

15.2.2 Communication

Communicating plans for fieldwork and applying for necessary permits in a timely manner avoids unnecessary complications. Essentially, the earlier researchers get in touch with site administrators, the better. Regulations vary, but obtaining necessary permits can take months, especially in areas with a high protection status. Often, research proposals are evaluated by a panel. However, it is good practice to get in touch with site administrators before writing a detailed proposal because local regulations might influence project plans, including changes to the location, timing, and duration of data collection; the equipment used; and the number of people on site. In addition, it is good practice to figure out logistics, such as transportation of people and equipment, early in the planning phase. Early communication provides time to understand the rationale behind a research plan and to adjust the plan appropriately. Often, it is not until details are questioned that it becomes clear which aspects of a research plan are critical and which can be handled somewhat more flexibly.

For research projects planned on or close to Indigenous lands, it is critical to inquire early in the planning process about local procedures and ethical guidelines. Generally, site administrators would be the first points of contact and should be aware of how to communicate the research objectives to the Indigenous communities and judge the level of involvement required (e.g., between short- and long-term studies). However, it is important for researchers to initiate this inquiry and to seek additional guidance if needed. For both short- and long-term projects, familiarizing oneself with ethical frameworks for research with Indigenous communities and/or on Indigenous lands (e.g., Claw et al. 2018) is good practice.

15.2.3 Planning Data Collection

Remote sensing of plant biodiversity typically includes the comparison of ecological or spectral data collected on the ground with remotely sensed images, often through a model that considers scale effects. Two aspects are critical to data collection: (i) sampling representative areas and/or individual plants that can be aligned with the imagery and (ii) sampling at high spatial accuracy and a level of precision that matches the sensor. With these two points in mind, the following sections discuss area selection (Sect. 15.2.3.1); value ranges (Sect. 15.2.3.2), which are important for model representativeness and thus also of interest for the validation of physical models; and sampling design (Sect. 15.2.3.3), with a particular focus on empirical-statistical analyses. Data collection is typically formalized in a sampling plan. A sampling plan describes data acquisition, recording, and processing (Domburg et al. 1997) and includes the first elements of the data life cycle (see Sect. 15.1.1). Consequently, the plan will likely include decisions about the area selection and variables measured, logistical constraints, sampling and analysis methods, sampling design, sampling protocols, estimation of measurement accuracy/precision, and operational costs.

Sampling is a method of selection from a larger population carried out to reduce the time and cost of examining the entire population (Justice and Townshend 1981). In the case of sampling plant biodiversity at a particular field site, we are generally selecting individual plants to represent local populations of a set of species that capture the range of functional and phylogenetic variation in a site or represent the dominant species. Data collection balances accuracy and representativeness against time and budget. Two questions are central to planning data collection: (i) Which population(s) is (are) best suited to answer the specific research question(s)? and (ii) Is (are) the population(s) adequately represented by the sampling scheme? During the early planning phase, it is important to get a sense of which environmental factors cause variation in the samples to be collected and variables to be measured (Johannsen and Daughtry 2009) and to choose research sites accordingly. This requires researchers to familiarize themselves with the conditions at the site. It is also critical to define at an early planning stage how the data will be analyzed and to choose the sampling design and sample size accordingly (see Sect. 15.2.3.3). Improper randomization in particular can lead to biased conclusions based on inappropriate assumptions (De Gruijter 1999).

15.2.3.1 Area Selection

It is rarely possible to describe all aspects of an ecological phenomenon of interest in a single study. Model boundaries help clarify the conditions under which a model is likely valid (see Sect. 15.1.2). Research areas represent conceptual model boundaries, including ecosystem(s), plant community(ies), species, and geographic location. Although it is good practice to formulate model boundaries first and pick research areas accordingly, in practice model boundaries usually need to be adjusted after selecting a research area to reflect the conditions at the site. Formulating the model first and refining it during the planning process help clarify a study's limitations; stating them clearly is important for the analysis and further synthesis work.

Once model boundaries have been formulated, it is important to investigate which environmental conditions influence the response and explanatory variables in the study system and how they are spatially distributed. Generally, it is critical to cover the range of environmental conditions, both biotic and abiotic, for which model inferences are being made, including the diversity and distribution of vegetation communities, plant species, successional gradients, soil types, soil moisture and nutrient gradients, aspects, slopes, land uses, microclimatic conditions, animal communities, pathogens, and other factors determining environmental heterogeneity in a study area. Accounting for the variation of every factor might not be possible, but many environmental factors are correlated. If possible, it is good practice to investigate the covariance structure of environmental factors based on previous studies and existing data and to focus on a few factors that are expected to have the most effect on the phenomenon of interest. Data collection might need to be limited to a smaller area than anticipated due to a high degree of environmental heterogeneity and/or time constraints, which affects the range of conditions for which conclusions can be drawn. However, it is generally advantageous to work with sound models for small areas with a limited degree of environmental variation than to work with weaker models for larger areas. Testing model predictions outside the model boundaries can provide important insights regarding model transferability, the comparability of ecological conditions, and differences and similarities in ecosystem function between areas.

Maps, local sources, and other research groups can provide important information regarding environmental variation in a research area. Again, it is helpful to contact site administrators early to gain access to resources and build connections to other research groups. Local administrators can often give advice regarding the timing and location of sample collection, including practical considerations such as accessibility. Visiting a potential research area can be extremely helpful during the

planning phase. Often, it is easier to discuss issues and concerns in person, and seeing the conditions on site facilitates decision-making. Joining another research
group for some time in the field or accompanying a person who knows the area well
provides a great way to get to know an area. Covering the heterogeneity of a research
area is important for all aspects of remote sensing of plant biodiversity, including
collecting spectral references for image processing and sampling ground data (such
as vegetation spectra and vegetation samples) for model calibration, validation, and
interpretation. For example, empirical line correction (ELC, Kruse et al. 1990,
Broge and Leblanc 2001), a common method for correcting atmospheric and instrument influences on remotely sensed data, utilizes ground measurements of invariant
surfaces (e.g., pavements, rocky outcrops, snow, water, calibration tarps) that can be
readily identified in the image and/or for which accurate location data has been
acquired. As for any empirical method, the performance of ELC depends on the
representativeness and accuracy of the input data, and model transferability is limited. Thus, target surfaces for ELC should ideally be distributed across the area of
interest (e.g., located across all flight lines) and cover differences in altitude, slope,
and aspect (see Sect. 15.3.3.2).

Sampling biodiversity in a way that fits remotely sensed data means incorporating the heterogeneity of a research area but also requires thinking about the size and
shape of sampling units. Generally, sampling units should be delineated to encompass areas with similar environmental conditions. Remotely sensed imagery are
usually raster data, so ground measurements need to represent areas rather than
points. The optimal size of the sampling units on the ground depends on the spatial
heterogeneity and spatial resolution of the imagery. Sampling units that are smaller
or the same size as the pixels in the imagery are usually unrepresentative (Justice
and Townshend 1981). Pixel shifts are a common consequence of image processing,
and averaging RS data across several pixels is common practice for noise reduction.
As a general guideline, the minimal dimensions of a representative sampling unit
can be calculated as $A = P * (1 + 2 L)$ (Justice and Townshend 1981), with P being
the pixel dimensions of the image and L the accuracy of image alignment in number
of pixels. For example, if the spatial resolution of an image is 3 m and a one-pixel
shift is expected to occur during image processing, the minimal size of an internally
homogeneous sampling area would be $3*(1 + 2 * 1) = 9$ m × 9 m. This makes it
possible to capture similar environmental conditions even when the image pixel that
should align with the sampling unit's center pixel has shifted one pixel in either
direction or when averaging the center pixel and its neighboring pixels (Fig. 15.1).
In this particular context, internal homogeneity does not mean that the sampling
unit can only consist of one particular feature but rather that all features should be
evenly distributed throughout the unit. In other words, to be considered internally
homogeneous, a sampling unit does not have to consist of a single plant species of
one particular age or size class; it can consist of different species and individuals as
long as their spatial distribution is comparable among the pixels within that unit. For
example, the center pixel in Fig. 15.1a is not representative for the sampling unit
because species abundance varies among the nine pixels; the sampling area is internally heterogeneous. In contrast, the center pixel in Fig. 15.1b is representative for

a   b

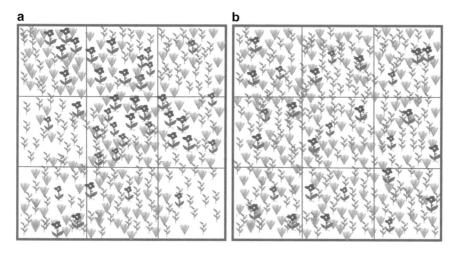

Fig. 15.1 (**a, b**) Internally heterogeneous and homogeneous sampling units. (Adapted from Justice and Townshend 1981)

the sampling unit because species abundance is similar among the nine pixels; the sampling area is internally homogeneous.

Another way to determine the minimum size of sampling units for RS studies are structural cells, which are defined as area units that are large enough to fully capture the variation within one particular feature on the ground, such as the variation in the spatial distribution of individual plant species within a plant community or the variation in the terrain characteristics within a particular topographic feature (Grabau and Rushing 1968). If the size of a structural cell is larger than the image pixel plus its accuracy buffer, structural cells should be preferred; their size can vary depending on the variation of the environmental feature of interest. It is also worth mentioning that pixels as seen by a sensor are not square but elliptic and that surrounding pixels contribute substantially to the signal detected per focus pixel (Inamdar et al. 2020). Theoretically, elliptic or hexagonal sampling units, representing shapes that are frequent in nature, should capture local environmental conditions better than square plots. However, since remotely sensed images are usually subsampled to make pixels quadratic, a case for square sampling plots can be made. As pointed out earlier (Sect. 15.1.2), drawing hard boundaries around any natural feature is notoriously flawed because gradients are the norm and abrupt changes the exception. Thus, it is good practice to specifically sample ecotones and other transition zones if possible—or, if not, to acknowledge that a model might not be representative for transition zones when they are not sampled. Generally, a sampling unit can be considered adequately described when measurements within that unit cover the variation of the characteristic of interest. Thus, it is not necessary to sample entire sampling units when they are internally homogeneous (Fig. 15.1b). For example, when the plant species composition in every 1 m^2 in a 9 m × 9 m research plot closely resembles that of every other 1 m^2, it is sufficient to conduct a species inventory within 1 m^2, ideally in the plot center. Likewise, it can be expected that the chemical composition of the biomass clipped in a 1 m × 20 cm strip in the central 1 m^2 is representative for the chemical composition of the vegetation in the entire 9 m × 9 m plot, given that the clip strip captures the variation in species composition, height, and age-class distribution within the central 1 m^2.

The accuracy and precision of the surveying equipment used for measuring plot coordinates is another aspect to consider when determining the minimal size of homogeneous sampling units; ideally, measurement errors should be estimated under field conditions and added to the minimal size of the sampling unit. Additionally, edge effects influence remotely sensed data. Sampling units should be placed sufficiently far from landscape features, such as open soil, gravel, snow, water, large rocks, roads, footpaths, bridges, and trees (when working in grasslands), that influence the spectral properties of adjacent areas. Depending on the time of day of image acquisition, shadow effects from tall objects such as trees, mountains, or buildings need to be taken into account as well.

15.2.3.2 Range of Values

For the remote sensing of biodiversity, the range of conditions (time of year, value range, species, and environmental context) used to calibrate a model should cover the range of conditions for which inferences or predictions should be made. For example, extrapolating beyond the range of values relies on the assumption that the estimated relationship holds beyond the investigated range. This cannot be assumed without additional information, because nonlinearities (e.g., saturating curves) are common in ecological data, especially when covering large areas and multiple environmental gradients and ecosystems.

For modeling continuous data with regression-style empirical-statistical approaches, the sampling design should cover the expected range of values in the area of interest with a sufficient number of evenly distributed samples. Predictions outside the calibrated range are not reliable because deviations from the 1:1 line between measured and predicted values (Fig. 15.2a) increase at the lower and upper ends of the distribution (Fig. 15.2b). During model validation the entire range of values should be covered as well, specifically paying attention to the value range most important for the research question(s). When the tails of the distribution are of interest for predictions, it is good practice to include a number of extreme values in the validation. These values can be used for updating the model, extending model validity beyond the previously calibrated range or beyond previously covered environmental contexts (Fig. 15.3c).

However, a larger range of values and environmental contexts is not automatically better. Empirical-statistical models are context specific. Transferability beyond the time of year, value range, species, and environmental context for which they are calibrated cannot be assumed without a test. The power of empirical-statistical methods lies in their ability to fit the data. Thus, empirical-statistical models should be calibrated for the range of values and the environmental conditions that are most

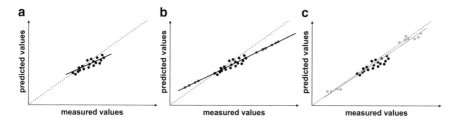

Fig. 15.2 (**a**) A calibrated model (solid black line) generally deviates to some degree from the ideal 1:1 relationship (dashed black line) between measured and predicted values; (**b**) the deviation becomes more pronounced when predicting samples outside the calibrated range of values (red dots); (**c**) using the measured values of these samples (green dots) for calibrating a new model extends the range of values for which the model is valid (green solid line; note that in this case the samples are not evenly distributed such that model performance in the gaps of the value range, i.e., between black and green points, is unknown)

relevant for answering a specific research question. Extending the calibrated range beyond the range of values for which predictions are being made generally decreases model accuracy for these values, as compared to a more narrowly defined model that fits that range.

For classification models, including empirical-statistical clustering and supervised classification methods, covering all classes of interest with a similar and sufficiently large number of samples is also important. However, the range of values/ environmental conditions question is more nuanced. On the one hand, it can be advantageous to include all major classes as end-members for classification, even when not all of them are of interest for prediction. For example, when the aim is to differentiate broadleaf and needleleaf forest using remotely sensed imagery, it makes sense to include other classes present in the image, such as grasslands, roads, and water bodies, as well. The reason is that when broadleaf and needleleaf forest are the only two classes used for model calibration, the model will, when applied to the full image, try to assign grassland, road, and water pixels to these two forest types, decreasing model accuracy. On the other hand, too many extra classes can make it difficult for a model to differentiate among the classes of interest. If one is interested in differentiating two forest types, it would probably not make sense to use tree species as input classes, because a species differentiation model is likely overall less accurate than a model trained on just the two forest types of interest. Stepwise approaches to such classification problems are often helpful. First, one could differentiate broader classes such as vegetation, roads, and water bodies from each other; then, forest from grassland within the vegetation class; and finally different forest types within the forest class (for more details see textbooks on remote sensing of vegetation, e.g., Jones and Vaughan 2010; Thenkabail et al. 2012, and specific topics, such as on deep learning approaches to image classification, e.g., Cholet and Allair 2018).

Areas from within the calibrated value range are usually prioritized during model interpretation. Nevertheless, visits to areas with vegetation or other site characteristics at the edge of or beyond the calibrated range can be insightful regarding the limits of a model's applicability. Field visits are ideal for investigating how environmental context and time of year influence model performance and for determining under which conditions a model performs well or poorly.

15.2.3.3 Sampling Design

A good approach to deciding on a sampling design and to planning data collection for remote sensing of plant biodiversity in general is to start at the end and think about (i) what type of modeling result or product would be most useful with respect to the research question (such as a hypothesis test at a given significance level or a map of a variable with a given accuracy), (ii) what kind of data analysis would lead to that result, (iii) what data properties are needed for the specific analysis, and (iv) how these data can be collected efficiently (De Gruijter 1999).

In a spatial context, a sampling design assigns a probability of selection to any set of points in a research area, while a sampling strategy is defined as the combination of sampling design and the estimator of the variable of interest, for which statistical quality measures (such as bias or variance) can be evaluated (De Gruijter 1999). Sampling designs can be model-based or design-based; the two approaches use different sources of randomness for sample selection and model inferences (Brus and DeGruijter 1993; Domburg et al. 1997). Harnessing as much information about spatial variation as possible, including maps of the study region and theory about spatial patterns, facilitates finding an efficient sampling design for both model- and design-based sampling strategies.

Model-based sampling is based on geostatistical theory and evaluates uncertainties by using a fixed set of sampling points while the pattern of the values of interest varies according to a defined random model of spatial variation (De Gruijter 1999). Model-based sampling strategies are, for example, used for kriging, a spatial interpolation method that uses measured point values to estimate unknown points on a surface. The ideal situation for using a model-based sampling scheme is when the desired result should be the prediction of values at individual points or of the entire spatial distribution of values in the research area (i.e., a map), when a large number of sample points can be afforded to calculate the variogram (~100–150 sample points, Webster and Oliver 1992), and, most importantly, when a reliable model of spatial variation is available, the spatial autocorrelation is high, and there is a strong association between the model of spatial variation and the variable of interest. The association between a geostatistical model and the variable of interest is particularly important, because the final inferences about the spatial distribution of the variable of interest are based on the model of spatial variation (De Gruijter 1999; Atkinson 1999). However, it is often difficult to decide if model assumptions are acceptable because several decisions for defining the spatial structure (e.g., about stationarity, isotropy, and the variogram) are subjective (Brus and DeGruijter 1993).

Ecological systems are often too complex to use model-based inference with much confidence (Theobald et al. 2007). Nevertheless, if a tight relationship between a geostatistical model and nature is expected, model-based sampling schemes are useful, for example, to find at a defined accuracy the optimal sampling grid orientation and spacing for kriging methods (see, e.g., Papritz and Stein 1999) and to define ideal locations for additional sampling points when some sampling points are predefined or spatially fixed. Moreover, model-based sampling encompasses methods such as convenience sampling (sampling at locations that are easy to reach) and purposive sampling (sampling sites chosen subjectively to represent "typical" conditions). Although no formal statement of representativeness can be made for these methods (Justice and Townshend 1981), and they are not appropriate for accuracy assessments (Stehman and Foody 2009), they can provide valuable information in a geostatistical context (for an introduction to geostatistics, see, e.g., Atkinson 1999, Chun and Griffith 2013).

Design-based sampling is based on classic sampling theory. It evaluates uncertainty by varying the sample points while the underlying values are unknown but fixed (De Gruijter 1999). Statistical inferences from design-based sampling are valid, regardless of spatial variation and patterns of spatial autocorrelation, because no assumptions about spatial structure are being made. Design-based sampling schemes can be classified depending on how randomizations are restricted. Two or more designs can be combined (De Gruijter 1999; Fig. 15.3):

- *Simple random sampling* (Fig. 15.3a): No restriction is placed on randomization; all sample points are selected with equal probability and independently from each other.

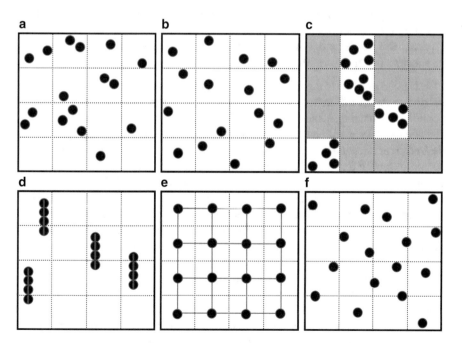

Fig. 15.3 Examples of sampling designs: (**a**) simple random, (**b**) stratified random, (**c**) two-stage, (**d**) cluster, (**e**) systematic, and (**f**) spatial systematic sampling. (Adapted from De Gruijter 1999)

- *Stratified random sampling* (Fig. 15.3b): The area is divided into subareas (strata; small squares), and simple random sampling is performed in each stratum. This reduces the variance at the same sampling effort or the sampling effort at the same variance. Strata can be based on maps of environmental parameters (soil types, vegetation types, aspect, etc.) and can have any shape. Cost functions can be included for determining sample size. Generally, more points are sampled in larger, more variable, or cheaper to sample strata.
- *Two-stage sampling* (Fig. 15.3c): The area is divided into subareas (also called principle units, PUs), but only a random subset of these subareas is sampled; within a subarea, sample points are selected with equal probability. This clustering of points is more time-efficient but less precise than simple random sampling.
- *Cluster sampling* (Fig. 15.3d): Predefined sets of points (clusters) are sampled. The starting point of each cluster is selected at random; the geometry of the cluster is independent of the starting point (e.g., transects with equidistant points extending in opposite, predefined directions from the starting point). The regularity of the clusters makes sampling more time-efficient but less precise than simple random sampling.
- *Systematic sampling* (Fig. 15.3e): Similar to cluster sampling, a predefined set of points is selected at random, but only one cluster is selected (e.g., a random grid); interference with periodic variations can be avoided by combining systematic sampling with a random element (e.g., two-stage sampling combined with cluster sampling).
- *Spatial systematic sampling* (Fig. 15.3f): Randomization restrictions are used at the coordinate level; the area is split into strata, and one point is selected at random. The points in the other strata are not selected independently but follow a specific model (e.g., a Markov chain).

It is good practice to conduct a sensitivity analysis for estimating the sample size needed to detect differences in the parameter of interest with the desired level of confidence (Johannsen and Daughtry 2009). The sample size needed to estimate a statistical property with a chosen probability depends on the sampling scheme, the desired error rate, and the variation of the ecosystem property of interest (which can be approximated from existing data or a pilot study or based on literature values and experience). Details for estimating sample sizes for the sampling designs mentioned above are given by De Gruijter (1999). However, error rates of spectroscopic models of vegetation characteristics also depend on the measurement accuracies of vegetation and spectral data and the tightness of the association between the property of interest and spectral data. For example, as a rule of thumb, the smaller the amount of the chemical compound of interest and the less precise the laboratory method used to determine that compound, the more samples will be needed for building a sound model. Similarly, for classification models, the number of samples needed to differentiate classes with a desired accuracy will depend on intra- and interclass variation or the distinctiveness of classes. In other words, when projecting samples from different classes into spectral (or more generally, feature) space, model accuracy for class differentiation depends on the number of classes, the spread of the distribution of values within classes, and the distance among the class centroids. As before, the strength of the relationship between classes and their spectral characteristics and measurement accuracies should be taken into account when deciding on sample sizes. As a rule of thumb, a minimum of 50 samples per class and 75–100 samples per class for more than 12 categories or areas larger than 4000 km^2 has been suggested (Congalton and Green 1999), but fewer samples can provide sufficient accuracy when classes are relatively dissimilar. Tracking error propagation is important for assessing the performance of spectral models (Singh et al. 2015, Wang et al. 2019, Serbin and Townsend, Chap. 3); ideally, the accuracy of laboratory analysis should be included in error assessments, as well.

For empirical-statistical models that combine vegetation characteristic and spectral measurements, stratified random sampling is often a good choice. Stratifying a research area based on ecologically relevant environmental variation helps cover the heterogeneity of a research area and the range of values of the vegetation characteristics of interest. A range of methods for automating sampling designs are available for R, including the packages spsurvey (Kincaid and Olsen 2016), spcosa (Walvoort et al. 2010), spatstat (Baddeley and Turner 2005), and spatialEco (Evans 2017), and ArcGIS, including the Geospatial Modeling Environment (Beyer 2010) and the Reverse Randomized Quadrant-Recursive Raster algorithm (RRQRR, Theobald et al. 2007). However, it can be difficult to automate sampling design completely, especially in natural ecosystems with limited accessibility. Moreover, for studies with a RS component, it can be difficult to select research plots that are internally homogeneous and located far enough from objects that influence the spectral signal of neighboring pixels (see Sect. 15.2.3.1) automatically. Under such circumstances, a mix of automated sampling based on GIS data and informed decision-making (convenience/purposive sampling) can be a good option. For example, information about environmental factors and gradients influencing the vegetation characteristics of interest and other relevant information about the study area, such as accessibility and travel time, can be used as input into a GIS and used as strata. Random points per stratum can be created automatically and used, for example, to define larger polygons within which the exact location of research plots is determined in the field. When vegetation characteristics are expected to vary along gradients, cluster sampling of plots at predefined intervals along these transects is a good choice—but again, it might be necessary to adjust these distances to avoid objects influencing the spectral signal of the plots or to find internally homogeneous areas. In this context, areas can be considered "internally homogeneous" when their biotic and abiotic characteristics are comparable, which means that they can actually show a high degree of small-scale heterogeneity (e.g., situations changing every 5 cm) as long as this small-scale heterogeneity creates a similar mosaic at the measurement scale (e.g., 1 m^2 is comparable to the adjacent 1 m^2; see Sect. 15.2.3.1). It is important to report the reasons for deviating from common sampling schemes in the methods.

When working at the level of individual plants, sampling random points within research plots makes it possible to capture interindividual variation, which can be important, for example, when scaling functional traits of individual plants to plot-level estimates (Wang et al. 2019; Serbin and Townsend, Chap. 3). Random sampling combined with species identification can also be used as an alternative to detailed botanical inventories because species frequencies approximate fractional cover when a sufficiently large number of points are sampled within a plot. For approximating fractional cover, it is good practice to choose random points (e.g., using the point frame method; Heady and Rader 1958; Jonasson 1988) and not random individuals within plots, to avoid overrepresenting species with more lateral growth. When botanical inventories are available, stratified random sampling within plots with plant species as strata followed by abundance weighting based on species fractional cover or biomass is a good way for capturing vegetation composition and for scaling traits of individuals to plot-level estimates. In plant communities where species abundances are unequally distributed, it is important to think about the pros and cons of sampling all species vs. sampling the most abundant species and of sampling all species at the same frequency vs. sampling more abundant species at higher frequencies (Table 15.1).

Table 15.1 Sampling options for several example situations

All species—same frequency	Most abundant species—same frequency	All species—depending on abundance	Random point method
• Rare species are expected to have a disproportionally large influence on the vegetation/ecosystem characteristic of interest • Investigating traits of species is an important part of the study • Vegetation community traits are calculated using species means multiplied by fractional cover or biomass, and the range of intraspecific variation is expected to be similar for all species • Botanical inventories are available	• Species with abundances above a certain threshold are expected to be the main source of variation influencing the vegetation/ecosystem characteristic of interest • The influence of rare species is expected to be negligible; their traits are not specifically studied • Vegetation community traits are calculated using species means multiplied by fractional cover or biomass, and the range of intraspecific variation is expected to be similar for all species • Botanical inventories are available	• The effect of species on the vegetation/ecosystem characteristic of interest is expected to depend on their abundance • No inventory data are available, but species can be differentiated in the field • Investigating traits of species or scaling traits of individuals to plot-level (community) estimates is not important for the study, or a minimum number of samples per species is being collected	• Species functional identity and/or intraspecific variation is expected to capture the influence of individual taxa on the vegetation/ecosystem characteristic of interest better than species mean traits • Taxonomic identity of individuals is not possible to determine, or not essential for the study • Investigating traits of species or scaling traits of individuals to plot-level (community) estimates is not important for the study, or species can be differentiated and random points are being used to approximate species fractional cover or biomass • No inventory data are available

15.3 Field Data Collection

Spectra of plants can be acquired across scales (see Gamon et al., Chap. 16), including at the leaf level, using proximal RS techniques (e.g., handheld spectrometers, robotic systems, UASs), airborne instruments, and satellite systems. For leaf-level studies, it is often of interest to collect information about taxonomic identity (species or clade), functional type (e.g., based on life form, growth form, dispersal type), functional traits (e.g., based on samples for chemical or structural analysis, growth measurements), developmental stage, and stress symptoms (e.g., signs of disease, herbivory, drought). For canopy-level studies, it is common to collect information about community composition and cover, spatial arrangement (or clustering), gap fractions, plant and canopy architecture (e.g., leaf area index, leaf angle distribution, branching structure, stem diameter, stratification), community biomass, and community traits. Additional data often collected together with vegetation spectra include soil characteristics (e.g., chemistry, water content), elevation, slope, and aspect. Important metadata include time and precise location, observer, nomenclature used, and photos, from which, for example, cover fractions can be estimated. Ideally data are recorded digitally to avoid the time and sources of error associated with transcriptions, and it is good practice to develop and test protocols for standardized data collection. Information should always be recorded as precisely as possible. For example, in grasslands it would be unnecessary to record vegetation height in classes because recording vegetation height at the cm level takes about the same amount of time, and classes can always be aggregated later if needed. Working together with other research groups can make it more efficient to collect additional data. This requires coordination at an early planning stage.

Offering educational opportunities might be part of the mission of a research area, and site administrators might be able to help with hiring students or technicians. However, it is advisable to focus on collecting the most ecologically relevant data, using well-trained personnel and sound methods, including appropriate sampling design and large enough sampling size, rather than collecting various kinds of data of poorer quality. For studies with a RS element, it is important to acquire accurate and precise coordinates of research plots and/or individuals to match their locations to the image data. Triangulation can be used to estimate plot coordinates from ground control points, and relative positions of individuals within plots can be estimated from plot coordinates. The level of accuracy and precision needed depends on the spatial resolution of the imagery, but professional surveying equipment can be needed. Again, early planning is important, because finding rental equipment can become difficult during peak season. Purchasing insurance for expensive equipment might be advisable. Research areas might have periodic surveying campaigns. Including research plots in such campaigns is a great option but requires marking plots temporarily; posts made out of a light but rot-resistant wood (e.g., larch, spruce) are well suited for this.

The following sections give some examples about spectral data acquisition at different levels of measurement; for details on the collection of ecological, non-spectral data, see textbooks on ecological methods (e.g., Sala et al. 2000; van der

Maarel and Franklin 2012). As mentioned in the introduction (Sect. 15.1), the choice of methods will depend on the research question, the site conditions, equipment, and personnel available, among other things. There are many good protocols available (see Sect 15.1); familiarizing oneself with a couple of options and their advantages and limitations and testing them under specific scenarios is generally good practice.

15.3.1   Leaf-Level Spectroscopy

A typical setup for leaf-level spectroscopy consists of a spectrometer, light source, fiber-optic cable, leaf clip or integrating sphere, and user interface. Leaf-level spectrometers can be classified into VNIR instruments, usually covering the visible to the beginning of the near-infrared (NIR) portion of the electromagnetic spectrum (~ 350–1000 nm), and full-range instruments, covering additional wavelengths in the NIR and the shortwave-infrared (~350–2400 nm). Generally, VNIR instruments use a silicon array detector, which does not require cooling, making VNIR instruments relatively light and easy to carry. Full-range instruments use additional indium gallium arsenide (InGaAs) photodiodes, which require cooling, to detect the longer wavelengths in the less energetic infrared part of the spectrum, making instruments heavier and less stable. The conditions at the field site should be kept in mind when choosing an instrument. If the spectrometer needs to be carried for longer times and does not come with its own backpack, some extra effort is required to figure out a good packing solution, especially for the fiber-optic cable, which can be easily damaged. It is good practice to check with the instrument companies if warranties are still valid when instruments are transported without their shipping cases; additional insurance might be worth considering.

A number of leaf clips are commercially available; some are easier for one person to handle and/or better suited to measure narrow leaves, such as conifer needles and grass blades, than others, and there is room for design improvements (e.g., using 3D printers). Leaf clip measurements can be used to calculate reflectance (the ratio of detected to incident light), which most instruments do internally, while measurements with integrating spheres can be used to calculate reflectance, transmittance, and absorptance. Leaf clip measurements are generally faster. Ideally, leaves should cover the entire field of view of the sensor; special protocols are available for narrow leaves (Noda et al. 2013). It is important to note that measurements with different setups and among different instruments cannot be directly compared (Hovi et al. 2017). One relatively laborious way to make measurements comparable (e.g., to include data from different instruments in one study) would be estimating empirical transfer functions. This requires measuring the same leaf samples with the instrument setups to be compared. Empirical transfer functions can be estimated for each wavelength and applied to transform measurements from one spectrometer and setup to the other, given that measurement conditions are comparable between model development and application.

Typically, measurements start with assembling the spectrometer, fiber-optic cable, light source, and leaf clip or integrating sphere, turning on the instrument and light source, and giving them some time (e.g., around 15–30 minutes) to warm up and stabilize. Meanwhile, the user interface can be connected to the spectrometer and folders can be organized. For instruments that are operated from a bench, it is important to find a stable position; ideally, neither the instrument nor the fiber-optic cable should be moved between measurements. The same applies to instruments operated from a backpack; the setup should be as consistent and stable as possible. To avoid damage to the fibers, it is important to avoid bending fiber-optic cables, including at connections between the fibers and the instrument. Generally, it is good practice to use and transport fiber-optic cables as stretched out as possible. A good option is coiling longer fiber-optic cables loosely while allowing for enough play at connections between cable and instruments and between cable and leaf clip/integrating sphere to keep angles around 180°. It is important to keep the fiber-optic cable away from branches when walking through vegetation, because it can be snagged and broken.

One way to test if the instrument has warmed up and is stabilized is to measure an invariant surface, such as a reflectance standard. Before starting sample measurements, it is time to take reference measurements, including measurements of so-called dark current (the background signal from the instrument), which some instruments take automatically, and white references (materials that approximate Lambertian surfaces, which reflect light at all angles equally or are perfectly diffuse). Dark current measurements correct for instrument noise, while white reference measurements determine the light entering an instrument and allow the calculation of reflectance (and transmittance). White references are usually made of polytetrafluoroethylene, better known by its commercial name Spectralon. They are available in different sizes and shapes; some leaf clips have built-in white references, but they are difficult to keep clean. It is important to keep white references as clean as possible, because even small traces of dirt and oil affect the spectral signal. Reference panels should only be held by their sides (touching the surface should be avoided), and they should be covered after each measurement. Depending upon usage and field conditions, frequent cleaning according to the manufacturer's instructions may be necessary.

Generally, a white reference reading should be made before the first measurement and whenever conditions (e.g., temperature, the arrangement of the fiber-optic cable, instrument, or lamp settings) change. However, it might be more practical or accurate to take white reference measurements at regular intervals, such as every 10 minutes, for each sample, or after a certain number of measurements, and to take additional measurements when needed. It is good practice to plot the reflectance spectrum of each white reference measurement and to save the spectrum. The reflectance of the white reference should be around 100% for all wavelengths except for the beginning and end of the spectrum, which are generally noisier; deviations from 100% or excessive noise can indicate a dirty panel, or issues with the cooling system, lamp, instrument setup, or a low battery. Measurements should be taken immediately after sample collection, because leaves dry out quickly. For most purposes, if samples need to be stored before measurement, they should be kept cool, moist, and dark and measured as soon as possible. It is good practice to keep the intensity of the light source at the minimum needed for a good signal-to-noise ratio, because leaves can get burned by the lamp. Defining a threshold for a "good" measurement, for example, based on the reflectance at the so-called NIR shoulder, which is the highest point of the reflectance curve at the beginning of the NIR, can be helpful. Spectral measurements should be made under dry conditions; instruments can be damaged by water, and water films alter the spectral characteristics of leaves.

Measurement protocols should specify how many leaves per plant to measure and which leaves to select. Generally, this means clarifying if a study deals with "ideal" or "average" plant individuals and if an entire plant should be characterized or only certain layers, such as the top canopy. Measuring mature, healthy, sunlit leaves is a good strategy for characterizing species or functional groups. For studies dealing with disease detection, asymptomatic and symptomatic leaves should be measured, ideally at different stages of the disease. Selecting leaves at random, including all ages, canopy layers, and stress levels, can make sense when aiming to relate leaf chemistry to ecosystem processes or when scaling leaf-level chemistry to plot-level estimates. If the aim is to characterize entire individuals, leaves from all canopy layers can be included, with the number of measurements per layer reflecting plant size, growth form, and architectural complexity. However, when leaf-level spectra are being matched to spectra acquired with RS, it makes sense to select only leaves from the layer that is captured by the sensor (i.e., from the top of the canopy). If possible, measuring the midvein should be avoided. Measuring different spots on the same leaf is also usually unnecessary, at least for small- to medium-sized leaves, because spectral variation at the leaf level is generally small. However, it is important to check the quality of every spectrum. Ideally, quality checks are done immediately after each measurement; bad measurement can be flagged for subsequent filtering, which considerably reduces preprocessing time.

15.3.2   Proximal Canopy-Level Spectroscopy

Proximal canopy spectra can be sampled with handheld spectrometers, robotic systems, and UASs. One important differentiation is between nonimaging and imaging systems. Nonimaging spectrometers integrate over a defined amount of time the spectral reflectance of an illuminated area; the output is one spectral curve per measurement. Instruments used for leaf-level spectroscopy (Sect. 15.3.1) fall into this category. The same instruments can be used in a handheld mode or mounted on a platform to sample spectra at the proximal canopy level, and the reach of instruments can be expanded using long fiber-optic cables attached to a beam. Creative solutions include mounting a spectrometer on a bike and using it as a mobile platform (see "reflectomobile" in Milton et al. 2009). Imaging spectrometers sample spectra in a spatially resolved fashion. The collected data are commonly represented

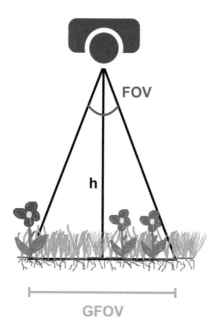

Fig. 15.4 Calculating the ground field of view (GFOV) based on the field of view (FOV) and height (h) of the sensor above the ground: GFOV = tan(FOV/2) ∗ h ∗ 2

as data cubes, with x- and y-axes representing the spatial extent of the imaged area and the z-axis representing the spectral response measured across the wavelength range (Vane and Goetz 1988). Commonly used systems are push-broom and whisk-broom imagers, which are usually operated from a moving platform. Alternatively, the imaged scene can move while the imaging spectrometer scans the samples, which is, for example, possible using conveyor-belt-like setups in the laboratory.

It is important to know the footprint, or ground field of view (GFOV), of remote sensing instruments. For nonimaging systems, the footprint equals the measured area on the ground; for imaging systems, it determines image and pixel size. The footprint depends on the field of view (FOV) of the sensor and the distance between sensor and measured object (h), and it is calculated as GFOV = tan(FOV/2) ∗ h ∗ 2 (Fig. 15.4). Foreoptic lenses can be used to narrow or expand an instrument's FOV, which is particularly relevant for handheld nonimaging systems. For spectrometers operated using robotic systems or UASs, the footprint is typically regulated by adjusting the height of the sensor above the ground; the farther away, the larger the GFOV, imaged area, and pixel size.

Handheld spectrometers and accessories need to be sturdy and easy to carry, particularly when collecting data over longer periods of time and in areas that cannot be reached by car. Spectrometers covering the VNIR range of the electromagnetic spectrum are usually small, such that neck straps securely attached to the instrument are often sufficient. Full-range spectrometers are heavier and typically need to be fit into a backpack, which means that cable connections have to be robust;

the entrance point of the fiber-optic cable in particular needs to be flexible, yet stable. Furthermore, it is important that the fiber-optic cable can be stored securely during transportation to avoid damaging the fibers and that the instrument is properly ventilated to avoid overheating. Changing the orientation of the fiber-optic cable changes the amount of light reaching the detector. Thus, it is good practice to ensure that the entire setup can be assembled easily in the same way every time. For spectrometers that are being used outside, the exterior should be made of materials that do not easily overheat when exposed to sunlight; polymeric surface films that provide radiative cooling (Zhai et al. 2017) could reduce the problem of overheating.

Often it is easier to collect data with handheld spectrometers in two-person teams, one person carrying the instrument and measuring and the other operating the computer and checking the data. However, it can be advantageous to have a system in place allowing one person to operate the spectrometer alone when needed. Vendor trays are a good option for carrying laptops while taking measurements. Small handheld devices can be very useful for collecting data, but their small screens make data checks difficult, and it can be impractical to name and rename files on small devices. Instrument software using voice control would be an advantage in this regard. Furthermore, when one person is operating a portable spectrometer, the white reference needs to be securely stored and in reach of the user. It is also important that backpacks fit comfortably, which means adjustable straps, cushioning, and ventilation. As with leaf-level spectrometers, instrument and light source should be switched on at least 15–30 minutes before data collection to allow the instrument to warm up and stabilize. Like other passive RS systems, handheld spectrometers should ideally be operated under stable illumination conditions, which is rarely possible. However, clear sky and no haze are a good place to start. Reference measurements (white reference and, if needed, dark current) should be taken before the first measurement and whenever illumination or temperature changes (after breaks, when adjusting the backpack, changing the sampling position, etc.). Again, it is good practice to take references at regular intervals, e.g., every 10 minutes, because of changing solar angle, ambient temperature, and sky conditions and to take additional references when needed. During measurements, the entrance optic of handheld systems should be positioned as far away from the body as possible (i.e., by stretching out the arm holding the fiber cable) to avoid measuring one's own shadow. In flat terrain, measurements should be taken in nadir position (i.e., with the fiber optic pointing directly down). In steep terrain, it is advantageous to point the fiber optic at a right angle toward the slope. It is good practice to wear nonreflective clothing and shoes in dark shades as stray reflected light off clothing can affect the spectral measurements.

At midlatitudes, the best time for measurements, given clear sky, is around solar noon (i.e., when the sun is at its highest point in the sky), whereas in the Arctic and the Tropics, time of day is usually less important than sky conditions. Generally, at midlatitudes, spectral measurements should be taken when the sun elevation angle (the angle measured from the horizon to the center of the sun) exceeds 45°; angles exceeding around 50° are better. For example, a good time window for canopy spectroscopy (sun angle >45°) for Minneapolis, MN, on June 21 would be from around

10:00 am to 4:30 pm. At midlatitudes, the longest time window for sun elevation angles greater than certain angles occurs around summer solstice, while around the equator, it is around the equinox. Calculators for sun position can be found online, such as from the National Oceanic and Atmospheric Administration (NOAA, https://www.esrl.noaa.gov/gmd/grad/solcalc/).

15.3.3 Airborne Campaigns

This section covers flight planning and some aspects of reference data collection for image processing. Remote sensing of plant biodiversity typically involves matching ecological and spectral data collected on the ground to remotely sensed images. Key aspects to match ground and remote sensing data include choosing vegetation plots that match the spatial resolution of the imagery (i.e., plots that are representative for at least one pixel) and collecting accurate coordinate information at a relevant precision for the remote sensor; this and more is covered in Sect. 15.2.3 and the beginning of Sect. 15.3.

15.3.3.1 Flight Planning

Flight planning for airborne imaging spectroscopy campaigns starts with deciding on the best time window(s) for the flight(s). The ideal time depends on the research question, but generally it is when the phenomena of interest are most pronounced. For example, for modeling and mapping aboveground productivity, peak biomass could be a good time for acquiring airborne images; for differentiating plant species, early or late growing season could be the times of year when certain species are most distinctive; for modeling and mapping plant disease or plant stress, different symptoms could be expressed at specific times of the year.

Schedules for flight crews are usually tight; thus communicating ideal flight windows early (i.e., at least several months in advance) is important. It is also critical to communicate flight windows to site administrators as soon as possible (see Sect. 15.2.2). On the one hand, it might be important to limit disturbance (e.g., trampling or destructive sampling) at the site during the week(s) leading up to a flight and to have no people and/or equipment on-site during the time of the overflight. On the other hand, other research groups might be interested in data collection around the time of the overflight. Ideally, airborne data are collected under clear sky conditions and low humidity. It is good practice to take the typical weather conditions at the site into account and plan flights at a time of year with generally good sky conditions, if possible.

Typically, the next steps of flight planning include determining the desired pixel size and drawing the flight lines. The ideal pixel size depends, again, on the goal of the study and the study system. Image pixels can be larger for modeling biomass and chemical composition at the plant community level than for predicting functional traits or species identity of individuals. The desired pixel size (within the limits of instrument and platform) determines flight altitude and swath width. Airplanes need to fly lower and slower to acquire images with smaller pixels resulting in narrower flight lines. Thus, there is a trade-off between pixel size and the area covered with a single flight line or during one flight. Generally, flight lines should overlap 15–25% to ensure full area coverage. Flight lines can be stitched together in a process called mosaicking, but spectra from the same pixels from different flight lines vary, because of changing sun angles and atmospheric conditions over the course of the day. In part, these differences can be accounted for during atmospheric correction, but not perfectly. Given inevitable constraints, covering a research area in fewer flight lines or a single line and working with larger pixels could be an option, depending on the phenomenon of interest.

At midlatitudes, flights approximately ±2 h from solar noon are often ideal; solar noon times can be looked up, for example, on the NOAA website (https://www.esrl. noaa.gov/gmd/grad/solcalc/). When images are acquired around solar noon, flight lines are often oriented in a north-south direction to align the flight direction with the principle plane of the sun and to prevent the plane from casting a shadow on the image. However, in areas such as the Tropics, it might be better to fly in the morning or afternoon because of cloud formation during the middle of the day. In the morning, flight lines oriented southeast-northwest are a good option; in the afternoon, flight lines oriented southwest-northeast may be preferable. The sun azimuth angle (i.e., the angle between the sun's position and north along the horizon: north equals 0°, east equals 90°, etc.) can be used as flight-line bearings. Additionally, it is seldom possible for flight crews to commit to specific dates and exact times, so the sun azimuth angle for the approximate time of overpass is generally fine as a bearing. Drawing separate sets of flight lines for different times of day is also an option. Again, information on sun azimuth angles for specific dates, times, and locations can be found online. During the time window of the light, it is often a judgment call if sky conditions are "good enough" for image acquisition; the costs of having the plane, air, and ground crews wait for a delay are important factors to consider. Generally, although cumulus clouds obscure parts of the image, they are preferable over cirrus or stratus clouds, which keep changing illumination conditions resulting in overall low-quality image data.

15.3.3.2   Reference Data Collection for Image Processing

In summary, the most important steps in image processing are:

- *Radiometric correction*: Sensors record electromagnetic radiation in digital numbers (DNs). During radiometric correction, DNs are converted to at-sensor radiance using sensor- and pixel-specific radiation sensitivity coefficients. Information about the sun's geometry, including Earth-sun distance and solar angle, is used to convert at-sensor radiance to top of the atmosphere (TOA) reflectance.

- *Spectral correction*: Every pixel contains spectral information recorded at slightly different wavelengths, which are standardized to a common wavelength interval in this step.
- *Atmospheric correction:* Atmospheric correction transforms TOA reflectance to surface reflectance using information about atmospheric conditions and aerosol properties at the time of image acquisition. This can be done with atmospheric radiative transfer models (e.g., MODTRAN, Berk et al. 1987), some of which are included in image processing software; dark subtraction, the subtraction of values from dark image pixels; or ELC (explained below).
- *Geometric correction (including geometric resampling, orthorectification, and georeferencing):* Geometric resampling makes pixels square (initially they are elliptic). Orthorectification corrects image distortions caused by sensor tilt, flight altitude, and changes in surface terrain, creating planimetric images, which allow direct measurement of distances, areas, and angles. Geometric resampling and orthorectification require information about camera properties, the sensor position recorded by the inertial measurement unit (IMU), and an accurate digital elevation model (DEM), which provides information about terrain height (elevation above sea level). Georeferencing aligns images to a specific coordinate system. This is frequently done using ground control points (GCPs).

Ground reference data collected for image processing frequently include GCPs for georeferencing and reference spectra for image calibration/validation procedures and atmospheric correction with ELC. Generally, GCPs need to be easily identifiable in the acquired images; road intersections, corners of buildings, and trees are good choices. Accurate and precise coordinates of GCPs can be either determined from existing georeferenced imagery, in which case it is good practice to check if the features still exist, or measured on the ground. For ELC, the reflectance of large calibration targets on the ground is measured with a handheld spectrometer during the overflight. At-sensor radiance is transformed to ground-level reflectance by applying bandwise transformation coefficients estimated based on the difference between sensor and calibration target spectra (Smith and Milton 1999). Essentially, this subtracts atmospheric influences from the spectra recorded by the sensor. Calibration targets should meet several requirements. They need to be identifiable in the images; when in doubt, location data should be collected. Furthermore, calibration targets should allow the extraction of several pixels from the image, so they should be sufficiently large (e.g., targets measuring 7×7 pixels allow using 5×5 pixels from the image after removing the pixels at the edges). In addition, calibration targets should be Lambertian as possible (i.e., they should reflect light equally in all directions, independent of viewing angle). Ideally, calibration targets should include the range of values in the image, which means including targets with low and high reflectance. Good options are calibration tarps in different shades (e.g., white, gray, and black) that are as spectrally invariant as possible. Tarps can be made of boat canvas (e.g., acrylic-coated woven polyester) and should include grommets, so that they can be tightly pulled and secured with pegs. Calibration tarps should be placed in flat areas with short vegetation. Ideally, one set of tarps is placed in the center of each flight line and measured exactly at the time of overflight, such that each flight line can be corrected separately. However, when multiple sets of tarps are not an option, a single set of tarps can be placed in a central area instead. Other surfaces can be used for ELC, including water bodies, road intersections, rooftops, snow, and ice. Ideally, these targets are distributed throughout the image and are measured at the time of the overpass. However, it might be difficult to find already existing calibration targets that are spectrally stable and uniform; for example, the reflectance of lakes can vary substantially depending on the distance from the shore and currents; snow and ice are often less spectrally uniform than expected due to surface irregularities, dust, and algae growth; and asphalt concrete varies spectrally depending on the aggregate composition (bitumen varies less). If possible, limiting movement and traffic around calibration targets during measurements and image acquisition is advantageous for reducing the amount of dust and dirt (e.g., when using calibration tarps or roads), as well as limiting surface disturbance and shadow cast (e.g., when using snow). However, prioritizing calibration target quality over quantity is good practice. For example, measuring one set of spectrally stable tarps in a flat area in the center of an image is preferable over measuring several natural calibration targets that are likely not as spectrally uniform and might be difficult to locate precisely.

15.4 Conclusions

As ecologists and remote sensing scientists are coming together to develop methods for the continuous assessment and monitoring of plant biodiversity, connecting the local to the global scale, studies of species to communities and ecosystems, and ecological resources to human needs and values become more and more feasible. Field campaigns are essential to this effect, because only (the repeated process of generating) ecological knowledge and data (including spectral measurements of plants) make it possible to understand better what is happening to the ecosystems and species we care about and why. Thanks to the ecological data revolution, remote sensing and organismal data as well as climate, land use, and socioeconomic data are becoming increasingly publicly available. At the same time, bioinformatics and cyberinfrastructure promote innovative ways for data handling, storage, and visualization and for integrating and analyzing these data across scales. Harnessing this amount of data requires developing and documenting data standards that facilitate collaborations across disciplines, data integration across sites and scales, data discovery for meta-analysis, and model re-calibrations. This makes the planning of data collection with consideration of the data life cycle as important as the data collection itself. It is important in the face of our current and future challenges and opens a wealth of opportunities in biodiversity science.

Acknowledgments AKS wishes to thank Susan L. Ustin and Shawn P. Serbin for reviewing this chapter and the National Institute for Mathematical Biology and Synthesis (NIMBioS) working group on remote sensing for stimulating discussions. This work was supported by the NSF/NASA

Dimensions of Biodiversity grant "Linking remotely sensed optical diversity to genetic, phylogenetic and functional diversity to predict ecosystem processes" (DEB-1342872) to Jeannine Cavender-Bares.

References

Anderson TM, Hopcraft JGC, Stephanie E, Ritchie M, Grace JB, Olff H (2010) Landscape-scale analyses suggest both nutrient and antipredator advantages to serengeti herbivore hotspots. Ecology 91:1519–1529

Atkinson PM (1999) Spatial statistics. In: Stein A, Van Der Meer F, Gorte B (eds) Spatial statistics for remote sensing. Kluwer Academic Publishers, Dortrecht

Baddeley A, Turner R (2005) Spatstat: an r package for analyzing spatial point patterns. J Stat Softw 12:1–42

Bartholomé E, Belward AS (2005) Glc2000: a new approach to global land cover mapping from earth observation data. Int J Remote Sens 26:1959–1977

Berk A, Bernstein LS, Robertson DC (1987) Modtran: a moderate resolution model for lowtran. Spectral Sciences Inc, Burlington

Beyer HL (2010) Geospatial modelling environ

Bicheron P, Huc M, Henry C, Bontemps S, Lacaux J (2008) Globcover products description manual. European Space Agency, Paris

Bonham CD (2013) Measurements for terrestrial vegetation. Wiley, Hoboken

Bonn A, Storch D, Gaston KJ (2004) Structure of the species–energy relationship. Proc R Soc Lond Ser B Biol Sci 271:1685–1691

Broge NH, Leblanc E (2001) Comparing prediction power and stability of broadband and hyperspectral vegetation indices for estimation of green leaf area index and canopy chlorophyll density. Remote Sens Environ 76:156–172

Brown JH (1981) Two decades of homage to Santa Rosalia: toward a general theory of diversity. Am Zool 21:877–888

Brus DJ, Degruijter JJ (1993) Design-based versus model-based estimates of spatial means: theory and application in environmental soil science. Environmetrics 4:123–152

Cavender-Bares J, Gamon JA, Hobbie SE, Madritch MD, Meireles JE, Schweiger AK, Townsend PA (2017) Harnessing plant spectra to integrate the biodiversity sciences across biological and 1391 spatial scales. Am J Bot 104:966–969

Chollet F, Allaire JJ (2018) Deep learning with R, 1st edn. Manning Publications, Shelter Island, New York

Chun Y, Griffith DA (2013) Spatial statistics and geostatistics: theory and applications for geo1393 graphic information science and technology. Sage Publications, London

Clark M, Roberts D, Clark D (2005) Hyperspectral discrimination of tropical rain forest tree spe1395 cies at leaf to crown scales. Remote Sens Environ 96(3 4):375 398

Claw KG, Anderson MZ, Begay RL, Tsosie KS, Fox K, Garrison NA (2018) A framework for enhancing ethical genomic research with indigenous communities. Nat Commun 9(1):2957

Combal B, Baret F, Weiss M, Trubuil A, Mace D, Pragnere A, Myneni R, Knyazikhin Y, Wang L (2003) Retrieval of canopy biophysical variables from bidirectional reflectance: using prior information to solve the ill-posed inverse problem. Remote Sens Environ 84:1–15

Congalton RG, Green K (1999) Assessing the accuracy of remotely sensed data: principles and 1402 applications. Lewis Publishers, Boca Raton

Congalton RG, Gu J, Yadav K, Thenkabail P, Ozdogan M (2014) Global land cover mapping: a review and uncertainty analysis. Remote Sens 6:12070–12093

Cook RB, Wei Y, Hook LA, Vannan SK, Mcnelis JJ (2018) Preserve: protecting data for long-term use. In: Ecological informatics. Springer, Berlin, Heidelberg

Coops NC, Kearney SP, Bolton DK, Radeloff VC (2018) Remotely-sensed productivity clusters capture global biodiversity patterns. Sci Rep 8:16261

Corti L, Van Den Eynden V, Bishop L, Woollard M (2014) Managing and sharing research data: a guide to good practice. Sage Publications, London

Couture JJ, Singh A, Rubert-Nason KF, Serbin SP, Lindroth RL, Townsend PA (2016) Spectroscopic determination of ecologically relevant plant secondary metabolites. Methods Ecol Evol 7:1402–1412

Dahabreh IJ, Chan J, Earley A, Moorthy D, Avendano E, Trikalinos T, Balk E, Wong J (2017) Modeling and simulation in the context of health technology assessment: review of existing guidance, future research needs, and validity assessment. Technical report, Rockville: Agency for Healthcare Research and Quality (US). Available from: https://www.ncbi.nlm.nih.gov/books/NBK424024/

De Gruijter J (1999) Spatial sampling schemes for remote sensing. In: Spatial statistics for remote sensing. Springer, Berlin, Heidelberg

Domburg P, De Gruijter JJ, Van Beek P (1997) Designing efficient soil survey schemes with a knowledge-based system using dynamic programming. Geoderma 75:183–201

Evans JS (2017) Spatialeco. R package version 2:0–0

Fassnacht FE, Latifi H, Stereńczak K, Modzelewska A, Lefsky M, Waser LT, Straub C, Ghosh A (2016) Review of studies on tree species classification from remotely sensed data. Remote Sens Environ 186:64–87

Féret J-B, Asner GP (2014) Mapping tropical forest canopy diversity using high-fidelity imaging spectroscopy. Ecol Appl 24:1289–1296

Fleishman E, Seto KC (2009) Applications of remote sensing to ecological modeling [online]. Available: https://ncep.amnh.org/index.php/Detail/objects/36. Accessed 26 Nov 2018

Gamon J, Penuelas J, Field C (1992) A narrow-waveband spectral index that tracks diurnal changes in photosynthetic efficiency. Remote Sens Environ 41:35–44

Gamon JA, Cheng Y, Claudio H, Mackinney L, Sims DA (2006) A mobile tram system for systematic sampling of ecosystem optical properties. Remote Sens Environ 103:246–254

Gastellu-Etchegorry J, Martin E, Gascon F (2004) Dart: a 3d model for simulating satellite images and studying surface radiation budget. Int J Remote Sens 25:73–96

Goetz AFH, Vane G, Solomon JE, Rock BN (1985) Imaging spectrometry for earth remote sensing. Science 228:1147–1153

Gould W (2000) Remote sensing of vegetation, plant species richness, and regional biodiversity hotspots. Ecol Appl 10:1861–1870

Grabau WE, Rushing WN (1968) A computer-compatible system for quantitatively describing the physiognomy of vegetation assemblages. In: Land evaluation. Macmillan of Australia, Melbourne, pp 263–275

Heady HF, Rader L (1958) Modifications of the point frame. Rangeland Ecol & Manag/J Range Manag Arch 11:95–96

Herrmann I, Vosberg S, Ravindran P, Singh A, Chang H-X, Chilvers M, Conley S, Townsend P (2018) Leaf and canopy level detection of fusarium virguliforme (sudden death syndrome) in soybean. Remote Sens 10:426

Horning N, Robinson JA, Sterling EJ, Turner W, Spector S (2010) Linking remote sensing with modeling. In: Remote sensing for ecology and conservation: a handbook of techniques. Oxford University Press, Oxford

Hovi A, Forsström P, Mõttus M, Rautiainen M (2017) Evaluation of accuracy and practical applicability of methods for measuring leaf reflectance and transmittance spectra. Remote Sens 10:25

Huemmrich K (2001) The geosail model: a simple addition to the sail model to describe discontinuous canopy reflectance. Remote Sens Environ 75:423–431

Huete AR (1988) A soil-adjusted vegetation index (savi). Remote Sens Environ 25:295–309

Inamdar D, Kalacska M, Leblanc G, Arroyo-Mora JP (2020) Characterizing and mitigating sensor generated spatial correlations in airborne hyperspectral imaging data. Remote Sens 12:641

Jacquemoud S, Baret F (1990) Prospect: a model of leaf optical properties spectra. Remote Sens Environ 34:75–91

Jacquemoud S, Verhoef W, Baret F, Bacour C, Zarco-Tejada PJ, Asner GP, François C, Ustin SL (2009) Prospect+ sail models: a review of use for vegetation characterization. Remote Sens Environ 113:S56–S66

Johannsen CJ, Daughtry CS (2009) Surface reference data collection. In: Warner T, Nellis M, Foody G (eds) The SAGE handbook of remote sensing. Sage Publications, London

Jonasson S (1988) Evaluation of the point intercept method for the estimation of plant biomass. Oikos 52:101–106

Jones R (2009) Categories, borders and boundaries. Prog Hum Geogr 33:174–189

Jones HG, Vaughan RA (2010) Remote sensing of vegetation: principles, techniques, and applications. Oxford University Press, Oxford

Justice CO, Townshend JR (1981) Integrating ground data with remote sensing. In: Townshend JRG (ed) Terrain analysis and remote sensing. Allen & Unwin, London

Kattenborn T, Eichel J, Fassnacht FE (2019) Convolutional neural networks enable efficient, accurate and fine-grained segmentation of plant species and communities from high-resolution UAV imagery. Sci Rep 9:17656

Kattge J, Diaz S, Lavorel S, Prentice I, Leadley P, Bönisch G, Garnier E, Westoby M, Reich PB, Wright I (2011) Try–a global database of plant traits. Glob Chang Biol 17:2905–2935

Kincaid T, Olsen A (2016) Spsurvey: spatial survey design and analysis. R package version 3:3

Kruse F, Kierein-Young K, Boardman J (1990) Mineral mapping at cuprite, Nevada with a 63-channel imaging spectrometer. Photogramm Eng Remote Sens 56:83–92

Lopatin J, Fassnacht FE, Kattenborn T, Schmidtlein S (2017) Mapping plant species in mixed grassland communities using close range imaging spectroscopy. Remote Sens Environ 201:12–23

Martens H (2001) Reliable and relevant modelling of real world data: a personal account of the development of pls regression. Chemom Intell Lab Syst 58:85–95

McCoy RM (2005) Field methods in remote sensing. The Guilford Press, New York

Milton EJ, Schaepman ME, Anderson K, Kneubühler M, Fox N (2009) Progress in field spectroscopy. Remote Sens Environ 113:S92–S109

Noda HM, Motohka T, Murakami K, Muraoka H, Nasahara KN (2013) Accurate measurement of optical properties of narrow leaves and conifer needles with a typical integrating sphere and spectroradiometer. Plant Cell Environ 36:1903–1909

Papritz A, Stein A (1999) Spatial prediction by linear kriging. In: Spatial statistics for remote sensing. Springer, Berlin, Heidelberg

Perez-Harguindeguy N, Diaz S, Garnier E, Lavorel S, Poorter H, Jaureguiberry P, Bret-Harte M (2013) New handbook for standardised measurement of plant functional traits worldwide. Australian J 800:167–234

Pettorelli N, Ryan S, Mueller T, Bunnefeld N, Jedrzejewska B, Lima M, Kausrud K (2011) The normalized difference vegetation index (ndvi): unforeseen successes in animal ecology. Clim Res 46:15–27

Pontius J, Hallett R, Martin M (2005) Using aviris to assess hemlock abundance and early decline in the catskills, New York. Remote Sens Environ 97:163–173

Rouse JW Jr, Haas RH, Deering DW, Schell JA, Harlan JC (1974) Monitoring the vernal advancement and retrogradation (green wave effect) of natural vegetation. NASA/GSFC type III final report, Greenbelt. Available from: https://ntrs.nasa.gov/search.jsp?R=19750020419

Sala OE, Jackson RB, Mooney HA, Howarth RW (2000) Methods in ecosystem science: progress, tradeoffs, and limitations. Springer, New York

Schaepman ME (2007) Spectrodirectional remote sensing: from pixels to processes. Int J Appl Earth Obs Geoinf 9:204–223

Schmidtlein S, Feilhauer H, Bruelheide H (2012) Mapping plant strategy types using remote sensing. J Veg Sci 23:395–405

Schweiger AK, Risch AC, Damm A, Kneubühler M, Haller R, Schaepman ME, Schütz M (2015a) Using imaging spectroscopy to predict above-ground plant biomass in alpine grasslands grazed by large ungulates. J Veg Sci 26:175–190

Schweiger AK, Schütz M, Anderwald P, Schaepman ME, Kneubühler M, Haller R, Risch AC (2015b) Foraging ecology of three sympatric ungulate species - behavioural and resource maps indicate differences between chamois, ibex and red deer. Mov Ecol 3:6

Schweiger AK, Schütz M, Risch AC, Kneubühler M, Haller R, Schaepman ME (2017) How to predict plant functional types using imaging spectroscopy: linking vegetation community traits, plant functional types and spectral response. Methods Ecol Evol 8:86–95

Serbin SP, Singh A, Mcneil BE, Kingdon CC, Townsend PA (2014) Spectroscopic determination of leaf morphological and biochemical traits for northern temperate and boreal tree species. Ecol Appl 24:1651–1669

Sims DA, Gamon JA (2002) Relationships between leaf pigment content and spectral reflectance across a wide range of species, leaf structures and developmental stages. Remote Sens Environ 81:337–354

Singh A, Serbin SP, Mcneil BE, Kingdon CC, Townsend PA (2015) Imaging spectroscopy algorithms for mapping canopy foliar chemical and morphological traits and their uncertainties. Ecol Appl 25:2180–2197

Smith GM, Milton EJ (1999) The use of the empirical line method to calibrate remotely sensed data to reflectance. Int J Remote Sens 20:2653–2662

Stehman SV (2001) Statistical rigor and practical utility in thematic map accuracy assessment. 1525. Photogramm Eng Remote Sens 67:727–734

Stehman S, Foody G (2009) Accuracy assessment. In: Warner T, Nellis M, Foody G (eds) The SAGE handbook of remote sensing. Sage Publications, London

Stein A, Van Der Meer FD, Gorte B (2002) Spatial statistics for remote sensing. Kluwer Academic Publishers, Dordrecht

Szary ALA (2015) Boundaries and borders. In: The Wiley Blackwell companion to political geography

Thenkabail PS, Lyon JG, Huete A (2012) Hyperspectral remote sensing of vegetation. CRC Press, Boca Raton

Theobald DM, Stevens DL, White D, Urquhart NS, Olsen AR, Norman JB (2007) Using gis to generate spatially balanced random survey designs for natural resource applications. Environ Manag 40:134–146

Tucker CJ (1979) Red and photographic infrared linear combinations for monitoring vegetation. Remote Sens Environ 8:127–150

Ustin SL, Gamon JA (2010) Remote sensing of plant functional types. New Phytol 186:795–816

Ustin SL, Gitelson AA, Jacquemoud S, Schaepman ME, Asner GP, Gamon JA, Zarco-Tejada P (2009) Retrieval of foliar information about plant pigment systems from high resolution spectroscopy. Remote Sens Environ 113:S67–S77

van der Maarel E, Franklin J (2012) Vegetation ecology. Wiley-Blackwell, Chichester

Vane G, Goetz AFH (1988) Terrestrial imaging spectroscopy. Remote Sens Environ 24:1–29

Verhoef W (1984) Light scattering by leaf layers with application to canopy reflectance modeling: the sail model. Remote Sens Environ 16:125–141

Verrelst J, Camps-Valls G, Muñoz-Marí J, Rivera JP, Veroustraete F, Clevers JGPW, Moreno J (2015) Optical remote sensing and the retrieval of terrestrial vegetation bio-geophysical properties – a review. ISPRS J Photogramm Remote Sens 108:273–290

Vohland M, Jarmer T (2008) Estimating structural and biochemical parameters for grassland from spectroradiometer data by radiative transfer modelling (prospect+ sail). Int J Remote Sens 29:191–209

Walvoort DJ, Brus D, De Gruijter J (2010) An r package for spatial coverage sampling and random sampling from compact geographical strata by k-means. Comput Geosci 36:1261–1267

Wang R, Gamon J, Montgomery R, Townsend P, Zygielbaum A, Bitan K, Tilman D, Cavender-Bares J (2016) Seasonal variation in the ndvi–species richness relationship in a prairie grassland experiment (cedar creek). Remote Sens 8:128

Wang Z, Townsend PA, Schweiger AK, Couture JJ, Singh A, Hobbie SE, Cavender-Bares J (2019) Mapping foliar functional traits and their uncertainties across three years in a grassland experiment. Remote Sens Environ 221:405–416

Warner TA, Nellis MD, Foody GM (2009) The SAGE handbook of remote sensing. Sage Publications, London

Webster R, Oliver MA (1992) Sample adequately to estimate variograms of soil properties. J Soil
     Sci 43:177–192
Weiss M, Baret F, Myneni R, Pragnère A, Knyazikhin Y (2000) Investigation of a model inversion
     technique to estimate canopy biophysical variables from spectral and directional reflectance
     data. Agronomie 20:3–22
Wold S, Martens H, Wold H (1983) The multivariate calibration problem in chemistry solved by
     the pls method. In: Ruhe A, Kagstrom B (eds) Matrix pencils, lecture notes in mathematics.
     Springer, Heidelberg, pp 286–293
Wright DH (1983) Species-energy theory: an extension of species-area theory. Oikos 41:496–506
Zhai Y, Ma Y, David SN, Zhao D, Lou R, Tan G, Yang R, Yin X (2017) Scalable-manufactured
     randomized glass-polymer hybrid metamaterial for daytime radiative cooling. Science
     355:1062–1066

Open Access   This chapter is licensed under the terms of the Creative Commons Attribution 4.0
International License (http://creativecommons.org/licenses/by/4.0/), which permits use, sharing,
adaptation, distribution and reproduction in any medium or format, as long as you give appropriate
credit to the original author(s) and the source, provide a link to the Creative Commons license and
indicate if changes were made.

   The images or other third party material in this chapter are included in the chapter's Creative
Commons license, unless indicated otherwise in a credit line to the material. If material is not
included in the chapter's Creative Commons license and your intended use is not permitted by
statutory regulation or exceeds the permitted use, you will need to obtain permission directly from
the copyright holder.

Chapter 16
Consideration of Scale in Remote Sensing of Biodiversity

John A. Gamon, Ran Wang, Hamed Gholizadeh, Brian Zutta, Phil A. Townsend, and Jeannine Cavender-Bares

16.1 Introduction

Biodiversity is critical to ecosystem function and provides many goods and services essential to human well-being (Hooper et al. 2012; Tilman et al. 2012). Despite centuries of effort, we lack a comprehensive account of global biodiversity, at a time the world is facing a sixth mass extinction due to human disturbance and climate change (Barnosky et al. 2011). Effective management of biological resources to preserve diversity and maintain ecosystem function in a rapidly changing world remains difficult, in part due to sampling challenges and lack of globally consistent data sets. Sampling biodiversity using traditional field methods alone simply cannot address this need, leading to recent calls for remote sensing (RS) as part of a global

J. A. Gamon (✉)
Department of Earth & Atmospheric Sciences, University of Alberta, Edmonton, AB, Canada

Department of Biological Sciences, University of Alberta, Edmonton, AB, Canada

CALMIT, School of Natural Resources, University of Nebraska – Lincoln, Lincoln, NE, USA

R. Wang
Department of Earth & Atmospheric Sciences, University of Alberta, Edmonton, AB, Canada

H. Gholizadeh
CALMIT, School of Natural Resources, University of Nebraska – Lincoln, Lincoln, NE, USA

B. Zutta
Vice-rectorate for Research and Technological Innovation, Universidad Alas Peruanas, Lima, Peru

P. A. Townsend
Department of Forest and Wildlife Ecology, University of Wisconsin, Madison, WI, USA

J. Cavender-Bares
Department of Ecology, Evolution & Behavior, University of Minnesota, Saint Paul, MN, USA

© The Author(s) 2020
J. Cavender-Bares et al. (eds.), *Remote Sensing of Plant Biodiversity*,
https://doi.org/10.1007/978-3-030-33157-3_16

biodiversity monitoring system (Scholes et al. 2012; Pereira et al. 2013; Turner 2014; Jetz et al. 2016; Geller et al. Chap. 20).

In response to this need for a more complete accounting of biodiversity, there have been several recent attempts to define Essential Biodiversity Variables (EBVs), many of which involve RS (Pereira et al. 2013; Turner 2014; Vihervaara et al. 2017; Fernández et al., Chap. 18). However, most of the appeals for a global biodiversity monitoring system involving RS have not fully addressed the topic of *how* RS would be used or what aspects of biodiversity would be measured. A review of the literature on biodiversity assessment via RS reveals a wide array of methods and definitions of biodiversity (Table 16.1), most of which are strongly scale-dependent in the measurements and/or in the definitions of biodiversity. Many of these RS studies do not directly address standard biological metrics of species diversity (e.g., alpha or beta diversity; Whittaker 1972; see also Chap. 1), but may be indirectly related to biodiversity through characterization of habitat, dominant vegetation, or vegetation functional traits, some of which can, in principle, be captured with proposed EBVs (Pereira et al. 2013; Kissling et al. 2018), but often involve mismatches between sampling scales and the biodiversity variables being sampled.

With the advent of hyperspectral sensors and imaging spectrometers, a growing number of studies have utilized optical diversity, or the variability in vegetation optical properties (also called spectral diversity) to assess species diversity (typically alpha or beta diversity), or to address plant traits related to functional diversity. These methods offer the opportunity to directly detect species and functional diversity, but also require close attention to scale (Asner et al. 2015). In this chapter, our primary focus is on these latter RS methods involving optical RS, with the understanding that other RS methods can also make important contributions to our understanding of biodiversity.

Table 16.1 Examples of biodiversity-related studies using different methods of optical remote sensing

Method	Reference(s)
Habitat assessment	Kerr et al. (2001), Nagendra et al. (2013)
Community composition (dominant species mapping)	Wang et al. (2004), Roth et al. (2015), Franklin and Ahmed (2018)
Productivity assessment	Gould (2000), Psomas et al. (2011), Gaitán et al. (2013)
Plant trait assessment	Asner and Martin (2009), Singh et al. (2015), Chadwick and Asner (2016)
Optical diversity assessment	Féret and Asner (2014), Schäfer et al. (2016)

16.1.1 Why a Chapter on Scale?

A key thesis of this chapter is that much of the uncertainty in the RS of biodiversity arises from the lack of attention to sampling scale, which affects both traditional biodiversity metrics and our ability to detect biodiversity remotely. In most cases, the sampling scales of typical satellite or airborne RS methods do not match those of our biological definitions or biodiversity measurements on the ground, confounding our interpretation of biodiversity from RS. In part because of scale mismatches, the interpretation of remotely sensed data from one time or place often cannot be applied to another, and we lack a universal, operational approach to RS of biodiversity. For RS of biodiversity to be meaningful, a careful consideration of scale is essential.

The purpose of this chapter is to address this need, with the goal of contributing to the design of an effective, operational global biodiversity monitoring system. Our primary focus is on *optical diversity* (a.k.a. "spectral diversity") using passive optical RS in the visible to shortwave-infrared (VIS-SWIR) range (400–2500 nm) because this approach allows species and functional diversity assessment. However, we acknowledge that other methods, including lidar (Asner et al. 2012; Lausch et al., Chap. 13, this volume), can also make important contributions to our understanding of biodiversity. Our key examples involve optical studies of terrestrial vegetation primarily at the level of alpha and beta diversity (Whittaker 1972), with the underlying assumption that vegetation diversity may be related to the diversity of other trophic levels via surrogacy (Magurran 2004) or to belowground diversity via biogeochemical cycling (Madritch et al. 2014; Madritch et al., Chap. 8). Similarly, in aquatic environments optical diversity (often expressed as "ocean color") can reveal dynamic structure related to the distribution of phytoplankton (Moses et al. 2016; Muller-Karger et al. 2018) and benthic organisms (Goodman and Ustin 2007), and scaling principles discussed here may apply in these cases. While a comprehensive assessment of all aspects of biodiversity in all environments is beyond the scope of this review, our hope is that the principles discussed here with a primary focus on terrestrial vegetation will enable progress toward an operational global biodiversity monitoring system involving RS.

16.1.2 What Is Optical (Spectral) Diversity?

Optical diversity can be defined and measured in many ways. It is often based on spectral reflectance of leaves and canopies, in which case the term "spectral diversity" is often applied. One definition is based on the number of different kinds of reflectance spectra ("spectral types," "spectral species," or "spectral signatures," Fig. 16.1) present in a given area, a direct analogy to the biological concept of species diversity (Féret and Asner 2014). The categorical *spectral type* concept presumes distinct and stable spectral patterns exist for a given species. However, this is

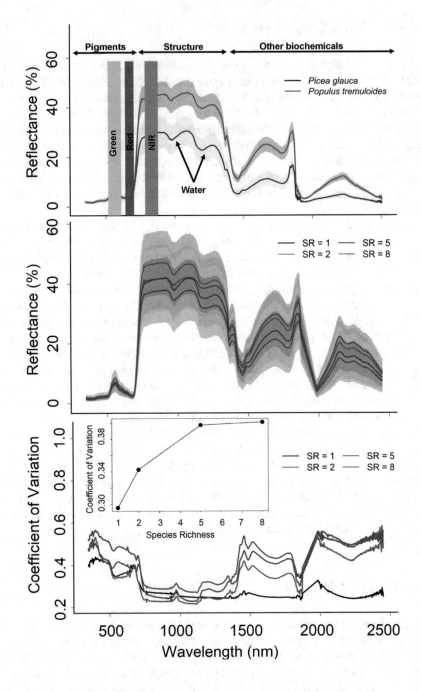

usually not the case, in part because species' spectra are dynamic and in part because intraspecific spectral variation may approach or exceed that of interspecific variation, particularly when the full range of environmental conditions is expressed (Roth et al. 2015). For these reasons, the number of distinguishable spectral types does not always match the number of species in a given area.

An alternate definition refers to the amount of *spectral variability* in a given area using statistical metrics of spectral information content, which can be measured many ways (Table 16.2). This concept has an early expression in the spectral variability hypothesis, which stated that variation in spectral characteristics scales with species richness (Palmer et al. 2000, 2002). In recent years, this concept has been further developed in many studies that explore the links between expressions of spectral variability and metrics of biodiversity, typically at the level of alpha or beta diversity (Baldeck et al. 2014; Féret and Asner 2014; Wang et al. 2016; Wang et al. 2018a; Gholizadeh et al. 2019).

Many spectral variability methods derive from information theory, which provides a rich array of methods for assessing the abstract "information content" or "entropy" in a given data set (Table 16.2). One simple method expresses spectral variation as the coefficient of variation (CV) spectrum for a given region, which can then be averaged into a single metric (Fig. 16.1). At this point, it is not entirely clear if there is a "best" method, because most of these methods work to some degree and their strength of correlation may vary with the circumstances (e.g., Gholizadeh et al. 2018).

The fundamental reason spectral patterns reveal underlying biological diversity is that plant reflectance spectra contain information on plant structure and chemical composition (Ustin and Jacquemoud, Chap. 14) that can differ slightly between species or functional types (Ustin and Gamon 2010) and can indicate different evolutionary histories (Schweiger et al. 2018; Meireles et al., Chap. 7, this volume). Thus, another way to utilize spectral information is to directly relate spectral patterns to plant functional traits (Serbin and Townsend, Chap. 3), providing a link between optical diversity and functional diversity (Cavender-Bares et al. 2017; Schweiger et al. 2018).

The topic of how to measure spectral diversity remains an active area of research, and the "best" metric is likely to vary depending upon the particular context and

Fig. 16.1 Top panel: contrasting canopy reflectance spectra of boreal tree seedlings (*Picea glauca*, an evergreen conifer, and *Populus tremuloides*, a deciduous angiosperm) illustrating spectral regions (arrows) influenced by leaf traits (categorized here as pigments, biochemicals, water, and structural features). Also shown are examples of green, red, and near-infrared (NIR) bands to illustrate the more limited spectral coverage provided by most airborne cameras and many satellite sensors. Middle panel: mean (+/− SD) spectral for experimental tree plots of varying species richness (SR: 1, 2, 5, and 8). Species include *Acer negundo, Fraxinus pennsylvanica, Picea glauca, Pinus contorta, Populus balsamifera, Populus tremuloides, Larix sibirica, and Prunus virginiana.* Bottom panel: coefficient of variation (CV) spectra for the same plots. Inset: average CV for each level of species richness (1, 2, 5, and 8). (Data from canopy reflectance spectra sampled in Edmonton, Alberta, (summer 2013) using a full-range spectrometer (PSR 3500, Spectral Evolution, North Andover, MA, USA.). From (DeLancey 2014))

Table 16.2 Examples of methods and metrics used to assess optical diversity, several of which are based on information content

Method (metric)	Reference(s)
Values or variance of vegetation indices (e.g., the Normalized Difference Vegetation Index, NDVI, or Enhanced Vegetation Index, EVI)	Gould (2000); Fairbanks and McGwire (2004); Gaitán et al. (2013); Tuanmu and Jetz (2015)
Measures of variance (e.g., coefficient of variation, CV)	Rey-Benayas and Pope (1995); Lucas and Carter (2008); Somers et al. (2015)
Measures of spectral angle (e.g., Spectral Angle Mapper, SAM)	Kruse et al. (1993); Gholizadeh et al. (2018)
Measures of area or volume in spectral or principle components space (e.g., convex hull area or volume)	Dahlin (2016); Gholizadeh et al. (2018)
Principal components analysis (PCA)	Oldeland et al. (2010); Rocchini et al. (2011); Asner et al. (2012)
Regression methods (e.g., partial least squares regression, PLSR)	Fava et al. (2010)
Spectral classification (clustering) methods	Schäfer et al. (2016); Paz-Kagan et al. (2017)

Some methods (e.g., principle components, partial least squares analysis, and spectral clustering methods) are often used as an initial step in a multistep procedure, and some studies combine more than one approach

purpose. Rather than provide a detailed summary of all the possible methods used, we present the concept of spectral information content as a viable proxy for multiple dimensions of biodiversity at the species, functional, or genetic and phylogenetic levels (Cavender-Bares et al. 2017; Schweiger et al. 2018; Cavender-Bares et al., Chap. 2, this volume) and note that recent reviews (e.g., Wang and Gamon 2019) consider this topic in more detail.

16.2 What Is "Scale" and Why Is It Important?

"Scale" has several definitions and is used as both a noun and a verb. As a noun, it refers to a level of observation and can have several dimensions, including spatial, temporal, spectral, and biological. As a verb, it refers to the act of examining a phenomenon at multiple levels, usually referring to transcending spatial scales, as in "upscaling" (extrapolation from fine-scale data to a coarser scale) or "downscaling" (interpreting underlying patterns or mechanisms from coarse-scale data).

Discussions of scale have a rich history in both biology and RS. In biology, scale typically refers to levels of organization (genetic, cellular, organismal, species or population, community or ecosystem, etc.). Biodiversity can be defined at many of these levels, requiring different study approaches (Bonar et al. 2011). Biological systems typically exhibit complex, nonlinear interactions, and feedbacks, resulting in emergent properties and thus requiring evaluation across multiple levels of organization (Heffernan et al. 2014). Scale can also refer to sampling scale, as in the grain size and extent of field sampling (Turner et al. 1989). There is an abundant literature on the importance of considering scale when exploring ecological phenomena (Ehleringer and Field 1993; Levin 1992) and a number of "scaling rules" have emerged. For example, Levin (1992) stated the importance of matching the scale of observation to the phenomenon (and grain size or patch size) in question, and specific rules of sample grain size have been developed (e.g., O'Neill et al. 1996). However, these rules are often violated when remotely sensing biodiversity.

In RS, scale has several aspects or dimensions (Malenovský et al. 2007). The *spatial* dimension typically refers to the pixel size (grain size) and spatial extent of a remotely sensed image. The *temporal* dimension can refer to the time of sampling, repeat frequency, or temporal extent of sampling. The *spectral* dimension includes band position, bandwidth (full width half maximum, FWHM, and sampling interval), and range. The *directional* or *angular* dimension, including the illumination or viewing angle, leading to variations in anisotropic reflectance and the bidirectional reflectance distribution function (BRDF), is also an important scale dimension in optical RS, as it strongly influences the ability to detect signals present in reflectance spectra (Schaepman-Strub et al. 2006; Malenovský et al. 2007; Gamon 2015). These scale dimensions often interact, with the effects of one influencing the effects of another, so it is often best to consider multiple scale dimensions together. A consideration of sampling scale in all these dimensions is relevant to a discussion of RS of biodiversity. Sampling across scales often reveals critical information that is not apparent from a single-scale observation alone. Below, we consider these scale effects in more detail by providing examples of how these dimensions impact biodiversity assessment from RS.

16.2.1   Biological Scale

Biodiversity exists across multiple scales of organization from genes to biomes. A detailed discussion of the various "dimensions" of biodiversity at different scales of organization is beyond the scope of this chapter and has been reviewed elsewhere (e.g., Magurran 2004; see also Cavender-Bares et al., Chap. 2, this volume). In the context of this chapter, a key challenge lies in matching the scale of the measurement approach to the biological scale of organization, a topic considered in the sections below.

16.2.2   Spatial Scale

Optical instruments used for sampling biodiversity range from laboratory spectrometers and proximal field spectrometers to airborne and satellite-based imaging spectrometers, with grain sizes (pixel sizes) covering approximately six orders of magnitude (Fig. 16.2). If we include the molecular cross-section of DNA (e.g.,

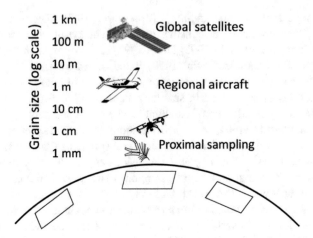

Fig. 16.2 Schematic of a proposed, multi-scale, global biodiversity monitoring system. Satellite imaging spectrometry would provide the context for understanding patterns in time and space, and regional and proximal sampling would provide sampling at progressively finer scales. This design would be replicated systematically around the world, using field sampling plots for different biomes (indicated by parallelograms). Note that spatial scale (sampling grain size, typically measured as optical cross-section or pixel size) associated with various optical sampling methods span roughly six orders of magnitude

when determining genetic diversity), our grain sizes span an even wider range, from roughly 2 nm (in the case of the DNA double helix; van Holde 1989) to roughly 1 km (for "Pando," a large clonal aspen stand (DeWoody et al. 2008), a range of approximately 12 orders of magnitude). A rule of thumb for distinguishing spectral differences is that the sampling grain size should be smaller than the cross-section of the target (e.g., leaf or individual canopy crown in the case of individual plants; Woodcock and Strahler 1987). These numbers imply that the sampling grain size needs to span an extremely wide range to properly match all our definitions of biodiversity. Clearly, this is not possible from a single instrument, but can be considered in multi-scale field campaigns employing multiple instruments and platforms. Additional challenges arise when designing a field campaign to validate remotely sensed biodiversity data. Many field sampling methods for species richness entail quadrat or transect sampling (Bonar et al. 2011), neither of which is well-matched to the size, shape, and location of typical airborne or satellite pixels. Moreover, our classical definitions of biodiversity at different scales (alpha, beta, or gamma) reflect relative rather than absolute spatial scales and often poorly match the scale of both field sampling and RS. These challenges of scale mismatch abound and require careful attention to definitions and sampling protocols.

Similar issues of scale mismatch arise when exploring plant traits with RS. It is unclear how well the ability to detect leaf traits transcends spatial scales, with some studies suggesting certain leaf traits (nitrogen content) cannot be detected at the sampling scale of a typical aircraft or satellite pixel (Knyazikhin et al. 2013). However, many leaf traits (if not all) can be detected at the canopy scale (i.e., the scale of an individual plant crown), particularly when appropriate sampling and scaling methods are followed, suggesting that scale-appropriate RS methods can often resolve trait differences associated with functional or species diversity (Townsend et al. 2013; Asner et al. 2015). This is particularly relevant to airborne data, where pixel sizes often approximate those of individual tree crowns, allowing individual traits to be distinguished (Asner et al. 2015; Singh et al. 2015), but calls into question the accuracy of trait retrievals when the pixel sizes are much larger than individual canopies, as is the case for most satellite RS.

Not surprisingly, a review of the recent literature suggests strong effects of spatial scale on the ability to detect biodiversity using optical RS. Most of this research has been conducted in North American prairie, with relatively short-statured vegetation, so may be biased by the relatively small plant crown sizes (roughly 10 cm). One set of studies, conducted using tallgrass prairie species at the Cedar Creek Ecosystem Science Reserve in Minnesota (USA), used pixel sizes ranging from 1 mm (sampled on the ground) to 1 m (sampled from aircraft) and found a significant correlation between optical diversity and alpha diversity at finer scales (1 mm to 10 cm), but most of the information on alpha diversity was lost at pixel sizes larger than about 10 cm, roughly the size of many plant crowns (Fig. 16.3, red line). Another study conducted at Wood River in Nebraska (USA) using airborne data found a strong relationship (R^2) between optical diversity and alpha diversity for prairie species at pixel sizes of 0.5–1 m and noted that the optical diversity-alpha diversity relationship was markedly weakened at progressively larger spatial scales (up to 6 m) (Fig. 16.3, black line). A third study, conducted at Mattheis Ranch in southern Alberta (Canada), found an intermediate relationship between the two other prairie sites (Fig. 16.3, blue square).

These studies have significant implications for attempts to sample optical diversity from aircraft or satellite sensors and illustrate the importance of matching sampling scale (pixel size) to crown size when designing airborne campaigns. Clearly, there is a strong scale-dependence of the optical diversity-biodiversity relationship, but this scale-dependence varies between study sites, even for the same biome (prairie grassland, in this case). These site differences have been attributed to a number of possible factors, including the degree of disturbance (fire regimes, invasion by exotic weeds, or subsequent weed removal), the size of the plot (sampling extent) relative to the pixel (grain) size, and the differences in species richness between studies. The Wood River site was less disturbed, with less bare ground and larger, more species-rich plots than the Cedar Creek site (Gholizadeh et al. 2019). It is likely that multiple features influence the scale dependence of the optical diversity-biodiversity relationship, illustrating that the larger context and experimental design of a study can matter.

We know less about the scale dependence of these relationships for other biomes where explicit scaling experiments involving the RS of biodiversity have not yet been conducted. Despite unanswered questions, these experiments in prairie ecosystems demonstrate the importance of spatial scale and support the idea that can-

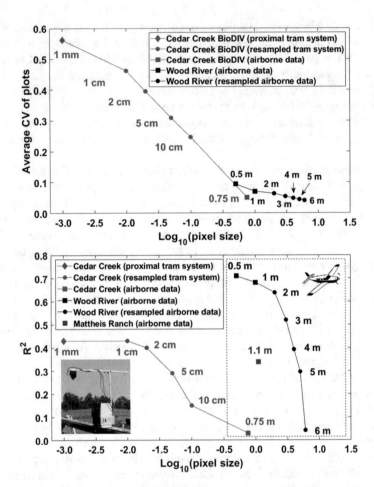

Fig. 16.3 Scale dependence of the optical diversity-biodiversity relationship from three experimental studies in prairie grasslands: Cedar Creek, Minnesota, USA (red), Wood River, Nebraska, USA (black), and Mattheis Ranch, Alberta, Canada (blue). (Data combined from Wang et al. (2016) and Gholizadeh et al. (2018, 2019))

opy crown size is an important factor in scale dependence. These findings have clear implications for studies using airborne and satellite platforms where the pixel sizes often exceed the crown sizes of common plant species.

16.2.3 Temporal Scale

The temporal dimension is rarely considered in most RS campaigns; many RS studies are based on a single overpass (e.g., a single aircraft image or satellite image), or at best a few overpasses, limiting the opportunities for examining temporal effects.

A typical field campaign might focus on an optimal time to collect data from a RS perspective (e.g., the dry season in the tropics or the summer growing season in higher latitudes). However, biological communities are inherently dynamic, and the visibility of different species changes with ontogeny, season, and over longer time spans due to disturbance, invasion, succession, climate change, and other processes. Consequently, our ability to detect species richness will vary with time, often in ways that are poorly understood. The few studies that have investigated the temporal aspect of biodiversity using RS or spectral reflectance demonstrate the importance of temporal dynamics when examining the optical diversity-biodiversity links.

A study of California chaparral (Zutta 2003) found a clear seasonal dependence in the ability to distinguish plant functional types using reflectance spectra (Fig. 16.4a). Different methods involving contrasting spectral indices and bands yielded distinct seasonal patterns, indicating the importance of the spectral dimension (further discussed below) and illustrating interactions between temporal and spectral dimensions. In that study, photosynthetic and flowering phenology contributed to the seasonal patterns observed. Similarly, when evaluating several functional leaf traits with spectral reflectance in tropical species, Chavana-Bryant et al. (2017) found a clear seasonal effect on the trait retrievals using PLSR, again emphasizing an interaction between temporal and spectral dimensions. A study of optical diversity for prairie species also revealed strong phenological effects that were different at the leaf and canopy scales, demonstrating an interaction between temporal and spatial scale (Fig. 16.4b). Clearly, the temporal dimension should be considered in any study of the RS of biodiversity, yet most studies have been limited to a single time frame, limiting the power to distinguish biodiversity. These examples also demonstrate that measurement technique (e.g., instrument foreoptics) and interacting scale effects can influence optical diversity.

16.2.4 Spectral Scale

The advent of hyperspectral sensors, both imaging and nonimaging, provides rich opportunities for exploring spectral features related to biodiversity. Approaches range from detection of species or functional traits to methods based on the information content of the spectra themselves (Fig. 16.1; Table 16.2). All of these methods require attention to spectral scale, including spectral resolution and range, which influence biodiversity detection in complex ways. Furthermore, our methods of analysis, ranging from simple vegetation indices to more complex full-spectral statistical methods (Table 16.2), explore the spectral dimension in different ways and to varying degrees. To date, relatively few studies have explicitly addressed spectral scale in the context of biodiversity detection, but most show that more spectral information is generally better than less (e.g., Asner et al. 2012). Consequently, hyperspectral sensors are more informative than multiband sensors, and full-range (VIS-SWIR) detectors are usually more useful than limited range (e.g., VIS-NIR) detectors for detecting plant traits or biodiversity. The importance of spectral scale

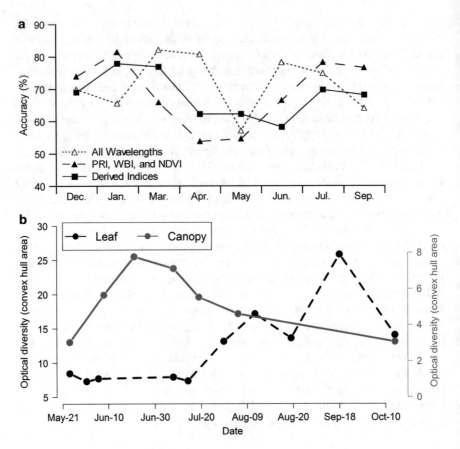

Fig. 16.4 (**a**) Ability (% accuracy) of spectral variability to distinguish plant phenological types (evergreen, winter deciduous, drought deciduous, and annual) in California chaparral (Santa Monica Mountains, California, USA, Dec 1998–Sept 1999). Input variables include reflectance at all wavelengths (450–1000 nm), three physiological indices derived from reflectance (Photochemical Reflectance Index, PRI; Water Band Index WBI, and Normalized Difference Vegetation Index, NDVI) and indices derived from the coefficients produced in discriminant analysis. (From Zutta (2003).) (**b**) Phenology of optical diversity (convex hull area in spectral space) at the leaf (black) and canopy (red) scale for prairie vegetation sampled at the Cedar Creek Ecosystem Science Reserve, Minnesota, USA, in summer 2014. Leaf-scale data sampled with a leaf clip and canopy-scale data sampled with a straight fiber yielding a field-of view of approximately 10 cm diameter. Canopy data are available as "Phenology Canopy Spectra Big Biodiversity Experiment Cedar Creek LTER 2014" on EcoSIS (doi: https://doi.org/10.21232/C2Z070)

in biodiversity detection can be readily seen when comparing multiband data (measured from a drone) to hyperspectral data (measured from a tram system) for Cedar Creek; in this case, multiband drone imagery failed to detect different alpha diversity levels, despite pixel sizes (2.3 cm) that were intermediate between those of the hyperspectral sensor (Fig. 16.5).

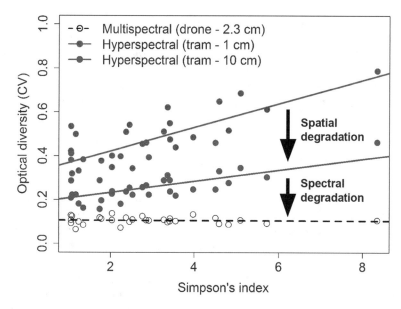

Fig. 16.5 Effect of spatial and spectral scale on the relationship between optical diversity (measured as the coefficient of variation) and Simpson Index, a measure of alpha diversity combining species richness and evenness (Simpson 1949). Data collected by a hyperspectral sensor on a tram, sampled at two resolutions (1 cm = red dots, 10 cm = blue dots) (for methods see Wang et al. 2018a) and a multispectral sensor on a drone (resolution = 2.3 cm, black circles) (Parrot Sequoia, Parrot Drones, Paris, France). All data measured at the Cedar Creek Ecosystem Science Reserve

A number of studies have demonstrated that some wavebands provide more information than others, and this information can vary seasonally (Zutta 2003; Chavana-Bryant et al. 2017; see also Fig. 16.4). Similarly, visible bands (revealing pigment composition) and NIR bands (revealing structural information) respond differently to spatial scale (Wang et al. 2018b) or sampling angle (Gamon 2015) illustrating the functionally distinct responses of these different spectral regions (Fig. 16.1) and providing further evidence of interactions between the spectral, spatial, and angular dimensions of scale.

16.2.5   Angular Scale

Illumination and view angle both interact with vegetation structure to influence the shape and intensity (brightness) of reflectance spectra. While these effects have been well-studied in RS and can be characterized by the BRDF for a particular surface (Malenovský et al. 2007), angular information has generally not been employed in detection of biodiversity. A few studies have noted that the BRDF response can help distinguish vegetation types (Gamon et al. 2004), suggesting that angular information may be useful in biodiversity detection. Sensor view angle affects different spectral regions differently, illustrating an interaction between angular scale and spectral scale (Gamon 2015). With the advent of lidar (Asner et al. 2012) and structure-from-motion (SFM; Wallace et al. 2016) to characterize plant 3-D structure, angular information can now be better understood and effectively integrated with optical RS to improve biodiversity or plant trait detection.

16.3 Implementing Scaling Approaches

Properly addressing scale effects in biodiversity detection requires explicit attention to scale in all its dimensions and designing a study approach that transcends scale limitations. Scaling methods can be empirical or theoretical. Typical empirical methods involve a multi-scale sampling strategy, often using multiple RS instruments operating at different scales, along with traditional field sampling for validation. Often, a goal of such field campaigns is to aggregate fine-scale data to be used as validation for coarse-scale data (e.g., Cohen and Justice 1999; Wehlage et al. 2016), yet aggregation tends to obscure spectral variability at the scale of individual leaves or plant canopies and undercuts the goal of detecting local (alpha) biodiversity with optical diversity. On the other hand, sampling at progressively larger scales often involves transitions from local (alpha) to regional (beta) diversity, creating the opposite effect of increasing optical diversity with increasing pixel size, and these transitions can themselves be scale-dependent and vary with vegetation type (Fig. 16.6).

The complexity of scaling effects involving the transition from alpha to beta diversity is briefly illustrated in Fig. 16.6, which illustrates optical diversity (CV in this case) calculated for different regions along a transect crossing several plant communities, including woodland, grassland, and experimental grassland plots at the Cedar Creek Ecosystem Science Reserve, Minnesota (USA). In this case, the highest optical diversity values occur at edges, points of abrupt landscape transitions marking the boundaries between adjacent communities (i.e., ecotones), an effect commonly seen in remotely sensed images (Paz-Kagan et al. 2017). In this complex landscape, (and in contrast to the plot-level patterns shown in Fig. 16.3), CV generally increases with spatial scale, reflecting a transition from alpha to beta diversity with increasing spatial lag (pixel sizes). Interestingly, this transition occurs at about 10 m × 10 m in the manipulated grassland (see arrow and blue curve, Fig. 16.6), matching the plot sizes in this experiment, but occurs more gradually in the natural grassland (black curve, Fig. 16.6). These patterns seem to contradict findings of declining optical diversity in grasslands with increasing pixel size obtained at finer scales (1 mm to 1 m in Wang et al. 2018a) (see also Fig. 16.3). This comparison illustrates that the patterns of scale dependence are themselves scale dependent and can differ within the same landscape or for different communities depending upon both individual crown size and larger landscape structure. These

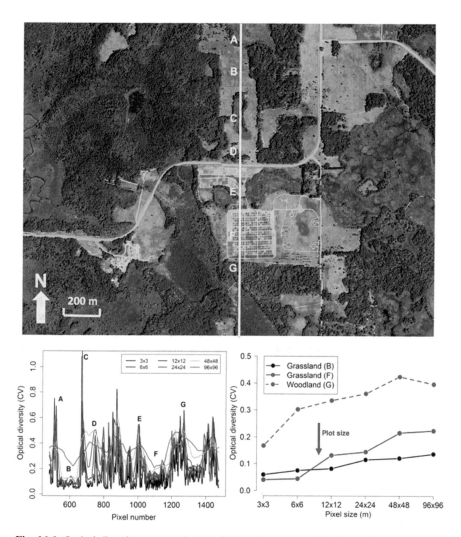

Fig. 16.6 Optical diversity, expressed as coefficient of variation (CV) derived from reflectance spectra, along a transect (yellow line, top panel) crossing forest and prairie communities at the Cedar Creek Ecosystem Science Reserve, Minnesota. This image cube (top panel) was collected on July 22, 2016, using an imaging spectrometer (AISA Eagle, Specim, Oulu, Finland) mounted on a fixed-wing aircraft (Piper Saratoga, Piper Aircraft, Vero Beach, Florida, USA) operated by the University of Nebraska Center for Advanced Land Management Information Technologies (CALMIT) Hyperspectral Airborne Monitoring Program (CHAMP). Images were collected from a height of approximately 1435 m and a speed of approximately 177 km/h, providing a ground pixel size of approximately 1 m^2. The imaging spectrometer provided hyperspectral images covering 400–970 nm with 10 nm spectral resolution (FWHM). Airborne data were collected and pre-processed by Rick Perk from CALMIT. A–G indicate particular points of interest, including transition points between communities (A, C, E, grassland (B), forest (G), and the Cedar Creek biodiversity experiment (F). CV has been calculated at various pixel sizes (3 m × 3 m to 96 m × 96 m); bottom panels to illustrate the effect of spatial resolution on retrieved optical diversity, illustrating a general increase in CV with pixel size, but this pattern varies between woodland (G) and natural (B) vs. experimental (F) grassland plots

observations agree with a body of RS literature that show wide variation in patterns of scale dependency for different landscapes using geostatistical approaches, with the local variance a function of the relative sizes of the pixels and the discrete targets themselves (e.g., Woodcock and Strahler 1987). For these reasons, the spatial aggregation approaches often used in other fields (e.g., plant productivity) may actually confound the detection of optical diversity at certain scales. Consequently, optical diversity (alpha or beta diversity) or variation in plant traits from RS requires particular attention to scaling methodologies (e.g., Asner et al. 2015).

By providing both fine-scale patterns (Fig. 16.3) and the larger context of landscape structure, imaging spectroscopy (Fig. 16.6) provides a valuable tool for further, more detailed studies, and can help define the proper scale at which alpha and beta diversity can be most effectively sampled. Remotely sensed data can also be used to define patterns of geodiversity, the physical template influencing biodiversity (Record et al., Chap. 10, this volume). Geostatistical approaches that examine optical diversity as a function of distance, analogous to the use of semivariograms (Curran 1988), can help design appropriate sampling methodologies by illustrating the influence of landscape features (e.g., ecotones or different vegetation types) on optical diversity (Fig. 16.6). Imaging spectrometry can also reveal temporal dynamics and disturbance patterns that can provide additional context for understanding both the drivers and the consequences of biodiversity changes. Furthermore, image spectrometry can be used to assess the relationships between optical diversity and ecosystem function (e.g., productivity) over large areas (Wang et al. 2016; Schweiger et al. 2018). For these reasons, satellite and airborne RS can be a powerful complement to local and regional field studies of biodiversity.

Other approaches to scaling involve the use of models for upscaling or downscaling, including radiative transfer models and statistical models (Knyazikhin et al. 2013; Malenovský et al. 2019; Verrelst et al. 2019). While potentially powerful, such models are generally limited by the lack of suitable data, bringing us back to the need for experimental studies incorporating empirical methods. Ultimately, a combined empirical and theoretical framework that explicitly considers multiple dimensions of scale is needed to advance our knowledge of biodiversity detection with RS.

16.4 Designing a Scale-Aware Biodiversity Monitoring System

To be truly robust, a global biodiversity monitoring system will need to consider scale in all the dimensions discussed here. Biological concepts (e.g., grain size and extent) often have a direct analog to RS concepts (e.g., pixel size and extent). Ideally, we should design our instrument and optical sampling protocols to match our sampling scale to a particular organizational scale of biodiversity. While specific sampling rules have been proposed (e.g., Justice and Townshend 1981), the general rule of thumb that the pixel size must be smaller than the size of the individual sampling unit (e.g., typical plant canopy) is a good starting point, and this has been supported by studies discussed above (see Fig. 16.3).

However, matching instrument to organizational scale remains a challenge; we typically have a poor match between the pixel size and sampling extent and the organism size and distribution on the ground, often due to the practical constraints of field sampling (in the case of biological studies) and instrument design (in the case of RS). Most remote sensing instruments are designed for a particular airborne or spaceborne platform with physics and engineering requirements in mind. The detector response is constrained by the amount of electromagnetic energy available, which in turn determines the sensor design, pixel size, and spectral resolution needed to achieve a given signal-to-noise ratio. Greater signal-to-noise ratios can be attained by reducing the spectral resolution (combining narrow bands into broad-bands, e.g., via spectral binning), or by reducing spatial resolution (e.g., pixel binning), but these choices limit the ability to distinguish individuals, species, and functional types due to the degradation of spectral and spatial information. Orbital and altitudinal considerations also determine the pixel size obtainable from a particular sensor platform. Together, these constraints often reduce the ability to properly distinguish individual organisms or vegetation functional types. Adding to this mismatch, field sampling (including plot size, transect size, and location) is often limited by practical considerations of personnel, time, and budget and is rarely designed with RS in mind.

Improvements in sensor design and novel sampling platforms can relax these impediments, but with trade-offs. For example, flexible airborne platforms, emerging unmanned aerial vehicles (UAV) systems, or robotic ground-based systems (Wang et al. 2018a, b) provide useful platforms for testing the effects of spatial scale, but may not always have the temporal or spectral coverage desired for biodiversity detection (Fig. 16.5). Plans to deploy global satellite imaging spectrometers with frequent repeat visits (Schimel et al., Chap. 19) offer new ways to explore the temporal dimension with high spectral resolution and wide spectral range, although at relatively coarse spatial resolution. Consequently, a key application of satellite RS can be to provide the larger context within which effective sampling regimes can then be defined at finer scales.

Due to these inherent limitations and trade-offs between scale dimensions, we suggest that the ideal global monitoring system would be an integrated, multi-component system, combining RS at different scales (satellite and aircraft sensors) with proximal sensing (ground optical sensors and field sampling of biodiversity) (Fig. 16.2). Such a system would operate within a clearly defined scaling framework, incorporating empirical and modeling approaches, with explicit attention to sampling scale in each of the dimensions mentioned above. Although rarely used in RS, explicit experimental approaches involving cross-scale and cross-instrument comparisons should be a key capability of an effective global biodiversity monitoring system. With a multi-scale system, it would be possible to express results (e.g., spectral variability) as a function of sampling scale (pixel or grain size; see Figs. 16.3

and 16.6), much in the way that semivariograms are used to express landscape structure (Curran 1988). These patterns could be compared to the driving variables, including the underling patterns of "geodiversity," (Record et al., Chap. 10, this volume). Such a system would reveal ideal spatial scales for sampling alpha and beta diversity and would help reveal the causes of biodiversity patterns at multiple scales, as illustrated in Figs. 16.3 and 16.6.

Such a scale-aware global biodiversity monitoring system could incorporate and integrate many aspects of existing networks such as NEON (Hopkin 2006), Forest GEO (Anderson-Teixeira et al. 2014), and many others (see Geller et al., Chap. 20), but would provide many benefits currently not provided by existing biodiversity monitoring efforts. The system would require global imaging spectrometry with repeat coverage, revealing global patterns in time and space and providing essential context for more detailed studies at higher spatial resolution (see Schimel et al. Chap. 19). More detailed resolution could be achieved by a fleet of regional aircraft carrying imaging spectrometers at a spatial resolution matching the crown sizes of many shrub and tree species (e.g., Kampe et al. 2010). For even more detailed sampling resolution needed for smaller statured vegetation or for resolving individual leaf traits, UAVs, robotic, or tower-mounted imaging spectrometers could be deployed at key sites. These methods would be coupled to systematic ground sampling of species composition (alpha and beta diversity) using traditional field sampling methods, along with leaf and canopy optical properties (using field spectrometry) for a detailed assessment of plant traits. Radiative transfer models and statistical scaling methods (Serbin and Townsend, Chap. 3) could provide a framework for integrating and analyzing data across scales.

A systematic global evaluation of optical diversity across multiple scales could readily detect dynamics in biodiversity, identify causes biodiversity changes, adapt to these changes, and help to identify monitoring and conservation priorities. By integrating diversity metrics with measures of ecosystem function, our understanding of the ecosystem impacts of biodiversity would be enhanced, allowing better management for resilience. On our rapidly changing planet, such a system would enable the monitoring required for sustainable management of ecosystems globally.

Acknowledgments The authors wish to acknowledge funding from a NSF and NASA Dimensions grant to JCB (DEB 1342872), PT (DEB 1342778), and JG (DEB 1342823), as well as the Cedar Creek NSF Long-Term Ecological Research grant (DEB 1234162) for supporting many of the examples cited here. We also thank NIMBioS and the Keck Institute of Space Studies (KISS) for sponsoring meetings that helped develop some of the ideas in this chapter. We also acknowledge additional funding support from NSERC, CFI, and from Rangeland Research Institute (U. Alberta) grants to JG. Thanks to Evan Delancey for spectral data (Fig. 16.1), Greg Crutsinger for access to the Parrot Drone data (Fig. 16.5), and to Rick Perk and Brian Leavitt (CALMIT, University of Nebraska-Lincoln) for technical assistance in airborne data collection and processing (Figs. 16.3 and 16.6).

References

Anderson-Teixeira KJ et al (2014) CTFS-ForestGEO: a worldwide network monitoring forests in an era of global change. Glob Chang Biol 21(2):528–549

Asner GP, Martin RE (2009) Airborne spectranomics: mapping canopy chemical and taxonomic diversity in tropical forests. Front Ecol Environ 7:269–276. https://doi.org/10.1890/070152

Asner GP et al (2012) Carnegie Airborne Observatory-2: increasing science data dimensionality via high-fidelity multi-sensor fusion. Remote Sens Environ 124:454–465. https://doi.org/10.1016/j.rse.2012.06.012

Asner GP, Martin RE, Anderson CB, Knapp DE (2015) Quantifying forest canopy traits: imaging spectroscopy versus field survey. Remote Sens Environ 158:15–27. https://doi.org/10.1016/j.rse.2014.11.011

Baldeck CA, Colgan MS, Feret JB, Levicx SR, Martin RE, Asner GP (2014) Landscape-scale variation in plant community composition of an African savanna from airborne species mapping. Ecol Appl 24:84–93. https://doi.org/10.1890/13-0307.1

Barnosky AD et al (2011) Has the Earth's sixth mass extinction already arrived? Nature 471:51–57. https://doi.org/10.1038/nature09678

Bonar S, Fehmi J, Mercado-Silva N (2011) An overview of sampling issues in species diversity and abundance surveys. In: Magurran AE, McGill BJ (eds) Biological diversity: Frontiers in measurement and assessment. Oxford University Press, Oxford, England, p 376

Cavender-Bares J, Gamon JA, Hobbie SE, Madritch MD, Meireles JE, Schweiger AK, Townsend PA (2017) Harnessing plant spectra to integrate the biodiversity sciences across biological and spatial scales. Am J Bot 104:966–969. https://doi.org/10.3732/ajb.1700061

Chadwick KD, Asner GP (2016) Organismic-scale remote sensing of canopy foliar traits in lowland tropical forests. Remote Sens 8. https://doi.org/10.3390/rs8020087

Chavana-Bryant C et al (2017) Leaf aging of Amazonian canopy trees as revealed by spectral and physiochemical measurements. New Phytol 214:1049–1063. https://doi.org/10.1111/nph.13853

Cohen WB, Justice CO (1999) Validating MODIS terrestrial ecology products: linking in situ and satellite measurements. Remote Sens Environ 70:1–3. https://doi.org/10.1016/s0034-4257(99)00053-x

Curran PJ (1988) The semivariogram in remote sensing: an introduction. Remote Sens Environ 24:493–507. https://doi.org/10.1016/0034-4257(88)90021-1

Dahlin KM (2016) Spectral diversity area relationships for assessing biodiversity in a wildland-agriculture matrix. Ecol Appl 26:2756–2766. https://doi.org/10.1002/eap.1390

DeLancey E (2014) Hyperspectal remote sensing of boreal forest tress diversity at multiple scales. Master's Thesis, University of Alberta

DeWoody J, Rowe CA, Hipkins VD, Mock KE (2008) "Pando" lives: molecular genetic evidence of a giant aspen clone in Central Utah. West N Am Nat 68:493–497. https://doi.org/10.3398/1527-0904-68.4.493

Ehleringer JR, Field CB (1993) Scaling physiological processes: leaf to globe, 1st edn. Academic Press, New York

Fairbanks DHK, McGwire KC (2004) Patterns of floristic richness in vegetation communities of California: regional scale analysis with multi-temporal NDVI. Glob Ecol Biogeogr 13:221–235. https://doi.org/10.1111/j.1466-822X.2004.00092.x

Fava F, Parolo G, Colombo R, Gusmeroli F, Della Marianna G, Monteiro AT, Bocchi S (2010) Fine-scale assessment of hay meadow productivity and plant diversity in the European Alps using field spectrometric data. Agric Ecosyst Environ 137:151–157. https://doi.org/10.1016/j.agee.2010.01.016

Féret JB, Asner GP (2014) Mapping tropical forest canopy diversity using high-fidelity imaging spectroscopy. Ecol Appl 24:1289–1296

Franklin SE, Ahmed OS (2018) Deciduous tree species classification using object-based analysis and machine learning with unmanned aerial vehicle multispectral data. Int J Remote Sens 39:5236–5245. https://doi.org/10.1080/01431161.2017.1363442

Gaitán JJ et al (2013) Evaluating the performance of multiple remote sensing indices to predict the spatial variability of ecosystem structure and functioning in Patagonian steppes. Ecol Indic 34:181–191. https://doi.org/10.1016/j.ecolind.2013.05.007

Gamon JA (2015) Reviews and syntheses: optical sampling of the flux tower footprint. Biogeosciences 12:4509–4523. https://doi.org/10.5194/bg-12-4509-2015

Gamon JA et al (2004) Remote sensing in BOREAS: lessons learned. Remote Sens Environ 89:139–162. https://doi.org/10.1016/j.rse.2003.08.017

Gholizadeh H, Gamon JA, Zygielbaum AI, Wang R, Schweiger AK, Cavender-Bares J (2018) Remote sensing of biodiversity: soil correction and data dimension reduction methods improve assessment of alpha-diversity (species richness) in prairie ecosystems. Remote Sens Environ 206:240–253. https://doi.org/10.1016/j.rse.2017.12.014

Gholizadeh H et al (2019) Detecting prairie biodiversity with airborne remote sensing. Remote Sens Environ 221:38–49. https://doi.org/10.1016/j.rse.2018.10.037

Goodman JA, Ustin SL (2007) Classification of benthic composition in a coral reef environment using spectral unmixing. J Appl Remote Sens 1:011501. https://doi.org/10.1117/1.2815907

Gould W (2000) Remote sensing of vegetation, plant species richness, and regional biodiversity hotspots. Ecol Appl 10:1861–1870. https://doi.org/10.2307/2641244

Heffernan JB et al (2014) Macrosystems ecology: understanding ecological patterns and processes at continental scales. Front Ecol Environ 12:5–14. https://doi.org/10.1890/130017

Hooper DU et al (2012) A global synthesis reveals biodiversity loss as a major driver of ecosystem change. Nature 486:105–108. https://doi.org/10.1038/nature11118

Hopkin M (2006) Spying on nature. Nature 444:420–421

Jetz W et al (2016) Monitoring plant functional diversity from space. Nature Plants 2. https://doi.org/10.1038/nplants.2016.24

Justice CO, Townshend JRG (1981) Integrating ground data with remote sensing. In: Townshend JRG (ed) Terrain analysis and remote sensing. Allen and Unwin, London, pp 38–101

Kampe TU, Johnson BR, Kuester M, Keller M (2010) NEON: the first continental-scale ecological observatory with airborne remote sensing of vegetation canopy biochemistry and structure. J Appl Remote Sens 4:043510. https://doi.org/10.1117/1.3361375

Kerr JT, Southwood TRE, Cihlar J (2001) Remotely sensed habitat diversity predicts butterfly species richness and community similarity in Canada. Proc Natl Acad Sci U S A 98:11365–11370. https://doi.org/10.1073/pnas.201398398

Kissling WD et al (2018) Towards global data products of Essential Biodiversity Variables on species traits. Nat Ecol Evol 2:1531–1540. https://doi.org/10.1038/s41559-018-0667-3

Knyazikhin Y et al (2013) Hyperspectral remote sensing of foliar nitrogen content. Proc Natl Acad Sci U S A 110:E185–E192. https://doi.org/10.1073/pnas.1210196109

Kruse FA, Lefkoff AB, Boardman JW, Heidebrecht KB, Shapiro AT, Barloon PJ, Goetz AFH (1993) The spectral image processing system (SIPS) - interactive visualization and analysis of imaging spectrometer data. Remote Sens Environ 44:145–163. https://doi.org/10.1016/0034-4257(93)90013-n

Levin SA (1992) The problem of pattern and scale in ecology. Ecology 73:1943–1967. https://doi.org/10.2307/1941447

Lucas KL, Carter GA (2008) The use of hyperspectral remote sensing to assess vascular plant species richness on Horn Island, Mississippi. Remote Sens Environ 112:3908–3915. https://doi.org/10.1016/j.rse.2008.06.009

Madritch MD, Kingdon CC, Singh A, Mock KE, Lindroth RL, Townsend PA (2014) Imaging spectroscopy links aspen genotype with below-ground processes at landscape scales. Philos Trans R Soc B Biol Sci 369:20130194. https://doi.org/10.1098/rstb.2013.0194

Magurran AE (2004) Measuring biological diversity. Blackwell Publishing, Malden

Malenovský Z, Bartholomeus HM, Acerbi-Junior FW, Schopfer JT, Painter TH, Epema GF, Bregt AK (2007) Scaling dimensions in spectroscopy of soil and vegetation. Int J Appl Earth Obs Geoinf 9:137–164. https://doi.org/10.1016/j.jag.2006.08.003

Malenovský Z et al. (2019) Variability and uncertainty challenges in upscaling imaging spectroscopy observations from leaves to vegetation canopies Surv Geophys 40:631–656

Moses WJ, Ackleson SG, Hair JW, Hostetler CA, Miller WD (2016) Spatial scales of optical variability in the coastal ocean: implications for remote sensing and in situ sampling. J Geophys Res Oceans 121:4194–4208. https://doi.org/10.1002/2016jc011767

Muller-Karger FE et al (2018) Satellite sensor requirements for monitoring essential biodiversity variables of coastal ecosystems. Ecol Appl 28:749–760. https://doi.org/10.1002/eap.1682

Nagendra H, Lucas R, Honrado JP, Jongman RHG, Tarantino C, Adamo M, Mairota P (2013) Remote sensing for conservation monitoring: assessing protected areas, habitat extent, habitat condition, species diversity, and threats. Ecol Indic 33:45–59. https://doi.org/10.1016/j.ecolind.2012.09.014

O'Neill RV, Hunsaker CT, Timmins SP, Jackson BL, Jones KB, Riitters KH, Wickham JD (1996) Scale problems in reporting landscape pattern at the regional scale. Landsc Ecol 11:169–180. https://doi.org/10.1007/bf02447515

Oldeland J, Wesuls D, Rocchini D, Schmidt M, Jurgens N (2010) Does using species abundance data improve estimates of species diversity from remotely sensed spectral heterogeneity? Ecol Indic 10:390–396. https://doi.org/10.1016/j.ecolind.2009.07.012

Palmer MW, Wohlgemuth T, Earls PG, Arévalo JR, Thompson SD (2000) Opportunities for long-term ecological research at the Tallgrass Prairie Preserve, Oklahoma. In: Lajtha K, Vanderbilt K (eds) Cooperation in long term ecological research in central and eastern Europe: proceedings of the ILTER Regional Workshop, Budapest, Hungary, 22–25 June 1999. Oregon State University, Corvallis, pp 123–128

Palmer MW, Earls PG, Hoagland BW, White PS, Wohlgemuth T (2002) Quantitative tools for perfecting species lists. Environmetrics 13:121–137. https://doi.org/10.1002/env.516

Paz-Kagan T, Caras T, Herrmann I, Shachak M, Karnieli A (2017) Multiscale mapping of species diversity under changed land use using imaging spectroscopy. Ecol Appl 27:1466–1484. https://doi.org/10.1002/eap.1540

Pereira HM et al (2013) Essential biodiversity variables. Science 339:277–278. https://doi.org/10.1126/science.1229931

Psomas A, Kneubuhler M, Huber S, Itten K, Zimmermann NE (2011) Hyperspectral remote sensing for estimating aboveground biomass and for exploring species richness patterns of grassland habitats. Int J Remote Sens 32:9007–9031. https://doi.org/10.1080/01431161.2010.532172

Rey-Benayas JM, Pope KO (1995) Landscape ecology and diversity patterns in the seasonal tropics from Landsat TM imagery. Ecol Appl 5:386–394. https://doi.org/10.2307/1942029

Rocchini D, McGlinn D, Ricotta C, Neteler M, Wohlgemuth T (2011) Landscape complexity and spatial scale influence the relationship between remotely sensed spectral diversity and survey-based plant species richness. J Vegetation Sci 22:688–698. https://doi.org/10.1111/j.1654-1103.2010.01250.x

Roth KL, Roberts DA, Dennison PE, Alonzo M, Peterson SH, Beland M (2015) Differentiating plant species within and across diverse ecosystems with imaging spectroscopy. Remote Sens Environ 167:135–151. https://doi.org/10.1016/j.rse.2015.05.007

Schaepman-Strub G, Schaepman ME, Painter TH, Dangel S, Martonchik JV (2006) Reflectance quantities in optical remote sensing-definitions and case studies. Remote Sens Environ 103:27–42. https://doi.org/10.1016/j.rse.2006.03.002

Schäfer E, Heiskanen J, Heikinheimo V, Pellikka P (2016) Mapping tree species diversity of a tropical montane forest by unsupervised clustering of airborne imaging spectroscopy data. Ecol Indic 64:49–58. https://doi.org/10.1016/j.ecolind.2015.12.026

Scholes RJ et al (2012) Building a global observing system for biodiversity. Curr Opin Environ Sustain 4:139–146. https://doi.org/10.1016/j.cosust.2011.12.005

Schweiger AK et al (2018) Plant spectral diversity integrates functional and phylogenetic components of biodiversity and predicts ecosystem function. Nat Ecol Evol 2:976–982. https://doi.org/10.1038/s41559-018-0551-1

Simpson EH (1949) Measurement of diversity. Nature 163:688–688. https://doi.org/10.1038/163688a0

Singh A, Serbin SP, McNeil BE, Kingdon CC, Townsend PA (2015) Imaging spectroscopy algorithms for mapping canopy foliar chemical and morphological traits and their uncertainties. Ecol Appl 25:2180–2197. https://doi.org/10.1890/14-2098.1.sm

Somers B, Asner GP, Martin RE, Anderson CB, Knapp DE, Wright SJ, Van De Kerchove R (2015) Mesoscale assessment of changes in tropical tree species richness across a bioclimatic gradient in Panama using airborne imaging spectroscopy. Remote Sens Environ 167:111–120. https://doi.org/10.1016/j.rse.2015.04.016

Tilman D, Reich PB, Isbell F (2012) Biodiversity impacts ecosystem productivity as much as resources, disturbance, or herbivory. Proc Natl Acad Sci U S A 109:10394–10397. https://doi.org/10.1073/pnas.1208240109

Townsend PA, Serbin SP, Kruger EL, Gamon JA (2013) Disentangling the contribution of biological and physical properties of leaves and canopies in imaging spectroscopy data. Proc Natl Acad Sci U S A 110:E1074. https://doi.org/10.1073/pnas.1300952110

Tuanmu MN, Jetz W (2015) A global, remote sensing-based characterization of terrestrial habitat heterogeneity for biodiversity and ecosystem modelling. Glob Ecol Biogeogr 24:1329–1339. https://doi.org/10.1111/geb.12365

Turner W (2014) Sensing biodiversity. Science 346:301–302. https://doi.org/10.1126/science.1256014

Turner MG, Dale VH, Gardner RH (1989) Predicting across scales: theory development and testing. Landsc Ecol 3:245–252. https://doi.org/10.1007/bf00131542

Ustin SL, Gamon JA (2010) Remote sensing of plant functional types. New Phytol 186:795–816. https://doi.org/10.1111/j.1469-8137.2010.03284.x

van Holde KE (1989) Chromatin. Springer, New York

Verrelst J et al (2019) Quantifying vegetation biophysical variables from imaging spectroscopy data: a review on retrieval methods. Surv Geophys 40:589–629. https://doi.org/10.1007/s10712-018-9478-y

Vihervaara P et al (2017) How Essential Biodiversity Variables and remote sensing can help national biodiversity monitoring. Glob Ecol Conserv 10:43–59. https://doi.org/10.1016/j.gecco.2017.01.007

Wallace L, Lucieer A, Malenovsky Z, Turner D, Vopenka P (2016) Assessment of forest structure using two UAV techniques: a comparison of airborne laser scanning and structure from motion (SfM) point clouds. Forests 7. https://doi.org/10.3390/f7030062

Wang R, Gamon JA (2019) Remote sensing of biodiversity. Remote Sens Environ 231:111218 https://doi.org/10.1016/j.rse.2019.111218

Wang L, Sousa WP, Gong P, Biging GS (2004) Comparison of IKONOS and QuickBird images for mapping mangrove species on the Caribbean coast of Panama. Remote Sens Environ 91:432–440. https://doi.org/10.1016/j.rse.2004.04.005

Wang R, Gamon JA, Emmerton CE, Hitao L, Nestola E, Pastorello G, Menzer O (2016) Integrated analysis of productivity and biodiversity in a Southern Alberta prairie. Remote Sens 8:214. https://doi.org/10.3390/rs8030214

Wang R, Gamon JA, Cavender-Bares J, Townsend PA, Zygielbaum AI (2018a) The spatial sensitivity of the spectral diversity-biodiversity relationship: an experimental test in a prairie grassland. Ecol Appl 28:541–556. https://doi.org/10.1002/eap.1669

Wang R, Gamon JA, Schweiger AK, Cavender-Bares J, Townsend PA, Zygielbaum AI, Kothari S (2018b) Influence of species richness, evenness, and composition on optical diversity: a simulation study. Remote Sens Environ 211:218–228. https://doi.org/10.1016/j.rse.2018.04.010

Wehlage DC, Gamon JA, Thayer D, Hildebrand DV (2016) Interannual variability in dry mixed-grass prairie yield: a comparison of MODIS, SPOT, and field measurements. Remote Sens 8. https://doi.org/10.3390/rs8100872

Whittaker RH (1972) Evolution and measurement of species diversity. Taxon 21:213–251. https://doi.org/10.2307/1218190

Woodcock CE, Strahler AH (1987) The factor of scale in remote sensing. Remote Sens Environ 21:311–332. https://doi.org/10.1016/0034-4257(87)90015-0

Zutta B (2003) Assessing vegetation functional type and biodiversity in Southern California using spectral reflectance. Master's Thesis, California State University, Los Angeles

Open Access   This chapter is licensed under the terms of the Creative Commons Attribution 4.0 International License (http://creativecommons.org/licenses/by/4.0/), which permits use, sharing, adaptation, distribution and reproduction in any medium or format, as long as you give appropriate credit to the original author(s) and the source, provide a link to the Creative Commons license and indicate if changes were made.

The images or other third party material in this chapter are included in the chapter's Creative Commons license, unless indicated otherwise in a credit line to the material. If material is not included in the chapter's Creative Commons license and your intended use is not permitted by statutory regulation or exceeds the permitted use, you will need to obtain permission directly from the copyright holder.

Chapter 17
Integrating Biodiversity, Remote Sensing, and Auxiliary Information for the Study of Ecosystem Functioning and Conservation at Large Spatial Scales

Franziska Schrodt, Betsabe de la Barreda Bautista, Christopher Williams, Doreen S. Boyd, Gabriela Schaepman-Strub, and Maria J. Santos

17.1 Introduction

In the face to accelerated environmental change, being able to assess different aspects of plant biodiversity, such as those related to, e.g., the productivity or health of an ecosystem, repeatedly at large spatial scales, is increasingly important. The recent decade has seen an explosion of in-situ databases necessary to assess such patterns and processes, often cover large parts of the Earth (e.g., plant functional traits, phenology (PhenoCam networks)), and integrating this data with remotely sensed products enables assessment at critical scales which would otherwise be impossible or extremely costly to do.

However, RS data comes with limitation of their own, and despite of the many opportunities offered by RS data, certain aspects and scales of biodiversity are currently not measurable using RS technology alone. Thus, the combination of RS, in-situ and other auxiliary data, provides the most powerful approach to assessing ecosystem functioning and conservation at large spatial scales.

F. Schrodt (✉) · B. de la Barreda Bautista · D. S. Boyd
School of Geography, University of Nottingham, Nottingham, UK
e-mail: lgzfs@nottingham.ac.uk; lgzbd1@nottingham.ac.uk; doreen.boyd@nottingham.ac.uk

C. Williams
GeoAnalytics and Modelling Directorate, British Geological Survey, Nottingham, UK
e-mail: chrwil@bgs.ac.uk

G. Schaepman-Strub
Department of Evolutionary Biology and Environmental Studies, University of Zürich, Zürich, CH, Switzerland
e-mail: gabriela.schaepman@geo.uzh.ch

M. J. Santos
University Research Priority Program in Global Change and Biodiversity
and Department of Geography, University of Zürich, Zürich, CH, Switzerland
e-mail: maria.j.santos@geo.uzh.ch

© The Author(s) 2020
J. Cavender-Bares et al. (eds.), *Remote Sensing of Plant Biodiversity*,
https://doi.org/10.1007/978-3-030-33157-3_17

The means and uses of RS to link biodiversity with other relevant ecosystem
metrics are manifold, and some are covered in detail in other chapters within this
book. A compilation of some of the major RS data types and sensors used for vegetation analyses at the landscape scale are provided in Fig. 17.1. Here, we discuss
strengths, weaknesses, and caveats of linking RS with in-situ data to address the
themes of ecosystem functioning and conservation of biodiversity.

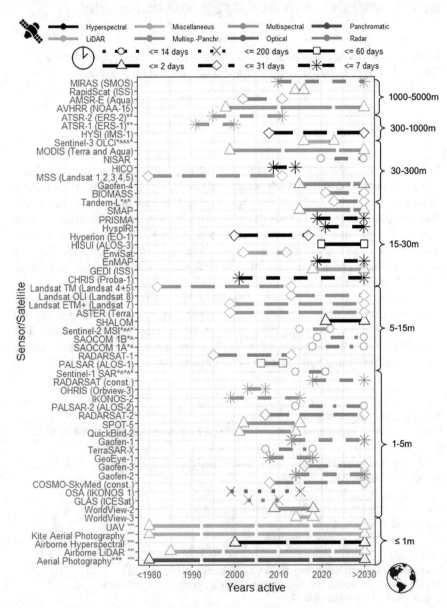

Fig. 17.1 Non-exhaustive compilation of some of the major RS data types and sensors used for
vegetation analyses at the landscape scale, revisit times (line style and symbols), spatial resolution
(from low resolution at the top to high resolution at the bottom), sensor types (colors), and years

17.2 Ecosystem Functioning

Pettorelli et al.'s (2017) framework for monitoring ecosystem functions at all scales lends itself well to the flexibilities and strengths provided by RS. We consider ecosystem functions as those attributes related to the performance of an ecosystem that are the consequence of one or more ecosystem processes (Lovett et al. 2005). With respect to plant diversity, the attributes underpinning functions that benefit plant species (and indirectly fauna and humans) are crucial. Such functions include pollination, water regulation, disturbance regulation, supporting habitats, and biological control. The measurement and monitoring of these ecosystem functions often relies on a remotely sensed proxy. For example, the ecosystem function of greenhouse gas regulation could be monitored with RS-based measurements of emissions from fires, as provided by Moderate Resolution Imaging Spectroradiometer (MODIS) and expected from missions such as Environmental Mapping and Analysis Program (EnMap) and the Surface Biology and Geology imaging spectrometer (currently under planning to replace the cancelled HyspIRI mission). The advantage of such sensors is their moderate to high spatial and temporal resolution that creates dense time series. Analyzing ecosystem functions over large scales can provide information on drivers of species diversity and abundance change, as well as aspects that affect human well-being such as those related to ecosystem services (ES). Ecosystems provide regulating, provisioning, and cultural and supporting services to society, such as nutrient regulation, the provision of food, and waste treatment (De Groot et al. 2002; Ma 2005). Detailed techniques for mapping ES at large scales (Englund et al. 2017) and rapid assessments (Meyer et al. 2015; Cerreta and Poli 2017) are discussed elsewhere. Although ES functions and processes are closely related, "services" implies inherent contributions to humans and an attached monetary or cultural value (De Groot et al. 2002). All ESs are under increasing threat due to pollution, overexploitation, land use change, and climate change, with concerns of overstepping the "safe operating space for humanity" (Rockström et al. 2009).

We choose to focus on a few contrasting ecosystem functions in this chapter, with particular focus on those that plants provide or require for survival and fitness;

Fig. 17.1 (continued) active (x-axis). For airborne data, it was assumed that theoretically, data can be available for the past although these data will be very sparse and not easily accessible. In all cases, highest possible spatial and temporal resolutions are shown. This might not be the case for all bands covered by these sensors. Where satellites are currently active or proposed, they are presented here as being active into the future; this is subject to change and should be reviewed frequently. Codes after sensor names: "not continuous, ✳✳ ATSR-1/2 also contained a Microwave (MW) sensor, ✳^ 16-day revisit with one satellite, 8 days if using data from both, ✳^✳ 2 satellites proposed—details specific to Biosphere observations, ✳✳✳ Film and digital. Green and red boxes refer to the examples given in the text, enclosing MODIS and ALOS PALSAR sensors. Other than Kite, UAV, and declared airborne methods, all instruments are either satellites or sensors carried by a satellite. Many satellites have a payload of a range of instruments; where this is the case, hyperspectral or multispectral units have been presented. Many satellites also carry panchromatic sensors, which are not represented in this figure. (For more information, see Toth and Jóźków et al. (2016) and Khorram et al. (2016))

further details on other ecosystem functions can be obtained in Pettorelli et al. (2017). The narrative below focuses on satellite RS measurements because they are widely accessible and offer a relatively inexpensive and verifiable means of deriving data with complete spatial coverage in a consistent manner over large areas with (near) appropriate temporal resolutions, thus offering great potential for tracking change in ecosystem functions (Cabello et al. 2012; Nagendra et al. 2013). However, these satellite measurements can (and should) be integrated with a host of other data sources that are remotely acquired; these are also discussed where appropriate.

17.2.1 Pollination

The transport of pollen between plants is crucial for reproduction. Different pollination types exist with varying distributions in space and time. The ecosystem function of pollination is under varying threats especially with declines in abundance and the loss of the organisms providing the service (e.g., bees) (Vanbergen et al. 2013; IPBES 2016). Thus, up-to-date information on pollination is extremely important.

Two distinct RS approaches can be used to study this function: (i) direct RS of different pollination types and (ii) remotely sensed indicators of pollination (i.e., vegetation phenology or biomass as an indicator). The former is challenging because many pollination traits cannot be directly measured with RS sensors due to the signal contribution of some pollination traits (nectar content, flower structure, etc.) being too low relative to other surface components. Alternatively, Feilhauer et al. (2016) posited that different pollination types might be inferred from leaf and canopy optical traits (leaf area index, leaf tilt angle, mean canopy height, cover, specific leaf area, leaf dry matter content, leaf dry mass; see also Olinger 2011), allowing for an indirect classification of plant pollination types. Using data acquired by an airborne hyperspectral sensor, pollination types were related to canopy reflectance in a way that allows their discrimination, opening the potential to expand this approach to other ecosystems and different phenological stages. Such an approach should benefit from upcoming satellite missions equipped with hyperspectral sensors, for example, the German EnMAP and the Chinese GF-5 Hyperspectral Imager (Fig. 17.1). Further, pollination traits such as floral display size may have too little of a signal to be discernible from other surface components with RS, although several studies have been able to detect flowers with hyperspectral airborne sensors and use the information to, for example, detect invasive species (see Bolch et al., Chap. 12, in this book).

Plant phenology is another ecosystem function directly linked to pollination, since any change in the phenological cycle may affect interactions such as competition between plant species and mutualism with pollinators (Buitenwerf et al. 2015). Any direct alteration in functional or taxonomic plant diversity as a result of change in vegetation phenology is further compounded by both short- and long-term climatic changes. RS has demonstrated capacity for measuring and monitoring vegetation phenology (Cleland et al. 2007) and thus indirectly detecting pollination (Neil and Wu 2006). In the next section, we describe how RS can be used to measure phenology.

17.2.2 Phenology

Recent climate change has shifted phenology and associated species distributions globally. This has in turn increased the risk of extinction for affected species through the alteration of development rates of species or by altering the timing of environmental cues that affect a species' presence in the community (Yang and Rudolf 2009). Ongoing climatic and phenological changes are expected to further increase this risk. Moreover, the extent to which changes in vegetation phenology will feed back into the climate system by modifying albedo and hence cloud formation is a major source of uncertainty in climate change projections (Zhao et al. 2013). Recent work on land surface phenology has focused on assessing changes in phenology globally (rather than regionally) over long time periods as is now afforded by the Earth Observation (EO) archive (Ganguly et al. 2010). Buitenwerf et al. (2015) used phenomes (83 phenologically similar zones) in their global study of 32 years (1980–2012) and showed via metrics extracted from the normalized difference vegetation index (NDVI) record that most of Earth's land surface has undergone some form of change in the seasonal pattern of vegetation activity. Other studies used alternative indices, such as the MERIS Global Vegetation Index (MGVI) and Terrestrial Chlorophyll Index (MTCI), from the now defunct Envisat platform and MODIS Enhanced Vegetation Index (EVI). Indeed, comparison between these indices for a temperate deciduous forest showed that MTCI corresponded more closely with vegetation phenology from ground observations and climatic proxies than any of the other indices (Boyd and Foody 2011). This finding suggests that the Envisat MTCI is best suited for monitoring vegetation phenology, advocated by its sensitivity to canopy chlorophyll content, a proxy for the canopy's physical and chemical alterations occurring during phenological cycles (Boyd et al. 2012). These studies also point to the value of increasing both the temporal and spatial sampling of vegetation phenology. Limitations to spatial resolution mean that satellite-derived data represent land surface phenology rather than direct vegetation phenology and are therefore too coarse to detect critical individual, species, or community-level responses. Further improving the temporal sampling means that those challenges to using satellite data, including high sensitivity to effects of clouds and atmospheric conditions, could be overcome. With the recent launch of the ESA Sentinel-2 and Sentinel-3, this improvement in data characteristics is assured to go forward.

Key to applying data from these or any satellite for the derivation and study of phenology, however, are accurate calibration and atmospheric correction to obtain surface reflectance data. National Oceanic and Atmospheric Administration (NOAA) satellite sensors have legacy calibrated data. For the Landsat and Terra/Aqua satellites, calibrated top-of-atmosphere radiance and surface reflectance products are delivered directly to the users. While Sentinel satellites are calibrated, users are provided with top-of-atmosphere reflectance and have to perform the atmospheric correction (using plug-ins such as sen2cor (http://step.esa.int/main/third-party-plugins-2/sen2cor/)). That said, Vuolo et al. (2016) have demonstrated very good agreement between calibrated Sentinel-2 and Landsat-8 data for six test sites. A Harmonized Landsat and Sentinel-2 land surface reflectance data set is now readily available (Claverie et al. 2018).

The requirement for high-temporal-resolution RS data has, since 1982, been provided by the NOAA Advanced Very High Resolution Radiometer (AVHRR) sensor and its successors, with much regional scale analyses undertaken using NDVI (Reed et al. 2009). Capturing the seasonal pattern of photosynthetically active radiation absorbed by the land surface, using repeated measures of vegetation indices such as NDVI throughout the year for an area of interest, allows depicting the cycle of events that drive the seasonal progression of vegetation through stages of dormancy, active growth, and senescence. Several phenological metrics can be extracted from the temporal sequence, relating to leaf-on and leaf-off, length of growing season, peak of growing season, trough, and measures of seasonal amplitude (integral, trough), and their patterns over time and space. Specially written and customizable open-source software, such as TIMESAT (Eklundh and Jonsson 2015) and QPhenoMetrics (Duarte et al. 2018), afford some robustness to the study of phenology.

With this emphasis on using satellite RS of land surface phenology to assess changes in vegetation, validation of extracted metrics is imperative. Three principal approaches are suitable, all of which use observations taken at ground level. The first relies on collaboration between experts to ensure suitable spatial and temporal coverage. The PEP725 ground phenology database generated as a result of the Pan European Phenology (PEP) project (a European infrastructure to promote and facilitate phenological research, education, and environmental monitoring) is one example (Templ et al. 2018). The second approach is an extension of the first and uses citizen science projects where the public exploits Web 2.0 technologies and contributes ground-based observations (Kosmala et al. 2016). The idea of citizens as sensors is not new, but full use of their input still requires effort. This is the focus of a current European Union (EU) Horizon 2020 project, LandSense (https://landsense.eu/). The third approach uses proximal sensing (often automated systems) to provide detailed information at particular locations (e.g., traffic cameras, Morris et al. 2013; archived TV video footage, De Frenne et al. 2018; and webcams, e.g., the PhenoCam project, Richardson et al. 2018). However, as Brown et al. (2016) pointed out, these technological advances present challenges with respect to data standards. These authors suggest that continental-scale ecological research networks, such as the US National Ecological Observatory Network (NEON) and the EU's Integrated Carbon Observation System (ICOS), can serve as templates for developing rigorous data standards and extending the utility of PhenoCam data through standardized protocols for ground-truthing.

17.2.3 Carbon Storage

Biomass estimates are fundamental to estimate carbon fluxes and stocks and link them to carbon credit initiatives. Active RS such as laser scanning provides fundamental data on canopy structure useful to include in allometric equations to estimate biomass (Chave et al. 2014). For example, the analysis of first returns from laser scanning is used to estimate canopy height (Bouvet et al. 2017). Recent Terrestrial Laser Scanning (TLS) technology shows promising advances in more accurate estimates of biomass because it allows for a first-order estimate of diameter at breast height. Further, TLS analysis is able to provide more precise representation of tree structure, allowing for moving beyond the assumed cylindrical shape of a tree trunk used for many biomass estimates. Novel methods also make it possible to fit geometric shapes along not only the trunk but also the branches and stems, giving a more precise estimate of the woody component of trees and producing more reliable estimates of biomass.

Another area of active research in carbon storage and credits is the estimation of tree cover and the number of trees per pixel, because even if the biomass per tree is correctly estimated, it is scaled to regional or global estimates by a multiplier of tree density. MODIS offers a tree cover product (as a layer within the Vegetation Continuous Fields product) systematically and repeatedly at a global scale, but this product makes some assumptions on the minimum cover of trees needed to be detected by the MODIS sensors. In addition, it is important to estimate the contribution of non-tree functional types such as shrubs and grasslands to the carbon storage and credits calculations. Current assessments of the performance of reducing emissions from deforestation and forest degradation (REDD+) programs have shown varied success; most are linked to the quality of the biomass estimates, which are fundamental to calculate the carbon potential of a given ecosystem as a fraction (often assumed around 1/2) of its biomass.

17.2.4 Challenges

Although there is a clear role for RS for monitoring ecosystem functions (see Serbin and Townsend, Chap. 3; Martin, Chap. 5), there are still many challenges to be overcome to ensure its full potential is realized. Many acknowledge the lack of an acceptable framework that brings together the many proxies for ecosystem functioning that can be directly remotely sensed (e.g., Asner and Olinger 2009; De Araujo Barbosa et al. 2015). But fundamentally there is a need to improve the RS estimates of the many proxies that are used to infer the ecosystem functions of interest. Developments in methodologies for processing, analyzing, and interpreting RS data will serve to improve the mapping accuracy and monitoring opportunities. However, those developments cannot exist in isolation; a dialogue between computational scientists and those concerned with ecosystem functions must occur for the full potential of RS to be realized (Cabello et al. 2012; Paganini et al. 2016). Coupled with this is the need for advances in sensor technologies that will enable accurate, timely data with the right thematic, i.e., where interpretation of the sensor's raw data can provide the information necessary for applications in ecosystem function and biodiversity. Further, data should be open access, maintained, and interoperable (Pettorelli et al. 2017), particularly given the plethora of existing local to regional scale data capture initiatives by airborne methods based on drones or airplanes (Cord et al. 2017). Finally, RS proxies will often need to be combined with field measurements to accurately represent the desired ecosystem function (e.g., Tong et al. 2004). Indeed, joint analyses of satellite data with in-situ measurements or process measurements in the lab may be essential steps to the refinement and increased capacity and utility of satellite-based indicators for ecosystem function monitoring (see also Meireles et al., Chap. 7, in this book). This is likely to be a nontrivial task, particularly in highly dynamic situations.

17.3   Conservation

Global environmental change has led to major losses, changes, and erosion of biodiversity and ecosystem function and counteracted to some extent by conservation action. Conservation science focuses on understanding the distribution of organisms, their rarity status, the viability of populations, drivers and disturbances, and current and future restoration need. RS has been increasingly used to answer conservation science questions and applications (Rose et al. 2015), namely, species mapping (see Chaps. 9, 10, and 11 in this book by Record et al.; Paz et al.; Pinto-Ledezma), biodiversity monitoring (Feret and Asner 2014; Rocchini et al. 2017), detecting invasive alien species (see Bolch et al., Chap. 12, in this book), assessing vegetation condition, monitoring carbon storage and credits, and assessing habitat extent and condition (see Record et al., Chap. 10, in this book), among others (Bustamante et al. 2016; Lawley et al. 2016; Niphadkar and Nagendra 2016; Reddy et al. 2017). Here, we focus on aspects of conservation related to the abovementioned measures of ecosystem function at large scales.

17.3.1   Biodiversity Monitoring

RS has long been recognized as useful for biodiversity measurement (Stoms and Estes 1993; Turner et al. 2003; Turner 2014), and more recently it has emerged prominently in biodiversity monitoring, through essential biodiversity variables (Pereira et al. 2013; Skidmore et al. 2015; Pereira et al. 2015). Rose et al. (2015) identified the top ten applications of RS in conservation, namely, for species distribution and abundance, movement and life stages, ecosystem processes, climate change, rapid response, protected areas, ecosystem services, conservation effectiveness, changes in land use/cover, and degradation and disturbance regimes.

Beyond land cover classification, one of the most extensive uses of RS data is to produce distribution maps of species, communities, and ecosystems (Kerr and Ostrovsky 2003). Most RS studies of biodiversity focused on mapping species using all kinds of information from optical, radar, and Light Detection and Ranging (LiDAR) data. Nagendra (2001) reviewed the potential of RS data for assessing biodiversity, namely, species mapping and species diversity and habitat mapping, and concluded that at the time the most feasible application of RS would be to map species distributions and habitat, the former at smaller spatial scales and the latter at larger scales. RS data such as those from Landsat and other multispectral sensors have been widely used to map species and vegetation communities (Xie et al. 2008). There is now a growing literature on identifying species in many case studies. However, use of imaging spectroscopy for species identification needs to be understood at a more fundamental level—especially the development of generalized methodologies and rules for detection and mapping, which is an area of active research today. Conceptually, we have yet to resolve how to identify unique spectral signatures for the estimated 400,000 extant plant species or groups of species (i.e., functional types or optical types; Ustin and Gamon 2010). In contrast to geologic minerals, which are often spectrally distinct, all land plants share a common basic metabolism and biochemistry. This fundamental similarity makes identification of plant species difficult. The interactions of a spectral signal with environmental conditions and shifts in spectral signatures through phenological stages contribute to spectral variation, in addition to the characteristic properties of individual species (Ustin and Jacquemoud, Chap. 14). In recent years, with the advent of hyperspectral sensors and the fusion of these data sets with other auxiliary data, novel avenues to map and monitor biodiversity have emerged. These new data sets make it possible to directly discriminate species in terrestrial and freshwater ecosystems (Jones and Vaughan 2010; Turak et al. 2017; Choa et al. 2012; Fassnacht et al. 2016) and assess relationships between the diversity of spectra and the diversity of species and the fundamentals of the spectral diversity hypothesis (i.e., that the diversity of spectral profiles generally predicts diversity of species, Nagendra 2001), which are presented in other chapters of this book (see especially Schweiger, Chap. 15; Cavender-Bares et al., Chap. 2). A recent study highlights the relation of spectral diversity to functional and phylogenetic components of biodiversity (Schweiger et al. 2018), and this topic is covered in more detail in Meireles et al. (Chap. 7).

Another avenue in which RS can contribute to biodiversity monitoring is through its use in species distribution models (SDMs) (see Pinto-Ledézma and Cavender-Bares, Chap. 9; Paz et al., Chap. 11). SDMs are empirical statistical approaches that predict the spatial distribution of species (Guisan and Zimmermann 2000), and the choice of environmental predictors is fundamental for SDM. RS measurements of vegetation condition (Turner et al. 2003), ecosystem productivity (Running et al. 2004), and seasonality (Reed et al. 1994), among others, are now available over time series (e.g., Landsat time series; Kennedy et al. 2014) and might be used in SDMs (Bradley and Fleishman 2008; He et al. 2015) making it possible to predict species distributions over time. Although the use of RS in SDMs is widely advocated and applied, it has yet to be scaled to most species, especially non-plant taxa. Promising progress toward the inclusion of RS products in SDMs includes responses to nutritional value (Sheppard et al. 2007), food resources (Coops et al. 2009), seasonal variation (Bischof et al. 2012), and combined effects of climate and land use changes (Santos et al. 2017). Upcoming sensors are expected to provide even better and more diverse measurements of ecosystem processes and other information that might be relevant to map species distributions at finer spatial, temporal, and spectral resolutions (e.g., Sentinel satellites; Berger and Aschbacher 2012). Further, through SDMs we can expand our capacity to monitor taxonomic groups beyond plants, providing a broader understanding of the dynamics and dimensions of biodiversity, its feedbacks and interactions, and its change.

Yet another way in which RS can be useful in biodiversity monitoring is to detect animals using unmanned aerial vehicles with visible and thermal sensors and LiDAR data. Nagendra et al. (2013) concluded that despite the potential for RS in monitoring habitat, the integration has not happened yet because of technical challenges of conducting and accurately interpreting image analyses, insufficient integration between in-situ data and expert knowledge RS data, and lack of funding and platforms that provide such services and capacity in an accessible way globally. However, progress in this domain includes the provision of environmental data layers from RS sources in Movebank, a major animal movement data repository. Environmental data are directly extracted at the recorded locations and interpolated to the date and time of each GPS fix (Dodge et al. 2013). This allows animal movement ecologists to easily extract environmental variables co-registered in space and time with their animal location data.

Current RS capabilities allow for improving species mapping and monitoring such as the local species pool (i.e., alpha diversity, Feret and Asner 2014) as well as to move beyond species (Jetz et al. 2016) to measure and monitor other components of diversity such as compositional turnover (i.e., beta diversity, Leitão et al. 2015; Schwieder et al. 2016; Rocchini et al. 2017). There were a few attempts to map species richness using Landsat TM multispectral data to calculate NDVI as a proxy for species richness with limited success (Gould 2000). The spectral resolution of imaging spectrometer data today is sufficiently fine to implement and test the spectral diversity hypothesis (Feret and Asner 2014) because plant species exhibit a set of traits that respond to light at different wavelengths (plant optical types, Ustin and Gamon 2010). Novel findings, however, show some limitations to the application of the spectral diversity principles at larger spatial resolutions (Schmidtlein and Fassnacht 2017), and more studies are needed to identify challenges and opportunities of this approach to mapping and monitoring biodiversity.

17.3.2   Vegetation Condition

There have been many efforts to move beyond species assessments toward functional aspects, which include vegetation condition. The current wealth of time series data from sensors like NOAA AVHRR, MODIS, or Landsat allows measurements of ecosystem phenology, seasonality, and changes in onset of seasons and assessment of ecosystem condition. Vegetation condition is the measurement of the vegetation response to stress (Liu and Kogan 1994). High or good condition corresponds to green, photosynthetically active vegetation, while stress such as water and nutrient limitations, pest outbreaks, and fire results in low or poor condition, after accounting for seasonal changes.

Physiologically, plants respond to stress by reducing chlorophyll activity and subsequently expressing other pigments. Plants also respond to stress by closing their stomata and reducing gas and water exchanges, which results in cells becoming turgid under water stress. These responses can be measured in the RS signal in both the visible and near infrared (NIR). In the visible the signal switches from a weak to a stronger reflectance signal due to chlorophyll absorption of red and blue wavelengths. Water stress can be measured in the NIR because as cells become turgid, they increase scattering of NIR radiation and therefore change the measured signal. One advance toward systematic measurements of vegetation condition is the Australian BioCondition (Lawley et al. 2016). This approach provides a framework to systematically assess terrestrial biodiversity condition—"[t]he similarity in key features of the regional ecosystem being assessed with those of the same regional ecosystem in its reference state"—using attributes like fraction of large trees, tree canopy height, recruitment of canopy species, native plant richness, size of patch, and connectivity. Thus, such in-situ measurements of vegetation condition can be linked to RS estimates to better provide an assessment of an ecosystem's stress level and ability to function and to provide services like habitat provisioning. In the next section, we will cover the later issue.

17.3.3   Habitat Intactness and Critical Transitions

Habitat intactness may be defined temporally or spatially as either (i) the degree to which the condition of the vegetation that forms habitat has not changed beyond what is expected from natural processes such as phenology and other dynamics or (ii) the spatial pattern of a given habitat, its degree of connectivity or fragmentation, and its edge extent. An active area of research on the potential of RS for conservation is the assessment of habitat intactness. Nagendra et al. (2013) employed RS to estimate how habitat has been changing in several regions of the Western Ghats in India. Coops et al. (2008, 2009) developed a dynamic habitat index using time series satellite data and showed its potential to monitor habitat condition in Canada. Later the approach was expanded to other regions, and dynamic habitat indices have been used as predictors of the richness of other taxa (Hobi et al. 2017). Coops et al. (2018) have now expanded it globally and have shown how these data sets mimic global biodiversity patterns. These data sets are currently available at http://silvis. forest.wisc.edu/data/DHIs-clusters/ and can be very useful to monitor protected area performance.

RS data are increasingly applied across large spatial scales to study stable state conditions of habitats and assess early warning signals for catastrophic shifts.

For example, the relationship between stable ecosystem states and rainfall can be inferred from the global probability density of forests extracted from remotely sensed forest cover and its relation to rainfall (Verbesselt et al. 2016). Temporal autocorrelation of NDVI time series and vegetation optical depth from radar over tropical forests indicated a reduced rate of recovery (critical slowing down) when tall canopy trees of intact forests under decreasing rainfall approached a tipping point, inducing high mortality. The rainfall threshold was at a similar level as the one indicated for stable state transitions from the spatial analysis. Similar to the temporal warning signals, the patch size distribution of spatially patterned ecosystems, such as arid ecosystems, showed a meltdown when approaching extinction (Dakos et al. 2011). High-resolution RS data across large spatial scales and patch size analysis can therefore be used to assess the extinction risk of these vulnerable ecosystems under decreasing rainfall conditions.

17.3.4 Protected Area Monitoring

RS plays an essential role in monitoring natural ecosystems, especially in protected areas. Human pressure over these areas has changed dramatically over the last decades (Geldmann et al. 2014), justifying a need for monitoring. Perhaps the two most influential papers that first demonstrated the usefulness of RS to monitor biodiversity in protected areas were Liu et al. (2001), which showed ecological degradation in protected areas designed to protect giant pandas (*Ailuropoda melanoleuca*), and Asner et al. (2005), which mapped deforestation in the Amazon with Landsat data and showed large rates of deforestation within legally designated protected areas. In 2007, the journal *Remote Sensing of Environment* published a special issue on monitoring protected areas in which a series of papers provided a framework for establishing monitoring programs, presented techniques and methods to make operational the use of remotely sensed data in protected area monitoring, and showcased a few examples linking remotely sensed data to models used to inform ecological assessments (Gross et al. 2008). RS can aid monitoring of many aspects of biodiversity (Cavender-Bares et al., Chap. 2; Gamon et al., Chap. 16) and ecosystem functioning within protected areas, including forest extent, land use/land cover change, local species pool and turnover, invasions (Bolch et al., Chap. 12), and carbon dynamics. The technical aspects of how RS can address some of these issues are presented in other chapters; here we review a few selected examples.

One of the major uses of RS in monitoring protected areas involves assessing land cover change and dynamics, for example, due to anthropogenic or natural disturbance. Liu et al. (2001) used Landsat data to estimate changes in forest cover and giant panda habitat before and after a reserve was created and showed how the Wolong Nature Reserve was becoming progressively more fragmented and how this resulted in crucial loss of habitat for the giant panda. Koltunov et al. (2009) showed how selective logging in the Amazon region led to different forest dynamics and to land cover change. Asner et al.'s (2005) seminal work was followed by global maps of deforestation and was also produced using the Landsat archive (Hansen et al. 2013), which have been used to track deforestation dynamics and development frontiers (Potapov et al. 2017). These data sets are now freely available for initiatives such as the Global Forest Watch (https://www.globalforestwatch.org/), in which the dynamics of deforestation can be monitored within and outside protected areas. These initiatives are fundamental to downscale global biodiversity goals to local scale action (Geijzendorffer et al. 2018). The quality of the forest classification in the Hansen et al. (2013) product is, however, limiting. In this data set, any pixel with >25% tree cover is considered a forest, and there is no distinction between naturally occurring forests and planted forests (e.g., eucalyptus or oil palm plantations) or forests planted for REDD+ programs. While these readily available products are fundamental to monitoring protected areas, they still rely on careful interpretation of their results on the ground. Asner and Tupayachi (2017) showed the extent of mining in the Amazon, and within this system, road development has long been shown to lead to deforestation and land cover changes within and outside protected areas. For example, Gude et al. (2007) showed how land use change around Yellowstone National Park could increase the risk to biodiversity inside the park. Svancara et al. (2009) showed areas surrounding US national parks are more protected and natural than areas farther away but had higher human population density and subsequently higher conversion risk for the parks' ecosystems and natural processes.

Such changes in land cover might result in changes in habitat availability and quality both outside and within protected areas. For example, Taubert et al. (2018) showed global patterns of tropical forest fragmentation follow a power-law distribution, which suggests that tropical forest fragmentation is close to a critical point. Santos et al. (2017) showed how the extent of habitat for small mammals in Yosemite National Park has changed in the last 100 years and that these habitat changes might in some cases counteract negative effects of climate change on species persistence. Platforms such as the Global Forest Watch (mentioned above) and the Global Surface Water Explorer from the EU Joint Research Centre show ways by which this integration may be achieved. Novel satellite configurations also show unexpected potential to monitor ecosystems and their responses to disturbance, providing potential new avenues for further integration at conservation-relevant scales.

17.3.5 Challenges

Two main challenges are more immediate in the conservation applications of RS. The first is that it is important to move beyond considering biodiversity as only number of species, and novel approaches looking at the four dimensions of biodiversity (genetic, species, function, and ecosystem structure) are necessary. Jetz et al. (2016) provide a framework for such an approach, and several chapters in this book (see Record et al., Chap. 10, in this book) already show the state of the art of the RS potential in these areas.

The second is that there is a growing need that conservation moves beyond protected areas to protected landscapes, which include livelihoods. Remotely sensed supporting services include habitat, nutrient cycling, terrestrial and aquatic primary production, soil formation, and provision of biological refugia, but there is an acute lack of RS applications in the study of cultural ES with the exception of cultural heritage and recreation (Andrew et al. 2014; de Araujo Barbosa et al. 2015; Cord et al. 2017). Similar to the concept of essential biodiversity variables, indicators have been developed to allow fast and efficient assessment of ES across large spatial scales. One of these indicators is the recently proposed Rapid Ecosystem Function Assessment (REFA, Meyer et al. 2015). REFA builds on a suite of core variables such as aboveground primary productivity, soil fertility, decomposition, and pollination. While many of these could theoretically be assessed using RS technology, the concept was developed for in-situ measured data. On the other hand, Cerreta and Poli (2017) propose a GIS-based framework with scalable and transferable methodology to rapidly assess multiple ecosystem functional features of a landscape using a multi-criteria spatial decision support system. While none of these approaches has been implemented for the use of RS technology at large spatial extents, there is potential for setting up near-real-time systems of fast ES assessment. One issue with extending temporal coverage and amount of ESs covered, despite increasing amounts and quality of remotely sensed data products, is the lack of ground data necessary to validate satellite outputs (Jones and Vaughan 2010). Alternatives to traditional validation approaches based on comparison with ground data are new methods such as the application of process models to test the consistency of time series of more complex satellite data products (Loew et al. 2017).

17.4 Data Availability and Issues

Large amounts of auxiliary data from RS and other sources are now freely available, together with the models and technology necessary to process disparate data in geospatial frameworks. To take advantage of the full range of aspects covered, increase the reliability of different data sets, and account for data uncertainty as much as possible, auxiliary data sources are often used in tandem with RS data. For example, satellite and census land use data can be integrated to scale administrative-level information to the globe (Ellis et al. 2013), and vegetation indices from satellites can be combined with local ecological knowledge to improve assessments of ecosystem degradation (Eddy et al. 2017). We provide an overview of the most frequently and widely integrated remotely sensed and auxiliary data products using approaches discussed at the end of this chapter. We focus on data sources that are useful in assessing plant diversity-related aspects at the landscape scale. In a recent review, Englund et al. (2017) found RS-related publications to use the term *landscape* rather loosely as describing studies at anything between 24 and 122 million ha. Here, we define landscape as referring to studies going beyond local, plot-level scales and generally not past country-level scales, although many methods and data products will be applicable to continental and global scales, too. Where the RS data products are described in other chapters, we only report on non-remotely sensed products; otherwise, references to both are given. Details of data, spatial, and temporal resolution as well as the formats available are given in Table 17.1.

17.4.1 In-Situ Biodiversity-Related Data

Decades of field work by dedicated researchers, their assistants, and students have resulted in collection of large amounts of biodiversity data, including plant species records on geographical location and abundance, traits of individual species, taxonomy, and phylogenetic data, as well as information on associated parameters such as pollination or dispersal and growth. This plethora of information is often scattered, hidden in scientific publications and a range of online (and offline) databases, herbaria, and agency reports. One means of retrieving relevant data semiautomatically from online sources is web scraping. Tools have been developed that make this a viable option for people who are not experts in languages routinely used for creating web pages and applications. These include several R packages (e.g., rvest, xml2, httr, TR8), Python libraries (e.g., Beautiful Soup), online tools (e.g., Nokogiri), and software assisting with identifying relevant CSS selectors on websites (e.g., Selector Gadget).

Recent efforts to cover this step and make dissemination of data more traceable, convenient, and standardized have resulted in large databases covering all the aspects discussed above. For example, large global databases exist for plant functional traits (e.g., TRY, Kattge et al. 2011), plant community data (species co-occurrences; sPlot, Dengler and sPlot Core Team 2014), plant phylogeny (e.g., TreeBASE, Smith and Brown 2018; Open Tree of Life), species distributions (e.g., Global Biodiversity Information Facility (GBIF), and botanical description and identification tools (e.g., JSTOR's Global Plants) to name just a few.

Some issues with using such large databases, however, are unavoidable. Regarding plant phylogeny data, one needs to be aware of the lack of molecular data associated with most species of plants resulting in many phylogenetic placements being based on data at the genus or even family level (Smith and Brown 2018). Where genetic data are available at the species level, large uncertainties with regard to the placement of many taxa remain (Smith and Brown 2018), and, increasingly, genetic sequences have not yet been linked to species names (so-called dark taxa). Species distribution data, on the other hand, are known to have an inherently large spatial sampling bias (see, e.g., Fig. 17.2) and generally lack absence data, which can inflate the effect of sampling bias even further (Barbet-Massin et al. 2012; Kramer-Schadt et al. 2013; Beck et al. 2014; Maldonado et al. 2015).

On the other hand, in the case of trait data, for example, TRY—a global database of plant functional traits—has been shown to be biased toward more extreme trait values, that is, frequently measured species consistently have higher or lower trait

Table 17.1 Non-exhaustive examples of key, global, open access auxiliary variables; some sources; their maximum spatial and temporal resolution; and the available data format

Variable		Sources	Maximum resolution		Data format
			Temporal	Spatial	
Soils	Physical/chemical	SoilGrids (Hengl et al. 2017)	Static	250 m	Raster
Geology	Lithology/mineralogy	GLIM (Hartmann and Moosdorf 2012)	Static	<10 km	Raster
	General	OneGeology	Static	Varied	Raster/vector
Topography	Land topography, ocean bathymetry	ETOPO1	Static	1 arc min	Raster
		SRTM	Static	1 arc sec	Raster
Hydrology	Vertical soil, groundwater reservoir	WaterGAP (Flörke et al. 2013)	Daily	5 arc min	Raster
	Permeability and porosity	GLHYMPS (Gleeson et al. 2014)	Static	100 km^2	Vector
	Global reservoirs/dams	GRanD (Lehner et al. 2011)	Static	Country	Vector
	Irrigated areas	Global irrigation maps (Siebert et al. 2013)	Static	5 arc min	Raster
	Water used for irrigation	FAOSTAT	Static	Country	Vector
	Rivers and lake centerlines	Natural Earth	Static	10 m	Vector
	Lakes and reservoirs	Natural Earth	Static	10 m	Vector
Climate	Temperature/ precipitation	CHELSA (Karger et al. 2017)	Monthly	30 arc s	Raster
		CRU TS v. 4.01 (Harris et al. 2014)	Monthly (1901–2016)	0.5°	Raster
	Precipitation	TRMM	3-hourly (1998–2016)	0.25°	Raster
		CHIRPS	Daily (1981–2016)	0.05°	Raster
	Evapotranspiration	MODIS, NASA	Yearly/8-day (2000–2010)	500 m	Raster

Category	Variable	Source	Temporal resolution	Spatial resolution	Data type
Socioeconomic	Population density	The World Bank	Yearly	Country	Vector
		NASA	5-yearly	30 arc s	Raster/vector
	Health/Demography	The World Bank	Yearly	Country	Vector
		Global Health Data Exchange/WHO	Yearly	Country	Vector
	Nutrition	The World Bank	Yearly	Country	Vector
	GDP	The World Bank	Yearly	Country	Vector
	Education	The World Bank	Yearly	Country	Vector
		UNESCO	Yearly to 5-yearly	Country	Vector
	Road networks	OpenStreetMap	Minutely-monthly	Variable	Vector
	Roads/railroad/airports/ports	Natural Earth	Static	10 m	Vector
Land use	Irrigated areas	International Water Management Institute	Static	10 km	Raster
	Land use	LandSense (Fritz et al. 2017)	Static (2011–2012)	300 m–1 km	Raster
	Wilderness and human impact	LandSense (Fritz et al. 2017)	Static (2011–2012)	300 m–1 km	Raster
	Global map of rainfed cropped areas	International Water Management Institute	Static	10 km	Raster
	Protection status	World database on protected areas (UNEP-WCMC 2017)	Monthly	Variable	Vector
Land cover	Land cover	USGS	Static (1992–1993)	1 km	Raster
		ESA GlobCover V2	Static	300 m	Raster
		LandSense (Fritz et al. 2017)	Static (2011–2012)	300 m–1 km	Raster
		MODIS12C1	Yearly (2001–2013)	500 m	Raster
		FAO SHARE	Static	1 km	Raster
		National Geomatics Center of China GlobeLand30	Static	30 m	Raster
		Corine (European Environment Agency)	Static (1990, 2000, 2006, 2012, 2018)	Country	Raster/vector
	Mosaic and forest/non-forest	Japan Aerospace Exploration Agency PALSAR	Static (2007–2010)	25 m	Raster
	Global forest change	University of Maryland (Hansen et al. 2013)	Yearly (2000–2017)	30 m	Raster

(continued)

Table 17.1 (continued)

Variable		Sources	Maximum resolution			Data format
			Temporal	Spatial		
Related to in-situ plant diversity	Functional traits	TRY database (Kattge et al. 2011)	Static	Varied		Vector, point
	Fine root traits	FRED (Iversen et al. 2018)	Static	Varied		Vector, point
	Co-occurrences	sPlot database (Dengler and sPlot Core Team 2014)	Static	Plot location		Vector
	Phylogeny (no branch lengths)	Open Tree of Life	Static	None		Point
	Seed plant phylogeny (with branch lengths)	Smith and Brown (2018)	Static	None		Point
	Taxonomy	iNaturalist (crowd sourced)	Static	None		ID
		Angiosperm Phylogeny Group	Static	None		ID
		Global Plants (JSTOR)	Static	None		ID
	Biodiversity change	Dornelas et al. (2018)	1874–2016	158 cm²– 100 km²		Vector
	Species distributions	Global Biodiversity Information Facility (GBIF)	Static	Varied		Vector, raster, polygon

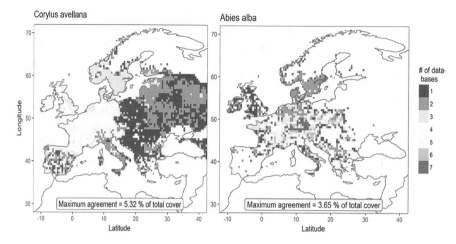

Fig. 17.2 Number of species distribution databases reporting the presence of two tree species: Abies alba *(right)* and Corylus avellana *(left)* across Europe. Color shows 0.5° raster including geo-references of A. alba or C. avellana presence, ordered from red (only one database reports presence of the species in this pixel) to dark blue (all seven examined databases report presence of the species in this pixel). Note the strong country border-related pattern for C. avellana. The number of pixels as a percentage of the total number of "presence pixel" where all seven databases agree is indicated in the plot (3.65% and 5.32%). Seven species distribution databases covering Europe were analyzed, including the Atlas Florae Europaeae, GBIF (status Nov. 2017), European Vegetation Archive, EUFORGEN (EU Forestry Commission), data from Brus et al. (2012), data collated by colleagues from the University of Leipzig (DE), and the FunDivEUROPE project (Baeten et al. 2013)

values than species missing in TRY (Sandel et al. 2015). Although plant functional traits are conventionally measured at the peak of the growing season and in full light conditions (top of canopy) (Pérez-Harguindeguy et al. 2013), one of the most commonly measured traits, specific leaf area (SLA) has been shown to have values in TRY that are typical of partial canopy shading (Keenan and Niinemets 2016). Due to the extremely diverse nature of studies contributing data to the TRY database, most entries have on average only three traits measured simultaneously, which makes multivariate analyses at the individual plant level extremely challenging (Schrodt et al. 2015).

These issues are mainly due to studies represented within these databases not necessarily following standardized protocols (e.g., Pérez-Harguindeguy et al. 2013), studies having different foci, data from opportunistic sampling (Maes et al. 2015) being mixed with data from directed approaches, and rare species, from a purely statistical viewpoint, being less likely to be measured. As such, avoiding them at the database level, especially where such a large number of data entries are managed in open access databases (e.g., Version 4 of TRY contained almost seven million trait records), is currently virtually impossible. An additional challenge when using trait data in tandem with RS is the lack of geo-referenced measurements within trait databases. For example, only about 60% of all data points within TRY are geo-referenced with variable levels of precision.

This leaves it to the data user to work around and with the data issues. Means of dealing with some of the aforementioned challenges regarding data availability and quality are (i) gap filling of missing trait data (Swenson 2014; Schrodt et al. 2015) and (ii) spatial extrapolation of plant traits (Butler et al. 2017), accounting for "dark diversity," i.e., the portion of species absent from species distribution data such as biodiversity maps (Ronk et al. 2015). Care must be taken to avoid circularity (e.g., using phylogeny to gap fill trait data when including an aspect of taxonomy or phylogeny in subsequent analyses). In addition, a possible lack of representativeness throughout analyses should be considering, as well as the fact that different approaches might require different data collection protocols (e.g., statistical versus process models).

17.4.2 In-Situ Abiotic Factors

Much abiotic information, including data from the lithosphere, atmosphere, hydrosphere, and cryosphere, can be assessed remotely (see Record et al., Chap. 10, in this book for a thorough discussion, including access to climatological data). However, many important aspects are only accessible from in-situ sources. These include, for example, soil chemical and physical characteristics, geomorphology, and subsurface hydrology (see Table 17.1). Many are available at static temporal but relatively high spatial resolutions with high associated uncertainties in geolocation, bias due to different sampling efforts depending on the location, etc. For example, Generalized Linear Interactive Modelling (GLIM), a lithology and mineralogy data source, has been shown to be highly biased by country boundaries—an issue that is perpetuated in other products using GLIM, such as the SoilGrids database (Hengl et al. 2017), resulting in error propagation to higher-level agglomerate analyses. Other challenges include breakdown of concepts and assumptions related to up- and downscaling of composite products (e.g., inter-cell redistribution of soil water at fine spatial resolution, which can be ignored at coarser resolutions) and a lack of knowledge about parameters and processes acting at different resolutions (Bierkens et al. 2015).

17.4.3 Socioeconomic Factors and Land Use

Socioeconomic aspects are often ignored in assessments of plant biodiversity at the landscape scale, despite the obvious imprint humans have left on most of the globe. For example, Abelleira Martínez et al. (2016) found that studies linking local plant trait measurements to environmental gradients without accounting for anthropogenic effects on these traits render them of limited use due to the multivariate nature of the processes governing observed patterns. The same applies to studies integrating in-situ plant trait variability with land cover types, e.g., for ES assessments without explicitly taking into account human modification of the landscape through management, engineered novel communities, land use history, and heterogeneous landscapes (Abelleira Martínez et al. 2016). Anthropogenic aspects that can be assessed easily and incorporated into landscape scale analyses include data on population density, socioeconomics, and pollution (see Table 17.1).

Land use is another important anthropogenic aspect but less easily assessed at the landscape scale using RS techniques alone. This is mainly due to the same land cover (which refers to the physical characteristics of a landscape) having the potential of belonging to different land use categories (referring to the human use of this landscape). For example, the land cover class *forest* could be within the land use category *natural primary forest* or *heavily managed degraded forest*. Consequently, large-scale land use mapping depends on auxiliary data such as that coming from open access crowdsourced land cover and land use data to improve ground-truthing and validation (Fritz et al. 2017) and a combination of RS data sources, such as fusion of spaceborne optical data with radar data (Joshi et al. 2016).

At the other end of the spectrum of land uses are the human modified, urbanized, and infrastructure types such as cities and roads. RS has demonstrated a great potential to map impervious surface and more limited success in detecting roads. There has been a growing interest in urban ecology because more than half of the global population now lives in cities, and there is a growing interest to increase healthy urban living that combines well-being and biodiversity (Botzat et al. 2016). The first step to reach this goal is to create urban green belts (Hostetler et al. 2011), which are expected to bring about increasing numbers of native species and increased connectivity (Aronson et al. 2017). However, urban areas are also linked to high richness (Gavier-Pizarro et al. 2010) and spread of invasive species (Hui et al. 2017), and small urban centers are sources of invasive plants into natural areas (McLean et al. 2017).

RS of urban (invasive) plant species is covered in another chapter in this book (see Bolch et al., Chap. 12). Roads are more difficult to retrieve with RS alone, although the fishbone patterns in the Brazilian Amazon are evident in Landsat data (Alves and Skole 1996). Initiatives like OpenStreetMap can provide auxiliary data to improve the accuracy of RS-only estimates. These data are fundamental to assess global roadless areas and fundamental for maintaining biodiversity processes and avoiding deleterious effects of fragmentation.

17.4.4 Land Cover

Among the most traditional applications of RS are those related to the estimation of biophysical variables (e.g., tree density, vegetation health). AVHRR, Landsat, Sentinel-2, and MODIS are the most widely used sensors for this purpose, but integration of optical RS with LiDAR technology significantly improves the estimation and assessment of vegetation structure due to added horizontal and vertical information of vegetation properties (e.g., canopy height) (Lim et al. 2003). Studies combining optical, LiDAR, and radar RS have been applied to study interactions between biotic (i.e., vegetation) and abiotic (i.e., soil, geomorphology) elements at landscape scales and to quantify the carbon cycle and biomass.

Land cover is another commonly analyzed measure (in their recent review, Ma et al. (2017) analyzed 254 experimental cases and 173 scientific papers on the subject) that has been shown to be highly sensitive to the classification method applied, with the optimal approach depending to a large extent on spatial resolution, differences between land cover types, and training set size. Land cover maps are often used to derive landscape structural features such as patch size, isolation, and perimeter-to-area ratio.

Such landscape metrics can be assessed using patch matrix models (PMM), which are most suitable for high-hemeroby (low naturalness and high anthropogenic pressure, e.g., urban) landscapes due to reduced spatiotemporal heterogeneity, while gradient models (GM) are recommended for low-hemeroby landscapes (e.g., undisturbed forest) (Lausch et al. 2015). While PMMs are relatively well established and easy to use, disadvantages include that heterogeneity information might be lost, patches tend to have sharp boundaries, and results are highly sensitive to misclassifications of land cover and use metrics (Lausch et al. 2015). GMs, on the other hand, are more complex to use and require more computing capacity and RS expertise while being less susceptible to loss of heterogeneity information and artificially sharp boundaries (Lausch et al. 2015). Both models use a variety of data as inputs, including hyperspectral and LiDAR RS as well as in-situ data, thereby taking full advantage of opportunities offered by each methodology.

17.5 Methods to Integrate Remotely Sensed Measures of Plant Biodiversity with In-Situ Plant Diversity, Abiotic, and Socioeconomic Data

Studies of plant diversity at the landscape level frequently require a mix of data sources from various sensors as well as in-situ data (Table 17.1, Fig. 17.3). There are thus three main reasons for integrating measures of biodiversity-related variables across different data sources: (i) combining data from different sensors to make use of different vegetation aspects measured (e.g., MODIS vs. Advanced Land Observation Satellite Phased Array type L-band Synthetic Aperture Radar (ALOS PALSAR)); (ii) combining different sensors to simulate higher spatiotemporal and spectral resolutions to save financial resources or account for gaps in available RS data (e.g., Zeng et al. 2017); and (iii) combining in-situ with RS data for upscaling and validation.

The process of integrating, combining, and correlating data from different sensors and data types of different temporal and spatial scales is not straightforward. Challenges are numerous and include sensor calibration, the propagation of uncertainties from individual data sets with inherent and variable uncertainty and impreciseness, outliers and spurious data, bias due to spatial autocorrelation, differences

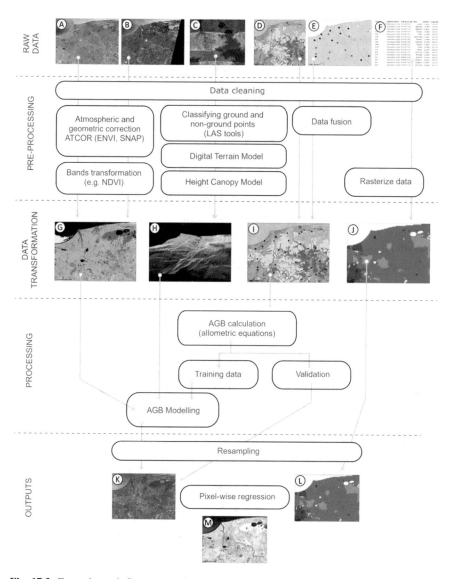

Fig. 17.3 Example work flow to correlate biomass (**k**) with human population density (**f**). Raw data from a variety of sources needs to be integrated, including Sentinel-2 (**a**), Landsat (**b**), and LiDAR (**c**) from RS as well as modeled land cover (**d**, INEGI 2013), plot level in-situ biomass measurements (**e**), and raw population density data (**f**, GPWv4 2016). After general data checking and cleaning (which is advisable for any data source), atmospheric and geometric corrections are performed on the remotely sensed data using software such as ENVI or SNAP, followed by transformation of the bands—in this case, calculation of the NDVI vegetation index (**g**). Radar data are classified into ground and nonground points using LAStools software, followed by application of a digital terrain and height canopy model to derive a canopy height map (**h**). Aboveground biomass (AGB, **k**) is calculated using the NDVI, canopy height, and (to validate the model) ground data and vegetation map (**i**). (**j**) Rasterized population density map, (**m**) pixelwise regression between (**k**) and (**l**). Please note no visible difference between (**j**) and (**l**) due to resampling, resulting in only small changes in pixel size

in geospatial data registration and alignment, and different processing frameworks (see Quattrochi et al. 2017 for a thorough discussion).

Figure 17.3 presents an example workflow that depicts the steps required to assess if there is a correlation between biomass and human population density in a coastal area of Mexico. In this case, since biomass cannot be inferred directly from RS sources (as described elsewhere in this book), several preprocessing and processing steps are required. Preprocessing steps include data preparation (data cleaning, atmospheric and geometric corrections), data transformation (e.g., from tabular into rasterized data) and fusion, running of auxiliary methods such as classification methods, and application of digital terrain models and height canopy models. Processing steps incorporate the use of allometric equations using in-situ plot level measurements of plant biomass in that area, which are also used as training and validation data to formulate the final aboveground biomass model fusing RS with in-situ data. Here, we present examples of techniques dealing with some of the abovementioned challenges in aligning different sensors, fusing data from these different sources across space and time, training fusion methods, and validating results.

17.5.1 Fusion

Data fusion is an invaluable tool to assess patterns and processes of biodiversity at large spatial scales and integrate data across different aspects of remotely sensed plant diversity, abiotic, and socioeconomic factors. Data fusion allows integration of data from different sensors and of diverging spatial, spectral, and temporal extent to produce outputs of increased fidelity and usefulness. Fusion is often performed to account for limitation in one data source, e.g., where single data rather than time series data are available in the sensor of interest (Carreiras et al. 2017) or to resample low-resolution data from one spaceborne RS channel using data from another, high-resolution channel on the same sensor.

In the example mentioned above (Fig. 17.3), the authors chose to use optical data from MODIS and radar data from ALOS PALSAR, and for their successful fusion, it is important to consider the different spatial resolutions, temporal data availability, and sensor characteristics (Fig. 17.1). This kind of constellation is frequently used to map a range of land cover and land use characteristics, including change, conversion, and modification where detailed information on both broad land cover classes from optical data and detailed surface roughness and moisture information from radar images are required (Pereira et al. 2013; Dusseux et al. 2014; Stefanski et al. 2014).

Different fusion techniques are applied to spaceborne or airborne sensors. For example, Sankey et al. (2018) described an approach to fuse unmanned aerial vehicle (UAV) LiDAR with hyperspectral data using a decision tree classification technique and found the combined use of these sensors provided more accurate assessments of 3D analyses of plant characteristics and plant species identification at submeter spatial resolutions.

A common application of fusion is an increase of spatial or spectral resolution, either accounting for limitations in the available data or imitating high-cost systems using low-cost alternatives, e.g., in precision agriculture. For example, Zeng et al. (2017) developed a system imitating very high spatial resolution hyperspectral measurements such as those required for the calculation of some vegetation indices (VI) using low-cost UAV-mounted sensors and fusing multispectral imagery with spectrometer data using Bayesian imputation and principal component analysis. Data fusion can also be applied when linking RS data to in-situ data. In their recent review, Lesiv et al. (2016) compared different algorithms fusing RS with crowd-sourced data for forest cover mapping, including geographically weighted logistic regression (GWR), naïve Bayes, nearest neighbor, logistic regression, and classification and regression trees (CART), finding GWR to perform slightly better where input data were disparate.

In its simplest form, fusion can be a basic overlay of high- (spectral/spatial/temporal) resolution data over low-resolution data. However, as Lesiv et al. (2016) and others have shown, it is worth comparing different fusion techniques. Several studies have performed such comparisons but mainly with respect to land cover classification and specific to certain sensors and spatiotemporal scales (e.g., Caruana and Niculescu-Mizil 2006, Clinton et al. 2015). Consequently, Liu et al. (2017) recommend routine use of statistical comparisons between different fusion techniques (e.g., a Wilcoxon signed-ranks test for two algorithms or a Friedman test with Iman and Davenport extension if more than five algorithms are compared) to detect the optimal solution for a given application.

17.5.2 Assimilation

In essence, data assimilation is an extension of data fusion, linking noisy RS measurements with the outputs from imperfect numerical models to optimize estimates of measures that are not directly observable from RS [e.g., for detailed, high spatiotemporal drought monitoring (Ahmadalipour et al. 2017) or, in the example given in Fig. 17.3, to derive biomass estimates using a combination of canopy height, digital terrain, and aboveground biomass (AGB) modeling]. Advantages of data assimilation include enhanced quality control, the ability to take into account errors and uncertainties in data and models simultaneously, gap filling in data-poor locations and where insufficient temporal information is available, and improved parameter estimation in models.

However, data assimilation can also result in circular and inconsistent analyses. The end user needs to be aware that many remotely sensed variables [e.g., leaf area index (LAI)] are based on models incorporating ancillary information and are thus not independently retrieved. In our example, land cover might already be used as an information layer to tune the RS LAI retrieval within the data assimilation step. Thus, using LAI as biodiversity variable and adding land cover as an explanatory variable could be problematic (inconsistent if from different sources or circular).

This illustrates the importance of carefully considering all input variables at different steps of the data fusion and assimilation process before using a RS product for further analyses.

A range of data assimilation techniques is available, from univariate (scalar) and multivariate (vector) 3D and 4D Kalman filter to ensemble methods that are particularly suitable where large sets of parameters are required and models are complex (for details see, e.g., Bouttier and Courtier 2002; Evensen 2002).

17.5.3 Validation

Like any other source of data, spaceborne remotely sensed products have errors and uncertainties associated with them. Validation is thus an ongoing challenge, although a number of guidelines and recommendations for best practice exist. For example, the NASA Land Product Validation Subgroup has published a framework for product validation and inter-comparison, as well as a "guide to the expression of uncertainty in measurement" (Schaepman-Strub et al. 2014). In the case of remotely sensed LAI, the list of auxiliary parameters with associated uncertainties that should be considered when validating LAI measures is long. It includes input data [land cover, radiometric calibration error, geometry, aerosol optical depth at 550 nm, canopy condition (chlorophyll, dry matter, and moisture content), understory reflectance and geolocation (sensitivity to terrain slope), sensor noise (especially for dark targets such as dense vegetation), clear sky top-of-atmosphere radiance, bidirectional reflectance distribution function (BRDF) modeling uncertainty, canopy and understory modeling uncertainty, and geometric considerations (where products are gridded in map projection systems of varying shape and area, Fernandes et al. 2014). A detailed overview of validation techniques used across different levels of RS data is given in Zeng et al. (2015), and guidelines on terminology, unified satellite validation metrics, and strategies, as well as explicit examples of RS validation techniques, including their mathematical basis, are provided in Loew et al. (2017). Luckily for the end user, many of these validation steps are performed by the respective satellite agencies (e.g., the European Space Agency (Dorigo et al. 2017) and NASA (Justice et al. 2013). However, being aware of the complexity of this endeavor and the importance of considering both the target variable (e.g., LAI) and its associated quality measure (e.g., uncertainty) as provided by the space agencies is of utmost importance to ensure appropriate use of RS products.

Apart from validating spaceborne RS data, validation techniques are also used to evaluate the quality of modeled secondary indices, as well as to assess uncertainty propagation after data fusion [e.g., accounting for uncertainty due to variable data quality of in-situ or crowdsourced data (see, e.g., Comber et al. 2016) or to validate downscaled RS products and airborne RS products]. For instance, crowdsourced data have been used to validate a high-resolution global land cover map (Fritz et al. 2017), and in-situ measurements of LAI collected simultaneously with airborne hyperspectral images were used to validate canopy radiative transfer models in agricultural landscapes (Haboudane et al. 2003). In its simplest form, validation is a pixelwise comparison of presumably high accuracy (often in-situ) data with the remotely sensed or modeled data, using an x-fold validation approach (splitting the in-situ data into training and validation data) if some of the in-situ data are needed for model development or downscaling. For an overview of more complex techniques see, for example, Montesano et al. (2016) for Landsat-derived tree cover, Lesiv et al. (2016) for crowdsourced forest cover, Joshi et al. (2016) for optical- and radar-derived land use, and Sun et al. (2017) for in-situ validated land cover.

With the rapidly growing availability of RS and auxiliary data, validation can become a time-consuming and complex task. Thus, increasingly, web-based validation systems are being developed that integrate big data access and storage, adjustment, and different intercomparison and validation techniques simultaneously (e.g. Sun et al. 2017).

One of the potential issues with the data fusion and assimilation methods described above is that they are often applied globally without testing whether variables and correlations remain stable in space and time (Comber et al. 2012). This is a recognized problem, and solutions have been proposed for over a decade (e.g., geographically distributed correspondence matrices, Foody 2005) with new approaches being continually proposed. Often these are specific for certain applications, such as net primary production (Wang et al. 2005), epidemiology (Khormi and Kumar 2011), biomass (Propastin 2012) or population segregation (Yu and Wu 2013). One recently proposed more generic approach is that of locally geographically weighted correspondence matrices, which combine categorical difference measures (Pontius and Milones 2011, Pontius and Santacruz 2014) with spatially distributed kappa coefficient, user, and producer accuracy estimates—with code to run these tests in R being available, e.g., see packages gwxtab, differ and RSLcode (available on github (https://github.com/lexcomber/RSLcode)) (Comber et al. 2017). All of these draw attention to the fact that, even after performing data cleaning, fusion, assimilation, and validation steps, local approaches, data, and techniques cannot necessarily be directly transferred from one location and spatiotemporal resolution to another.

17.6   Conclusions

We are living in an increasingly data-rich world, in which not only more but also more accurate and reliable data are available on many aspects related to plant biodiversity, both from remotely sensed as well as in-situ measurements. Increasingly, limitations and potential circularities inherent to these data are acknowledged, often aided by the provision of associated estimates of uncertainties and dedicated intercomparison studies. Techniques are being developed that enable even nonexperts to account for and learn from these. Even in the case of data limitations, however, our ability to map all aspects of biodiversity over large spatial and temporal scales has increased exponentially over the last decade, and monitoring and understanding ecosystem functions is easier than ever before. Nevertheless, significant challenges remain, including scaling mismatches, misuse of data and techniques, insufficiently high spatiotemporal resolution of RS data, biases in in-situ data, and many more. It is imperative that we acknowledge and work with these challenges to devise even more accurate and suitable approaches to assessing biodiversity for the study of ecosystem function, conservation, and other applications at large spatial scales.

Acknowledgments FS was supported by an Anne McLaren fellowship by the University of Nottingham (UK). We thank Nathalie Pettorelli for useful discussions and suggestions. CW acknowledges support from the British Geological Survey. MJS acknowledges the support of the University Research Priority Program in Global Change and Biodiversity and the Department of Geography at the University of Zurich. We also thank the NIMBioS Working Group on Remote Sensing of Biodiversity.

References

Abelleira Martínez OJ et al (2016) Scaling up functional traits for ecosystem services with remote sensing: concepts and methods. Ecol Evol 613:4359–4371. https://doi.org/10.1002/ece3.2201

De Araujo Barbosa CC, Atkinson PM, Dearing JA (2015) Remote sensing of ecosystem services: a systematic review, ecological indicators. Elsevier Ltd 52:430–443. https://doi.org/10.1016/j.ecolind.2015.01.007

Alves DS, Skole LD (1996) Characterizing land cover dynamics using multi-temporal imagery. Int J Remote Sens 17:835–839

Ahmadalipour A, Moradkhani H, Yan H, Zarekarizi M (2017) Remote sensing of drought: vegetation, soil moisture, and data assimilation. In: Remote sensing of hydrological extremes, pp 121–149

Andrew ME, Wulder MA, Nelson TA, Coops NC (2014) Spatial data, analysis approaches, and information needs for spatial ecosystem service assessments: a review. GIScience Remote Sens 52:344–373

Aronson MFJ, Patel MV, ONeill KM, Ehrenfeld JG (2017) Urban riparian systems function as corridors for both native and invasive plant species. Biol Invasions 19:3645–3657

Asner GP et al (2005) Selective logging in the Brazilian Amazon. Science 310(5747):480–482

Asner GA, Olinger SV (2009) Remote sensing for terrestrial biogeochemical modelling. In: Warner TA, Nellis MD, Foody GM (eds) The SAGE handbook of remote sensing. SAGE, London, pp 411–422

Asner GP, Tupayachi R (2017) Accelerated losses of protected forests from gold mining in the Peruvian Amazon. Environ Res Lett 12:094004

Baeten L et al (2013) A novel comparative research platform designed to determine the functional significance of tree species diversity in European forests. Persp Pl Ecol Evol Syst 155:281–291. https://doi.org/10.1016/j.ppees.2013.07.002

Barbet-Massin M et al (2012) Selecting pseudo-absences for species distribution models : how, where and how many? Methods Ecol Evol 3:327–338. https://doi.org/10.1111/j.2041-210X.2011.00172.x

Beck J et al (2014) Spatial bias in the GBIF database and its effect on modeling species geographic distributions. Eco Inform 19:10–15

Berger M, Aschbacher J (2012) Preface: The Sentinel missions—new opportunities for science. Remote Sens Environ 120:1–2

Bierkens MFP et al (2015) Hyper-resolution global hydrological modelling: what is next?: "Everywhere and locally relevant" M. F. P. Bierkens et al. Invited Commentary. Hydrol Process 292:310–320. https://doi.org/10.1002/hyp.10391

Bischof R, Loe LE, Meisingset EL, Zimmentmann B, van Moorter B, Mysterud A (2012) A migratory northern ungulate in the pursuit of Spring: jumping or surfing the green wave? Am Nat 180:407–424

Botzat A, Fischer LK, Kowarik I (2016) Unexploited opportunities in understanding liveable and biodiverse cities. A review on urban biodiversity perception and valuation. Glob Environ Chang 39:220–233

Bouttier F, Courtier P (2002) Data assimilation concepts and methods. Meteorol Train Course Lect Ser 1–58

Bouvet A, Mermoz S, Le Toan T, Villard L, Mathieu R, Naidoo L, Asner GP (2017) An aboveground biomass map of African savannahs and woodlands at 25 m resolution derived from ALOS PALSAR. Remote Sens Environ 206:156–173

Boyd DS et al (2012) Evaluation of Envisat MERIS terrestrial chlorophyll index-based models for the estimation of terrestrial gross primary productivity. IEEE Geosci Remote Sens Lett 93:457–461. https://doi.org/10.1109/LGRS.2011.2170810

Boyd DS, Foody GM (2011) An overview of recent remote sensing and GIS based research in ecological informatics, Ecological Informatics. Elsevier BV 61:25–36. https://doi.org/10.1016/j.ecoinf.2010.07.007

Bradley BA, Fleishman E (2008) Can remote sensing of land cover improve species distribution modelling? J Biogeogr 35:1158–1159

Brown TB et al (2016) Using phenocams to monitor our changing earth: toward a global phenocam network. Front Ecol Environ 142:84–93. https://doi.org/10.1002/fee.1222

Brus DJ et al (2012) Statistical mapping of tree species over Europe. Eur J For Res 1311:145–157. https://doi.org/10.1007/s10342-011-0513-5

Buitenwerf R, Rose L, Higgins SI (2015) Three decades of multi-dimensional change in global leaf phenology. Nat Clim Chang 54:364–368. https://doi.org/10.1038/nclimate2533

Bustamante MMC et al (2016) Toward an integrated monitoring framework to assess the effects of tropical forest degradation and recovery on carbon stocks and biodiversity. Glob Chang Biol 221:92–109. https://doi.org/10.1111/gcb.13087

Butler EE et al (2017) Mapping local and global variability in plant trait distributions. Proc Natl Acad Sci 114(51):201708984. https://doi.org/10.1073/pnas.1708984114

Cabello J et al (2012) The ecosystem functioning dimension in conservation: insights from remote sensing. Biodivers Conserv 2113:3287–3305. https://doi.org/10.1007/s10531-012-0370-7

Carreiras JMB, Jones J, Lucas RM, Shimabukuro YE (2017) Mapping major land cover types and retrieving the age of secondary forests in the Brazilian Amazon by combining single-date optical and radar remote sensing data. Remote Sens Environ 194:16–32

Caruana R, Niculescu-Mizil A (2006) An empirical comparison of supervised learning algorithms. In: Proceedings of the 23rd International Conference on Machine Learning, Pittsburgh, pp 161–168

Cerreta M, Poli G (2017) Landscape services assessment: a hybrid Multi-Criteria Spatial Decision Support System (MC-SDSS). Sustainability 9:1310–1328

Choa MA, Mathieu R, Asner GP, Naidoo L, Aardt J, Ramoelo A, Debba P, Wessels K, Main R, Smit IPJ, Erasmus B (2012) Mapping tree species composition in south African savannas using an integrated airborne spectral and LiDAR system. Remote Sens Environ 125:214–226

Chave J et al (2014) Improved allometric models to estimate the aboveground biomass of tropical trees. Glob Chang Biol 20:3177–3190

Claverie M, Ju J, Masek JG, Dungan JL, Vermote EF, Roger J-C, Skakun SV, Justice C (2018) The harmonized landsat and sentinel-2 surface reflectance dataset. Remote Sens Environ 219:145–161

Cleland EE, Chuine I, Menzel A, Mooney HA, Schwartz MD (2007) Shifting plant phenology in response to global change. Trends Ecol Evol 22:357–365

Clinton N, Yu L, Gong P (2015) Geographic stacking: decision fusion to increase global land cover map accuracy. Glob L Cover Mapp Monit 103:57–65

Comber A, Brunsdon C, Charlton M, Harris P (2017) Geographically weighted correspondence matrices for local error reporting and change analyses: mapping the spatial distribution of errors and change. Remote Sens Lett 8:234–243

Comber A, Fisher P, Brunsdon C, Khmag A (2012) Spatial analysis of remote sensing image classification accuracy. Remote Sens Environ 127:237–246

Comber A, Mooney P, Purves R, Rocchini D, Walz A (2016) Crowdsourcing: it matters who the crowd are. The impacts of between group variations in recording land cover. PLoS One 11:e0158329

Coops NC, Wulder MA, Duro DC, Han T, Berry S (2008) The development of a Canadian dynamic habitat index using multi-temporal satellite estimates of canopy light absorbance. Ecol Indic 8:754–766

Coops NC, Waring RH, Wulder MA, Pidgeon AM, Radeloff VC (2009) Bird diversity: a predictable function of satellite-derived estimates of seasonal variation in canopy light absorbance across the United States. J Biogeogr 365:905–918

Cord AF et al (2017) Priorities to advance monitoring of ecosystem services using earth observation, trends in ecology & evolution. Elsevier Ltd 326:416–428. https://doi.org/10.1016/j.tree.2017.03.003

Dakos V, Kefi S, Rietkerk M, van Nes EH, Scheffer M (2011) Slowing down in spatially patterned ecosystems at the brink of collapse. Am Nat 177:E153–E166

De Frenne P et al (2018) No title. Methods Ecol Evol 00:1–9

De Groot RS, Wilson MA, Boumans RMJ (2002) A typology for the classification, description and valuation of ecosystem functions, goods and services. Ecol Econ 41:393

Dengler J, sPlot Core Team (2014) sPlot: the first global vegetation-plot database and opportunities to contribute. IAVS Bulletin 2:34–37

Dodge S, Bohrer G, Weinzierl R, Davidson SC, Kays R, Douglas D, Cruz S, Han J, Brandes D, Wikelski M (2013) The environmental-data automated track annotation Env-DATA system: linking animal tracks with environmental data. Mov Ecol 1:3. https://doi.org/10.1186/2051-3933-1-3

Dorigo E, Wagner W, Albergel C, Albrecht F, Balsamo G, Brocca L, Chung D, Ertl M, Forkel M, Gruber A, Haas E, Hamer P, Hirschi M, Ikonen J, de Jeu R, Kidd R, Lahoz W, Liu Y, Miralles D, Mistelbauer T, Nicolai-Shaw N, Parinussa R, Pratola C, Reimer C, van der Schlie R, Senebiratne S, Smolander T, Lecomte P (2017) ESA CCI Soil Moisture for improved Earth system understanding: state-of-the art and future directions. Remote Sens Environ 15:185–215

Dornelas M et al (2018) BioTIME: a database of biodiversity time series for the anthropocene. Glob Ecol Biogeogr 27:760. https://doi.org/10.1111/geb.12729

Duarte L et al (2018) QPhenoMetrics: an open source software application to assess vegetation phenology metrics, computers and electronics in agriculture. Elsevier 148May 2016:82–94. https://doi.org/10.1016/j.compag.2018.03.007

Dusseux P et al (2014) Combined use of multi-temporal optical and radar satellite images for grassland monitoring. Remote Sens 6:6163–6182

Eddy IMS et al (2017) Integrating remote sensing and local ecological knowledge to monitor rangeland dynamics. Ecol Indic 82May:106–116. https://doi.org/10.1016/j.ecolind.2017.06.033

Eklundh L, Jonsson P (2015) TIMESAT: a software package for time-series processing and assessment of vegetation dynamics. In: Remote sensing time series. Springer, New York, pp 141–158. https://doi.org/10.1007/978-3-319-15967-6_7

Ellis EC et al (2013) Used planet: a global history. Proc Natl Acad Sci 11020:7978–7985. https://doi.org/10.1073/pnas.1217241110

Englund O, Berndes G, Cederberg C (2017) How to analyse ecosystem services in landscapes – a systematic review, ecological indicators. Elsevier Ltd 73:492–504. https://doi.org/10.1016/j.ecolind.2016.10.009

Evensen G (2002) Data assimilation: the ensemble Kalman filter. Springer-Verlage, Berlin

Fassnacht FE, Latifi H, Stereńczak K, Modzelewska A, Lefsky M, Waser LT, Straub C, Ghosh A (2016) Review of studies on tree species classification from remotely sensed data. Remote Sens Environ 186:64–87

Feilhauer H et al (2016) Mapping pollination types with remote sensing. J Veg Sci 275:999–1011. https://doi.org/10.1111/jvs.12421

Feret JB, Asner GP (2014) Mapping tropical forest canopy diversity using high-fidelity imaging spectroscopy. Ecol Appl 24:1289–1296

Fernandes R, Plummer S, Nightingale J, Baret F, Camacho F, Fang H, Garrigues S, Gobron N, Lang M, Lacaze R, LeBlanc S, Meroni M, Martinez B, Nilson T, Pinty B, Pisek J, Sonnentag O, Verger A, Welles J, Weiss M, Widlowski J, Schaepman-Strub G, Roman M, Nicheson J (2014) Global leaf area index product validation good practices. Best Pract Satell L Prod Validation L Prod Valid Subgr:1–78

Flörke M, Kynast E, Bärlund I, Eisner S, Wimmer F, Alcamo J (2013) Domestic and industrial water uses of the past 60 years as a mirror of socio-economic development: a global simulation study. Glob Environ Chang 23:144–156

Foody GM (2005) Local characterization of thematic classification accuracy through spatially constrained confusion matrices. Int J Remote Sens 26:1217–1228

Fritz S et al (2017) A global data set of crowdsourced land cover and land use reference data. Sci Data 4:1–8. https://doi.org/10.1038/sdata.2017.75

Ganguly S, Friedl MA, Tan B, Zhang X, Verma M (2010) Land surface phenology from MODIS: characterization of the collection 5 global land cover dynamics product. Remote Sens Environ 114:1805–1816

Gavier-Pizarro GI, Radeloff VC, Stewart SI, Huebner CD, Keuler NS (2010) Housing is positively associated with invasive exotic plant species richness in New England, USA. Ecol Appl 20:1913–1925

Geijzendorffer I, van Teeffelen A, Allison H, Braun D, Horgan K, Iturrate-Garcia M, Santos MJ, Pellissier L, Prieur-Richard A-H, Quatrini S, Sakai S, Zuppinger-Dingley D (2018) How can global targets for biodiversity and ecosystem services guide local conservation actions? Curr Opin Environ Sustain 29:145–150

Geldmann J, Joppa LN, Burgess ND (2014) Mapping change in human pressure globally on land and within protected areas. Conserv Biol 28:1604

Gleeson T, Moosdorf N, Hartmann J, van Beek LP (2014) A glimpse beneath earth's surface: GLobal HYdrogeology MaPS (GLHYMPS) of permeability and porosity. Geophys Res Lett 41

Gould W (2000) Remote sensing of vegetation, plant species richness, and regional biodiversity hotspots. Ecol Appl 10:1861–1870

GPWv4 (2016) Center for International Earth Science Information Network Gridded population of the world version 4: Population density (2016), https://doi.org/10.7927/H4NP22DQ

Gross JE, Goetz SJ, Cihlar J (2008) Monitoring protected areas: introduction to the special issue. Remote Sens Environ 113:1343–1345

Gude PH, Hansen AJ, Jones DA (2007) Biodiversity consequences of alternative future land use scenarios in greater Yellowstone. Ecol Appl 17:1004–1018

Guisan A, Zimmermann NE (2000) Predictive habitat distribution models in ecology. Ecol Model 1352–3:147–186

Haboudane D, Miller JR, Pattey E, Zarco-Tejada PJ, Strachan IB (2003) Hyperspectral vegetation indices and novel algorithms for predicting green LAI of crop canopies: modelling and validation in the context of precision agriculture. Remote Sens Environ 90:337–352

Hansen MC, Potapov PV, Moore R, Hancher M, Turubanova SA, Tyukavina A, Thau D, Stehman SV, Goetz SJ, Loveland TR, Kommareddy A, Egorov A, Chini L, Justice CO, Townshend JRG (2013) High-resolution global Maps of. Science 850:2011–2014

Harris I, Jones P, Osborn T, Lister D (2014) Updated high-resolution grids of monthly climatic observations – the CRU TS3.10 Dataset. Int J Climatol 34:623–642

Hartmann J, Moosdorf N (2012) The new global lithological map database GLiM: a representation of rock properties at the Earth surface. 13:1–37

He KS, Bradley BA, Cord AF, Rocchini D, Tuanmu M-N, Schmidtlein S et al (2015) Will remote sensing shape the next generation of species distribution models? Remote Sens Ecol Conserv 11:4–18

Hengl T et al (2017) SoilGrids250m: global gridded soil information based on machine learning. PLoS One 12(2):e0169748. https://doi.org/10.1371/journal.pone.0169748

Hobi ML, Dubinin M, Graham CH, Coops NC, Clayton MK, Pidgeon AM, Radeloff VC (2017) A comparison of dynamic habitat indices derived from different MODIS products as predictors of avian species richness. Remote Sens Environ 194:142–152

Hostetler M, Allen W, Meurk C (2011) Conserving urban biodiversity? Creating green infrastructure is only the first step. Landsc Urban Plan 100:369–371

Hui C, Richardson DM, Visser V (2017) Ranking of invasive spread through urban green areas in the world's 100 most populous cities. Biol Invasions 19:3527–3539

IPBES (2016) In: Potts SG, Imperatriz-Fonseca VL, Ngo HT (eds) The assessment report of the intergovernmental science-policy platform on biodiversity and ecosystem services on pollinators, pollination and food production. Secretariat of the Intergovernmental Science-Policy Platform on Biodiversity and Ecosystem Services, Bonn, p 552

INEGI (2013) National Institute for Statistics and Geography Mexico Land Cover Map. https://www.inegi.org.mx/

Iversen C et al (2018) Fine-root ecology database FRED: a global collection of root trait data with coincident site, vegetation, edaphic, and climatic data, version 2, Oak Ridge National Laboratory, TES SFA, U.S. Department of Energy, Oak Ridge. Available at: https://doi.org/10.25581/ornlsfa.012/1417481

Jetz W, Cavender-Bares J, Pavlick R, Schimel D, Davis FW, Asner GP, Guralnick R, Kattge J, Latimer AM, Moorcroft P, Schaepman ME, Schildhauer MP, Schneider FD, Schrodt F, Stahl U, Ustin SL (2016) Monitoring plant functional diversity from space. Nat Plants 2

Jones HG, Vaughan RA (2010) Remote sensing of vegetation: principles, techniques and applications. Oxford University Press, Oxford

Joshi N et al (2016) A review of the application of optical and radar remote sensing data fusion to land use mapping and monitoring. Remote Sens 81:1–23. https://doi.org/10.3390/rs8010070

Justice C, Roman M, Csiszar I, Vermonte E, Wolfe R, Hook S, Friedl M, Wang Z, Schaaf C, Miura T, Tschudi M, Riggs F, Hall D, Lyapustin A, Devadiga S, Davidson C, Masuoka E (2013) Land and cryosphere products from Suomi NPP VIIRS: overview and status. J Geophys Res Atmos 118:9753–9765

Karger DN, Conrad O, Böhner J, Kawohl T, Kreft H, Soria-Auza RW, Zimmermann NE, Linder HP, Kessler M (2017) Climatologies at high resolution for the earth's land surface areas. Sci Data 4:170122

Kattge J et al (2011) TRY – a global database of plant traits. Glob Chang Biol 17:2905–2935. https://doi.org/10.1111/j.1365-2486.2011.02451.x

Keenan TF, Niinemets Ü (2016) Global leaf trait estimates biased due to plasticity in the shade, Nature Plants. Nat Publ Group 3:16201. https://doi.org/10.1038/nplants.2016.201

Kennedy RE, Andrefouet S, Cohen WB, Gomez C, Griffiths P, Hais M et al (2014) Bringing an ecological view of change to Landsat-based remote sensing. Front Ecol Environ 126:339–346. https://doi.org/10.1890/130066

Kerr J, Ostrovsky M (2003) From space to species: ecological applications for remote sensing. Trends Ecol Evol 18:299–305

Khormi HM, Kumar L (2011) Modeling dengue fever risk based on socioeconomic parameters, nationality and age groups: GIS and remote sensing based case study. Sci Total Environ 409:4713–4719

Khorram S, van der Wiele CF, Koch FH, Nelson SAC, Potts MD (2016) Principles of applied remote sensing, pp 21–31

Koltunov A, Ustin SL, Asner GP, Fung I (2009) Selective logging changes forest phenology in the Brazilian Amazon: evidence from MODIS images time series analysis. Remote Sens Environ 113:2431–2440

Kosmala M et al (2016) Season spotter: using citizen science to validate and scale plant phenology from near-surface remote sensing. Remote Sens 89:1–22. https://doi.org/10.3390/rs8090726

Kramer-Schadt S et al (2013) The importance of correcting for sampling bias in MaxEnt species distribution models. Divers Distrib 19:1366–1379. https://doi.org/10.1111/ddi.12096

Lausch A et al (2015) Understanding and quantifying landscape structure – A review on relevant process characteristics, data models and landscape metrics, ecological modelling. Elsevier BV 295:31–41. https://doi.org/10.1016/j.ecolmodel.2014.08.018

Lawley V et al (2016) Site-based and remote sensing methods for monitoring indicators of vegetation condition: an Australian review, ecological indicators. Elsevier Ltd 60:1273–1283. https://doi.org/10.1016/j.ecolind.2015.03.021

Leitão PJ, Schwieder M, Suess S, Catry I, Milton EJ, Moreira F, Osborne PE, Pinto MJ, van der Linden S, Hostert P (2015) Mapping beta diversity from space: sparse generalised dissimilarity modelling SGDM for analysing high-dimensional data. Methods Ecol Evol 6:764–771

Lehner B, Liermann Reidy C, Revenga C, Vörösmarty C, Fekete B, Crouzet P, Döll P, Endejan M, Frenken K, Magome J, Nilsson C, Robertson JC, Rodel R, Sindorf N, Wisser D (2011) High-resolution mapping of the World's reservoirs and dams for Sustainable River-Flow management. Front Ecol Environ 9:494–502

Lesiv M, Moltchanova E, Schepaschenko D, See L, Shvidenko A, Comber A, Fritz S (2016) Comparison of data fusion methods using crowdsourced data in creating a hybrid forest cover map. Remote Sens 8

Lim K et al (2003) LiDAR remote sensing of forest structure. Prog Phys Geog 27(1):88–106

Liu WT, Kogan FN (1994) Monitoring regional drought using the vegetation condition index. Int J Remote Sens 17:2761–2782

Liu J, Linderman M, Ouyang Z, An L, Yang J, Zhang H (2001) Ecological degradation in protected areas: the case of Wolong nature reserve for giant Pandas. Science 292:98–101

Liu Z, Blasch E, John V (2017) Statistical comparison of image fusion algorithms: recommendations. Inf Fusion 36:251–260

Loew A, Bell W, Brocca L, Bulgin CE, Burdanowitz J, Calbet X, Donner RV, Ghent D, Gruber A, Kaminski T, Kinzel J, Klepp C, Lambert J-C, Schaepman-Strub G, Schröder M, Verhoelst T (2017) Validation practices for satellite-based earth observation data across communities. Rev Geophys 39:779–817

Lovett GM et al (2005) Ecosystem function in heterogeneous landscapes, ecosystem function in heterogeneous landscapes. Springer, New York

MA [Millennium Ecosystem Assessment] (2005) Ecosystems and human wellbeing: synthesis. Island Press, Washington, DC

Maes D, Isaac NJ, Harrower CA, Coleen B, van Strien AJ, Roy DB (2015) The use of opportunistic data for IUCN red list assessments [in special issue: fifty years of the biological records Centre]. Biol J Linn Soc 115:690–706

Ma L et al (2017) A review of supervised object-based land-cover image classification. ISPRS J Photogramm Remote Sens 130:277–293. https://doi.org/10.1016/j.isprsjprs.2017.06.001

Maldonado C et al (2015) Estimating species diversity and distribution in the era of Big Data : to what extent can we trust public databases? Glob Ecol Biogeogr 24:973–984. https://doi.org/10.1111/geb.12326

McLean P, Gallien L, Wilson JRU, Gaertner M, Richardson DM (2017) Small urban centres as launching sites for plant invasions in natural areas: insights from South Africa. Biol Invasions 19:3541–3555

Meyer ST, Koch C, Weisser WW (2015) Towards a standardized rapid ecosystem function assessment REFA. Trends Ecol Evol 30:390–397

Montesano PM, Neigh CSR, Sexton J, Feng M, Channan S, Ranson KJ, Townshend JR (2016) Calibration and validation of Landsat tree cover in the Taiga-Tundra Ecotone. Remote Sens 8:5–7

Morris DE et al (2013) Exploring the potential for automatic extraction of vegetation phenological metrics from traffic webcams. Remote Sens 55:2200–2218. https://doi.org/10.3390/rs5052200

Nagendra H (2001) Using remote sensing to assess biodiversity. Int J Remote Sens 22:2377–2400

Nagendra H et al (2013) Remote sensing for conservation monitoring: assessing protected areas, habitat extent, habitat condition, species diversity, and threats, ecological indicators. Elsevier Ltd 33:45–59. https://doi.org/10.1016/j.ecolind.2012.09.014

Neil K, Wu J (2006) Effects of urbanization on plant flowering phenology: a review. Urban Ecosyst 9:243–257

Niphadkar M, Nagendra H (2016) Remote sensing of invasive plants: incorporating functional traits into the picture. Int J Remote Sens. Taylor & Francis 3713:3074–3085. https://doi.org/1 0.1080/01431161.2016.1193795

Olinger SV (2011) Sources of variability in canopy reflectance and the convergent properties of plants. New Phytol 189:375–394

Paganini M et al (2016) The role of space agencies in remotely sensed essential biodiversity variables. Remote Sens Ecol Conserv 23:132–140. https://doi.org/10.1002/rse2.29

Pereira HM et al (2013) Essential biodiversity variables. Science 3396117:277–278. https://doi.org/10.1126/science.1229931

Pereira HM, Belnap J, Böhm M, Brummitt N, Garcia-Moreno J, Gregory R, Martin L, Peng C, Proença V, Schmeller D, van Swaay C (2015) Monitoring essential biodiversity variables at the species level. In: Walters M, Scholes R (eds) The GEO handbook on biodiversity observation networks. Springer, Cham

Pérez-Harguindeguy N et al (2013) New Handbook for standardized measurement of plant functional traits worldwide. Aust J Bot 6134:167–234. https://doi.org/10.1071/BT12225

Pettorelli N et al (2017) Satellite remote sensing of ecosystem functions: opportunities, challenges and way forward. Remote Sens Ecol Conserv 4:1–23. https://doi.org/10.1002/rse2.59

Pontius RG, Millones M (2011) Death to Kappa: birth of quantity disagreement and allocation disagreement for accuracy assessment. Int J Remote Sens 32:4407–4429

Pontius RG, Santacruz A (2014) Quantity, exchange, and shift components of difference in a square contingency table. Int J Remote Sens 35:7543–7554

Potapov P, Hansen MC, Laestadius L, Turubanova S, Yaroshenko A, Thies C, Smith W, Zhuravleva I, Komarova A, Minnemeyer S, Esipova E (2017) The last frontiers of wilderness: tracking loss of intact forest landscapes from 2000 to 2013. Sci Adv 3:e1600821

Propastin P (2012) Modifying geographically weighted regression for estimating aboveground biomass in tropical rainforests by multispectral remote sensing data. Int J Appl Earth Obs Geoinf 18:82–90

Quattrochi DA, Wentz EA, Siu-Ngam Lam N, Emerson CW (2017) Integrating scale in remote sensing and GIS. Routledge, New York

Reed BC, Brown JF, Vanderzee D, Loveland TR, Merchant JW, Ohlen DO (1994) Measuring phenological variability from satellite imagery. J Veg Sci 55:703–714

Reddy CS et al (2017) Earth observation data for habitat monitoring in protected areas of India, remote sensing applications: society and environment. Elsevier BV 8May:114–125. https://doi.org/10.1016/j.rsase.2017.08.004

Reed B et al (2009) Integration of MODIS-derived metrics to assess interannual variability in snowpack, lake ice, and NDVI in southwest Alaska, Remote Sensing of Environment. Elsevier BV 1137:1443–1452. https://doi.org/10.1016/j.rse.2008.07.020

Richardson AD et al (2018) Intercomparison of phenological transition dates derived from the PhenoCam Dataset V1.0 and MODIS satellite remote sensing. Sci Rep 81:1–12. https://doi.org/10.1038/s41598-018-23804-6

Rocchini D, Luque S, Pettorelli N, Bastin L, Doktor D, Faedi N, Feilhauer H, Feret J-B, Foody GM, Gavish Y, Godinho S, Kunin WE, Lausch A, Leitao PJ, Marcantonio M, Neteler M, Ricotta C, Schmidtlein S, Vihervaara P, Wegmann M, Nagendra H (2017) Measuring beta-diversity by remote sensing: a challenge for biodiversity monitoring. Methods Ecol Evol 9(8):1787–1798

Rockström J et al (2009) A safe operating space for humanity. Nature 461:472–478

Ronk A, Szava-kovats R, Pärtel M (2015) Applying the dark diversity concept to plants at the European scale. Ecography 38:1015–1025. https://doi.org/10.1111/ecog.01236

Rose RA et al (2015) Ten ways remote sensing can contribute to conservation. Conserv Biol 292:350–359. https://doi.org/10.1111/cobi.12397.

Running SW, Nemani RR, Heinsch FA, Zhao MS, Reeves M, Hashimoto H (2004) A continuous satellite-derived measure of global terrestrial primary production. Bioscience 546:547–560

Sandel B et al (2015) Estimating the missing species bias in plant trait measurements. J Veg Sci 265:828. https://doi.org/10.1111/jvs.12292.

Santos MJ, Smith AB, Thorne JH, Moritz C (2017) The relative roles of changing vegetation and climate on elevation range dynamics of small mammals. Clim Chang Res 4:7

Sankey TT et al (2018) UAV hyperspectral and LiDAR data and their fusion for arid and semi-arid land vegetation monitoring. Remote Sens Ecol Conserv, Early View 4:20

Schmidtlein S, Fassnacht FE (2017) The spectral variability hypothesis does not hold across landscapes. Remote Sens Environ 192:114–125

Schrodt F et al (2015) BHPMF – a hierarchical Bayesian approach to gap-filling and trait prediction for macroecology and functional biogeography. Glob Ecol Biogeogr 2412:1510–1521. https://doi.org/10.1111/geb.12335

Schweiger AK, Cavender-Bares J, Townsend PA, Hobbie SE, Madritch MD, Wang R, Tilman D, Gamon JA (2018) Plant spectral diversity integrates functional and phylogenetic components of biodiversity and predicts ecosystem function. Nat Ecol Evol 2:976–982

Smith SA, Brown JW (2018) Constructing a broadly inclusive seed plant phylogeny. Am J Bot 1053:302–314. https://doi.org/10.1002/ajb2.1019.

Stefanski J et al (2014) Mapping land management regimes in western Ukraine using optical and SAR data. Remote Sens 6:5279–5305

Stoms DM, Estes JE (1993) A remote sensing research agenda for mapping and monitoring biodiversity. Int J Remote Sensing 1:14

Sun B, Chen X, Zhou Q (2017) Analyzing the uncertainties of ground validation for remote sensing land cover mapping in the era of big geographic data. In: Zhou C (ed) Advances in geographic information science. Springer Nature, Singapore, pp 31–38

Svancara LK, Scott JM, Loveland TR, Pidgorna AB (2009) Assessing the landscape context and conversion risk of protected areas using satellite data products. Remote Sens Environ 113:1357–1369

Swenson NG (2014) Phylogenetic imputation of plant functional trait databases. Ecography 37:105–110. https://doi.org/10.1111/j.1600-0587.2013.00528.x

Schwieder M, Leitão PJ, Bustamante MMC, Ferreira LG, Rabe A, Hostert P (2016) Mapping Brazilian savanna vegetation gradients with Landsat time series. Int J Appl Earth Obs Geoinf 52:361–370

Sheppard JK, Lawler IR, Marsh H (2007) Seagrass as pasture for seacows: landscape-level dugong habitat evaluation. Estuar Coast Shelf Sci 711–2:117–132

Siebert S, Henrich V, Frenken K, Burke J (2013) Global map of irrigation areas version 5. Rheinische Friedrich-Wilhelms-University, Bonn, Germany/Food and Agriculture Organization of the United Nations, Rome

Skidmore AK, Pettorelli N, Coops NC, Geller GN, Hansen M, Lucas R, Mücher CA, OConnor B, Paganini M, Pereira HM, Schaepman ME, Turner W, Wang T, Wegmann M (2015) Environmental science: agree on biodiversity metrics to track from space. Nature 523:403–405

Taubert F, Fischer R, Groeneveld J, Lehmann S, Müller MS, Rödig E, Wiegand T, Huth A (2018) Global patterns of tropical forest fragmentation. Nature 554:519–522

Templ B et al (2018) Pan European phenological database PEP725: a single point of access for European data. Int J Biometeorol 626:1109–1113. https://doi.org/10.1007/s00484-018-1512-8

Tong C et al (2004) A landscape-scale assessment of steppe degradation in the Xilin River Basin, Inner Mongolia, China. J Arid Environ 591:133–149. https://doi.org/10.1016/j.jaridenv.2004.01.004

Toth C, Jóźków G (2016) Remote sensing platforms and sensors: a survey. ISPRS J Photogramm Remote Sens 115:22–36

Turak E et al (2017) Essential biodiversity variables for measuring change in global freshwater biodiversity, biological conservation. Elsevier Ltd 213:272–279. https://doi.org/10.1016/j.biocon.2016.09.005

Turner W, Spectro S, Gardiner N, Fladeland M, Sterling E, Steininger M (2003) Remote sensing for biodiversity science and conservation. Trends Ecol Evol 18:306–314

Turner W (2014) Sensing biodiversity. Science 346:301–302

Ustin SL, Gamon JA (2010) Remote sensing of plant functional types. New Phytol 186:795–816

Vanbergen AJ et al (2013) Threats to an ecosystem service: pressures on pollinators. Front Ecol Environ 115:251–259. https://doi.org/10.1890/120126

Verbesselt J, Umlauf N, Hirota M, Holmgren M, Van Nes EH, Herold M, Zeileis A, Scheffer M (2016) Remotely sensed resilience of tropical forests. Nat Clim Chang 6:1028

Vuolo F et al (2016) Data service platform for sentinel-2 surface reflectance and value-added products: system use and examples. Remote Sens 811:938. https://doi.org/10.3390/rs8110938.

Wang Q, Ni J, Tenhunen J (2005) Application of a geographically-weighted regression analysis to estimate net primary production of Chinese forest ecosystems. Glob Ecol Biogeogr 14:379–393

Xie Y, Sha Z, Yu M (2008) Remote sensing imagery in vegetation mapping: a review. J Plant Ecol 1:9–23

Yang LH, Rudolf VHW (2009) Phenology, ontogeny and the effects of climate change on the timing of species interactions. Ecol Lett 13:1–10

Yu D, Wu C (2013) Understanding population segregation from Landsat ETM+ Imagery: a geographically weighted regression approach. GIScience Remote Sens 41:187–206

Zeng C et al (2017) Fusion of multispectral imagery and spectrometer data in UAV remote sensing. Remote Sens 97:696–716. https://doi.org/10.3390/rs9070696

Zeng Y, Su Z, Calvet J-C, Manninen T, Swinnen E, Schulz J, Roebeling R, Poli P, Tan D, Riihela A, Tanis C-M, Arslan A-N, Obregon A, Kaiser-Weiss A, John V, Timmermans W, Timmermans J, Kaspar F, Gregow H, Barbu A-L, Fairbairn D, Gelati E, Meurey C (2015) Analysis of current validation practices in Europe for space-based climate data records of essential climate variables. Int J Appl Earth Obs Geoinformatics 42:150–161

Zhao M et al (2013) Plant phenological modelling and its application in global climate change research: overview and future challenges. Environ Rev 211:1–14. https://doi.org/10.1139/er-2012-0036

Open Access   This chapter is licensed under the terms of the Creative Commons Attribution 4.0 International License (http://creativecommons.org/licenses/by/4.0/), which permits use, sharing, adaptation, distribution and reproduction in any medium or format, as long as you give appropriate credit to the original author(s) and the source, provide a link to the Creative Commons license and indicate if changes were made.

The images or other third party material in this chapter are included in the chapter's Creative Commons license, unless indicated otherwise in a credit line to the material. If material is not included in the chapter's Creative Commons license and your intended use is not permitted by statutory regulation or exceeds the permitted use, you will need to obtain permission directly from the copyright holder.

Chapter 18
Essential Biodiversity Variables: Integrating In-Situ Observations and Remote Sensing Through Modeling

Néstor Fernández, Simon Ferrier, Laetitia M. Navarro, and Henrique M. Pereira

18.1 Introduction

Which facets of biodiversity are changing, and what is the magnitude and direction of these changes? How is biodiversity responding to the variety of human pressures? Are the management policies put into place effective to tackle the impact of those pressures? While the scientific community has been addressing these questions for decades, the information gap in biodiversity science remains a major obstacle for reducing the large uncertainties associated with answering those questions. Technological advances, collection of data by an increasing number of scientists and volunteer citizens, and increased access to Earth observations (EO) should help reduce this gap. Yet quantitative information is still limited, as has been its ability to inform important international commitments such as the Aichi Biodiversity Targets in response to the global biodiversity crisis (Tittensor et al. 2014). Data collection and monitoring protocols are often adopted by scientists, public administrations, and environmental organizations with no effective international coordination, and

N. Fernández (✉) · L. M. Navarro
German Centre for Integrative Biodiversity Research (iDiv) Halle-Jena-Leipzig, Leipzig, Germany

Institute of Biology, Martin Luther University Halle-Wittenberg, Halle (Saale), Germany
e-mail: nestor.fernandez@idiv.de

S. Ferrier
CSIRO Land and Water, Australian Capital Territory, Canberra, Australia

H. M. Pereira
German Centre for Integrative Biodiversity Research (iDiv) Halle-Jena-Leipzig, Leipzig, Germany

Institute of Biology, Martin Luther University Halle-Wittenberg, Halle (Saale), Germany

InBio - Research Network in Biodiversity and Evolutionary Biology, Universidade do Porto, Vairão, Portugal

© The Author(s) 2020
J. Cavender-Bares et al. (eds.), *Remote Sensing of Plant Biodiversity*,
https://doi.org/10.1007/978-3-030-33157-3_18

there is no consensus on adopting priority metrics to quantify biodiversity change. While strengthening efforts to reduce the multiple biases present in biodiversity data remains critical (including spatial, temporal, and taxonomic biases, among others; Meyer et al. 2016; Proença et al. 2017), parallel efforts are needed to consolidate data from in-situ and remote sensing (RS) EO so as to increase their usability and information value.

The concept of *essential biodiversity variables* (EBVs) was proposed in 2013 as a framework to prioritize, integrate, and consolidate biodiversity observations and monitoring programs worldwide (Pereira et al. 2013). Since then, EBVs have gained acceptance among scientists, along with the interest and endorsement of the policy-making community, including the Convention on Biological Diversity (e.g., Decision XI/3 in UNEP/CBD/COP/11/35) and the Intergovernmental Science-Policy Platform on Biodiversity and Ecosystem Services (IPBES). By providing an integrative framework for quantifying biodiversity change in time, EBVs also hold great potential for advancing research on biodiversity and responses to pressures and conservation actions. However, the concept is still evolving, and divergent viewpoints have emerged on what actually constitutes an EBV. Here we discuss recent progress in defining an operational EBV framework and the importance of this framework for biodiversity data integration. We start with discussing key attributes of EBVs. We then describe recent conceptual developments in support of their implementation. Finally, we illustrate the role of biodiversity models as a cornerstone for integrating data obtained from in-situ and satellite RS EO to support global assessments of biodiversity change and as a critical component of a global biodiversity monitoring system (Geller et al., Chap. 20).

18.2 The EBV Framework

18.2.1 Definition of Essential Biodiversity Variables

EBVs are defined as a minimum set of complementary measurements needed to detect and document biodiversity change across all levels of biodiversity, from genes to species and ecosystems (Pereira et al. 2013). EBVs are part of a larger family of Essential Variables (EVs) that was first conceptualized by the climate community with the Essential Climate Variables (Box 18.1).

Like all EVs, EBVs must meet the criteria of feasibility, cost-effectiveness, and scientific and policy relevance. Additional characteristics that might be specific to the EBVs are generalization (to the best extent possible) across terrestrial, marine, and freshwater realms and scalability. Importantly, the EBVs evolved from initially covering multiple aspects of the Driver-Pressure-State-Impact-Response (DPSIR) framework to focusing exclusively on biological state variables (i.e., EBVs describe the condition or the status of a particular biological entity). This is not to say that nonbiological variables are irrelevant for EBVs. On the contrary, some of these variables, such as temperature, fire occurrence, or elevation, may be extremely important (e.g., as covariates in biodiversity models); however, they do not constitute EBVs themselves. Furthermore, EBVs can be analyzed in relation to other variables to attribute biodiversity change to specific pressures and drivers (Pereira et al. 2012), to predict how different biodiversity metrics might behave with different scenarios of change (Kim et al. 2018), and to assess the effectiveness of management policies for biodiversity and ecosystem services (Geijzendorffer et al. 2016).

EBVs are best understood as the level of integration between primary observations, including in-situ and RS EO, and indicators of biodiversity change, calculated for a given spatial reporting unit (country, set of protected areas, etc.; Fig. 18.1). The power of EBVs emerges from their flexibility to incorporate new data as technology evolves and/or more exhaustive primary data are collected. This is already the case with the advent of citizen science and the technical progress made with, for instance, metagenomics, metabarcoding, field sensor networks, and RS (Turner 2014; Bush et al. 2017; Haase et al. 2018; Muller-Karger et al. 2018a). This means that the underlying measurement, coverage, and frequency of primary observations are likely to change (Fig. 18.2). Likewise, the needs of end users in terms of biodiversity

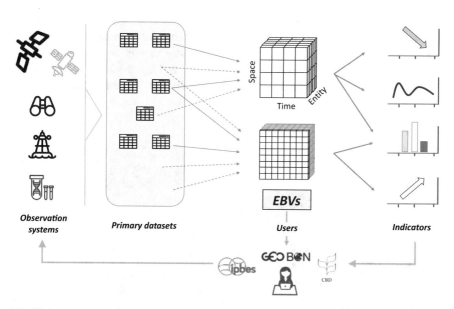

Fig. 18.1 EBVs are intermediate products between primary observations and biodiversity change indicators. Observations obtained with different methods and protocols require different levels of integration, often with the use of biodiversity models, to consolidate the information in an EBV. The EBV cube typically structures biological measurements in a space defined by geographic and temporal references and a biological entity, such as species or ecosystem class. While end users (including scientists, managers, public administrations, and international policy forums and bodies) determine the need for indicators, they also influence the implementation of observation systems. However, the EBV remains the same so that it is complemented with new primary data, e.g., from repeated in-situ surveys or future satellite missions

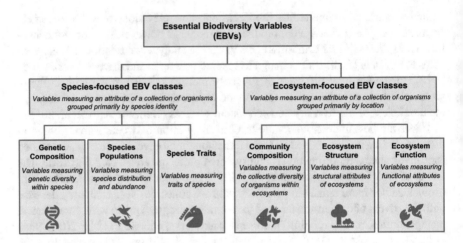

Fig. 18.2 Framework of the six classes of EBVs grouped by species-focused and ecosystem-focused approaches

change indicators have evolved in the past and will continue to do so. However, the EBVs are designed to remain conceptually stable, making them adaptable to different and unforeseen users' needs. For example, even though the methods used to acquire and integrate data on *species occurrence*, and the indicators it can inform on, are likely to change in the future, the *species distribution* variable remains *essential*.

Box 18.1 Essential Variables

Essential variables (EVs) emerged from the need for openly available data sets with transparent production processes that offer an appropriate spatial and temporal coverage to allow their use in policy- and decision-making (Bojinski et al. 2014). As a result, EVs are meant to allow the development of indicators that can support dynamic users' needs while being resilient to changing and/or evolving observation systems (Reyers et al. 2017). From a pool of candidate variables, both science and technology will determine which are feasible, cost-effective, and, most importantly, relevant, and thus essential (Bojinski et al. 2014). In practice, although EVs can be interpreted and adapted differently among disciplines, the process of their development and endorsement remains similar, with a community of practice that self-organizes to provide both the scientific foundation (research, data, monitoring) and technical guidance to produce those EVs.

EVs were first adopted by the climate community as the Essential Climate Variables (ECVs) in the early 1990s, to respond to the needs of Parties of the United Nations Framework Convention on Climate Change and the

Intergovernmental Panel on Climate Change, but the concept has since been expanded to go beyond climate science, including with the Essential Biodiversity Variables (EBVs) and the Essential Ocean Variables (EOVs, Miloslavich et al. 2018). While there is value in increasingly expanding the concept to other domains, a coordinated approach within disciplines to define and prioritize the EVs and avoid the duplication of efforts is currently being discussed within the Group on Earth Observations (GEO). One example is the joint effort by the Marine Biodiversity Observation Network of the Group on Earth Observations Biodiversity Observation Network (GEO BON) and the Global Ocean Observing System (GOOS) of the Intergovernmental Oceanographic Commission (IOC) to streamline the marine observations that underpin the EBVs and EOVs (Muller-Karger et al. 2018b). The discussion on EVs is also permeating other domains, such as agriculture, health, and disaster risk reduction (Reyers et al. 2017).

18.2.2   A Space-Time-Biology Cube

The data structure of an EBV can be described as a hypercube and has analogy to a multidimensional data array in computer programming. The first two dimensions of the hypercube are space (latitude and longitude) and time, while the third dimension represents biological entities (Fig. 18.1). The latter dimension can, for instance, describe taxonomy in a species-centered EBV (see below), and values will inform the presence/absence or population abundance (e.g., Kissling et al. 2018a). Unlike the Essential Climate Variables (ECVs), the biological dimension of the EBVs makes them especially challenging in terms of developing the conceptual framework and producing the EBV data products. For ecosystem-level EBVs, this dimension can also inform ecosystem structure metrics (e.g., extent of different habitat types) or functions (e.g., primary productivity) in the time-space coordinates. The hypercube thus provides an intuitive representation of the EBV concept and at the same time has a direct translation in data computing language that suits implementation. Other EBVs are also more challenging to represent with three dimensions, even more so when considering that their value is likely to change depending on the spatial scale and extent, as is the case for the community spatial turnover.

18.2.3   Six EBV Classes

Each EBV measures a particular attribute (property) of a given entity (object). EBVs are grouped into six broad classes based on similarities and differences in the attributes and entities they address (Fig. 18.2). These classes are sets of variables describing the structure, composition, and function of biodiversity across its hierar-chical levels (Noss 1990). The entity addressed by an EBV can be of two broad types, distinguished by the approach used to define the set of organisms forming this entity.

In the first approach, entities are formed by grouping organisms primarily on the basis of their species identity. In other words, EBVs of this broad type measure particular attributes of species—i.e., genetic diversity within a species in the case of the Genetic Composition Class; distribution and abundance of a species in the Species Populations Class; and traits of a species in the Species Traits Class.

The second approach to forming entities involves grouping organisms primarily on the basis of where they occur. EBVs of this broad type measure collective attri-butes of the entire ecosystem formed by all of the organisms occurring within a defined area (most typically an individual cell within a regular grid)—i.e., struc-tural attributes of the ecosystem in the case of the Ecosystem Structure Class; func-tional attributes of the ecosystem in the Ecosystem Function Class; and various dimensions of compositional diversity (e.g., taxonomic, genetic/phylogenetic, functional) of organisms occurring within the ecosystem in the Community Composition Class.

The relationships between these six EBV classes is depicted in Fig. 18.2. A few key aspects of this overall typology are worth noting. First, the two broad approaches to defining the entity addressed by an EBV, species-focused and ecosystem-focused, essentially work with the same pool of individual organisms but view these organisms from two different perspectives—one grouping organisms accord-ing to species identity and the other according to location. While the entity employed in species-focused EBVs will typically be defined primarily on the basis of species identity, this could in some instances be qualified to focus, for example, on the population of a species occurring in a particular area. Likewise, ecosystem-focused EBVs might, in some instances, focus on measuring collective attributes of a particular subset of organisms occurring in an ecosystem rather than all organ-isms, with this subset defined in terms of taxonomy (e.g., all birds) or any other trait of interest (e.g., all pollinators). Finally, it is important to caution against directly equating the species-focused versus ecosystem-focused typology with major sources of in-situ versus RS observation. Many different sources and types of data can, and should, contribute to the population of EBVs across this entire framework. Any given EBV class can typically be populated using data from mul-tiple sources of in-situ and remote-sensing observation, and any given source will typically contribute data to more than one EBV class. For example, EBVs in the Community Composition Class could be populated with data both from RS of compositional diversity (e.g., Morsdorf et al., Chap. 4) and from aggregation of in-situ species observations and models (e.g., Pinto-Ledezma and Cavender-Bares, Chap. 9), with the latter also contributing data simultaneously to the Species Populations Class.

18.3   Production Workflows for EBVs

The estimation of EBV information products typically involves multiple levels of data integration, from the collection of raw observations to the production of a final, consistent information set that provides comparable measurements in space and time. Data integration procedures need to be customized for almost every EBV, since they need to accommodate highly diverse biological quantities that are often specific to a particular EBV. Designing open, consistent, and fully reproducible workflows is key to support the full operationalization process, from data collection to publication of an EBV product that is ready to use for multiple science and policy purposes.

18.3.1   The Need for Open EBV Workflows

Workflows are defined as precise descriptions of data processing from one analytical step to another in a formal language. In recent years a multiplication of biodiversity data availability, novel analytical capabilities, and virtual infrastructures have laid the foundations for producing better integrated and more detailed information for measuring biodiversity change (e.g., Jetz et al. 2012; Hansen et al. 2013). However, the increasing variety of analytical procedures and project-specific designs also means that analytical standards are difficult to establish (Borregaard and Hart 2016). Open workflows benefit the preservation of processing steps and support data *interoperability* and the automation of biological and environmental data integration (e.g., via virtual biodiversity e-infrastructures; La Salle et al. 2016). These workflows require provenance of derived products to be also recorded so others can understand the relationships among data, processing, and results (Michener and Jones 2012) and thus facilitate product updating as new data and better processing algorithms become available. All these aspects are critical in the EBV framework since the production of EBVs depends on large research collaborations built on the basis of knowledge transfer and open access to data and production protocols.

At present, fully operational workflows that facilitate the automated and widespread production of EBVs are missing. However, recent efforts have identified critical steps and bottlenecks for the definition of workflows in support of the production of specific EBVs.

18.3.2   From Data Collection to Biodiversity Models

Generic workflows have been outlined so far for the production of a few species-centered EBVs, including species distributions, population abundances, and species traits (Kissling et al. 2018a, b). For example, 11 steps have been identified to build

spatially continuous and temporally consistent EBV products for species distributions, from the integration of multiple data sources, including traditional direct species observations collected in many different ways, automated records from sensor
networks—such as camera traps and sound detection—and emerging uses of satellite remote sensing (RS) for detecting species (Kissling et al. 2018a). These workflows pay special attention to the integration among in-situ observations and RS
data. Other approaches may use in-situ observations only as ground-truth data,
while the rest of the process is dominated by image processing (e.g., mapping vegetation cover; Hansen et al. 2013). However, traceability of the ground-truth sampling and processing remains equally important and therefore also applies to the
entire process similar principles of annotation, uncertainty reporting, and conformance with data management guidelines (see below).

The key workflow steps can be summarized into three main groups (Fig. 18.3):

1. *Standardization of primary biodiversity observations.* At the core of the EBV
   concept is the aggregation of primary observations from multiple sources into a
   harmonized product that provides more comprehensive and richer information
   than each individual data set. Before this aggregation can take place, primary
   data must be curated, standardized, and annotated with appropriate metadata that
   record characteristics such as location, time, measurement units, and, ideally,
   sampling designs, collection procedures, and data quality control procedures
   (Rüegg et al. 2014). For example, for Species Populations EBVs, harmonized
   observations would consist of sets of species occurrence and abundance data
   expressed in appropriate units (such as species occupancies and number of individuals per unit area, respectively) complemented with metadata in appropriate
   standards such as the "Darwin Core Standard" with the "Event Core" extension
   (Wieczorek et al. 2012), which makes it possible to capture monitoring protocols

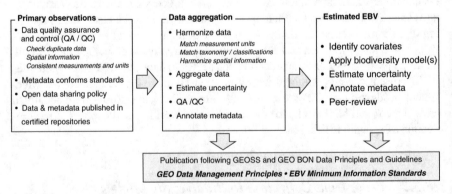

Fig. 18.3 Outline of an EBV production workflow for the integration of in-situ and RS data from
disparate primary sources of data to final modeled information and publishing. Some authors consider the result of the intermediate data integration level also as an EBV-ready data set from which
some indicators can be calculated, even from sparse observations in space and time (Kissling
2018a), while fully continuous coverage in the spatial and temporal dimensions is typically
obtained only in the last level of integration

and sampling efforts together with the data (https://www.gbif.org/darwin-core). Full documentation of sampling events using adequate metadata standards not only is critical for facilitating reuse of data by secondary users but also provides important information for quantifying the associated uncertainty and eventually applying correction techniques in subsequent steps. In practice, for a decade or so, the critical importance of annotating data with standard metadata has guided data management practices (e.g., in the context of long-term ecological research networks; Michener et al. 2011). However, poor data practices that ignore the annotation of metadata or that fail to adopt interoperable formats are still common for many biodiversity data sets, including those accessible through public data archives of scientific journals (Roche et al. 2015). These deficiencies constitute a major bottleneck for building EBVs (Hugo et al. 2017).

2. *Primary data aggregation.* A second set of steps leads to the production of consolidated data products that typically conform to all or most of the following characteristics: They contain consistent biological quantities expressed in the same measurement unit; other relevant biological attributes have been checked and harmonized (e.g., into a harmonized taxonomy or a consistent typology of traits or of ecosystem types); spatial and temporal references are matched; and data uncertainties have been quantified. Standardized observations need to be checked at this stage using quality control (QC) mechanisms that are documented transparently (Rüegg et al. 2014), for example, looking at outliers to ensure data quality. Collation of data in support of user requirements will ideally be automated using virtual infrastructures that are able to map the different (standardized) data sets with metadata into fully interoperable formats (Hugo et al. 2017). As detailed in Kissling et al. (2018b), an excellent example of this for the Species Traits EBVs is the Global Plant Phenology Data Portal (https://www.plantphenology.org), a platform that integrates phenology observations from three different networks using disparate data frameworks (Stucky et al. 2018). Key for this integration was the design of a new "Plant Phenology Ontology" that was able to provide a semantic framework as a basis to overcome interoperability problems produced by network-specific terminologies for data recording. Finally, data integration needs to deal with, and report on, uncertainties resulting from errors that may propagate throughout the different EBV production steps, including uncertain geographic locations of in-situ data, heterogeneous sampling methods and efforts (Proença et al. 2017), and measurement errors.

3. *Model-based estimation.* Final EBV products ideally provide continuous information in space and at different time periods so biodiversity change can be measured throughout the entire spatial domain. This is the case for EBVs that can be directly estimated using algorithms applied to satellite RS imagery with complete area coverage. On the contrary, for many EBVs that are primarily estimated from in-situ data, an additional level of integration is required to overcome the sparsity of data. Biodiversity models provide this level of integration by combining the strengths of in-situ observations and state-of-the-art RS products based on correlative or deductive approaches (Jetz et al. 2012; Ferrier et al. 2017). For instance, species distribution models are often based on a correlative relationship

between environmental variables and the probability of the occurrence of a species. These models are calibrated or trained using species occurrence and sometimes absence data as response variables and environmental variables as predictors. The probability of occurrence of a species can be spatially interpolated between the observation points because environmental variables are available as continuous surfaces (i.e., wall-to-wall), which are themselves generated from models using in-situ and EO data. In deductive habitat modeling, expert-based assessment of the habitat preference and environmental constraints of species is used to refine the potential species distribution. When habitat predictors are available in high resolution, this makes it possible to go from coarse potential species distributions to fine-grain species distributions because species also respond more locally to habitat variables than to, for instance, climate (Triviño et al. 2011; Martins et al. 2014). Other EBVs can also be projected with models that integrate in-situ observations with RS data and other environmental data. Community composition variables such as the beta diversity between two sites can be projected from climate and other variables using generalized dissimilarity models (Ferrier et al. 2007), while alpha and gamma diversity can be projected from land-use using the countryside species-area model (Pereira and Borda-de-Água 2013). Hence, environmental predictors derived from RS constitute the backbone of higher-resolution EBV products that are consistent in space and time. However, it is important to note that such model-based EBVs provide information that is fundamentally different from the aggregated data sets described in the preceding steps and that while it improves the spatial and temporal coverage of the data set, it also introduces additional uncertainties that need to be documented.

Massive integration of biodiversity data based on the EBV framework and workflows requires implementation via interoperable informatics infrastructures (Hugo et al. 2017). Projects aligned with the mission and concepts of EBVs, such as Map of Life (www.mol.org) or the Biogeographic Infrastructure for Large-scaled Biodiversity Indicators (BILBI) (Hoskins et al. 2018), already constitute a proof of concept of the potential of virtual infrastructures for developing a biodiversity-modeling framework that delivers global information from multi-sourced EO data integration. While the technological implementation of these infrastructures should not constitute a major limitation, redoubled efforts are needed, first, on making the large amounts of in-situ data being collected available and interoperable and, second, on developing and adapting biodiversity models that are able to ingest massive and novel sources of data, both in-situ (e.g., eDNA data) and RS (e.g., imaging spectroscopy).

18.3.3   Access Principles

The open publication of intermediate and final processed products and the adherence to open data-sharing principles is key to maximize scientific and policy benefits of the EBV framework. The Group on Earth Observations (GEO) has established a set of Data Management Principles to support publication of information using open standards and to ensure discoverability and accessibility through GEOSS, the Global Earth Observation System of Systems (Fig. 18.4). These principles allow full traceability, ensuring accessible information on data sources and processing history via provenance information. All of these management principles are directly applicable to EBVs. For example, traceability is critical for facilitating the updating of the information contained in an EBV product with new data (e.g., from new monitoring and/or observation systems) and the timely incorporation of new biodiversity model developments.

In addition, GEO BON is developing an "Essential Biodiversity Variables Portal" that supports this process and enhances accessibility to EBV products. Open distribution of these products is complemented by reporting on their compliance with a set of "EBV Minimum Information Standards". Besides ensuring good data management practices, these information standards aim to provide a guideline for the standardized description of EBV products. The purpose is to ensure consistent information about the EBV hypercube (i.e., the attributes of space, time, biological entity, and uncertainties) among the different EBV classes so that final users can easily access the relevant information (e.g., when searching for suitable EBVs for specific indicators).

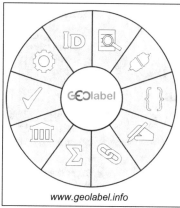

1. **Discoverable** data and metadata with access conditions clearly indicated
2. **Accessible** online, preferably with services for access
3. **Data encoding** following community standards
4. **Data documented** via metadata
5. **Data traceable** with provenance metadata
6. **Data quality control** and results indicated in metadata
7. **Data preservation** planned for future use
8. **Data and metadata** verified to ensure integrity
9. **Data reviewed** and updated, including
10. **Persistent identifiers** are assigned to the data

www.geolabel.info

Fig. 18.4 The ten GEOSS Data Management Principles promote the practical implementation of openness in scientific data and best practices ensuring that data are easily discoverable, accessible, and (re)usable. Data providers may assess conformance with each of the principles, in which case a labeling system helps the user to recognize such conformance. For detailed guidelines on the implementation of these principles, see www.geolabel.info.

18.4 Seamless Integration of Past Trends to Future Scenarios Using EBVs

Besides providing spatial interpolation of EBVs, biodiversity models can project changes in EBVs over time based on the relationship between drivers of biodiversity change and state variables of biodiversity. This means that, when historical data on drivers is available, past trends for an EBV can be backcast. In other words, a single snapshot of biodiversity and driver data at a given moment in time can be used to establish the relationship between driver variables and biodiversity variables across points in space (Fig. 18.5). Then, in order to project for other moments in time, these spatially inferred relationships are assumed to also hold when drivers evolve over time, using space-for-time replacement. When scenarios exist for the future trajectories of the drivers, the future trends in the EBV can be forecast as well (Fig.18.5; Ferrier et al. 2017). Estimated EBVs allow for seamless comparison of historical trends of biodiversity to future scenarios of biodiversity change. Indicators aggregating spatial information can be easily calculated from the spatially explicit EBV and plotted in time for any spatial unit of interest, such as a country or region (GEO BON 2015; Navarro et al. 2017).

Recently, a set of EBVs was historically reconstructed and projected into the future in an inter-model comparison study carried out by the Expert Group on

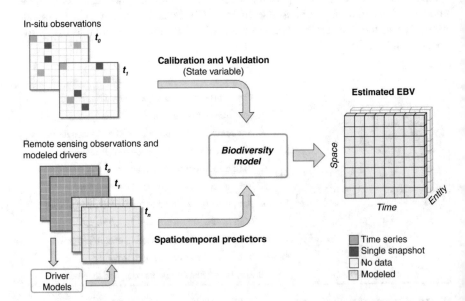

Fig. 18.5 Estimation of EBVs using biodiversity models. In-situ observations of an EBV often are sparse in space, and only a few time series exist. Drivers are often modeled continuously in space for a given moment in time and can be used by biodiversity models to project an EBV continuously in space after calibration and validation with the in-situ observations. When driver layers exist for other moments in time, either from RS observations or from scenario projections using models, the EBV can be estimated over time

Scenarios and Models of the Intergovernmental Platform on Biodiversity and Ecosystem Services (Kim et al. 2018). Species distribution, community composition, ecosystem function variables, and ecosystem services variables were reconstructed since 1900 and projected to 2050 globally, at a 0.5° resolution, using harmonized land-use data sets developed for the shared-socioeconomic pathways and climate data sets for the relative concentration pathway scenarios. In this exercise, a space-for-time substitution was used in the biodiversity models. In other words, no explicit time series biodiversity data were used to calibrate the models. Instead, current spatial patterns of biodiversity and drivers were used to infer how biodiversity changes over time when driver variables change. A future challenge for biodiversity modelers is to use biodiversity time series to fully model biodiversity across space and time (Ferrier et al. 2017).

18.5   Concluding Remarks

Since EBVs were first defined, there have been significant advances in the consolidation of the framework, substantial conceptual work on implementation, and increasing enthusiasm about their many potential applications in biodiversity science and policy. Now the scientific community needs to face the challenge of delivering EBV products and workflows that provide estimates of changes for the main facets of biodiversity and support our understanding of the driving mechanisms and the consequences of such changes. There are important opportunities for mobilizing primary data offered by the open-access movement, which continue to permeate the biodiversity community. These include public institutions responsible for promoting scientific and technological advancement. Data gaps will be covered by combining technological development with appropriate biodiversity models. For example, spaceborne sensors such as the Global Ecosystem Dynamics Investigation (GEDI) Lidar launched in 2018 are providing unprecedented global coverage in vertical measurements of vegetation and topography and will most likely support model-based integration of information for biodiversity variables in unforeseen ways.

Fulfilling the EBV vision requires renewed efforts, first, in continued scientific and technological support for the mobilization of in-situ data and for designing more comprehensive and better coordinated monitoring schemes and, second, in the implementation of workflows and interoperable infrastructures that support wall-to-wall integration of biodiversity data. GEO BON as a network defined at multiple levels, from scientific to institutional and infrastructure support, is instrumental for this endeavor (Hugo et al. 2017; Navarro et al. 2017). Key priorities are the implementation of mechanisms that enhance data mobilization as exemplified by the Darwin Event Core; a common understanding between the biodiversity research community and the space agencies of the processes to develop the technology required for detecting biodiversity change (Paganini et al. 2016); global informatics infrastructure support that meets the requirements for operationalizing EBVs (Hardisty et al. 2019); and broad scientific cooperation in implementing and enhancing biodiversity models that integrate all types of Earth observations.

References

Bojinski S, Verstraete M, Peterson TC, Richter C, Simmons A, Zemp M (2014) The concept of essential climate variables in support of climate research, applications, and policy. Bull Am Meteorol Soc 95:1431–1443

Borregaard MK, Hart EM (2016) Towards a more reproducible ecology. Ecography 39: 349–353

Bush A, Sollmann R, Wilting A, Bohmann K, Cole B, Balzter H, Martius C, Zlinszky A, Calvignac-Spencer S, Cobbold CA, Dawson TP, Emerson BC, Ferrier S, Gilbert MTP, Herold M, Jones L, Leendertz FH, Matthews L, Millington JDA, Olson JR, Ovaskainen O, Raffaelli D, Reeve R, Rödel M-O, Rodgers TW, Snape S, Visseren-Hamakers I, Vogler AP, White PCL, Wooster MJ, Yu DW (2017) Connecting earth observation to high-throughput biodiversity data. Nat Ecol Evol 1:0176

ConnectinGEO (2015) Deliverable D2.2 EVs current status in different communities and way to move forward. CREAF

Ferrier S, Manion G, Elith J, Richardson K (2007) Using generalized dissimilarity modelling to analyse and predict patterns of beta diversity in regional biodiversity assessment. Divers Distrib 13:252–264

Ferrier S, Jetz W, Scharlemann J (2017) Biodiversity modelling as part of an observation system. In: Walters M, Scholes RJ (eds) The GEO handbook on biodiversity observation networks. Springer International Publishing, Cham, pp 239–257

Geijzendorffer IR, Regan EC, Pereira HM, Brotons L, Brummitt N, Gavish Y, Haase P, Martin CS, Mihoub J-B, Secades C, Schmeller DS, Stoll S, Wetzel FT, Walters M (2016) Bridging the gap between biodiversity data and policy reporting needs: an essential biodiversity variables perspective. J Appl Ecol 53:1341–1350

GEO BON (2015) Global biodiversity change indicators: model-based integration of remote-sensing & in situ observations that enables dynamic updates and transparency at low cost. GEO BON Secretariat, Leipzig

Haase P, Tonkin JD, Stoll S, Burkhard B, Frenzel M, Geijzendorffer IR, Häuser C, Klotz S, Kühn I, McDowell WH, Mirtl M, Müller F, Musche M, Penner J, Zacharias S, Schmeller DS (2018) The next generation of site-based long-term ecological monitoring: linking essential biodiversity variables and ecosystem integrity. Sci Total Environ 613–614:1376–1384

Hansen MC, Potapov PV, Moore R, Hancher M, Turubanova SA, Tyukavina A, Thau D, Stehman SV, Goetz SJ, Loveland TR, Kommareddy A, Egorov A, Chini L, Justice CO, Townshend JRG (2013) High-resolution global maps of 21st-century forest cover change. Science 342: 850–853

Hardisty AR, Michener WK, Agosti D, Alonso García E, Bastin L, Belbin L, Bowser A, Buttigieg PL, Canhos DAL, Egloff W, De Giovanni R, Figueira R, Groom Q, Guralnick RP, Hobern D, Hugo W, Koureas D, Ji L, Los W, Manuel J, Manset D, Poelen J, Saarenmaa H, Schigel D, Uhlir PF, Kissling WD (2019) The Bari manifesto: an interoperability framework for essential biodiversity variables. Eco Inform 49:22–31

Hoskins AJ, Harwood TD, Ware C, Williams KJ, Perry JJ, Ota N, Croft JR, Yeates DK, Jetz W, Golebiewski M, Purvis A, Ferrier S (2018) Supporting global biodiversity assessment through high-resolution macroecological modelling: Methodological underpinnings of the BILBI framework

Hugo W, Hobern D, Kõljalg U, Tuama ÉÓ, Saarenmaa H (2017) Global infrastructures for biodiversity data and services. In: Walters M, Scholes RJ (eds) The GEO handbook on biodiversity observation networks. Springer International Publishing, Cham, pp 259–291

Jetz W, McPherson JM, Guralnick RP (2012) Integrating biodiversity distribution knowledge: toward a global map of life. Trends Ecol Evol 27:151–159

Kim H, Rosa IMD, Alkemade R, Leadley P, Hurtt G, Popp A, van Vuuren DP, Anthoni P, Arneth A, Baisero D, Caton E, Chaplin-Kramer R, Chini L, De Palma A, Di Fulvio F, Di Marco M, Espinoza F, Ferrier S, Fujimori S, Gonzalez RE, Gueguen M, Guerra C, Harfoot M, Harwood TD, Hasegawa T, Haverd V, Havlík P, Hellweg S, Hill SLL, Hirata A, Hoskins AJ, Janse JH, Jetz W, Johnson JA, Krause A, Leclère D, Martins IS, Matsui T, Merow C, Obersteiner M, Ohashi H, Poulter B, Purvis A, Quesada B, Rondinini C, Schipper A, Sharp R, Takahashi K, Thuiller W, Titeux N, Visconti P, Ware C, Wolf F, Pereira HM (2018) A protocol for an inter-comparison of biodiversity and ecosystem services models using harmonized land-use and climate scenarios. Geosci Model Dev Discuss:1–37

Kissling WD, Ahumada JA, Bowser A, Fernandez M, Fernández N, García EA, Guralnick RP, Isaac NJB, Kelling S, Los W, McRae L, Mihoub J-B, Obst M, Santamaria M, Skidmore AK, Williams KJ, Agosti D, Amariles D, Arvanitidis C, Bastin L, De Leo F, Egloff W, Elith J, Hobern D, Martin D, Pereira HM, Pesole G, Peterseil J, Saarenmaa H, Schigel D, Schmeller DS, Segata N, Turak E, Uhlir PF, Wee B, Hardisty AR (2018a) Building essential biodiversity variables (EBVs) of species distribution and abundance at a global scale: building global EBVs. Biol Rev 93:600–625

Kissling WD, Walls R, Bowser A, Jones MO, Kattge J, Agosti D, Amengual J, Basset A, van Bodegom PM, Cornelissen JHC, Denny EG, Deudero S, Egloff W, Elmendorf SC, Alonso García E, Jones KD, Jones OR, Lavorel S, Lear D, Navarro LM, Pawar S, Pirzl R, Rüger N, Sal S, Salguero-Gómez R, Schigel D, Schulz K-S, Skidmore A, Guralnick RP (2018b) Towards global data products of essential biodiversity variables on species traits. Nat Ecol Evol 2:1531–1540

La Salle J, Williams KJ, Moritz C (2016) Biodiversity analysis in the digital era. Philos Trans R Soc B 371:20150337

Martins IS, Proença V, Pereira HM (2014) The unusual suspect: land use is a key predictor of biodiversity patterns in the Iberian Peninsula. Acta Oecol 61:41–50

Meyer C, Weigelt P, Kreft H (2016) Multidimensional biases, gaps and uncertainties in global plant occurrence information. Ecol Lett 19:992–1006

Michener WK, Jones MB (2012) Ecoinformatics: supporting ecology as a data-intensive science. Trends Ecol Evol 27:85–93

Michener WK, Porter J, Servilla M, Vanderbilt K (2011) Long term ecological research and information management. Eco Inform 6:13–24

Miloslavich P, Bax NJ, Simmons SE, Klein E, Appeltans W, Aburto-Oropeza O, Garcia MA, Batten SD, Benedetti-Cecchi L, Checkley DM, Chiba S, Duffy JE, Dunn DC, Fischer A, Gunn J, Kudela R, Marsac F, Muller-Karger FE, Obura D, Shin Y-J (2018) Essential Ocean variables for global sustained observations of biodiversity and ecosystem changes. Glob Chang Biol 24:2416–2433

Muller-Karger FE, Hestir E, Ade C, Turpie K, Roberts DA, Siegel D, Miller RJ, Humm D, Izenberg N, Keller M, Morgan F, Frouin R, Dekker AG, Gardner R, Goodman J, Schaeffer B, Franz BA, Pahlevan N, Mannino AG, Concha JA, Ackleson SG, Cavanaugh KC, Romanou A, Tzortziou M, Boss ES, Pavlick R, Freeman A, Rousseaux CS, Dunne J, Long MC, Klein E, McKinley GA, Goes J, Letelier R, Kavanaugh M, Roffer M, Bracher A, Arrigo KR, Dierssen H, Zhang X, Davis FW, Best B, Guralnick R, Moisan J, Sosik HM, Kudela R, Mouw CB, Barnard AH, Palacios S, Roesler C, Drakou EG, Appeltans W, Jetz W (2018a) Satellite sensor requirements for monitoring essential biodiversity variables of coastal ecosystems. Ecol Appl 28:749–760

Muller-Karger FE, Miloslavich P, Bax NJ, Simmons S, Costello MJ, Sousa Pinto I, Canonico G, Turner W, Gill M, Montes E, Best BD, Pearlman J, Halpin P, Dunn D, Benson A, Martin CS, Weatherdon LV, Appeltans W, Provoost P, Klein E, Kelble CR, Miller RJ, Chavez FP, Iken K, Chiba S, Obura D, Navarro LM, Pereira HM, Allain V, Batten S, Benedetti-Checchi L, Duffy JE, Kudela RM, Rebelo L-M, Shin Y, Geller G (2018b) Advancing marine biological observations and data requirements of the complementary essential ocean variables (EOVs) and essential biodiversity variables (EBVs) frameworks. Front Mar Sci 5:211

Navarro LM, Fernández N, Guerra C, Guralnick R, Kissling WD, Londoño MC, Muller-Karger F, Turak E, Balvanera P, Costello MJ, Delavaud A, El Serafy G, Ferrier S, Geijzendorffer I, Geller GN, Jetz W, Kim E-S, Kim H, Martin CS, McGeoch MA, Mwampamba TH, Nel JL, Nicholson E, Pettorelli N, Schaepman ME, Skidmore A, Sousa Pinto I, Vergara S, Vihervaara P, Xu H, Yahara T, Gill M, Pereira HM (2017) Monitoring biodiversity change through effective global coordination. Curr Opin Environ Sustain 29:158–169

Noss RF (1990) Indicators for monitoring biodiversity: a hierarchical approach. Conserv Biol 4:355–364

Paganini M, Leidner AK, Geller G, Turner W, Wegmann M (2016) The role of space agencies in remotely sensed essential biodiversity variables. Remote Sens Ecol Conserv 2:132–140

Pereira HM, Borda-de-Água L (2013) Modeling biodiversity dynamics in countryside and native habitats. In: Encyclopedia of biodiversity. Elsevier, San Diego, pp 321–328

Pereira HM, Navarro LM, Martins IS (2012) Global biodiversity change: the bad, the good, and the unknown. Annu Rev Environ Resour 37:25–50

Pereira HM, Ferrier S, Walters M, Geller GN, Jongman RHG, Scholes RJ, Bruford MW, Brummitt N, Butchart SHM, Cardoso AC, Coops NC, Dulloo E, Faith DP, Freyhof J, Gregory RD, Heip C, Hoft R, Hurtt G, Jetz W, Karp DS, McGeoch MA, Obura D, Onoda Y, Pettorelli N, Reyers B, Sayre R, Scharlemann JPW, Stuart SN, Turak E, Walpole M, Wegmann M (2013) Essential biodiversity variables. Science 339:277–278

Proença V, Martin LJ, Pereira HM, Fernandez M, McRae L, Belnap J, Böhm M, Brummitt N, García-Moreno J, Gregory RD, Honrado JP, Jürgens N, Opige M, Schmeller DS, Tiago P, van Swaay CAM (2017) Global biodiversity monitoring: from data sources to essential biodiversity variables. Biol Conserv 213:256–263

Reyers B, Stafford-Smith M, Erb K-H, Scholes RJ, Selomane O (2017) Essential variables help to focus sustainable development goals monitoring. Curr Opin Environ Sustain 26–27:97–105

Roche DG, Kruuk LEB, Lanfear R, Binning SA (2015) Public data archiving in ecology and evolution: how well are we doing? PLoS Biol 13:e1002295

Rüegg J, Gries C, Bond-Lamberty B, Bowen GJ, Felzer BS, McIntyre NE, Soranno PA, Vanderbilt KL, Weathers KC (2014) Completing the data life cycle: using information management in macrosystems ecology research. Front Ecol Environ 12:24–30

Stucky BJ, Guralnick R, Deck J, Denny EG, Bolmgren K, Walls R (2018) The plant phenology ontology: a new informatics resource for large-scale integration of plant phenology data. Front Plant Sci 9:517

Tittensor DP, Walpole M, Hill SLL, Boyce DG, Britten GL, Burgess ND, Butchart SHM, Leadley PW, Regan EC, Alkemade R, Baumung R, Bellard C, Bouwman L, Bowles-Newark NJ, Chenery AM, Cheung WWL, Christensen V, Cooper HD, Crowther AR, Dixon MJR, Galli A, Gaveau V, Gregory RD, Gutierrez NL, Hirsch TL, Hoft R, Januchowski-Hartley SR, Karmann M, Krug CB, Leverington FJ, Loh J, Lojenga RK, Malsch K, Marques A, Morgan DHW, Mumby PJ, Newbold T, Noonan-Mooney K, Pagad SN, Parks BC, Pereira HM, Robertson T, Rondinini C, Santini L, Scharlemann JPW, Schindler S, Sumaila UR, Teh LSL, van Kolck J, Visconti P, Ye Y (2014) A mid-term analysis of progress toward international biodiversity targets. Science 346:241–244

Triviño M, Thuiller W, Cabeza M, Hickler T, Araújo MB (2011) The contribution of vegetation and landscape configuration for predicting environmental change impacts on Iberian birds. PLoS One 6:e29373

Turner W (2014) Sensing biodiversity. Science 346:301–302

Wieczorek J, Bloom D, Guralnick R, Blum S, Döring M, Giovanni R, Robertson T, Vieglais D (2012) Darwin core: an evolving community-developed biodiversity data standard. PLoS One 7:e29715

Open Access This chapter is licensed under the terms of the Creative Commons Attribution 4.0 International License (http://creativecommons.org/licenses/by/4.0/), which permits use, sharing, adaptation, distribution and reproduction in any medium or format, as long as you give appropriate credit to the original author(s) and the source, provide a link to the Creative Commons license and indicate if changes were made.

The images or other third party material in this chapter are included in the chapter's Creative Commons license, unless indicated otherwise in a credit line to the material. If material is not included in the chapter's Creative Commons license and your intended use is not permitted by statutory regulation or exceeds the permitted use, you will need to obtain permission directly from the copyright holder.

Chapter 19
Prospects and Pitfalls for Spectroscopic Remote Sensing of Biodiversity at the Global Scale

David Schimel, Philip A. Townsend, and Ryan Pavlick

19.1 Introduction

Understanding how Earth's ecosystems will respond to drivers of change requires quantifying the composition and diversity of functions (e.g., growth, nutrient uptake, decomposition) of the taxa present in those ecosystems. Biodiversity is critical to maintenance of ecosystem health, which plays a critical role in global biogeochemical cycling. In turn, human well-being is dependent on ecosystems for services ranging from food and fiber to water and air quality.

Biodiversity can change in time and space as environmental conditions change (e.g., seasonally or due to natural disturbances, human activities, or climate variability) and is changing rapidly due to climate change and human activities. Even as it does, we have remarkably little information on biodiversity worldwide, with major spatial gaps in global coverage and very limited ability to detect changes over time (Jetz et al. 2016). Ground-based repeat data are needed to track changes in biodiversity and function but are particularly sparse. Our lack of understanding of species and functional diversity in both time and space leads to great uncertainty in predicting impacts of future changes in terrestrial and aquatic coastal marine ecosystems, making this an urgent target for improved understanding.

Functional biodiversity in terrestrial and aquatic coastal marine ecosystems is controlled by environmental factors, such as soils and climate, as well as by the evolutionary history and environmental filtering of the species pool. Changes can occur at a range of temporal scales, due to both ephemeral and catastrophic disturbances as well as to long-term processes (climate change, tectonics). Even among

D. Schimel (✉) · R. Pavlick
Jet Propulsion Laboratory, Pasadena, CA, USA
e-mail: David.Schimel@jpl.nasa.gov

P. A. Townsend
Department of Forest and Wildlife Ecology, University of Wisconsin, Madison, WI, USA

© The Author(s) 2020
J. Cavender-Bares et al. (eds.), *Remote Sensing of Plant Biodiversity*,
https://doi.org/10.1007/978-3-030-33157-3_19

similar ecosystems (e.g., tropical rainforests in South America vs. Africa vs. SE Asia), species and functional diversity can vary widely, and differential responses of those ecosystems to change depend on their functional composition and diversity. Current ecosystem models do not have access to data representing functional diversity and need better spatially explicit and temporally resolved measurements to more accurately characterize responses to change than can be obtained from ground-based and airborne observations.

Space-based measurements add a unique dimension to biodiversity studies. No single measurement can fully characterize biodiversity, and any single measurement only captures one or a subset of the dimensions of the diversity of life on Earth. Species diversity can be captured by taxonomic identification quantified by richness and turnover, while genetic analyses can capture additional levels of variation, including within-species variation. Traits are often used to characterize diversity of function, which may or may not parallel either species or within-species diversity. Remote sensing (RS) has long been used to characterize controls or correlates of diversity such as land use, fragmentation, productivity, habitat, and habitat structure. More recently, RS has been used to measure functional diversity (Schneider et al. 2017) and spectral diversity that predicts species diversity (Gholizadeh et al. 2019) and to characterize species diversity itself (Féret and Asner 2014). RS observations are indirect in the sense that inference must be made from the interaction of electromagnetic radiation with matter, but they provide far more sampling in time and space than most in-situ techniques and so are a valuable complement for inherently variable and even ephemeral aspects of the distribution and abundance of organisms (Saatchi et al. 2015).

For decades, scientists have proposed spaceborne imaging spectrometers as a means of obtaining information about vegetation in more detail than provided by multispectral sensors (Ustin et al. 1991), but only limited actual deployments have occurred (Pearlman et al. 2003) so ecologists have had access primarily to data collected from aircraft, over relatively limited areas. While global access is valuable (i.e., the ability to sample over all or most of the world while only collecting data over particular places), global coverage has proved enormously valuable for other measurements such as greenness, and global spectroscopy will enable studies of global biogeography, biogeochemical cycles, and biodiversity. As of 2020, this capability is in planning, and global assessment of plant functional diversity is likely to become a reality by the latter part of the 2020s. This paper will serve as an introduction to the likely system so ecologists can become aware of the general characteristics of the data that are likely to become available.

In 2018, the US National Research Council released the second Decadal Survey for the Earth Sciences (National Academies of Sciences, Engineering, and Medicine 2018). This document, following the first such survey, released in 2007 (National Research Council 2007), presented a clear vision for the role of RS in addressing some of the many dimensions of biodiversity. That report notes that "[k]ey fluxes within ecosystems are mediated by the composition and functional traits of the organisms present. Imaging spectroscopy is a tool for determining global terrestrial and marine plant functional traits, functional types and in some cases, provides taxonomic composition. Traits, types, and taxonomic composition, as well as their variability and how they are changing, are poorly understood globally. Nor is there a comprehensive understanding of how they feed back to the climate system via altered biogeochemical fluxes."

The report notes that physiological, chemical and morphological traits influence the functioning of ecosystems with respect to matter and energy fluxes and emphasizes that environmental change can change the distributions of the functional characteristics of organisms, thereby potentially feeding back to the climate system and other drivers of environmental change.

The concept that the unique characteristics of organisms mediate the functioning of ecosystems, and that changes to the distribution of functional traits can feed back to the environment and even the planetary climate, leads to a key question in the report:

- What are the structures, functions, and biodiversity of Earth's ecosystems, and how and why are they changing in time and space (National Academies of Sciences, Engineering, and Medicine 2018)?

and an objective

- Quantify the distribution of the functional traits, functional types, and composition of terrestrial and shallow aquatic vegetation ... spatially and over time (National Academies 2018).

The report identifies imaging spectroscopy (IS) as "the *only* technology that can provide the detailed spectral data to allow identification and quantification of major biochemical and structural components of plant canopies" and calls for an instrument in orbit optimized for the retrieval of a set of plant functional traits globally and able to track their changes over time. The report identifies a set of traits for terrestrial vegetation that are priorities and that are known to be observable based on the current literature, reflecting extensive field, lab, and airborne research. These traits, shown in Table 19.1, include plant traits that are linked to growth rate, photosynthetic capacity, longevity, and the decomposition rate of nonliving plant material (Serbin and Townsend, Chap. 3).

Table 19.1 Terrestrial plant functional traits identified by the National Academies report as currently observable by remote sensing (National Academies of Sciences, Engineering, and Medicine 2018). This list includes important remotely observable plant traits, but many other traits have been identified as candidates for remote measurement

Plant functional trait	Controls
Chlorophyll	Photosynthesis
Xanthophyll	Stress response of photosynthesis
Lignin	Structure rigidity and decomposability of dead plant material
Cellulose	Plant cell wall structure
Leaf nitrogen content	Photosynthesis, digestibility by herbivores, and decomposability of dead plant material
Leaf mass per area (LMA)	Foliar morphology: ratio of foliar dry mass to leaf area (reciprocal of specific leaf area, SLA)

The functional traits are used to estimate parameters and state variables in terrestrial ecosystem models controlling growth, turnover, and microbial respiration. As such they are key to Earth system prediction and simulation of future carbon cycling. They also define specific ecosystem characteristics. For example, chlorophyll is linked to absorption of sunlight and so affects albedo and the surface energy balance (Bonan 2016). Lignin and nitrogen content affect the palatability of forage for herbivores (Hobbs et al. 1991), thus influencing higher trophic levels, and xanthophyll indices have been proposed as leading to improved models of carbon flux (Garbulsky et al. 2010). Nitrogen additionally scales with the photosynthetic enzyme concentration of plants and therefore is a widely used indicator of photosynthetic capacity (Chapin 2003). LMA is indicative of relative investment by a plant in photosynthesis, leaf structure and/or leaf longevity and is strongly correlated with photosynthetic capacity (Poorter et al. 2009). The local diversity of plant functional traits has also been used as a predictor of plant species richness (Féret and Asner 2014) as well as for mapping functionally different plant communities (Asner et al. 2017).

The set of traits called for in the National Academies (2018) report includes plant characteristics long known to be useful for ecosystem studies (Schimel et al. 1991a), ecosystem modeling (Melillo et al. 1982), and studies of higher trophic levels (Hobbs et al. 1991). More recently, these same functional traits have been used in Earth system models (Verheijen et al. 2015) and studies of global biogeography (Butler et al. 2017; Moreno-Martínez et al. 2018). In many cases, particular species with distinct spectral characteristics can be mapped (see National Academies 2018 for a review), and phylogenetic relationships have been mapped based on spectral similarity (Cavender-Bares et al. 2016).

As a result of the studies mentioned above and many others, the National Academies (2018) report identified an imaging spectrometer with global coverage as a high priority and designated it for inclusion in NASA's next decadal program (2017–2027). This instrument, provisionally dubbed Surface Biology and Geology (SBG), is the first instrument planned for flight in the 2020s as a result of the 2018 decadal survey and could fly mid-decade. As well, the European Space Agency is proceeding with planning for the similar CHIME (Copernicus Hyperspectral Imaging Mission for the Environment) global imaging spectrometer (Rast et al. 2019).

19.2   Characteristics and Objectives for a Global Imaging Spectrometer

A number of IS instruments have been used for ecological studies over the years, starting with the AIS (the Airborne Imaging Spectrometer, Wessman et al. 1988) and including a number of airborne and spaceborne instruments. These instruments have all covered some part of the visible and shortwave infrared spectrum between 350 and 2500 nm, usually with spectral resolutions of 10–40 nm. Over the evolution of this technology, the signal-to-noise performance has systematically improved, so that ever-subtler spectral features could be resolved. Many other performance characteristics have also improved with time, leading to instruments that are better and better suited for ecological applications (Asner et al. 2012).

A diversity of airborne instruments exist today. In addition, several spaceborne imaging spectrometers are likely to fly in the next decade (see below), including NASA's Earth Surface Mineral Dust Source Investigation (EMIT), aimed at studying atmospheric dust; the Hyperspectral Imager Suite (HISUI) from Japan (Ogawa et al. 2017), a general purpose instrument deployed to the International Space Station (ISS) in 2019; and EnMAP from Germany. Nevertheless, the SBG instrument is needed to meet rigorous and community-determined objectives to quantify plant functional traits globally. SBG will be aimed at studying plant functional diversity, along with aspects of geology and hydrology, and will be paired with a thermal imager designed to study hydrological and geological processes as well as thermal drivers of ecosystem processes, such as evapotranspiration. The following discussion covers only those aspects of the instrument that directly address plant functional diversity. SBG's objectives for observing plant functional diversity are summarized in Table 19.2, excerpted from the National Academies (2018) document.

The Science and Applications Traceability Matrix (SATM) for the plant functional trait designated observable of SBG (Table 19.2) allows us to see the shape of the mission likely to fly in the mid-2020s. An SATM links science of a mission to the measurements necessary for the mission's objectives, as well as providing metrics for assessment of instrument performance for those measurements. What will the data from this mission look like? The matrix defines a number of key parameters, and those parameters allow us to see the opportunities and challenges this instrument will raise, once flown.

The data, radiance measurements of reflected sunlight, should cover the spectral region from 400 to 2500 nm, with approximately 10 nm spectral discretization, leading to an instrument with >200 spectral channels, meaning the data volume from this instrument will be very high. The information content would likewise be high, as work by Thompson et al. (2017) and Asner et al. (2012) shows that data like these collected over vegetated surfaces have as many as 30 significant principal components, meaning that pixels can be described in 30-dimensional space, while by contrast LANDSAT has 3 dimensions (Schimel et al. 2013).

The National Academies report calls for SBG's pixels to be of order 30–45 m, meaning each pixel is about 900–2000 m^2, collected nominally at least every 16, which could be halved if paired with the CHIME mission. The terrestrial Earth surface is about 2/3 covered with clouds at any given time, though that varies geographically and with time of day and season, so not every acquisition succeeds in capturing the land surface. A nominal 16-day revisit means most sites will be imaged, on the average, about once a month, but in some very cloudy regions and ecosystems—for example, humid tropical ecosystems—a 16-day revisit might mean one acquisition every few years. In the National Academies (2018) report, more frequent but less spectral resolved measurements are suggested as a complement to SGB-type measurements, providing less functional information though more phenological or time series information.

Signal-to-noise ratio is one of several spectroscopic parameters that determine the information content of IS data. Gross features of plant composition emerge with

Table 19.2 Science and Applications Traceability Matrix (SATM) for the SBG plant functional trait designated observable. This matrix relates the science question and objective to its targeted observable and then defines the objectives for the instrument (e.g., spectral resolution) and mission (e.g., revisit time) needed to achieve the targeted observable

Question	Earth science objective	Priority	Geophysical observable	Measurement parameters	Method	Data source	Targeted observable
What are the structure, function, and biodiversity of Earth's ecosystems, and how and why are they changing in time and space?	Quantify the distribution of the functional traits, functional types, and composition of terrestrial and shallow aquatic vegetation … spatially and over time	Very important DESIGNATED program element	Chemical and morphological properties of vegetation	Spectral radiance (10 nm) 380–2500 nm. Pixel size 30–45 m Revisit <15 days Signal to noise >400:1	High-fidelity imaging spectrometer, 150–200 km swath from low Earth orbit	SBG	Functional traits of terrestrial vegetation

SNRs of 100–200, but modern airborne IS instruments have SNRs of >1000:1, depending on the fraction of incoming sunlight reflected. The brighter the surface, the more light is reflected and the more light is incident on the instrument's detector, increasing SNR. This is also true spectrally, so even over a particular location, parts of the spectrum will have higher or lower SNR, depending on the spectral reflectance of the surface. With modern instruments (Mouroulis et al. 2011), SNR is not expected to be limiting for plant functional trait estimates.

Several other objectives for an SBG mission may be inferred from the wording of the SATM but require additional discussion to fully quantify. For example, the defining question for the plant functional trait designated observable asks why traits are changing in time and space. This implies broad coverage, ideally global, to observe spatial changes (e.g., along environmental gradients) and a long-enough time series to see at least some timescales of change.

While in many ecosystems, an SBG instrument would observe phenological change, several years or more on orbit would be required to see other types of change, such as successional change or change after disturbance. The report discusses three types of disturbance that could cause change, each with a different timescale. Wildfire causes extremely rapid change but may require decades or longer for full recovery. Pine beetle infestations, such as those have occurred in the Western United States and Canada, emerge more slowly, over several years, but may likewise trigger decades of recovery. Permafrost thaw is the slowest disturbance to emerge of the three considered; it does not lead to recovery but rather may initiate a long-term cascade of change to a new state. A global imaging spectroscopy mission would see some aspects of these disturbance processes as expressed through changing plant functional traits, but likely only a sequence of missions could see the slower aspects of recovery and change.

The formal mission study for SBG began in 2018. Some changes from the above description could occur between the National Academies (2018) definition of the observables and objectives identified for construction, so the above descriptions should be considered provisional.

19.3   EMIT, HISUI, and EnMAP

Several other imaging spectrometers with global access, although more limited coverage, are likely to fly before SBG. All will prove valuable in gathering imaging spectroscopy data in previously unsampled or inaccessible regions to serve as baselines for change in areas surveyed later by SBG (Schimel et al. 2013). While several other imaging spectrometers may fly, the three described below are most likely to provide data of interest to ecologists.

EMIT is a NASA Earth Venture Instrument investigation selected competitively and planned for flight on the ISS in 2021. It will be an advanced spectrometer, spanning 380–2500 nm, with 30 m pixels and a swath of ~40 km, aimed at studying the arid, dust source regions of the world to better understand the interaction of dust with the climate system. EMIT has a specific disciplinary focus in geology but will provide significant information about vegetation, especially in arid lands.

HISUI is a Japanese spaceborne hyperspectral instrument being developed by the Ministry of Economy, Trade and Industry. HISUI was launched in December, 2019 onboard ISS. HISUI has 185 spectral bands from 0.4 to 2.5 μm with 20 m × 30 m spatial resolution with swath of 20 km (note the contrast in mission characteristics—SBG will have a swath of 150,200 km). This narrow swath is limiting; observations have to be requested for specific targets during the planned HISUI mission lifetime of 3 years, with data availability starting in 2021.

EnMAP will orbit in a sun-synchronous orbit 653 km above the Earth. The satellite will be a high-resolution hyperspectral imager capable of resolving 230 spectral bands from 420 to 2450 nm with a ground resolution of 30 m × 30 m. The swath width will be 30 km at a maximum swath length of up to 5000 km/day due to data storage and downlink limits. The off-nadir (+/− 30°) pointing feature enables fast target revisits of 4 days and so can be used to study phenology.

Other instruments that may be of interest include DESIS, launched in 2018 to the ISS for the German aerospace agency (with 235 bands between 400 and 1000 nm and 30 m ground resolution), and SHALOM, planned by Italy and Israel. Together, all of these instruments will begin a revolution in observation of vegetation for research and applications in agriculture and natural resources.

19.4   Pitfalls and Opportunities in Remote Sensing of Global Plant Diversity

Every methodology for studying biodiversity has advantages and limitations. Identifying organisms and their locations provides foundational information but only captures biological function inferentially, for example, through correlation of ranges with climate. Genomic information is extremely informative, but it can only provide inferential insight into ecological function. Functional and/or physiognomic type (e.g., graminoid, tree, evergreen, deciduous, needleleaf, broadleaf) provide broad insights into ecological niches but obscure variation of other properties within groups. Each type of observation contributes some aspect or dimension of biological diversity in ancestry, form, function, distribution, and abundance, and remote observation is no exception. Spectroscopic measurement has its own unique characteristics, which depend partly on spectral region and resolution but also on the scale and frequency of the observation. Below, we discuss three aspects of measurement and how they affect the interpretation and use of IS data.

19.5   Vegetation Structure

The quantitative mapping of biodiversity in ecosystems may differ by physiognomic or functional type due to (i) differences in plant size within the field of view (e.g., trees vs. grasses; see below), (ii) phenology (e.g., grasslands in which dominant species and number of species presenting to a sensor may vary across the growing season), and (iii) vegetation physical structure, including overall plant form and leaf shape and longevity. At the leaf level, the interaction between a ray of light and foliar tissue is generally consistent among taxa, with biochemistry (e.g., pigments), water content, and leaf structure as the primary drivers of absorption and scattering (Jacquemoud et al. 2009). Scaled up to the canopy—i.e., as viewed by a satellite—the interactions change due to both the structure and ecology of vegetation. Variations in leaf area and layering, leaf orientation distribution and clumping, multiple scattering, and canopy heterogeneity add complexity to the signal that varies based on gross vegetation structure (Jacquemoud et al. 2009). However, ecological differences among vegetation structural types may be as significant: In some cases, the structural variability itself may be important to characterizing ecological dimensions of biodiversity (Townsend et al. 2013) or may confound the detection of variables driving diversity (Knyazikhin et al. 2013). Most significantly, however, the interpretation of variability (diversity) in retrieved traits may differ among ecosystem types.

Because of vegetation structure, the algorithms to map components of biodiversity (e.g., functional traits; see Serbin and Townsend, Chap. 3) differ. For example, a key ecological trait to characterize ecosystems is leaf mass per area (LMA), which is frequently used as a basis to estimate photosynthetic capacity and is also representative of environmental or evolutionary trade-offs between leaf construction costs and carbon uptake, i.e., longevity (thick leaves or needles, high LMA) vs. fast growth (thin leaves, high leaf area, low LMA) (Díaz et al. 2016; Reich et al. 1992; Shipley et al. 2006; Wright et al. 2004).

Differences in algorithms among functional types are illustrated by the retrieval algorithms to predict LMA and nitrogen in grasslands vs. forests (Fig. 19.2), which exhibit considerable differences due to variations in bidirectional reflectance. Standardized partial least squares regression (PLSR) coefficients indicate the relative contribution of a wavelength to prediction of a trait, and PLSR is one among multiple methods for mapping traits from imaging spectroscopy (discussed in more detail in Serbin and Townsend, Chap. 3). PLSR coefficients are especially useful for diagnostic purposes because they can be interpreted with respect to known foliar features. Figure 19.1 illustrates the differences in PLSR retrievals between ecosystem types but also that combined equations are possible. Also, LMA and nitrogen are generally negatively correlated among ecosystems worldwide (Wright et al. 2004), and as a consequence their PLSR coefficients are inverse of each other at some wavelengths (e.g., the grassland and forest model at 1730 nm).

The difference in structure among ecosystem types is driven in large part by several components of vegetation variation within a pixel. This has implications for both the derivation of biodiversity parameters and their interpretation. Leaf area index and canopy layering greatly influence the detection of vegetation characteristics. Elements of the canopy not directly exposed to the sun will be greatly reduced in significance for the calculation of any canopy characteristic. In forests, this means that a rich understory is often obscured, in contrast to a grassland in which many more components of the canopy are in the sunlit portion of the canopy exposed to a sensor. In addition to having technical consequences for the detection of functional or taxonomic diversity of a whole canopy, this also means that diversity metrics of

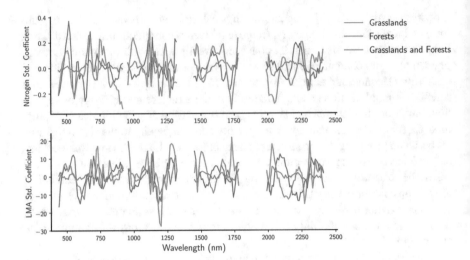

Fig. 19.1 Standardized partial least squares regression (PLSR) coefficients from three separate studies for predicting foliar nitrogen concentration and leaf mass per area (LMA). The standardized coefficients can be interpreted as the relative importance of each wavelength to retrieval of the trait. Differences in PLSR models between physiognomic types are generally related to canopy structure/physiognomic rather than leaf composition characteristics. The importance of different wavelengths varies due to both differences in absorption characteristics as a function of leaf structure and differences in canopy characteristics due to leaf shape, orientation, vertical layering, etc. Missing wavelengths are in spectral regions that are not used due to atmospheric absorption. (Grassland data from Wang et al. (2019), forest data from Singh et al. (2015), and grassland and forest data from Wang et al. (in revision). Model performance (R^2) for validation data: forest LMA = 0.88, %N = 0.84; grassland LMA = 0.83, %N = 0.57; together, LMA = 0.78, %N = 0.55)

forests may not be fully informative compared to lower stature canopies. This can be mitigated in seasonal vegetation by imaging prior to leaf emergence or following canopy senescence, but this could be challenging given cloud cover and timing of 15-day repeat cycles. This forms a critical area for new research.

Canopy height and its variation are related confounding factors, as shadowing within canopies creates significant effects. Leaf physiognomy and shadowing add further complexity (e.g., consider the contrast between herbaceous vegetation with only vertically distributed graminoids and those with mixed grasses and broadleaf forbs). Likewise, broadleaf and conifer foliage in forests exhibit contrasting spectral characteristics at the leaf and canopy levels due to shading caused by angular distribution of needles. At a basic level, a structurally continuous canopy will exhibit lower diversity than a heterogeneous one; while this effect can be leveraged to characterize diversity, it is important to understand that the resulting spectral diversity is a consequence of the combination of biological, chemical, and physical characteristics of vegetation.

It is conceivable that an acceptable cross-ecotype model is possible for mapping some elements of biodiversity in different physiognomic vegetation types (see green lines in Fig. 19.1), but it is equally possible that better, or more precise, models will be achieved using stratification by a few biome types (grasslands/tundra, savannas, shrublands/forests). Indeed, this approach has been adopted for many global RS products, such as multiple MODIS variables (e.g., LAI, GPP). Much work is now required to develop practical, broadly applicable yet acceptably accurate and well-characterized retrievals for global application. More specific and computational methods, such as radiative transfer model inversion, may be used for local applications (Morsdorf et al., Chap. 4). The history of imaging spectroscopy has been of relatively small data collections from aircraft and from space with Hyperion (Pearlman et al. 2003), and as a result, investigators have not been greatly constrained by computation and have used algorithms that require a fair amount of manual intervention. The likely volume of data from a global spectrometer is so great that algorithms will have to run largely unattended and simple, robust approaches will be needed!

19.6   Pixel Size

Much of the extant literature on IS and ecology has been collected using airborne imagers flying at comparatively low altitudes, so that the pixels are quite small; 1–2 m^2 is common (Asner et al. 2012; Kampe et al. 2010), and even data collected using NASA's research aircraft, the ER-2, flying at >20,000 m, are often 15 m^2, smaller than the expected 30–45 m^2 spaceborne pixels. Data on most of the dimensions of biodiversity can be tied to measurements or samples from not only a particular species but often an individual of that species. Most field sampling involves taxonomic identification followed by some sort of measurement or sampling process, increasingly tied to a geographic coordinate as well (Fig. 19.2).

By contrast, a 30 m pixel, 900 square meters, is not consistent with delineating a single organism or species in most cases. In fact, not only is there a scale mismatch between 30 m pixels and plants, but the magnitude of the mismatch depends on the ecosystem (see Gamon et al., Chap. 16). However conceptualized, the reality is that ecologists will have to cope with this range of mismatch and may only be able to inject intermediate scales at specific field sites, efforts which are likely to prove essential (Gamon et al., Chap. 16, Barnett et al. 2019). The range in size differences between plants and pixels is extraordinary. For example, to take one extreme, there is a factor of 10^{10} difference in size between a single phytoplankter and a 30 m pixel. A grassland represents an intermediate case, as a single grass plant (neglecting the fact that large swards of genetically identical individuals are known) is about 10^{-6} of a 30 m pixel, a million times smaller, while a large tree may cover a tenth or more of a 30 m pixel. There is thus not only a huge scale mismatch between plants and pixels, but the degree of mismatch may vary by 10^9 between ecosystems.

The scale mismatch implies that the plant functional trait measurement from space characterizes a collection of organisms, a huge number for phytoplankton, and quite a large one for herbaceous plants, but possibly just a few individuals in a forest. The interpretation of the functional trait measurements will thus depend

Fig. 19.2 The scale gap between organisms and the standard 30 m pixel ranges from 10^{10} to 10^1, depending on whether phytoplankton or large canopy trees are being imaged. When spectroscopic data are applied to address ecological and phylogenetic questions, this scale gap will have to explicitly factor into the analysis

somewhat on the ecosystem type or the gradient of ecosystem types. If, for example, the diversity of plant functional traits estimated within a region (Asner et al. 2017) were to be used to look at some aspect of plant competition, assuming that plants with different functional traits are interacting competitively (as in many individual-based forest models), how would pixel-level diversity be used in such a framework? This requires research and validation to determine, since these models are usually built around data collected in association with taxonomic identification, or highly aggregated into generalized plant functional types. The pixel-level data could be intermediate between these current practices.

The scale mismatch also raises challenges with calibration and validation of plant functional trait estimates. In situ, these measurements are normally made on individual leaves or other tissue, or small composites of leaves, samples typically $1/10^7$ the size of a pixel. Within a pixel, distributed either horizontally or vertically within a canopy, trait values may vary dramatically (Schimel et al. 1991b; Serbin et al. 2014), and careful extrapolation is required to even compare with remote observations (Singh et al. 2015). This extrapolation itself has to be done differently, for example, in forests where vertically sun and shade leaves differ, compared to grasslands, where hundreds of species may partition the environment, varying in traits within this pixel and sampling representatively can be very challenging.

19.7 Phenology and Temporal Aggregation

Functional traits and biodiversity are normally presented as stable characteristics of an ecosystem and are often mapped at broad scales (Butler et al. 2017; Moreno-Martínez et al. 2018) as a snapshot and implemented within models as static data. When functional traits are observed in situ, they are typically based on species-level sampling, sometimes repeated over a phenological sequence, though most commonly near-peak biomass. By contrast, a satellite comes over at a fixed schedule and collects data when cloud and aerosol conditions permit. Data may be selected to capture maximum "greenness" or some other criterion (e.g., Kampe et al. 2010), or data may be collected multiple times each year.

In many ecosystems, the species visible to a sensor in space vary (Sherry et al. 2011) over the course of a growing season. What is the best way to characterize that system's functional diversity? The simple sum of traits over a seasonal cycle might double-count some species, while a peak greenness assessment might underestimate functional diversity and neglect functionally important species specialized to early or late season. A global imaging spectrometer with consistent repeat imaging will enable us to both (i) address the effects of vegetation phenology on the retrieval of foliar traits from imaging spectroscopy data and (ii) assess the phenological drivers of variation in functional diversity at unprecedented spatial and temporal scales. Existing studies have largely been conducted in periods of peak greenness, and neither the seasonal variability of vegetation traits nor the effects of seasonal variability on retrieval algorithms have been addressed. These will be needed for an instrument such as SBG, with a ~15-day return interval. Addressing this problem, like the spatial scaling problem, requires considering the measurement in the context of ecological theory as well as considering pragmatic reality.

19.8 Conclusion

The likely flight of a spaceborne imaging spectrometer, with global repeat coverage, represents a transformative moment in global ecology. For the first time, a global measurement will address not only the function of life on Earth, as do greenness and other older measures, but at least one dimension of the diversity of life on Earth. SBG, CHIME and precursor missions like HISUI and EnMAP will provide a unique view of the range of plant functions and will no doubt produce new insights about the range of function, the extent and distribution of functional diversity of plants, and new aspects of habitat for other trophic levels. In addition, global spectroscopy is likely to become available at a time when many other structural and functional dimensions of biodiversity and ecosystem function are also available, reviewed in Schimel, Schneider et al. (2019). The parallel emergence of new ecological observatories will provide a wealth of systematic ground data to aid in developing theory and practical algorithms (Barnett et al. 2019). A number of challenges need to be confronted in order to consistently integrate IS with ecological theory and with the other dimensions of biodiversity, and the upcoming years will be the critical time for ecologists to address these challenges.

Acknowledgments PT received support from NSF Dimensions of Biodiversity grant 1342778, NSF Macrosystems Biology and Early Neon Science grant 1638720, NASA/JPL contracts 1579654 and 1590148, and USDA McIntire-Stennis/Hatch awards WIS01809 and WIS01874. This research, carried out at the Jet Propulsion Laboratory, California Institute of Technology, was under a contract with the National Aeronautics and Space Administration, copyright 2019 California Institute of Technology. We also thank the NIMBioS working group on remote sensing of biodiversity.

References

Asner GP, Knapp DE, Boardman J, Green RO, Kennedy-Bowdoin T, Eastwood M, Martin RE, Anderson C, Field CB (2012/9) Carnegie airborne observatory-2: increasing science data dimensionality via high-fidelity multi-sensor fusion. Remote Sens Environ 124:454–465

Asner GP, Martin RE, Suhaili AB (2012) Sources of canopy chemical and spectral diversity in lowland Bornean Forest. Ecosystems 15(3):504–517

Asner GP, Martin RE, Knapp DE, Tupayachi R, Anderson CB, Sinca F, Vaughn NR, Llactayo W (2017) Airborne laser-guided imaging spectroscopy to map forest trait diversity and guide conservation. Science 355(6323):385–389

Barnett DT, Adler PB, Chemel BR, Duffy PA, Enquist BJ, Grace JB, Harrison S, Peet RK, Schimel DS, Stohlgren TJ, Vellend M (2019) The plant diversity sampling design for the National Ecological Observatory Network. Ecosphere 10(2):e02603. https://doi.org/10.1002/ecs2.2603

Bonan GB (2016) Ecological climatology: concepts and applications. 3rd edition. Cambridge University Press, Cambridge, p 692

Butler EE, Datta A, Flores-Moreno H, Chen M, Wythers KR, Fazayeli F, Banerjee A, Atkin OK, Kattge J, Amiaud B, Blonder B, Boenisch G, Bond-Lamberty B, Brown KA, Byun C, Campetella G, Cerabolini BEL, Cornelissen JHC, Craine JM, Craven D, de Vries FT, Díaz S, Domingues TF, Forey E, González-Melo A, Gross N, Han W, Hattingh WN, Hickler T, Jansen S, Kramer K, Kraft NJB, Kurokawa H, Laughlin DC, Meir P, Minden V, Niinemets Ü, Onoda Y, Peñuelas J, Read Q, Sack L, Schamp B, Soudzilovskaia NA, Spasojevic MJ, Sosinski E, Thornton PE, Valladares F, van Bodegom PM, Williams M, Wirth C, Reich PB (2017) Mapping local and global variability in plant trait distributions. Proc Natl Acad Sci U S A 114(51):E10937–E10946

Cavender-Bares J, Meireles J, Couture J, Kaproth M, Kingdon C, Singh A, Serbin S, Center A, Zuniga E, Pilz G, Townsend P (2016) Associations of leaf spectra with genetic and phylogenetic variation in oaks: prospects for remote detection of biodiversity. Remote Sens 8(3):221

Chapin FS 3rd (2003) Effects of plant traits on ecosystem and regional processes: a conceptual framework for predicting the consequences of global change. Ann Bot 91(4):455–463

Díaz S, Kattge J, Cornelissen JHC, Wright IJ, Lavorel S, Dray S, Reu B, Kleyer M, Wirth C, Prentice IC, Garnier E, Bönisch G, Westoby M, Poorter H, Reich PB, Moles AT, Dickie J, Gillison AN, Zanne AE, Chave J, Wright SJ, Sheremet'ev SN, Jactel H, Baraloto C, Cerabolini B, Pierce S, Shipley B, Kirkup D, Casanoves F, Joswig JS, Günther A, Falczuk V, Rüger N, Mahecha MD, Gorné LD (2016) The global spectrum of plant form and function. Nature 529(7585):167–171

Féret J-B, Asner GP (2014) Mapping tropical forest canopy diversity using high-fidelity imaging spectroscopy. Ecol Appl 24(6):1289–1296

Gholizadeh H, Gamon JA, Townsend PA, Zygielbaum AI, Helzer CJ, Hmimina GY, Yu R, Moore RM, Schweiger AK, Cavender-Bares J (2019) Detecting prairie biodiversity with airborne remote sensing. Remote Sens Environ 221:38–49

Hobbs NT, Schimel DS, Owensby CE, Ojima DS (1991) Fire and grazing in the tallgrass prairie: contingent effects on nitrogen budgets. Ecology 72(4):1374–1382

Jacquemoud S, Verhoef W, Baret F, Bacour C, Zarco-Tejada PJ, Asner GP, François C, Ustin SL (2009) PROSPECT+ SAIL models: a review of use for vegetation characterization. Remote Sens Environ 113:S56–S66

Jetz W, Cavender-Bares J, Pavlick R, Schimel D, Davis FW, Asner GP, Guralnick R, Kattge J, Latimer AM, Moorcroft P, Schaepman ME, Schildhauer MP, Schneider FD, Schrodt F, Stahl U, Ustin SL (2016) Monitoring plant functional diversity from space. Nature Plants 2:16024

Kampe TU, Johnson BR, Kuester MA, Keller M (2010) NEON: the first continental-scale ecological observatory with airborne remote sensing of vegetation canopy biochemistry and structure. JARS 4(1):043510

Knyazikhin Y, Schull MA, Stenberg P, Mõttus M, Rautiainen M, Yang Y, Marshak A, Latorre Carmona P, Kaufmann RK, Lewis P, Disney MI, Vanderbilt V, Davis AB, Baret F, Jacquemoud S, Lyapustin A, Myneni RB (2013) Hyperspectral remote sensing of foliar nitrogen content. Proc Natl Acad Sci U S A 110(3):E185–E192

Melillo JM, Aber JD, Muratore JF (1982) Nitrogen and lignin control of hardwood leaf litter decomposition dynamics. Ecology 63(3):621–626

Moreno-Martínez Á, Camps-Valls G, Kattge J, Robinson N, Reichstein M, van Bodegom P, Kramer K, Cornelissen JHC, Reich P, Bahn M, Niinemets Ü, Peñuelas J, Craine JM, Cerabolini BEL, Minden V, Laughlin DC, Sack L, Allred B, Baraloto C, Byun C, Soudzilovskaia NA, Running SW (2018) A methodology to derive global maps of leaf traits using remote sensing and climate data. Remote Sens Environ 218:69–88

Mouroulis P, Van Gorp BE, White VE, Mumolo JM, Hebert D, Feldman M (2011) A compact, fast, wide-field imaging spectrometer system, in next-generation spectroscopic technologies IV. Int Soc Opt Photon 8032:80320U

National Academies of Sciences (2018) Engineering, and medicine: thriving on our changing planet: a decadal strategy for earth observation from space. The National Academies Press, Washington, DC

National Research Council (2007) Earth science and applications from space: national imperatives for the next decade and beyond. National Academies Press, Washington, DC

Ogawa K, Konno Y, Yamamoto S, Matsunaga T, Tachikawa T, Komoda M (2017) Observation planning algorithm of a Japanese space-borne sensor: hyperspectral imager SUIte (HISUI) onboard international Space Station (ISS) as platform, in sensors, systems, and next-generation satellites XXI. Int Soc Opt Photon 10423:104230R

Pearlman JS, Barry PS, Segal CC, Shepanski J, Beiso D, Carman SL (2003) Hyperion, a space-based imaging spectrometer. IEEE Trans Geosci Remote Sens 41(6):1160–1173

Poorter H, Niinemets U, Poorter L, Wright IJ, Villar R (2009) Causes and consequences of variation in leaf mass per area (LMA): a meta-analysis. New Phytologist 182:565–588

Rast M, Ananasso C, Bach H, Dor EB, Chabrillat S, Colombo R, Bello UD, Feret J-B, Giardino C, Green RO, Guanter L, Marsh S, Mieke J, Ong C, Rum G, Schaepman M, Schlerf M, Skidmore AK, Strobi P (2019) Copernicus hyperspectral imaging mission for the environment (CHIME) mission requirements document. 2019. European Space Agency. ESA-EOPSM-CHIM-MRD-3216. http://esamultimedia.esa.int/docs/EarthObservation/Copernicus_CHIME_MRD_v2.1_Issued20190723.pdf

Reich PB, Walters MB, Ellsworth DS (1992) Leaf life-span in relation to leaf, plant, and stand characteristics among diverse ecosystems. Ecol Monogr 62(3):365–392

Saatchi S, Mascaro J, Xu L, Keller M, Yang Y, Duffy P, Espírito-Santo F, Baccini A, Chambers J, Schimel D (2015) Seeing the forest beyond the trees. Glob Ecol Biogeogr 24(5):606–610

Schimel DS, Kittel TGF, Knapp AK, Seastedt TR, Parton WJ, Brown VB (1991a) Physiological interactions along resource gradients in a tallgrass prairie. Ecology 72(2):672–684

Schimel DS, Kittel TGF, Parton WJ (1991b) Terrestrial biogeochemical cycles: global interactions with the atmosphere and hydrology. Tellus A 43(4):188–203

Schimel DS, Asner GP, Moorcroft P (2013) Observing changing ecological diversity in the Anthropocene. Front Ecol Environ 11(3):129–137

Schimel D, Schneider F, JPL Carbon and Ecosystem Participants (2019) Flux towers in the sky: global ecology from space. New Phytologist (2019) 224:570–584. https://doi.org/10.1111/nph.15934

Schneider FD, Morsdorf F, Schmid B, Petchey OL, Hueni A, Schimel DS, Schaepman ME (2017) Mapping functional diversity from remotely sensed morphological and physiological forest traits. Nat Commun 8(1):1441

Serbin SP, Singh A, McNeil BE, Kingdon CC, Townsend PA (2014) Spectroscopic determination of leaf morphological and biochemical traits for northern temperate and boreal tree species. Ecol Appl 24(7):1651–1669

Sherry RA, Zhou X, Gu S, Arnone JA, Johnson DW, Schimel DS, Verburg PSJ, Wallace LL, Luo Y (2011) Changes in duration of reproductive phases and lagged phenological response to experimental climate warming. Plant Ecol Divers 4(1):23–35

Shipley B, Lechowicz MJ, Wright I, Reich PB (2006) Fundamental trade-offs generating the worldwide leaf economics spectrum. Ecology 87(3):535–541

Singh A, Serbin SP, McNeil BE, Kingdon CC, Townsend PA (2015) Imaging spectroscopy algorithms for mapping canopy foliar chemical and morphological traits and their uncertainties. https://doi.org/10.1890/14-2098.1

Thompson DR, Boardman JW, Eastwood ML, Green RO (2017) A large airborne survey of Earth's visible-infrared spectral dimensionality. Opt Express 25(8):9186–9195

Townsend PA, Serbin SP, Kruger EL, Gamon JA (2013) Disentangling the contribution of biological and physical properties of leaves and canopies in imaging spectroscopy data. Proc Natl Acad Sci U S A 110(12):E1074

Ustin SL, Wessman CA, Curtis B, Kasischke E, Way J, Vanderbilt VC (1991) Opportunities for using the EOS imaging spectrometers and synthetic aperture radar in ecological models. Ecology 72(6):1934–1945

Verheijen LM, Aerts R, Brovkin V, Cavender-Bares J, Cornelissen JHC, Kattge J, van Bodegom PM (2015) Inclusion of ecologically based trait variation in plant functional types reduces the projected land carbon sink in an earth system model. Glob Chang Biol 21(8):3074–3086

Wang Z, Townsend PA, Schweiger AK, Couture JJ, Singh A, Hobbie SE, Cavender-Bares J (2019) Mapping foliar functional traits and their uncertainties across three years in a grassland experiment. Remote Sens Environ 221:405–416

Wang Z, Chlus A, Geygan R, Singh A, Couture JJ, Kruger EL, Townsend PA In prep. A database of foliar functional trait maps across NEON domains in eastern U.S. Ecology

Wessman CA, Aber JD, Peterson DL, Melillo JM (1988) Remote sensing of canopy chemistry and nitrogen cycling in temperate forest ecosystems. Nature 335(6186):154–156

Wright IJ, Reich PB, Westoby M, Ackerly DD, Baruch Z, Bongers F, Cavender-Bares J, Chapin T, Cornelissen JHC, Diemer M, Flexas J, Garnier E, Groom PK, Gulias J, Hikosaka K, Lamont BB, Lee T, Lee W, Lusk C, Midgley JJ, Navas M-L, Niinemets U, Oleksyn J, Osada N, Poorter H, Poot P, Prior L, Pyankov VI, Roumet C, Thomas SC, Tjoelker MG, Veneklaas EJ, Villar R (2004) The worldwide leaf economics spectrum. Nature 428(6985):821–827

Open Access This chapter is licensed under the terms of the Creative Commons Attribution 4.0 International License (http://creativecommons.org/licenses/by/4.0/), which permits use, sharing, adaptation, distribution and reproduction in any medium or format, as long as you give appropriate credit to the original author(s) and the source, provide a link to the Creative Commons license and indicate if changes were made.

The images or other third party material in this chapter are included in the chapter's Creative Commons license, unless indicated otherwise in a credit line to the material. If material is not included in the chapter's Creative Commons license and your intended use is not permitted by statutory regulation or exceeds the permitted use, you will need to obtain permission directly from the copyright holder.

Chapter 20
Epilogue: Toward a Global Biodiversity Monitoring System

Gary N. Geller, Jeannine Cavender-Bares, John A. Gamon, Kyle McDonald, Erika Podest, Phil A. Townsend, and Susan Ustin

The loss of biological diversity in our era is occurring at a faster rate than at any time in the recent past. To effectively manage losses and avoid catastrophic outcomes, it is imperative to advance the understanding of how ecosystems are changing, what is being lost, and the fundamental causes driving extinction. With this in mind, the Group on Earth Observations Biodiversity Observation Network (GEO BON) was started in 2008 to begin building a global observation network and support improved management of the world's biodiversity and ecosystem services (Scholes et al. 2008). Despite good progress, much work remains to fully realize this vision, and RS has an important role to play – its potential to contribute to monitoring biodiversity and ecosystems has long been discussed in the literature, and it has proven to be extremely valuable (e.g., Stoms and Estes 1993; Nagendra 2001; Cash et al. 2003;

G. N. Geller (✉) · E. Podest
Jet Propulsion Laboratory, California Institute of Technology, Pasadena, CA, USA
e-mail: gary.n.geller@jpl.nasa.gov

J. Cavender-Bares
Department of Ecology, Evolution and Behavior, University of Minnesota,
Saint Paul, MN, USA

J. A. Gamon
Department of Earth & Atmospheric Sciences, University of Alberta, Edmonton, AB, Canada

CALMIT, School of Natural Resources, University of Nebraska – Lincoln, Lincoln, NE, USA

K. McDonald
Jet Propulsion Laboratory, California Institute of Technology, Pasadena, CA, USA

The City College of New York, City University of New York, New York, NY, USA

P. A. Townsend
Department of Forest and Wildlife Ecology, University of Wisconsin, Madison, WI, USA

S. Ustin
Department of Land, Air and Water Resources, University of California Davis,
Davis, CA, USA

© The Author(s) 2020
J. Cavender-Bares et al. (eds.), *Remote Sensing of Plant Biodiversity*,
https://doi.org/10.1007/978-3-030-33157-3_20

Turner et al. 2003; Jetz et al. 2016; Luque et al. 2018). The chapters in this book examine the use of RS to characterize and monitor biodiversity, focused largely on plant diversity. The authors have collectively explained the technologies involved and the analytical and conceptual approaches available for applying RS to monitor changes in the multiple dimensions of biodiversity and for evaluating ecosystem condition and function. The techniques discussed reflect recent advances in RS technologies, which have enabled massive increases in the amount of data available, and data science, which has developed new methods for applying these data to the detection and prediction of biodiversity. These advances are occurring in parallel with those in computing technology and statistical and analytical methods. Integrating all of these advances creates new opportunities to monitor biodiversity change and ecosystem condition and function at a global scale. Coincidentally, these opportunities arise at a time when the climate is changing rapidly and human population continues to grow, both of which further increase pressures on the world's natural systems and its biodiversity, accelerating declines that began decades ago. The emergence of international efforts to develop a global biodiversity monitoring system that can integrate RS with traditional field methods offers some hope that the ability to assess change and enhance management efforts can be improved.

Spaceborne RS has special value for monitoring these biodiversity changes because it is global, consistent, periodic, and, for Landsat, has a retrospective record going back more than 40 years. A system that combines the techniques, approaches, and lessons learned described in this book's 19 chapters would support monitoring Earth's biodiversity globally and at unprecedented levels of spatial and temporal resolution. Such a system would facilitate breakthroughs in scientific knowledge as well as provide previously unavailable information to manage biodiversity and natural resources. But spaceborne RS, by itself, is not enough. Utilizing it requires a suite of additional data, including airborne RS and a wide variety of in-situ measurements, so these must also be part of any global monitoring system.

How can these many pieces be brought together? It seems a fitting way to end this book by briefly exploring ideas for a system that can do that.

20.1 Current Situation

Most RS technology and instrumentation are developed and operated by government space agencies such as the National Aeronautics and Space Administration (NASA, USA) and the European Space Agency (ESA, Europe); while there are others, these organizations currently provide the bulk of the world's freely accessible data. Agencies such as these acquire, store, manage, and process data; they distribute data that support development of the algorithms used to generate these products. While some recent projects take a different approach, traditionally, development was assigned to specific funded teams to develop algorithms, which were then used to generate a mostly static list of standard products over the life of a particular satellite program. These products are available for download by individual users who can then process them further and apply them to problems of interest.

This approach was compatible with the computing environment available at the time; since then, however, huge advances in computing technology, including greatly decreased cost and the availability and flexibility of the cloud (which, in turn, enables new levels of co-development), have enabled alternative approaches. In fact, these approaches are necessitated by the increase in the amount and diversity of RS data available (e.g., downloading huge volumes of data for local processing is impractical).

20.2   Remote Sensing for Global Biodiversity Monitoring: Building on GEO BON

Taking advantage of these new opportunities requires not only a new level of agility in algorithm and product development but also new approaches for data management and processing. For example, collaboration both within and across disciplines is crucial because advances in science often require multidisciplinary collaboration. Effective use of data from different sources or collected at different times and scales requires a greater degree of data sharing and collaboration than is currently common. Not only have advances in understanding ecosystems and the interaction of their many components made the science increasingly multidisciplinary, but technology, particularly its move toward cloud computing, makes collaboration both easier and more natural. The workspaces the cloud provides also facilitate experimentation with and exploration of both algorithms and data, and the cloud addresses the data volume issue by "bringing the user to the data."

Recognition of the benefits of multidisciplinary collaboration, shared workspaces, and a shift toward cloud computing is not completely new. The NASA Earth Exchange, for example, first became available in 2012, and Google Earth Engine started around the same time. The ESA Thematic Exploitation Platforms, which focus on specific themes, began in 2014, and the European Commission-funded Data and Information Access Services started to go online in 2016. Most recently, development began on the Multi-mission Algorithm and Analysis Platform (MAAP), a cloud-based, joint NASA-ESA activity that will support several new missions. Key features of MAAP include support for collaborative algorithm development as well as flexibility in terms of which algorithms are developed or products generated. However, despite the critical importance to society, no system has yet emerged that provides these and the suite of other capabilities and data needed to monitor biodiversity or related ecosystem functions.

A concept for such a system is summarized as a high-level architecture in Box 20.1. It makes extensive use of RS data, particularly satellite RS, because that is the only practical way to obtain the periodic coverage needed for regular global monitoring. As such, it augments GEO BON's global monitoring work with a focus on RS and cloud-based processing that can take advantage of a variety of new technology-enabled opportunities. It also anticipates several new sensors and sensor types, particularly imaging spectroscopy, lidar, and radar; these technologies and their application to biodiversity monitoring are discussed in many of the book's

Box 20.1 System Concept Overview

The system concept consists of several basic components that reside in the cloud. At the bottom is a large data store that feeds the processing area and also acts as a repository for published products. The RS feedstock consists largely of analysis-ready data (ARD), which are data that have been preprocessed to simplify further processing. A variety of other data (in-situ, airborne, and ancillary data such as DEMs) also reside there or are accessed directly from the provider's site during processing. The "sandbox" serves two purposes. First, it supports the development of algorithms by providing a space where code can be developed, shared, and tested; published algorithms can be stored in an "algorithm warehouse." The sandbox also provides a space for experimentation by scientists that need, for example, to run an existing algorithm with nonstandard parameters, to run their own models, or to combine data and algorithms in new ways in support of their research. A variety of tools are available to support both types of sandbox users. The processing area is where algorithms are staged and then run to generate products; depending on the product, processing may involve a chain of steps that produce intermediate products that may or may not be published. On the far right is a toolbox with tools to find and access the products in the data store as well as to interact with, understand, and utilize the data; these latter tools are of particular importance for applied users.

chapters. Importantly, the architecture is inherently flexible, providing a suite of basic capabilities that can be utilized in a variety of ways; for example, which algorithms are developed and how they are assessed and published depends upon the data the system hosts and how the system is governed. Box 20.2 summarizes some

Box 20.2 Some Key Characteristics of a Global Biodiversity Monitoring System

- Easy collaboration and facilitation of cross-discipline interactions. This enhances algorithm development, scientific experimentation, and applications.
- Agility and flexibility in algorithm development and data processing approaches. This facilitates a broad range of algorithms and products, makes the system responsive to the needs of users of all types, and can increase product quality.
- Integration – including fusion among different sensors, such as optical and radar, as well as among RS and in-situ data. Integration also involves addressing the challenge of data interpretation across spatial and temporal scales. The data already available provides opportunities that have not yet been fully utilized, in part because different types of data are handled by different communities.
- Simplified processing. Development of analysis-ready data will save both algorithm development time and computer resources. Analysis-ready data are standardized data for which some key processing steps, such as atmospheric correction, have already been executed.
- Utilization of advancing technologies. These include those related to sensors, such as imaging spectroscopy, thermal, radar, and lidar, but also of genomic technology, and those related to processing huge volumes of time series data in the cloud.
- Derived products and tools to increase usability and understanding. To extract the full value from data acquired and the products derived from it, many users, particularly decision-makers such as land managers, will need more highly derived products as well as tools to help them understand what the data mean and the problems they can address. A cloud-based system can facilitate development, generation, analysis, and sharing of derived products.
- Increased access to in-situ, flux tower, and airborne or UAV RS data. These data are absolutely essential because they tie the spaceborne data to what is happening on the ground. Although a tremendous amount of this type of data has been collected, only a fraction of it is accessible for integration with satellite RS data. These data will be needed to take full advantage of the opportunities discussed, but accessing them is a challenge.

of the key characteristics that such a system should have. Most of these explicitly take advantage of advancing science and technology, but there are some that technology alone cannot enable. Addressing these is important because they limit the value that a system can extract from RS data and thus the value of those data to society. By addressing the challenges of integrating vastly different data types across a range of spatial and temporal scales – and particularly when combined with the new and forthcoming spaceborne sensors – such a system will enable a new era for biodiversity monitoring, and it will be global.

A system like that in Box 20.1, of course, is only part of the picture – the other part, upon which it completely depends, acquires the data it utilizes. While many of these data are collected by the spaceborne instruments operated by several space agencies, utilizing them to understand biodiversity on the ground depends critically on in-situ data. Integrating the two is a key challenge at hand, as discussed in the introduction to the book (Chap. 1), and is what enables inferences to be made from space about the biodiversity on the ground. Airborne data is often used as an intermediary – basically, a "scaling tool" for understanding the scale dependence and process-level understanding of signals related to biodiversity (Gamon et al., Chap. 16). Thus, off the bottom edge of the figure in Box 20.1, there exists a huge suite of data collection activities that are not shown. While some of those data are made widely available – the Global Biodiversity Information Facility (GBIF) is an excellent example – a tremendous amount of in-situ and airborne data remains inaccessible or difficult to locate and utilize (see Fernández et al., Chap. 18). The data may not be published online, but even if published the variables collected, the methods used, and the formats of the data are often specific to each activity because they are operated by independent projects or organizations.

Lack of standardization is one of the challenges involved in developing an integrated system for biodiversity monitoring, though this issue is starting to be addressed. GEO BON and its parent organization (the Group on Earth Observations, GEO), the International Long-Term Ecological Research (ILTER) site network, the US National Ecological Observatory Network (NEON), and a variety of sponsors and other organizations are working to enhance coordination and to develop guidelines and standards. Many of these activities are sponsored by governments, and in fact it is government agencies that can best facilitate and develop coordinated, operational observation systems. Thus, one of GEO BON's focal areas is development of national and regional Biodiversity Observation Networks (BONs), and as discussed in Fernández et al. (Chap. 18), GEO BON is developing a suite of Essential Biodiversity Variables to provide top-level guidance on what data these BONs should collect and to facilitate development of standards. For spaceborne RS data, the Committee on Earth Observation Satellites (CEOS), which includes most national space agencies, facilitates the coordination of missions as well as of data standards.

Even a brief overview of a system concept like this should discuss how to ensure that the products it generates meet the needs of its target users. The "usability" of any product depends on who the user is and their level of expertise. Historically, most RS-based datasets and products have been oriented toward scientific users who have the resources and expertise to process them further. However, other users such as land managers or decision-makers, whose expertise lies elsewhere, require more specialized and more highly derived products – as well as user-friendly tools that enable them to explore and understand the meaning of those products for efficient application. These tools appear toward the right edge of the figure in Box 20.1.

As GEO BON has demonstrated, a global biodiversity monitoring system must be a coordinated effort among many international and national organizations and user communities and be built upon the vast amount of existing knowledge and data that has already been acquired. Taking full advantage of RS data and several advancing technologies will provide new insights into the status and trends in biodiversity, ecosystem functions, and ecosystem services and support operational monitoring at new scales. This will vastly enhance the understanding of our biological systems, how they are changing, and how society should respond.

Acknowledgments Many of the ideas presented in this Epilogue reflect discussions within the workshop on "Unlocking a New Era in Biodiversity Science" at the Keck Institute for Space Studies. Participants, in addition to the chapter authors, included Laura Bertola, Elizabeth Blood, Ana Carnaval, Patrick Comer, David Coomes, Néstor Fernández, Christian Frankenberg, Robert Guralnick, Walter Jetz, Troy Magney, José Eduardo Meireles, Charles Miller, Enrique Montes, Frank Muller-Karger, Helene Muller-Landau, Ruth Oliver, Monica Papeş, Ryan Pavlick, Naiara Pinto, David Schimel, Anna Schweiger, Gary Spiers, Derek Tesser, David Thompson, Woody Turner, and Maria Tzortziou. The research described herein was in part carried out at the Jet Propulsion Laboratory, California Institute of Technology, under contract with NASA; government sponsorship acknowledged.

References

Cash DW, Clark WC, Alcock F, Dickson NM, Eckley N, Guston DH, Jäger J, Mitchell RB (2003) Knowledge systems for sustainable development. Proc Natl Acad Sci 100:8086
Jetz W, Cavender-Bares J, Pavlik R, Schimel D, Davis FW, Asner GP, Guralnick R, Kattge J, Latimer AM, Moorcroft P, Schaepman ME, Schildhauer MP, Schneider FD, Schrodt F, Stahl U, Ustin SL (2016) Monitoring plant functional diversity from space. Nat Plants 2:16024. https://doi.org/10.1038/NPLANTS.2016.24
Luque S, Pettorelli N, Vihervaara P, Wegmann M (2018) Improving biodiversity monitoring using satellite remote sensing to provide solutions towards the 2020 conservation targets. Methods Ecol Evol 9:1784–1786. https://doi.org/10.1111/2041-210X.13057
Nagendra H (2001) Using remote sensing to assess biodiversity. Int J Remote Sens 22:2377–2400
Scholes RJ, Mace GM, Turner W, Geller GN, Jurgens N, Larigauderie A, Muchoney D, Walther BA, Mooney HA (2008) Toward a global biodiversity observing system. Science 321:1044–1045
Social Learning Group (2001) Learning to manage global environmental risks. A functional analysis of social responses to climate change, ozone depletion, and acid rain. MIT (Massachusetts Institute of Technology) Press, Cambridge, MA
Stoms DM, Estes JE (1993) A remote sensing research agenda for mapping and monitoring biodiversity. Int J Remote Sens 14:1839–1860
Turner W, Spector S, Gardiner N, Fladeland M, Sterling E, Steininger M (2003) Remote sensing for biodiversity science and conservation. Trends Ecol Evol 18:306–314

Open Access This chapter is licensed under the terms of the Creative Commons Attribution 4.0 International License (http://creativecommons.org/licenses/by/4.0/), which permits use, sharing, adaptation, distribution and reproduction in any medium or format, as long as you give appropriate credit to the original author(s) and the source, provide a link to the Creative Commons license and indicate if changes were made.

The images or other third party material in this chapter are included in the chapter's Creative Commons license, unless indicated otherwise in a credit line to the material. If material is not included in the chapter's Creative Commons license and your intended use is not permitted by statutory regulation or exceeds the permitted use, you will need to obtain permission directly from the copyright holder.

Correction to: A Range of Earth Observation Techniques for Assessing Plant Diversity

Angela Lausch, Marco Heurich, Paul Magdon, Duccio Rocchini, Karsten Schulz, Jan Bumberger, and Doug J. King

Correction to:
Chapter 13 in: J. Cavender-Bares et al. (eds.),
Remote Sensing of Plant Biodiversity,
https://doi.org/10.1007/978-3-030-33157-3_13

The original version of this book was inadvertently published with an incorrect affiliation "Humboldt University to Berlin, Geography Department, Leipzig, Germany, Humboldt Universität zu Berlin, Geography Department, Berlin, Germany, e-mail: angela.lausch@ufz.de" of the 13th chapter author Angela Lausch.

The correct affiliation "Department Computational Landscape Ecology, Helmholtz Centre for Environmental Research–UFZ, Leipzig, Germany, Geography Department, Humboldt University Berlin, Berlin, Germany, e-mail: angela.lausch@ufz.de" has been updated in 13th chapter.

The updated online version of this chapter can be found at
https://doi.org/10.1007/978-3-030-33157-3_13

© The Author(s) 2020
J. Cavender-Bares et al. (eds.), *Remote Sensing of Plant Biodiversity*,
https://doi.org/10.1007/978-3-030-33157-3_21

Glossary

3-D segmentation or 3-D image segmentation object categorization A specific case of spectral clustering applied to image segmentation.

Abaxial The surface of a leaf facing away from the stem (particularly during early development), usually the lower surface of a leaf or petiole.

Abiotic Non-biological physical environmental variables, such as net radiation, temperature, and precipitation.

Abiotic niche The characteristic physical attributes, such as weather and climate, of a species fundamental niche.

Absorptance The portion of electromagnetic radiation incident upon a leaf that is absorbed, expressed as a percentage or decimal fraction. Absorptance (A) is typically derived from integrating sphere measurements of reflectance (R) and transmittance (T), where A + R + T = 1 (100%). Note that absorptance differs both from *absorbance* (the log of the ratio of the transmitted to incident light) and *absorption* (the process of absorbing light).

Absorption feature An absorption pattern associated with particular wavelengths that can frequently be attributed to a particular compound that absorbs at those wavelengths.

Abundance refers to the evenness or number of individuals per species in a community and can be calculated as absolute or relative abundance.

Accuracy A measure of distance of a measurement from the true value.

Active radar A remote sensing method that employs an antenna to send and receive microwave frequency/wavelength pulses.

Active sensor One that emits a signal and detects the amount returned.

Adaxial The surface of a leaf facing toward the stem axis (during early development), typically the upper surface of a leaf or petiole.

Airborne platform A plane, drone (defined as an unmanned aerial system (UAS); see below), balloon, blimp, or other flying object that carries a remote sensing instrument.

Airborne remote sensing Remote sensing performed with sensors installed on an airborne platform (e.g., drone, copter, or fixed-wing aircraft).

© The Author(s) 2020

J. Cavender-Bares et al. (eds.), *Remote Sensing of Plant Biodiversity*,
https://doi.org/10.1007/978-3-030-33157-3

Aichi targets   The 20 global biodiversity targets identified in the Strategic Plan for Biodiversity 2011–2020 of the Convention on Biological Diversity.

Albedo   The reflected incident radiation of a surface, typically reported as the fraction of reflected incident radiation over the full range of the solar spectrum.

Alpha diversity   Diversity within a site or community, calculated based on any dimension of biodiversity.

Amplicon   A fragment of DNA or RNA that is amplified in the laboratory often using polymerase chain reactions (PCR) or ligase chain reactions (LCR) or naturally via gene duplication.

Analysis ready data (ARD)   Remote sensing data products that have been consistently processed to a level required for direct use in monitoring and assessing change. Products are held to high scientific standards and typically involve derived products such as top of atmosphere reflectance, surface reflectance, and pixel quality assessment.

Angular scale   in optical remote sensing, often defined as the incident and sampling angle of reflected radiation (i.e., the sampling geometry). Sometimes called directional scale.

Anisotropic reflectance   Scattering of photons off a surface in a directional manner due to the structure and optical properties of the surface.

Anthocyanin   A type of flavonoid pigment (a family of more than 6000 polyphenolic compounds) that is water-soluble. The colors range from red to blue or violet, depending on the pH. Anthocyanins can be found in all plant organs. In leaves they are non-photosynthetic, vacuolar pigments that provide protection from excess light or herbivory or serve as attractants.

Anthropocene   The proposed current geological epoch, categorized by substantial human impact on the world. [Note that as of this writing, the International Commission on Stratigraphy has not yet finalized approval of this term, but it is widely accepted by the scientific community.]

Aspect   The direction a slope is facing.

Atmosphere   The layer of gases, particles, and aerosols surrounding a planet that is maintained by gravity.

Atmospheric correction   The process by which the effects of scattering and absorption of photons by the atmosphere are accounted for and removed to generate the apparent surface reflectance values of imagery acquired by satellite or airborne sensors.

Aufeis   Accumulations of large masses of ice in many Arctic rivers.

Azimuth angle   The angle along the horizon between the sun's position and north.

Backscatter   The electromagnetic signal that is scattered by the target back to the detector.

BAM diagram   A graphical representation of the geographical space for understanding the distribution of species. The diagram is composed of three determinants of species distributions: the biotic niche (B), the abiotic niche (A), and areas accessible through migration or colonization (M).

Band   A defined wavelength interval of electromagnetic radiation, sometimes determined by passing radiation through a filter or other device. See also "spectral band," "FWHM."

Band center The central wavelength of a wavelength interval, both of which define a spectral band.

Band collinearity The degree of linear correlation between spectral bands.

Band-depth analysis characterizing the intensity of reflectance at an absorption feature of a particular spectral region, usually in comparison with the reflectance at the shoulders of the absorption feature or reflectance in the absence of the feature. See also continuum removal.

Basal area refers to the cross-sectional area of tree stems in a given sampling area. Typically, stem diameters are measured at breast height (dbh), approximately 1.3 m from the ground, and basal area is reported as total stem surface area per unit land area.

Bathymetry The study of or measurements of underwater depth in oceans, lakes, or seas.

Bayesian statistical methods based on Bayes theorem in which parameters of a probability distribution or statistical model are estimated using prior information about the parameters (priors), observed data, and a model. Posterior probabilities of the parameters are determined from the prior probability and a likelihood function based on the model, given the observed data.

Belowground processes processes that occur in the soil, including both biological and physical processes. Biological processes include, for example, microbial processes and processes conducted or mediated by plant roots and soil invertebrates, such as nutrient uptake, root respiration, root exudation, root growth and death, microbial respiration, nutrient mineralization and immobilization, litter breakdown and decomposition, nitrification, denitrification, water transport, solute transport, and cation exchange. Physical processes include, for example, mineral weathering and bulk flow of water and dissolved nutrients.

Beta diversity The spatial differentiation of biodiversity, the variation in diversity, and composition among sampling units (nondirectional beta diversity) or along spatial, temporal, or environmental gradients (directional beta diversity).

Bidirectional reflectance The two-directional path (incidence and reflectance) of photons that interact with a surface and are reflected. See also angular scale.

Bidirectional reflectance distribution function A mathematical description of surface reflectance patterns consisting of both azimuthal and zenith angle information. In remote sensing of vegetation, this considers both the shape/density of a canopy and the sun-sensor-target viewing geometry.[Our variables influence the pattern of light reflected from an opaque surface, including incoming and outgoing light directions, which are defined in terms of their azimuth and zenith angle directions.]

Bignones Members of the plant family Bignoniaceae.

Biochemical refers to chemical processes and substances that occur within living organisms. Biochemicals of interest in plants are typically those related to a particular function, such as carbon fixation, defense, metabolism, stress tolerance, structural support, or genetic coding.

Bioclimatic variable A climatic variable that is believed to have biological significance and represents temperature or precipitation data derived from interpolated monthly values. Such variables are interpolated across the globe from weather station data and provide annual trends, extremes, and seasonality in both temperature and precipitation.

BioCondition   A measure of the capacity of the terrestrial ecosystem to maintain biodiversity values at a particular scale.

Biodiversity   The variability among living organisms and the ecosystems in which they live, including genetic and phenotypic diversity within species, variation among species across the tree of life, and variation among ecosystems. The Convention on Biodiversity defines it as "the variability among living organisms from all sources including, inter alia, terrestrial, marine and other aquatic ecosystems and the ecological complexes of which they are part; this includes diversity within species, between species and of ecosystems."

Biodiversity ecosystem function (BEF) relationships   The nature and mechanisms by which biodiversity influences (or is influenced by) ecosystem functions and processes, such as productivity of the ecosystem, decomposition, nutrient cycling, resistance to drought, stability of biomass through time, etc.

Biogeographic assembly   The distribution, movement, and grouping of species in a given geographical area over evolutionary time scales.

Biological scale   The hierarchical nature of biological organization. For example, individual organisms are nested within populations, which are nested within species and within increasingly deeper clades.

Biotic homogenization/differentiation   A change in species composition, structure, and/or function where different ecosystems, ecological communities, and/or land areas become more similar (homogenized) or less similar (differentiated) through time.

Biological species concept   Species defined as a group of individuals that interbreed (or can potentially interbreed). Applies to sexually reproducing organisms.

Biome   A large-scale assemblage of species occupying a particular region of the planet, typically defined by climate space. Examples include the boreal forest, the Arctic tundra, deserts, grasslands, and tropical rainforests.

Bio-optical   Optical properties resulting from the presence of biological matter.

Biophysical   Any interactions between a biological organism and its physical environment. In the context of remote sensing of plant biodiversity, it involves characterizing how vegetation properties interact with photons.

Biophysiology   The study of physiological processes of whole plants or animals and/or their organs, including the study of structure, growth, and morphology.

Bootstrap   to generate new datasets based on a current dataset. This term often implies *nonparametric bootstrapping*: generating samples with replacement from an existing dataset. Another approach, less common, is *parametric bootstrapping*: simulating new datasets with a model based on your existing dataset. Bootstrap approaches are typically used to estimate uncertainty in parameter estimates from data.

Broadleaf   (broad-leaf; broadleaved) an angiosperm plant, typically a dicotyledonous species. It is usually contrasted with needle-leaf plants, like conifers, or narrow leaves of graminoids.

Bromeliads   Informal name for members of the plant family Bromeliaceae, also known as the pineapple family.

Brownian motion model of evolution A model that depicts the evolution of a trait as the accumulation of many random changes over time in a diffusion-like process. At any point in time, a trait value is equally like to increase or decrease. The pace at which those random changes accumulate is referred to as the rate of evolution.

Calibration Comparison of an instrument or measurement to a known standard, often with known units (e.g., referenced to the US National Institute of Standards and Technology).

Calibration data Data used to train models from imagery, including calibration for atmospheric effects, georegistration, spectral resolution, and any retrieval of interest.

Canopy The upper, aerial region of a plant or group of plants (e.g., tree crown or crowns), often including other biological organisms (epiphytes, lianas, arboreal animals, etc.). In ecology, the canopy often refers to the crown of a single plant, whereas in remote sensing, it often refers collectively to the crowns of many plants in a stand (e.g., trees in a forest).

Canopy cover Projection of tree crowns onto the ground divided by ground surface area.

Canopy structure The architecture of the aerial portion of a plant or plant stand that influences the way plants interact with light and the reflectance patterns they generate. Typical metrics include leaf area index, plant area index, leaf angle distribution, branching structure, crown shape, etc.

Carbon cycling A critical biogeochemical cycle on Earth that governs how carbon moves from living to nonliving chemical forms and involves transfers between various "spheres" (biosphere, atmosphere, hydrosphere, lithosphere). The fast carbon cycle (involving the biosphere) can be distinguished from the slow carbon cycle (involving the deep lithosphere and long-term geological storage).

Carbon flux Movement of carbon between Earth spheres (e.g., atmosphere, biosphere, and hydrosphere). Typically expressed as unit carbon per cross-sectional area per unit time.

Carotenoids Plant tetraterpenoid pigments (a family of more than 1100 molecules), with hydrocarbon chains of conjugated double bonds ending in benzene rings. They form red, yellow, and orange colors in leaves, fruit, and flowers (e.g., autumn leaf colors). They are grouped into two types: those without oxygen (the carotenes) and those with oxygen (the xanthophylls). In plants, these molecules help capture light energy for photosynthesis, or shield against excess energy, and also act as antioxidants for photoprotection.

Cation-exchange capacity A measure of how many cations (atoms or molecules with a positive charge) can be held on soil surface particles (typically negatively charged) and thus are made available to plants.

Chemical phylogeny The evolutionary relationship of chemical traits.

Chemometrics A subdiscipline of chemistry aimed at extracting information from chemical measurements in mixed environments using mathematical and statistical procedures that can be applied to spectroscopic data and remotely sensed imagery.

Chlorenchyma A type of parenchyma cell in leaf mesophyll tissue that contains chloroplasts and is photosynthetic.

Chlorophyll Green pigment used by plants to absorb energy from sunlight to generate the energy necessary for plants to fix carbon via photosynthesis.

Ciliate A single-celled protozoan characterized by hair-like organelles called cilia. Ciliates are typically found in water.

Citizen science The collection and analysis of data by members of the general public, frequently led by and generally in collaboration with professional scientists. Crowdsourcing is a form of citizen science in which datasets are built continuously through the contribution of groups of citizen scientists.

Clade A branch of a phylogenetic tree composed of an ancestor and all of its descendants. Also known as lineage. *Or* a group of taxa containing all the descendants of a single ancestor.

Classification/image classification A statistical procedure to identify different classes of objects or features in an image. It can be unsupervised (based on the variance in the data) or supervised (based on the characteristics of the training data).

Classifier (i.e., in species detection) A set of rules or a mathematical function that uses pixel data to assign or predict class membership. Also known as a classification model.

Close-range remote sensing Remote sensing using sensors installed on (or integrated into) platforms at a close distance from the target being measured. See also "proximal remote sensing" and "near-surface remote sensing."

Cluster A predefined set of points for sampling.

Coefficient of variation The ratio of the standard deviation to the mean, sometimes used as a measure of spectral diversity.

Collinearity A situation in which variables are highly correlated to each other.

Colonizers New arrivals that successfully establish in a community.

Community A set of interacting or potentially interacting species within a spatial area defined by the investigator.

Community assembly The processes by which species from a regional species pool arrive in and populate local neighborhoods and form ecological assemblages, shaping their composition and diversity. Important processes include dispersal, colonization and establishment, sorting based on the biotic and abiotic environment, environmental filtering, species interactions—such as competition, facilitation, predation, pest and pathogen dynamics, etc.— diversification, gene flow, reproduction, and others.

Community ecology The branch of ecology that focuses on the interaction of species and the mechanisms that influence assembly of species into natural communities.

Complementarity in the context of species or organismal interactions within ecological communities, refers to the positive effects attributable to mixing of organisms or species that differ; the extent to which two or more biological entities fit together minimizing overlap in resource use; related to niche or resource partitioning.

Conditional autoregressive (CAR) model A category of spatial regression models derived from standard linear regression models yet incorporating additional terms that describe the spatial autocorrelation of the data in the form of specified error structure.

Conservation science Interdisciplinary study of the protection and maintenance of biodiversity.

Constellation in the context of remote sensing, a group of satellites operating in concert with coordinated ground coverage under shared control and synchronized to optimize overlap and coverage.

Contingency An event whose occurrence depends on the existence of other events. In evolution, this refers to the different ways evolution could have shaped current life based on the random occurrences of mutations and the inheritance of those across the history of life.

Continuum removal Normalization of reflectance spectra by their local maxima, allowing comparison of individual absorption features. This is done by fitting a straight line segment to connect local spectral maxima (forming a "continuum") and then using those values to assess the individual spectral features (absorptions). The "removal" is performed by subtracting the spectral feature value from the line value at each wavelength, thus estimating the area of the enclosed feature. This procedure highlights a spectral feature of interest against a changing spectral background.

Convention on Biodiversity (CBD) A multilateral international treaty signed by 150 nations and entered into force on December 29, 1993. The treaty has three main objectives: 1) conservation of biological diversity, 2) the sustainable use of the components of biodiversity, and 3) the fair and equitable sharing of the benefits arising out of the utilization of genetic resources. *Or* a treaty between nations with the objective of developing national strategies for conservation and sustainable use of biological diversity, as well as equitable sharing of benefits arising from the use of genetic resources.

Convergent evolution When the same traits evolve independently in different clades, typically in response to the same selective pressure. For example, the multiple independent evolution of red, tubular flowers across flowering plants that are pollinated by hummingbirds is convergent.

Convex hull volume Volume enclosed by the smallest convex set or envelope that contains a group of points, a commonly used metric for functional diversity.

Critical zone The part of the Earth where biotic and abiotic processes support life on Earth's surface.

Cross-validation A method for internally validating a model, often used in studies where data are repeatedly jackknifed (split) to compare differences in model output among subsets of data.

Cryosphere The component of the Earth's geosphere made up of frozen water. The cryosphere is composed of glaciers, snow, floating ice on the sea and on lakes and rivers, and frozen ground (permafrost).

CSR strategy Plant functional type classification according to Philip Grime in which plant traits are related to their stressors and to disturbance regimes (C, competitor; S, stress tolerator; R, ruderal).

Cuticle Outer waterproof covering of the plant epidermis synthesized by the epidermal cells and consisting of hydrocarbon polymers and various lipids (including wax).

Dark current   The apparent signal of an instrument (e.g., spectrometer) in the absence of any external stimulus (e.g., light). Typically, dark current is a function of temperature and must be subtracted from the desired signal to yield the true value of the external signal.

Dark subtraction   The subtraction of the dark current values to yield the true external signal detected; typically done for spectra or images.

Data augmentation   Image transformations that improve machine learning performance for image classification by using the data available to create additional data.

Data dimensionality   The number of orthogonal (independent) dimensions (attributes) in a dataset, e.g., defined by principal components analysis, singular value decomposition, or discriminant analysis.

Data latency   The time it takes for an end user to retrieve source data from a data collection such as a field campaign, flight campaign, or orbital overpass.

Deciduous   Shedding leaves, so that in part of the year the plant lacks leaves. Contrast with evergreen. Many high latitude oak trees shed leaves in the autumn and are thus deciduous, but in lower latitudes some keep leaves all year and so are evergreen.

Decomposition   An important process in nutrient cycling by which plant and animal litter, detritus, and other organic substances are broken down into more simple organic or mineral forms.

Decomposition analysis   in the context of processing radar data, the separation of polarimetric radar backscatter into components representing different scattering mechanisms, such as volume scattering, surface scattering, and double bounce reflection.

Dehiscence   Splitting along a built-in line of weakness in plant structure in order to release its contents, such as in the rupture of seed capsules for the release and dispersal of seeds.

Demography   The study of births, deaths, incidence of disease, etc., to understand the changing structure, distribution, and size of populations.

Diameter at breast height (DBH)   Tree stem diameter at breast height, approximately 1.3 m above the ground; a common measurement in forest inventories.

Diatoms   Single-celled algae (or phytoplankton) in the class Bacillariophyceae with cell walls composed of silica. They are found in oceans, freshwater bodies, and soils and considered to be one of the most diverse and ecologically important groups of phytoplankton. Globally, there are an estimated 20,000 to 2 million species of diatoms.

Dicot   Short for *dicotyledon.* These are the type of flowering plants (angiosperms) that have two cotyledons (seed leaves) at germination.

Diffuse light   Radiation scattered from a surface in all directions. A surface that scatters perfectly in all directions is termed a Lambertian (isotropic) surface. Contrast to anisotropic.

Digital surface model (DSM)   A gridded dataset containing the absolute elevations including natural and built features of the Earth's surface.

Digital terrain model (DTM) A gridded dataset containing the absolute elevations of the terrain with built objects and vegetation *removed or* data describing the bare terrain surface (without buildings, vegetation, etc.).

Dimensions of biodiversity A term referring to the different aspects of biodiversity—including taxonomic diversity, phylogenetic diversity, functional diversity, and spectral diversity—that can be applied at different scales (including alpha, beta, and gamma diversity).

Direct/indirect species detection directly observing species presence or composition/inferring species presence or composition based on correlated measurements.

Directional scale See angular scale.

Discrete return system in the context of LiDAR measurements, instrumentation that records individual points (representing peaks in the leading edge) in the returning signal from a transmitted laser pulse from (c.f. full waveform).

Dispersal ability The capacity of organisms to move themselves or their propagules and offspring from one location to another.

(Dis)similarity coefficient A measure or index of the similarity (or difference)—using any number of distance-based or other methods—between two objects, variables, or samples, for example, ecological communities. Dissimilarity and similarity coefficients are often used in multivariate analyses, including ordination methods, cluster analysis, multidimensional scaling, principal components analysis, etc. Common dissimilarity metrics include Euclidean distance, Manhattan distance, covariance, correlation, and Bray-Curtis.

Distance decay Describes—often mathematically—the effect of distance on interactions of the components in a study system, such that the decline in the interaction of components between two locations is a function of the distance between them.

Diversity Variation in characteristic or feature of interest. See also phylogenetic diversity, structural diversity, taxonomic diversity, functional diversity, geodiversity, optical diversity, spectral diversity, phylogenetic diversity, structural diversity, taxonomic diversity, alpha diversity, beta diversity, and gamma diversity.

Diversity metrics A broad array of mathematical formulae for calculating biodiversity within a given spatial extent based on biological entities (individuals, species, functions, groups) that variously consider numbers, abundance, and distribution. Many are related to metrics of information content or entropy.

Dominant species The most abundant in terms of biomass, cover, or stem density in a community, ecosystem, or landscape. Developed most clearly by E. Lucy Braun for use in characterizing forest types in North America.

Double bounce Reflection of a radar signal from one surface and then immediately from a second perpendicular surface such that most or all of the signal returns to the antenna.

Downscaling A general procedure for taking information at broad spatial scales to make predictions, or to infer states and processes, at finer scales (c.f. upscaling), e.g., interpolating from coarse to fine pixel or grain sizes.

Drone See unmanned aerial vehicle (UAV).

Earth observation (EO) platforms Surfaces on which remote sensors are mounted, including aircraft or satellites; also known as remote sensing (RS) platforms.

Earth system model A coupled climate model that simulates the atmosphere, hydrosphere, cryosphere, lithosphere, and biosphere by including physical, chemical, and biological processes.

Ecological community A group of species that co-occur and interact with one another.

Ecological heterogeneity The variability in ecological entities—such as local neighborhoods, communities, ecosystems, or landscapes—in space and time, often measured at multiple scales.

Ecological niche modeling (ENM) Modeling of species niches (an approximation to observed or realized niches) or estimation of abiotically suitable areas based primarily on environmental conditions or scenopoetic variables such as climate or topography.

Ecosystem The biotic and abiotic components of an ecological community and their interactions. The boundaries of an ecosystem are variously defined in practice and sometimes circumscribed by those of a watershed. The Convention on Biodiversity defines an ecosystem as "a dynamic complex of plant, animal and micro-organism communities and their nonliving environment interacting as a functional unit."

Ecosystem engineers Species that directly or indirectly alter their physiochemical environment to become more favorable for their continued success often changing the availability of resources to other species.

Ecosystem function Processes and emergent properties of a whole ecosystem, including its annual productivity and growth, decomposition rates, nutrient cycling, stability, resistance to invasion, etc.

Ecosystem services The benefits to human well-being provided by ecosystems. These include regulating services, such as filtering of water, prevention of erosion, carbon sequestration, air pollution removal, etc.; provisioning services, such as food and fiber; and aesthetic and cultural services. Equivalent to Nature's Contributions to People defined by IPBES as the many and varied benefits that humans freely gain from the natural environment and from properly functioning ecosystems.

Ecotron A controlled environmental facility for the investigation of plant and animal populations and ecosystem processes under near-natural conditions using noninvasive methods.

Ecotype A distinct form or genetic race of a plant or animal species occupying a particular habitat.

Eddy covariance A micrometeorological method of estimating turbulent vertical fluxes of matter and energy within the atmospheric boundary layer, especially carbon dioxide and water. Also known as eddy correlation and eddy flux, eddy covariance is widely used to estimate net ecosystem exchange of CO_2, from which gross primary production (GPP) can be modeled.

Elementary surface unit The smallest unit of a surface that is homogeneous.

Elevational range The absolute difference (maximum minus minimum) between elevation at two sites or sample units.

Ellenberg indicator values A classification procedure for Central European plants according to their ecological "behavior" and botanical properties, first described in detail by Heinz Ellenberg in the mid-1970s.

Emissivity A measure of the degree to which an object's surface emits thermal energy (expressed relative to a blackbody of the same temperature).

Empirical line correction A method for correcting atmospheric influences on remotely sensed data to estimate surface reflectance using calibrated or stable ground targets (a form of vicarious calibration).

Empirical model A mathematical model derived directly from observable data, often using statistical methods (as opposed to first principles or conceptual mechanisms) to derive a generalized representation of a process (c.f. mechanistic model or process model).

Empirical-statistical model a model derived from applying statistics to data (c.f. mechanistic model).

Endmembers Spectral signature "types" selected to represent different categories of surface features, such as different vegetation classes or scene components (e.g., forest, grassland, soil, standing water), phylogenetic groups (e.g., plant families, species), or vegetation states (e.g., alive, dead). Can also represent spectral types determined mathematically, e.g., from imaging spectrometry.

Endemism Degree of spatial restriction of a species. A species is endemic to a place if it only exists in that location (e.g., an island, mountain, country, region).

Ensemble classification A classification method using multiple learning algorithms to obtain better performance than use of a single learning algorithm.

Environmental filtering The process by which organisms cannot persist in a local environment because they lack the necessary physiological tolerances to survive or reproduce there. In the strict sense, environmental filtering refers only to abiotic factors that prevent persistence rather than biotic factors, such as competition or pest or pathogen pressure. In contrast, environmental sorting encompasses both abiotic and biotic factors.

Environmental heterogeneity The variability of environmental entities—such as topography, hydrology, soils, and climate—in space and time.

Environmental sorting The process by which species assemble into particular ecological niches, defined by biotic and abiotic variables, based on their functional characteristics and environmental tolerances. Environmental sorting is a consequence of both abiotic factors (such as light, nutrient and water availability, or temperature) and biotic factors (such as herbivory, pathogens, competition, pollination, etc.).

Environmental space The suite of environmental conditions at a given space and/or time.

Epidermis The outermost layer of cells on plant organs that have not undergone secondary growth, e.g., the "skin" of the surfaces of the leaves or other primary plant organs.

Epigenetic Heritable changes in gene expression patterns that do not involve alterations in the DNA sequence. An example is the addition of methyl groups to the DNA, which changes its three-dimensional structure, influencing which regions of the genome are expressed (active).

Epiphyte An organism (typically a plant, fungus, or bacterium) that grows on the surface of a plant, or sometimes multiple plants, often obtaining its moisture and nutrients from the air or from debris that accumulates in structural features of the supporting plant(s). Epiphytes are physically supported by their host and can affect its function. The term is frequently used specifically to refer to plants that grow on other plants, such as Spanish moss or many orchids.

Erectophilic A vegetation canopy structural orientation that is primarily vertical.

Essential biodiversity variable (EBV) A minimum set of measurements, complementary to one another, required for the study, reporting, and management of changes in biodiversity (e.g., allelic diversity, population abundance and distribution, etc.) based on the framework established by Pereira et al. (2013).

ETOPO1 Global relief model used to calculate the proportion of Earth's surface at different elevations—or hypsographic curve; used for *Volumes of the World's Oceans* and to derive the *Hypsographic Curve of Earth's Surface.*

Evapotranspiration The combined movement of water from the Earth's surface (e.g., soil) to the atmosphere through evaporation and plant transpiration. Plant transpiration involves the movement of liquid water from locations of higher water potential—typically the rhizosphere—to areas of lower (more negative) water potential, such as the leaves, through the vascular system. Evaporation refers to the phase change (liquid to gas) which can occur from any surface (living or nonliving) and includes the transition of transpired liquid water to water vapor through intercellular spaces and stomatal pores on leaf surfaces.

Evolutionary legacy effects Functional trait and ecosystem consequences of evolutionary history due to shared ancestry and adaptation to past environments and/or biogeographic and historical processes that influenced the course of evolution and thus current ecological processes. The concept does not indicate that future evolution is constrained by past evolution, but rather that current organismal or ecosystem structure and function reflect past evolutionary processes.

Evolutionary tree see phylogeny.

Evenness The similarity in abundance among species in an area; how equal species within a community are numerically, i.e., a measure of heterogeneity. Mathematically defined as a diversity index.

Extent Attribute of scale that defines the area covered in time or space.

Extinction coefficient (attenuation coefficient; molar absorption coefficient) an index or measure of how much energy a chemical element or compound absorbs at a particular wavelength of the electromagnetic spectrum.

Extracellular enzymes Enzymes operating outside of the cell; microorganisms often produce and secrete extracellular enzymes, which function outside of their own cells, to degrade substances such as proteins, lipids, starches, and other organic molecules.

Faith's phylogenetic diversity index An index of phylogenetic diversity that sums the distances of all the branches of a phylogeny that includes the species present in the observed community.

Field campaign A coordinated data collection effort usually in a particular outdoor location.

Field spectrometry The outdoor, in situ, measurement of spectral properties using a spectrometer.

Flavonoids Widely distributed polyphenolic plant compounds (anthocyanin pigments and flavones) in flowers, fruits, and leaves. Typically these range from near colorless to a range of reds and blues. They have diverse functions as antioxidants, pollinator attractants, seed and fruit colors for dispersal, and more. The red colors of leaves are often due to a type of flavonoid.

Flight time Time or duration of an aircraft campaign or the time it takes a signal, such as a light or sound wave, to go from an emitter, bounce off something, and go to a receiver.

Flux in optics, the rate of electromagnetic radiation contacting a surface of given area. In general, it is used to describe the rate of flow through a surface of given area, such as light transmission through a leaf or carbon, water, or oxygen flow into or out of the leaf.

Flux tower or eddy flux tower A structure, often a metal scaffold, supporting the instruments used to measure eddy covariance.

Focal taxa The set of organisms, species, or lineages that are the subject of study and/or analysis.

Foreoptics The component attached to an optical sensor that transfers and modifies the signal between the target and detector by attenuating the light or changing the field of view. Can consist of lenses, filters, fiber optics, cosine heads, or field-of-view restrictors (e.g., tubes).

Foundation species in an ecological community, a species that plays a critical role in structuring the community and maintaining the integrity of the food web, thereby influencing and stabilizing ecosystem properties and processes.

Fragmented landscapes Landscapes characterized by a dispersion of smaller patches, e.g., by terrain features or more frequently by disturbance, including roads, paths, or clearings.

Full waveform in the context of LiDAR, full laser echo intensity recorded and digitized over time (c.f. discrete return systems).

Functional diversity The variability of life as measured by functional traits. Includes ecosystem functional diversity and species functional diversity.

Functional trait A morpho-physio-phenological or phenotypic attribute expressed by organisms in their environment that represents or is linked to organismal performance. Examples for plants include photosynthetic pigments, leaf chemical composition, water content, dry leaf mass, leaf mass per area, leaf economic spectrum traits, canopy structure, plant height, flower color, wood density, seed size and production, etc.

Functional type See plant functional type.

Fusion The process of integrating multiple data sources, such as different satellite products or environmental data layers, with the goal of producing more useful information than individual data sources provide alone.

Gamma diversity Total diversity within a region. It can be defined in various ways that incorporate the alpha diversity of individual communities or habitats within a region and the beta diversity across communities or habitats, or variation among these.

Generalized additive model (GAM)   an extension of the generalized linear model in which the linear predictor is replaced by a sum of unspecified smooth functions.

Genetic algorithm   A computational method for solving optimization problems that is based on natural selection.

Genotype   The genetic makeup of an individual or group of individuals; also refers to individuals of a particular genetic background, such as a recombinant inbred line, a seed family, or known genetic variants. Contrasts with phenotype.

Geodiversity   The variety of abiotic features and processes of nature (e.g., landforms, soils, and hydrological patterns) that influences the development and maintenance of biodiversity.

Geographical space   The extent of a particular region or study area.

Geometric changes   in the context of remote sensing, typically refers to changes in the geometric resolution (ground sampling distance or pixel resolution) of satellite data. It can also refer changes in sampling geometry.

Geometric correction   Resampling of spectral data from the pixels in an image (which are originally not square and vary in size) to a common pixel grid, often using nearest neighbor interpolation, to reduce image distortion.

Geomorphological   Pertaining to the study of the physical features of the Earth's surface, including the origin and evolution of topographic and underwater features that result from physical, chemical, or biological processes.

Geosphere   The Earth's abiotic environment. The geosphere is made up of four subcomponents: the atmosphere, lithosphere, hydrosphere, and cryosphere.

Geostatistical   considering the spatial positioning or scale of study objects.

Georeferenced data   Data accompanied by locality information in the form of geographically explicit coordinates that can be mapped.

Glabrous   A condition that defines a smooth leaf surface without hairs, bristles, or rough cuticle surface, often shiny in appearance.

Global biodiversity monitoring system   An emerging concept that integrates tools and approaches to sampling, monitoring, and understanding temporal and spatial patterns in biodiversity in multiple dimensions and at multiple scales. Often involves satellite, aircraft, and in situ sampling and ranges from the molecular to the global scale.

Global mapping of biodiversity   Depicting spatial patterns of biodiversity at a global scale, often using satellite data, and typically at a relatively coarse spatial resolution. Can also entail mapping of "essential biodiversity variables" to infer changing patterns of biodiversity.

Global navigation satellite system   A constellation of satellites that provide geospatial positioning and timing information to networks of ground control stations for calculation of ground positions used in computer networks, air traffic control, power grids, etc.

Global spectroscopy data   High spectral resolution data collected by an instrument measuring the entire globe, presumably from satellite(s).

GLOPNET   The Global Plant Trait Network – a group of scientists who contributed trait data across the plant tree of life that was used to establish the leaf economic spectrum. The GLOPNET database was a precursor to the TRY database.

Gradient   A continuous change in a property (such as temperature, rainfall, elevation) in space or time.

Grain size   The dimensions of a sample unit. In remote sensing, analogous to pixel size.

Graminoid   A herbaceous plant with a grass-like morphology, such as a grass or sedge.

Ground control point   A coordinate that is measured on the ground or that can be easily identified from preexisting imagery, such as road intersections, building corners, and trees, and used to apply geometric correction to aircraft or satellite image data.

Habitat   The ecological or environmental area or conditions in which a particular species or population lives, comprised of the biological and physical properties and including the suite of resources, on which it depends.

Habitat heterogeneity   The variability of habitat structure or composition in space and time.

Hardwood   The wood of dicotyledonous angiosperm trees.

Hemeroby   refers to the degree of human influence on the environment in contrast to the degree of naturalness of the environment in the absence of human influence. It can be measured in terms of the magnitude of the deviation from the potential natural vegetation caused by human activities.

Heterogeneity   Variability in space and/or time.

Hill numbers   A unified standardization method for quantifying and comparing species diversity across samples, originally presented by Mark Hill. These are generalizable to all of the dimensions of diversity and consider the number of species and their relative abundances within a local community.

Hydraulic traits   Traits in plants related to water transport through vascular tissues (e.g., xylem), such as leaf water potential.

Hydrosphere   A collective term for all water below, on, and above the Earth's surface.

Hyperspectral data   Measurements in a large number of concurrent, narrow spectral bands (>20 to hundreds) across a wavelength region in a single spectrum or within a pixel in a spectroscopic image (image cube).

Hyperspectral imaging   See imaging spectroscopy.

Hyperspectral remote sensing   Collection and processing of a wide range of data in contiguous wavelengths with fine spectral resolution. Commonly used as a synonym for imaging spectroscopy but can also include non-imaging measurements.

Hyperspectral sensor or spectrometer   An instrument for measuring many spectral bands. It is often used interchangeably with imaging spectrometer but can also indicate a non-imaging spectrometer.

Hyperspatial data   Spectral data or imagery having very small pixel sizes, generally less than 1 m ground sampling resolution. There are several satellites with

panchromatic bands of this size. Airplane and UAS-collected imagery at submeter scale are considered hyperspatial.

Image band   refers to the range of wavelengths measured by a single sensor channel, often named according to the color or region of the EMR spectrum (e.g., red
or near IR) or by the wavelength(s) of the energy being recorded. Multispectral
data, including color digital photos, are made up of multiple image bands (also
called channels or layers).

Image campaign   A remote sensing campaign for the acquisition of image data.

Imaging spectrometer   A spectrometer that captures an image where each pixel
is a full spectrum covering many narrow adjacent spectral bands. "Pushbroom"
spectrometers consist of a two-dimensional array in which the x-dimension has
spatial pixels and the y-dimensions are co-registered wavelengths. A spatial
image is constructed through the forward movement of the spectrometer.

Imaging spectroscopy   The acquisition and analysis of data from an imaging spectrometer, involving simultaneous acquisition of spatially co-registered
images in many spectrally contiguous bands, defined by their wavelength centroid and full width half maximum describing the spectral response centered on
that wavelength. Also known as hyperspectral remote sensing.

Immobilization   The process by which microorganisms consume inorganic compounds, converting them to organic compounds, making them unavailable to
plants. Immobilization is the reverse of mineralization, but the two processes
occur contemporaneously.

Incumbents   in the context of biodiversity, the resident organisms or species in a
system prior to colonization or invasion by new organisms or species.

Independent validation   See model validation.

Index of refraction   a value calculated by the speed of light in a vacuum relative
to the speed in a denser medium. The difference causes the light ray to bend
toward the denser medium. It is the real part of the complex refractive index, an
intrinsic property of the medium. The imaginary part describes the attenuation,
often called the extinction coefficient (absorption coefficient).

Individual tree crown (ITC)   Delineation of individual tree crown polygons from
remote sensing data, typically provided by airborne laser scanning or high-resolution multispectral sensors.

Informatics   The scientific discipline that involves studying and designing the
means of storage, transformation, and representation of large volumes of information of diverse types and provenance in computational systems.

Infrared (IR)   Electromagnetic radiation with wavelengths longer than those of
visible light, i.e., from about 700 up to about 1050 nanometers (in the case of
near infrared), 700 to 2500 nanometers (in the case of shortwave infrared), and
from 3 microns to 1 mm (in the case of thermal infrared).

In situ   of or relating to an actual location or habitat, for example, an intact leaf on
a plant or plants in their natural environment.

Intactness (of habitat)   (i) The degree to which the condition of vegetation that
forms habitat has not changed beyond what is expected from natural processes
such as phenology and other dynamics, or (ii) the spatial pattern of a given habi-

tat, often described in terms of the degree of connectivity or fragmentation, and edge extent.

Interferometric SAR Radar remote sensing that employs two signals from different antennas or from the same antenna in a repeated pass, where the signals are emitted at different angles. The phase difference of the two backscattered signals from the same location on the terrain surface is related to the relative elevation of the surface or changes in that elevation.

Interferometry A group of methods for extracting information based on the interference of two sets of superimposed electromagnetic waves. In remote sensing, generally used with synthetic aperture radar to characterize surface change.

Intergovernmental Science-Policy Platform on Biodiversity and Ecosystem Services (IPBES) The intergovernmental body, established by the United Nations member states in 2012, which assesses the state of scientific knowledge regarding Earth's biodiversity and the ecosystem services that nature contributes to people.

Internal validation See model validation.

Interpolated climate surfaces Interpolation of climatic data from weather stations into climate surfaces, generally using spline smoothing surface fitting techniques.

Interspecific Comparison of observations between different species.

Intraspecific Comparison of observations within individuals of the same species.

Invasive alien species (IAS) An exotic species prone to rapid dispersal, often displacing native species, altering ecosystems, and reducing biological diversity. The Convention on Biodiversity defines an IAS as "a species that is established outside of its natural past or present distribution, whose introduction and/ or spread threaten biological diversity."

Inverse model A model that uses outputs to infer or derive inputs, often by mathematical inversion. For example, in radiative transfer (RT) modeling, the ingesting of vegetation reflectance and transmittance data to predict biophysical and biochemical characteristics (traits). The forward model would predict a spectrum, given the traits as inputs.

Isotropy Directional equality in reflection of electromagnetic radiation. An isotropic surface reflects radiation equally in all directions. See also "Lambertian." Contrast to anisotropy.

Janzen-Connell type mechanisms The processes by which host-specific herbivores, pathogens, or other natural enemies make the areas near a parent inhospitable for the survival of offspring. They were proposed as an explanation for the maintenance of high species diversity in different ecological systems, including tropical forests.

Kappa coefficient A measure that indicates the overall accuracy of a classification analysis compared to the expected accuracy, controlling for chance agreement.

Kernel dependent Relying on data not just from an individual point but from surrounding points as well.

Keystone species An ecologically important species that has a disproportionate impact on an ecological community and the ecosystem in which it is embedded, such that if it were removed the community structure and ecosystem function would change drastically.

Kriging   A class of methods for spatial interpolation that use statistical approaches to estimate a continuous surface from known points with measured values.

Lambertian   Reflecting light equally in all directions, independent of viewing angle. See also isotropy.

Land cover stratification   A process that segments and groups land based on land cover types, land use types, and/or percent cover; usually a combination of two or all three. Stratification is often used to identify land classes (e.g., managed/ non-managed forest areas or cultivated/uncultivated crop areas) for subsequent analyses such as biomass or biodiversity inventories.

Landscape   A geographically defined land area that can be viewed at one time from one place that may encompass multiple ecosystems interacting through the movement of species, energy, and matter. A landscape is generally described by physical and biological features such as topography and patterns of vegetation cover.

Laser echo   A reflection of light as recorded by a LiDAR instrument, the ranging part of a laser scanning system. It has the attribute range (i.e., distance of object from LiDAR instrument) and can have additional attributes such as energy and echo width, which are derived from the recorded full-waveform information.

Laser light sheet   A laser beam focused in a single direction.

Laser pulse   The pulse emitted by a LiDAR instrument. It has the properties of length (i.e., duration), shape, and energy.

Laser scanning, airborne   LiDAR mounted on an airborne platform with a scanner deflecting the beam across the swath, while differential GPS and inertial navigation systems provide location and orientation of the measurement platform. It is capable of providing several hundred thousand 3-D locations of reflecting objects per second with decimetric accuracy.

Laser scanning, terrestrial   LiDAR on a nonmoving terrestrial platform. Scanning is performed by deflecting the laser pulse through rotation of the instrument along two axes (azimuth and elevation), covering a (hemi)sphere.

Latent heat flux   Flux of energy due to phase changes, i.e., evapotranspiration and condensation (contrast to sensible heat flux).

Leaf anatomy   The microscopic structure of a leaf, including the arrangement and size of cells, air spaces, and the intra- and extracellular components of the leaf.

Leaf area index (LAI)   The one-sided area of leaves divided by the subtending area on the ground surface.

Leaf economic spectrum (LES)   The concept formalized in Wright et al. (2004) that leaf traits of plants are coordinated and correlated—given biophysical and ecological trade-offs in how resources can be used and deployed for plant function—such that they constrain the variation in leaves among species to a single axis of variation that ranges from "fast" acquisitive traits to "slow" stress tolerance traits associated with leaf construction costs. Leaf mass per area (LMA) is an easy-to-measure proxy for the entire spectrum that represents the variation in leaf life span, the original theoretical basis for trade-offs in carbon and nutrient allocation across the spectrum.

Leaf mass per area (LMA)   Dry weight per one-sided leaf area (typically g/m^2). Inverse of specific leaf area (SLA) (based on unit conventions, LMA = 1000/SLA).

Leaf morphology   The shape and structural characteristics of leaves.

Leaf traits   Characteristics that describe physiochemical attributes of leaves, such as pigment concentration, water content, dry matter, and structure, all of which have a functional role in plant growth and resource allocation.

Liana   A woody plant that is rooted in the soil and grows by climbing up trees or other substrates by various means (tendrils, adventitious roots, twining stems, etc.) in order to reach the top of the canopy.

LiDAR   Acronym for light detection and ranging. In LiDAR measurements, time of transmission of a laser pulse between the instrument and a reflecting object is measured and converted to a distance measurement. LiDAR can also provide detailed 3-D information of the landscape, trees in a forest, and other objects.

Light   An imprecise term that when used in relation to spectroscopy, remote sensing, or plant photosynthesis requires precise definition in terms of the portion of the sun's electromagnetic spectrum that is being referred to. Visible light is considered to be the range of wavelengths perceivable to the human eye—approximately 380–740 nm, which is similar to the range active in photosynthesis (400–700 nm).

Lignin   A large polymorphous molecule with repeating complex polyphenol units that is a common component in secondary plant cell walls that provides structural strength, enables water transport, and resists decomposition in soil organic matter.

Lineage   Typically means the same as clade (see definition); more rarely, the set of ancestors leading to a specified taxon.

Lithology   The study of rocks or the character of a rock formation.

Lithosphere   The solid Earth, from its surface to its core. The lithosphere is one subcomponent of the geosphere.

Lysimeter   An apparatus for the measurement of water within a soil profile that when combined with rain gauge data enables inference of soil drainage rates, plant water use, and evapotranspiration. They are also used to estimate soil water nutrient dynamics to understand nutrient availability and leaching.

Macrophyte   A large aquatic plant—as opposed to phytoplankton (algae)—that grows in or near water.

Maximum likelihood estimation   A supervised classification method for estimating parameter values of statistical model in which observed data are most probable given the process described by the assumed statistical model. Parameter values are estimated by a likelihood function that maximizes the probability that the observations are true given the parameters.

Mean normalization   Standardizing a vector to the mean of the components, which changes the magnitude but not the direction of the vector; a useful transformation for spectral reflectance data from plants when the magnitude of reflectance varies among samples (see also vector normalization).

Mean phylogenetic distance (MPD)   The average evolutionary distance between each pair of species in a given community, where evolutionary distance is measured as the sum of the branch lengths between two species in the phylogeny.

Mechanistic model A model that describes—often mathematically—a process in terms of its component physical, chemical, and biological processes, akin to a process model (c.f. empirical, statistical, phenomenological model).

Melastomes Members of the plant family Melastomataceae, also known as the princess flower family.

Mesophyll The tissue within a leaf, comprised of photosynthetic parenchyma cells.

Metabarcoding A taxonomic method that uses a designated portion of a specific gene or genes (proposed to be analogous to a barcode) to identify an organism to species. These "barcodes" are sometimes used in an effort to identify unknown species, parts of an organism, or simply to catalog as many extant taxa as possible. Also known as DNA barcoding.

Microendemic Species whose geographical distribution is restricted to a very small location (e.g., a single locality or a single mountain top).

Microhabitat The local environmental space in which an organism or species lives—and is restricted to—as a consequence of its evolved tolerances to the biotic and abiotic environment and ability to persist long-term.

Mineral nitrogen Soil nitrogen that is directly available to plants as nitrate or ammonium.

Mineralization The process by which organic compounds, such as nitrogen, are converted to plant-available inorganic forms. For nitrogen, these include nitrate and ammonium. Mineralization is the inverse process to immobilization, but the two work in conjunction within the soil-plant system.

Minimum noise fraction A linear transformation consisting of two principal component rotations used to decrease noise in spectral data by redistributing it across all channels (whitening the bands) and transforming the original bands into orthogonal bands that combine information from different channels. Transformed bands are ordered from the band containing most of the information to those with least information. Also used to determine the data dimensionality.

Mission in the context of remote sensing, the full process of instrument deployment including planned design, implementation, and launch of a spaceborne instrument on the part of a space agency. Alternatively, it can refer to the deployment of an airborne instrument.

Model A simplified description, usually a visual, conceptual, or mathematical depiction, of some aspect of the world and often how it works; a set of rules or mathematical function(s).

Model of trait evolution A mathematical description of how a trait—such as leaf size or chlorophyll content—changes over time due to evolutionary processes.

Model calibration The process of adjusting model parameter values to match the model output to observed data.

Model selection Choosing the best model using some criterion, e.g., the Akaike Information Criterion.

Model validation The process of comparing model outputs with observed data. Internal validation refers to testing a model's ability to explain the data used to populate the model. Independent validation refers to testing the accuracy of model outputs with new or withheld observed data.

Monte Carlo ray-tracing A statistical, sampling-based way of producing a physically accurate representation of the light field in a given environment.

Morphological related to the form of organisms and their structural features, including shape, structure, pattern, size, etc. See also physiognomic.

Morphological species concept Characterization of a species based on its body shape or other structural features.

Morphometric Quantitative measurement(s) of a shape or form (e.g., size).

Multispectral refers to an instrument or data having a few (more than one) reflectance bands but generally less than 20. Multispectral bands typically have gaps between them, and therefore do not measure all contiguous bands in a spectral interval. For comparison: panchromatic (1 band), multispectral (2–20 bands), hyperspectral (>20 several hundred), ultraspectral (several thousand).

Multispectral satellite data Satellite imagery capturing specific, typically broad, noncontiguous wave bands, across the electromagnetic spectrum.

Multiband See multispectral.

Nadir Direction aligned with the direction of the force of gravity at a location often used to describe the direction of looking straight down at a 90° angle perpendicular to a flat, level surface.

National Ecological Observatory Network (NEON) A major US National Science Foundation-funded program involving a group of sites—distributed across 20 domains of the United States, including Alaska, Hawaii, and Puerto Rico—representing the ecoclimatic variability of the United States. NEON is connected through cyberinfrastructure that delivers standardized ecological datasets based on ground sampling, remote sensing, and flux towers.

Non-imaging sensor technology A remote sensing technology in which no image data are generated, e.g., leaf-level spectroscopy or canopy temperature.

Normalized Difference Vegetation Index (NDVI) A normalized expression of red and NIR spectral bands created by subtracting the difference in value between the two bands and dividing this by the sum of the values in the two bands. It generally indicates greenness and often closely relates to several measures of green canopy material such as green leaf area index, green biomass, and the amount of radiation absorbed by green canopy material.

Net biodiversity effect See overyielding.

Neural network in the context of computing systems, a machine learning technique inspired by the neural network in the brain that can adapt to changing input without redesign. It can be described as a framework for coordinating many different machine learning algorithms that process complex data inputs for a defined purpose in accordance with a set of output criteria.

Niche The range of biotic and abiotic conditions a species requires for persistence or, alternatively, a species' ecological role in an ecosystem. Evelyn Hutchinson defined it as the multidimensional hypervolume circumscribed by an indefinite number of biotic and abiotic axes which describe the variation in resources or environmental conditions affecting the performance of an organism or species. Hutchinson distinguished between the *fundamental niche* (pre-interactive), the hypervolume in which a species can live and reproduce absent competition from others, and the *realized niche* (post-interactive), the hypervolume in which it actually lives.

Niche conservatism   The tendency of a species to remain in the same ecological niche as its ancestors and hence to share similar ecological niches with its close relatives.

Niche model   A predictive model of species potential distribution along multiple environmental dimensions (or niche axes) based on current occurrence information of a species in relation to its current environment. See environmental niche model.

Near-infrared shoulder   The sharp rise in a vegetation reflectance spectrum beyond 700 nm. Also called the red edge. Spectral position (e.g., inflection point) of the red edge can indicate the relative health or phenological status of vegetation, with red edges farther into the near infrared indicating more vigorous growth.

Near-surface remote sensing   See close-range remote sensing and proximal remote sensing.

Nitrogen   An element whose molecular form (N_2) makes up most of the troposphere. In this book, it most commonly refers to nitrogen concentration (measured by unit mass) or content (measured by unit area) of soil or plant tissues, which reflects nitrogen incorporated in proteins (including enzymes like RuBisCO), chlorophyll, and nucleic acids in DNA and RNA. Nitrogen comprises 0.5–4% of the dry biomass of a leaf.

Noise (in spectra)   Random variation of a signal, e.g., often an inherent product of an electronic instrument and source of error.

Normalization of band centers   See spectral correction.

Nutrient cycling   The movement and exchange of organic and inorganic matter back into the production of matter. Microorganisms degrade organic matter, which allows the release of nutrients, which can in turn be incorporated into the bodies of living organisms.

Ontogenetic   related to the development of an organism or its organs. For example, in vascular plants it refers to developmental shifts from the embryo to the seedling stage to the mature adult form. For plant leaves it refers to the developmental shifts that can occur seasonally from various stages of bud growth and budbreak to the early, heterotrophic leaf stage, to a fully autotrophic leaf, and finally to a senescing leaf. Ontogeny in a community of plants refers the processes affecting the origin, structure, and composition of plants interacting in an ecosystem.

Open source   Software for which the original source code is made freely available and may be redistributed and modified. Contrast with proprietary or closed source code.

OpenStreetMap   A voluntary, collaborative effort to generate an editable world map.

Optical detection   Detection of phenomena (such as chemical properties, physiological functions, structural characteristics, or taxonomic identities of plants) based on spectral patterns, spectral indices, or other optical metrics. In remote sensing, "optical" refers to wavelengths from the near-UV to the SWIR region (including the visible).

Optical diversity The variation in optical properties, typically measured using reflectance spectra but can also be measured via spectral indices or other optical metrics (e.g., NDVI, albedo, or chlorophyll fluorescence). See spectral diversity.

Optical imaging system A passive sensor, including any remote sensing system, for recording electromagnetic radiation in the wavelength range between 0.3 and 3 μm (300–3000 nm). Also known as optical sensor.

Optical properties Spectral or other properties of a surface or biological entity; typically related to the entity's absorbing or scattering properties (e.g., leaf chemical constituents and surface structure) and derived from the pattern of absorption and scattering of different wavelengths of light but can also include fluoresced or emitted radiation.

Optical remote sensing Remote sensing technology using optical sensors in the wavelength range between 0.3 and 3 μm (300–3000 nm).

Optical signals See optical properties.

Optical spectrum The spectrum in the wavelength region between 0.3 and 3 μm (300–3000 nm). See spectrum.

Optical type A class of organisms (or objects) distinguishable from optical properties, particularly via remote sensing. See also spectral type.

Organic nitrogen A nitrogen compound that has its origin in living material, for example, as a component of amino acids, which are the building blocks of all proteins.

Ornstein-Uhlenbeck model of evolution A model of evolution that extends the Brownian motion model by describing the evolution of a trait being pulled toward some optimum value.

Ornstein-Uhlenbeck process A process that has stochastic components as well as an attraction to an optimum.

Orthorectification The correction of image distortions caused by factors such as sensor tilt, flight altitude, and changes in surface terrain, creating an image that is geographically registered to surface coordinates irrespective of topographic variation.

Overall classification accuracy The probability (often expressed as a percentage) that a classification model will correctly classify an unknown sample.

Overyielding Producing more than expected. Refers to the synergistic effects on plant growth, where mixtures of species yield more biomass than the same set of species are predicted to yield based on their growth in separate monocultures.

Palisade parenchyma A specialized chloroplast-containing cell or tissue in the mesophyll of dicotyledonous angiosperms. These cells are typically tightly packed and elongated perpendicular to the epidermis and located on the adaxial side of the leaf.

Panchromatic refers to a single band extending across all or most of the visible bands and, often, part of the near infrared. By collecting light across a wide band, the energy collected per unit of time is much greater than narrow wavelength bands can acquire in the same time interval. This allows data to be collected in smaller pixels with high signal-to-noise characteristics than is possible with narrower bands.

Parameter   A value (generally held constant in a given model or equation) describing a phenomenon of interest. Contrast to variable.

Parameterization   The process of defining or selecting parameters for equations or sets of equations (models) to explain or describe phenomena of interest.

Parenchyma   Living thin-walled, relatively unspecialized (undifferentiated) cells or tissues. In leaves, parenchyma cells comprise the mesophyll and can be modified to perform more specific functions. Parenchyma tissue may be compact, as in the palisade parenchyma, or have large spaces between the cells, as in the spongy mesophyll.

Partial least squares regression (PLSR)   A predictive statistical regression method that is used to find the fundamental relations between a predictor variable matrix (X) and dependent variable matrix (Y) based on identifying an optimal set of latent vectors descriptive of the variable of interest. PLSR is widely used in chemometrics and is especially useful when the number of predictor variables (e.g., spectral measurements) is large compared to the number of observations.

Passive optical remote sensing   The measurement of optical signatures or reflectance from an object or phenomenon by a sensor that is not in physical contact with the object. Passive remote sensing systems depend on solar illumination or thermal emission (i.e., they do not generate their own energy for measurement).

Passive sensor   A sensor that does not require active or pulsed light emission but uses solar irradiance or thermal emission to detect the feature or objects of interest.

Petiole   The specialized stalk that attaches the plant leaf blade to the stem, containing vascular tissues and providing structural support.

PhenoCam   A webcam for monitoring vegetation phenology. Also, the name for a vegetation phenology network based on webcams.

Phenogram   Depiction of a phylogenetic tree where one axis represents trait values and the other axis represents time.

Phenol   Organic compound containing the basic structural unit C_6H_5OH.

Phenology   The study of periodic, seasonal processes such as budbreak, flowering, seed maturation, leaf senescence, and leaf fall in plants. Although most typically used for visible events or processes, the term is also increasingly applied to less visible, seasonally changing processes, as in "photosynthetic phenology."

Phenological types   Species groupings based on regularity, date of onset, and duration of phenological cycles.

Phenome   A phenologically similar terrestrial *zones or* the set of phenotypes (physical and biochemical traits) that can be produced by a given organism over the course of development and in response to genetic mutation and environmental influences.

Phenomics   The study or measurement of phenotypes that often uses proximal remote sensing (typically cameras, imaging spectrometers, or other devices) to sample and distinguish plant features.

Phenotype   The observable aspects of an organism, population, or biological entity. Contrast with genotype. While phenotype depends on genotype, the interaction of the genotype with the environment determines the phenotype.

Phenotypic plasticity When organisms with the same genotype display different phenotypes in response to different environmental conditions.

Phenotyping The process of determining or measuring an organism's phenotype, often involving automated sensors or imaging systems (proximal remote sensing) and computing power (e.g., high-throughput phenotyping).

Phloem The vascular tissue that conducts sugars and metabolites between organs in the plant. The tissue is composed of several types of specialized cells for conducting, support, or regulation and other control functions.

Photochemical reflectance index (PRI) A narrowband normalized difference index, typically using a band at or near 531 nm (that responds to changes in the relative amounts of xanthophyll cycle pigments), referenced to a band at or near 570 nm that does not respond to these changes. Under some conditions over short time periods (e.g., diurnally varying illumination), PRI can provide an indicator of xanthophyll cycle activity and changing photosynthetic light-use efficiency. Under longer (seasonal) periods or across individual plants within a landscape, it is often correlated with the relative concentrations of chlorophylls and carotenoids or the amount of green canopy material (e.g., leaf area index), both of which can be indicative of relative photosynthetic activity.

Photogrammetry The discipline of making maps, Earth surface measurements, or 3-D models from photographs.

Photon An elementary particle describing a quantum of electromagnetic radiation (e.g., light) that is described by both particle and wave properties.

Photosynthetic capacity The maximum photosynthetic rate for a specific species under a set of environmental conditions, often defined as full sunlight, optimal temperature, and ambient carbon dioxide concentration. Is often standardized to a specific temperature such as 25 °C for comparison among taxa. For describing enzyme kinetic processes, is often broken down into the carboxylation (RuBisCO)-limited rate Vc_{max} and maximum electron transport rate J . Correlates strongly with leaf nitrogen, phosphorus, and specific leaf area (SLA).

Phycocyanins Accessory blue photosynthetic pigments found in cyanobacteria, including those comprising potentially harmful algal blooms.

Phylogenetic diversity The evolutionary distances between species or individuals, represented in terms of millions of years since divergence from a common ancestor or molecular distances based on accumulated mutations since divergence.

Phylogenetic clustering The tendency of close relatives to occur together in an area (or within a community) more often than expected by chance.

Phylogenetic conservatism The tendency of close relatives to be more similar to each other in traits or niches than expected by chance as a consequence of shared ancestry.

Phylogenetic diversity (PD) The amount of evolutionary history represented by a set of taxa: typically the total length of branches on a phylogenetic tree connecting them, though the number of speciation events is also sometimes used. See also Faith's phylogenetic diversity index.

Phylogenetic endemism The amount of evolutionary history (represented by the amount of branch length, measured in a genealogy containing all species present in an area) that is uniquely represented in that specific geographical location.

Phylogenetic overdispersion   The tendency of distant relatives to occur together in an area (or within a community) more often than expected by chance.

Phylogenetic signal   The degree to which closely related organisms tend to resemble each other. A high phylogenetic signal suggests that there is a strong relationship between species' traits values and their shared evolutionary history.

Phylogenetic species concept   The concept of a species as a group of individuals descended from a common ancestor that share a set of derived traits that define the group.

Phylogenetic tree   See phylogeny.

Phylogenetics   Discipline that investigates the evolutionary relationships between species, and the methods to uncover such relationships.

Phylogeny   A history of shared ancestry between species, often depicted as a branched tree. Each tip is typically a species and internal branching points represent speciation events. Also known as evolutionary tree, cladogram, or chronogram. *Or (phylogenetic tree, evolutionary tree)* a reconstruction of the estimated evolutionary relationship between species.

Physiognomic   related to external attributes (of plants) visible to a human observer; related to the external appearance (phenotype) of plants or vegetation. See also morphological.

Phytoplankton   Microscopic, photosynthesizing organisms that inhabit almost all water bodies on Earth. Phytoplankton consist largely of single-celled bacteria or protists, although some are multi-celled.

Phytosociological   refers the composition of plant communities and the system of analyzing and classifying them according to their interactions.

Pixel   The smallest element in a detector array, or the smallest spatially resolved unit in a digital image. Sometimes used to refer to the ground sampling unit (grain size) detected or represented in an image.

Pixel binning   The process of combining the signal from multiple adjacent pixels into a single signal. This is often done to improve the signal-to-noise ratio of a sensor. Sometimes also called co-adding. Binning can be spatial (spatially adjacent pixels) or spectral (binning adjacent spectral bands).

Pixel shift   Spatial displacement of image pixels; occurs when the location of a pixel is spatially misregistered.

Planimetric   related to geographic elements, maps, or images that are independent of elevation and that can be used to determine distances, areas, and angles within a two-dimensional plane.

Plant area index   similar to leaf area index but including both leafy and woody (e.g., stem) components to express their combined area projected to a given ground area.

Plant biodiversity   The variation among plants, detected at any of a number of levels (e.g., within plant species, between plant species, or between vegetation types).

Plant functional type   A group of plant species that share critical functional traits or trait values related to performance; they are expected to respond similarly to environmental conditions and/or have similar effects on ecosystem processes.

Plant functional types are often used in Earth system models or dynamic global vegetation models to simplify the surface representation of vegetation into functional classes. Plant functional types defined for models are usually physiognomic vegetation types, defined by their structure, shape, stature, and leaf shape/habit, e.g., graminoids, forbs, shrubs, broadleaf/conifer, and evergreen/deciduous.

Plant litter Dead plant material—leaves, stems, bark, etc.—on the ground that can form the surface layer of the soil, sometimes called the "duff" layer.

Platform in remote sensing, a structure or system for mounting or supporting a measuring instrument. Examples include unmanned aerial vehicle (UAV) or system (UAS) (also called a drone), airplane, satellite, or tower.

Point cloud A set (cloud) of points representing the three-dimensional structure of an object or set of objects (e.g., tree or forest canopy), typically derived from LiDAR or structure from motion.

Polarization parameters used in radar remote sensing to represent polarizations (e.g., horizontal, vertical, or other configurations) of the microwave energy emitted from the antenna and the signal backscattered from the target.

Polarimetric in radar remote sensing, horizontal and vertical phase information from both transmit and receive data is recorded, usually reported as HH, HV, VH, and VV.

Pollination The process that brings pollen, which produces the male gametes (sperm), into contact with the female organs of a plant. This process often results in the fertilization of an ovule by the sperm and the development of a seed.

Polyphenolics A large category of secondary metabolites found in plant tissues and soil that contain many phenol (C_6H_5OH) groups.

Polytomy When more than two branches descend directly from the same ancestral node. It could mean that one species instantly became three or more (this is sometimes called a hard polytomy) – more commonly, it represents uncertainty in not knowing the relative order of branching of the descendants (called a soft polytomy).

Post-processing in remote sensing, further modification of data beyond initial processing steps to a standard output, often with a particular objective in mind, e.g., filtering or smoothing.

Precision Refinement in a measurement, calculation, or specification, represented by the number of significant digits used. Also, a measure of the measurement repeatability, related to the variability in values or predictions resulting from applying models or measurements multiple times, with low variability corresponding to high precision, or high reproducibility; not to be confused with accuracy, which is a measure of measurement error or measurement, or model bias, with low bias corresponding to a high degree of correspondence between predictions and observations/measurements.

Principal components analysis A linear transformation method that maximizes the variance of the data through transformation (rotation) of the data axes using the covariance matrix. When applied to spectra, it produces a series of orthogonal components that correspond to linear combinations of the original bands aligned to represent the main axes of variation within the original data.

Process model A conceptual or mathematical representation of a process that attempts to represent component steps or mechanisms (c.f. empirical or statistical model).

Producer's accuracy The fraction of correctly classified pixels in the reference dataset used to train a model. Contrast to user's accuracy.

Propagule Plant material that can become detached and develop into a new organism (e.g., seeds, spores, corms, tubers, stems or rhizomes, etc.).

Provisioning ecosystem services The material or energy outputs from ecosystems that benefit—and are generally consumed by—humans, including fuel, fiber, food, clean water, and other resources. Also called Nature's Contributions to People (NCPs) within the IPBES framework.

Proximal remote sensing Observations from a close distance, usually in the range of centimeters to meters. Also known as close-range remote sensing or near-surface remote sensing. Typically uses trams, platforms, tractors, and UASs for deployment or is even handheld.

Proximal sensors Sensors mounted on platforms on or near the ground that collect information in close proximity to the target of interest.

Quadrat Square frame used to define a small study area for ecological studies.

Radar Acronym for radio detection and ranging. It involves methods, systems, techniques, and equipment for using the timing of transmitted, reflected, and detected electromagnetic radiation to detect, locate, and track objects or to measure abiotic and biotic surface and subsurface traits. In remote sensing, usually encompasses imaging synthetic aperture radar (SAR) wavelengths 3–70 cm (frequencies 435 MHz to 8 GHz).

Radiance Radiant flux density emanating from a surface per unit solid angle. Units are often given in watts per steradian per square meter, $W \cdot sr^{-1} \cdot m^{-2}$. *Spectral* radiance units would also include wavelength (e.g., nm^{-1}).

Radiative transfer The transfer and propagation of electromagnetic energy through a medium (e.g., the atmosphere, a forest canopy, the water column), involving absorption, scattering, and emission of radiation as described by the fundamental radiative transfer equation. These processes are modeled to describe the relationship between the characteristics of the medium and reflectance.

Radiative transfer modeling in remote sensing, radiative transfer modeling (RTM) pertains to the use of computer programs of varying complexity to simulate the reflectance, transmittance, and absorption of solar radiation within different media such as the atmosphere (air, clouds, aerosols), snow, water, and vegetation (leaves and canopies). In each of these domains, physical laws govern the scattering and absorbing behavior of the radiation. RTMs are used to predict the shape of the spectral response from the characteristics of a specific material (forward mode), or to predict material characteristics from spectra (backward mode). Specific RTMs are developed using a wide range of assumptions, different model representations, and a range of approaches to simplify the solution to the full radiative transfer equation, resulting in a range of model complexities with trade-offs between computational demand, accuracy, and scalability. All RTMs require inputs in the form of external conditions (e.g., incident solar radiation for canopy modeling) and parameters (e.g., leaf angle distribution, leaf area index) used to solve the radiative transfer equation in the shortwave, optical, or longwave (i.e., thermal) domains.

Radiometric quality refers to the degree or accuracy of radiometric characterization, including resolution (see Radiometric resolution), typically determined by radiometric calibration, providing an instrument response in absolute radiometric units (e.g., $W\ m^{-2}\ sr^{-1}nm^{-1}$).

Radiometric resolution The smallest increment in spectral radiance that can be detected by a sensor. This is determined by the dynamic range (the bit quantization level) and signal-to-noise ratio of the sensor.

Random forests A supervised machine learning algorithm that constructs many decision trees and utilizes their outputs to get an accurate value of a variable or class prediction based upon training data.

Random noise Noise component of a signal that does not contain information about the property being measured. In remote sensing, random noise is often a function of instrument electronics and temperature and is typically removed by "dark correction" (subtracting the signal obtained when no external energy is being measured, e.g., shutter is closed).

Random-walk process A stochastic process that describes a sequence of random steps. The Brownian motion process used in models of evolution is an example of a random walk. Note that a random walk of a species trait through time does not necessarily mean that there is no selection: if the optimum value moves due to many random factors (including randomly shifting forces of selection), and the trait follows this closely, it is still approximated by a random walk.

Rate of evolution How fast trait changes accumulate over evolutionary time. In the context of the evolution of quantitative traits, it typically refers to the pace of evolution in a Brownian motion model.

Red edge The region of rapid change in a vegetation electromagnetic reflectance spectrum that occurs at the transition from visible red to near infrared (between 650 and 800 nm) and indicates the boundary between chlorophyll absorption (in the visible red) and vegetation scattering (in the near infrared).

Reflectance The ratio or percent of the radiance scattered backward from the surface relative to the irradiance incident on the surface. Reflectance is a primary product of remote sensing optical data because it largely normalizes for illumination, avoids the need for detailed radiometric calibration (e.g., calibration in energy units), and provides unique spectral shapes for different materials. Contrast to absorptance and transmittance.

Reflectance signature the characteristic shape or pattern of reflectance across wavelengths observed for a specific object (e.g., plant species) or condition. Sometimes referred to as a spectral fingerprint.

Refractive index See index of refraction.

Regeneration Formation or regrowth of tissue, organ, organism or assemblage of organisms (e.g., community or ecosystem).

Regional pool The set of species, populations, or organisms found within a defined region that have the potential to disperse over time.

Regulating ecosystem services Benefits from ecosystem processes that help maintain the Earth's life support systems; examples include maintenance of the quality of air and soil, flood and disease control, and crop pollination; also called Nature's Contributions to People (NCPs) within the IPBES framework.

Remote sensing Measurement of an object from a distance without physical contact. Usually involves detectors that measure energy in the visible to short-wave infrared or the thermal infrared region. Both passive instruments and active instruments (like LiDAR and radar sensors) are considered remote sensing instruments, as are various "sounders" that probe the atmosphere measuring in the optical or thermal bands, or certain acoustic instruments that detect and measure sound.

Remote sensing products Data or information obtained from remote sensing, often in the form of spectral information or maps of quantities of interest expressed a raster grid.

Reproducible research Research where the data and methods are available and sufficiently explained so that others can, without help from the original researchers, perform the same analysis and get the same results. Research where datasets are only available upon request from authors (who may lose them, leave the field, etc.) or where methods are opaque or not available (outliers removed by hand, unavailable scripts used to process data) are not reproducible.

Resilience The ability to return to an original functional state, e.g., the capacity of an ecosystem to regain its function after disturbance.

Resistance The ability of an ecosystem (or any system) to maintain function and not be adversely affected by a perturbation or stress (such as drought).

Resolution The smallest interval an instrument can measure (e.g., spectral, radiometric, spatial, or temporal resolution) based on detector characteristics or principles of measurement.

Retrieval Obtaining information or values of interest. In remote sensing, retrieval implies that values of interest are derived from measurements and *models*, i.e., the process of finding the inverse of a forward function which describes the relation between the measurements and the values of the quantities of interest (e.g., vegetation properties, atmospheric concentrations, water column values).

Retrieval algorithm The computer calculation used to make a remote sensing estimation of a property (see retrieval).

Ribulose-1,5-bisphosphate carboxylase/oxygenase (RuBisCO) A nitrogen-containing enzyme that catalyzes the reaction that fixes carbon dioxide to ribulose-1,5-bisphosphate (RuBP)to form three-carbon sugars in the "dark reactions" of photosynthesis. RuBisCO also catalyzes the reaction that binds RuBP with oxygen and releases carbon dioxide in photorespiration. It is the most abundant enzyme on Earth.

Root exudates Secretions from plant roots that alter local soil conditions and soil chemistry, which influences microbial processes.

Root mean square error (RMSE) A measure of the differences between values (sample or population values) predicted by a model or an estimator and the values observed. Also known as root-mean-square deviation (RMSD) or root-mean-squared error.

Ruderal A "weedy" plant species that tends to colonize disturbed lands.

Scale (noun) Level of observation, including spatial, temporal, spectral, and biological. Can also refer to the extent and grain of observation units.

Scale (verb) to change the resolution of a dataset. *Upscaling* involves extrapolation to larger areas (increasing extent), for example, through data aggregation or forward modeling; *downscaling* interpolates values or inverts a model to create a higher-resolution dataset or to infer component contributions at finer scales (increasing grain size). From a biological perspective, upscaling might involve considering emergent properties, whereas downscaling might involve looking at underlying mechanisms or properties.

Scale dimensions The different types of scale (e.g., spatial, temporal, spectral, biological).

Scaling The act of examining a phenomenon at multiple levels, usually referring to transcending spatial scales, as in "upscaling" (increasing extent by extrapolation from fine-scale data to a coarser, usually spatially broader scale) or "downscaling" (interpreting the variation in underlying patterns or mechanisms from coarse-scale data to finer resolution patterns or grain size).

Scattering (scatter, scattered) in remote sensing, reflecting and transmitting radiation. Radiation that is not absorbed is scattered either backward (reflected) or forward (transmitted).

Scattering phase height center The approximate height above the ground within a vegetation canopy where most of a radar signal interacts with scattering elements (leaves, branches) and is backscattered to the sensor.

Sclerophyllous having a hard, leathery texture, as in "sclerophyllous" leaf characteristic of Mediterranean or desert climates to maximize water retention.

Secondary metabolites Plant metabolites not directly associated with primary metabolism (e.g., photosynthesis or respiration) tied to plant growth, development, or reproduction, often associated with plant defense or signaling.

Semantic web An effort to make the World Wide Web machine readable by encoding semantics with data (e.g., through provision of metadata).

Semivariogram A plot of semivariance in some property vs. lag (distance between sampling points), used in geostatistics to compare similarity between sampling points and evaluate the scale dependence of features in a landscape.

Senescence The process of tissue or plant death.

Shortwave infrared (SWIR) The range of the electromagnetic spectrum between 1100 and 3000 nm.

Signal-to-noise ratio A ratio comparing the desired signal to the background noise of the sensor; sometimes calculated as the mean signal value divided by the standard deviation of the noise.

Simpson Index One of multiple metrics of alpha diversity that combines species richness and evenness (relative abundance).

Singular value decomposition (SVD) Matrix factorization; decomposes a matrix into an orthonormal basis transformation multiplied by a diagonal matrix followed by another orthonormal change of bases. Contrast with principal components analysis (PCA), which finds a new representation for a dataset in terms of a (typically lower rank) orthogonal basis that preserves the variance of the original projection. The two methods are related as SVD is regularly used to find the eigenvalues of a dataset's covariance matrix needed by PCA.

Solar spectrum   The light emitted by the sun—or solar radiation measured as irradiance upon a surface—expressed as a function of wavelength. These wavelengths extend from the UV to mid-infrared or from 200 nm to 3.0 µm. Optical remote sensing is primarily interested in the wavelengths of solar light that penetrate the atmosphere and reach the ground surface, which range from about 370 nm to 2.5 µm.

Spaceborne   Measurement sensors mounted on satellites or space stations and deployed in orbit (as in "spaceborne remote sensing").

Spatial domain   The realm of measurements or properties existing in space.

Spatial grain   The spatial resolution (pixel size) of a remote sensing dataset. See spatial resolution.

Spatial resolution   Generally, the size of the smallest measurement unit (pixel) in an image.

Spatial scale   The grain size and spatial extent at which data about a phenomenon are sampled or expressed.

Spatially explicit   Exhibiting spatial properties such as geographic position and/ or extent.

Species   A group of organisms of a type, distinct from other such groups. Defining what constitutes a species remains a contentious discussion in taxonomy, and several common definitions exist: the *biological species concept* involves a group of interbreeding (or potentially interbreeding) organisms; the *morphological species concept* is based on morphology; the *phylogenetic species concept* is based on evolutionary history.

Species distribution model   An empirical statistical approach that predicts the spatial and temporal distribution of species, usually as a function of climatic and other environmental (soils, topography) variables. See species distribution modeling.

Species distribution modeling (SDM)   A modeling procedure with the purpose of predicting the occurrence of taxa in geographical space and time as a function of gridded geographic variables related to climate and environment. See environmental niche model and niche model.

Species diversity   Taxonomic diversity at the species level, as distinct from functional diversity, phylogenetic diversity, etc.

Species evenness   The relative abundance of species relative to other species in a given area.

Species richness   The number of species found in a given area.

Species turnover   Changes in species composition from one community to another, over space (geographic distance), or time. Often characterized as a rate, one aspect of beta diversity.

Specific leaf area (SLA)   Leaf area per unit dry mass, usually expressed in $m^2 \, kg^{-1}$; it is the reciprocal of leaf mass per area (based on unit conventions, SLA = 1000/LMA).

Spectral band   A defined region of the electromagnetic spectrum measured by a detector. See band.

Spectral correction   Resampling to a true wavelength array; spectral calibration.

Spectral centroid   The center point of a spectrum or many spectra, sometimes as viewed in multidimensional space.

Spectral characteristics   Characteristic properties or features related to spectra collected from a target of interest (e.g., vegetation).

Spectral database (or spectral library)   An Internet-based (generally) repository of spectra tied to metadata including measurement techniques and—for plants—species identity and/or concurrent trait measurements. For plant spectral databases, entries include leaf- and canopy-level spectra or image-derived spectra and have specific metadata standards (e.g., EcoSIS.org).

Spectral diversity   Variation in spectra, typically measured as reflectance spectra in the optical range (VIS-SWIR), among a group of plant species, functionally distinct vegetation types (e.g., grassland vs. forest), or among pixels in a spectroscopic image. See also optical diversity.

Spectral diversity hypothesis   The hypothesis that the diversity of spectral profiles predicts some aspect or dimension of biodiversity (e.g., species richness, or functional diversity). See also optical diversity.

Spectral distance   The distance in multidimensional spectral space between spectra. Also a measure of dissimilarity for two or more species or functional types based on spectral reflectance.

Spectral distortion   Systematic inaccuracies in spectra, possibly arising from several sources (e.g., atmosphere or stray light in an instrument).

Spectral domain   referring to spectral properties across specific portions of the electromagnetic spectrum, involving information such as wavelength or spectral bandwidth.

Spectral feature   Pattern in a spectrum due to absorption or scattering characteristics of a material, often involving only a few adjacent wavelengths.

Spectral fingerprint   See spectral signature.

Spectral heterogeneity   The heterogeneity in spectral characteristics; expected to be positively related to environmental heterogeneity and also to certain metrics of biodiversity, but also often a consequence of canopy or landscape variability.

Spectral imaging   See imaging spectroscopy.

Spectral index (plural: spectral indices)   A mathematical expression, often a ratio or normalized difference, of measured reflectance or radiance from two or more spectral bands, with or without a constant or scaling factor, that highlights a particular feature in the spectral signal.

Spectral laboratory   A laboratory using spectrometers, indoors or outdoors, mobile or stationary.

Spectral mixture analysis   A mathematical procedure for estimating the fraction of subpixel elements in mixed pixels, based on distinct spectral signatures, also called "endmembers." The analysis treats pixels as mixtures of these endmembers, which are resolved by calculating the best fitting fractional composition of all endmembers. Mathematically the fraction of each endmember is multiplied by the endmember's spectrum, and all fractions are added linearly until the mixed spectrum is approximated sufficiently, or this process is inverted to identify relative contributions of each endmember to the mixture. Can be used to estimate the distribution of types of targets (e.g., dominant plant cover types, or most commonly soil, green vegetation, and non-photosynthetic vegetation fraction) in a landscape.

Spectral reflectance   Reflectance expressed as a function of wavelength (i.e., as a spectrum).

Spectral reflectance profile   See spectral reflectance.

Spectral resolution   related to a sensor's ability to resolve the features of an electromagnetic spectrum. The resolution depends on the number and width of spectral bands measured by an instrument and is often defined as the bandwidth at half the maximum amplitude of the spectral response per band ("full width half maximum").

Spectral response function   The mathematical and/or graphical description of a spectral pattern, typically associated with a spectral band. In remote sensing, generally is used to describe the response across wavelengths measured in a specific band (e.g., in imaging spectroscopy, the reflectance at different wavelengths surrounding the center wavelength, or "full width half maximum," of the wave band).

Spectral scale   Spectral resolution and wavelength range.

Spectral signals   features in a spectrum determined by specific absorbing and scattering patterns.

Spectral signature   The specific (in theory, unique) shape of the spectrum of a specific vegetation type, plant species, or any particular material or combination of materials. Also known as spectral fingerprint.

Spectral space   A multidimensional space with axes consisting of wavelength bands, or combinations of wavelengths bands (e.g., principal component axes, spectral indices).

Spectral species   Types (categories or groupings) of spectra, typically determined using unsupervised classification that may or may not match to actual biological (taxonomically defined) species. They can be used to estimate alpha diversity, beta diversity, and gamma diversity.

Spectral traits (ST)   features in the reflectance or absorption spectra of plant compounds, leaves, canopies, ecosystems, or landscapes that can be meaningfully interpreted and directly or indirectly associated with underlying sources of variation in plants—at the relevant scale of detection—such as their anatomical, chemical, morphological, biophysical, physiological, structural, phenological, or functional characteristics that are influenced by phylogenetic, taxonomic, population, community, environmental, ecosystem, and/or landscape-level properties.

Spectral trait variation (STV)   changes to spectral traits that can be directly or indirectly recorded by spectroscopic techniques in space, over time, among samples or within a sample.

Spectral type   A group of organisms or species that share similar spectral profiles (see also "spectral species," "endmember," and "optical type"); can also refer to a characteristic spectral pattern associated with a particular group of organisms or species.

Spectral variability   See spectral diversity.

Spectral variation hypothesis (SVH) A hypothesis that states that the spatial variability in the remotely sensed signal, i.e., the spectral heterogeneity, is expected to be positively related to environmental heterogeneity, or to variation in plant traits, and could therefore be used as a proxy of biodiversity.

Spectral vegetation index A mathematical combination of reflectance from wavelength regions that together are relevant to vegetation properties or processes (see spectral index).

Spectranomics Term introduced by Asner and Martin (2008) to describe an approach to link plants, canopies, and community as well as their functional traits to their spectral properties with the objective of providing time-varying, scalable methods for remote sensing of ecosystems and biodiversity.

Spectrophotometer An instrument for measuring the spectral transmittance of light to infer the spectral absorbance. Spectrometers are typically used for determining absorbance in a solution to quantify the concentration of particular compounds. See also spectrometer.

Spectrometer A device that is designed to measure electromagnetic radiation as a function of wavelength; not necessarily radiometrically calibrated (c.f. spectroradiometer). See also spectrophotometer and spectroradiometer.

Spectrometry Measurement with a spectrometer.

Spectroradiometer A device that is designed to measure electromagnetic radiation across a range of wavelengths; a radiometrically calibrated spectrometer (c.f. spectrometer).

Spectroscopy An area of study focused on the interactions between electromagnetic radiation and matter. For imaging spectroscopy of plant biodiversity, the term is used to describe the measurement of electromagnetic radiation in a pixel of an image to study the properties of leaves, canopies, ecosystems, and landscapes and their variation in time and space.

Spectrum A depiction of the intensity or distribution of electromagnetic radiation, typically the emittance, radiance, reflectance, absorptance, or transmittance over the wavelength interval measured and expressed in digital numbers, radiance (watts per steradian per square meter, $W \cdot sr^{-1} \cdot m^{-2}$), or as a fraction or percentage of outgoing to incoming radiance (reflectance) or emissivity. Can also be used to express absorbance, fluorescence, or emittance as a function of wavelength. Plural: spectra.

Specular A type of electromagnetic reflection that is scattered away from a very smooth, mirror-like surface. When light is specularly reflected from a surface, it is reflected at the same angle as the incident ray but on the opposite side of the plane normal to the surface (c.f. Lambertian or isotropic).

Spongy mesophyll Photosynthetic parenchyma tissue in plant leaves where the cells are loosely arranged. Typically located on the abaxial side (normally the lower surface) of the leaf and specialized to facilitate the transport and exchange of CO_2, H_2O, and O_2.

Stationarity Equality or constancy of parameter values or statistical descriptors (e.g., mean, variance) of equidistant points in space or time.

Stomata The turgor-controlled valves, comprised of specialized epidermal cells (guard cells), typically located on the abaxial (lower) side of the leaf, that open and close, regulating gas exchange, i.e., movement of water, carbon, and oxygen into and out of the leaf. Also called stomates. Singular: stoma or stomate.

Stress Conditions such as drought, temperature extremes, nutrient limitation, pest/pathogen exposure, and fire that produce sub-optimal growth, reproduction, or survival conditions for living organisms.

Structural diversity Variability or heterogeneity in the arrangement and distribution of physical features on the landscape (e.g., topography, vertical and horizontal canopy structure, or vertical and horizontal stand structure).

Structural complexity in the context of biodiversity, the display of organization in the components of biological systems, particularly in their horizontal and vertical structure, distribution, morphology, and/or anatomy; it is also a science of applied mathematics that describes the morphological, structural, and/or anatomical intricacies of a complex system.

Structure from motion techniques (SfM) A technique for estimating 3-D structure from a series of 2-D images that can be integrated using motion that provides multiple viewpoints via parallax. These techniques can be used to produce 3-D models based on point clouds similar to LiDAR.

Successional change Change in a community over time; for example, after a fire, grasses may grow first and then gradually be replaced by shrubs and trees species that are increasingly shade-tolerant. Ecologists distinguish primary succession—initial establishment of an ecological community on bare substrate after an extreme disturbance, e.g., glaciation or volcanic eruption—and secondary succession, community change over time as a consequence of less extreme disturbance, such as fire or treefall.

Supervised classification methods Models or classifiers in which reference samples, or training data, are used to define classes.

Support vector machine (SVM) A supervised learning algorithm for classification and regression problems. Given labeled training data, the algorithm outputs an optimal hyperplane that separates classes in multidimensional space.

Surrogacy in relation to biodiversity, the idea that diversity at one trophic level (plant diversity) may be related to diversity at other trophic levels, or the idea that one aspect of biodiversity can serve as a proxy of another.

Surface energy balance The formal relationship between incoming and outgoing energy that describes the energy striking a surface (e.g., leaf or Earth surface) and the energy leaving that surface. The energy is typically split into several components such as shortwave (incoming solar), longwave (outgoing thermal), sensible (temperature), and latent (e.g., phase change) energy. To maintain a constant temperature over time, these components must be in equilibrium or the object would heat or cool.

Sustainable Development Goals A collection of 17 global goals set by the United Nations General Assembly in 2015 for the year 2030. They cover social, economic, environmental, and development issues including poverty, hunger, health, education, gender equality, clean water, sanitation, affordable energy, decent work, urbanization, global warming, environment, social justice, and peace.

Synonymy   Multiple scientific names are applied to the same taxon; this can happen when multiple taxonomists describe the same taxon or when phylogenetic relationships are newly resolved. When a scientific name is updated, the prior names are called synonyms.

Surface emissivity   The effectiveness of the surface of a body in emitting energy as electromagnetic radiation. Ranges from 0 to 1 (blackbody that emits perfectly).

Swath   The width of Earth surface that an airborne or satellite sensor samples as it moves.

Tachymeter, geodetic   Terrestrial surveying instrument providing locations with millimeter precision.

Taxon (plural taxa)   A general term for a named set of organisms. The term typically applies to a species (e.g., *Quercus alba*) or a higher group, such as a genus (e.g., *Quercus*), family (e.g., Fagaceae), or major lineage (e.g., angiosperms).

Taxonomic diversity   Measurements that incorporate the number of distinct species or higher-level taxa and/or abundances of those taxa. The variability in species or taxonomic groups present. See also species richness.

Taxonomic group   A named clade: a set of organisms all descended from one ancestor.

Taxonomy   A system of naming; typically in this volume, the way organisms are named. Much of biology uses a taxonomy derived from the system developed by Linnaeus that groups organisms hierarchically according to species, genera, families, and higher-order classifications.

Temporal domain   related to time or timing of a process or measurement. See also temporal scale.

Temporal scale   refers to data or measurement in the time domain and characterized by properties such as the timespan, duration, period, or repeat frequency of a measurement.

Terra/aqua   NASA's Earth-observing satellites launched in 1999 and 2002 that carry a variety of instruments and having different equatorial crossing times (morning and afternoon, respectively). For example, the MODIS instrument on each satellite platform images the entire Earth every 1–2 days and is the basis for long-term records of vegetation process at broad scales (250–1000 m pixels).

Texture   Smoothness or roughness of the surface of an object.

Thematic classification   The process of assigning discrete categories, types, or classes to each pixel in a continuous raster image.

Thermal remote sensing   Remote sensing of temperature effects carried out by sensing radiation emitted from materials in the thermal infrared region of the spectrum. Most thermal sensing of solids and liquids occurs in two atmospheric windows where absorption is a minimum. The windows normally used are in the 3–5 μm (middle infrared, important for detection of fire) and 8–14 μm wavelength regions.

TIMESAT   A software package for analyzing time series of satellite data.

Tomography   Imaging by sections or sectioning through an object or scene by the use of any kind of penetrating wave.

Topographic illumination Differential shading and brightness of reflectance caused by land surface relief, i.e., relative slope exposure to direct sunlight.

Topology in phylogenetics, the structure of a phylogeny (dendrogram of evolutionary relationships among taxa).

Tracheid Specialized support and water transport cell in the xylem tissue of plants. Tracheids have thick and rigid cell walls and are nonliving at maturity. Conifers use tracheids for both water transport and mechanical support, whereas angiosperms primarily use vessels for water transport and tracheids for mechanical support.

Trait A biochemical, physiological, morphological, structural, phenological, or functional characteristic of a plant, population of plants, or community. Traits exist on all levels of biological organization. See also functional trait.

Trait diversity Variation of traits on all levels of biological organization, generally calculated as a metric of the volume or breadth of the multidimensional scatter of those traits. Trait diversity can be linked to phylogenetic, structural, taxonomic, and functional diversity.

Transect A linear sampling approach used for quantifying number, size, species composition, and other attributes of vegetation based on intersection with line of fixed length.

Transmittance The fraction of light entering a leaf that is scattered out through the opposite surface.

Transpiration The process by which liquid water is transported through plants from roots to mesophyll tissue in leaves, where it evaporates and is released through stomata (small pores on the underside of leaves) and to the atmosphere.

Tree of life The phylogeny—or inferred evolutionary relationships—of all species on Earth.

Trophic levels Positions of organisms on a food chain: organisms at higher trophic levels consume organisms at lower trophic levels.

TRY database Online repository for plant functional traits (https://www.try-db.org).

Uncertainty The probability (or value) of error regarding the data, algorithms, or model outputs; also known as associated error.

Understory The vegetation below the canopy.

Unmanned aerial vehicle (UAV) or system (UAS) or device (UAD) An airborne vehicle (drone) that can be flown without a human on board, including multicopters and fixed-wing aircraft.

Unsupervised classification The classification of remote sensing data without previously classified training or reference samples; mapped categories are labeled or merged post hoc.

Upscaling Extrapolation from comparatively fine scales to larger areas, for example, through data aggregation or forward modeling. Contrast to downscaling.

User's accuracy The fraction of correctly classified pixels with regard to the classes present on the ground. Contrast to producer's accuracy.

Vacuole A membrane-bound organelle within the plant cell that stores water, enzymes, and other organic molecules, sometimes including toxins or waste materials. As the largest organelle in the cell, the vacuole maintains turgor pressure within the plant cell and provides structure and support for the growing cell.

Validation (internal and external validation) Quantitative assessment of the accuracy and precision of a measurement, model output, or model assessment. Internal validation: based on withholding test samples repeatedly from the calibration process (e.g., through permutation, n-fold cross-validation, leave-one-out cross-validation) to predict the data withheld (validation is not independent). External or independent validation: model assessment based on samples that were never part of the modeling process (e.g., samples split from the population or samples collected independently).

Validation pixels Pixels not used during the training phase of a supervised classifier or prediction model that are later used to assess the accuracy of the classifier or prediction.

Variogram Function or plot describing the degree of spatial dependence or patterns of variation with distance. See also semivariogram.

Vector normalization Standardizing a vector to unit length, which changes the magnitude but not the direction; a useful transformation for spectral reflectance data from plants when the magnitude of reflectance varies among samples (see also mean normalization).

Vegetation Plants; plants aggregated in space without reference to individual species.

Vegetation structure Constitutes the three-dimensional vertical and horizontal components of vegetation.

Voxel Unit of space formed by using grids to subdivide a 3-D space. While a pixel includes x- and y-coordinates in two-dimensional space, a voxel also has a z-coordinate. Most commonly for vegetation studies, a voxel provides characterization of the vertical distribution of plant components within a pixel, e.g., leaf area index by height. In the context of imaging spectroscopy, a voxel is an image pixel in coordinate space with spectral bands as the z-dimension.

Water content The amount of water in a tissue, organ, or organism, typically expressed as the mass of unbound liquid water per unit area (content) or unit fresh or dry weight (usually termed concentration). In remote sensing, water content is often expressed as equivalent water thickness (thickness of water, assuming all water is present as a single layer).

Wave band See band.

Wavelength range The spectral interval sampled by an instrument or covered by a spectrum.

Wavelet transform A linear transformation of a continuous function of one variable into a continuous function of two variables, translation and scale, which are often used to remove noise in digital image processing, for image compression, or to identify the resolution important to variation in a signal.

Waveform decomposition Methods used to derive statistical parameters of full waveform light detection and ranging (LiDAR) signals for characterizing vegetation three-dimensional structure and complexity.

Whole plant economic spectrum   The concept of a single "fast-slow" axis of plant ecological strategies put forth by Peter Reich that integrates across traits of leaves, stems, and roots and explains resource use, growth potential, stress, and pest/pathogen tolerance and competitive advantage of plants in different environments. Using an "economic" framework following the "leaf economic spectrum" (LES), the concept is based on coordinated function and trade-offs among traits and is relevant to explaining community assembly and ecosystem function.

Wireless sensor networks (WSN)   A group of sensors for monitoring and recording physical or environmental conditions—such as temperature, sound, electromagnetic radiation, pollutants, pressure, humidity, and wind speed. The network may contain hundreds of thousands of sensor nodes that are spatially dispersed. The data are collected via wireless communication at a central location that serves as an interface between users and the network.

Wood density   The wood dry mass in a unit volume of wood.

Workflow   Sequence of tasks (often substantially automated but not necessarily) that process input data to a desired output; precise description of data processing from one analytical step to another in a formal language.

WorldClim variables   19 gridded bioclimatic variables that reflect spatial and temporal (annual, seasonal) differences in precipitation and temperature in the WorldClim climatic database, generally intended for species distribution modeling (www.worldclim.org).

Xylem   The vascular tissue that transports water and nutrients from the roots to all parts of the plant. It includes vessels (angiosperms only), tracheids, parenchyma, and fiber cells.

Z-dimension   Typically the third dimension in a three-dimensional space or dataset, such as the vertical element (e.g., elevation above or below sea level), or the spectral dimension in an image cube.

Index

© The Author(s) 2020
J. Cavender-Bares et al. (eds.), *Remote Sensing of Plant Biodiversity*,
https://doi.org/10.1007/978-3-030-33157-3

Printed in the United States
By Bookmasters